AF353099

Natural Compounds in Plant-Based Food

Natural Compounds in Plant-Based Food

Editors

Andreas Eisenreich
Bernd Schaefer

Basel • Beijing • Wuhan • Barcelona • Belgrade • Novi Sad • Cluj • Manchester

Editors

Andreas Eisenreich
Department Food Safety
German Federal Institute for
Risk Assessment
Berlin
Germany

Bernd Schaefer
Department Food Safety
German Federal Institute for
Risk Assessment
Berlin
Germany

Editorial Office
MDPI
St. Alban-Anlage 66
4052 Basel, Switzerland

This is a reprint of articles from the Special Issue published online in the open access journal *Foods* (ISSN 2304-8158) (available at: www.mdpi.com/journal/foods/special_issues/Natural_Compounds_Plant-Based_Food).

For citation purposes, cite each article independently as indicated on the article page online and as indicated below:

Lastname, A.A.; Lastname, B.B. Article Title. *Journal Name* **Year**, *Volume Number*, Page Range.

ISBN 978-3-0365-8479-9 (Hbk)
ISBN 978-3-0365-8478-2 (PDF)
doi.org/10.3390/books978-3-0365-8478-2

Contents

Preface

Culinary herbs and spices confer characteristic flavors or coloring to food. In this Special Issue of *Foods*, we aim to focus on culinary herbs and spices from a health perspective. Substances of plant origin are often regarded as harmless. The intention of this Special Issue is to provide a science-based view of the health impact of naturally occuring ingredients in plants with a particular focus on the health impact of plants used as culinary herbs and spices. In this context, different aspects will be addressed, including toxicological, analytical, and regulatory issues regarding naturally occurring ingredients in culinary herbs and spices as well as in products, such as essential oils and the food supplements based on them. For this purpose, some selected examples of these compounds will be utilized to illustrate potential effects on human health, together with knowledge gaps related to hazard and exposure assessment. Moreover, safety issues related to products, such as essential oils or food supplements containing extracts of culinary herbs and spices that have become widely available to consumers through several distribution channels in the EU, will be covered.

Andreas Eisenreich and Bernd Schaefer
Editors

Editorial

Natural Compounds in Plant-Based Food

Andreas Eisenreich *[image] and Bernd Schäfer

Department of Food Safety, German Federal Institute for Risk Assessment (BfR), Max-Dohrn-Str. 8-10, 10589 Berlin, Germany
* Correspondence: andreas.eisenreich@bfr.bund.de; Tel.: +49-30-18412-25202

Plant-based foods include a wide range of products, such as fruits, vegetables, herbs and spices, as well as food products based on them, such as sauces, soups, or beverages. Particularly, culinary herbs and spices are used to confer characteristic flavor or coloring to those food products due to their aromatizing and color-giving properties. This, in turn, is based on the occurrence of various compounds, naturally generated as secondary plant metabolites in those herbs and spices [1]. In this context, those substances of plant origin are often regarded as harmless. However, several secondary plant metabolites, originally generated to deter pests or herbivores (e.g., phenylpropanoids and pyrrolizidine alkaloids), may also exhibit toxic properties to consumers [2–4].

In this Special Issue, several expert contributions addressed aspects regarding "Natural Compounds in Plant-Based Food". These include the occurrence, toxicological relevance, and potential health benefits of different compounds, such as alkenylbenzenes, pyrrolizidine alkaloids, glucosides, curcumin, and piperine in culinary spices and herbs, as well as in essential oils and processed foods made up of these.

Alkenylbenzenes represent an important group of naturally occurring substances in plants used as food or for food production [1]. Therefore, a part of this Issue is devoted to these compounds and the associated safety issues. In this context, different aspects regarding the occurrence (e.g., in herbs and spices), toxicokinetics, toxicity, and analytics, as well as the corresponding uncertainties, were discussed in detail [2,5]. Along with this, experimental data regarding state-of-the-art analytics of alkenylbenzenes were also presented in this compilation [6].

Another complex of this Issue dwelled on the toxicology of pyrrolizidine alkaloids [7]. In this context, different studies dealing with the regulation and associated molecular processes induced by a pyrrolizidine alkaloid in experimental liver settings were also included here [8,9].

Human cancer risk is an issue of particularly high importance in the context of an exposure to potentially genotoxic and carcinogenic naturally occurring substances in plant-based food, such as alkenylbenzenes and pyrrolizidine alkaloids [2,3]. To address this topic, a review giving a detailed overview of relevant food-borne chemical carcinogens and the evidence for human cancer risk was also included in this Issue [10].

In addition, other articles published in this Issue deal with the further issues of naturally occurring compounds (e.g., curcuminoids and piperine) in different herbs and spices, such as dill, tarragon, black pepper, or turmeric [11–14]. In this context, different relevant topics were addressed in more detail, such as analytical aspects but also the current safety issues as well as potential health benefits associated with those compounds.

The scientific publications curated in this compilation certainly do not represent a comprehensive and conclusive summary of the abovementioned field of science. However, the present Special Issue offers a balanced view on the issue "Natural Compounds in Plant-Based Food", thereby attempting to shed more light on the different relevant facets of this topic with a particular focus on the potential health effects of the ingredients present in culinary herbs and spices as well as in essential oils and other food products made of them as part of the human diet.

Citation: Eisenreich, A.; Schäfer, B. Natural Compounds in Plant-Based Food. *Foods* 2023, *12*, 857. https://doi.org/10.3390/foods12040857

Received: 13 February 2023
Accepted: 15 February 2023
Published: 17 February 2023

Author Contributions: Conceptualization, A.E.; writing—original draft preparation, A.E. and B.S.; writing—review and editing, A.E. and B.S. All authors have read and agreed to the published version of the manuscript.

Acknowledgments: We thank Janine Frenzel and Vivian Kümpel for technical and administrative support.

Conflicts of Interest: The authors declare no conflict of interest.

References

1. Kristanc, L.; Kreft, S. European medicinal and edible plants associated with subacute and chronic toxicity part I: Plants with carcinogenic, teratogenic and endocrine-disrupting effects. *Food Chem. Toxicol.* **2016**, *92*, 150–164. [CrossRef] [PubMed]
2. Eisenreich, A.; Götz, M.E.; Sachse, B.; Monien, B.H.; Herrmann, K.; Schäfer, B. Alkenylbenzenes in Foods: Aspects Impeding the Evaluation of Adverse Health Effects. *Foods* **2021**, *10*, 2139. [CrossRef] [PubMed]
3. Schramm, S.; Kohler, N.; Rozhon, W. Pyrrolizidine Alkaloids: Biosynthesis, Biological Activities and Occurrence in Crop Plants. *Molecules* **2019**, *24*, 498. [CrossRef] [PubMed]
4. War, A.R.; Sharma, H.C.; Paulraj, M.G.; War, M.Y.; Ignacimuthu, S. Herbivore induced plant volatiles: Their role in plant defense for pest management. *Plant Signal. Behav.* **2011**, *6*, 1973–1978. [CrossRef] [PubMed]
5. Gotz, M.E.; Sachse, B.; Schafer, B.; Eisenreich, A. Myristicin and Elemicin: Potentially Toxic Alkenylbenzenes in Food. *Foods* **2022**, *11*, 1988. [CrossRef]
6. Hermes, L.; Römermann, J.; Cramer, B.; Esselen, M. Phase II Metabolism of Asarone Isomers In Vitro and in Humans Using HPLC-MS/MS and HPLC-qToF/MS. *Foods* **2021**, *10*, 2032. [CrossRef]
7. Chmit, M.S.; Horn, G.; Dübecke, A.; Beuerle, T. Pyrrolizidine Alkaloids in the Food Chain: Is Horizontal Transfer of Natural Products of Relevance? *Foods* **2021**, *10*, 1827. [CrossRef]
8. Enge, A.-M.; Sprenger, H.; Braeuning, A.; Hessel-Pras, S. Identification of microRNAs Implicated in Modulating Senecionine-Induced Liver Toxicity in HepaRG Cells. *Foods* **2022**, *11*, 532. [CrossRef] [PubMed]
9. Glück, J.; Henricsson, M.; Braeuning, A.; Hessel-Pras, S. The Food Contaminants Pyrrolizidine Alkaloids Disturb Bile Acid Homeostasis Structure-Dependently in the Human Hepatoma Cell Line HepaRG. *Foods* **2021**, *10*, 1114. [CrossRef] [PubMed]
10. Kobets, T.; Smith, B.P.C.; Williams, G.M. Food-Borne Chemical Carcinogens and the Evidence for Human Cancer Risk. *Foods* **2022**, *11*, 2828. [CrossRef] [PubMed]
11. Newerli-Guz, J.; Śmiechowska, M. Health Benefits and Risks of Consuming Spices on the Example of Black Pepper and Cinnamon. *Foods* **2022**, *11*, 2746. [CrossRef] [PubMed]
12. Setzer, W.N.; Duong, L.; Poudel, A.; Mentreddy, S.R. Variation in the Chemical Composition of Five Varieties of Curcuma longa Rhizome Essential Oils Cultivated in North Alabama. *Foods* **2021**, *10*, 212. [CrossRef]
13. Ziegenhagen, R.; Heimberg, K.; Lampen, A.; Hirsch-Ernst, K.I. Safety Aspects of the Use of Isolated Piperine Ingested as a Bolus. *Foods* **2021**, *10*, 2121. [CrossRef]
14. Saini, R.K.; Assefa, A.D.; Keum, Y.-S. Spices in the Apiaceae Family Represent the Healthiest Fatty Acid Profile: A Systematic Comparison of 34 Widely Used Spices and Herbs. *Foods* **2021**, *10*, 854. [CrossRef] [PubMed]

Food-Borne Chemical Carcinogens and the Evidence for Human Cancer Risk

Tetyana Kobets [1],*, Benjamin P. C. Smith [2] and Gary M. Williams [1]

1   Department of Pathology, Microbiology and Immunology, New York Medical College, Valhalla, NY 10595, USA
2   Future Ready Food Safety Hub, Nanyang Technological University, Singapore 639798, Singapore
*   Correspondence: tetyana_kobets@nymc.edu; Tel.: +1-914-594-3105; Fax: +1-914-594-4163

Abstract: Commonly consumed foods and beverages can contain chemicals with reported carcinogenic activity in rodent models. Moreover, exposures to some of these substances have been associated with increased cancer risks in humans. Food-borne carcinogens span a range of chemical classes and can arise from natural or anthropogenic sources, as well as form endogenously. Important considerations include the mechanism(s) of action (MoA), their relevance to human biology, and the level of exposure in diet. The MoAs of carcinogens have been classified as either DNA-reactive (genotoxic), involving covalent reaction with nuclear DNA, or epigenetic, involving molecular and cellular effects other than DNA reactivity. Carcinogens are generally present in food at low levels, resulting in low daily intakes, although there are some exceptions. Carcinogens of the DNA-reactive type produce effects at lower dosages than epigenetic carcinogens. Several food-related DNA-reactive carcinogens, including aflatoxins, aristolochic acid, benzene, benzo[a]pyrene and ethylene oxide, are recognized by the International Agency for Research on Cancer (IARC) as causes of human cancer. Of the epigenetic type, the only carcinogen considered to be associated with increased cancer in humans, although not from low-level food exposure, is dioxin (TCDD). Thus, DNA-reactive carcinogens in food represent a much greater risk than epigenetic carcinogens.

Keywords: carcinogens; food; DNA-reactants; epigenetic; risk assessment

Citation: Kobets, T.; Smith, B.P.C.; Williams, G.M. Food-Borne Chemical Carcinogens and the Evidence for Human Cancer Risk. *Foods* **2022**, *11*, 2828. https://doi.org/10.3390/foods11182828

Academic Editors: Andreas Eisenreich and Bernd Schaefer

Received: 9 August 2022
Accepted: 8 September 2022
Published: 13 September 2022

Publisher's Note: MDPI stays neutral with regard to jurisdictional claims in published maps and institutional affiliations.

1. Introduction

Foods and beverages are essentially complex mixtures of chemicals consumed for either sustenance or pleasure. The diversity of chemicals found in food is vast, as are their varying properties. It has long been known that chemicals with carcinogenic activity in rodent models can be found in many commonly consumed foods [1–5] from a variety of sources including plants, microorganisms, contaminations, additive uses and reactions which occur during storage, processing and cooking [2] (Table 1). In addition, carcinogens can be formed endogenously, from food materials [6–8]. This review focuses mainly on carcinogens, both rodent and human, present in foods and beverages at low concentrations which are imperceptible, and a few components present at levels associated with adverse effects. It does not address drinking water contaminants, such as arsenic, or the contributions of caloric content and macro components such as fat content, or the excess consumption of alcoholic beverages, all of which, nevertheless, have been implicated in increased cancer risks in humans [9–16].

Chemical carcinogens exert their effects through two distinct types of mechanism of action (MoA), which have been characterized as DNA-reactive (genotoxic) and epigenetic (non-genotoxic) [17–20], as discussed below. Chemical structure determines the carcinogenic MoAs; DNA-reactive carcinogens have structures that form reactive electrophiles, either directly or following bioactivation, whereas epigenetic carcinogens lack such properties, but have structures that exert other molecular and cellular effects leading to cancer [17,18]. These differences in MoA underly the nature of human cancer risks from exposures [21,22].

Table 1. Sources of detectable carcinogens in food.

Source	Examples [a]	
1. Naturally occurring		
Plant:	alkenylbenzene derivatives aristolochic acid cycasin ptaquiloside d-limonene	psoralen pyrrolizidine alkaloids pulegone β-myrcene
Microbial/Fungal:	various mycotoxins	
2. Contaminants		
Introduced before processing:	daminozide dioxins	DDT flumequine
Introduced during processing:	trichloroethylene	methylene chloride
Food contact materials:	plastics (polyolefins, polyesters, polystyrene, polyamides, etc.) polymeric coatings	monomers (vinyl chloride, styrene, acrylonitrile)
3. Additives		
Anthropogenic:	α,β-aldehydes butylated hydroxyanisole and butylated hydroxytoluene	hexenal saccharin
4. Formed from food components		
During processing:	acrylamide chloropropanols ethyl carbamate (urethane)	furan various nitrosamines alkylated imidazoles
During packaging:	bisphenol A furan	phthalates
During storage:	benzene	
During cooking:	acrylamide benzo[a]pyrene	various heterocyclic amines
In the body:	nitrosamines and nitrosamides α,β-aldehydes	ethylene oxide

[a] Many of the agents listed are detectable only at minute levels by highly sensitive analytical techniques.

1.1. Mechanisms of Carcinogenicity of DNA-Reactive Carcinogens

DNA-reactive carcinogens have structures that permit formation of electrophilic reactants that covalently bind (adduct) to nucleophilic sites in nuclear DNA, as well as in other macromolecules, including RNA and proteins, in the target tissue(s) of carcinogenicity [23–25]. In target tissue(s), a single DNA reactant can form different DNA adducts on various nucleophilic sites either on a single base or on different bases. Each adduct can undergo different rates of repair depending upon its location in the genome. For example, adducts in transcriptionally active regions are repaired by a transcription-coupled repair system whereas adducts in transcriptionally silent regions are repaired by a global repair system [26]. The levels of DNA adducts resulting from exposures are a function of several metrics including dose levels, the frequency of exposure, and rates of DNA repair for specific adducts. Each adduct has a characteristic efficiency with which it gives rise to mutations, with those at sites of base pairing being more mutagenic.

Pro-mutagenic DNA alterations are converted to mutations during cell replication [27–29]. Mutations in critical growth control genes lead to neoplastic conversion, and subsequent neoplastic development [28,30]. DNA-reactive carcinogens can also exert other cellular effects, such as cytotoxicity, leading to enhanced cell proliferation, which can contribute to their carcinogenic activity [31,32]. DNA-reactive carcinogens can have additive effects with one another in their target organ(s).

Some DNA adducts evidently do not lead to carcinogenicity, since some adducts can be found in tissues where no tumors are induced following administration of a carcinogen [33–36]. For example, acrylamide, which is discussed below, forms adducts in target and non-target tissues [37]. It could also be the case that epigenetic effects are required to enable neoplastic conversion resulting from some adducts [38,39].

As a result of DNA interactions, DNA-reactive carcinogens are typically genotoxic in assay systems in which appropriate bioactivation is represented [17,18,24,40,41]. Moreover, DNA-reactive carcinogens often produce tumors at multiple sites and with a short duration of exposure, even after administration of a single dose for some. This property underlies their activity in limited short-term bioassays [18].

Some DNA-reactive carcinogens have been demonstrated to exhibit no-observed-adverse-effect-levels (NOAELs) for carcinogenic effects in animal models [25,31,42–46], although conflicting data have been reported. Based on the steps for tumorigenesis, it is evident that biological thresholds that may influence the likelihood of cancer progression for genotoxic carcinogens exist. Nevertheless, currently, thresholds are not generally accepted for DNA-reactive carcinogens from a risk assessment and management perspective [47]. It is acknowledged that the derivation of NOAELs can be dependent on the study design, and more research is needed in this space. It is outside the scope of this paper to discuss thresholds for carcinogens in detail; however, this topic is reviewed elsewhere [25,31,42–46].

1.2. Mechanisms of Carcinogenicity of Epigenetic Carcinogens

Epigenetic carcinogens do not chemically react with DNA [17,20,48–52]. In the target tissue(s) of carcinogenicity, MoAs of these types of carcinogens involve molecular or cellular effects, which through secondary mechanisms, can either indirectly result in modification of DNA function or cell behavior [17,48]. For example, epigenetic carcinogens can induce oxidative stress, resulting in oxidative DNA damage [53–55], leading to either neoplastic conversion or stimulation of cell proliferation, thereby facilitating neoplastic development, often from cryptogenic pre-neoplastic cells. Epigenetic carcinogens can also affect gene expression [56,57], leading to neoplastic conversion. Such effects are often specific for rodents (e.g., d-limonene). Epigenetic carcinogens can enhance carcinogenicity of DNA-reactive carcinogens through interactive effects such as neoplasm promotion (e.g., butylated hydroxyanisole).

Due to their lack of direct DNA reactivity, epigenetic carcinogens, in contrast to DNA-reactive agents, are typically negative in genotoxicity assays, even in the presence of bioactivation, unless some artifact, such as extreme cytotoxicity, mediates mutagenicity. To exert their carcinogenicity, epigenetic agents often require prolonged high-level exposures. Their MoA underlies the fact that in limited bioassays they are negative for initiating activity, but may be positive for promoting activity [18].

Epigenetic carcinogens are well established to exhibit NOAELs for the cellular effect underlying their carcinogenicity in animal models [17,19], as discussed for several of the food-borne carcinogens reviewed herein. Accordingly, thresholds are generally accepted for DNA-reactive carcinogens from a risk assessment perspective [47].

2. Risk Assessment of Food-Derived Carcinogens

2.1. Application of Carcinogenicity Data to Human Risk

Two types of carcinogenicity data are used in the assessment of risk: human epidemiologic data and tumor data obtained in testing in rodent models [58]. The former is considered more relevant for a variety of reasons [59–62], although such data are often limited in human exposure information and can be poorly controlled [63].

Animal data are usually more robust, but frequently involve findings whose relevance to humans is uncertain [18,64,65], because the tumorigenic effect involves MoAs operational only in rodents. In addition, rodent studies do not mimic real life human exposures with respect to both the concentration and frequency of exposure. The human diet is also composed of mixture of components, which can both enhance and inhibit carcinogenicity.

Thus, in assessing human risk, two considerations are critical, i.e., the MoA of carcinogenicity and human exposure dose [21,25].

Once a chemical has been identified in a food product and its structure determined, it is possible to undertake an in silico analysis to determine, based on structure-activity relationships, the potential for DNA reactivity [66]. While this works well for relatively simple compounds, with the complexity of many natural products, the subtleties of metabolic activation become increasingly difficult to predict. If sufficient material is available, direct testing for DNA reactivity is the preferred approach [18].

This review focuses primarily on chemicals present in food that have sufficient evidence of carcinogenicity in either humans or experimental animals and which were classified by the International Agency for Research on Cancer (IARC) as either carcinogenic to humans (Group 1), probably (Group 2A) or possibly (Group 2B) carcinogenic to humans [58,67]. IARC also recognizes a third group of substances (Group 3) which lack sufficient evidence to be classified as carcinogenic to humans but nonetheless can have the potential to cause carcinogenicity in animals. Moreover, a variety of chemicals has not yet been characterized as to their carcinogenic risk to humans. Where available, evaluations by other expert groups are cited. Data on classification of carcinogens by government agencies and their carcinogenic potencies (TD_{50}) calculated based on the tumorigenicity findings in rodents are provided in Table 2.

Table 2. Classifications and characteristics of food-borne carcinogens.

Chemical Name	CAS Registry Number	Classification		Carcinogenic Potency (TD_{50}, mg/kg/d) [c]	MoA
		IARC [a]	NTP [b]		
1. Human carcinogens					
Aflatoxins		1	1	0.343 (mouse) 0.0032 (rat)	GTX
Aristolochic acid I	313-67-7	1	1	N/A	GTX
Benzene	71-43-2	1	1	77.5 (mouse) 169 (rat)	GTX
Benzo[a]pyrene	50-32-8	1	2	3.47 (mouse) 0.956 (rat)	GTX
Dioxin (TCDD)	1746-01-6	1	1	0.000156 (mouse) 0.0000235 (rat)	EPI
Dioxin-like compounds (PBCs)		1	N/L	N/A	EPI
Ethylene oxide	75-21-8	1	1	63.7 (mouse) 21.3 (rat)	GTX
Methoxsalen with UV A radiation	298-81-7	1	1	32.4 (rat)	GTX
Processed meat		1	N/L	N/A	GTX
Salted fish		1	N/L	N/A	GTX
2. Likely to be human carcinogens					
Acrylamide	79-06-1	2A	2	3.75 (rat)	GTX
2-Amino-3-methylimidazo[4,5-*f*]quinoline	76180-96-6	2A	2	19.6 (mouse) 0.812 (rat)	GTX
p,p′-Dichlorodiphenyl-trichloroethane (DDT)	50-29-3	2A	2	12.8 (mouse) 84.7 (rat)	EPI
Ethyl carbamate (urethane)	51-79-6	2A	2	16.9 (mouse) 41.3 (rat)	GTX
5-Methoxypsoralen	484-20-8	2A	N/L	N/A	GTX
N-nitrosodiethylamine	55-18-5	2A	2	0.0265 (rat)	GTX
Red meat		2A	N/L	N/A	GTX
2-Amino-3,4-dimethylimidazo[4,5-*f*]quinoline	77094-11-2	2B	2	15.5 (mouse)	GTX
2-Amino-3,8-dimethylimidazo[4,5-*f*]quinoline	77500-04-0	2B	2	24.3 (mouse) 1.66 (rat)	GTX
2-Amino-1-methyl-6-phenylimidazo[4,5-*b*]pyridine	105650-23-5	2B	2	33.2 (mouse) 1.78 (rat)	GTX
Benzophenone	119-61-9	2B	N/L	152 (rat) 379 (mouse)	EPI

Table 2. *Cont.*

Chemical Name	CAS Registry Number	Classification		Carcinogenic Potency (TD$_{50}$, mg/kg/d) [c]	MoA
		IARC [a]	NTP [b]		
Bracken fern		2B	N/L	N/A	GTX
Butylated hydroxyanisole	25013-16-5	2B	2	5530 (mouse) 405 (rat)	EPI
3-Chloro-1,2-propanediol	96-24-2	2B	N/L	117 (rat)	Uncertain
Crotonaldehyde	4170-30-3	2B	N/L	4.2 (rat)	GTX
Cycasin	14901-08-7	2B	N/L	N/A	GTX
1,3-Dichloro-2-propanol	96-23-1	2B	N/L	46.4 (rat)	GTX
Di(2-ethylhexyl) phthalate	117-81-7	2B	2	476 (rat) 484 (mouse)	EPI
1,4-Dioxane	123-91-1	2B	2	204 (mouse) 267 (rat)	Uncertain/EPI
Fumonisin B$_1$	116355-83-0	2B	N/L	6.79 (mouse) 5.75 (rat)	Uncertain/EPI
Fusarin C	79748-81-5	2B	N/L	N/A	Uncertain/EPI
Furan	110-00-9	2B	2	2.72 (mouse) 0.396 (rat)	EPI
Lasiocarpine	303-34-4	2B	N/L	0.389 (rat)	GTX
Methyl eugenol	93-15-2	2B	2	19.3 (mouse) 19.7 (rat)	GTX
Methylazoxymethanol	592-62-1	2B	N/L	N/A	GTX
2-Methylimidazole	693-98-1	2B	N/L	782 (mouse) 868 (rat)	EPI
4-Methylimidazole	822-36-6	2B	N/L	387 (mouse) 317 (rat)	EPI
Methyl isobutyl ketone	108-10-1	2B	N/L	612 (rat)	EPI
Monocrotaline	315-22-0	2B	N/L	0.94 (rat)	GTX
β-Myrcene	123-35-3	2B	N/L	15,400 (rat)	EPI
N-nitrosodiethanolamine	1116-54-7	2B	2	3.17 (rat)	GTX
Ochratoxin A	303-47-9	2B	2	6.41 (mouse) 0.136 (rat)	GTX/EPI
Pickled vegetables		2B	N/L	N/A	GTX
Pulegone	89-82-7	2B	N/L	232 (mouse) 156 (rat)	EPI
Riddelliine	23246-96-0	2B	2	1.97 (mouse) 0.119 (rat)	GTX
Safrole	94-59-7	2B	2	51.3 (mouse) 441 (rat)	GTX
trans,trans-2,4-Hexadienal	142-83-6	2B	N/L	176 (mouse) 62.2 (rat)	GTX
3. *Unknown carcinogenic potential*					
Agaritine [d]	2757-90-6	3	N/L	N/A	GTX
Butylated hydroxytoluene	128-37-0	3	N/L	653 (mouse)	EPI
Carrageenan (native) [d]	9000-07-1	3	N/L	N/A	
Chlorate (sodium salt) [d]	7775-09-9	3	N/L	69.1 (mouse) 0.865 (rat)	EPI
Eugenol [d]	97-53-0	3	N/L	N/A	
Furfural [d]	98-01-1	3	N/L	197 (mouse) 683 (rat)	Uncertain
Hydroquinone	123-31-9	3	N/L	225 (mouse) 82.8 (rat)	EPI
Isatidine [d]	15503-86-3	3	N/L	0.716 (rat)	GTX
d-Limonene [d]	5989-27-5	3	N/L	204 (rat)	EPI
Malondialdehyde	24382-04-5	3	N/L	14.1 (mouse) 122 (rat)	GTX

Table 2. *Cont.*

Chemical Name	CAS Registry Number	Classification		Carcinogenic Potency (TD$_{50}$, mg/kg/d) [c]	MoA
		IARC [a]	NTP [b]		
Patulin [d]	149-29-1	3	N/L	N/A	Uncertain
Ptaquiloside	87625-62-5	3	N/L	N/A	GTX
Quercetin [d]	117-39-5	3	N/L	10.1 (rat)	EPI
Retrorsine [d]	480-54-6	3	N/L	0.862 (rat)	GTX
Senkirkine [d]	2318-18-5	3	N/L	1.7 (rat)	GTX
Sodium saccharin [d]	128-44-9	3	N/L	2140 (rat)	EPI
Symphytine [d]	22571-95-5	3	N/L	1.91	GTX
Zearalenone [d]	17924-92-4	3	N/L	39 (mouse)	EPI
4. Not classified by IARC/NTP					
Daminozide [d]	1596-84-5	N/L	N/L	1030 (mouse) 2500 (rat)	EPI
Estragole	140-67-0	N/L	N/L	51.8 (mouse)	GTX
Genistein [d]	446-72-0	N/L	N/L	27.1 (rat)	EPI
N-methyl-*N*-formylhydrazine [d]	758-17-8	N/L	N/L	1.37 (mouse)	GTX

[a] IARC group 1—carcinogenic to humans; group 2A—probably carcinogenic to humans; group 2B—possibly carcinogenic to humans; group 3—not classifiable as to its carcinogenicity to humans; group 4—probably not carcinogenic to humans. Source—Agents Classified by the IARC Monographs, Volumes 1–131 [67]. [b] 1—known to be a human carcinogen; 2—reasonably anticipated to be a human carcinogen. Source—NTP Report on Carcinogens, 15th Edition [68]. [c] Only rodent data was included for comparison; Source—Lhasa Carcinogenicity Database, https://carcdb.lhasalimited.org/ (accessed on 9 July 2022). [d] Not discussed in this review. EPI, epigenetic modifications; GTX, genotoxicity; N/A, not available N/L, not listed.

In this review, the evidence for human cancer risk from intake of food borne carcinogens of both the DNA-reactive and epigenetic types is discussed. In the assessment of risk from experimental studies, the greatest weight is given to studies with oral administration since that route of intake is most relevant to human consumption. The demonstration of human carcinogenicity is made in epidemiologic studies, although, the absence of an effect can be due to inadequacy of the studies.

2.2. Risk Assessment of DNA-Reactive Rodent Carcinogens

In order to evaluate possible safety concerns arising from presence of carcinogens with DNA-reactive MoA in the diet, many regulatory and advisory agencies, including the European Food Safety Authority Panel on Contaminants in the Food Chain (EFSA CONTAM) and the Joint Food and Agriculture Organization of the United Nations (FAO)/World Health Organization (WHO) the Expert Committee on Food Additives (JECFA) use a margin of exposure (MoE) approach [69,70]. MoE is calculated as a ratio between an appropriate Point of Departure for a tumor response, such as NOAELs obtained from animal studies, and a predicted or estimated human exposure level. A number of considerations should be taken into account when a MoE is derived, including the biological relevance of carcinogenic MoAs to humans [65].

Among DNA-reactive rodent carcinogens, only aflatoxins, aristolochic acid I, benzene, benzo[a]pyrene and ethylene oxide, have been found to be associated with cancer causation in humans (Table 2). Nevertheless, all materials in this class are genotoxic, indicating an MoA that represents human risk [22].

2.3. Risk Assessment of Epigenetic Carcinogens

The contribution and relevance of epigenetic mechanisms produced by dietary factors leading to the development of cancer in humans is uncertain [71], and the best approach to risk assessment of such carcinogens remains a topic of a debate [72]. Nevertheless, at low intermittent exposures (less than 1 mg/day) epigenetic carcinogens are not considered to pose cancer risks to humans [21]. This may reflect the absence in humans of the processes involved in the MoAs in rodents, e.g., d-limonene alpha $2\mu(\alpha_{2\mu})$-globulin nephropathy in male rats leading to kidney tumors [73], or the much lower exposures of humans, e.g., forestomach irritation in rats caused by butylated hydroxyanisole leading to squamous cell carcinoma [74]. Additionally, the fact that epigenetic changes can be reversible could contribute to lack of human risk. Hence, for epigenetic carcinogens NOAELs are used to derive safety values, such as tolerable daily intake (TDI) [21].

3. DNA-Reactive Carcinogens and Related Chemicals Present in Food

This section provides an overview of food-derived carcinogens that typically produce genotoxic and mutagenic effects in vitro and/or in vivo, in particular with appropriate bioactivation. Chemical structures of DNA-reactive carcinogens and related chemicals discussed in this section are provided in Figures 1–5.

3.1. Phytotoxins

A recent inventory of botanical ingredients that are of possible concern for human health because of their genotoxic and carcinogenic properties revealed that the majority of the compounds identified belong to the group of alkenylbenzenes or the group of unsaturated pyrrolizidine alkaloids [75].

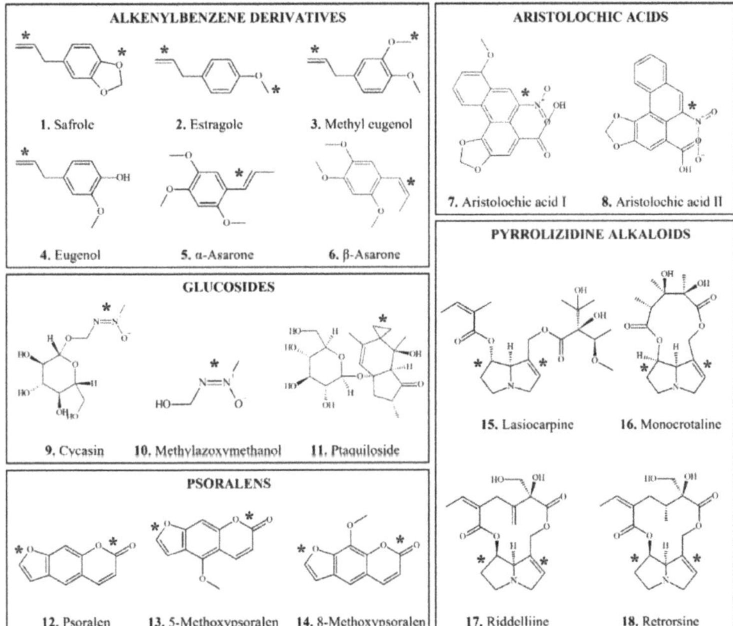

Figure 1. Chemical structures of DNA-reactive carcinogenic phytochemicals and related chemicals present in foods. Asterisks indicate sites of activation.

3.1.1. Alkenylbenzene Derivatives

Alkenylbenzene (AB) compounds (Figure 1(1–6)) are important constituents of herbs and spices such as nutmeg (*Myristica fragrans*), cinnamon (*Cinnamomum verum*), anise star (*Illicium verum*), tarragon (*Artemisia dracunculus*), sweet basil (*Ocimum basilicum*), and sweet fennel (*Foeniculum vulgare*) which are present in the modern food chain mainly as a result of use of these herbs and spices and the use of their essential oils as flavorings [76]. There are two general types of ABs, methylenedioxyallylbenzenes and methoxyallylbenzenes [77,78], with different potential for bioactivation.

ABs are well absorbed following oral intake [79]. Biotransformation pathways are influenced by dose; at lower doses, ring oxidation occurs, whereas, at higher doses, the allyl side chain is also oxidized ultimately through sulfate ester formation to chemically reactive intermediates [78–82]. Polymorphisms in metabolism and lifestyle differences are likely to influence metabolism of these compounds [83].

Some ABs are DNA-reactive, as shown in a study in mice in which several cola beverages were administered in place of drinking water leading to formation of significant levels of DNA adducts in the livers [84]. These adducts were detected in mice treated with extracts of nutmeg or mace, or myristicin (1-allyl-5-methoxy-3,4-methylenedioxybenzene), the major spice constituent of nutmeg and smaller amounts of adducts derived from safrole, a minor constituent of nutmeg.

The ABs discussed in this section have exhibited carcinogenic activity in rodents. Other ABs, such as eugenol (Figure 1(4)) and methyl isoeugenol, with structures not conducive to formation of an electrophile have not been found to be carcinogenic under conditions in which related ABs with structures that form electrophiles were [78,82,85,86].

3.1.1.1. Safrole

Occurrence: Safrole (SAF) (4-allyl-1,2-methylenedioxy-benzene) (Figure 1(1)), the prototype compound of the AB group, is present in *Sassafras*, nutmeg, cinnamon, sweet basil and star anise [77,82,87]. Until 1960, the beverage root beer contained approximately 30 ppm safrole it being made from the root of *Sassafras albidum* which contains about 85% safrole in the essential oil from its root bark [88].

Carcinogenicity: Dietary administration of SAF at up to 5000 mg/kg body weight (bw) to mice and rats caused increases in the incidences of hepatocellular carcinoma or cholangiocarcinoma [78,89,90]. SAF carcinogenicity in mice was strain specific [91]. The hepatocarcinogenicity of 1'-hydroxy SAF metabolites has been also demonstrated [91,92].

Genotoxicity/DNA Binding (Adducts): Genotoxicity tests yielded inconclusive results, being generally negative in vitro, although, some genotoxicity was observed in vivo [78,93,94]. SAF also induced in vitro chromosomal aberrations sister chromatid exchange (SCE), unscheduled DNA synthesis (UDS) and DNA damage [90]. Nevertheless, guanine derivative SAF-DNA adducts were isolated from the livers of multiple species, including rats, mice [82,90,95,96], chicken and turkey [85,86] and humans [97,98].

Biotransformation: SAF undergoes bioactivation primarily on its side chain (Figure 1(1)) to form a hydroxy metabolite which is subsequently sulfated [99,100]. These reactions involve several cytochrome P450 (CYP) enzymes, especially CYP1A2 [83] and sulfotransferase [101]. Genotoxic effects of SAF are likely mediated by metabolites, 1'-hydroxysafrole and 1'-sulfoxysafrole [77]. A number of hydroxylated metabolites have been isolated from human urine [102].

MoA: SAF was considered to be a genotoxic carcinogen, based on its ability to induce formation of DNA adducts [90].

Human Exposure: Humans may ingest SAF with edible spices, such as sassafras, cinnamon, nutmeg, mace, star anise, ginger, black and white pepper, and from chewing betel quid [68]. An Estimated Daily Intake (EDI) for SAF was reported to be 300 µg/person/day [82,90]. JECFA [78] estimated the intake of safrole to be around 879 µg/person/day.

Human Effects: Most of the evidence that SAF may be carcinogenic to humans comes from studies of individuals who chew betel quid, which is known to contain up to 15 mg/g SAF. Thus, SAF-like DNA adducts have been reported in oral squamous cell cancers [97,103] and hepatocellular carcinoma [104] isolated from users of betel quid. Betel quid users are known to have an increased risk for oral cancer development [103].

Risk: SAF was classified by the IARC [87] as possibly carcinogenic to humans (Group 2B) (Table 2). Reflecting these concerns, JECFA did not allocate an Acceptable Daily Intake (ADI) [78]. The direct addition of SAF to food is prohibited in the USA (21 CFR § 189.180) [105] and Europe (Regulation EC No. 1334/2008) [106]. Nevertheless, exposure to SAF continues to occur [107].

3.1.1.2. Estragole

Occurrence: Estragole (1-methoxy-4-(2-propenyl)-benzene) (Figure 1(2)) is a natural constituent of a number of aromatic plants and their essential oil fractions including among others tarragon, sweet basil, sweet fennel and anise star [77,78,82,108]. As a flavoring agent it is used at maximum levels of 50 ppm [79].

Carcinogenicity: Estragole and its 1′-hydroxy metabolite were hepatocarcinogenic in mice when administered in diet at doses up to 600 mg/kg bw for 12 months [91,92]. In mice susceptibility to estragole carcinogenicity was strain specific [91]. In rats, estragole administered by gavage up to 600 mg/kg bw, 5 days/week for 3 months showed evidence of carcinogenic activity, increasing incidences of cholagiocarcinomas and hepatocellular adenomas [109].

Genotoxicity/DNA Binding (Adducts): Genotoxicity and DNA binding of estragole has been reported [78,79,85,96,98,110–112]. However, it was primarily negative in in vitro tests [109], likely due to inadequate bioactivation [25].

Biotransformation: With regard to metabolism, studies in rats indicate that the proximate carcinogen, the 1′-hydroxy metabolite, was produced in minimal amounts at doses in the range of 1–10 mg/kg bw/day [79]. In humans, this metabolite appears to be produced at an even lower rate [113]. These considerations would argue for the existence of a practical threshold for carcinogenic risk in human population [114].

MoA: Formation of DNA adducts and genotoxicity are considered to underly carcinogenicity of estragole [79,108].

Human Exposure: Based on the annual production volume for flavoring substances, the per capita intake of estragole in the US is 5 μg/day [78], while other sources estimated average baseline exposures to estragole from food intake to range from 500 to 5000 μg/day, with an average exposure of 1000 μg/person/day [108].

Human Effects: No evidence for human carcinogenicity of estragole is available [108].

Risk: The Expert Panel of the Flavor and Extract Manufacturers' Association (FEMA), concluded that based on the fact that genotoxic and carcinogenic effects of estragole are dose dependent, present dietary exposures to estragole do not pose a significant cancer risk to humans [79]. However, JECFA indicated that further research is required to assess potential human risk from low-level exposures [78]. Analyses of cancer responses in rodents demonstrated that thresholds for estragole carcinogenicity were well above the levels normally associated with human consumption [114]. Based on the carcinogenic potency, the European Medical Agency (EMA) [108] calculated an ADI for adults of 52 μg/person/day.

3.1.1.3. Methyl Eugenol

Occurrence: Methyl eugenol (ME) (1,2-dimethoxy-4-(2-propenyl)benzene) (Figure 1(3)) occurs in a variety of plants, including nutmeg, sweet basil, tarragon, allspice and pimento [77–79,82,115]. Both ME and eugenol (Figure 1(3,4)) were found in juice from oranges treated on the tree with rind-injuring abscission agents used to loosen the fruit for mechanical harvesting [116]. As a flavoring agent, ME was used in the past at a maximum level of 50 ppm [79]; however, since 2008, ME has been banned for direct addition to foods in Europe (Regulation EC No. 1334/2008) [106].

Carcinogenicity: In a 2-year study, with ME administered to rats and mice of both sexes at doses up to 150 mg/kg bw by gavage, 5 days/week for 105 weeks, chemical-related increases in liver neoplasms occurred in all dosed groups of rats [79,115,117]. In the glandular stomach, mucosal atrophy, an early indication of potential neoplasia, was increased at all doses in rats and malignant gastric neuroendocrine tumors were observed in high dose group in male mice. In rats, gastric neuroendocrine cell hyperplasia was evident at 6 months and neuroendocrine tumors occurred in the high dose group. Other neoplasms with increased incidence included forestomach squamous cell papilloma or carcinoma, renal tubule adenomas, malignant mesotheliomas, mammary gland fibroadenomas and fibromas of the subcutaneous tissue [115,117].

Genotoxicity/DNA Binding (Adducts): ME tested generally negative in genotoxicity tests in vitro and in vivo [115,117]. However, it induced chromosomal aberrations and UDS in vitro [82,115] and formed DNA adducts in human hepatocytes [98] and the livers of rats [118], turkey and chicken fetuses [85,86]. Moreover, correlation between formation of DNA adducts and tumor formation has been shown for ME, and a threshold for tumors was calculated at $10^{20.1}$ molecules/kg/day [119]. Results of PBPK modelling for rats and humans support validity of linear extrapolation of ME tumor data from rodents to humans [120]. However, the application of this log/linear plot for extrapolation is not uniformly accepted [121].

Biotransformation: Similar to other ABs discussed above, ME is bioactivated by CYP1A2 through hydroxylation at the $1'$ position (Figure 1(3)) to produce reactive $1'$-hydroxymethyle ugenol, followed by sulfation. Other metabolic pathways include oxidation of the $2',3'$-double bond to form ME-2,3-oxide and O-demethylation followed by spontaneous rearrangement to form eugenol quinone methide [68,79,115].

MoA: DNA-binding of $1'$-hydroxy ME metabolite most likely underlies MoA for the several types of ME-induced neoplasms [115,122]. In rat liver, ME rapidly induced preneoplastic lesions indicating tumor initiating activity [118]. In addition, based on mechanistic studies of other chemicals that have induced gastric neuroendocrine tumors [123], the mucosal atrophy may have produced decreased hydrochloric acid production which stimulates gastrin production leading to neuroendocrine cell proliferation, and eventually to neuroendocrine neoplasia.

Human Exposure: The overall EDI of ME in US from dietary sources was estimated to be 0.77 μg/kg bw/day, with basil, nutmeg and allspice being primary sources of exposure [79]. JECFA calculated mean per-capita dietary exposure to ME of 80.5 μg/day in US and 9.6 μg/day in Europe [78]. The total dietary intake of food containing ME was calculated to be 66 μg/kg bw/day for regular consumers [122].

Human Effects: No epidemiological studies evaluating evidence of human carcinogenicity from ME are available [68,115].

Risk: ME has been classified by IARC [115] as possibly carcinogenic to humans (Group 2B) (Table 2) based on sufficient evidence for carcinogenicity in animals. While FEMA concluded that present exposures to ME do not pose significant risk to human health [79], estimated MoE based on the dose–response modelling ranges from 100 to 800, suggesting that the dietary intake of ME is of high concern [122]. In 2018, the FEMA Expert Panel removed ME from the FEMA Generally Recognized as Safe (GRAS) list, citing the need for additional data to clarify the relevance of DNA adducts formed by ME in humans [124].

3.1.1.4. α- and β-asarone

Occurrence: Propenylic phenylpropenes, α- and β-asarone ((E)-/(Z)-1,2,4-trimethoxy-5-prop-1-enylbenzene) (Figure 1(5,6)), are constituents of essential oils (e.g., calamus oil) which are present in certain plants such as *Acorus* spp. and *Aarum* spp. and are used as flavoring agents [125,126]. *β-asarone* content varies with the source of the plant; Indian plant oil is approximately 75–95% β-asarone, whereas European is 5–10% [127,128].

Carcinogenicity: When fed to rats for 2 years at doses up to 2000 mg/kg bw, *β-asarone* induced leiomyosarcomas of the small intestine of males but not females [126–128]. Feeding Indian calamus oil at 0.05% and greater produced intestinal tumors in male and female rats, while feeding European calamus oil induced leiomyosarcomas and additionally, liver neoplasms at 1% and greater. Hepatocarcinogenicity of α- and *β-asarone* was also reported following oral administration or intraperitoneal injections to mice [91,129].

Genotoxicity/DNA Binding (Adducts): In the in vitro genotoxicity assays, α- and *β-asarone* produced conflicting results, while in vivo mutagenicity data is limited [128]. Nevertheless, positive results in the in vitro mutagenicity assays were obtained in the presence of bioactivating systems or in metabolically competent cell lines, including human Hepa-G2 cells [94,126,129–131]. Asarones also induced SCE, UDS and DNA breaks in vitro [126,132]. Both isomers produced DNA adducts in rat hepatocytes [133] and in avian embryos [86].

Biotransformation: In rat hepatocytes, the major metabolite of asarones was 2,4,5-trimethoxycinnamic acid, which was not genotoxic [131]. In rat and human liver microsomes epoxide-derived side-chain diols were the major metabolites, and the major bioactivation pathway for *α-asarone* was considered to be 3′-hydroxylation of propenylic side chain by CYP1A2, while for β-asarone, epoxidation by CYP3A4 prevails [126,134–136]. O-demethylation catalyzed by CYP1A1, 2A6, 2B6, and 2C19 was a minor reaction.

MoA: The mutagenicity and DNA binding of side chain epoxides formed during bioactivation of asarones suggests that this intermediate is responsible for carcinogenic effects, at least in the liver [129,136]. The MoA for induction of the intestinal tumors remains undetermined.

Human Exposure: The primary source of human exposure to asarones is through the consumption of alcoholic beverages such as bitters, liqueurs and vermouths, in which levels of calamus oil have been detected up to 0.35 mg/kg [128]. While no regulations for the use of *α-asarone* are currently in place, limits of 0.1 and 1 mg/kg are set for *β-asarone* in food and alcoholic beverages, respectively [126]. Nevertheless, some alcoholic drinks can contain up to 4.96 mg/kg of *β-asarone* [128]. Based on limited British data, maximum EDI for *β-asarone* is approximately 115 µg/day or 2 µg/kg bw/day [127,128].

Human Effects: No epidemiological studies investigating association of asarones with human cancer risk has been reported; however, some in vitro studies indicate anticarcinogenic properties of *β-asarone* [137,138].

Risk: JECFA and the Scientific Committee on Food (SCF) concluded that the existence of a threshold cannot be assumed for *β-asarone* due to its genotoxicity [127,128]. Accordingly, an ADI for nutritional exposure could not be derived. Committees recommended that calamus oil used in foods should have the lowest practicable levels of *β-asarone*. Calamus oil and its extracts are prohibited from use in the USA (21 CFR § 189.110) [139].

3.1.2. Aristolochic Acids

Occurrence: Aristolochic acid I (AAI) (8-methoxy-6-nitrophenanthro[3,4-d]-1,3-dioxole-5-carboxylic acid) (Figure 1(7)) is one of a group of about 14 AAs known to be present in plants belonging to the family *Aristolochiaceae* (Birthwort family). Species known to contain AAs include *A. contorta, A. debilis, A. fangchi,* and *A. manshuriensis* [15,68,140–142].

Carcinogenicity: AAI, either purified or as a mixture with AAII (Figure 1(8)), was carcinogenic in rats and mice after oral exposure producing tumors predominantly in the forestomach and in the kidneys [15,68,140,143,144]. Other target organs of carcinogenicity include lung, uterus and lymphatic system in female mice and urinary bladder, thymus, small intestine and pancreas in rats. In addition, extracts from *Aristolochia* plants, *A. manshuriensis* and *A. fructus* induced forestomach and kidney tumors in rats when administered orally [15].

Genotoxicity/DNA Binding (Adducts): AAI and AAII, have been found to be genotoxic in vitro and in vivo [141,145,146] and to form DNA adducts in vitro and in rodent tissues [141,147–151], as well as in humans urothelial tissues of patients with Chinese herb nephropathy, Balkan endemic nephropathy or urothelial cancer [152,153]. The major AA-specific DNA adducts were 7-(deoxyadenosin-*N*6-yl)aristolactam and 7-(deoxyguanosin-*N*2-yl)aristolactam [141]. Adducts of deoxyguanosine and deoxyadenosine were found in animal studies in both target (forestomach) and non-target tissues (glandular stomach, liver, lung, and bladder). In addition, AAs can bind to codon 61 of the *ras* oncogene and to purines in the *p53* tumor suppressor gene [68,141,153].

Biotransformation: Bioactivation of AAI occurs by nitro reduction in the presence of NAD(P)H quinone oxidoreductase and CYP1A2 [154] leading to formation of a nitrenium ion which, by rearrangement reactions, forms adducts on both deoxyguanosine and deoxyadenosine, the latter being biologically more stable [155].

MoA: Covalent binding to DNA and resulting mutagenicity is the predominant MoA of AAI carcinogenicity [15,68]. The most frequently observed mutation is a single *TP53* mutation (A to T transversion), consistent with the presence of persistent AAI-adenine adducts in DNA of exposed patients [141,153,156].

Human Exposure: AAs are present in herbal products and several teas made from *Aristolochia* plants [68,157] and in wild ginger used by North American Indians [158]. A combined EDI for AAI and AAII was calculated to be 1.7×10^{-3}–30 μg/kg bw/day [142].

Human Effects: Consumption of herbal supplements containing AAs has been linked to nephropathy [159] and cases of urothelial cancer [160,161]. Among patients with AA nephropathy, the rate of urothelial cancer is much higher compared to the prevalence of transitional-cell carcinoma of the urinary tract [68].

Risk: Based on the evidence that AA-specific DNA adducts and *TP53* mutations have been described in humans, IARC [15] upgraded classification of AAI from probable human carcinogen (Group 2A) to human carcinogen (Group 1) (Table 2). MoEs for kidney tumor formation calculated based on the rodent data were below 10,000 indicating risk to humans [142]. The US Food and Drug Administration (FDA) advised consumers in 2001 to discontinue use of botanical products that contain AA; however, exposure to AA continues despite its known hazards [162].

3.1.3. Glucosides

3.1.3.1. Cycasin

Occurrence: *Cycasin* (methylazoxymethanol-D-glucoside) (Figure 1(9)) is a glucoside produced by the cycad nut, which grows in most tropical climates [89,163]. The amount of cycasin ranges from 0.02% to 2.3% [89].

Carcinogenicity: With oral administration, cycasin induced neoplasia in mice, rats, hamsters, guinea pigs, and monkeys mainly in liver, kidney and colon [89,164]. A metabolite of cycasin, methylazoxymethanol (MAM) (Figure 1(10)), has also been shown to induce hepatocellular carcinomas and tumors in other organs, including kidneys and intestinal tract, in nonhuman primates [164], and colon carcinogenesis in rodents [165–168].

Genotoxicity/DNA Binding (Adducts): Cycasin was genotoxic after removal of a sugar residue to yield the aglycone, MAM, which is an alkylating agent [89,169–173]. MAM produced DNA adducts, specifically O^6-methylguanine and $N7$-methylguanine, in vitro and in vivo in rats and guinea pigs [174–178].

Biotransformation: Bioactivation of cycasin to MAM occurs by hydroxylation of the methyl group, a reaction which is catalyzed by CYP2E1 [179]. Interspecies differences in metabolic bioactivation of cycasin to MAM was suggested to underly different susceptibility to its carcinogenicity [178].

MoA: The genotoxic metabolite MAM was shown to target cellular processes involved in carcinogenesis [180].

Human Exposure: Human exposure to cycasin is limited since cycad nuts are no longer used as a source of starch. Cycasin can, however, contaminate improperly prepared flour, as has occurred in Guam, where concentrations of cycasin ranged from 0.004 to 75.93 μg/g [175].

Human Effects: Human ingestion of cycad plant toxins has been associated with neurodegenerative disorders in inhabitants of Guam [181,182], but no appreciable increase in cancer mortality was evident at 2 to 7 years after heavy intake [89]. Cases of acute toxicity from high exposures have been reported but all with complete initial recovery [183].

Risk: IARC [89] classified both, cycasin and MAM, as possibly carcinogenic to humans (Group 2B) (Table 2).

3.1.3.2. Ptaquiloside and Bracken Fern

Occurrence: *Ptaquiloside* (Figure 1(11)) is an unstable norsesquiterpene glucoside of the illudane type [184,185]. It is present in bracken fern (*Pteridium aquilinum*), in wild species and in products made from fronds at concentration ranges of 6300 ± 520 and 44 ± 3 μg/g, respectively [186]. High quantities of ptaquiloside, in various studies ranging from 0.0006 to 0.0058 µg/mL, were found in the milk from farm animals that consume diet containing bracken fern [187–190].

Carcinogenicity: Ingestion of bracken fern by cattle and sheep has been reported to cause cancers of the esophagus and urinary bladder [191,192]. Feeding of bracken fern to rats and mice induced intestinal and bladder cancers [193], which was initially attributed to the content of quercetin [194], but ptaquiloside was subsequently demonstrated to be the carcinogenic constituent [190,195]. With oral administration, ptaquiloside induced tumors of mammary glands, ileum and urinary bladder in female rats [196,197] and oral squamous cell carcinomas in HPV16-transgenic mice [198].

Genotoxicity/DNA Binding (Adducts): Ptaquiloside was genotoxic in bacterial mutagenicity assays and in the rat hepatocyte primary culture DNA-repair assay [199–201]. In addition, it produced chromosomal aberrations in Chinese hamster lung cells and human mononuclear blood cells [202,203], and DNA damage in human gastric epithelial cells [204]. Formation of DNA adducts was reported in upper gastrointestinal tract of mice that were fed bracken fern [205,206] and in target tissue of carcinogenicity, ileum, in rats injected with ptaquiloside intravenously [207]. DNA adducts formed after exposure to bracken fern were distinctly different from the adducts formed by ptaquiloside [206].

Biotransformation: Bioactivation of ptaquiloside is not enzyme mediated, and involves conversion to aglycone, ptaquilosin, which, under alkaline conditions undergoes aromatization resulting in a reactive metabolite, bracken dienone [185]. Dienone has an ability to alkylate DNA, forming adducts primarily on *N3* position of adenine and *N7* position of guanine [185].

MoA: DNA alkylation of adenine bases with subsequent DNA depurination and breakage leading to induction of mutations, in particular to activation of *H-ras* proto-oncogenes and frameshift mutations of *p53* gene [192,204,208,209] is thought to be the main mechanism underlying ptaquiloside-related carcinogenicity. Other potential MoAs, including clastogenicity and aneugenicity, as well as alteration of monocyte function, TNFα expression and cell proliferation, cannot be excluded [202,207,210].

Human Exposure: Estimation of human consumption of ptaquiloside with cow's milk resulted in intake levels ranging from 1.75 to 13.4 mg/day [211]. Some populations in Japan, Brazil and Canada can also consume cooked or salted bracken crosiers [185,186].

Human Effects: No study on human carcinogenicity of ptaquiloside is available. However, in areas where bracken fern is consumed, there is a correlation between the consumption of ptaquiloside-contaminated milk and increased risk of esophageal or stomach cancer [184,212,213].

Risk: While IARC recognizes bracken fern as possibly carcinogenic to humans (Group 2B), it considers ptaquiloside to be unclassifiable as to its carcinogenicity (Group 3) (Table 2) based on limited evidence [193]. Nevertheless, genotoxicity and mutagenicity of ptaquiloside, as well as some epidemiological evidence of potential carcinogenicity raises concerns for human safety [184,190,202].

3.1.4. Psoralens

Occurrence: *Psoralen* (7H-furo[3,2-g][1]benzopyran-7-one) (Figure 1(12)) is a furocoumarin which is naturally present in several plants, notably *Psoralea corylifolia*, celery, parsley and in all citrus fruits, including bergamot orange peel, whose oils are used as flavors [214–217]. In citrus-flavored beverages, the highest levels of psoralens, 29 and 24 mg/L, were found in bergamot juice and home-made limoncello, respectively [218]. Levels of psoralens in celery varied, depending on when it was harvested, from 26 to 84 µg/g fresh weight [216]. Psoralens are widely used in the photochemotherapy of various skin conditions in humans [219].

Carcinogenicity: Psoralen derivatives, 5- and 8-methoxypsoralen (methoxsalen) (Figure 1(13,14)), produced skin tumors in mice in the presence of UV A light, even with oral administration [15,68,193,220,221]. In male rats, tumors of Zymbal glands and kidneys were also reported after oral gavage with methoxsalen [222].

Genotoxicity/DNA Binding (Adducts): Psoralen can be photoactivated to DNA cross linking reactant which exhibit genotoxicity and photomutagenicity [15,193,222–224]. Intercalation occurs predominantly on pyrimidine bases of DNA, mainly with thymine, which leads to inhibition of DNA synthesis, in addition, psoralens have high affinity for uridine bases on RNA [219].

Biotransformation: Metabolism of psoralen involves hydroxylation of phenyl ring, hydrogenation and hydrolysis of the unsaturated lactone ester, and oxidation of the furan ring to generate epoxide or/and γ-ketoenal intermediates [225,226]. These reactions are catalyzed by CYP3A4, CYP1A1, CYP1A2 and CYP2B6 [226–230]. Incubation of psoralen with liver microsomes from different species, including humans, dogs, non-human primates and rodents, demonstrated similarity of metabolites produced by humans and dogs, while metabolic capabilities of rat and monkey microsomes were the closest to those of human microsomes [225].

MoA: Photochemical genotoxicity and mutagenicity are most likely responsible mechanism of psoralens carcinogenicity [15,68,193]. Other potential MoAs may involve oxidative damage [219].

Human Exposure: Dietary exposure to psoralens occurs mainly from either limes, with estimated per capita exposure of 1300 μg/day [231], or grapefruit juice, with dietary exposure re-estimated to be in the range of 548 to 2237 μg/day [232].

Human Effects: Human exposures have so far been mainly associated with photodermatitis due to occupational contact [216,233,234] with only one report of phototoxicity following ingestion [235]. Human carcinogenicity studies relating to oral psoralens have only been made with patients receiving photodynamic therapy (PUVA) [15,68,236,237] and no attempt to extrapolate these positive results to normal populations has been undertaken. One study identified an association between high citrus consumption and melanoma [238]. This study, however, did not specifically assess risk from psoralen consumption.

Risk: IARC classifies methoxsalen with UV A radiation as a human carcinogen (Group 1) and 5-methoxypsoralen as probable human carcinogen (Group 2A) [15,193] (Table 2). Further investigation to establish potential health risks of dietary intake of psoralens in humans is warranted [219].

3.1.5. Pyrrolizidine Alkaloids

Occurrence: *Pyrrolizidine alkaloids* (PAs) are heterocyclic compounds (Figure 1(15–18)), most of which derive from esters of basic alcohols known as necine bases [239–243]. Close to 500 PAs have been identified [244]. They occur widely in flowering plants, and consequently in honey, and are present in herbal teas from many countries [241,244–246].

Carcinogenicity: Over 20 PAs are established to be carcinogenic in experimental animals [89,240,247–250]. Oral administration of lasiocarpine (Figure 1(15)), monocrotaline (Figure 1(16)), riddelliine (Figure 1(17)), and retrorsine (Figure 1(18)) produced tumors primarily in the liver of rats [68,242,251,252]. Other target organs of PA carcinogenicity include lung, kidney, skin, bladder, brain and spinal cord, pancreatic islets and adrenal gland.

Genotoxicity/DNA Binding (Adducts): Many PAs are genotoxic and mutagenic in vivo and in vitro following metabolic activation [240,242,245,248,250,252–256]. Several PAs, including retronecine-type PAs riddelliine [257] and monocrotaline [258], are known to form DNA crosslinking and DNA adducts in vivo [250,253,259]. Levels of DNA adducts was reported to closely correlate with the carcinogenic potency of some PAs [25,250,253,257].

Biotransformation: The bioactivation of PAs is mediated by CYPs, in particular, CYP3A4, which catalyze hydroxylation of the necine base, followed by dehydration to form the corresponding dehydropyrrolizidine derivatives [240,245,246,249,260]. The dehydropyrrolizidine derivatives (i.e., pyrrole metabolites) have been reported to be strong alkylating agents and have been linked to tumor initiation [261,262]. Similarities have been observed between metabolic activation of several PAs in vitro by human and rat microsomes [260,263]; however, certain differences in formed metabolites were reported [264,265].

MoA: Genotoxicity and acute toxicity of PAs are the most likely mechanisms involved in the carcinogenicity of these compounds [250,253].

Human Exposure: In the majority of developed countries, human exposure to PAs, which mainly occurs from consumption of contaminated foods of animal origin, grains and plant-derived foods, including herbs, spices and teas, is low, ranging from 0.035 to 0.214 µg/kg bw/day [240,249,266]. Mean total dietary intakes of PAs were estimated to be 0.019 µg/kg bw/day for children and to 0.026 µg/kg bw/day for adults [246], with the highest dietary exposure, ranging from 0.0013 to 0.26 µg/kg bw/day, resulting from herbal tea consumption, while consumption of honey has been calculated to result in chronic dietary exposure ranging between 0.0001 and 0.027 µg/kg bw/day [240,266]. In Europe, levels of PAs is various foods is limited up to 400 µg/kg for herbal infusions [249].

Human Effects: In humans, PAs are known to be teratogenic and to act as abortafacients, and exposure can be potentially lethal [267]. Hepatotoxicity of PAs in humans has been also reported [268]. There is a need for epidemiologic studies on acute and long-term effects of PAs.

Risk: IARC [89,247] classified lasiocarpine, monocrotaline and riddelliine as possibly carcinogenic to humans (Group 2B) (Table 2), even though there is no epidemiological evidence to indicate that intake of these substances, even at toxic levels, present a carcinogenic risk [246,269]. Other PAs, namely hydroxysenkirkine, isatidine, jacobine, retrorsine, seneciphylline, senkirkine and symphytine, were not classifiable as to their carcinogenicity to humans (Group 3) (Table 2) [89,247]. EFSA and JECFA concluded that based on calculated MoEs, there is a potential concern for human health, in particular for high-level long-term consumers [240,246,248,270]. Genotoxic and carcinogenic potentials of PAs indicates priority for risk management and warrants effort to continue reduction of PAs content in herbal products [271].

3.2. Mycotoxins

Mycotoxins are produced by fungi that can contaminate a variety of crops pre- and post-harvest, and which are associated with several diseases in animals and humans. Mycotoxins cannot be completely eliminated from food by food processing procedures, including thermal processing [272]. Of major concern are the mycotoxins aflatoxins, ochratoxin A and fumonisins [273,274].

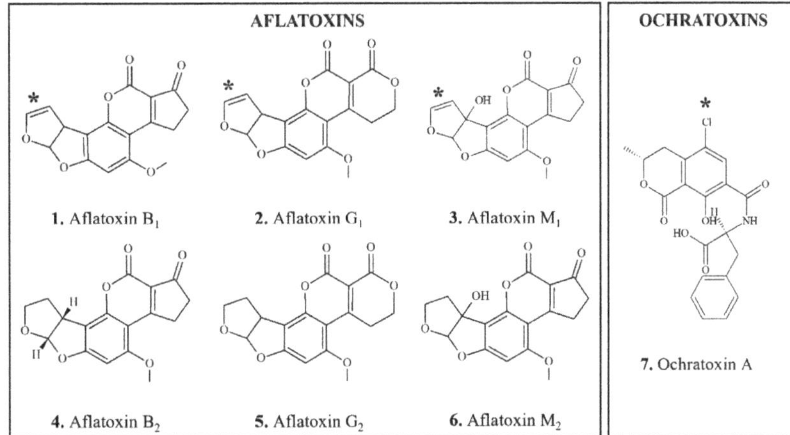

Figure 2. Chemical structures of DNA-reactive carcinogenic mycotoxins and related chemicals present in foods. Asterisks indicate sites of activation.

3.2.1. Aflatoxins

Occurrence: Aflatoxins (AFs) (Figure 2(1–6)) are mycotoxins formed by various strains of the fungus, *Aspergillus flavus*, and are present in contaminated foods, particularly corn and peanuts [68,275,276]. Food levels of AFs are often expressed as total AFs [276], which is useful for monitoring purposes. AFB_1 has a tetrahydrocyclopenta[c]-furo [3′,2′:4,5]-furo [2,3-h]chromene skeleton with oxygen functionality at positions 1, 4 and 11 (Figure 2(1)).

Carcinogenicity: AFB_1 is the most highly carcinogenic AF [277] and one of the most potent carcinogens [278,279] (Table 2). Oral administration of AFB_1, including as AFs mixtures, produced sufficient evidence for carcinogenicity in multiple species [275]. Specifically, AFB_1-induced increases in the incidences of hepatocellular or cholangiocellular carcinomas were observed in rats, hamsters, marmosets, tree shrews, and monkeys; in addition, increase were observed in renal cell carcinomas and colon tumors in rats, lung adenomas in mice as well as osteogenic sarcoma, gallbladder tumors and adenocarcinoma of the pancreas in monkeys [68,275]. AFB_2 (Figure 2(4)), AFG_1 (Figure 2(2)), and AFM_1 (Figure 2(3)) also produced liver tumors in experimental animals, but their potency was significantly lower compared to that of AFB_1 [68,280]. No evidence for carcinogenicity of AFG_2 have been reported [275].

Genotoxicity/DNA Binding (Adducts): AFB_1 is genotoxic in vitro and in vivo, producing mutagenic, aneugenic and clastogenic effects [275,281,282], as well as DNA adducts in multiple species [275,283–286], with the AFB_1-$N7$-guanine adduct being assumed to be pro-mutagenic and pro-carcinogenic [278,287–289]. The initial AFB_1-DNA adduct is unstable in vivo; it either depurinates to give an AFB_1-guanine residue which can be detected in the urine, or forms a more stable ring opened formamidopyrimidine derivative measurable in cellular DNA. AFB_1-DNA adducts show high correlation with tumor incidence, but no threshold for hepatic DNA adduct formation was reported [25]. AFB_1 also elicited DNA repair synthesis in cultured human hepatocytes [290] and γH2AX induction in human cell lines derived from hepatoblastoma, renal cell adenocarcinoma, and epithelial colorectal adenocarcinoma [291]. DNA adduct formation has been also reported after AFG_1 and AFM_1 exposures [292,293].

Biotransformation: The genotoxic and carcinogenic AF, AFB_1, is metabolically activated predominantly by CYP3A4 oxidation at the 8–9 positions (Figure 2(1)) to form an AFB_1-8,9-epoxide, which is highly reactive and binds to the $N7$ position of guanine residues in DNA [287–289,294]. There is abundant evidence that in humans AFB_1 is bioactivated by CYP1A2, 2B6, 3A4, 3A5, 3A7 and GSTM1 enzymes [281]. Ramsdell and Eaton [279] reported that mouse and monkey microsomes formed AFB_1-8,9-epoxide at higher rates compared to rat and human; however, at lower substrate concentrations, conversion to AFB_1-8,9-epoxide increased with rat and human microsomes, but not with mouse of monkey microsomes. Thus, the authors attributed interspecies differences in carcinogenic potency of AFB_1 to differences in patterns of epoxide formation. AFG_1 and AFM_1, which also have a double bond at the 8,9-position (Figure 2(3,4)), can form epoxides; however, they are less DNA-reactive compared to AFB_1-8,9-epoxide [281]. Non- or weakly carcinogenic AFs, e.g., AFB_2, AFG_2, and AFM_2, lack the double bond in the 8–9 position (Figure 2(4–6)) [276] and, except in the duck, are not metabolized to detectable levels of AFB_1 [295]. CYP3A4 and CYP1A2 can also metabolize AFB_1 to hydroxylated metabolites, AFM_1 and AFQ_1. Roebuck and Wogan [296] reported that AFQ_1 was the principal metabolite produced by monkey, human, and rat liver, whereas duck liver produced mainly chloroform-insoluble derivatives. Monkey, human, and mouse liver also produced AFP_1, which was not observed in duck and rat. The authors noticed that duck, monkey, and human livers were most active, each metabolizing approximately 80% of available substrate in half an hour. In comparison, activity of rat and mouse livers was lower, each metabolizing from 15 to 20% of substrate. No consistent pattern of metabolism that could explain interspecies differences in susceptibility to AFB_1 carcinogenicity was detected. Detoxication of AFB_1 occurs predominantly via conjugation with glutathione (GSH), and extent of this reaction differs among species, with mouse showing the highest and humans having the lowest conjugation rates [281].

MoA: Covalent binding of AFB_1-8,9-epoxide to *N7* of guanine in DNA is considered to be the primary MoA of AFB_1 carcinogenicity [275,278,281,289,297]. The adduct is believed to induce mutations of *TP53* gene in humans [275,276]. In addition, AFB_1 epoxide reacts with serum proteins, including albumin. All have been used as biomarkers to assess AFB_1 exposure [298,299]. Such studies have led to the clear association of AFB_1 exposure and hepatocellular carcinoma, particularly in those infected with hepatitis B virus [278,281,300]. This is believed to be due to enhanced liver cell proliferation with hepatitis [300]. A strong correlation of urinary adducts indicative of AFB_1 exposure, notably AFB_1-*N7*-guanine, serological markers of hepatitis B infection, and liver cancer risk exists [301]. Induction of oxidative stress, immunomodulation and epigenetic modification also play a role in carcinogenicity of AFB_1 [278,281].

Human Exposure: Overall, exposure to AFB_1 results from ingestion of foods contaminated with *Aspergillus flavus*. Total EDI to AFs ranges from 0.0001 to 0.049 µg/kg bw/day in developing countries and is generally less than 0.001 µg/kg bw/day in developed countries [276]. In parts of the worlds where *Aspergillus* contamination of food is prevalent, AFB_1 occurs in such foods at significant levels [278,302]. In the United States, consumption of food contaminated with up to 20 ppb AFB_1, mainly corn and peanuts, is permitted [303], with the exception of milk, which is required to contain less than 0.5 ppb [304], corresponding to about 30 µg/day for a 70 kg adult. Obviously, high exposures are occurring in parts of the world where crop contaminations are not well controlled and accordingly, the cancer risk is much higher.

Human Effects: In humans, exposure to AFs is associated with increased risk of liver cancer, particularly in association with concurrent hepatitis B [68,275,281,299,300].

Risk: IARC [275] considers AFs to be carcinogenic to humans (Group 1) (Table 2). JECFA estimated the cancer potency for exposure to AFB_1 per 100,000 population at 0.001 µg/kg bw/day, and recommended that efforts to reduce aflatoxin exposure continue [276]. The Committee also noticed that AFM_1 will generally make a negligible (<1%) contribution to aflatoxin-induced cancer risk for the general population. EFSA estimated that MoEs, which range from 5000 to 29 for AFB_1 and from 100,000 to 508 for AFM_1 exposures, respectively, raise a concern for human health [281].

3.2.2. Ochratoxin A

Occurrence: Ochratoxin A (OTA) (*N* [(3R)-5-chloro-8-hydroxy-3-methyl-1-oxo-3,4-dihydro-1H-isochromen-7-yl]carbonyl-L-phenylalanine) (Figure 2(7)), is a mycotoxin produced by a single *Penicillium* and several *Aspergillus* fungal species [305–307]. The ochratoxins are pentaketides, consisting of a dihydroisocoumarin coupled to 8-phenylalanine (Figure 2(7)) and, unusually for natural products, OTA is chlorinated. OTA is formed in improperly stored foods which have been produced mainly in Europe and Canada, including cereals, beans, ground nuts, oleaginous seeds, meat and wine [306].

Carcinogenicity: In several strains of mice, OTA fed in the diet induced kidney neoplasms, including carcinomas, at a concentration of 40 mg/kg bw, and liver neoplasms at 1 mg/kg bw. When administered by gavage to rats it induced renal tumors starting at 70 µg/kg bw, [305,308]. Male rats were considerably more susceptible than females. A feature of the renal toxicity of OTA is formation of karyomegalic nuclei in the tubular epithelia, predominantly in the corticomedullary zone [309].

Genotoxicity/DNA Binding (Adducts): OTA was consistently negative in studies assessing mutagenicity in *Salmonella typhimurium*, both with and without exogenous metabolic activation. In contrast to bacteria, however, overall results from genotoxicity tests in mammalian cell systems provide some evidence for a weak genotoxic activity of OTA [305,307,310–312]. It elicited DNA repair synthesis in cultured rat and mouse hepatocytes at cytotoxin doses, increased DNA strand breaks levels and mutagenicity in target tissue, kidney, in rodents [307,312]. Controversy exits over whether OTA reacts directly with DNA. OTA did not form DNA adducts in the kidneys of male rats when measured using radiolabeled OTA and accelerator mass spectrometry [313], while others obtained positive results in isolated DNA and cell culture by dehalogenation and redox reactions analyzed by nucleotide ^{32}P-postlabeling (NPL) [314]. Mantle et al. [315] was able to identify a small amount (20–70 adducts per 10^9 normal nucleotides) of a single DNA adduct in the kidneys of rats using refined NPL methodology. These conflicting data have been reviewed [306,310,312]. EFSA concluded that while formation of covalent OTA-DNA adducts remains controversial, OTA mutagenicity cannot not simply be a consequence of oxidative DNA damage [307].

Biotransformation: OTA is characterized by rapid absorption and distribution, but slow elimination and excretion [307]. The major metabolite of OTA forms as a result of hydrolyses of amide bond between phenylalanine and dihydroisocoumaric acid. OTA also undergoes oxidative dichlorination in the presence of CYPs, generating electrophilic quinone, which can be further reduced to hydroquinone metabolite and excreted in urine, as has been shown in rats and humans [307,312]. In addition, peroxidase enzymes are involved in oxidation of OTA to electrophilic phenolic radical, which is believed to cause oxidative stress. Radical and benzoquinone intermediates formed during metabolism of OTA can covalently bind to DNA, generating C-bound C8-dG adducts [306].

MoA: The definitive MoA for carcinogenicity of OTA remains unclear, and most likely involves a combination of mutagenicity and increased reactive oxygen species (ROS) level leading to oxidative DNA damage [306,310,312,316,317]. Alternatively, an epigenetic MoA for renal carcinogenicity has been postulated to be a combination of inhibition of histone acetyltransferase, producing mitotic disruption leading to increased cell proliferation and genetic instability [310]. The mitotic disruption may be the basis for karyomegaly observed in rodent kidneys [318]. Thus, MoA other than DNA reactivity are possible for OTA. The pathogenesis of the renal tumors in male rats probably does not involve an α2u-globulin MoA [319]. Moreover, sex and strain differences are suggestive of biotransformation being important [320].

Human Exposure: In European Union, dietary exposures range between 0.00064 to 0.0178 μg/kg bw/day across all age groups [307]. The EDI values for OTA calculated from food products range from 1×10^{-7} to 0.0252 μg/kg bw/day [306]. In areas where contamination occurs, biomarkers of OTA exposure are measurable in human blood, urine and milk [306].

Human Effects: OTA is suspected to be the main etiologic agent for human Balkan endemic nephropathy and the associated urinary tract tumors [311,321].

Risk: IARC [305] classified OTA as possibly carcinogenic to humans (Group 2B) based on sufficient evidence for carcinogenicity in experimental animals (Table 2). JECFA concluded that maximum levels of at 5 or 20 μg/kg in contaminated cereal grains would be unlikely to have an impact on dietary exposure to OTA, and established a provisional tolerable weekly intake of 0.112 μg/kg bw [322]. EFSA estimated that MoE for chronic neoplastic effects ranged from 22,615 to 815, indicating possible health concern for high-level consumers and breastfed infants [307].

3.3. Carcinogens Formed during Food Processing

While many carcinogens associated with food processing are generated during heating (see next section), some can be formed through nonthermal process or during storage [5]. Such chemicals include benzene, cholopropanols and ethyl carbamate.

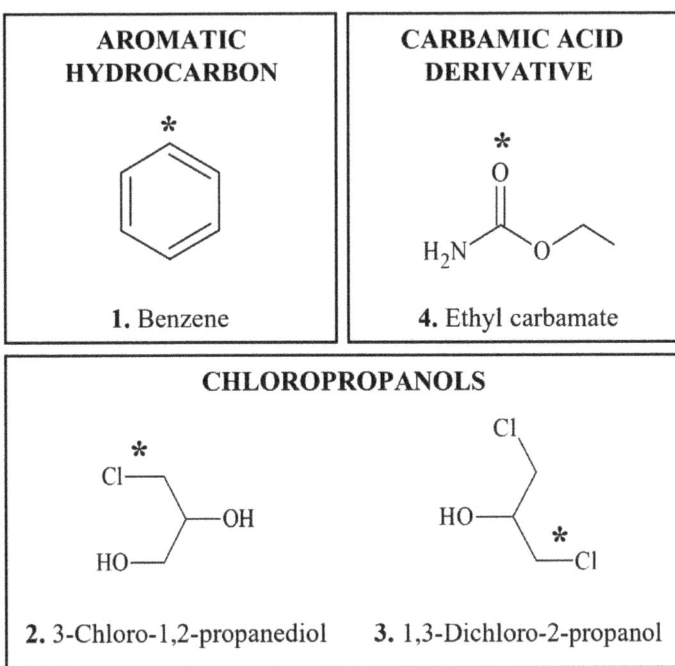

Figure 3. Chemical structures of DNA-reactive carcinogens formed during food processing and related chemicals present in foods. Asterisks indicate sites of activation.

3.3.1. Benzene

Occurrence: Benzene (BZ) (Figure 3(1)) is present at low levels in a wide variety of foods [323], in particular processed products, as well as in drinking water and soft beverages [324,325]. Highest concentrations of BZ, up to 2100 ppb, have been reported in eggs, haddock, beef and butter [326]. It is formed from a reaction between sodium or potassium benzoate and ascorbic acid, which are often used as food preservatives [327], this can result in BZ concentrations ranging from 0.001 to 0.038 µg/g in some products such as eggs [323]. In addition, BZ can be introduced to foods from packaging materials [328]. In soft beverages BZ is present in small amounts, below 5 µg/kg [329,330].

Carcinogenicity: BZ produced leukemia, and other neoplasms, in rats and in mice with inhalation exposure [68,324,331,332]. With oral administration, BZ at doses up to 200 mg/kg bw is reported to produce oral cavity and skin tumors in rats, Zymbal gland carcinoma in rats and mice, malignant lymphoma, lung cancer, preputial gland carcinoma and cancer of the mammary gland in mice [324,333].

Genotoxicity/DNA Binding (Adducts): The genotoxicity data are mixed, although, DNA damage and chromosomal aberrations were often reported in animals and occupationally exposed humans [324,331,334–339]. In addition, BZ metabolites form DNA adducts in vitro, in mice and human hematopoietic cells [324,340–344]. Importantly, hydroquinone (HQ), a major metabolite of BZ and a food component, does not form DNA adducts in vivo [343,345,346]. *p*-Benzoquinone is a possible candidate for the genotoxic metabolite of BZ [347–349], although other metabolites have been considered, including indirect mechanisms involving oxidative DNA damage [324,331,334,350].

Biotransformation: BZ is biotransformed by CYPs, mainly CYP2E1, to benzene oxide, which is further metabolized by various pathways to phenol, HQ, catechol and *trans,trans*-muconic acid in experimental animals and humans [324,331,351].

MoA: The exact molecular mechanism which BZ exerts its carcinogenicity remains to be elucidated, especially by the oral route. Oxidative DNA damage, genotoxicity, aneugenicity and clastogenicity, as well as interference with cell cycle and immunosuppression may be involved [324,334,336,350].

Human Exposure: Predominant exposure to BZ in general population occurs through air, rather than foods [68,324,326,352,353]. Similarly, while detectable BZ levels are present in human milk, infant exposure occurs predominantly from the air [354]. A dietary exposure to BZ through various sources was estimated to be in the range of 0.003 to 0.05 µg/kg bw/day [326,355]. In Canada, intake level of BZ from food and water was estimated at approximately 10 µg/day [356]. Currently, no limits for BZ are established in foods and beverages, while in water it ranges from 1 to 10 ppb in different countries [325].

Human Effects: In humans, BZ is associated with increased risk of leukemia, myelodysplastic syndrome and other hematopoietic disorders with airborne occupational exposures [324,351,357]. Some studies also report association between BZ exposure in occupational settings and cancers in other organs, including respiratory, gastrointestinal, urinary, central nervous systems and skin [324]. No data on carcinogenicity via food exposure are available [326].

Risk: BZ is recognized by IARC [324] as carcinogenic to humans (Group 1) (Table 2). JECFA concluded that based on known effects, BZ in not suitable for use as an additive in food [358]. Using probabilistic modelling, Cheasley et al. [356] estimated that lifetime excess cancer risk associated with BZ dietary intake was 35 per million. Nevertheless, MoEs calculated based on the estimated dietary intake did not indicate human risk from dietary exposures [325,326,355,359], however more studies are warranted.

3.3.2. Chloropropanols

Occurrence: *3-Chloro-1,2-propanediol* (CP) (Figure 3(2)) and *1,3-dichloro-2-propanol* (DP) (Figure 3(3)) are formed during the acid hydrolysis of vegetable proteins through the reaction of chloride ions with triglycerides [115,360–362]. Several are present at low levels ranging from 9.6–82.7 µg/kg, in various foods, most notably refined oils, acid-hydrolyzed proteins and soy sauces; however, some sauces contain as high as 18 mg/kg or 876 ppm [115,276,361,363–365]. Chloropropanols can be also found in paper-based food contact materials [366,367].

Carcinogenicity: In a two-year bioassay in rats, CP produced increases in kidney, Leydig cell, and mammary neoplasms administered at doses up to 400 ppm (29.5 mg/kg bw/day) in drinking water [115,276]. DP produced increases in neoplasms in the liver, kidney and tongue in rats at doses up to 30 mg/kg bw/day [115,365,368,369].

Genotoxicity/DNA Binding (Adducts): CP was genotoxic in some in vitro assays, but not in vivo [115,370,371]. In contrast, DP was clearly genotoxic in vivo and in vitro, with or without metabolic activation [115,276,361,365,368]; however, formation of DNA adducts has not been reported. Genotoxicity of DP was attributed to formation of epoxide intermediate [368].

Biotransformation: CP is metabolized by alcohol dehydrogenase to chlorolactic acid, while DP is metabolized by CYP2E1, to cytotoxic metabolites, including 1,3-dichloroacetone [115,361].

MoA: No clear MoA has been established for carcinogenicity of CP and DP. There is evidence that CP induces tumors by a hormonally mediated and/or cytotoxic MoA [362,372]. Oxidative damage has been also implicated [373]. Nevertheless, genotoxic MoA cannot be excluded for CP and DP [115].

Human Exposure: Mean EDI for CP was calculated to range from 0.2 to 3.8 µg/kg bw/day in adults and to be 1.3 µg/kg bw/day in children [276,362]. Mean EDI for DP was estimated to be 7 µg/person/day from soy sauce consumption, and 0.1 µg/person/day from dietary sources other than soy sauce [365,369].

Human Effects: No adequate data are currently available to assess the potential carcinogenicity of the chloropropanols in humans [115].

Risk: IARC [115] classifies CP and DP as possibly carcinogenic to humans (Group 2B) (Table 2). JECFA [276] set a provisional maximum tolerable daily intake (PMTDI) of 4 μg/kg bw/day for CP, while EFSA [362] established a much lower TDI of 2 μg/kg bw/day. JECFA concluded that no TDI can be estimated for DP based on its effects [365,369]; however, based on calculated MoE, DP in diet was considered to be of low concern for human health [361].

3.3.3. Ethyl Carbamate (Urethane)

Occurrence: *Ethyl Carbamate* (EC) (Figure 3(4)), also referred to as urethane, is a fermentation product formed from the reaction of ethanol and carbamyl phosphate [374,375]. It is present as a natural trace constituent in various alcoholic beverages and in fermented food items, including cheese, bread, yogurt, soy sauce and fermented soybean products [374,376–379]. Mean concentrations of EC in some spirits, particularly in stonefruit brandies, have been measured in a range of 4 to 122 μg/kg (or 0.1 to 1400 μg/L), while in foods lower concentrations, ranging from 0.2 to 16 μg/kg, were observed [374,376–378]. EC content in foods can also increase with thermal processing [380].

Carcinogenicity: With oral administration to mice, EC up to 600 ppm induced mainly liver, lung, harderian gland, skin, mammary gland, ovaries, blood vessels and forestomach neoplasms [377,381,382]. In rats, oral administration of EC resulted in an increased incidence of Zymbal and mammary gland carcinomas [68,377].

Genotoxicity/DNA Binding (Adducts): Genotoxicity and clastogenicity of EC has been demonstrated in vitro and in vivo [377,382–384]. The formation of etheno DNA adducts was reported in the liver [385], lung [386] and other organs [387] in rats and mice.

Biotransformation: EC is metabolized predominantly by CYP2E1 to reactive metabolites, vinyl carbamate and vinyl carbamate epoxide [377,382,388–390]. Formation of vinyl carbamate was also reported after incubation of human liver and lung microsomes with EC [391,392], suggesting similarities in metabolism of EC between humans and rodents.

MoA: Formation of reactive metabolite and consequent transition mutations in *Kras* oncogene is considered as major mechanism involved in tumorigenesis of EC [377,388,393]. Other potential MoAs may involve proinflammatory signaling, mitochondrial dysfunction and ROS formation [376].

Human Exposure: Under normal dietary habits, excluding alcoholic beverages, the EDI was in the range of 0.01 to 0.02 μg/kg bw/day, or an average of 0.015 μg/kg bw/day [378,394]. Mean EC intake from diet and alcoholic beverages rages from 0.015 to 0.065 μg/kg bw/day [374].

Human Effects: No epidemiologic studies are currently available to assess hazardous effects of EC in humans [377,378].

Risk: IARC [377] evaluated EC as probably carcinogenic to humans (Group 2A) (Table 2). JECFA [378] and EFSA [379] concluded that exposure to EC in food, excluding alcoholic beverages, poses low concern for human health. Nevertheless, health concern exists for consumers of alcoholic beverages, and mitigation measures should the implemented to reduce levels of EC in certain spirits. Schlatter and Lutz [395] calculated a virtually safe dose for EC of 0.02 to 0.08 μg/kg bw/day, which represents negligible risk to human health. Nevertheless, consumption of alcoholic beverages, in particular stone-fruit distillates, increases cancer risk to approximately 0.01%.

3.4. Heat-Generated Carcinogens

Heating and combustion of organic materials (e.g., tobacco) is well established to generate carcinogens. With respect to foods, three major types are acrylamide, heterocyclic amines and polycyclic aromatic hydrocarbons [396].

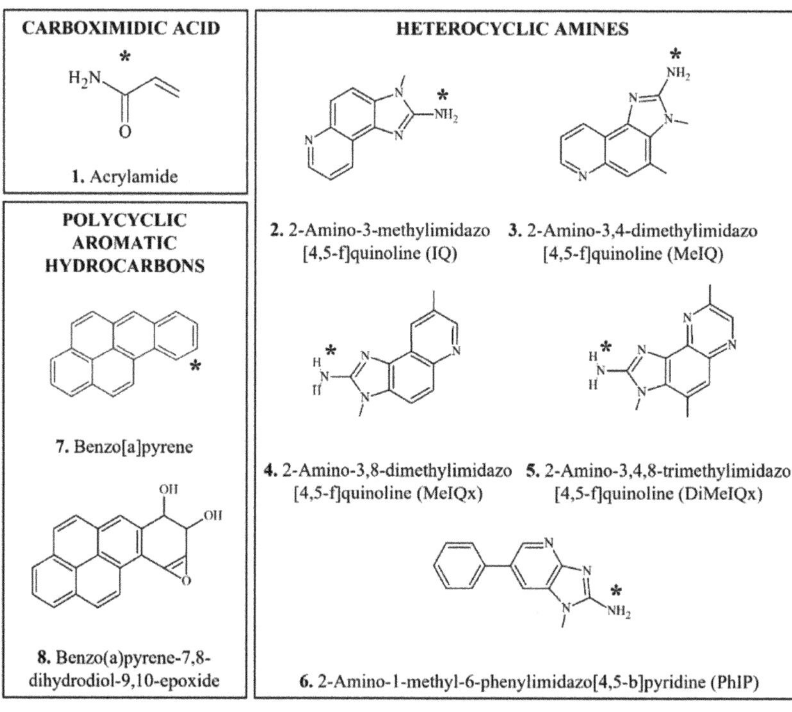

Figure 4. Chemical structures of DNA-reactive heat-generated carcinogens and related chemicals present in foods. Asterisks indicate sites of activation.

3.4.1. Acrylamide

Occurrence: Acrylamide (AC) (propen-2-amide) is an unsaturated amide (Figure 4(1)), which is formed in heated foods, especially those rich in carbohydrates, by a reaction of reducing sugars with asparagine [397–399] and consequently is present in a variety of food products, notably baked or fried foods [68,400–403]. Efforts to reduce AC formation in foods have been active. For example, asparaginase has been proposed for use in food manufacture to convert asparagine to aspartic acid, thereby depleting one of the precursors of AC formation [403–405].

Carcinogenicity: AC was tested for carcinogenicity in rats by oral administration. In males, it increased the incidences of peritoneal mesotheliomas found in the region of the testis and of follicular adenomas of the thyroid gland. In females, thyroid follicular tumors, mammary gland tumors, glial tumors of the central nervous system, oral cavity papillomas, uterine adenocarcinomas and clitoral gland adenomas were increased. In four screening bioassays in mice, AC, given either orally or intraperitoneally, increased both the incidence and multiplicity of lung tumors in all experiments [402,406,407]. In a two-year rodent carcinogenicity bioassay [408], AC produced clear evidence of carcinogenic activity in rats and mice. Specifically, administration of up to 50 ppm AC in drinking water resulted in increased incidences of thyroid gland and heart tumors in rats of both sexes, of malignant mesotheliomas and cancer in the pancreatic islets in male rats and of cancers in the clitoral gland, liver, mammary gland, skin, and mouth or tongue in female rats. Increased incidences of cancer in the harderian gland, lung, and stomach were observed in male and female mice, in addition, female mice also had increased incidences of cancer in the mammary gland, skin, and ovary.

Genotoxicity/DNA Binding (Adducts): AC is genotoxic and forms DNA adducts in target as well as non-target tissues of carcinogenicity, including the liver, lung, kidney, leucocytes and, testes in mice and in the liver, brain, thyroid, leukocytes and testes or mammary gland in rats [37,409]. In a dose–response study, a 0.1 µg/kg bw was established as a NOAEL for DNA adduct formation [410], suggesting plausibility of thresholds for carcinogenicity. In rats dosed with AC at 2 or 15 mg/kg bw for up to 28 days, DNA synthesis was increased in target tissues, but not in a non-target tissue, the liver [411]. In mice, administration of AC at 7.5, 15 and 30 mg/kg bw/day by gavage for 28 days produced a significant increase in micronuclei formation in the peripheral blood and an increase in *gpt* mutation frequencies in testes and lungs [412]. In addition, AC induced DNA strand breaks in various tissues in rats and mice [413–415].

Biotransformation: AC has two potentially reactive sites, a conjugated double bond and an amide group (Figure 4(1)) [399]. In vivo, AC is epoxidized at its double bond to glycidamide (GA) [416]. Both, AC and GA, are reactive and while there is some evidence for genotoxicity of AC [417], GA appears to be the DNA-reactive metabolite of AC [418]. GA reacts readily with DNA [407,419,420], forming purine adducts [410,421]. In rats, following administration of AC at 3 mg/kg bw, the majority of metabolites excreted in urine were AC-GSH conjugates, while a substantial proportion of the remainder consisted of two GA-derived mercapturic acids [422]. GA and dihydroxypropionamide were not detected at this dose level. The metabolism of AC in humans was investigated in a study in which male volunteers were administered 3 mg/kg AC orally. At 24 h, urine contained a third of the administered dose, and the majority of metabolites was derived from direct conjugation of AC with GSH [422]. GA, dihydroxypropionamide and one unidentified metabolite were also detected in urine. This study indicated both similarities and differences in the metabolism of AC between humans and rodents.

MoA: The carcinogenicity of AC may result from a combination of DNA reactivity and increased cell proliferation in target tissues. However, non-genotoxic MoA, such as alterations of calcium signaling, might be more relevant for tumorigenicity of AC, since evidence of its genotoxicity are weak [25]. In addition, some studies provide evidence of oxidative DNA damage by AC, as well as epigenetic modifications which might be involved in tumorigenesis [423–425]. Several possible MoAs have been reviewed [409].

Human Exposure: Dietary exposure to AC results from consumption of foods preparation of which involves cooking at high temperatures (e.g., French fries and potato chips), other exposure routes may involve dermal contact and inhalation. EDI of AC in several European populations was estimated to range from 0.4 to 1.9 µg/kg bw/day [425], with average dietary intake of 0.5 µg/kg bw/day in adults worldwide [426].

Human Effects: Numerous epidemiologic studies have examined the relationship of dietary consumption of AC and risk for cancers of the kidney, large bowel, urinary bladder, oral cavity and pharynx, esophagus, larynx, breast, and ovary [406,427]. No evidence of any association was found. Individual susceptibility, however, may be related to genetic polymorphisms in enzymes involved in activation and detoxification of AC [428]. Two cohort mortality studies were conducted among workers exposed to AC. The first showed no significant excess of cancer but was of small size, short duration of exposure and short latency. In the other study, in one plant in the Netherlands and three in the US, a nonsignificant increase was found in deaths from pancreatic cancer, but there was no trend with increasing exposure. In a prospective study, increased risks were found for postmenopausal endometrial and ovarian cancer with increasing dietary AC intake, particularly among never-smokers. Risk of breast cancer was not associated with AC intake [429].

Risk: Based on sufficient evidence for carcinogenicity in experimental animals and inadequate evidence in humans, IARC [402] classified AC as "probably carcinogenic to humans" (Group 2A) (Table 2). JECFA [403] concluded that estimated MoE for cancer events of 310 for general population and 78 for population with high exposure indicates a human health concern. EFSA [425] also concluded that although the epidemiological studies have not demonstrated AC to be a human carcinogen, MoE indicates a concern for neoplastic effects based on animal evidence. However, based on the dietary intakes [400,425], exposures to AC are mainly at or below those considered acceptable for AFB_1, which is a more potent carcinogen in animal models (Table 2). It would therefore seem unlikely that a significant risk exists for the general population. Furthermore, an expert panel convened by the German Federal Agency of Consumer Health Protection and Veterinary Medicine opined that while AC was a genotoxic carcinogen, it was likely to show a non-linear dose–response curve with respect to carcinogenic effect [430]. In support of this, Baum et al. [431] have shown that at concentrations added to human blood which are comparable to those achieved by intake from food, AC preferentially reacts with protein components of blood, and is 'quenched' without affecting DNA in lymphocytes.

3.4.2. Heterocyclic Amines

Occurrence: Heterocyclic Amines (HCAs) are generated in meats cooked at high temperatures which produce protein decomposition [4,305,396,432]. A variety of different HCAs have been identified, representing several structural types reflecting the chemistry of their formation. The major subclass of HCAs, aminoimidazoazaarenes, which is the most abundant in food, includes *2-amino-3-methylimidazo [4,5-f]quinoline* (IQ) (Figure 4(2)), *2-amino-3,4-dimethylimidazo [4,5-f]quinoline* (MeIQ) (Figure 4(3)), *2-amino-3,8-dimethylimidazo [4,5-f]quinoline* (MeIQx) (Figure 4(4)), *2-amino-3,4,8-trimethylimidazo[4,5-f]quinoline* (diMeIQx) (Figure 4(5)), and *2-amino-1-methyl-6-phenylimidazo [4,5-b]pyridine* (PhIP) (Figure 4(6)) [433].

Carcinogenicity: HCAs are potent multisite carcinogens in several species [68,305]. Specifically, oral administration of IQ, MeIQ and MeIQx to rats or mice caused increases in the incidences of tumors in the liver, small and large intestine, forestomach, lung, Zymbal gland, skin, mammary and clitoral glands, as well as lymphomas and leukemias. In the case of MeIQx, hepatic neoplastic lesions were observed only at high doses, indicating possibility of thresholds [45]. In rats, carcinogenicity of IQ in several target tissues was potentiated with high fat diet [434]. IQ was also shown to induce hepatocellular carcinomas in cynomolgus monkeys after chronic dosage of 10 or 20 mg/kg for 5 days/week. PhIP administered orally caused lymphoma in male rats and in mice of both sexes. Moreover, several cancers associated with the Western diet, specifically carcinoma of prostate gland, adenocarcinoma of the small intestine and colon and mammary gland carcinoma were observed in rats orally exposed to PhIP. Neonatal mice are also extremely sensitive to carcinogenic HCAs [68,305]. The order of carcinogenic potencies of selected HCAs is as follows: IQ > MeIQ > MeIQx > PhIP (Table 2).

Genotoxicity/DNA Binding (Adducts): HCAs are potent genotoxic mutagens in various systems in vitro and in vivo [68,305,435,436], including human cells [437]. IQ, MeIQ, MeIQx and PhIP were shown to induce DNA damage and chromosomal aberrations, SCE, micronucleus formation and UDS. HCAs have been shown to form DNA and protein adducts in vitro and in vivo in various species, including humans [305,438–443]. Data from studies investigating formation of PhIP DNA adducts reported that in human tissues at dietarily relevant exposures DNA adducts form with greater efficiency compared to rodents [436]. There was a liner correlation between the dose and DNA-binding of some HCAs in the liver [25].

Biotransformation: The bioactivation of HCAs involves mainly *N*-hydroxylation, usually by CYP1A2 [444], and subsequent acetylation by *N*-acetyltransferase type 2 [445], leading to formation of a reactive nitrenium ion, as with other aromatic amines. Nitrenium ion primarily binds to C8 atom of guanine bases [432,436]. Genetic polymorphism of these enzymes in humans might play a role in susceptibility to genotoxicity and carcinogenicity of HCAs [446]. For example, high levels of DNA damage were observed in cell cultures with rapid acetylation [447] and individuals with rapid acetylator phenotype are believed to have higher risk of certain cancers after exposure to HCAs compared to slow-acetylators [432,445].

MoA: Carcinogenicity of HCA most likely results from formation of DNA adducts which lead to mutations in proto-oncogenes and tumor-suppressor genes, including *K-ras*, *Ha-ras*, *Apc*, *β-catenin*, and *TP53* [68,432].

Human Exposure: Human intake of HCAs is estimated to range from 0.001 to 0.017 µg/kg bw/day [448] with some intakes as high as 1900 ng [449]. The average lifetime time-weighted consumption of HCAs for US population is estimated to be approximately 0.009 µg/kg/day, with PhIP comprising two thirds of the intake [450].

Human Effects: HCAs are reasonably anticipated to be human carcinogens [68,432]. They have been implicated in causing cancers of the breast [451], colon and rectum [449], stomach and esophagus [452], and lung [453]. Estimated consumption by humans of at least one HCA, PhIP, was associated with increased levels of DNA adducts in breast [454] and prostate tissues [455] of patients with cancers at those sites. While consumption of cooked or grilled meat has been associated with various types of cancers, the data do not definitively implicate HCAs as the causative component of these associations [449].

Risk: IARC [305] classifies HCAs as either possible (Group 2A) (e.g., IQ) or probable (Group 2B) (e.g., MeIQ, MeIQx, PhIP) human carcinogens (Table 2). An upper-bound risk for US population from dietary exposures to HCAs was estimated to be 28,000 cancers, with PhIP accounting for almost half (46%) of the total risk [448]. The consumption of cooked meat and fish was the primary contributor to cancer risk in humans. Nevertheless, currently no regulations targeting reduction of exposure to HCAs exist [68].

3.4.3. Polycyclic Aromatic Hydrocarbons

Occurrence: Polycyclic Aromatic Hydrocarbons (PAHs) is a group of compounds composed of two or more fused aromatic rings, which are present in many foods, either from deposition from air pollution or formed during cooking processes such as with char broiling of meats [275,378,456,457]. Heating of food above 350–400 °C leads to formation of PAHs, notably, the prototypical PAH, *benzo[a]pyrene* (BaP) (Figure 4(7)). BaP is found in smoked foods, charcoal-broiled steaks and ground meats [458–460]. The highest levels of BaP are found in grilled meats, at up to 4 ng BaP/g of cooked meat, [460].

Carcinogenicity: A variety of PAHs produced sufficient evidence of carcinogenicity in experimental animals [68,461]. In particular, BaP produced tumors in multiple species, including mouse, rat, hamster, guinea pig, rabbit, duck, newt, and monkey, following exposure by many different routes [68,275]. When administered orally, either via gavage or with diet, BaP at dosages up to 14 mg/kg bw/day increased incidences of tumors in lymphoid and haematopoeitic systems and in several organs in mice, including the lung, forestomach, liver, oesophagus and tongue [68,275,378]. Administration of BaP to rats by gavage for two years, produced liver tumors and tumors of forestomach at the lowest dose of 10 mg/kg bw and higher [462]. PAH mixtures, in particular creosote oils, coal-tar pitches, shale oils, anthracene oils, and certain bitumens have been shown to induce skin tumors in mice upon topical application [461,463].

Genotoxicity/DNA Binding (Adducts): PAHs, including BaP, are mutagenic and genotoxic in a variety of test systems, both in vitro and in vivo [47,275,378]. Reactive metabolites of PAHs can covalently bind to DNA, predominantly at the *N2* position of desoxyguanosine [378]. A linear correlation between DNA adduct formation and mutagenicity was reported, providing evidence against the existence of thresholds for BaP effects [25].

Biotransformation: BaP and other PAHs with appropriate structures are bioactivated to oxides and dihydrodiols, which in turn are oxidized to diol epoxides, in a multi-step, inducible pathway involving CYP and epoxide hydrolase microsomal enzyme systems [464,465]. The dihydrodiol epoxide intermediate(s) (e.g., benzo(a)pyrene-7,8-dihydrodiol-9,10-epoxide (Figure 4(8))) form stable and depurinating DNA adducts, which are primarily responsible for the mutagenic and carcinogenic action of BaP and other PAHs [47,275]. PAHs lacking the structural basis for formation of epoxides which can open and generate relatively stable carbonium ions with appropriate conformations of their hydroxyl groups are at most weakly carcinogenic [466]. For PAHs with lower ionization potential, the one-electron oxidation pathway, which results in the formation of unstable DNA adducts, might also be important [47].

MoA: The formation of DNA adducts by reactive metabolites, oxides and diol epoxides, is considered to be an initiating event in the development of tumors caused by BaP and some other PAHs [47,68]. These adducts were shown to induce mutations in oncogenes and tumor suppressor genes, such as *K-ras* and *TP53*, in humans and rodents [275]. However, due to poor quantitative relationship between levels of DNA adduct in target tissues and tumor formation, other factors involved in MoA should be considered. For example, induction of oxidative stress [467], immunosuppression [468], alterations of cell cycle [469] and epigenetic modifications [470] might also contribute to carcinogenic effects of PAHs.

Human Exposure: Food contaminated with PAHs, either from environmental sources or during processing and cooking, is the major source of exposure in non-smokers [47,457]. JECFA [378] reported EDI for BaP to be in a range of 0.0006 to 2.04 μg/kg bw/day, and for other PAHs EDI varies from 0.0001 to 0.015 μg/kg bw/day. Intake of PAHs in children is approximately double of the intake in adults [378]. EFSA [47] identified cereals and cereal products as well as seafood as the two highest contributors to the dietary exposure to PAHs. The European Union legislation (Regulations EC No. 835/2011 and No. 2020/1255) provides specific regulations for maximum levels of PAHs in various foods, which, depending on the product, ranges from 1 to 10 μg/kg for BaP and from 1 to 50 μg/kg for all PAHs [457,471,472]. In US, no maximum limits for PAHs in foods has been established, with exception of maximum permissible level of BaP in bottled water of 0.0002 mg/L [68,457].

Human Effects: No epidemiological studies on association between exposure to the individual PAHs and human cancers have been conducted, and data on the carcinogenic effects of PAHs in humans is available only for mixtures [68,275,461,463]. Thus, studies of smokers and consumers of certain meat products uncovered evidence of the carcinogenicity of BaP and other PAHs in humans. Lung cancer has been shown to be induced in humans by cigarette smoke, and by exposures to roofing tar and coke oven emissions, all of which contain mixtures of PAH [68]. A recent case control study reported an association between exposure to BaP in the diet and an increased risk for colorectal adenoma [473].

Risk: IARC [275] concluded that BaP is a human carcinogen (Group 1) based on the biological plausibility of mechanism of carcinogenicity in humans (Table 2). In a study of human intake in Korea, a possible excess cancer risk ascribed to PAHs using the cancer potency of BaP was calculated to be 2.3 cases per 100,000 persons [474]. Based on the MoE of 25,000 (mean) and 10,000 (high), and a human exposure estimate of 0.004 (mean) to 0.01 (high) μg/kg bw/day, JECFA [378] concluded that the estimated oral intakes of PAHs were of low concern for human health. Similarly, EFSA [47] established that MoE of 17,900 for BaP is indicative of low concern for consumer health at the average estimated dietary exposures; however, for high-level consumers potential concern exists.

3.5. Carcinogens Formed Exogenously and Endogenously

Consumer exposure to carcinogens is multifactorial and, in some cases, in addition to exogenous sources might involve endogenous formation of hazardous chemicals as a result of metabolic reactions [6]. Such endogenous exposures can significantly contribute to total exposures and complicate risk assessment.

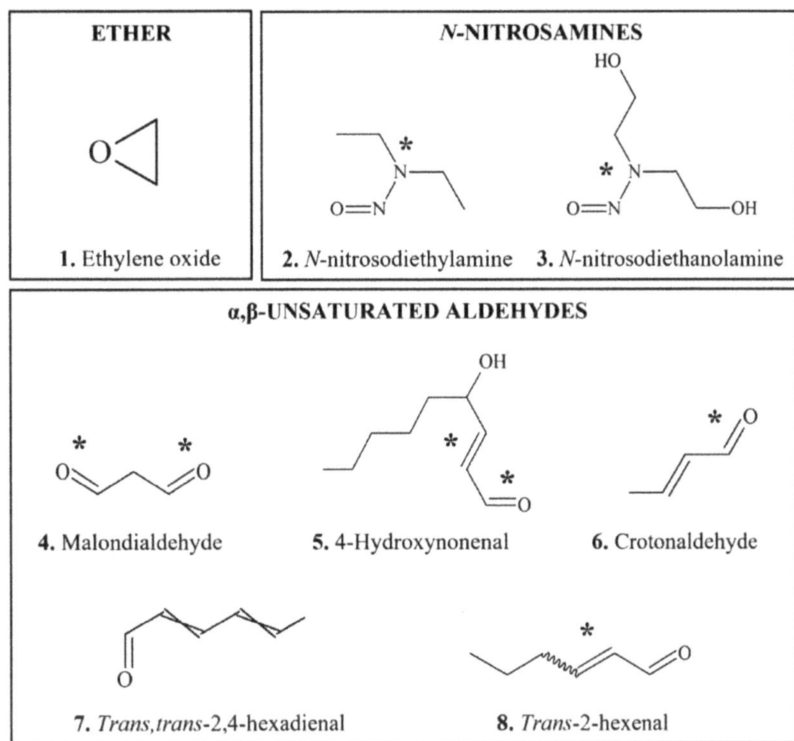

Figure 5. Chemical structures of DNA-reactive carcinogens and related chemicals formed endogenously and exogenously present in foods. Asterisks indicate sites of activation.

3.5.1. Ethylene Oxide

Occurrence: *Ethylene oxide* (EtO) (Figure 5(1)) is primarily used as an intermediate in the production of ethylene glycol [275]. However, in some countries, including USA, Canada and India, EtO has been also used as a fumigant for its disinfectant properties, and hence it may be present as a residue on contaminated foods including spices, nuts, sesame seeds, dry fruits and vegetables, milk powder and cereal, at various concentrations exceeding 0.05 mg/kg and even reaching 1800 mg/kg in some herbal teas [475–477]. Due to known hazardous effects, use of EtO in food production in Europe is prohibited (Regulation (EC) No 1107/2009) [478], and maximum residue levels are established between 0.02 and 0.1 mg/kg (Regulation (EC) No 396/2005) [479]. Such regulations led to a recent recall of a variety of products containing EtO contaminated sesame seeds or locust bean gum, including bread, sauces, ice cream and other fermented milk products [480,481]. In addition to exogenous sources, EtO can be also formed endogenously as a result of lipid peroxidation reactions, metabolic activity of microbiota or following metabolism of ethylene [482–484]. Endogenous levels of EtO in humans were estimated to range from 0.13 to 6.9 ppb [485].

Carcinogenicity: EtO is a multisite carcinogen in rodents, with target organs including the hematopoietic system in mice and rats, the lung, Harderian gland, mammary gland, and uterus in mice, and the brain and mesothelium in rats [68,275,486–489]. Oral administration of up to 30.5 mg/kg bw of EtO through gavage twice a week produced an increase in the forestomach squamous cell carcinoma in rats [490].

Genotoxicity/DNA Binding (Adducts): EtO exhibited genotoxic and mutagenic effects, albeit weak, in experimental systems in vitro and in vivo, moreover, cytogenic alterations and DNA damage, including chromosomal aberrations, SCE, *hprt* mutations, micronucleus formation, and DNA single-strand breaks were reported in peripheral blood of humans with occupational exposures [275,487–489,491–495]. As a direct alkylating agent, EtO covalently binds to DNA, predominantly at the *N7* position of guanine [487,496–499].

Biotransformation: EtO can be either hydrolyzed, spontaneously or in the presence of microsomal epoxide hydrolase, to ethylene glycol with subsequent conversion to oxalic acid, formic acid and carbon dioxide, or conjugate with GSH to form mercapturic and thiodiacetic acids [275,488,500]. Several studies implied that polymorphism in human GST genotype, in particular GSTT1, might underly the difference in susceptibility to adverse effects of EtO [501–504].

MoA: While carcinogenicity of EtO is attributed to its genotoxicity and mutagenicity [275,489], formation of *N7*-guanine DNA adducts are not likely to contribute to the carcinogenic MoA. These adducts are not pro-mutagenic and are steadily repaired not leading to accumulation of abasic sites [497,505]. Hence, mutagenicity of EtO was attributed to minor adducts, $N3$-2-hydroxyethyladenine and O^6-2-hydroxyethylguanine [506]. Due to high repair rate of DNA adducts formed by alkylating agents, existence of thresholds for EtO genotoxicity it plausible [489,507]. Several studies also attempted to use dose–response data for genotoxicity endpoints to estimate safe exposure levels to EtO [508,509].

Human Exposure: Exposures to EtO occur predominantly through inhalation, while dietary exposures are negligible [275]. Thus, EDI from all food sources amounts to 10 µg/person/day, which is lower than endogenous production of EtO by bacteria (15–20 µg/day) [510]. Average per capita consumption of EtO with spices was estimated to range from 0.21 µg/person/day in New Zealand to 1.6 µg/person/day in US [511].

Human Effects: There is no evidence of adverse health effects related to the consumption contaminated foods, mainly due to the rapid breakdown of EtO. While some epidemiological studies report association between occupational exposure (primarily through inhalation) to EtO and higher risk of lympho-haematopoietic [512] and breast cancers [513], the evidence is limited and not supported by meta-analysis studies [68,275,487,489,514–516].

Risk: Despite only limited evidence of carcinogenicity in humans, IARC [275] classified EtO as human carcinogen (Group 1) (Table 2), based on the mechanistic evidence of its genotoxicity in workers. While the NTP and the US EPA [68,517] reached the same conclusion, systemic analyses of carcinogenicity and toxicity data conducted by several authors suggests that such classification grossly overestimates the risk of EtO [489,516]. One study estimated that cancer risk from consumption of EtO with contaminated spices was negligible [511]. The German Federal Institute for Risk Assessment also established that EtO intake at or below 0.037 µg/kg bw/day should be considered of low concern [518].

3.5.2. *N*-Nitroso Compounds

Occurrence: *N-Nitroso compounds* (Figure 5(2,3)) are formed in food and in vivo at acidic pH by nitrosation of secondary and tertiary amino compounds [68,519–521]. Their formation also occurs in vivo at neutral pH by nitric oxide generated by bacteria converting nitrates and nitrites, or macrophages or endothelial cells metabolizing arginine [522–524]. *N*-Nitroso compounds have been found in over 200 foods, including fruits, vegetables, beverages, meats and cereals, and can be present in drinking water [456,519,521,525,526]. Concentrations of nitrosamines in food depend on the method, time and temperature of cooking or fat composition [527]. Thus, mean levels of *N*-nitroso compounds were 0.5 µmol/kg of fresh meat and over 5.5 µmol/kg of frankfurters and salted, dried fish [526]. Nevertheless, the levels in most foods are generally below 10 ppb [519,528] or less than 10 µg/kg, with exception of fired food, such as bacon, in which an average concentration is 36 µg/kg [521].

Carcinogenicity: Various dialkyl and cyclic nitrosamines have been found to produce tumors at multiple sites in a range of species, including rats, mice and hamsters. Target organs of carcinogenicity include the liver, esophagus, nasal and oral mucosa, kidney, pancreas, urinary bladder, lung and thyroid [8,68,521,529–532]. *N-nitrosodiethylamine* (NDEA)

(Figure 5(2)) is considered to be an exceptionally potent carcinogen, compared to other nitrosamines [533] (Table 2).

Genotoxicity/DNA Binding (Adducts): As a class, nitrosamines, particularly the volatile nitrosamines, are potent mutagens/genotoxins, both in vitro and in vivo [68,521,529,530]. Many of the nitrosamines act as alkylating agents [534,535] leading to pro-mutagenic lesions, including alkylation at the $N7$ and O^6 positions of guanine or O^4 position of thymidine [536–538]. For example, NDEA (Figure 5(2)) and *N-nitrosodiethanolamine* (Figure 5(3)) form poorly repaired O^4-ethyldeoxythymidine and O^6-2-hydroxyethyl-deoxyguanosine DNA adducts [536,539]. Some organ specificity of nitrosamines may arise because, unlike the liver, sensitive tissues such as the brain lack a DNA-repair enzyme, alkyltransferase, that regenerates guanine from O^6-alkylguanine [540].

Biotransformation: N-Nitroso compounds are bioactivated to diazonium ions by hydroxylation involving several CYP isozymes [521,541], in particular CYP2E1 [536,542]. The organ specificity probably stems from tissue specific CYPs that activate the nitrosamines which alkylate DNA in the organ where they are activated.

MoA: Formation of alkylated DNA adducts and consequent mutagenesis and genomic instability are likely the most prevalent cancer-related mechanism of nitrosamines [529,533,536,543].

Human Exposure: In Europe, EDI of volatile *N*-nitrosamines ranges from 0.001 to 0.02 µg/kg bw/day [521]. In US the average daily intake was calculated to be 1.8 µg/day in a vegetarian diet and 1.9 µg/day in a Western diet [544]. The highest values of nitrosamines, up to 0.531 µg/serving, were found in alcohol, meat and dairy products [525].

Human Effects: In humans, indirect evidence exists of the carcinogenic action of nitrosamines through reported associations between gastrointestinal (esophageal, gastric, colorectal), pancreatic, bladder cancers and the consumption of foods containing relatively high concentrations of nitrosamines, nitrites and nitrates [68,521,529,530]. Additionally, nasopharyngeal cancers associated with consumption of salted fish have been attributed to *N*-nitroso compounds (see below) [545].

Risk: IARC classified the majority of food-borne nitrosamines as either probable (Group 2A) or possible (Group 2B) human carcinogens [67], although, certain practices known to result in increased cancer risks, including consumption of processed meat and fish, smoking, and betel quid chewing, and certain occupations in the rubber industry, result in exposures to various nitrosamines. The lack of identification of nitrosamines as "known human carcinogens" is largely a consequence of the low levels of human exposure to these compounds. Using the benchmark approach, permissible daily exposures (PDE) for cancer and mutagenicity were calculated to be 6.2 and 0.6 µg/person/day for *N*-nitrosodimethylamine and 2.2 and 0.04 µg/person/day for NDEA, respectively [543].

3.5.3. α,β-Unsaturated Aldehydes

α,β-Unsaturated aldehydes (Figure 5(4–8)) compose a wide ranging class of aldehydes which naturally occur in a variety of foods and can be added as flavor ingredients. In addition, they can be formed endogenously through lipid peroxidation [546–549]. They are formed from the polyunsaturated fatty acids (PUFA) in triglycerides, as well as from any free fatty acids, which are susceptible to auto oxidation [548,550].

3.5.3.1. Malondialdehyde, 4-Hydroxynonenal, Crotonaldehyde, *trans,trans*-2,4-hexadienal

Occurrence: Several α,β-unsaturated aldehydes, including *malondialdehyde* (MDA) (Figure 5(4)), in its enolic form, *4-hydroxynonenal* (HNE) (Figure 5(5)) and *crotonaldehyde* (CA) (Figure 5(6)), occur as contaminants in food, especially edible oils [548,551,552]. *Trans,trans-2,4-hexadienal* (Figure 5(7)) is used as a flavoring agent [548,549,553] and was detected in a variety of food products, including olives, caviar, chicken and beef [115].

Carcinogenicity: The α,β-unsaturated aldehydes that have been tested in standard 2-year bioassays in rodents include CA [554], MDA [555], and *trans,trans*-2,4-hexadienal [556,557]. In the rat, CA administered at 0.6 and 6.0 mM/L in drinking water for 113 weeks was associated with development of neoplastic nodules of the liver [554] in conjunction with overt hepatotoxicity (necrosis, fibrosis, cholestasis, and inflammation). In the same strain of rats, the incidence of thyroid follicular cell neoplasms was increased following 103 weeks of administration of MDA at 100 mg/kg bw/day (5 days/week) by oral gavage, in addition, pancreatic islet cell adenomas were observed in male rats in the group that received 50 mg MDA/kg [555,558]. Dosing of rats and mice with *trans,trans*-2,4-hexadienal by gavage in com oil at dosages greater than 45 mg/kg, 5 days/week, for up to 105 weeks resulted in an increased incidence of squamous-cell papillomas and carcinomas of the forestomach in both species [115,556]. In a neonatal mouse model, no tumors were observed after administration of CA, MDA and HNE via intraperitoneal injections up to 3000 nmol [559].

Genotoxicity/DNA Binding (Adducts): CA [552,560,561], HNE [562], MDA [551,563], and *trans,trans*-2,4-hexadienal [115,557,560,561] have been shown to be genotoxic, especially in vitro [548]. Unsaturated aldehydes are considered to be strong alkylating agents, and as such they can covalently bind to DNA and proteins. In particular, formation of DNA adducts was detected in vitro and in vivo in multiple tissues of rats and mice after exposure to CA [552], MDA [551], and *trans,trans*-2,4-hexadienal [115,556]. DNA adducts of CA have been also detected in exposed humans [552,564]. Another lipid peroxidation product, HNE, reacts with DNA chemically and can form DNA adducts [565,566], however it may be too reactive with proteins for DNA adducts to be formed in vivo if administered directly. Thus, in serum-containing medium HNE was not mutagenic to cultured cells, whereas a protected form was [567]. Depletion of GSH and resulting oxidative stress are thought to be prerequisites for formation of DNA adducts by unsaturated aldehydes [548].

Biotransformation: The oxidation of fatty acids leads to formation of hydroperoxides, which in turn, decompose in a terminal reaction to form aldehydes from the methyl terminus of the fatty acid chain [548]. The levels of hydroperoxides formed are often estimated by assay of thiobarbituric acid-reactive substances [568,569], but more precise measurement techniques are available [570]. Detoxication of unsaturated aldehydes occurs primarily through reactions with GSH, yielding metabolites that are excreted in the urine of rats and humans [548,552].

MoA: In addition to direct DNA-reactivity and mutagenicity, oxidative stress and immunomodulation might also play a role in the carcinogenic MoA of α,β-unsaturated aldehydes [115,548,552]. Based on the findings in vitro and in vivo, EFSA ruled out genotoxicity concern for *trans,trans*-2,4-hexadienal [557].

Human Exposure: Humans are exposed to α,β-unsaturated aldehydes from food and alcoholic beverages, as well as endogenously, particularly in some disease states [548,571]. Levels of dietary exposure are low, particularly in the case of the flavoring agent *trans,trans*-2,4-hexadienal whose per capita intake is estimated at 100 µg/kg of bw/day [549].

Human Effects: Several epidemiological studies provided little evidence of a positive association between CA exposure with the lung [572,573], oral cavity, stomach, and colon [574] cancer risks in humans [552].

Risk: IARC [115,552] classified CA and *trans,trans*-2,4-hexadienal as possibly carcinogenic to humans (Group 2B), while MDA was considered to be not classifiable as to its carcinogenicity to humans (Group 3) [558] (Table 2). In a risk assessment of CA [575], a conclusion was made that based on an analysis of the doses which produced DNA adducts, use of the hepatocellular tumor data from the rat drinking water carcinogenicity study is likely to overestimate human cancer risk. This is indicative of a practical threshold for the genotoxic and carcinogenic effect of CA. The JECFA evaluation [549] of the NOAEL for *trans,trans*-2,4-hexadienal of 15 mg/kg bw/day was based on the NTP Report [556] and was estimated to be >100,000 times its current EDI when used as a flavoring agent, thus JECFA concluded that this flavoring agent does not pose a safety concern. FEMA reaffirmed α,β-unsaturated aldehydes as GRAS, based on the lack of evidence of potential hazard to human health at concentrations present in food [548].

3.5.3.2. Trans-2 hexenal

Occurrence: *Trans-2-hexenal* (2-HEX) is a 6-carbon aliphatic unsaturated aldehyde (Figure 5(8)), which accounts for over 65% of total annual volume of unsaturated aldehydes used as flavor ingredients [548,549]. 2-HEX has been identified in a variety of plant species, including peppers, tomatoes and potatoes, and is referred to as leaf aldehyde [576]. The highest content of 2-HEX was reported in bananas, which contain approximately 32 ppm or 76 mg/kg [548,577].

Carcinogenicity: 2-HEX has not been tested for carcinogenicity in a 2-year bioassay [578]; however, based on structural similarities with *trans,trans*-2,4-hexadienal (Figure 5(7,8)), it can be expected to produce forestomach tumors in rodents when administered by oral gavage [548]. Some evidence of tumorigenicity was found in rats and mice that received three intraperitoneal injections of 2-HEX at the total dose of 150 mg/kg bw 18-month after the exposure [579]. Specifically, higher incidences of leukemia, liver and kidney tumors were described in mice, while in rats, tumors of parotid gland and lungs were reported.

Genotoxicity/DNA Binding (Adducts): 2-HEX was genotoxic in several assays in vitro producing DNA damage, mutagenicity, clastogenicity and aneugenicity [548,553,578,580]. No induction of gene mutations, direct DNA strand breaks or micronucleus formation was reported in rodents in vivo [553,579]. In humans, non-smoking volunteers that consumed three to six bananas per day for 3 days showed at least a doubling of micronuclei in exfoliated buccal cells [581]. Rinsing the oral cavity with water containing 10 ppm 2-HEX produced a more pronounced effect. Subsequent experiments in rats [575,582] concluded that such exposures posed a negligible risk except in very special situations. 2-HEX has been also shown to form adducts with DNA and proteins [548,553]. Specifically, cyclic 1,*N*2-propanodeoxyguanosine adducts were detected in vitro and in several tissues of rats, including forestomach, esophagus, liver and kidneys, following oral doses of up to 500 mg/kg bw; however, the covalent binding index was calculated to be extremely low (0.06) [583–586]. Using a physiologically based in silico model developed in rats, formation of 2-HEX DNA adducts in humans from current levels of dietary intake was predicted to be three orders of magnitude lower compared to endogenous DNA adduct levels [587].

Biotransformation: 2-HEX is readily oxidized to *trans*-2-hexenoic acid in vitro by mouse cytosolic fraction and isoenzymes of rat aldehyde dehydrogenase [548,553]. Conjunction with GSH is a major mechanism involved in detoxication [548,549,587].

MoA: While DNA adduct formation can be related to mutagenicity of 2-HEX in vitro, EFSA concluded that based on the in vivo findings, concern for genotoxicity for this compound can be ruled out [553]. Induction of oxidative damage, exacerbated by GSH depletion, can potentially play an important role in carcinogenicity [588].

Human Exposure: Combined daily per capita intake of 2-HEX from foods was calculated to be 2390 μg/person per day or 31–165 μg/kg bw/day, with majority of exposures occurring with consumption of bananas [548,575]. An EDI based on the maximized survey-derived daily intake of 2-HEX as a flavoring substance was calculated to be 409 and 2800 μg/capita per day (or 0.007 and 0.05 mg/kg bw per day) in US and Europe, respectively [578].

Human Effects: Data assessing association of human cancer risk with 2-HEX exposure are currently lacking.

Risk: JECFA [549] and EFSA [578] concluded that 2-HEX would not pose a safety concern at the current levels of intake as a flavoring substance. Based on low covalent binding of 2-HEX, estimated cancer risk of 1–5 per 10^7 lives was considered to be negligible; however, under certain circumstances, utilization of 2-HEX as a flavoring agent or fungicide can increase cancer risk to 2–6 per 10^4 lives [575,582].

3.6. Carcinogenicity of Preserved and Processed Foods

Smoking and pickling of foods have long been suspected of leading to formation of carcinogens, particularly *nitrosamines* [519,589,590].

3.6.1. Preserved Vegetables

Consumption of pickled vegetables, prepared with or without salting, which is predominant in some regions of Asia, such as China, Japan and Korea, showed some association with higher risk of stomach, nasophageal or esophageal cancers; however, the evidence is not consistent [305,591]. In a meta-analysis of 16 case-control studies, the highest versus lowest preserved vegetable intake was associated with a 2-fold increase in the risk of nasopharyngeal cancer, whereas consumption of non-preserved vegetables was associated with reduced risk [592]. The increased risk was attributed to the content of nitrates and nitrosamines. Based on the limited evidence for carcinogenicity of pickled vegetables in humans and inadequate evidence for carcinogenicity in experimental animals, IARC [305] classifies pickled vegetables as possibly carcinogenic to humans (Group 2B) (Table 2).

3.6.2. Red and Processed Meat

In many countries, consumption of red meat, usually cooked, and processed meats that have been prepared thorough salting, curing, fermentation and smoking varies from 50 to 200 g/day [593]. While the evidence of carcinogenicity of consumption of either red or processed meat in experimental animals is inadequate, a variety of studies reported an association between the consumption of such meats with higher risk of colorectal, pancreatic, prostate, breast, endometrial, liver and gastric cancers in humans [593–596]. Cooking and processing of meat results in formation of various genotoxic carcinogens, including N-nitroso compounds, HACs and PAHs (discussed above), which are capable of inducing pro-mutagenic DNA damage, contributing to carcinogenesis [593,597]. In addition, consumption of processed meat can produce oxidative stress and formation of lipid peroxidation products that could contribute to carcinogenic MoA [593]. Based on the available data, IARC [593] classified consumption of red meat as probably carcinogenic to humans (Group 2A), while consumption of processed meat was considered to be carcinogenic to humans (Group 1) (Table 2).

3.6.3. Salted Fish

Salted fish is produced and consumed primarily in Southeast Asia and northern Europe. Low levels of several volatile nitrosamines (discussed above) have been detected in Chinese-style salted fish [598], which is prepared by treating with dry salt or an aqueous salt solution followed by drying in the sun, and high levels of N-nitrosodimethylamine have been reported in some samples [599–601].

Several experiments have demonstrated that feeding of high concentrations (i.e., >5%) of Chinese-style salted fish in the diet induced nasal cavity tumors in rats [602–604], a site for carcinogenicity of nitrosamines [531]. Administration of an extract of nitrate-treated fish induced glandular stomach cancer in rats [605]. In addition, N-nitrosamines-specific DNA adducts were detected in the livers and kidneys of rats which were fed Chinese salted fish [606].

In humans, association between consumption of salted fish and cancer incidences has been demonstrated [599]. A population-based case-control study [607] showed that individuals with the highest intake of salted fish had an 80% increase in risk of nasopharyngeal carcinoma and a linear trend with respect to protein-containing preserved foods. In addition, a meta-analysis of cohort studies in Korea, China and Japan supported the evidence that consumption of salted fish is associated with a 1.2-fold increase in the risk of gastric cancer [591], 1.2- to 1.45-fold increased risk for nasopharyngeal carcinoma [608], as well as higher risk of stomach and colon cancer [609]. Potential association of nasopharyngeal tumors in endemic areas with Epstein-Barr virus cannot be excluded [599]. IARC [599] concluded that there was sufficient evidence in humans for causation of nasopharyngeal carcinomas by Chinese-style salted fish, and classified it as carcinogenic to humans (Group 1) (Table 2).

4. Epigenetic Carcinogens or Carcinogens with Uncertain Mode of Action and Related Chemicals Present in Food

This section provides an overview of food-derived carcinogens that are typically negative in genotoxicity assays in vitro and in vivo, and which facilitate neoplastic development through molecular and cellular mechanisms other than direct DNA reactivity. This section also includes carcinogens that do not have enough mechanistic data for classification. Chemical structures of carcinogens and related chemicals discussed in this section are provided in Figures 6–10.

4.1. Phytotoxins

In 2018, FDA announced a ban on seven synthetically derived agents, including methyl eugenol, myrcene, pulegone, benzophenone, ethyl acrylate, pyridine and styrene for use as flavoring substances [610]. The majority of these substances have natural counterparts, discussed in this manuscript.

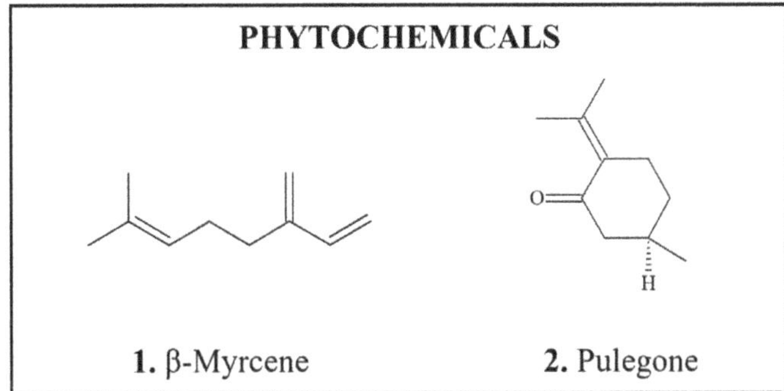

Figure 6. Chemical structures of non-DNA-reactive carcinogenic phytochemicals present in foods.

4.1.1. β-Myrcene

Occurrence: β-*myrcene* (Figure 6(1)) is an acyclic monoterpene, which occurs naturally in a variety of plants, including verbena, lemongrass, bay, rosemary, basil, cardamon and is a constituent in many fruits, vegetables and beverages, such as citrus peel oils and juices, pineapple, celery, carrot, beer, white wine, and many others [611–613]. The highest levels of β-myrcene, up to 10 g/kg dry weight, were reported in hops [611]. It is also widely used as a flavor and fragrance material.

Carcinogenicity: Oral administration of β-myrcene by gavage up to 1000 mg/kg bw, 5 days a week, induced a significant increase in liver tumors (adenomas and carcinomas) in male and female mice. In rats, increased incidences of renal tubular adenomas and carcinomas were reported [611,614–616].

Genotoxicity/DNA Binding (Adducts): β-myrcene lacks genotoxicity and mutagenicity in vitro and in vivo [611,613,614,616–619], accordingly, no covalent DNA binding was reported.

Metabolism: Metabolism of β-myrcene involves oxidation of the carbon–carbon double bond (Figure 6(1)) to an epoxide intermediate, which after hydrolysis gives rise to diol conjugates, 10-hydroxylinalool and 7-methyl-3-methyleneoct-6-ene-1,2-diol, that were detected in the urine of rabbits and rats [611,620–622]. Diols are further oxidized to corresponding aldehydes and hydroxy acids [613]. These reactions are likely metabolized by CYPs [619], and β-myrcene was shown to inhibit activity of CYP2B1 in vitro and induce CYP2B1/B2 in vivo [623,624].

MoA: The mechanism of tumor induction by β-myrcene remains largely unknown [612]. Analyses of cancer data by FEMA [619,622] suggested that hepatocarcinogenesis in mice and renal tumors in rats are secondary to cytotoxicity of β-myrcene at high carcinogenic doses, and in the kidney are related to the chronic progressive nephropathy and possibly unusual nephrosis. While it is structurally similar to another terpene, d-limonene, which is known to bind to α_{2u}-urinary globulin producing nephropathy, IARC [611] concluded that β-myrcene did not meet all of the criteria to explain its carcinogenicity by a α_{2u}-globulin-associated mechanism. Histopathologic assessment of the kidneys from rats chronically dosed with β-myrcene confirmed that due to complex renal pathology, α_{2u}-globulin nephropathy cannot be the sole MoA of carcinogenicity [625].

Human Exposure: Daily per capita intake of β-myrcene in US was calculated to be 3 µg/kg bw/day [622,626]. Another, more recent FDA estimation suggested an EDI of 1.23 µg/kg bw/day [612]. In Europe, estimated per capita intake was calculated to be 4.8 µg/kg bw/day [616].

Human Effects: No findings on human carcinogenicity are available [611].

Risk: IARC [611] classifies β-myrcene as possibly carcinogenic to humans (Group 2B) (Table 2). Safety assessment of β-myrcene by FEMA concluded that MoA of carcinogenicity in rodents is not relevant to humans and rodent carcinogenicity is not indicative of a health risk [619,622]. JECFA and EFSA [616,626] concluded that at estimated current dietary intake, β-myrcene would not pose a safety concern. Despite these conclusions, FDA recently withdrew authorization for use of a synthetic form of myrcene as a food additive due to its carcinogenicity in accordance with the Delaney Clause [610,612,627].

4.1.2. Pulegone

Occurrence: Pulegone (PUL) ((R)-5-methyl-2-(1-methylethylidine)cyclohexanone) (Figure 6(2)) is a naturally occurring monoterpene ketone found in a variety of plants, in particular mint species, such as *Nepeta cataria* (catnip), *Mentha piperita*, and pennyroyal and is used as a flavoring agent [628–631].

Carcinogenicity: Oral administration of PUL to mice at up to 150 mg/kg bw in corn oil by gavage, 5 days/week for 105 weeks, significantly increased incidences of hepatocellular adenoma in both sexes and incidences of hepatoblastoma in male mice [628,632]. In rats, increased incidences of urinary bladder neoplasms were observed in females only, while no evidence of carcinogenic activity was observed in males.

Genotoxicity/DNA Binding (Adducts): PUL was not genotoxic or mutagenic in vitro and in vivo [628,629,631–633]. Genotoxicity studies with herbal preparations containing PUL, such as peppermint oil, also yielded negative result [631]. No covalent DNA binding has been reported, although reactive metabolites of PUL can covalently bind to proteins [634].

Metabolism: PUL is metabolized by different pathways, including hydroxylation in the 9-position to toxic metabolite menthofuran or in the 5-position to piperitenone; reduction of the carbon-carbon double bond, which results in formation of menthone and isomenthone; or conjugation with GSH [631,635,636]. Hydroxylation to menthofuran involves multiple CYPs, including human CYP2E1, CYP1A2, CYP2C19 and CYP3A4 [637–639]. The major metabolites of PUL detected in humans were 10-hydroxypulegone, 8- and 1-hydroxymenthone, and menthol [640]. Metabolism of PUL to menthofuran can result in formation of reactive metabolites, in particular, epoxide pulegone 8-aldehyde (γ-ketoenal) and *p*-cresol, that can bind to proteins and deplete GSH levels [628,631,634,639,641].

MoA: An epigenetic MoA for PUL-induced urinary bladder tumors in female rats was proposed to involve chronic exposure to high concentrations resulting in excretion and accumulation of PUL and its cytotoxic metabolites, particularly piperitenone, in the urine, leading to urothelial cytotoxicity and sustained regenerative urothelial cell proliferation eventually resulting in development of urothelial tumors [642]. In addition, toxicity of menthofuran and covalent binding of its metabolites to proteins can lead to chronic regenerative cell proliferation, which can contribute to liver and urinary bladder carcinogenesis [628,630,631].

Human Exposure: Dietary exposure to PUL results primarily from ingestion of products flavored with spearmint or peppermint oil, such as confectionery, chewing gum, as well as alcoholic and non-alcoholic beverages [630,631]. JECFA [629] estimated an intake for PUL of approximately 2 µg/person/day or 0.04 µg/kg bw/day in Europe and 12 µg/person/day or 0.03 µg/kg bw per day in USA. The European Commission (Regulation EC No. 1334/2008) [106] set a limit of 20 mg/kg for PUL and menthofuran in foods and beverages.

Human Effects: No epidemiological studies linking PUL to human cancer risk have been conducted [628].

Risk: IARC [628] concluded that PUL was possibly carcinogenic to humans (Group 2B) (Table 2) based on sufficient evidence for carcinogenicity in experimental animals but inadequate evidence in humans. JECFA [629] found no safety concern when PUL is used as a flavoring agent. EMA [631] suggested that MoA for tumor induction in rodents is not relevant for carcinogenicity risk in humans, and recommended an acceptable exposure limit of 0.75 mg/kg bw/day.

4.2. Mycotoxins

Fumonisin B_1 and Fusarin C are the major toxins derived from *Fusarium* fungi species, *Fusarium verticilloides* (also known as *moniliforme*) and *proliferatum*, which are common contaminants on crops, in particular corn [247,276,305,643].

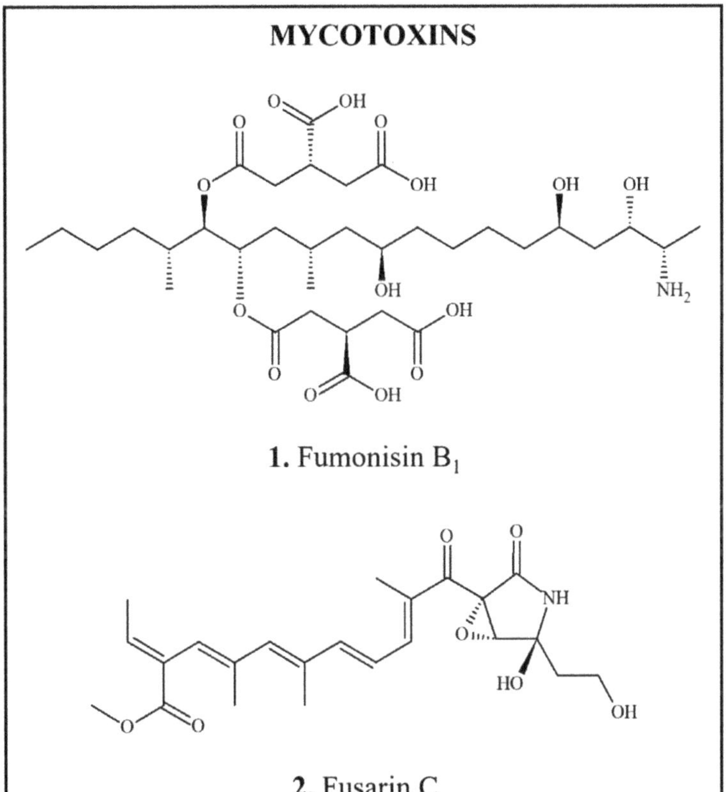

Figure 7. Chemical structures of non-DNA-reactive carcinogenic mycotoxins and related chemicals present in foods.

4.2.1. Fumonisins

Occurrence: Fumonisin B_1 (FB$_1$) (Figure 7(1)), is the most prevalent member of fumonisins class. It has the chemical structure of a substituted 2-amino-icosane diester, which has features in common with the sphingoid base backbone of sphingolipids [644,645]. The highest concentrations of FB_1, which range from 310 to up to 23,800 μg/kg, were reported in maize and maize-based cereal products [276].

Carcinogenicity: In female mice, oral administration of FB_1 caused an increase in hepatocellular adenomas and carcinomas. In the study in male rats, an increase in cholangocarcinomas and hepatocellular carcinomas were observed, while in the other rat study, FB_1 induced renal tubule carcinomas in males exposed to up to 100 ppm [247,646,647]. FB_1 was reported to have liver cancer initiating activity, as evidenced by induction of preneoplastic foci in rats by 7 weeks of dosing [648]. Some studies suggest that it also has tumor promoting activity [649,650]. For example, FB_1 administered in diet had promotional activity on liver tumors initiated by AFB_1 and *N*-methyl-*N'*-nitro-*N*-nitrosoguanidine in trout [651].

Genotoxicity/DNA Binding (Adducts): The structure of FB_1 (Figure 7(1)) lacks features that confer DNA reactivity. Accordingly, it was not mutagenic in bacteria; however, a positive result was reported in a luminescence induction assay in the absence of metabolic activation [247,647]. The compound did not induce DNA repair synthesis in the liver cells of rats in vitro or in vivo and no evidence for DNA adduct formation with oligonucleotides in vitro was found [247]. However, evidence for induction of DNA damage by FB_1 was reported in rat brain glioma cells and human fibroblasts in vitro, and in spleen and liver cells isolated from exposed rats [652–654]. In addition, FB_1 caused DNA fragmentation in rat liver and kidney [655]. Positive results were obtained in micronucleus assays in vitro with human-derived hepatoma (HepG2) cells but not with rat hepatocytes [656]. In bone marrow of mice, an increase in formation of micronuclei was found after intraperitoneal injection of FB_1 [657]. Positive results were obtained in chromosomal aberration assays with rat hepatocytes.

Metabolism: There is little or no evidence that fumonisins are metabolized in vivo or in vitro [247,646,647]. Nevertheless, FB_1 induced CYP1A activity in hepatoma cell line, and CYP2E activity in rats, while inhibiting CYP2C11 and CYP1A2 enzymes [658,659]. Liver and kidney retain most absorbed material [660]. Hydrolyzed FB_1 is more toxic compared to the parental form, and recently, hydrolyzed metabolites were detected in the kidney and liver of rats administered FB_1 by intraperitoneal injections [661].

MoA: One postulated MoA for FB_1 carcinogenicity involves disruption of sphingolipid metabolism, either through inhibition of ceramide synthesis [662] or due to changes in polyunsaturated fatty acid and phospholipid pools [649], leading to alteration of signaling pathways that control cell behavior and DNA synthesis [247,645,660,663–666]. Such perturbations produce alterations in cell turnover. Another proposed MoA involves oxidative stress, which is likely to mediate DNA damage observed in some assays [276,291,652,654,660]. In support of this hypothesis, FB_1 was demonstrated to increase lipid peroxidation in rat kidney and liver and decrease levels of antioxidant enzymes [667,668]. In addition, dosing of rats with 100 μg/kg bw for 12 weeks resulted in downregulation of hepatic antioxidant genes [655].

Human Exposure: In Europe and North America, EDI to FB_1 ranges from 0.01 to 0.2 μg/kg bw/day, while in other countries with different climate, cultivation practices and higher consumption of maize and maze-based products, EDI levels are much higher, reaching up to 354.9 μg/kg bw/day in South America and Africa, and up to 740 μg/kg bw/day in China [660]. The highest levels of chronic dietary exposure, ranging from 0.18 to 3.9 μg/kg bw/day for FB_1 and from 0.27 to 6.4 μg/kg bw/day for total fumonisins were reported in children, with cakes, cookies and pies, cereal-based foods and cereal grain being the main contributors [276]. JECFA reported that in adults, mean chronic exposures to FB_1 did not exceed 0.56 μg/kg bw/day for FB_1 and 0.82 μg/kg bw/day for total fumonisins, respectively.

Human Effects: Epidemiological evidence shows a link between exposure to *F. moniliforme* contaminated corn and esophageal and hepatocellular cancer [247,305,669–672] but these reports do not indicate the specific compounds involved. Others [673,674] investigated FB_1 specifically, but were not able to find significant association between exposure and cancer risk [276].

Risk: IARC [247] evaluated FB_1 as possibly carcinogenic to humans (Group 2B) (Table 2) based on sufficient evidence for carcinogenicity in experimental animals and inadequate evidence in humans. JECFA [276,647] established a PMTDI of 2 µg/kg bw/day for FB_1 alone or in combination with other fumonisins, and recommended to reduce exposures to fumonisins, especially in the areas where maize is consumed at higher levels.

4.2.2. Fusarin C

Occurrence: *Fusarin C* (FC) (Figure 7(2)) belongs to 2-pyrrolidinone metabolites produced by various species of the fungus *Fusarium*, including *Fusarium moniliforme* and *oxysporum* [305,675]. FC has been detected in corn and maize grain in concentrations ranging from 28 to 83 mg/kg [305,676,677]. While unstable to heat, FC may survive cooking process [678].

Carcinogenicity: FC induced papillomas and carcinomas of the oesophagus and forestomach in mice and rats when administered by oral gavage at 0.5 or 2 mg twice a week to mice or rats, respectively [305]. FC did not act as a promoter in rat liver [679].

Genotoxicity/DNA Binding (Adducts): FC was genotoxic in vitro in the presence of exogenous bioactivation, producing mutagenicity, SCE, chromosomal aberration and micronuclei formation [305,680,681]. Only marginal effect was observed in UDS assay in rat hepatocytes [682]. Currently, no in vivo genotoxicity studies with FC were reported. While crude extracts of *Fusarium moniliforme* produced direct mutagenicity in bacteria as well as positive results in NPL assays, no DNA adducts were measured by NPL assay with pure FC [683,684].

Metabolism: FC accumulates mainly in the intestines, stomach and liver after administration by gavage to rats. Studies utilizing rat liver microsomal enzymes showed that FC is metabolized by carboxyesterase to water-soluble fusarin PM_1, while monooxygenase is involved in the FC conversion to a mutagenic metabolite [305,685,686]. Hydroxylation at the 1-position resulted in production of two genotoxic metabolites, fusarin Z, which was the most potent mutagen in vitro, and fusarin X [687].

MoA: The role of mutagenic effects of FC in the development of cancer is not clear. One study suggested that FC may act as an estrogenic agonist in vitro [688]; however, no effects on mammary glands were detected in carcinogenicity studies [305].

Human Exposure: Major source of exposure to FC is maize and maize grain [305,676,677]. Currently, no data on dietary intake levels of FC in humans have been reported.

Human Effects: FC has been suggested to be responsible for the high incidence of esophageal cancer in China [680] and South Africa [689]. Changes in the staple diet of Black South Africans from sorghum to maize (corn), on which the fungus grows more easily, has been associated with the epidemic of squamous carcinoma of the esophagus in that area [671].

Risk: IARC [305] classifies FC, similar to other toxins derived from *Fusarium moniliforme* as possibly carcinogenic to humans (Group 2B) (Table 2).

4.3. Environmental, Agricultural and Industrial Contaminants

A variety of industrial contaminants and chemicals used in crop protection and production have caused cancers in experimental models and have been considered to be likely human carcinogens [67,275,690,691]. Traces of these chemicals can contaminate food, albeit at extremely low levels. The cancer risk that such exposures might pose has been a matter of debate.

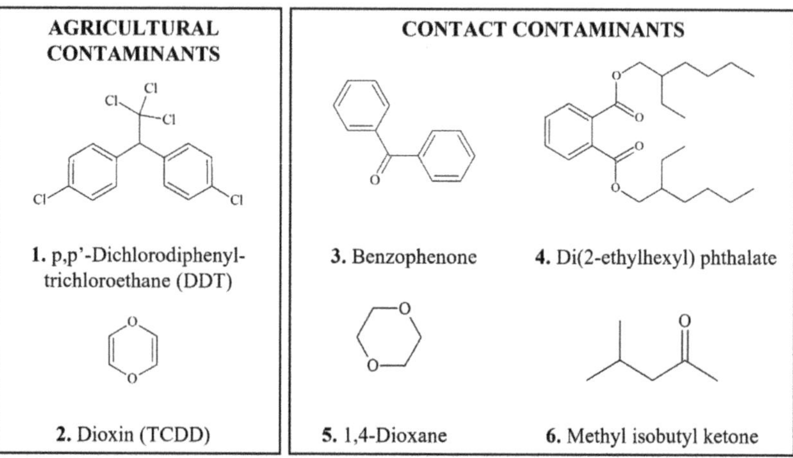

Figure 8. Chemical structures of non-DNA-reactive carcinogenic contaminants in food.

4.3.1. Agricultural Contaminants

4.3.1.1. *p,p′*-dichlorodiphenyltrichloroethane

Occurrence: Technical grade *p,p′-dichlorodiphenyltrichloroethane* (DDT) (Figure 8(1)) is a complex mixture of DDT, its isomers and related compounds. As an organochlorine insecticide, DDT, had a major impact on the incidence of malaria and typhus as a cheap and effective method of killing the female *Anopheles* mosquito, which is the malaria parasite vector, and lice, which spread *Rickettsia prowazekii*, the cause of epidemic typhus [691]. Cost-effectiveness analyses shows that DDT is the least expensive, yet effective insecticide for prevention of malaria that kills thousands of people each day [692,693]. Nevertheless, owing to the extensive use of DDT in agriculture to control insects, such as the pink boll worm (*Pectinophora gossypiella*) on cotton, codling moth (*Cydia pomonella*) on deciduous fruit, Colorado potato beetle (*Leptinotarsa decemlineata*), and the European corn borer (*Ostrinia nubilalis*), and DDT's resistance to degradation or metabolism, which results in bioaccumulation in the food chain, it has been largely banned in the early 1970s with the expectation that all use would be stopped [694]. In 2006, however, the World Health Organization (WHO) reversed a 30-year policy by endorsing the use of DDT for malaria control [695]. DDT and its metabolites and degradation products, *p,p′*-dichlorodiphenyldichloroethylene (DDE) and *p,p′*-dichlorodiphenyldichloroethane (DDD), have been found in human breast milk [696,697], as well as some of the foods, including American cheese, butter, catfish, carrots, summer squash, celery, and salmon [694]. IARC [691] and Smith [696] reported that the mean concentrations of DDT in population have declined in much of the world: from 5000–10,000 µg/kg to around 1000 µg/kg of milk fat or even lower over the last three decades. Although different concentrations are found in different regions, the declines seen in various countries correspond to their restrictions on use of DDT.

Carcinogenicity: In some studies, DDT produced liver tumors in rats and mice at doses exceeding 46 mg/kg bw by gavage or 250 mg/kg in diet, as well as increases in incidences of malignant lymphomas and lung neoplasms in mice [68,691,694]. In contrast, US National Cancer Institute [698] bioassays at up to 642 ppm in male rats and 175 ppm in female mice detected no evidence for carcinogenicity of DDT. The DDT metabolites, DDE and DDD, also were hepatocarcinogenic in mice [699].

Genotoxicity/DNA Binding (Adducts): The genotoxicity data on DDT and related compounds were overwhelmingly negative; however, some evidence of DNA damage, chromosome aberrations, and micronuclei formation was reported in human lymphocytes exposed to DDT in vitro [691,694,699,700]. No covalent DNA binding has been reported.

Metabolism: Due to its high lipophilicity, DDT, DDE and DDD tend to accumulate in the adipose tissue [694,701,702]. In mammals, including humans, DDT is primarily dehydrochlorinated to DDD, which is further metabolized to easily excreted 2,2-bis-chlorophenyl acetic acid isomers. To a lesser degree, DDT is also converted to DDE, which tends to bioaccumulate in lipid-rich tissues [691,700,702]. In rats, DDT and its metabolites has been shown to induce several CYPs, including CYP2B and CYP3A [703]. DDT can be also biotransformed to methylsulfonyl intermediates, which exhibit toxicity, in particular in adrenal gland [704].

MoA: DDT was shown to have a liver tumor promoting effect in mice [705], which, based on mechanistic studies [706], was attributed to its accumulation in the lipid layer of liver cell membranes and reduction of cell-cell communication, thereby diminishing tissue homeostatic control of incipient neoplastic cells. This MoA implies a requirement for a sufficient exposure over time to maintain the interference with intercellular communication throughout the liver. IARC also found strong evidence that DDT acts as endocrine disruptor, is immunosuppressive and can induce oxidative stress, all of these MoAs are operable in humans [691].

Human Exposure: Numerous studies have investigated human exposure to organochloride pesticides such as DDT, specifically due to concerns over its ability to bioaccumulate in the body and persist in the environment. However, due to the ban of DDT and declining levels of DDT and its metabolites in humans, more recent exposure data are scarce. It has been estimated that over 90% of the DDT detected in the general population is derived from food, particularly from meat, fish, poultry, and root and leafy vegetables [691,694]. The highest average daily intake ranging from 24.2 to 27.8 µg/day, was observed in Arctic populations, that consume foods such as seal or whale [694]. Nevertheless, most countries have seen a significant decline in DDT intake, ranging from 20 to 40% [700]. For example, in Europe total dietary exposure to DDT and its metabolites decreased from 0.00627 µg/kg bw/day in 1997 to 0.0051 µg/kg bw/day in 2005, and in US a decline from 0.0213 to 0.0056 µg/kg bw/day was observed from 1984 to 1991 [700]. EFSA also concluded that in most European countries, current EDI values for DDT, which range from 0.005 to 0.03 µg/kg bw/day in adult and children and up to 1 µg/kg bw in breastfed infants, are below the established provisional TDI of 0.01 mg/kg bw.

Human Effects: While some positive associations between DDT and cancers of the liver and testis, and non-Hodgkin lymphoma were reported, there seems to be limited evidence that DDT and related compounds from any source increase cancer rates in humans [691,699,707–710], even in agricultural workers [711]. This absence of carcinogenicity may be due to insufficient exposures, although some occupational exposures have been substantial, or it may reflect the fact that most human populations do not display rates of spontaneous liver tumor development that are as high as sensitive rodent models, indicating a low background of initiation available for promotion to tumor formation.

Risk: In 2018, IARC [691] upgraded classification for DDT from possibly (Group 2B) to probably carcinogenic to humans (Group 2A) (Table 2), based on sufficient evidence of carcinogenicity in experimental animals and strong mechanistic evidence that MoA for DDT carcinogenicity can operate in humans.

4.3.1.2. Dioxins and Dioxin-Like Compounds

Occurrence: Dioxins and related *Dioxin-Like-Compounds* (DLCs) refer to a complex family of chlorinated compounds with similar structures and biological effects. 2,3,7,8-Tetrachloro-*p*-dioxin (TCDD) (Figure 8(2)) is one of the most potent and prominent dioxins in the environment and is often referred synonymously as "dioxin". Dioxin and DLCs, including polychlorinated dibenzo-para-dioxins (PCDDs), dibenzofurans (PCDFs) and the polychlorinated biphenyls (PCBs), are formed by dimerization of chlorophenols produced during the synthesis of chlorophenoxy acetic acid herbicides [275]. The dimerization of 2,4,5-trichlorophenol yields TCDD while the heterodimerization of 2,4,5-trichlorophenol with related phenols such as 2,4-dichlorophenol yields tri- through heptachlorinated dibenzodioxins and dibenzofurans. Other sources of TCDD include the use of chlorophenol as wood preservatives, use of chlorine in pulp bleaching, incineration of halogen containing materials [712,713]. DLCs usually occur as mixtures and, in order to express the expected biological activity of mixtures by a common dose metric, toxic equivalency factors (TEFs) relative to the activity of TCDD have been developed [275,365,690,714]. Using TEFs and mass concentrations, dioxin toxic equivalents (TEQs) for a source can be calculated. Based on these values, there are at least 7 PCDDs, 10 PCDFs and 12 PCBs that have dioxin-like activity [715,716]. Food-mediated human exposure to TCDD and DLCs occurs when contaminants from the above-described sources are ingested by animals, including fish, which in turn are used as human foods [275,690,717]. Dioxins have also been detected in human milk, ranging from 5 to 15 ng TEQ/kg lipid [717–719]. Levels of TCDD and DCLs in the environment, and consequently in food, have been declining since the late 1970s because of reduced industrial emissions [720].

Carcinogenicity: In rodents, several DLCs, including TCDD, induced neoplasia, mainly of the liver [275,690,721–725]. Other target organs and tissues included thyroid gland, lungs and oral mucosa. TCDD acted as tumor promoter when administered with potent tumor initiators, such as nitrosamines [275,365,724].

Genotoxicity/DNA Binding (Adducts): TCDD is not DNA-reactive in vitro or in vivo and does not covalently bind to DNA [365,723,726–728]. Similarly, PCBs are mainly not DNA-reactive, although some evidence of DNA damage, SCE and chromosomal aberrations were observed in human lymphocytes [690,729]. PCBs can be metabolically activated to electrophilic quinoid intermediates, and can produce DNA adducts in vitro; however, no DNA adducts were observed in vivo [730].

Metabolism: Similar to DDT described in the section above, TCDD and DLCs are highly lipophilic and thus, tend to accumulate in the adipose-rich tissues, liver has been also shown as a primary site of TCDD accumulation in rodents [365,731–735]. TCDD metabolism is very slow and limited and the compound is eliminated mainly unchanged in the feces [734,735], although it induces activities of CYP1A1, CYP1A2, and CYP1B1 enzymes, which are also involved in the hydroxylation of PCDDs and PCBs [736], in mice and rats [365,734]. Rat hepatocytes show greater rate of TCDD metabolism compared to that in guinea pigs, this feature may underly the intraspecies differences in susceptibility to the toxicity of TCDD [737].

MoA: Dioxins are not DNA-reactive, but enhance liver tumor development through epigenetic mechanisms mediated by binding to the aryl hydrocarbon receptor leading to toxicity and enhanced cell proliferation [275,724,738–740]. Accordingly, for these events a threshold can be established. For example, NOAELs for hepatocyte proliferation in the NTP bioassays were 0.003 µg/kg at the 14-week interim evaluation and 0.022 µg/kg at 53 weeks for TCDD [721] and between 10 µg and 100 µg for 3,3′,4,4′-tetrachloroazoxybenzene (PCB 126) [722]. The hepatocarcinogenicity of TCDD in rats was greater in females than in males apparently due to the influence of estrogenic hormones [741], although the specific mechanism(s) has not been elucidated. Induction of oxidative damage can also play a role in carcinogenicity of dioxins. These MoAs are considered to be operational in humans [275].

Human Exposure: Mean dietary exposure to all dioxins in adults occurs primarily through consumption of food of animal origin, such as meat, dairy products, eggs and some fish, and is estimated to be 0.3–3 pg/kg bw day [275,365,716,717,742]. In 2010–2021 EDI for PCDDs and PCDFs varied from 0.001 pg TEQ/kg bw/day to 74.31 pg WHO-TEQ/day [743] depending on the country and method used for estimation of intake. Per capita intake of dioxins in US population is estimated to be lower (17 to 24 pg per capita) compared to that of European population (29 to 97 pg per capita) [365]. In nursing infants, dietary intake of dioxins can reach up to 53 pg TEQ/kg bw/day for TCDD [742] and over 150 WHO-TEQ/kg bw/day for PCDDs and PCBs [719]. Due to limitations in use, intake of dioxins has substantially reduced over the years.

Human Effects: There is no epidemiological evidence that implicates consumption of low-level DCLs-containing foods in human cancer causation [713,744]. Nevertheless, continuing evaluation of highly exposed individuals is strengthening the observations of increased cancer risk with dioxin exposure, in particular for lung cancer, soft-tissue sarcoma and non-Hodgkin lymphoma, although the increases are small for these relatively high exposures [745,746]. In addition to carcinogenicity, exposure to dioxins is associated with a variety of adverse effects, including dermatological effects (chloracne), cardiovascular diseases, endocrine disorders (diabetes, affected thyroid function), reproductive effects, neurological disorders and an increase in hepatic enzymes [365].

Risk: TCDD and DLCs have been classified by IARC [275,690] to be human carcinogens (Group 1) (Table 2) based on sufficient epidemiological information, animal carcinogenicity data and strong mechanistic considerations. In 1998, WHO modified a previously established TDI for TCDD from 10 pg/kg bw to a range of 1–4 pg TEQs kg bw/day [747,748], while the SCF [749] and JECFA [365] established TDI for dioxins of 2 and 2.3 pg/kg bw, respectively. In contrast, the US EPA [750] has proposed that dioxin doses in the range of 1 pg/kg might represent a cancer risk. This assessment was criticized as overly conservative [751–754], and the EPA has yet to issue a reanalysis of the cancer TCDD dose response reassessment.

4.3.2. Food Contact Materials

4.3.2.1. Benzophenone

Occurrence: Benzophenone (diphenylketone) (Figure 8(3)) is an aryl ketone that can occur in foods naturally or due to migration from packaging or its use as a food additive [115,365,755,756]. Naturally, benzophenone mainly occurs in grapes at concentrations up to 0.13 mg/kg, it is also a constituent of vanilla (up to 0.48 mg/kg), passion fruit (0.045 mg/kg) and papaya (less than 0.01 mg/kg). Benzophenone can migrate into foodstuff from paperboard packaging when used as photoinitiator for UV printing inks, or from plastic food packaging when used as a UV filter [757–760]. Concentrations of benzophonone residues migrated into food ranged from 0.01 to over 5 mg/kg, with the highest levels, 7.3 mg/kg, detected in confectionery products with high fat content [115,758]. As a flavoring agent, benzophenone is used at 0.5 to 1.28 mg/kg in non-alcoholic beverages and at 2 mg/kg in foods in general [115].

Carcinogenicity: Long-term oral administration of benzophenone in diet up to 1250 ppm (equivalent to 60 mg/kg bw in rats and 160 mg/kg bw in mice) produced some evidence of carcinogenic activity evident by increases in the incidences of mononuclear cell leukemia and renal tubular adenoma in male rats as well as liver tumors in male mice and histiocytic sarcoma in female mice [115,755,756,761–764].

Genotoxicity/DNA Binding (Adducts): Results of in vitro and in vivo genotoxicity testing for benzophenone were mainly negative [755,756,763,765]. However, in the presence of recombinant human CYP2A6 and NADPH-CYP reductase, benzophenone induced *umu* gene expression in *S. typhimurium*, which is an indicator of DNA damage [766]. Photoactivated benzophenone has been reported to react with DNA in vitro, producing single strand breaks, DNA-protein cross-links and abasic sites [767,768].

Metabolism: In rats, benzophenone is metabolized by reduction to benzhydrol or by oxidation to 4-hydroxybenzophenone, these metabolites and a sulphate conjugate of 4-hydroxybenzophenone were also detected in vitro [769,770].

MoA: Carcinogenic MoA of benzophenone is not well understood and likely involves multiple mechanisms, including endocrine-disrupting effects and oxidative damage [115]. Thus, benzophenone and its metabolite, 4-hydroxybenzophenone, exhibited estrogenic effects in vitro and in vivo [771–773]. In the subchronic and chronic rodent studies, oral administration of benzophenone induced CYP enzymes and consequent hepatocellular hypertrophy [761–763], indicating that these changes can be involved in hepatocarcinogenesis. Renal tumors in male rats were associated with the exacerbation of ageing chronic nephropathy, suggesting that this MoA largely contributes to renal tubular proliferation induced by benzophenone [764]. This MoA is not relevant to human renal carcinogenesis [764].

Human Exposure: Combined dietary exposures to benzophenone range from 8.5 µg/kg bw/day in adults to 22 µg/kg bw/day in children [755]. Similar findings were made in a study involving Taiwan population, where an average daily doses of benzophenone from dietary exposures were estimated to range from 4.54 to 25.8 µg/kg bw/day [774]. Daily per capita intakes of benzophenone based on its use as a flavoring ingredient were estimated to be 0.2 µg/kg bw/day in US and 0.4 µg/kg bw/day in Europe [365,756]. IARC estimated that dietary exposure to benzophenone from consumption of muscat grapes is approximately 0.3 µg/kg bw/day [115].

Human Effects: No data linking benzophenone and increased human cancer risk are currently available [115].

Risk: IARC [115] classified benzophenone as possibly carcinogenic to humans (Group 2B) (Table 2), stating that while evidence of rodent carcinogenicity is weak, relevance of carcinogenic MoA to humans cannot be excluded. EFSA estimated TDI for benzophenone of 0.03 mg/kg bw [755], and the current migration limit from packaging into foods is 0.6 mg/kg. Despite conclusions made by JECFA [365] and EFSA [755] that benzophenone poses no safety concerns at current levels of intake when used as a flavoring agent, FDA no longer allows use of synthetic benzophenone as a flavoring substance under the Delaney clause [610].

4.3.2.2. Di(2-ethylhexyl) Phthalate

Occurrence: Di(2-ethylhexyl) phthalate (DEHP) (Figure 8(4)) is produced by reaction of 2-ethylhexanol with phthalic anhydride and is primarily used as a plasticizer in the production of polyvinyl chloride [115]. Due to its wide presence in packaging materials, DEHP mainly contaminates food by leaching [775]. Concentrations of DEHP in food range from 0.001 to 7.5 mg/kg, with the highest levels of DEHP, exceeding 0.3 mg/kg and, in some cases, reaching 17 mg/kg, reported in foods with high fat content, namely oils, milk, butter, cheese, mayonnaise, fresh meat and fish products [775–781]. In soft drinks, DEHP occurs at concentrations ranging from 0.00003 to 0.0035 µg/L [115,782].

Carcinogenicity: Administration of DEHP in the diet up to 6000 ppm (equivalent to over 350 mg/kg/day) resulted primarily in development of hepatocellular adenomas and carcinomas in rats and mice of both sexes [68,115,777,783–787]. In addition, higher incidences of benign Leydig cell tumors and pancreatic adenomas were observed after DEHP administration. DEHP also showed tumor promoting activity on hepatocellular lesions induced by NDEA and skin tumors induced by 7,12-dimethylbenz[a]anthracene (DMBA) in mice [788,789].

Genotoxicity/DNA Binding (Adducts): DEHP and its major metabolite, mono(2-ethylhexyl) phthalate, produced mainly negative results in the in vitro genotoxicity tests with and without exogenous metabolic activation system; however, some positive responses were observed in cell transformation and DNA damage assays [115,777,780,784,790,791]. In vivo results were mixed [115,339,792,793] although genotoxic effects were likely secondary to oxidative stress [777,794–798]. DEHP did not covalently bind to liver DNA in mice and rats [799–801].

Metabolism: In rodents and humans, DEHP is hydrolyzed in the presence of carboxyesterases, in particular pancreatic lipases, to mono(2-ethylhexyl) phthalate, which is further metabolized to phthalic acid and oxidative metabolites, such as mono-(2-ethyl-5-hydroxyhexyl)phthalate, mono-(2-ethyl-5-oxohexyl)phthalate, mono-(2-ethyl-5-carboxypentyl)phthalate and mono-[2-(carboxymethyl)hexyl]phthalate, that can be detected in urine and feces either as glucuronide conjugates or unconjugated [115,802–804]. Ito and colleagues [805] reported species differences in the activities of enzymes that participate in the metabolism of DEHP, specifically lipase, UDP-glucuronyltransferase, alcohol dehydrogenase and aldehyde dehydrogenase in various tissues of rats, mice and marmosets. When comparing metabolic activity of human and mouse microsomes, activity of most DEHP-metabolizing enzymes was significantly higher in mice compared to that in humans; however, inter-individual variability varied from 10- to 26-fold [806].

MoA: Activation of peroxisome proliferator-activated receptor alpha (PPARα) pathway and downstream events are likely the major MoA involved in the hepatocarcinogenesis produced by phthalates, including DEHP [115,787,807,808]. While PPARα-dependent pathway is not relevant to humans [809], other molecular pathways might be involved in carcinogenicity of DEHP, including activation of nuclear receptors, nuclear factor kappa B (NFκB) and constitutive androstane receptor (CAR) [791,807,810]. Oxidative stress may also play a role. Production of benign testicular tumors is most likely caused by reproductive effects of DEHP, specifically, reduction of testosterone production [616].

Human Exposure: Exposure of general population to DEHP occurs mainly through consumption of contaminated foods, including dairy products, meat, cereal, fish and seafood [115,777]. EDI ranges from 0.45 to 3.5 µg/kg bw per day in Europe [780] and from 1 to 30 µg/kg bw per day in USA, with an average of 0.673 µg/kg/day [777,811,812]. Worldwide exposures to DEHP have declined over the years, from 4.40 µg/kg bw/day in years prior to 2000 to 2.23 µg/kg bw/day in 2015–2017, however children still have the highest levels of exposure, reaching 5.50 µg/kg bw/day [813].

Human Effects: Only limited data assessing association between human cancer, in particular breast, prostate and thyroid cancers, and exposure specifically to DEHP are available [68,115]. No significant associations between dietary exposure to phthalates and breast cancer was found in a recent population-based study [814].

Risk: IARC [115] classified DEHP as possibly carcinogenic to humans (Group 2B) (Table 2). EFSA derived a TDI for phthalates of 0.05 mg/kg bw per day [780].

4.3.2.3. 1,4-dioxane

Occurrence: 1,4-dioxane (1,4-diethylene dioxide) is an oxygen-containing single ring molecule (Figure 8(5)), which is mainly used as a solvent and stabilizer [558]. It can occur in some foods, including meat, tomatoes, shrimp and coffee as a natural constituent, as a contact material from food packaging or contaminated water, or as an impurity in food additives, such as polyethylene glycol and polysorbate [815–817]. Analysis of food products in Japan, revealed that content of 1,4-dioxane ranged from 3 to 13 µg/kg [818].

Carcinogenicity: Administration of 1,4-dioxane in drinking water induced hepato-cellular adenomas or carcinomas in rats and mice of both sexes and in male guinea pigs [68,558,815,816,819,820]. Other target organs of carcinogenicity included nasal cavity in rats of both sexes, mammary gland in female and abdominal cavity in male rats, gallbladder in male guinea pigs. Administration of 1,4-dioxane in drinking water at 25,000 ppm to rats of both sexes in a 13-week study induced glutathione S-transferase (GST) placental form-positive hepatocellular foci, which are known preneoplastic lesions [821]. In addition, 1,4-dioxane promoted hepatocellular foci produced by administration of diethyl-nitrosamine [822].

Genotoxicity/DNA Binding (Adducts): 1,4-dioxane was non-genotoxic in vitro, in vivo genotoxicity studies in rodents were also mainly negative although some positive results suggesting weak genotoxicity were observed at high cytotoxic doses exceeding 1500 mg/kg [815,816,823–826]. More recent studies provide evidence that 1,4-dioxane induced chromosomal breaks, DNA damage and mutagenicity in the liver of rats or mice [827–830]; however, these studies also used high dose levels of 1,4-dioxane (above 1000 mg/kg) and thus, no clear conclusion concerning genotoxicity of 1,4-dioxane can be made. DNA adductome analysis detected several DNA adducts with unidentified chemical structure after administration of 1,4-dioxane to male rats in the drinking water at 200 and 5000 ppm; however, these adducts could have resulted from oxidative damage, rather than direct covalent binding [831].

Metabolism: 1,4-dioxane is mainly metabolized in the presence of mixed-function oxidases to 1,4-dioxane-2-one and then to β-hydroxyethoxyacetic acid, which is excreted in urine in rats and humans [558,815,816,832]. Induction of CYP2B1/2, CYP2C11, and CYP2E1 in the liver and only CYP2E1 in the kidney and nasal mucosa was observed in rats exposed to 1,4-dioxane in the drinking water, while dosing by gavage induced CYP3A activity in the liver [833].

MoA: The mechanism(s) of carcinogenicity of 1,4-dioxane has not been elucidated but is unlikely to involve genotoxicity. Studies suggested that hepatocarcinogenicity of 1,4-dioxane likely results from cytotoxicity followed by regenerative hyperplasia, in addition, mitogenic response was suggested as a key initiating event [815,834–837]. Such effects are threshold-dependent. Oxidative damage might also play a role [838]. Tumors of nasal passages were attributed to inhalation of drinking water containing 1,4-dioxane [815,839]. SCF [817] concluded that since 1,4-dioxane is likely to exert its carcinogenic effects by non-genotoxic mechanisms, use of a threshold approach to determine acceptable levels of exposure is justified.

Human Exposure: Dietary exposures to 1,4-dioxane is a minor exposure route, in contrast to inhalation. FDA [840] estimated per capita dietary intake of 1,4-dioxane to be low, averaging at 0.6 µg/person/day. Analyses of Japanese foods revealed an EDI of 1,4-dioxane averaging from 0.44 to 4.5 µg/kg bw/day [818,841]. SCF established that an estimated maximum exposure to 1,4-dioxane as a constituent of food additives, polysorbates, in bread ranges between 0.008 to 0.05 µg/kg bw/day [817].

Human Effects: No epidemiological studies investigating association of oral exposure of humans to 1,4-dioxane and cancer are currently available [68,558,816]. Limited occupational studies found no excess of death from cancer associated with 1,4-dioxane exposure [815].

Risk: IARC [558] assigned 1,4-dioxane to a group of chemicals which are possibly carcinogenic to human (Group 2B) (Table 2). SCF [817] established that exposure to 1,4-dioxane in food additives is significantly lower than the established NOAEL of 10 mg/kg bw/day, and thus is of no toxicological concern.

4.3.2.4. Methyl Isobutyl Ketone

Occurrence: Methyl isobutyl ketone (MIBK) (4-methylpentan-2-one) (Figure 8(5)) is produced from acetone by aldol condensation and is used primarily as denaturant and solvent [115,842,843]. It is also a natural constituent of many foods including orange and lemon juice, grapes, papaya, ginger, cooked eggs, meat, milk and cheeses, beer, mushrooms, coffee and tea, as at concentrations ranging from 0.008 to 6.5 mg/kg, and as a flavoring agent in meat products, dairy and non-alcoholic beverages, baked goods and puddings at maximum reported level of 25 mg/kg [115,844]. As a component of adhesive, MIBK can also migrate into foods from packaging at levels around 10 to 12 mg/kg.

Carcinogenicity: Currently, no studies assessed carcinogenicity of MIBK following oral exposure. Some evidence of carcinogenic activity were observed in inhalation studies, which reported increased incidences of renal tubule neoplasms in rats and hepatocellular neoplasms in mice at the highest tested dose of 1800 ppm (equivalent to 1725 mg/kg/day for rats and 3171 mg/kg/day for mice) [115,842,844–846].

Genotoxicity/DNA Binding (Adducts): MIBK produced overwhelmingly negative results in genotoxicity testing battery in vitro and in vivo [115,843,846–849] and thus, is not considered to be of concern for genotoxicity [842,844].

Metabolism: MIBK is metabolized in vivo by reduction in the presence of alcohol dehydrogenases to 4-hydroxyMIBK and by oxidation in the presence of CYP-dependent monooxygenase to 4-methyl-2-pentanol [843,850–853]. The letter metabolite was not detected with oral administration [852,854]. MIBK has been shown to induce liver and renal CYPs, potentiating hepato- and nephrotoxicity produced by chloroform and carbon tetrachloride [853,855,856].

MoA: In the carcinogenicity bioassay, histopathologic changes observed in the kidneys of rats were characteristic of α_{2u}-globulin nephropathy [846], suggesting that α_{2u}-globulin-mediated MoA is involved in renal carcinogenesis [275,845,857,858]. This MoA is not considered relevant to humans [73]. MoA underlying hepatocarcinogenicity of MIBK in mice is not well understood. While it potentiated hepatotoxicity and cholestasis produced by other chemicals [853,855,856,859,860] no evidence of hepatotoxicity was observed when

MIBK was administered alone [275,843]. A study by Hughes and colleagues [861] suggested involvement of receptor-mediated mechanism, specifically, activation of the CAR/PXR nuclear receptors, which results in hepatocellular proliferation consequently leading to tumor development.

Human Exposure: Dietary per capita exposure to MIBK was calculated to be 7 µg/person/day in Europe and 2 µg/person/day in USA [849]. More recent estimations suggest lower levels of intake of 0.02 µg/kg/day [115].

Human Effects: No data on human carcinogenicity of MIBK are currently available [115]. Long term exposure in occupational settings was reported to cause cognitive impairment [862].

Risk: IARC [275] classified MIBK as possibly carcinogenic to humans (Group 2B) (Table 2). JECFA [849] concluded that at the current levels of intake as a flavoring agent MIBK is unlikely to pose any hazard to human health.

4.4. Carcinogens Formed during Processing, Packaging and Storage of Food

Food processing contaminants are generated through cooking practices or as a result of food packaging and storage. Some of the carcinogens belonging to this type were discussed in the section on DNA-reactive carcinogens. Examples of epigenetic processing carcinogens in food include alkylated imidazoles and furan.

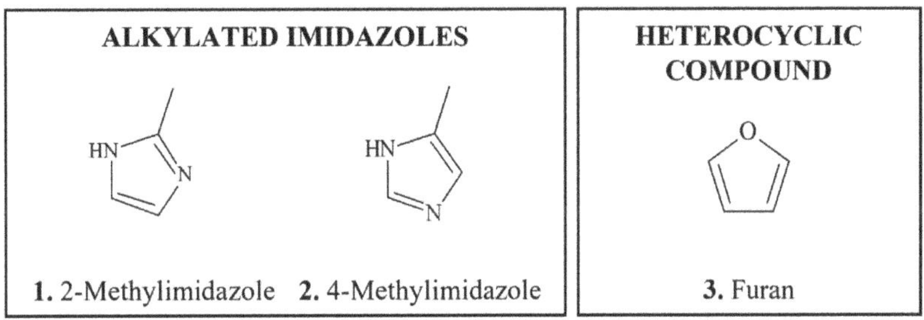

Figure 9. Chemical structures of non-DNA-reactive carcinogens formed during processing, packaging and storage of food.

4.4.1. Alkylated Imidazoles

Occurrence: *2-methylimidazole* (2-MI) (Figure 9(1)) and *4-methylimidazole* (4-MI) (Figure 9(2)) are formed during fermentation and cooking by ammoniation of simple sugars [115,863–865]. They have been identified as by-products in foods including caramel coloring (Classes III and IV) and caramel-colored syrups, cola, ammoniated molasses, wine, Worcestershire sauce, and soy sauce [115,863,866–871]. 4-MI has been also detected in the milk from cows fed ammoniated forage [872,873]. Alkylated imidazoles can be also formed during thermal processing of natural constituents not containing caramel coloring, thus up to 466 µg/kg of 4-MI and up to 135 µg/kg of 2-MI were detected in roasted barley, malt and cocoa powder [874].

Carcinogenicity: Both, 2-MI and 4-MI, were carcinogenic in rodent studies [115,875–878]. Specifically, 2-MI induced thyroid follicular cell hypertrophy in mice and hyperplasia in rats by 15 days [879]. In a 2-year feed study, there was some evidence of carcinogenic activity of 2-MI in male rats based on increased incidences of thyroid gland follicular cell neoplasms and clear evidence of carcinogenic activity in female rats based on increased incidences of thyroid gland follicular cell neoplasms [876,878]. In addition, increased incidences of hepatocellular adenoma in male and female rats may have been related to exposure. In mice, there was some evidence of carcinogenic activity of 2-MI, based on increased incidences of thyroid gland follicular cell adenoma and hepatocellular neoplasms in males and increased incidences of hepatocellular adenoma in females [876,878]. In NTP bioassays, 4-MI fed to groups of male and female mice in diet containing 312 ppm (equivalent to

80 mg/kg bw/day) and greater for 106 weeks, increased incidences of pulmonary alveolar/bronchiolar adenoma in all dosed groups of females, alveolar/bronchiolar carcinoma in males given 1250 ppm, and alveolar/bronchiolar adenoma or carcinoma (combined) in males fed 1250 ppm and in females fed 625 and 1250 ppm [875,877]. In male and female rats fed diets containing 4-MI at up to 2500 ppm to males or 5000 ppm to females (115 and 260 mg/kg bw/day, respectively) for 106 weeks, there was no evidence of carcinogenic activity in males and only equivocal evidence in females based on modest increases in the incidences of mononuclear cell leukemia [875,877].

Genotoxicity/DNA Binding (Adducts): 2-MI and 4-MI were negative in bacterial mutation assays when tested either with or without an exogenous bioactivation system [115,865,880,881]. 2-MI yielded mixed results in vivo for induction of chromosomal damage, as measured by micronucleated erythrocyte frequency and was negative in bone marrow micronucleus tests in rats and mice when administered intraperitoneally three times at 24-h intervals [115]. In a 14-week study of 2-MI; however, a significant exposure-related increase in the frequency of micronucleated erythrocytes was noted in peripheral blood of male and female mice [865,878]. While 4-MI produced SCE, chromosome aberrations and micronuclei induction in human peripheral lymphocytes in vitro [882], no increase in the frequencies of micronucleated erythrocytes was seen in the bone marrow of male rats or mice which were administered 4-MI by intraperitoneal injection, or in peripheral blood samples from male and female mice dosed in feed for 14 weeks [115,865,875,881].

Metabolism: 2- and 4-MI are rapidly absorbed and quickly eliminated in the urine mainly unchanged [871,883,884]. Mice cleared 2-MI faster than rats [878]. In rats, 2-MI is distributed to several tissues, including the thyroid [885], while 4-MI is mainly distributed to intestines, liver and kidney [115]. The principal urinary metabolite of 2-MI is the ring oxidized 2-MI, which possesses nucleophilicity [886,887]. Metabolism of 4-MI in rats and mice is similar, and the major metabolite detected in both species was hydroxylated 4-MI [884].

MoA: The MoA of 2-MI for induction of thyroid neoplasms likely involves interference with thyroid homeostasis, as described for several other chemicals [888]. 2-MI produced exposure-related reduction in thyroxine (T4) and increases in thyroid-stimulating hormone (TSH) in rats and had a lesser effect on T4 in mice [878]. The decrease in T4 can be attributed to increased hepatic UDP-glucuronyl transferase activity [876], which would lead to increased conjugation and excretion of T4. In response to T4 reduction, pituitary production of TSH increases which stimulates function and growth of the thyroid [888]. Likewise, the MoA of 2-MI for the liver tumors appears to involve a trophic effect on the liver reflected by liver weight and enzyme increases [876]. Both of these MoAs represent adaptive effects [889], which would be anticipated to be reversible. Nevertheless, IARC concluded that relevance of such tumor response in animals to humans cannot be excluded [115]. The MoA of 4-MI in induction of lung tumors in mice remains unclear, but does not involve genotoxicity, cytotoxicity or mitogenicity [115,890,891].

Human Exposure: The overall EDI for 4-MI for US population ranges from 0.13 to 0.51 µg/kg bw/day, with cola-type carbonated beverages being the highest contributor [892]. The average dietary intake for 4-MI in Europe was estimated to be between 0.4 to 3.7 µg/kg bw/day [893]. EDI of 4-MI from caramel colors ranges from 6 to 11 µg/kg bw/day for Class III and 7 to 9 µg/kg bw/day for Class IV [871].

Human Effects: No epidemiologic studies assessing human cancer risk of 2-MI or 4-MI were found [115].

Risk: IARC classifies 2-MI and 4-MI as possibly carcinogenic to humans (Group 2B) (Table 2). JECFA limits level of 4-MI to 200 and 250 mg/kg in caramel colors Classes III and IV, respectively [894]. EFSA suggest ADI of 300 mg/kg bw/day for all classes of caramel color [870]. FDA and EFSA concluded that at levels present in caramel colors 4-MI is not expected to be a concern to human health [871].

4.4.2. Furan

Occurrence: *Furan* (oxacyclopentadiene) (Figure 9(3)) is a volatile contaminant formed in some foods during heat treatment techniques such as canning and jarring where the furan cannot escape [895]. The sources for the formation of furan include the oxidation of polyunsaturated fatty acids or the decomposition of carbohydrates or amino acids, but the relative contributions of these processes in actual foods is not known [896,897]. Analysis of approximately 300 food samples found furan levels ranging from nondetectable (below the limits of detection of the method) to 175 ppb [898]. Particularly high levels were found in foods that are roasted (e.g., coffee, cocoa, nuts, toasted bread, popcorn) or heated in closed containers (e.g., canned food, ready meals and baby food) [896–899].

Carcinogenicity: In rodent carcinogenicity studies [68,900,901], furan, administered to rats by gavage at 8 mg/kg bw, 5 days/week, induced a high incidence of cholangio-carcinomas in both males and females, at lower doses these tumors were reclassified as cholangiofibrosis [896,902]. In addition, incidences of mononuclear cell leukemia were increased in both sexes, and in males, a high incidence of hepatocellular neoplasms was also produced, while in females the incidence was moderate. In mice, hepatocellular neo-plasms were induced at 8 mg/kg bw/day, 5 days/week [900]. In female mice, increased incidence of hepatocellular altered foci and hepatocellular tumors were preceded by a dose-dependent increase in cell proliferation [903]. At lower dosages (up to 2 mg/kg bw), furan administered to male rats by gavage induced malignant mesotheliomas in the epididymis and testicular tunics, dose-related increases in the incidence of mononuclear cell leukemia and dose-related increasing trend in the incidence of hepatocellular adenomas [902].

Genotoxicity/DNA Binding (Adducts): Furan produced some genotoxicity in bacterial and mammalian cells in vitro [896,897,901], and chromosomal aberrations in mice, but was negative in SCE, in the mouse bone marrow erythrocyte micronucleus assay, and did not induce UDS in rats or mice [900,904,905]. In Big Blue rats, furan produced mainly negative responses in micronucleus or mutagenicity assays; however, some DNA damage was reported in the comet assay only at cytotoxic doses [906]. DNA strand breaks reported in rat liver but not in the bone marrow were associated with oxidative stress [907]. DNA-protein crosslinking were observed in turkey embryos after dosing with furan [908]. A major metabolite of furan, cis-2-butene-1,4-dial, is a direct acting mutagen [909], which can covalently bind to DNA in vitro [910], but not in vivo [911]. Thus, some DNA binding observed in liver and kidney DNA was attributed to other furan metabolites [912].

Metabolism: Furan undergoes oxidation by CYP2E1 resulting in ring scission and formation of the α-unsaturated dicarbonyl cis-2-butene-1,4-dial [886,913,914], which is likely the toxic metabolite [915].

MoA: The MoA of furan carcinogenicity is uncertain, but likely involves chronic toxicity with increased cell proliferation, which results from binding of furan and cis-2-butene-1,4-dial to GSH and proteins [896,897]. Nevertheless, in vivo DNA reactivity has not been rigorously excluded. In addition, there is evidence that oxidative stress and epigenetic alterations play a role [896,916–918].

Human Exposure: As mentioned above, furan is formed in a variety of heat-treated foods by thermal degradation of natural food constituents. Mean dietary exposure to furan in Europe may be as high as 1.23 and 1.01 µg/kg bw/day for adults and 3- to 12-month-old infants, respectively [896]. In the US, FDA [919] calculations estimated that mean daily furan exposures ranged from 0.26 µg/kg bw/day for adults to 0.41 µg/kg/day for infants consuming baby food and 0.9 µg/kg/day for those consuming infant formula. EDI calculated based on the Dortmund Nutritional and Anthropometric Longitudinally Designed (DONALD) study for consumers of commercially jarred foods ranged between 0.182 and 0.688 µg/kg/day [920].

Human Effects: Currently, data on effects if furan in humans is limited, and no association with carcinogenicity of furan in humans has been investigated.

Risk: IARC [901] classified furan as possibly carcinogenic to humans (Group 2B) (Table 2). EFSA [896] concluded that exposure to furan indicates health concern, due to uncertainties regarding the MoA of furan carcinogenicity; however, current furan exposures are lower than established MoE of concern.

4.5. Food Additives

Food additives are added to food in order to improve or maintain certain characteristics, such as taste, texture, appearance or safety. Some are added to foods directly, while others migrate into foods in trace amounts during packaging, storage or handling [921]. According to FDA regulations established in late 1950s, any direct food additive with a carcinogenic potential should not be added to food; however, current advanced understanding of different mechanisms involved in chemical carcinogenesis puts relevance of such regulations under scrutiny [922,923].

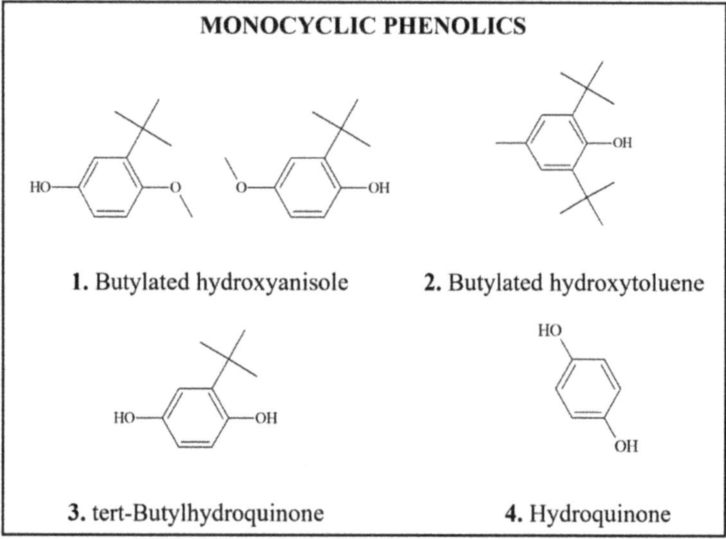

Figure 10. Chemical structures of non-DNA-reactive carcinogenic food additives.

Monocyclic Phenolics, Synthetic and Natural

Occurrence: The synthetic phenolic antioxidants *butylated hydroxyanisole* (BHA) (Figure 10(1)), *butylated hydroxytoluene* (BHT) (Figure 10(2)), *tert-butylhydroquinone* (BHQ) (Figure 10(3)), and *hydroquinone* (HQ) (Figure 10(4)) are widely used as food additives to prevent oxidation of lipids [193,849,924–928]. HQ (Figure 10(4)) can also occurs naturally in food, mainly as a glucose conjugate, 4-hydroxyphenyl-P-glucopyranoside, known as arbutin, but can be also found in the free form [558,929]. Foods rich in arbutin include wheat cereal, bread, coffee and pears, with wheat products and pears having the highest levels of arbutin, 10 and 15 ppm, respectively [929].

Carcinogenicity: When fed in the diet, BHA elicited increased forestomach neoplasms in rats, hamsters and mice at doses greater than 2% [193,926,930,931]. BHT produced an increase in mouse lung neoplasms at doses of 0.75% [193,927,931]. BHQ produced no neoplasms in rats or mice when fed up to 5000 ppm, in spite of the positive genotoxicity findings; however, in a 6-week feeding study, preneoplastic lesions and papillomas were observed in forestomach of rats [932,933]. In two-year rodent bioassays, HQ administration by gavage at 25 or 50 mg/kg in water led at both doses to increases in renal tubular adenomas in male rats and mononuclear cell leukemias in female rats [558,934]. While there was no evidence of carcinogenic activity in male mice administered 50 or 100 mg HQ/kg bw in water, 5 days/week by gavage, there was some evidence of carcinogenic activity in female mice, as shown by increases in hepatocellular neoplasms, mainly adenomas, and thyroid follicular cell adenoma at the same doses [558,934].

Genotoxicity/DNA Binding (Adducts): The synthetic phenolics are non-genotoxic [924,931,935,936], although BHQ has yielded some positive findings in vitro, but not in vivo [937,938]. No DNA binding was detected in the target tissue, forestomach, of rats administered BHA or its metabolites up to 1000 mg/kg [939]. HQ tested positive in some in vitro and in vivo genotoxicity assays [344,558,940], but did not form adducts with DNA in vivo [345,346].

Metabolism: BHA is mainly metabolized to glucuronide and sulphate conjugates or is demethylated to free phenols, including BHQ [926]. BHQ can further undergo either oxidation to a quinone metabolite or GSH conjugation, as has been shown in rats [941]. Metabolism of BHT in vitro, using mouse microsomes, produced quinone methides, while in vivo, it is oxidized at one or both *tert*-butyl group(s) by microsomal oxygenase, followed by conjugation with glucuronide [927]. HQ is metabolized mainly to sulfate and glucuronide conjugates, but a small percentage may be converted to 1,4-benzoquinone, which can be either conjugated with GSH or form DNA adducts in vitro [558].

MoA: The MoAs of BHA and BHT involve epigenetic mechanisms ultimately leading to promotion of background neoplasia [924,930,933,942]. A lifetime dose–response study of tumor promoting effect of BHA in the rat forestomach identified positive effects at 6000 ppm and above, and a NOAEL at 3000 ppm [943]. In the case of BHT, its MoA appears to involve infiltration of monocytes into the pulmonary alveoli and stimulation of proliferation of type II pneumocytes [944,945]. The MoA for the kidney tumor induction in male rats by HQ has been suggested to involve cytotoxicity leading to increased cell proliferation and exacerbation of chronic progressive nephropathy [946–949].

Human Exposure: The mean daily intake of BHA varies from 2 to 300 µg/kg bw/day, depending on consumer age, region and estimate methods, and the main sources of exposure, similar to other phenolics, include baked goods, snacks and processed potato products [849,924,926]. EFSA reports [927] a mean dietary exposure to BHT is in the range of 10 to 30 µg/kg bw/day for adults and 10 to 90 µg/kg bw/day for children, while JECFA [849] calculated EDI to be between 700 and 990 µg/kg bw. The EDI for BHQ ranges from 4 to 140 µg/kg bw/day based on poundage data and from 370 to 690 µg/kg bw/day, based on model diets [849]. EFSA [928] determined that exposure to BHQ as a food additive averages to approximately 5 µg/kg bw/day in adults and 257 µg/kg bw/day in children.

Human Effects: No epidemiological study has implicated BHA or BHT as human carcinogens [193]. To the contrary, studies has suggested an inverse relationship with cancers of gastrointestinal tract [950,951], which is consistent with demonstrated anticarcinogenicity of monocyclic phenolics against a variety of DNA-reactive carcinogens in animal models [345,951,952]. No evidence for human carcinogenicity was found in studies of occupational exposures or dermatological use of HQ [558,953]. No study of cancer risk with dietary exposures was reported.

Risk: IARC [193,558] classified BHA as possibly carcinogenic to humans (Group 2B), while BHT and HQ were considered as not classifiable as to human carcinogenicity (Group 3) (Table 2). BHA and BHT are not considered to pose carcinogenic risks to humans based on a threshold-dependent MoAs that are not relevant to potential human exposure levels [930]. Human exposures are not expected to exceed the ADI values of 0.5, 0.3 and 0.7 mg/kg bw, which have been allocated by JECFA to BHA, BHT and BHQ, respectively [849,954]. EFSA [926–928] arrived at a similar conclusion after re-evaluating ADIs for BHA, BHT and BHQ to be 1, 0.25 and 0.7 mg/kg bw/day, respectively, and comparing them to the mean dietary intakes in children and adult populations.

5. Food-Borne Chemopreventive Agents

A growing body of evidence suggests that a variety of food-derived constituents may exhibit properties that are preventive and/or protective of cancer formation [1,955–959]. Since 1940, almost half of antitumor drugs have been developed from natural products or their derivatives [960]. In contrast, use of naturally occurring chemicals present in diet for cancer prevention and suppression (also referred to as chemoprevention) remains a highly debatable topic due to the lack of success in clinical trials [961–964].

The MoA underlying chemopreventive properties of naturally occurring agents varies [955]. Some compounds block tumor induction during the initiation stage, others exhibit inhibitory effects during promotion stage, and certain anticarcinogenic substances affect process of carcinogenesis at multiple points [963,965].

Inhibition of preneoplastic and neoplastic effects produced by established chemical carcinogens (including chemicals discussed in this review), has been described in experimental settings for several food constituents. For example, naturally occurring coumarins found in citrus fruits, tonka beans, parsnip, parsley, cinnamon bark oil and peppermint oil, including simple coumarins (coumarin, limettin) and linear furanocoumarins (imperatorin and isopimpinellin) inhibited formation of pro-mutagenic DNA adducts by DMBA in mouse mammary gland and skin [966,967] and formation of BaP DNA adducts and skin tumors in mice [966]. Significant inhibition of DNA adduct levels produced by the heterocyclic amine, PhIP, in rat colonic tissue has been reported after administration of black tea, constituent of mustard plant and papaya seeds, benzylisothiocyanate, and diterpenes, kahweol and cafestol, found in coffee beans [968]. In addition, kahweol and cafestol prevented covalent DNA binding of AFB_1 in rat livers [969] and restricted tumor formation and growth caused by DMBA in hamster buccal pouch [970]. The flavonoid, nevadensin, which is present in basil and peppermint, significantly inhibited formation of ME specific DNA adducts and incidences of preneoplastic hepatocellular altered foci in the rat liver [971]. Naturally occurring in turmeric, curcumin has been shown to inhibit initiation and promotion of BaP-induced forestomach tumors and DMBA-induced skin tumors in mice [972] and stomach tumors induced by N-methyl-N'-nitro-N-nitrosoguanidine in rats [973]. Synthetic phenolic antioxidant BHT, which is often added to processed food, and its oxidative metabolites, 2,6-di-tert-butyl-4-hydroxymethylphenol and 2,6-di-tert-butyl-1,4-benzoquinone, inhibited DMBA-DNA adducts formation and tumorigenesis in mammary gland of rats [974]. BHT and BHA were also shown to inhibit the hepatocarcinogenicity of AFB_1 in rats [952,975,976]. Monocyclic phenolics, HQ, inhibited cancer-initiating effects of 2-acetylaminofluorene, including formation of DNA adducts, cell proliferation and formation of preneoplastic foci, in rat liver [345]. Chlorophyll [977], which is available in green vegetables, as well as constituents of cruciferous vegetables, 5-(2-pyrazinyl)-4-methyl-1,2-dithiol-3-thione (oltipraz) and related 1,2-dithiol-3-thiones and 1,2-dithiol-3-ones [978], inhibited experimental hepatocarcinogenicity of AFB_1 in rats. Other indoles, including indole-3-carbinol and 3,3'-diindolylmethane, also inhibited DMBA-induced mammary gland tumors in rats and BaP-induced forestomach neoplasms in mice [979]. A number of carotenoids, which are pigments present in yellow-orange vegetables, with some representatives of this class also serving as vitamin A precursors, mitigated bacterial mutagenicity of AFB_1, BaP, IQ, and cyclophosphamide in *Salmonella typhimurium*, and clastogenicity of BaP and cyclophosphamide in mouse bone marrow micronucleus assay [980], and were shown to decrease risk of cancers in different sites, including skin, lung, liver and colon [981].

One of the major mechanisms believed to be involved in cancer prevention by chemicals present in diet is modification in the activity of enzymes which are involved in xenobiotic metabolism. Thus, inhibition of phase I enzymes blocks bioactivation of procarcinogens to reactive metabolites, while induction of phase II enzymes leads to increased detoxication and excretion of carcinogens. For example, the inhibitory effect of naturally occurring coumarins on formation DMBA and BaP DNA adducts, is suggested to be a result of CYP1A1/1B1 inhibition and induction of GST activities [966,967]. Inhibition of tumorigenic effects of a wide array of carcinogens has been also linked to modification of phase I and phase II enzyme activities. Such activity was suggested for constituents of cruciferous vegetables, isothiocyanates (sodium cyanate, tert-butyl isocyanate, phenethyl isothiocyanate, benzyl isothiocyanate and sulforaphane), which inhibited carcinogenicity of ethionine, 2-acetylaminofluorene, 3,3'-diaminobenzidine, m-toluylenediamine and N-

butyl-*N*-(4-hydroxybutyl)nitrosamine in rat liver, DMBA in rat mammary gland, BaP and 4-(methylnitrosamino)-1-(3-pyridyl)-1-butanone in mouse lung and 1,2-dimethylhydrazine in large intestine of mice [951,982–984]. Moreover, consumption of cruciferous vegetables, including broccoli, brussels sprouts and cauliflower, was demonstrated to significantly enhance detoxication and excretion of PhIP present in cooked meat in humans, due to induction of phase I (CYP1A2) and phase II (glucuronidation) metabolism [985]. In addition, isothiocyanate compound, sulforaphane, exhibits anti-inflammatory and pro-apoptotic properties, thus, contributing to chemoprevention [986]. It should be noted, however, that persistent enzyme induction may be undesirable for humans.

In addition to the effects mentioned above, many plant components (often referred to as phytochemicals), such as flavone derivatives, isoflavones, catechins, coumarins, phenylpropanoids, polyfunctional organic acids, phosphatides, tocopherols, ascorbic acid, and carotenes act as antioxidants counteracting formation of ROS and thereby, preventing oxidative stress and oxidative DNA damage [987–989]. For example, antioxidant effects of resveratrol were associated with inhibition of formation and promotion of skin and breast tumors induced by DMBA in mice or rats [990]. Evidence suggests that food-derived phytochemicals can potentiate antioxidant effects of each other, emphasizing the importance of whole food diets rich in fruits and vegetables [988].

Elucidation of mechanisms of tumor prevention on the molecular level, led to discovery of various signaling pathways and molecular targets that are affected by chemopreventive agents [955,963,965]. For example, butyrate-containing structured lipids and tributyrin, which can be found in wholegrains, vegetables, fruits, nuts and beans, have been shown to prevent and/or inhibit activation of major oncogenes and induce apoptosis at early stages of hepatocarcinogenesis [991]. Tumor-suppressive activity of tributyrin was enhanced by concurrent administration of folic acid [992]. Resveratrol (3,4′,5-trihydroxy-trans-stilbene), which is present in the grape skins, peanuts and red wine, in addition to antioxidant effects, has been shown to modulate cell-cycle regulating pathways, such as MAPKs and NF-κB/AP-1, inducing apoptosis in carcinoma cell lines [993].

It should be noted, however, that some of the compounds described in this section can demonstrate dual effects in experimental settings. Thus, indole-3-carbinol, BHT and coumarin in addition to chemopreventive properties, under certain conditions, usually involving high exposures, can act as tumor promoters [935,994,995]. Nevertheless, presence of chemopreventive agents in diet and their beneficial effects on cancer prophylaxis in humans warrants further investigation.

6. Discussion and Conclusions

A wide variety of carcinogens is present in foods and beverages consumed by humans. Categorization of these as operating through MoAs that are either DNA-reactive or epigenetic reveals important differences. Dietary carcinogens of the DNA-reactive type often produce hazardous effects at much lower doses compared to the epigenetic carcinogens. In addition, they frequently affect multiple target tissues, whereas epigenetic carcinogens often affect no more than two sites, where their MoA is exerted. Most importantly, several DNA-reactive carcinogens found in diet are recognized as causes of human cancer, including aflatoxins, aristolochic acid, benzene, benzo[a]pyrene, ethylene oxide, and preserved food components. Other food-derived DNA-reactive chemicals that are considered likely to contribute to human cancer include nitrosamines from several sources, as well as polycyclic aromatic hydrocarbons and heterocyclic amines formed during processing or cooking of food. In contrast, the only food-borne epigenetic carcinogen considered by some authorities to be associated with increased cancer in humans, although not from low-level food exposure, is dioxin (TCDD). Accordingly, DNA-reactive carcinogens represent a much greater risk than epigenetic carcinogens.

Author Contributions: Conceptualization, G.M.W.; writing—original draft preparation, T.K. and G.M.W.; writing—review and editing, T.K., B.P.C.S. and G.M.W.; funding acquisition, T.K., B.P.C.S. and G.M.W. All authors have read and agreed to the published version of the manuscript.

Funding: This work was supported by the Chemical Risk Assessment Fund (New York Medical College Grant no. 161140) and the National Research Foundation Singapore Whitespace grant (Grant no. W20W3D0002) administered by the Agency for Science, Technology & Research.

Data Availability Statement: Data sharing not applicable. No new data were created or analyzed in this study. Data sharing is not applicable to this article.

Conflicts of Interest: The authors report no competing interests to declare.

Abbreviations

2-HEX, *Trans*-2-hexenal; 2-MI, 2-methylimidazole; 4-MI, 4-methylimidazole; AC, Acrylamide; AA, Aristolochic acid; AB, Alkenylbenzenes; ADI, Acceptable daily intake; AF, Aflatoxin; BaP, Benzo[a]pyrene; BHA, Butylated hydroxyanisole; BHQ, tert-Butylhydroquinone; BHT, Butylated hydroxytoluene; bw, body weight; BZ, Benzene; CA, Crotonaldehyde; CAR, Constitutive androstane receptor; CP, 3-chloro-1,2-propanediol; CONTAM, Panel on Contaminants in the Food Chain; CYP, Cytochrome P450; DDD, p,p'-dichlorodiphenyl-dichloroethane; DDE, p,p'-dichlorodiphenyl-dichloroethylene; DDT, p,p'-dichlorodiphenyl-trichloroethane; diMeIQx, 2-amino-3,4,8-trimethylimidazo[4,5-f]quinoline; DEHP, Di(2-ethylhexyl) phthalate; DMBA, 7,12-dimethylbenz[a]anthracene; DLC, Dioxin-like compound; DP, 1,3-dichloro-2-propanol; EC, Ethyl carbamate; EDI, estimated daily intake; EFSA, European Food Safety Authority; EMA, European Medical Agency; EtO, Ethylene oxide; FAO, Food and Agriculture Organization; FB_1, Fumonisin B_1; FC, Fusarin C; FDA, US Food and Drug Administration; FEMA, Flavor and Extract Manufacturers Association; GA, Glycidamide; GSH, Glutathione; GST, Glutathione S-transferase; HCA, Heterocyclic amine; HNE, 4-hydroxynonenal; HQ, Hydroquinone; IARC, International Agency for Research on Cancer; IQ, 2-amino-3-methylimidazo[4,5-f]quinoline; JECFA, WHO/FAO Joint Expert Committee of Food Additives; MAM, Methylazoxymethanol; MDA, Malondialdehyde; ME, Methyl eugenol; MeIQ, 2-amino-3,4-dimethylimidazo[4,5-f]quinoline; MelQx, 2-amino-3,8-dimethylimidazo[4,5-f]quinoline; MIBK, Methyl isobutyl ketone; MoA, Mechanism of action; MoE, Margins of exposure; NDEA, *N*-nitrosodiethylamine; NOAEL, No-observed-adverse-effect-level; NPL, 32P-nucleotide postlabeling; OTA, Ochratoxin A; PA, Pyrrolizidine alkaloid; PAH, Polycyclic aromatic hydrocarbon; PCB, polychlorinated biphenyl; PCDD, polychlorinated dibenzo-para-dioxin; PCDF, polychlorinated dibenzofuran; PhIP, 2-amino-1-methyl-6-phenylimidazo[4,5-b]pyridine; PMTDI, Provisional maximum tolerable daily intake; PPARα, peroxisome proliferator-activated receptor alpha; PUFA, Polyunsaturated fatty acids; PUL, Pulegone; ROS, reactive oxygen species; SAF, Safrole; SCE, Sister chromatid exchange; SCF, Scientific Committee on Food; T4, Thyroxine; TCDD, Dioxin; TD_{50}, Carcinogenic potency; TDI, Tolerable daily intake; TEF, Toxic equivalency factor; TSH, Thyroid stimulating hormone; UDS, Unscheduled DNA synthesis; WHO, World Health Organization.

References

1. National Research Council Committee, on Comparative Toxicity of Naturally Occurring Carcinogens. *Carcinogens and Anticarcinogens in the Human Diet: A Comparison of Naturally Occurring and Synthetic Substances*; The National Academies Collection: Reports Funded by National Institutes of Health; National Academies Press (US): Washington, DC, USA, 1996.
2. Williams, G.M. Food-borne carcinogens. *Prog. Clin. Biol. Res.* **1986**, *206*, 73–81.
3. Abnet, C.C. Carcinogenic food contaminants. *Cancer Investig.* **2007**, *25*, 189–196. [CrossRef] [PubMed]
4. Sugimura, T. Nutrition and dietary carcinogens. *Carcinogenesis* **2000**, *21*, 387–395. [CrossRef] [PubMed]
5. Jackson, L.S. Chemical food safety issues in the United States: Past, present, and future. *J. Agric. Food Chem.* **2009**, *57*, 8161–8170. [CrossRef] [PubMed]
6. Rietjens, I.M.C.M.; Michael, A.; Bolt, H.M.; Siméon, B.; Andrea, H.; Nils, H.; Christine, K.; Angela, M.; Gloria, P.; Daniel, R.; et al. The role of endogenous versus exogenous sources in the exposome of putative genotoxins and consequences for risk assessment. *Arch. Toxicol.* **2022**, *96*, 1297–1352. [CrossRef] [PubMed]
7. Hecht, S.S.; Hoffmann, D. N-nitroso compounds and man: Sources of exposure, endogenous formation and occurrence in body fluids. *Eur. J. Cancer Prev.* **1998**, *7*, 165–166.

8. Tricker, A.R.; Preussmann, R. Carcinogenic N-nitrosamines in the diet: Occurrence, formation, mechanisms and carcinogenic potential. *Mutat. Res./Genet. Toxicol.* **1991**, *259*, 277–289. [CrossRef]
9. Key, T.J.; Bradbury, K.E.; Perez-Cornago, A.; Sinha, R.; Tsilidis, K.K.; Tsugane, S. Diet, nutrition, and cancer risk: What do we know and what is the way forward? *BMJ* **2020**, *368*, m511. [CrossRef]
10. Reddy, B.S.; Cohen, L.A.; David McCoy, G.; Hill, P.; Weisburger, J.H.; Wynder, E.L. Nutrition and its relationship to cancer. *Adv. Cancer Res.* **1980**, *32*, 237–345. [CrossRef]
11. Clapp, R.W.; Howe, G.K.; Jacobs, M.M. Environmental and occupational causes of cancer: A call to act on what we know. *Biomed. Pharm.* **2007**, *61*, 631–639. [CrossRef]
12. Doll, R.; Peto, R. The causes of cancer: Quantitative estimates of avoidable risks of cancer in the United States today. *JNCI J. Natl. Cancer Inst.* **1981**, *66*, 1192–1308. [CrossRef]
13. Rumgay, H.; Shield, K.; Charvat, H.; Ferrari, P.; Sornpaisarn, B.; Obot, I.; Islami, F.; Lemmens, V.; Rehm, J.; Soerjomataram, I. Global burden of cancer in 2020 attributable to alcohol consumption: A population-based study. *Lancet Oncol.* **2021**, *22*, 1071–1080. [CrossRef]
14. Cogliano, V.J.; Baan, R.; Straif, K.; Grosse, Y.; Lauby-Secretan, B.; El Ghissassi, F.; Bouvard, V.; Benbrahim-Tallaa, L.; Guha, N.; Freeman, C.; et al. Preventable exposures associated with human cancers. *J. Natl. Cancer Inst.* **2011**, *103*, 1827–1839. [CrossRef] [PubMed]
15. IARC, International Agency for Research on Cancer. Pharmaceuticals. Volume 100 A. A review of human carcinogens. *IARC Monogr. Eval. Carcinog. Risks Hum.* **2012**, *100*, 1–401.
16. Pflaum, T.; Hausler, T.; Baumung, C.; Ackermann, S.; Kuballa, T.; Rehm, J.; Lachenmeier, D.W. Carcinogenic compounds in alcoholic beverages: An update. *Arch. Toxicol.* **2016**, *90*, 2349–2367. [CrossRef]
17. Kobets, T.; Iatropoulos, M.J.; Williams, G.M. Mechanisms of DNA-reactive and epigenetic chemical carcinogens: Applications to carcinogenicity testing and risk assessment. *Toxicol. Res.* **2019**, *8*, 123–145. [CrossRef]
18. Williams, G.M.; Iatropoulos, M.J.; Enzmann, H.G.; Deschl, U. Carcinogenicity of chemicals: Assessment and human extrapolation. In *Hayes' Principles and Methods of Toxicology*, 6th ed.; Hayes, A., Kruger, C.L., Eds.; Taylor and Francis: Philadelphia, PA, USA, 2014; pp. 1251–1303.
19. Williams, G.M. Mechanisms of chemical carcinogenesis and application to human cancer risk assessment. *Toxicology* **2001**, *166*, 3–10. [CrossRef]
20. Weisburger, J.H.; Williams, G.M. The distinction between genotoxic and epigenetic carcinogens and implication for cancer risk. *Toxicol. Sci.* **2000**, *57*, 4–5. [CrossRef]
21. Williams, G.M. Application of mode-of-action considerations in human cancer risk assessment. *Toxicol. Lett.* **2008**, *180*, 75–80. [CrossRef]
22. Williams, G.M. Chemicals with carcinogenic activity in the rodent liver; mechanistic evaluation of human risk. *Cancer Lett.* **1997**, *117*, 175–188. [CrossRef]
23. Miller, E.C.; Miller, J.A. Biochemical mechanisms of chemical carcinogenesis. *Mol. Biol. Cancer* **1974**, 377–402. [CrossRef]
24. Preston, R.J.; Williams, G.M. DNA-reactive carcinogens: Mode of action and human cancer hazard. *Crit. Rev. Toxicol.* **2005**, *35*, 673–683. [CrossRef]
25. Hartwig, A.; Arand, M.; Epe, B.; Guth, S.; Jahnke, G.; Lampen, A.; Martus, H.-J.; Monien, B.; Rietjens, I.M.C.M.; Schmitz-Spanke, S.; et al. Mode of action-based risk assessment of genotoxic carcinogens. *Arch. Toxicol.* **2020**, *94*, 1787–1877. [CrossRef] [PubMed]
26. Hanawalt, P. Functional characterization of global genomic DNA repair and its implications for cancer. *Mutat. Res./Rev. Mutat. Res.* **2003**, *544*, 107–114. [CrossRef] [PubMed]
27. Tong, C.; Fazio, M.; Williams, G.M. Cell cycle-specific mutagenesis at the hypoxanthine phosphoribosyltransferase locus in adult rat liver epithelial cells. *Proc. Natl. Acad. Sci. USA* **1980**, *77*, 7377–7379. [CrossRef] [PubMed]
28. Kaufmann, W.K.; Kaufman, D.G. Cell cycle control, DNA repair and initiation of carcinogenesis. *FASEB J.* **1993**, *7*, 1188–1191. [CrossRef] [PubMed]
29. Pagès, V.; Fuchs, R.P. How DNA lesions are turned into mutations within cells? *Oncogene* **2002**, *21*, 8957–8966. [CrossRef] [PubMed]
30. Vogelstein, B.; Papadopoulos, N.; Velculescu, V.E.; Zhou, S.; Diaz, L.A., Jr.; Kinzler, K.W. Cancer genome landscapes. *Science* **2013**, *339*, 1546–1558. [CrossRef]
31. Williams, G.M.; Iatropoulos, M.J.; Jeffrey, A.M. Mechanistic basis for nonlinearities and thresholds in rat liver carcinogenesis by the DNA-reactive carcinogens 2-acetylaminofluorene and diethylnitrosamine. *Toxicol. Pathol.* **2000**, *28*, 388–395. [CrossRef]
32. Cohen, S.M.; Arnold, L.L. Chemical carcinogenesis. *Toxicol. Sci.* **2011**, *120* (Suppl. S1), S76–S92. [CrossRef]
33. Poirier, M.C. Chemical-induced DNA damage and human cancer risk. *Discov. Med.* **2012**, *14*, 283–288. [CrossRef] [PubMed]
34. Paini, A.; Scholz, G.; Marin-Kuan, M.; Schilter, B.; O'Brien, J.; van Bladeren, P.J.; Rietjens, I.M.C.M. Quantitative comparison between in vivo DNA adduct formation from exposure to selected DNA-reactive carcinogens, natural background levels of DNA adduct formation and tumour incidence in rodent bioassays. *Mutagenesis* **2011**, *26*, 605–618. [CrossRef] [PubMed]
35. Hwa Yun, B.; Guo, J.; Bellamri, M.; Turesky, R.J. DNA adducts: Formation, biological effects, and new biospecimens for mass spectrometric measurements in humans. *Mass Spectrom. Rev.* **2020**, *39*, 55–82. [CrossRef] [PubMed]
36. Poirier, M.C.; Beland, F.A. DNA adduct measurements and tumor incidence during chronic carcinogen exposure in rodents. *Environ. Health Perspect.* **1994**, *102* (Suppl. S6), 161–165. [CrossRef]

37. Doerge, D.R.; Gamboa da Costa, G.; McDaniel, L.P.; Churchwell, M.I.; Twaddle, N.C.; Beland, F.A. DNA adducts derived from administration of acrylamide and glycidamide to mice and rats. *Mutat. Res.* **2005**, *580*, 131–141. [CrossRef]
38. Lafferty, J.S.; Kamendulis, L.M.; Kaster, J.; Jiang, J.; Klaunig, J.E. Subchronic acrylamide treatment induces a tissue-specific increase in DNA synthesis in the rat. *Toxicol. Lett.* **2004**, *154*, 95–103. [CrossRef]
39. Pavanello, S.; Bollati, V.; Pesatori, A.C.; Kapka, L.; Bolognesi, C.; Bertazzi, P.A.; Baccarelli, A. Global and gene-specific promoter methylation changes are related to anti-B[a]PDE-DNA adduct levels and influence micronuclei levels in polycyclic aromatic hydrocarbon-exposed individuals. *Int. J. Cancer* **2009**, *125*, 1692–1697. [CrossRef]
40. Williams, G.M.; Iatropoulos, M.J.; Weisburger, J.H. Chemical carcinogen mechanisms of action and implications for testing methodology. *Exp. Toxicol. Pathol.* **1996**, *48*, 101–111. [CrossRef]
41. Phillips, D.H.; Arlt, V.M. Genotoxicity: Damage to DNA and its consequences. *EXS* **2009**, *99*, 87–110. [CrossRef]
42. Neumann, H.-G. Risk assessment of chemical carcinogens and thresholds. *Crit. Rev. Toxicol.* **2009**, *39*, 449–461. [CrossRef]
43. Kobets, T.; Williams, G.M. Thresholds for hepatocarcinogenicity of DNA-reactive compounds. In *Thresholds of Genotoxic Carcinogens*; Academic Press: Cambridge, MA, USA, 2016; pp. 19–36. [CrossRef]
44. Nohmi, T. Thresholds of genotoxic and non-genotoxic carcinogens. *Toxicol. Res.* **2018**, *34*, 281–290. [CrossRef] [PubMed]
45. Kobets, T.; Williams, G.M. Review of the evidence for thresholds for DNA-reactive and epigenetic experimental chemical carcinogens. *Chem. Biol. Interact.* **2019**, *301*, 88–111. [CrossRef] [PubMed]
46. Williams, G.M.; Iatropoulos, M.J.; Jeffrey, A.M. Dose-effect relationships for DNA-reactive liver carcinogens. In *Cellular Response to the Genotoxic Insult: The Question of Threshold for Genotoxic Carcinogens*; Oxford Academic: Oxford, UK, 2012; pp. 33–51. [CrossRef]
47. EFSA CONTAM Panel, European Food Safety Authority, Panel on Contaminants in the Food Chain. *Scientific Opinion of the Panel on Contaminants in the Food Chain on a Request from the European Commission on Polycyclic Aromatic Hydrocarbons in Food*; The EFSA Journal Series, No. 724; European Food Safety Authority: Parma, Italy, 2008; 114p.
48. Williams, G.M. DNA reactive and epigenetic carcinogens. *Exp. Toxicol. Pathol.* **1992**, *44*, 457–463. [CrossRef]
49. Klaunig, J.E.; Kamendulis, L.M.; Xu, Y. Epigenetic mechanisms of chemical carcinogenesis. *Hum. Exp. Toxicol.* **2000**, *19*, 543–555. [CrossRef]
50. Sawan, C.; Vaissière, T.; Murr, R.; Herceg, Z. Epigenetic drivers and genetic passengers on the road to cancer. *Mutat. Res./Fundam. Mol. Mech. Mutagen.* **2008**, *642*, 1–13. [CrossRef]
51. Pogribny, I.P.; Rusyn, I.; Beland, F.A. Epigenetic aspects of genotoxic and non-genotoxic hepatocarcinogenesis: Studies in rodents. *Environ. Mol. Mutagen.* **2008**, *49*, 9–15. [CrossRef]
52. Pogribny, I.P.; Rusyn, I. Environmental toxicants, epigenetics, and cancer. *Adv. Exp. Med. Biol.* **2013**, *754*, 215–232. [CrossRef]
53. Williams, G.M.; Jeffrey, A.M. Oxidative DNA damage: Endogenous and chemically induced. *Regul. Toxicol. Pharmacol.* **2000**, *32*, 283–292. [CrossRef]
54. Cooke, M.S.; Evans, M.D.; Dizdaroglu, M.; Lunec, J. Oxidative DNA damage: Mechanisms, mutation, and disease. *FASEB J.* **2003**, *17*, 1195–1214. [CrossRef]
55. Klaunig, J.E.; Kamendulis, L.M. The role of oxidative stress in carcinogenesis. *Annu. Rev. Pharmacol. Toxicol.* **2004**, *44*, 239–267. [CrossRef] [PubMed]
56. Jones, P.A.; Baylin, S.B. The epigenomics of cancer. *Cell* **2007**, *128*, 683–692. [CrossRef] [PubMed]
57. Baylin, S.B.; Jones, P.A. Epigenetic determinants of cancer. *Cold Spring Harb. Perspect. Biol.* **2016**, *8*, a019505. [CrossRef] [PubMed]
58. IARC, International Agency for Research on Cancer. *Preamble to the IARC Monographs (Amended January 2019)*; International Agency for Research on Cancer: Lyon, France, 2019.
59. Wiltse, J.; Dellarco, V.L. U.S. Environmental Protection Agency guidelines for carcinogen risk assessment: Past and future. *Mutat. Res./Rev. Genet. Toxicol.* **1996**, *365*, 3–15. [CrossRef]
60. Jeffrey, A.M.; Williams, G.M. Risk assessment of DNA-reactive carcinogens in food. *Toxicol. Appl. Pharm.* **2005**, *207*, 628–635. [CrossRef]
61. Felter, S.P.; Bhat, V.S.; Botham, P.A.; Bussard, D.A.; Casey, W.; Hayes, A.W.; Hilton, G.M.; Magurany, K.A.; Sauer, U.G.; Ohanian, E.V. Assessing chemical carcinogenicity: Hazard identification, classification, and risk assessment. Insight from a Toxicology Forum state-of-the-science workshop. *Crit. Rev. Toxicol.* 2021; 51, 653–694. [CrossRef]
62. Barlow, S.; Schlatter, J. Risk assessment of carcinogens in food. *Toxicol. Appl. Pharm.* **2010**, *243*, 180–190. [CrossRef]
63. Raffaele, K.; Vulimiri, S.; Bateson, T. Benefits and barriers to using epidemiology data in environmental risk assessment. *Open Epidemiol. J.* **2011**, *411*, 99–105. [CrossRef]
64. Kobets, T.; Williams, G.M. Chemicals with carcinogenic activity primarily in rodent liver. In *Comprehensive Toxicology*; Elsevier: Amsterdam, The Netherlands, 2018; pp. 409–442. [CrossRef]
65. Edler, L.; Hart, A.; Greaves, P.; Carthew, P.; Coulet, M.; Boobis, A.; Williams, G.M.; Smith, B. Selection of appropriate tumour data sets for Benchmark Dose Modelling (BMD) and derivation of a Margin of Exposure (MoE) for substances that are genotoxic and carcinogenic: Considerations of biological relevance of tumour type, data quality and uncertainty assessment. *Food Chem. Toxicol.* **2014**, *70*, 264–289. [CrossRef]
66. Rosenkranz, H.S. SAR modeling of genotoxic phenomena: The consequence on predictive performance of deviation from a unity ratio of genotoxicants/non-genotoxicants. *Mutat. Res./Genet. Toxicol. Environ. Mutagen.* **2004**, *559*, 67–71. [CrossRef]
67. IARC, International Agency for Research on Cancer. *Agents Classified by the IARC Monographs*; International Agency for Research on Cancer: Lyon, France, 2022; Volume 1–131.

68. NTP, National Toxicology Program. *Report on Carcinogens, (RoC)*, 15th ed.; National Toxicology Program: Research Triangle, NC, USA, 2021.
69. O'Brien, J.; Renwick, A.G.; Constable, A.; Dybing, E.; Müller, D.J.G.; Schlatter, J.; Slob, W.; Tueting, W.; van Benthem, J.; Williams, G.M.; et al. Approaches to the risk assessment of genotoxic carcinogens in food: A critical appraisal. *Food Chem. Toxicol.* **2006**, *44*, 1613–1635. [CrossRef]
70. Benford, D.; Bolger, P.M.; Carthew, P.; Coulet, M.; DiNovi, M.; Leblanc, J.-C.; Renwick, A.G.; Setzer, W.; Schlatter, J.; Smith, B.; et al. Application of the Margin of Exposure (MOE) approach to substances in food that are genotoxic and carcinogenic. *Food Chem. Toxicol.* **2010**, *48*, S2–S24. [CrossRef]
71. Herceg, Z. Epigenetics and cancer: Towards an evaluation of the impact of environmental and dietary factors. *Mutagenesis* **2007**, *22*, 91–103. [CrossRef] [PubMed]
72. Braakhuis, H.M.; Slob, W.; Olthof, E.D.; Wolterink, G.; Zwart, E.P.; Gremmer, E.R.; Rorije, E.; van Benthem, J.; Woutersen, R.; van der Laan, J.W.; et al. Is current risk assessment of non-genotoxic carcinogens protective? *Crit. Rev. Toxicol.* **2018**, *48*, 500–511. [CrossRef]
73. Swenberg, J.A.; Lehman-McKeeman, L.D. alpha 2-Urinary globulin-associated nephropathy as a mechanism of renal tubule cell carcinogenesis in male rats. *IARC Sci. Publ.* **1999**, *147*, 95–118.
74. Williams, G.M.; Whysner, J. Epigenetic carcinogens: Evaluation and risk assessment. *Exp. Toxicol. Pathol.* **1996**, *48*, 189–195. [CrossRef]
75. van den Berg, S.; Restani, P.; Boersma, M.; Delmulle, L.; Rietjens, I. Levels of genotoxic and carcinogenic compounds in plant food supplements and associated risk assessment. *Food Nutr. Sci.* **2011**, *2*, 989–1010. [CrossRef]
76. Rietjens, I.M.; Cohen, S.M.; Fukushima, S.; Gooderham, N.J.; Hecht, S.; Marnett, L.J.; Smith, R.L.; Adams, T.B.; Bastaki, M.; Harman, C.G.; et al. Impact of structural and metabolic variations on the toxicity and carcinogenicity of hydroxy- and alkoxy-substituted allyl- and propenylbenzenes. *Chem. Res. Toxicol.* **2014**, *27*, 1092–1103. [CrossRef] [PubMed]
77. Eisenreich, A.; Götz, M.E.; Sachse, B.; Monien, B.H.; Herrmann, K.; Schäfer, B. Alkenylbenzenes in foods: Aspects impeding the evaluation of adverse health effects. *Foods* **2021**, *10*, 2139. [CrossRef]
78. JECFA, Joint FAO/WHO Expert Committee on Food Additives. *Safety Evaluation of Certain Food Additives. Prepared by the Sixty-Ninth Meeting of the Joint FAO/WHO Expert Committee on Food Additives*; World Health Organization: Geneva, Switzerland, 2009; Volume 952, pp. 9–21.
79. Smith, R.L.; Adams, T.B.; Doull, J.; Feron, V.J.; Goodman, J.I.; Marnett, L.J.; Portoghese, P.S.; Waddell, W.J.; Wagner, B.M.; Rogers, A.E.; et al. Safety assessment of allylalkoxybenzene derivatives used as flavouring substances—Methyl eugenol and estragole. *Food Chem. Toxicol.* **2002**, *40*, 851–870. [CrossRef]
80. Al-Subeihi, A.A.; Spenkelink, B.; Rachmawati, N.; Boersma, M.G.; Punt, A.; Vervoort, J.; van Bladeren, P.J.; Rietjens, I.M. Physiologically based biokinetic model of bioactivation and detoxification of the alkenylbenzene methyleugenol in rat. *Toxicol. Vitr.* **2011**, *25*, 267–285. [CrossRef]
81. Martati, E.; Boersma, M.G.; Spenkelink, A.; Khadka, D.B.; van Bladeren, P.J.; Rietjens, I.M.; Punt, A. Physiologically based biokinetic (PBBK) modeling of safrole bioactivation and detoxification in humans as compared with rats. *Toxicol. Sci.* **2012**, *128*, 301–316. [CrossRef]
82. Rietjens, I.M.C.M.; Boersma, M.G.; van der Woude, H.; Jeurissen, S.M.F.; Schutte, M.E.; Alink, G.M. Flavonoids and alkenylbenzenes: Mechanisms of mutagenic action and carcinogenic risk. *Mutat. Res./Fundam. Mol. Mech. Mutagen.* **2005**, *574*, 124–138. [CrossRef] [PubMed]
83. Jeurissen, S.M.F.; Punt, A.; Boersma, M.G.; Bogaards, J.J.P.; Fiamegos, Y.C.; Schilter, B.; van Bladeren, P.J.; Cnubben, N.H.P.; Rietjens, I.M.C.M. Human cytochrome P450 enzyme specificity for the bioactivation of estragole and related alkenylbenzenes. *Chem. Res. Toxicol.* **2007**, *20*, 798–806. [CrossRef] [PubMed]
84. Randerath, K.; Putman, K.L.; Randerath, E. Flavor constituents in cola drinks induce hepatic DNA adducts in adult and fetal mice. *Biochem. Biophys. Res. Commun.* **1993**, *192*, 61–68. [CrossRef]
85. Kobets, T.; Duan, J.-D.; Brunnemann, K.D.; Etter, S.; Smith, B.; Williams, G.M. Structure-activity relationships for DNA damage by alkenylbenzenes in turkey egg fetal liver. *Toxicol. Sci.* **2016**, *150*, 301–311. [CrossRef]
86. Kobets, T.; Cartus, A.T.; Fuhlbrueck, J.A.; Brengel, A.; Stegmüller, S.; Duan, J.-D.; Brunnemann, K.D.; Williams, G.M. Assessment and characterization of DNA adducts produced by alkenylbenzenes in fetal turkey and chicken livers. *Food Chem. Toxicol.* **2019**, *129*, 424–433. [CrossRef]
87. IARC, International Agency for Research on Cancer. *Overall Evaluations of Carcinogenicity: An Updating of IARC Monographs Volumes 1 to 42*; IARC Monographs Supplement 7; World Health Organization: Geneva, Switzerland, 1987.
88. Kamdem, D.; Gage, D. Chemical composition of essential oil from the root bark of Sassafras albidum. *Planta Med.* **1995**, *61*, 574–575. [CrossRef] [PubMed]
89. IARC, International Agency for Research on Cancer. Some naturally occurring substances. IARC Monographs on the Evaluation of the Carcinogenic Risk of Chemicals to Man. *Environ. Pollut.* **1976**, *10*, 1–353. [CrossRef]
90. SCF, Scientific Committee on Food. *Opinion of the Scientific Committee on Food on the Safety of the Presence of Safrole (1-allyl-3,4-Methylene Dioxy Benzene) in Flavourings and Other Food Ingredients with Flavouring Properties*; SCF/CS/FLAV/FLAVOUR/6; Scientific Committee on Food: Brussels, Belgium, 2002.

91. Wiseman, R.W.; Miller, E.C.; Miller, J.A.; Liem, A. Structure-activity studies of the hepatocarcinogenicities of alkenylbenzene derivatives related to estragole and safrole on administration to preweanling male C57BL/6J × C3H/HeJ F1 mice. *J. Ethnopharmacol.* **1987**, *22*, 319. [CrossRef]

92. Miller, E.C.; Swanson, A.B.; Phillips, D.H.; Fletcher, T.L.; Liem, A.; Miller, J.A. Structure-activity studies of the carcinogenicities in the mouse and rat of some naturally occurring and synthetic alkenylbenzene derivatives related to safrole and estragole. *Cancer Res.* **1983**, *43*, 1124–1134.

93. Daimon, H.; Sawada, S.; Asakura, S.; Sagami, F. In vivo genotoxicity and DNA adduct levels in the liver of rats treated with safrole. *Carcinogenesis* **1998**, *19*, 141–146. [CrossRef]

94. Kevekordes, S.; Spielberger, J.; Burghaus, C.M.; Birkenkamp, P.; Zietz, B.; Paufler, P.; Diez, M.; Bolten, C.; Dunkelberg, H. Micronucleus formation in human lymphocytes and in the metabolically competent human hepatoma cell line Hep-G2: Results with 15 naturally occurring substances. *Anticancer Res.* **2001**, *21*, 461–469. [PubMed]

95. Gupta, K.P.; van Golen, K.L.; Putman, K.L.; Randerath, K. Formation and persistence of safrole-DNA adducts over a 10,000-fold dose range in mouse liver. *Carcinogenesis* **1993**, *14*, 1517–1521. [CrossRef] [PubMed]

96. Randerath, K.; Haglund, R.E.; Phillips, D.H.; Reddy, M.V. 32P-post-labelling analysis of DNA adducts formed in the livers of animals treated with safrole, estragole and other naturally-occurring alkenylbenzenes. I. Adult female CD-1 mice. *Carcinogenesis* **1984**, *5*, 1613–1622. [CrossRef]

97. Lee, J.-M.; Liu, T.-Y.; Wu, D.-C.; Tang, H.-C.; Leh, J.; Wu, M.-T.; Hsu, H.-H.; Huang, P.-M.; Chen, J.-S.; Lee, C.-J.; et al. Safrole–DNA adducts in tissues from esophageal cancer patients: Clues to areca-related esophageal carcinogenesis. *Mutat. Res. /Genet. Toxicol. Environ. Mutagen.* **2005**, *565*, 121–128. [CrossRef] [PubMed]

98. Zhou, G.D.; Moorthy, B.; Bi, J.; Donnelly, K.C.; Randerath, K. DNA adducts from alkoxyallylbenzene herb and spice constituents in cultured human (HepG2) cells. *Environ. Mol. Mutagen.* **2007**, *48*, 715–721. [CrossRef]

99. Boberg, E.W.; Miller, E.C.; Miller, J.A.; Poland, A.; Liem, A. Strong evidence from studies with brachymorphic mice and pentachlorophenol that 1′-sulfoöxysafrole is the major ultimate electrophilic and carcinogenic metabolite of 1′-hydroxysafrole in mouse liver. *Cancer Res.* **1983**, *43*, 5163–5173.

100. Boberg, E.W.; Miller, E.C.; Miller, J.A. The metabolic sulfonation and side-chain oxidation of 3′-hydroxyisosafrole in the mouse and its inactivity as a hepatocarcinogen relative to 1′-hydroxysafrole. *Chem. Biol. Interact.* **1986**, *59*, 73–97. [CrossRef]

101. Daimon, H.; Sawada, S.; Asakura, S.; Sagami, F. Inhibition of sulfotransferase affecting in vivo genotoxicity and DNA adducts induced by safrole in rat liver. *Teratog. Carcinog. Mutagen.* **1997**, *17*, 327–337. [CrossRef]

102. Beyer, J.; Ehlers, D.; Maurer, H.H. Abuse of nutmeg (Myristica Fragrans Houtt.): Studies on the metabolism and the toxicologic detection of its ingredients elemicin, myristicin, and safrole in rat and human urine using gas chromatography/mass spectrometry. *Ther. Drug Monit.* **2006**, *28*, 568–575. [CrossRef]

103. Chen, C.L.; Chi, C.W.; Chang, K.W.; Liu, T.Y. Safrole-like DNA adducts in oral tissue from oral cancer patients with a betel quid chewing history. *Carcinogenesis* **1999**, *20*, 2331–2334. [CrossRef] [PubMed]

104. Chung, Y.T.; Chen, C.L.; Wu, C.C.; Chan, S.A.; Chi, C.W.; Liu, T.Y. Safrole-DNA adduct in hepatocellular carcinoma associated with betel quid chewing. *Toxicol. Lett.* **2008**, *183*, 21–27. [CrossRef] [PubMed]

105. FDA, Food and Drug Administration. *Title 21-Food and Drugs. Chapter I-Food and Drug Administration Department of Health and Human Services, Subchapter B-Food for Human Consumption (Continued). Part 189-Substances Prohibited from Use in Human Food. Subpart C-Substances Generally Prohibited from Direct Addition or Use as Human Food. Sec. 189.180 Safrole*; Food and Drug Administration: Silber Spring, MD, USA, 2017.

106. Commission Regulation, (EU). Regulation (EC) No 1334/2008 of the European Parliament and of the Council of 16 December 2008 on flavourings and certain food ingredients with flavouring properties for use in and on foods and amending Council Regulation (EEC) No 1601/91, Regulations (EC) No 2232/96 and (EC) No 110/2008 and Directive 2000/13/EC. *Off. J. Eur. Union* **2008**, *354*, 218–234.

107. Liu, T.Y.; Chung, Y.T.; Wang, P.F.; Chi, C.W.; Hsieh, L.L. Safrole-DNA adducts in human peripheral blood—An association with areca quid chewing and CYP2E1 polymorphisms. *Mutat. Res.* **2004**, *559*, 59–66. [CrossRef]

108. EMA, European Medicines Agency. *Public Statement on the Use of Herbal Medicinal Products Containing Estragole*; 2nd Draft, Revision 1, MA/HMPC/137212/2005 Rev 1; European Medicines Agency: Amsterdam, The Netherlands, 2019.

109. NTP, National Toxicology Program. *Technical Report on the 3-Month Toxicity Studies of Estragole (CAS No. 140-67-0) Administered by Gavage to F344/N Rats and B6C3F1 Mice*; 1521-4621 (Print); National Toxicology Program: Research Triangle, NC, USA, 2011; pp. 1–111.

110. Müller, L.; Kasper, P.; Müller-Tegethoff, K.; Petr, T. The genotoxic potential in vitro and in vivo of the allyl benzene etheric oils estragole, basil oil and trans-anethole. *Mutat. Res. Lett.* **1994**, *325*, 129–136. [CrossRef]

111. Ding, W.; Levy, D.D.; Bishop, M.E.; Pearce, M.G.; Davis, K.J.; Jeffrey, A.M.; Duan, J.D.; Williams, G.M.; White, G.A.; Lyn-Cook, L.E.; et al. In vivo genotoxicity of estragole in male F344 rats. *Environ. Mol. Mutagen.* **2015**, *56*, 356–365. [CrossRef] [PubMed]

112. Ishii, Y.; Suzuki, Y.; Hibi, D.; Jin, M.; Fukuhara, K.; Umemura, T.; Nishikawa, A. Detection and quantification of specific DNA adducts by liquid chromatography-tandem mass spectrometry in the livers of rats given estragole at the carcinogenic dose. *Chem. Res. Toxicol.* **2011**, *24*, 532–541. [CrossRef]

113. Anthony, A.; Caldwell, J.; Hutt, A.J.; Smith, R.L. Metabolism of estragole in rat and mouse and influence of dose size on excretion of the proximate carcinogen 1′-hydroxyestragole. *Food Chem. Toxicol.* **1987**, *25*, 799–806. [CrossRef]

114. Waddell, W.J. Thresholds of carcinogenicity of flavors. *Toxicol. Sci.* **2002**, *68*, 275–279. [CrossRef]
115. IARC, International Agency for Research on Cancer. *Some Chemicals Present in Industrial and Consumer Products, Food and Drinking Water*; IARC Monograhs on the Evaluation of Carcinogenic Risks to Humans; International Agency for Research on Cancer: Lyon, France, 2013; Volume 101.
116. Moshonas, M.G.; Shaw, P.E. Compounds new to essential orange oil from fruit treated with abscission chemicals. *J. Agric. Food Chem.* **1978**, *26*, 1288–1290. [CrossRef]
117. NTP, National Toxicology Program. *Technical Report on the Toxicology and Carcinogenesis Studies of Methyleugenol (CAS NO. 93-15-2) in F344/N Rats and B6C3F1 Mice (Gavage Studies)*; 0888-8051 (Print); National Toxicology Program: Research Triangle, NC, USA, 2000; pp. 1–412.
118. Williams, G.M.; Iatropoulos, M.J.; Jeffrey, A.M.; Duan, J.-D. Methyleugenol hepatocellular cancer initiating effects in rat liver. *Food Chem. Toxicol.* **2013**, *53*, 187–196. [CrossRef] [PubMed]
119. Waddell, W.J. Correlation of tumors with DNA adducts from methyl eugenol and tamoxifen in rats. *Toxicol. Sci.* **2004**, *79*, 38–40. [CrossRef] [PubMed]
120. Al-Subeihi, A.A.; Spenkelink, B.; Punt, A.; Boersma, M.G.; van Bladeren, P.J.; Rietjens, I.M. Physiologically based kinetic modeling of bioactivation and detoxification of the alkenylbenzene methyleugenol in human as compared with rat. *Toxicol. Appl. Pharm.* **2012**, *260*, 271–284. [CrossRef]
121. Lutz, W.; Gaylor, D.; Conolly, R.; Lutz, R. Nonlinearity and thresholds in dose–response relationships for carcinogenicity due to sampling variation, logarithmic dose scaling, or small differences in individual susceptibility. *Toxicol. Appl. Pharmacol.* **2005**, *207*, 565–569. [CrossRef]
122. Smith, B.; Cadby, P.; Leblanc, J.C.; Setzer, R.W. Application of the Margin of Exposure (MoE) approach to substances in food that are genotoxic and carcinogenic: Example: Methyleugenol, CASRN: 93-15-2. *Food Chem. Toxicol.* **2010**, *48* (Suppl. S1), S89–S97. [CrossRef]
123. Chandra, S.A.; Nolan, M.W.; Malarkey, D.E. Chemical carcinogenesis of the gastrointestinal tract in rodents: An overview with emphasis on NTP carcinogenesis bioassays. *Toxicol. Pathol.* **2010**, *38*, 188–197. [CrossRef] [PubMed]
124. Cohen, S.; Eisenbrand, G.; Fukushima, S.; Gooderham, N.; Guengerich, F.; Hecht, S.; Rietjens, I.; Harman, C.; Taylor, S. GRAS 28 flavoring substances. *Food Technol.* **2018**, *72*, 62–77.
125. Hermes, L.; Römermann, J.; Cramer, B.; Esselen, M. Quantitative analysis of β-asarone derivatives in Acorus calamus and herbal food products by HPLC-MS/MS. *J. Agric. Food Chem.* **2021**, *69*, 776–782. [CrossRef]
126. Uebel, T.; Hermes, L.; Haupenthal, S.; Müller, L.; Esselen, M. α-Asarone, β-asarone, and γ-asarone: Current status of toxicological evaluation. *J. Appl. Toxicol.* **2021**, *41*, 1166–1179. [CrossRef]
127. JECFA, Joint FAO/WHO Expert Committee on Food Additives. *Toxicological Evaluation of Certain Food Additives*; Monograph on β-asarone: WHO Food Additive Series No. 16; World Health Organization: Geneva, Switzerland, 1981; Volume 16.
128. SCF, Scientific Committee on Food. *Opinion of the Scientific Committee on Food on the Presence of β-Asarone in Flavourings and Other Food Ingredients with Flavouring Properties*; SCF/CS/FLAV/FLAVOUR/9 ADD1 Final 2002; European Commission Health & Consumer Protection Directorate-General: Brussels, Belgium, 2002.
129. Kim, S.G.; Liem, A.; Stewart, B.C.; Miller, J.A. New studies on trans-anethole oxide and trans-asarone oxide. *Carcinogenesis* **1999**, *20*, 1303–1307. [CrossRef]
130. Berg, K.; Bischoff, R.; Stegmüller, S.; Cartus, A.; Schrenk, D. Comparative investigation of the mutagenicity of propenylic and allylic asarone isomers in the Ames fluctuation assay. *Mutagenesis* **2016**, *31*, 443–451. [CrossRef] [PubMed]
131. Hasheminejad, G.; Caldwell, J. Genotoxicity of the alkenylbenzenes α− and β-asarone, myristicin and elemicin as determined by the UDS assay in cultured rat hepatocytes. *Food Chem. Toxicol.* **1994**, *32*, 223–231. [CrossRef]
132. Haupenthal, S.; Berg, K.; Gründken, M.; Vallicotti, S.; Hemgesberg, M.; Sak, K.; Schrenk, D.; Esselen, M. In vitro genotoxicity of carcinogenic asarone isomers. *Food Funct.* **2017**, *8*, 1227–1234. [CrossRef]
133. Stegmüller, S.; Schrenk, D.; Cartus, A.T. Formation and fate of DNA adducts of alpha- and beta-asarone in rat hepatocytes. *Food Chem. Toxicol.* **2018**, *116*, 138–146. [CrossRef] [PubMed]
134. Cartus, A.T.; Stegmüller, S.; Simson, N.; Wahl, A.; Neef, S.; Kelm, H.; Schrenk, D. Hepatic metabolism of carcinogenic β-asarone. *Chem. Res. Toxicol.* **2015**, *28*, 1760–1773. [CrossRef] [PubMed]
135. Cartus, A.T.; Schrenk, D. Metabolism of the carcinogen alpha-asarone in liver microsomes. *Food Chem. Toxicol.* **2016**, *87*, 103–112. [CrossRef]
136. Cartus, A.T.; Schrenk, D. Metabolism of carcinogenic alpha-asarone by human cytochrome P450 enzymes. *Naunyn Schmiedebergs Arch. Pharm.* **2020**, *393*, 213–223. [CrossRef]
137. Wu, J.; Zhang, X.X.; Sun, Q.M.; Chen, M.; Liu, S.L.; Zhang, X.; Zhou, J.Y.; Zou, X. β-Asarone inhibits gastric cancer cell proliferation. *Oncol. Rep.* **2015**, *34*, 3043–3050. [CrossRef]
138. Chen, M.; Zhuang, Y.W.; Wu, C.E.; Peng, H.Y.; Qian, J.; Zhou, J.Y. β-Asarone suppresses HCT116 colon cancer cell proliferation and liver metastasis in part by activating the innate immune system. *Oncol. Lett.* **2021**, *21*, 435. [CrossRef]
139. FDA, Food and Drug Administration. *Title 21-Food and Drugs. Chapter I-Food and Drug Administration Department of Health and Human Services, Subchapter B-Food for Human Consumption (Continued). Part 189-Substances Prohibited from Use in Human Food. Subpart C-Substances Generally Prohibited from Direct Addition or Use as Human Food. Sec. 189.110 Calamus and Its Derivatives*; Food and Drug Administration: Silver Springs, MD, USA, 2013.

140. EMA, European Medicines Agency. *Public Statement on the Risks Associated with the Use of Herbal Products Containing Aristolochia Species*; EMEA/HMPC/138381/2005; European Medicines Agency: Amsterdam, The Netherlands, 2005.

141. Arlt, V.M.; Stiborova, M.; Schmeiser, H.H. Aristolochic acid as a probable human cancer hazard in herbal remedies: A review. *Mutagenesis* **2002**, *17*, 265–277. [CrossRef]

142. Abdullah, R.; Diaz, L.N.; Wesseling, S.; Rietjens, I.M. Risk assessment of plant food supplements and other herbal products containing aristolochic acids using the margin of exposure (MOE) approach. *Food Addit. Contam. Part A Chem. Anal. Control Expo. Risk Assess.* **2017**, *34*, 135–144. [CrossRef] [PubMed]

143. Cui, M.; Liu, Z.H.; Qiu, Q.; Li, H.; Li, L.S. Tumour induction in rats following exposure to short-term high dose aristolochic acid I. *Mutagenesis* **2005**, *20*, 45–49. [CrossRef] [PubMed]

144. Mengs, U. Tumour induction in mice following exposure to aristolochic acid. *Arch. Toxicol.* **1988**, *61*, 504–505. [CrossRef]

145. Zhang, H.; Cifone, M.A.; Murli, H.; Erexson, G.L.; Mecchi, M.S.; Lawlor, T.E. Application of simplified in vitro screening tests to detect genotoxicity of aristolochic acid. *Food Chem. Toxicol.* **2004**, *42*, 2021–2028. [CrossRef] [PubMed]

146. Chen, R.; Zhou, C.; Cao, Y.; Xi, J.; Ohira, T.; He, L.; Huang, P.; You, X.; Liu, W.; Zhang, X.; et al. Assessment of Pig-a, micronucleus, and comet assay endpoints in Tg.RasH2 mice carcinogenicity study of aristolochic acid I. *Environ. Mol. Mutagen.* **2020**, *61*, 266–275. [CrossRef] [PubMed]

147. Bárta, F.; Dedíková, A.; Bebová, M.; Dušková, Š.; Mráz, J.; Schmeiser, H.H.; Arlt, V.M.; Hodek, P.; Stiborová, M. Co-exposure to aristolochic acids I and II increases DNA adduct formation responsible for aristolochic acid I-mediated carcinogenicity in rats. *Int. J. Mol. Sci.* **2021**, *22*, 479. [CrossRef]

148. Abdullah, R.; Wesseling, S.; Spenkelink, B.; Louisse, J.; Punt, A.; Rietjens, I. Defining in vivo dose-response curves for kidney DNA adduct formation of aristolochic acid I in rat, mouse and human by an in vitro and physiologically based kinetic modeling approach. *J. Appl. Toxicol.* **2020**, *40*, 1647–1660. [CrossRef]

149. Schmeiser, H.H.; Schoepe, K.B.; Wiessler, M. DNA adduct formation of aristolochic acid I and II in vitro and in vivo. *Carcinogenesis* **1988**, *9*, 297–303. [CrossRef]

150. McDaniel, L.P.; Elander, E.R.; Guo, X.; Chen, T.; Arlt, V.M.; Mei, N. Mutagenicity and DNA adduct formation by aristolochic acid in the spleen of Big Blue® rats. *Environ. Mol. Mutagen.* **2012**, *53*, 358–368. [CrossRef]

151. Mei, N.; Arlt, V.M.; Phillips, D.H.; Heflich, R.H.; Chen, T. DNA adduct formation and mutation induction by aristolochic acid in rat kidney and liver. *Mutat. Res.* **2006**, *602*, 83–91. [CrossRef]

152. Rebhan, K.; Ertl, I.E.; Shariat, S.F.; Grollman, A.P.; Rosenquist, T. Aristolochic acid and its effect on different cancers in uro-oncology. *Curr. Opin. Urol.* **2020**, *30*, 689–695. [CrossRef] [PubMed]

153. Grollman, A.P. Aristolochic acid nephropathy: Harbinger of a global iatrogenic disease. *Environ. Mol. Mutagen.* **2013**, *54*, 1–7. [CrossRef] [PubMed]

154. Stiborová, M.; Frei, E.; Arlt, V.M.; Schmeiser, H.H. Metabolic activation of carcinogenic aristolochic acid, a risk factor for Balkan endemic nephropathy. *Mutat. Res.* **2008**, *658*, 55–67. [CrossRef] [PubMed]

155. Fernando, R.C.; Schmeiser, H.H.; Scherf, H.R.; Wiessler, M. Formation and persistence of specific purine DNA adducts by 32P-postlabelling in target and non-target organs of rats treated with aristolochic acid I. *IARC Sci. Publ.* **1993**, *124*, 167–171.

156. Feldmeyer, N.; Schmeiser, H.H.; Muehlbauer, K.R.; Belharazem, D.; Knyazev, Y.; Nedelko, T.; Hollstein, M. Further studies with a cell immortalization assay to investigate the mutation signature of aristolochic acid in human p53 sequences. *Mutat. Res./Genet. Toxicol. Environ. Mutagen.* **2006**, *608*, 163–168. [CrossRef]

157. Levi, M.; Guchelaar, H.J.; Woerdenbag, H.J.; Zhu, Y.P. Acute hepatitis in a patient using a Chinese herbal tea—A case report. *Pharm. World Sci.* **1998**, *20*, 43–44. [CrossRef]

158. Schaneberg, B.T.; Applequist, W.L.; Khan, I.A. Determination of aristolochic acid I and II in North American species of Asarum and Aristolochia. *Pharmazie* **2002**, *57*, 686–689. [PubMed]

159. Gökmen, M.R.; Lord, G.M. Aristolochic acid nephropathy. *BMJ* **2012**, *344*, e4000. [CrossRef]

160. Nortier, J.L.; Martinez, M.C.; Schmeiser, H.H.; Arlt, V.M.; Bieler, C.A.; Petein, M.; Depierreux, M.F.; De Pauw, L.; Abramowicz, D.; Vereerstraeten, P.; et al. Urothelial carcinoma associated with the use of a Chinese herb (Aristolochia fangchi). *N. Engl. J. Med.* **2000**, *342*, 1686–1692. [CrossRef] [PubMed]

161. Nortier, J.L.; Vanherweghem, J.L. Renal interstitial fibrosis and urothelial carcinoma associated with the use of a Chinese herb (*Aristolochia fangchi*). *Toxicology* **2002**, *181–182*, 577–580. [CrossRef]

162. Martena, M.J.; van der Wielen, J.C.A.; van de Laak, L.F.J.; Konings, E.J.M.; de Groot, H.N.; Rietjens, I.M.C.M. Enforcement of the ban on aristolochic acids in Chinese traditional herbal preparations on the Dutch market. *Anal. Bioanal. Chem.* **2007**, *389*, 263–275. [CrossRef] [PubMed]

163. Bode, A.M.; Dong, Z. Toxic phytochemicals and their potential risks for human cancer. *Cancer Prev. Res.* **2015**, *8*, 1–8. [CrossRef] [PubMed]

164. Sieber, S.M.; Correa, P.; Dalgard, D.W.; McIntire, K.R.; Adamson, R.H. Carcinogenicity and hepatotoxicity of cycasin and its aglycone methylazoxymethanol acetate in nonhuman primates. *J. Natl. Cancer Inst.* **1980**, *65*, 177–189. [PubMed]

165. Kuniyasu, T.; Tanaka, T.; Shima, H.; Sugie, S.; Mori, H.; Takahashi, M. Enhancing effect of cholecystectomy on colon carcinogenesis induced by methylazoxymethanol acetate in hamsters. *Dis. Colon Rectum* **1986**, *29*, 492–494. [CrossRef] [PubMed]

166. Reddy, B.S.; Maeura, Y. Dose-response studies of the effect of dietary butylated hydroxyanisole on colon carcinogenesis induced by methylazoxymethanol acetate in female CF1 mice. *J. Natl. Cancer Inst.* **1984**, *72*, 1181–1187. [PubMed]

167. Tanaka, T.; Shinoda, T.; Yoshimi, N.; Niwa, K.; Iwata, H.; Mori, H. Inhibitory effect of magnesium hydroxide on methylazoxymethanol acetate-induced large bowel carcinogenesis in male F344 rats. *Carcinogenesis* **1989**, *10*, 613–616. [CrossRef]
168. Lijinsky, W.; Saavedra, J.E.; Reuber, M.D. Organ-specific carcinogenesis in rats by methyl- and ethylazoxyalkanes. *Cancer Res.* **1985**, *45*, 76–79.
169. Hoffmann, G.R.; Morgan, R.W. Review: Putative mutagens and carcinogens in foods. V. Cycad azoxyglycosides. *Environ. Mutagen.* **1984**, *6*, 103–116. [CrossRef]
170. Williams, G.M.; Laspia, M.F.; Mori, H.; Hirono, I. Genotoxicity of cycasin in the hepatocyte primary culture/DNA repair test supplemented with beta-glucosidase. *Cancer Lett.* **1981**, *12*, 329–333. [CrossRef]
171. Kawai, K.; Furukawa, H.; Hirono, I. Genotoxic activity in vivo of the naturally occurring glucoside, cycasin, in the Drosophila wing spot test. *Mutat. Res.* **1995**, *346*, 145–149. [CrossRef]
172. Matsushima, T.; Matsumoto, H.; Shirai, A.; Sawamura, M.; Sugimura, T. Mutagenicity of the naturally occurring carcinogen cycasin and synthetic methylazoxymethanol conjugates in Salmonella typhimurium. *Cancer Res.* **1979**, *39*, 3780–3782. [PubMed]
173. Cavanna, M.; Parodi, S.; Taningher, M.; Bolognesi, C.; Sciabà, L.; Brambilla, G. DNA fragmentation in some organs of rats and mice treated with cycasin. *Br. J. Cancer* **1979**, *39*, 383–390. [CrossRef] [PubMed]
174. Klaus, V.; Bastek, H.; Damme, K.; Collins, L.B.; Frötschl, R.; Benda, N.; Lutter, D.; Ellinger-Ziegelbauer, H.; Swenberg, J.A.; Dietrich, D.R.; et al. Time-matched analysis of DNA adduct formation and early gene expression as predictive tool for renal carcinogenesis in methylazoxymethanol acetate treated Eker rats. *Arch. Toxicol.* **2017**, *91*, 3427–3438. [CrossRef] [PubMed]
175. Kisby, G.E.; Ellison, M.; Spencer, P.S. Content of the neurotoxins cycasin (methylazoxymethanol beta-D-glucoside) and BMAA (beta-N-methylamino-L-alanine) in cycad flour prepared by Guam Chamorros. *Neurology* **1992**, *42*, 1336–1340. [CrossRef]
176. Esclaire, F.; Kisby, G.; Spencer, P.; Milne, J.; Lesort, M.; Hugon, J. The Guam cycad toxin methylazoxymethanol damages neuronal DNA and modulates tau mRNA expression and excitotoxicity. *Exp. Neurol.* **1999**, *155*, 11–21. [CrossRef]
177. Fiala, E.S.; Sohn, O.S.; Hamilton, S.R. Effects of chronic dietary ethanol on in vivo and in vitro metabolism of methylazoxymethanol and on methylazoxymethanol-induced DNA methylation in rat colon and liver. *Cancer Res.* **1987**, *47*, 5939–5943.
178. Sohn, O.S.; Puz, C.; Caswell, N.; Fiala, E.S. Differential susceptibility of rat and guinea pig colon mucosa DNA to methylation by methylazoxymethyl acetate in vivo. *Cancer Lett.* **1985**, *29*, 293–300. [CrossRef]
179. Sohn, O.S.; Fiala, E.S.; Requeijo, S.P.; Weisburger, J.H.; Gonzalez, F.J. Differential effects of CYP2E1 status on the metabolic activation of the colon carcinogens azoxymethane and methylazoxymethanol. *Cancer Res.* **2001**, *61*, 8435–8440.
180. Kisby, G.E.; Fry, R.C.; Lasarev, M.R.; Bammler, T.K.; Beyer, R.P.; Churchwell, M.; Doerge, D.R.; Meira, L.B.; Palmer, V.S.; Ramos-Crawford, A.L.; et al. The cycad genotoxin MAM modulates brain cellular pathways involved in neurodegenerative disease and cancer in a DNA damage-linked manner. *PLoS ONE* **2011**, *6*, e20911. [CrossRef]
181. Zhang, Z.X.; Anderson, D.W.; Mantel, N.; Román, G.C. Motor neuron disease on Guam: Geographic and familial occurrence, 1956–1985. *Acta Neurol. Scand.* **1996**, *94*, 51–59. [CrossRef]
182. Borenstein, A.R.; Mortimer, J.A.; Schofield, E.; Wu, Y.; Salmon, D.P.; Gamst, A.; Olichney, J.; Thal, L.J.; Silbert, L.; Kaye, J.; et al. Cycad exposure and risk of dementia, MCI, and PDC in the Chamorro population of Guam. *Neurology* **2007**, *68*, 1764–1771. [CrossRef] [PubMed]
183. Chang, S.S.; Chan, Y.L.; Wu, M.L.; Deng, J.F.; Chiu, T.F.; Chen, J.C.; Wang, F.L.; Tseng, C.P. Acute cycas seed poisoning in Taiwan. *J. Toxicol. Clin. Toxicol.* **2004**, *42*, 49–54. [CrossRef] [PubMed]
184. Alonso-Amelot, M.E.; Avendaño, M. Human carcinogenesis and bracken fern: A review of the evidence. *Curr. Med. Chem.* **2002**, *9*, 675–686. [CrossRef]
185. Gil da Costa, R.M.; Bastos, M.M.; Oliveira, P.A.; Lopes, C. Bracken-associated human and animal health hazards: Chemical, biological and pathological evidence. *J. Hazard. Mater.* **2012**, *203–204*, 1–12. [CrossRef] [PubMed]
186. Rasmussen, L.H. Presence of the carcinogen ptaquiloside in fern-based food products and traditional medicine: Four cases of human exposure. *Curr. Res. Food Sci.* **2021**, *4*, 557–564. [CrossRef]
187. Virgilio, A.; Sinisi, A.; Russo, V.; Gerardo, S.; Santoro, A.; Galeone, A.; Taglialatela-Scafati, O.; Roperto, F. Ptaquiloside, the major carcinogen of bracken fern, in the pooled raw milk of healthy sheep and goats: An underestimated, global concern of food safety. *J. Agric. Food Chem.* **2015**, *63*, 4886–4892. [CrossRef]
188. Aranha, P.C.; Hansen, H.C.; Rasmussen, L.H.; Strobel, B.W.; Friis, C. Determination of ptaquiloside and pterosin B derived from bracken (Pteridium aquilinum) in cattle plasma, urine and milk. *J. Chromatogr. B Anal. Technol. Biomed. Life Sci.* **2014**, *951–952*, 44–51. [CrossRef]
189. Francesco, B.; Giorgio, B.; Rosario, N.; Saverio, R.F.; Francesco, G.; Romano, M.; Adriano, S.; Cinzia, R.; Antonio, T.; Franco, R.; et al. A new, very sensitive method of assessment of ptaquiloside, the major bracken carcinogen in the milk of farm animals. *Food Chem.* **2011**, *124*, 660–665. [CrossRef]
190. Shahin, M.; Smith, B.L.; Prakash, A.S. Bracken carcinogens in the human diet. *Mutat. Res.* **1999**, *443*, 69–79. [CrossRef]
191. Potter, D.M.; Baird, M.S. Carcinogenic effects of ptaquiloside in bracken fern and related compounds. *Br. J. Cancer* **2000**, *83*, 914–920. [CrossRef]
192. Prakash, A.S.; Pereira, T.N.; Smith, B.L.; Shaw, G.; Seawright, A.A. Mechanism of bracken fern carcinogenesis: Evidence for H-ras activation via initial adenine alkylation by ptaquiloside. *Nat. Toxins* **1996**, *4*, 221–227. [CrossRef]

193. IARC, International Agency for Research on Cancer. Some Naturally Occurring and Synthetic Food Components, Furocoumarins and Ultraviolet Radiation. In *IARC Monographs on the Evaluation of the Carcinogenic Risk of Chemicals to Humans*; International Agency for Research on Cancer: Lyon, France, 1986; Volume 40.
194. Pamukcu, A.M.; Yalçiner, S.; Hatcher, J.F.; Bryan, G.T. Quercetin, a rat intestinal and bladder carcinogen present in bracken fern (*Pteridium aquilinum*). *Cancer Res.* **1980**, *40*, 3468–3472. [PubMed]
195. Hirono, I. Carcinogenic principles isolated from bracken fern. *Crit. Rev. Toxicol.* **1986**, *17*, 1–22. [CrossRef] [PubMed]
196. Hirono, I.; Aiso, S.; Yamaji, T.; Mori, H.; Yamada, K.; Niwa, H.; Ojika, M.; Wakamatsu, K.; Kigoshi, H.; Niiyama, K.; et al. Carcinogenicity in rats of ptaquiloside isolated from bracken. *Gan* **1984**, *75*, 833–836.
197. Hirono, I.; Ogino, H.; Fujimoto, M.; Yamada, K.; Yoshida, Y.; Ikagawa, M.; Okumura, M. Induction of tumors in ACI rats given a diet containing ptaquiloside, a bracken carcinogen. *J. Natl. Cancer Inst.* **1987**, *79*, 1143–1149.
198. Gil da Costa, R.M.; Neto, T.; Estêvão, D.; Moutinho, M.; Félix, A.; Medeiros, R.; Lopes, C.; Bastos, M.; Oliveira, P.A. Ptaquiloside from bracken (*Pteridium* spp.) promotes oral carcinogenesis initiated by HPV16 in transgenic mice. *Food Funct.* **2020**, *11*, 3298–3305. [CrossRef]
199. Mori, H.; Sugie, S.; Hirono, I.; Yamada, K.; Niwa, H.; Ojika, M. Genotoxicity of ptaquiloside, a bracken carcinogen, in the hepatocyte primary culture/DNA-repair test. *Mutat. Res.* **1985**, *143*, 75–78. [CrossRef]
200. Tourchi-Roudsari, M. Multiple effects of bracken fern under in vivo and in vitro conditions. *Asian Pac. J. Cancer Prev.* **2014**, *15*, 7505–7513. [CrossRef]
201. Nagao, T.; Saito, K.; Hirayama, E.; Uchikoshi, K.; Koyama, K.; Natori, S.; Morisaki, N.; Iwasaki, S.; Matsushima, T. Mutagenicity of ptaquiloside, the carcinogen in bracken, and its related illudane-type sesquiterpenes. I. Mutagenicity in *Salmonella typhimurium*. *Mutat. Res.* **1989**, *215*, 173–178. [CrossRef]
202. Gil da Costa, R.M.; Coelho, P.; Sousa, R.; Bastos, M.; Porto, B.; Teixeira, J.P.; Malheiro, I.; Lopes, C. Multiple genotoxic activities of ptaquiloside in human lymphocytes: Aneugenesis, clastogenesis and induction of sister chromatid exchange. *Mutat. Res.* **2012**, *747*, 77–81. [CrossRef]
203. Matsuoka, A.; Hirosawa, A.; Natori, S.; Iwasaki, S.; Sofuni, T.; Ishidate, M., Jr. Mutagenicity of ptaquiloside, the carcinogen in bracken, and its related illudane-type sesquiterpenes. II. Chromosomal aberration tests with cultured mammalian cells. *Mutat. Res.* **1989**, *215*, 179–185. [CrossRef]
204. Gomes, J.; Magalhães, A.; Michel, V.; Amado, I.F.; Aranha, P.; Ovesen, R.G.; Hansen, H.C.; Gärtner, F.; Reis, C.A.; Touati, E. Pteridium aquilinum and its ptaquiloside toxin induce DNA damage response in gastric epithelial cells, a link with gastric carcinogenesis. *Toxicol. Sci.* **2012**, *126*, 60–71. [CrossRef] [PubMed]
205. Povey, A.C.; Potter, D.; O'Connor, P.J. 32P-post-labelling analysis of DNA adducts formed in the upper gastrointestinal tissue of mice fed bracken extract or bracken spores. *Br. J. Cancer* **1996**, *74*, 1342–1348. [CrossRef] [PubMed]
206. Freitas, R.N.; O'Connor, P.J.; Prakash, A.S.; Shahin, M.; Povey, A.C. Bracken (Pteridium aquilinum)-induced DNA adducts in mouse tissues are different from the adduct induced by the activated form of the Bracken carcinogen ptaquiloside. *Biochem. Biophys. Res. Commun.* **2001**, *281*, 589–594. [CrossRef]
207. Shahin, M.; Smith, B.L.; Worral, S.; Moore, M.R.; Seawright, A.A.; Prakash, A.S. Bracken fern carcinogenesis: Multiple intravenous doses of activated ptaquiloside induce DNA adducts, monocytosis, increased TNF alpha levels, and mammary gland carcinoma in rats. *Biochem. Biophys. Res. Commun.* **1998**, *244*, 192–197. [CrossRef] [PubMed]
208. Sardon, D.; de la Fuente, I.; Calonge, E.; Perez-Alenza, M.D.; Castaño, M.; Dunner, S.; Peña, L. H-ras immunohistochemical expression and molecular analysis of urinary bladder lesions in grazing adult cattle exposed to bracken fern. *J. Comp. Pathol.* **2005**, *132*, 195–201. [CrossRef]
209. Shahin, M.; Moore, M.R.; Worrall, S.; Smith, B.L.; Seawright, A.A.; Prakash, A.S. H-ras activation is an early event in the ptaquiloside-induced carcinogenesis: Comparison of acute and chronic toxicity in rats. *Biochem. Biophys. Res. Commun.* **1998**, *250*, 491–497. [CrossRef]
210. Gil da Costa, R.M.; Oliveira, P.A.; Bastos, M.M.; Lopes, C.C.; Lopes, C. Ptaquiloside-induced early-stage urothelial lesions show increased cell proliferation and intact β-catenin and E-cadherin expression. *Environ. Toxicol.* **2014**, *29*, 763–769. [CrossRef]
211. Alonso-amelot, M.E.; Castillo, U.F.; Smith, B.L.; Lauren, D.R. Excretion, through milk, of ptaquiloside in bracken-fed cows. A quantitative assessment. *Lait* **1998**, *78*, 413–423. [CrossRef]
212. Alonso-Amelot, M.E. The link between bracken fern and stomach cancer: Milk. *Nutrition* **1997**, *13*, 694–696. [CrossRef]
213. Alonso-Amelot, M.E.; Avendaño, M. Possible association between gastric cancer and bracken fern in Venezuela: An epidemiologic study. *Int. J. Cancer* **2001**, *91*, 252–259. [CrossRef]
214. Liu, R.; Li, A.; Sun, A.; Kong, L. Preparative isolation and purification of psoralen and isopsoralen from Psoralea corylifolia by high-speed counter-current chromatography. *J. Chromatogr. A* **2004**, *1057*, 225–228. [CrossRef] [PubMed]
215. Siskos, E.P.; Mazomenos, B.E.; Konstantopoulou, M.A. Isolation and identification of insecticidal components from Citrus aurantium fruit peel extract. *J. Agric. Food Chem.* **2008**, *56*, 5577–5581. [CrossRef]
216. Finkelstein, E.V.E.; Afek, U.Z.I.; Gross, E.; Aharoni, N.; Rosenberg, L.; Halevy, S. An outbreak of phytophotodermatitis due to celery. *Int. J. Dermatol.* **1994**, *33*, 116–118. [CrossRef]
217. McCloud, E.S.; Berenbaum, M.R.; Tuveson, R.W. Furanocoumarin content and phototoxicity of rough lemon (Citrus jambhiri) foliage exposed to enhanced ultraviolet-B (UVB) irradiation. *J. Chem. Ecol.* **1992**, *18*, 1125–1137. [CrossRef]

218. Arigò, A.; Rigano, F.; Russo, M.; Trovato, E.; Dugo, P.; Mondello, L. Dietary intake of coumarins and furocoumarins through citrus beverages: A detailed estimation by a HPLC-MS/MS method combined with the Linear Retention Index System. *Foods* **2021**, *10*, 1533. [CrossRef] [PubMed]

219. Melough, M.M.; Chun, O.K. Dietary furocoumarins and skin cancer: A review of current biological evidence. *Food Chem. Toxicol.* **2018**, *122*, 163–171. [CrossRef] [PubMed]

220. Nagayo, K.; Way, B.H.; Tran, R.M.; Song, P.S. Photocarcinogenicity of 8-methoxypsoralen and aflatoxin B1 with longwave ultraviolet light. *Cancer Lett.* **1983**, *18*, 191–198. [CrossRef]

221. Forbes, P.D.; Davies, R.E.; Urbach, F.; Dunnick, J.K. Long-term toxicity of oral 8-methoxypsoralen plus ultraviolet radiation in mice. *J. Toxicol. Cutan. Ocul. Toxicol.* **1990**, *9*, 237–250. [CrossRef]

222. NTP, National Toxicology Program. *Toxicology and Carcinogenesis Studies of 8-Methoxypsoralen (CAS No. 298-81-7) in F344/N Rats (Gavage Studies)*; National Toxicology Program: Research Triangle, NC, USA, 1989; pp. 1–130.

223. Müller, L.; Kasper, P.; Kersten, B.; Zhang, J. Photochemical genotoxicity and photochemical carcinogenesis—Two sides of a coin? *Toxicol. Lett.* **1998**, *102–103*, 383–387. [CrossRef]

224. Chételat, A.A.; Albertini, S.; Gocke, E. The photomutagenicity of fluoroquinolones in tests for gene mutation, chromosomal aberration, gene conversion and DNA breakage (Comet assay). *Mutagenesis* **1996**, *11*, 497–504. [CrossRef]

225. Yang, A.; Chen, J.; Ma, Y.; Wang, L.; Fan, Y.; He, X. Studies on the metabolites difference of psoralen/isopsoralen in human and six mammalian liver microsomes in vitro by UHPLC-MS/MS. *J. Pharm. Biomed. Anal.* **2017**, *141*, 200–209. [CrossRef] [PubMed]

226. Ji, L.; Lu, D.; Cao, J.; Zheng, L.; Peng, Y.; Zheng, J. Psoralen, a mechanism-based inactivator of CYP2B6. *Chem. Biol. Interact.* **2015**, *240*, 346–352. [CrossRef]

227. Girennavar, B.; Poulose, S.M.; Jayaprakasha, G.K.; Bhat, N.G.; Patil, B.S. Furocoumarins from grapefruit juice and their effect on human CYP 3A4 and CYP 1B1 isoenzymes. *Bioorg. Med. Chem.* **2006**, *14*, 2606–2612. [CrossRef]

228. Santes-Palacios, R.; Romo-Mancillas, A.; Camacho-Carranza, R.; Espinosa-Aguirre, J.J. Inhibition of human and rat CYP1A1 enzyme by grapefruit juice compounds. *Toxicol. Lett.* **2016**, *258*, 267–275. [CrossRef] [PubMed]

229. Zhuang, X.M.; Zhong, Y.H.; Xiao, W.B.; Li, H.; Lu, C. Identification and characterization of psoralen and isopsoralen as potent CYP1A2 reversible and time-dependent inhibitors in human and rat preclinical studies. *Drug Metab. Dispos.* **2013**, *41*, 1914–1922. [CrossRef] [PubMed]

230. Bickers, D.R.; Pathak, M.A. Psoralen pharmacology: Studies on metabolism and enzyme induction. *Natl. Cancer Inst. Monogr.* **1984**, *66*, 77–84.

231. Wagstaff, D.J. Dietary exposure to furocoumarins. *Regul. Toxicol. Pharm.* **1991**, *14*, 261–272. [CrossRef]

232. Gorgus, E.; Lohr, C.; Raquet, N.; Guth, S.; Schrenk, D. Limettin and furocoumarins in beverages containing citrus juices or extracts. *Food Chem. Toxicol.* **2010**, *48*, 93–98. [CrossRef]

233. Ellis, C.R.; Elston, D.M. Psoralen-induced phytophotodermatitis. *Dermatitis* **2021**, *32*, 140–143. [CrossRef]

234. Berkley, S.F. Dermatitis in grocery workers associated with high natural concentrations of furanocoumarins in celery. *Ann. Intern. Med.* **1986**, *105*, 351. [CrossRef] [PubMed]

235. Ljunggren, B. Severe phototoxic burn following celery ingestion. *Arch. Derm.* **1990**, *126*, 1334–1336. [CrossRef]

236. Archier, E.; Devaux, S.; Castela, E.; Gallini, A.; Aubin, F.; Le Maître, M.; Aractingi, S.; Bachelez, H.; Cribier, B.; Joly, P.; et al. Carcinogenic risks of psoralen UV-A therapy and narrowband UV-B therapy in chronic plaque psoriasis: A systematic literature review. *J. Eur. Acad. Derm. Venereol.* **2012**, *26* (Suppl. S3), 22–31. [CrossRef] [PubMed]

237. Stern, R.S.; Liebman, E.J.; Väkevä, L. Oral psoralen and ultraviolet-A light (PUVA) treatment of psoriasis and persistent risk of nonmelanoma skin cancer. PUVA Follow-up Study. *J. Natl. Cancer Inst.* **1998**, *90*, 1278–1284. [CrossRef] [PubMed]

238. Marley, A.R.; Li, M.; Champion, V.L.; Song, Y.; Han, J.; Li, X. The association between citrus consumption and melanoma risk in the UK Biobank. *Br. J. Derm.* **2021**, *185*, 353–362. [CrossRef]

239. Moreira, R.; Pereira, D.M.; Valentão, P.; Andrade, P.B. Pyrrolizidine alkaloids: Chemistry, pharmacology, toxicology and food safety. *Int. J. Mol. Sci.* **2018**, *19*, 1668. [CrossRef]

240. JECFA, Joint FAO/WHO Expert Committee on Food Additives. *Safety Evaluation of Certain Food Additives and Contaminants: Prepared by the Eightieth Meeting of the Joint FAO/WHO Expert Committee on Food Additives (JECFA)*; Supplement 2: Pyrrolizidine alkaloids; Who Food Additives Series: 71-S2; World Health Organization: Geneva, Switzerland, 2020. [CrossRef]

241. Robertson, J.; Stevens, K. Pyrrolizidine alkaloids: Occurrence, biology, and chemical synthesis. *Nat. Prod. Rep.* **2017**, *34*, 62–89. [CrossRef]

242. EFSA CONTAM Panel, European Food Safety Authority, Panel on Contaminants in the Food Chain. Scientific Opinion on pyrrolizidine alkaloids in food and feed. *EFSA J.* **2011**, *9*, 2406. [CrossRef]

243. Robertson, J.; Stevens, K. Pyrrolizidine alkaloids. *Nat. Prod. Rep.* **2014**, *31*, 1721–1788. [CrossRef]

244. Kopp, T.; Abdel-Tawab, M.; Mizaikoff, B. Extracting and analyzing pyrrolizidine alkaloids in medicinal plants: A review. *Toxins* **2020**, *12*, 320. [CrossRef]

245. Fu, P.P.; Xia, Q.; Lin, G.; Chou, M.W. Pyrrolizidine alkaloids—Genotoxicity, metabolism enzymes, metabolic activation, and mechanisms. *Drug Metab. Rev.* **2004**, *36*, 1–55. [CrossRef] [PubMed]

246. Dusemund, B.; Nowak, N.; Sommerfeld, C.; Lindtner, O.; Schäfer, B.; Lampen, A. Risk assessment of pyrrolizidine alkaloids in food of plant and animal origin. *Food Chem. Toxicol.* **2018**, *115*, 63–72. [CrossRef] [PubMed]

247. IARC, International Agency for Research on Cancer. Some Traditional Herbal Medicines, Some Mycotoxins, Naphthalene and Styrene. In *IARC Monographs on the Evaluation of Carcinogenic Risks to Humans*; International Agency for Research on Cancer: Lyon, France, 2002; Volume 82.

248. JECFA, Joint FAO/WHO Expert Committee on Food Additives. *Evaluation of Certain Food Additives and Contaminants*; Prepared for the Eightieth Report of the Joint FAO/WHO Expert Committee on Food Additives; World Health Organization: Geneva, Switzerland, 2016.

249. EMA, European Medicines Agency. *Public Statement on the Use of Herbal Medicinal Products Containing Toxic, Unsaturated Pyrrolizidine Alkaloids (PAs) Including Recommendations Regarding Contamination of Herbal Medicinal Products with Pas*; EMA/HMPC/893108/2011; European Medicines Agency: Amsterdam, The Netherlands, 2021.

250. Fu, P.P. Pyrrolizidine alkaloids: Metabolic activation pathways leading to liver tumor initiation. *Chem. Res. Toxicol.* **2017**, *30*, 81–93. [CrossRef] [PubMed]

251. NTP, National Toxicology Program. *Bioassay of Lasiocarpine for Possible Carcinogenicity*; 0163-7185 (Print); National Toxicology Program: Research Triangle, NC, USA, 1978; pp. 1–66.

252. NTP, National Toxicology Program. *Toxicology and Carcinogenesis Studies of Riddelliine (CAS No. 23246-96-0) in F344/N Rats and B6C3F1 Mice (Gavage Studies)*; 0888-8051 (Print); National Toxicology Program: Research Triangle, NC, USA, 2003; pp. 1–280.

253. Chen, T.; Mei, N.; Fu, P.P. Genotoxicity of pyrrolizidine alkaloids. *J. Appl. Toxicol.* **2010**, *30*, 183–196. [CrossRef]

254. Allemang, A.; Mahony, C.; Lester, C.; Pfuhler, S. Relative potency of fifteen pyrrolizidine alkaloids to induce DNA damage as measured by micronucleus induction in HepaRG human liver cells. *Food Chem. Toxicol.* **2018**, *121*, 72–81. [CrossRef]

255. Williams, G.M.; Mori, H.; Hirono, I.; Nagao, M. Genotoxity of pyrrolizidine alkaloids in the hepatocyte primary culture/DNA-repair test. *Mutat. Res.* **1980**, *79*, 1–5. [CrossRef]

256. Mori, H.; Sugie, S.; Yoshimi, N.; Asada, Y.; Furuya, T.; Williams, G.M. Genotoxicity of a variety of pyrrolizidine alkaloids in the hepatocyte primary culture-DNA repair test using rat, mouse, and hamster hepatocytes. *Cancer Res.* **1985**, *45*, 3125–3129.

257. Chou, M.W.; Yan, J.; Nichols, J.; Xia, Q.; Beland, F.A.; Chan, P.C.; Fu, P.P. Correlation of DNA adduct formation and riddelliine-induced liver tumorigenesis in F344 rats and B6C3F1 mice. *Cancer Lett.* **2004**, *207*, 119–125. [CrossRef]

258. Wang, Y.P.; Yan, J.; Beger, R.D.; Fu, P.P.; Chou, M.W. Metabolic activation of the tumorigenic pyrrolizidine alkaloid, monocrotaline, leading to DNA adduct formation in vivo. *Cancer Lett.* **2005**, *226*, 27–35. [CrossRef]

259. Xia, Q.; Zhao, Y.; Von Tungeln, L.S.; Doerge, D.R.; Lin, G.; Cai, L.; Fu, P.P. Pyrrolizidine alkaloid-derived DNA adducts as a common biological biomarker of pyrrolizidine alkaloid-induced tumorigenicity. *Chem. Res. Toxicol.* **2013**, *26*, 1384–1396. [CrossRef]

260. Ebmeyer, J.; Braeuning, A.; Glatt, H.; These, A.; Hessel-Pras, S.; Lampen, A. Human CYP3A4-mediated toxification of the pyrrolizidine alkaloid lasiocarpine. *Food Chem. Toxicol.* **2019**, *130*, 79–88. [CrossRef] [PubMed]

261. Schoch, T.K.; Gardner, D.R.; Stegelmeier, B.L. GC/MS/MS detection of pyrrolic metabolites in animals poisoned with the pyrrolizidine alkaloid riddelliine. *J. Nat. Toxins* **2000**, *9*, 197–206. [PubMed]

262. Yang, Y.C.; Yan, J.; Doerge, D.R.; Chan, P.C.; Fu, P.P.; Chou, M.W. Metabolic activation of the tumorigenic pyrrolizidine alkaloid, riddelliine, leading to DNA adduct formation in vivo. *Chem. Res. Toxicol.* **2001**, *14*, 101–109. [CrossRef] [PubMed]

263. Wang, Y.P.; Yan, J.; Fu, P.P.; Chou, M.W. Human liver microsomal reduction of pyrrolizidine alkaloid N-oxides to form the corresponding carcinogenic parent alkaloid. *Toxicol. Lett.* **2005**, *155*, 411–420. [CrossRef] [PubMed]

264. Geburek, I.; Rutz, L.; Gao, L.; Küpper, J.H.; These, A.; Schrenk, D. Metabolic pattern of hepatotoxic pyrrolizidine alkaloids in liver cells. *Chem. Res. Toxicol.* **2021**, *34*, 1101–1113. [CrossRef] [PubMed]

265. Fashe, M.M.; Juvonen, R.O.; Petsalo, A.; Räsänen, J.; Pasanen, M. Species-specific differences in the in vitro metabolism of lasiocarpine. *Chem. Res. Toxicol.* **2015**, *28*, 2034–2044. [CrossRef]

266. EFSA, European Food Safety Authority. Dietary exposure assessment to pyrrolizidine alkaloids in the European population. *EFSA J.* **2016**, *14*, 4572. [CrossRef]

267. Rasenack, R.; Müller, C.; Kleinschmidt, M.; Rasenack, J.; Wiedenfeld, H. Veno-occlusive disease in a fetus caused by pyrrolizidine alkaloids of food origin. *Fetal Diagn. Ther.* **2003**, *18*, 223–225. [CrossRef]

268. Neuman, M.G.; Cohen, L.; Opris, M.; Nanau, R.M.; Hyunjin, J. Hepatotoxicity of pyrrolizidine alkaloids. *J. Pharm. Pharm. Sci.* **2015**, *18*, 825–843. [CrossRef]

269. Habs, M.; Binder, K.; Krauss, S.; Müller, K.; Ernst, B.; Valentini, L.; Koller, M. A balanced risk-benefit analysis to determine human risks associated with pyrrolizidine alkaloids (PA)-The case of tea and herbal infusions. *Nutrients* **2017**, *9*, 717. [CrossRef]

270. EFSA CONTAM Panel, European Food Safety Authority, Panel on Contaminants in the Food Chain. Statement on the risks for human health related to the presence of pyrrolizidine alkaloids in honey, tea, herbal infusions and food supplements. *EFSA J.* **2017**, *15*, 4908. [CrossRef]

271. Chen, L.; Mulder, P.P.J.; Louisse, J.; Peijnenburg, A.; Wesseling, S.; Rietjens, I. Risk assessment for pyrrolizidine alkaloids detected in (herbal) teas and plant food supplements. *Regul. Toxicol. Pharm.* **2017**, *86*, 292–302. [CrossRef]

272. Bullerman, L.B.; Bianchini, A. Stability of mycotoxins during food processing. *Int. J. Food Microbiol.* **2007**, *119*, 140–146. [CrossRef] [PubMed]

273. Marin, S.; Ramos, A.J.; Cano-Sancho, G.; Sanchis, V. Mycotoxins: Occurrence, toxicology, and exposure assessment. *Food Chem. Toxicol.* **2013**, *60*, 218–237. [CrossRef] [PubMed]

274. Bryden, W.L. Mycotoxins in the food chain: Human health implications. *Asia Pac. J. Clin. Nutr.* **2007**, *16* (Suppl. S1), 95–101. [PubMed]

275. IARC, International Agency for Research on Cancer. Chemical Agents and Related Occupations. Review of Human Carcinogens. In *IARC Monographs on the Evaluation of Carcinogenic Risks to Humans*; International Agency for Research on Cancer: Lyon, France, 2012; Volume 100F.

276. JECFA, Joint FAO/WHO Expert Committee on Food Additives. *Evaluation of Certain Contaminants in Food*; Prepared for the Eighty-Third Report of the Joint FAO/WHO Expert Committee on Food Additives; WHO: Geneva, Switzerland, 2017.

277. Wogan, G.N.; Hecht, S.S.; Felton, J.S.; Conney, A.H.; Loeb, L.A. Environmental and chemical carcinogenesis. *Semin. Cancer Biol.* **2004**, *14*, 473–486. [CrossRef]

278. Rushing, B.R.; Selim, M.I. Aflatoxin B1: A review on metabolism, toxicity, occurrence in food, occupational exposure, and detoxification methods. *Food Chem. Toxicol.* **2019**, *124*, 81–100. [CrossRef] [PubMed]

279. Ramsdell, H.S.; Eaton, D.L. Species susceptibility to aflatoxin B1 carcinogenesis: Comparative kinetics of microsomal biotransformation. *Cancer Res.* **1990**, *50*, 615–620.

280. Wogan, G.N.; Edwards, G.S.; Newberne, P.M. Structure-activity relationships in toxicity and carcinogenicity of aflatoxins and analogs. *Cancer Res.* **1971**, *31*, 1936–1942.

281. EFSA CONTAM Panel, European Food Safety Authority, Panel on Contaminants in the Food Chain. Scientific Opinion on risk assessment of aflatoxins in food. *EFSA J.* **2020**, *18*, 6040. [CrossRef]

282. Robens, J.F.; Richard, J.L. Aflatoxins in animal and human health. *Rev. Environ. Contam. Toxicol.* **1992**, *127*, 69–94. [CrossRef] [PubMed]

283. Choy, W.N. A review of the dose-response induction of DNA adducts by aflatoxin B1 and its implications to quantitative cancer-risk assessment. *Mutat. Res.* **1993**, *296*, 181–198. [CrossRef]

284. Williams, G.M.; Duan, J.-D.; Brunnemann, K.D.; Iatropoulos, M.J.; Vock, E.; Deschl, U. Chicken fetal liver DNA damage and adduct formation by activation-dependent DNA-reactive carcinogens and related compounds of several structural classes. *Toxicol. Sci.* **2014**, *141*, 18–28. [CrossRef] [PubMed]

285. Zhang, Y.J.; Chen, C.J.; Haghighi, B.; Yang, G.Y.; Hsieh, L.L.; Wang, L.W.; Santella, R.M. Quantitation of aflatoxin B1-DNA adducts in woodchuck hepatocytes and rat liver tissue by indirect immunofluorescence analysis. *Cancer Res.* **1991**, *51*, 1720–1725. [PubMed]

286. Coskun, E.; Jaruga, P.; Vartanian, V.; Erdem, O.; Egner, P.A.; Groopman, J.D.; Lloyd, R.S.; Dizdaroglu, M. Aflatoxin-guanine DNA adducts and oxidatively induced DNA damage in aflatoxin-treated mice in vivo as measured by Liquid Chromatography-Tandem Mass Spectrometry with Isotope Dilution. *Chem. Res. Toxicol.* **2019**, *32*, 80–89. [CrossRef]

287. Bedard, L.L.; Massey, T.E. Aflatoxin B1-induced DNA damage and its repair. *Cancer Lett.* **2006**, *241*, 174–183. [CrossRef]

288. Wang, J.S.; Groopman, J.D. DNA damage by mycotoxins. *Mutat. Res.* **1999**, *424*, 167–181. [CrossRef]

289. Eaton, D.L.; Gallagher, E.P. Mechanisms of aflatoxin carcinogenesis. *Annu. Rev. Pharm. Toxicol.* **1994**, *34*, 135–172. [CrossRef]

290. McQueen, C.A.; Way, B.M.; Williams, G.M. Genotoxicity of carcinogens in human hepatocytes: Application in hazard assessment. *Toxicol. Appl. Pharmacol.* **1988**, *96*, 360–366. [CrossRef]

291. Theumer, M.G.; Henneb, Y.; Khoury, L.; Snini, S.P.; Tadrist, S.; Canlet, C.; Puel, O.; Oswald, I.P.; Audebert, M. Genotoxicity of aflatoxins and their precursors in human cells. *Toxicol. Lett.* **2018**, *287*, 100–107. [CrossRef]

292. Marchese, S.; Polo, A.; Ariano, A.; Velotto, S.; Costantini, S.; Severino, L. Aflatoxin B1 and M1: Biological properties and their involvement in cancer development. *Toxins* **2018**, *10*, 214.

293. Garner, R.C.; Martin, C.N.; Smith, J.R.; Coles, B.F.; Tolson, M.R. Comparison of aflatoxin B1 and aflatoxin G1 binding to cellular macromolecules in vitro, in vivo and after peracid oxidation; characterisation of the major nucleic acid adducts. *Chem. Biol. Interact.* **1979**, *26*, 57–73. [CrossRef]

294. Guengerich, F.P.; Johnson, W.W.; Shimada, T.; Ueng, Y.F.; Yamazaki, H.; Langouët, S. Activation and detoxication of aflatoxin B1. *Mutat. Res.* **1998**, *402*, 121–128. [CrossRef] [PubMed]

295. Roebuck, B.D.; Siegel, W.G.; Wogan, G.N. In vitro metabolism of aflatoxin B2 by animal and human liver. *Cancer Res.* **1978**, *38*, 999–1002. [PubMed]

296. Roebuck, B.D.; Wogan, G.N. Species comparison of in vitro metabolism of aflatoxin B1. *Cancer Res.* **1977**, *37*, 1649–1656.

297. Benkerroum, N. Chronic and acute toxicities of aflatoxins: Mechanisms of action. *Int. J. Environ. Res. Public Health* **2020**, *17*, 423. [CrossRef]

298. Groopman, J.D.; Wild, C.P.; Hasler, J.; Junshi, C.; Wogan, G.N.; Kensler, T.W. Molecular epidemiology of aflatoxin exposures: Validation of aflatoxin-N7-guanine levels in urine as a biomarker in experimental rat models and humans. *Environ. Health Perspect.* **1993**, *99*, 107–113. [CrossRef]

299. Wogan, G.N.; Kensler, T.W.; Groopman, J.D. Present and future directions of translational research on aflatoxin and hepatocellular carcinoma. A review. *Food Addit. Contam. Part A Chem. Anal. Control. Expo. Risk Assess.* **2012**, *29*, 249–257. [CrossRef]

300. Kew, M.C. Synergistic interaction between aflatoxin B1 and hepatitis B virus in hepatocarcinogenesis. *Liver Int.* **2003**, *23*, 405–409. [CrossRef]

301. Ross, R.K.; Yuan, J.M.; Yu, M.C.; Wogan, G.N.; Qian, G.S.; Tu, J.T.; Groopman, J.D.; Gao, Y.T.; Henderson, B.E. Urinary aflatoxin biomarkers and risk of hepatocellular carcinoma. *Lancet* **1992**, *339*, 943–946. [CrossRef]

302. Fang, L.; Zhao, B.; Zhang, R.; Wu, P.; Zhao, D.; Chen, J.; Pan, X.; Wang, J.; Wu, X.; Zhang, H.; et al. Occurrence and exposure assessment of aflatoxins in Zhejiang province, China. *Environ. Toxicol. Pharmacol.* **2022**, *92*, 103847. [CrossRef] [PubMed]

303. FDA, Food and Drug Administration. *Compliance Policy Guide Sec. 555.400. Aflatoxins in Human Food*; Guidance for FDA Staff; Food and Drug Administration: Silver Spring, MD, USA, 2021.

304. FDA, Food and Drug Administration. *Compliance Policy Guide (CPG) Sec 527.400 Whole Milk, Lowfat Milk, Skim Milk-Aflatoxin M1*; Food and Drug Administration: Silver Spring, MD, USA, 2005.

305. IARC, International Agency for Research on Cancer. Some Naturally Occurring Substances: Food Items and Constituents, Heterocyclic Aromatic Amines and Mycotoxins. In *IARC Monographs on the Evaluation of Carcinogenic Risk to Humans*; International Agency for Research on Cancer: Lyon, France, 1993; Volume 56.

306. Malir, F.; Ostry, V.; Pfohl-Leszkowicz, A.; Malir, J.; Toman, J. Ochratoxin A: 50 years of research. *Toxins* **2016**, *8*, 191. [CrossRef] [PubMed]

307. EFSA CONTAM Panel, European Food Safety Authority, Panel on Contaminants in the Food Chain. Scientific Opinion on the risk assessment of ochratoxin A in food. *EFSA J.* **2020**, *18*, 6113. [CrossRef]

308. NTP, National Toxicology Program. *Toxicology and Carcinogenesis Studies of Ochratoxin A (CAS No. 303-47-9) in F344/N Rats (Gavage Studies)*; 0888-8051 (Print); National Toxicology Program: Research Triangle, NC, USA, 1989; pp. 1–142.

309. Mantle, P.; Kulinskaya, E.; Nestler, S. Renal tumourigenesis in male rats in response to chronic dietary ochratoxin A. *Food Addit. Contam.* **2005**, *22*, 58–64. [CrossRef] [PubMed]

310. Mally, A. Ochratoxin A and mitotic disruption: Mode of action analysis of renal tumor formation by Ochratoxin A. *Toxicol. Sci.* **2012**, *127*, 315–330. [CrossRef]

311. Pfohl-Leszkowicz, A.; Manderville, R.A. Ochratoxin A: An overview on toxicity and carcinogenicity in animals and humans. *Mol. Nutr. Food Res.* **2007**, *51*, 61–99. [CrossRef]

312. Pfohl-Leszkowicz, A.; Manderville, R.A. An update on direct genotoxicity as a molecular mechanism of ochratoxin a carcinogenicity. *Chem. Res. Toxicol.* **2012**, *25*, 252–262. [CrossRef]

313. Mally, A.; Dekant, W. DNA adduct formation by ochratoxin A: Review of the available evidence. *Food Addit. Contam.* **2005**, *22*, 65–74. [CrossRef]

314. Tozlovanu, M.; Faucet-Marquis, V.; Pfohl-Leszkowicz, A.; Manderville, R.A. Genotoxicity of the hydroquinone metabolite of ochratoxin A: structure-activity relationships for covalent DNA adduction. *Chem. Res. Toxicol.* **2006**, *19*, 1241–1247. [CrossRef]

315. Mantle, P.G.; Faucet-Marquis, V.; Manderville, R.A.; Squillaci, B.; Pfohl-Leszkowicz, A. Structures of covalent adducts between DNA and ochratoxin a: A new factor in debate about genotoxicity and human risk assessment. *Chem. Res. Toxicol.* **2010**, *23*, 89–98. [CrossRef]

316. Kamp, H.G.; Eisenbrand, G.; Janzowski, C.; Kiossev, J.; Latendresse, J.R.; Schlatter, J.; Turesky, R.J. Ochratoxin A induces oxidative DNA damage in liver and kidney after oral dosing to rats. *Mol. Nutr. Food Res.* **2005**, *49*, 1160–1167. [CrossRef] [PubMed]

317. Arbillaga, L.; Azqueta, A.; van Delft, J.H.M.; López de Cerain, A. In vitro gene expression data supporting a DNA non-reactive genotoxic mechanism for ochratoxin A. *Toxicol. Appl. Pharmacol.* **2007**, *220*, 216–224. [CrossRef] [PubMed]

318. Czakai, K.; Müller, K.; Mosesso, P.; Pepe, G.; Schulze, M.; Gohla, A.; Patnaik, D.; Dekant, W.; Higgins, J.M.G.; Mally, A. Perturbation of mitosis through inhibition of histone acetyltransferases: The key to Ochratoxin A toxicity and carcinogenicity? *Toxicol. Sci.* **2011**, *122*, 317–329. [CrossRef]

319. Rásonyi, T.; Schlatter, J.; Dietrich, D.R. The role of α2u-globulin in ochratoxin A induced renal toxicity and tumors in F344 rats. *Toxicol. Lett.* **1999**, *104*, 83–92. [CrossRef]

320. Pfohl-Leszkowicz, A.; Pinelli, E.; Bartsch, H.; Mohr, U.; Castegnaro, M. Sex- and strain-specific expression of cytochrome P450s in Ochratoxin A-induced genotoxicity and carcinogenicity in rats. *Mol. Carcinog.* **1998**, *23*, 76–85. [CrossRef]

321. Heussner, A.H.; Bingle, L.E. Comparative ochratoxin toxicity: A review of the available data. *Toxins* **2015**, *7*, 4253–4282. [CrossRef] [PubMed]

322. JECFA, Joint FAO/WHO Expert Committee on Food Additives. *Evaluation of Certain Food Additives and Contaminants*; Prepared by the Sixty-Eight Meeting of the Joint FAO/WHO Expert Committee on Food Additives (JECFA); WHO: Geneva, Switzerland, 2007.

323. McNeal, T.P.; Nyman, P.J.; Diachenko, G.W.; Hollifield, H.C. Survey of benzene in foods by using headspace concentration techniques and capillary gas chromatography. *J. AOAC Int.* **1993**, *76*, 1213–1219. [CrossRef]

324. IARC, International Agency for Research on Cancer. Benzene. *IARC Monographs on the Evaluation of Carcinogenic Risks to Humans*; International Agency for Research on Cancer: Lyon, France, 2018; Volume 120.

325. Salviano Dos Santos, V.P.; Medeiros Salgado, A.; Guedes Torres, A.; Signori Pereira, K. Benzene as a chemical hazard in processed foods. *Int. J. Food Sci.* **2015**, *2015*, 545640. [CrossRef]

326. Smith, B.; Cadby, P.; DiNovi, M.; Setzer, R.W. Application of the Margin of Exposure (MoE) approach to substances in food that are genotoxic and carcinogenic: Example: Benzene, CAS: 71-43-2. *Food Chem. Toxicol.* **2010**, *48* (Suppl. S1), S49–S56. [CrossRef]

327. Gardner, L.K.; Lawrence, G.D. Benzene production from decarboxylation of benzoic acid in the presence of ascorbic acid and a transition-metal catalyst. *J. Agric. Food Chem.* **1993**, *41*, 693–695. [CrossRef]

328. Jickells, S.M.; Crews, C.; Castle, L.; Gilbert, J. Headspace analysis of benzene in food contact materials and its migration into foods from plastics cookware. *Food Addit. Contam.* **1990**, *7*, 197–205. [CrossRef] [PubMed]

329. Meadows, M. Benzene in beverages. *FDA Consum.* **2006**, *40*, 9–10. [PubMed]

330. Nyman, P.J.; Diachenko, G.W.; Perfetti, G.A.; McNeal, T.P.; Hiatt, M.H.; Morehouse, K.M. Survey results of benzene in soft drinks and other beverages by headspace gas chromatography/mass spectrometry. *J. Agric. Food Chem.* **2008**, *56*, 571–576. [CrossRef] [PubMed]

331. Snyder, R.; Witz, G.; Goldstein, B.D. The toxicology of benzene. *Environ. Health Perspect.* **1993**, *100*, 293–306. [CrossRef]

332. Maltoni, C.; Cotti, G.; Valgimigli, L.; Mandrioli, A. Zymbal gland carcinomas in rats following exposure to benzene by inhalation. *Am. J. Ind. Med.* **1982**, *3*, 11–16. [CrossRef]

333. NTP, National Toxicology Program. *Toxicology and Carcinogenesis Studies of Benzene (CAS No. 71-43-2) in F344/N Rats and B6C3F1 Mice (Gavage Studies)*; 0888-8051 (Print); National Toxicology Program: Research Triangle, NC, USA, 1986; pp. 1–277.

334. Whysner, J.; Reddy, M.V.; Ross, P.M.; Mohan, M.; Lax, E.A. Genotoxicity of benzene and its metabolites. *Mutat. Res.* **2004**, *566*, 99–130. [CrossRef]

335. Wetmore, B.A.; Struve, M.F.; Gao, P.; Sharma, S.; Allison, N.; Roberts, K.C.; Letinski, D.J.; Nicolich, M.J.; Bird, M.G.; Dorman, D.C. Genotoxicity of intermittent co-exposure to benzene and toluene in male CD-1 mice. *Chem. Biol. Interact.* **2008**, *173*, 166–178. [CrossRef]

336. Tuo, J.; Loft, S.; Thomsen, M.S.; Poulsen, H.E. Benzene-induced genotoxicity in mice in vivo detected by the alkaline comet assay: Reduction by CYP2E1 inhibition. *Mutat. Res.* **1996**, *368*, 213–219. [CrossRef]

337. Provost, G.S.; Mirsalis, J.C.; Rogers, B.J.; Short, J.M. Mutagenic response to benzene and tris(2,3-dibromopropyl)-phosphate in the lambda lacI transgenic mouse mutation assay: A standardized approach to in vivo mutation analysis. *Environ. Mol. Mutagen.* **1996**, *28*, 342–347. [CrossRef]

338. Salem, E.; El-Garawani, I.; Allam, H.; El-Aal, B.A.; Hegazy, M. Genotoxic effects of occupational exposure to benzene in gasoline station workers. *Ind. Health* **2018**, *56*, 132–140. [CrossRef]

339. Kitamoto, S.; Matsuyama, R.; Uematsu, Y.; Ogata, K.; Ota, M.; Yamada, T.; Miyata, K.; Kimura, J.; Funabashi, H.; Saito, K. Genotoxicity evaluation of benzene, di(2-ethylhexyl) phthalate, and trisodium ethylenediamine tetraacetic acid monohydrate using a combined rat comet/micronucleus assays. *Mutat. Res. Genet. Toxicol. Environ. Mutagen.* **2015**, *786–788*, 137–143. [CrossRef] [PubMed]

340. Grigoryan, H.; Edmands, W.M.B.; Lan, Q.; Carlsson, H.; Vermeulen, R.; Zhang, L.; Yin, S.N.; Li, G.L.; Smith, M.T.; Rothman, N.; et al. Adductomic signatures of benzene exposure provide insights into cancer induction. *Carcinogenesis* **2018**, *39*, 661–668. [CrossRef] [PubMed]

341. Li, G.; Wang, C.; Xin, W.; Yin, S. Tissue distribution of DNA adducts and their persistence in blood of mice exposed to benzene. *Environ. Health Perspect.* **1996**, *104* (Suppl. S6), 1337–1338. [CrossRef] [PubMed]

342. Bodell, W.J.; Pathak, D.N.; Lévay, G.; Ye, Q.; Pongracz, K. Investigation of the DNA adducts formed in B6C3F1 mice treated with benzene: Implications for molecular dosimetry. *Environ. Health Perspect.* **1996**, *104* (Suppl. S6), 1189–1193. [CrossRef]

343. Reddy, M.V.; Blackburn, G.R.; Schreiner, C.A.; Mehlman, M.A.; Mackerer, C.R. 32P analysis of DNA adducts in tissues of benzene-treated rats. *Environ. Health Perspect.* **1989**, *82*, 253–257. [CrossRef]

344. Gaskell, M.; McLuckie, K.I.E.; Farmer, P.B. Genotoxicity of the benzene metabolites para-benzoquinone and hydroquinone. *Chem. Biol. Interact.* **2005**, *153–154*, 267–270. [CrossRef]

345. Williams, G.M.; Iatropoulos, M.J.; Jeffrey, A.M.; Duan, J.-D. Inhibition by dietary hydroquinone of acetylaminofluorene induction of initiation of rat liver carcinogenesis. *Food Chem. Toxicol.* **2007**, *45*, 1620–1625. [CrossRef]

346. English, J.C.; Hill, T.; O'Donoghue, J.L.; Reddy, M.V. Measurement of nuclear DNA Modification by 32P-postlabeling in the kidneys of male and female Fischer 344 rats after multiple gavage doses of hydroquinone. *Toxicol. Sci.* **1994**, *23*, 391–396. [CrossRef]

347. Gaskell, M.; McLuckie, K.I.; Farmer, P.B. Comparison of the mutagenic activity of the benzene metabolites, hydroquinone and para-benzoquinone in the supF forward mutation assay: A role for minor DNA adducts formed from hydroquinone in benzene mutagenicity. *Mutat. Res.* **2004**, *554*, 387–398. [CrossRef]

348. Linhart, I.; Mikes, P.; Frantík, E.; Mráz, J. DNA adducts formed from p-benzoquinone, an electrophilic metabolite of benzene, are extensively metabolized in vivo. *Chem. Res. Toxicol.* **2011**, *24*, 383–391. [CrossRef]

349. Xie, Z.; Zhang, Y.; Guliaev, A.B.; Shen, H.; Hang, B.; Singer, B.; Wang, Z. The p-benzoquinone DNA adducts derived from benzene are highly mutagenic. *DNA Repair.* **2005**, *4*, 1399–1409. [CrossRef] [PubMed]

350. Kolachana, P.; Subrahmanyam, V.V.; Meyer, K.B.; Zhang, L.; Smith, M.T. Benzene and its phenolic metabolites produce oxidative DNA damage in HL60 cells in vitro and in the bone marrow in vivo. *Cancer Res.* **1993**, *53*, 1023–1026. [PubMed]

351. Snyder, R. Leukemia and benzene. *Int. J. Environ. Res. Public Health* **2012**, *9*, 2875–2893. [CrossRef] [PubMed]

352. Wallace, L. Environmental exposure to benzene: An update. *Environ. Health Perspect.* **1996**, *104* (Suppl. S6), 1129–1136. [CrossRef] [PubMed]

353. Weisel, C.P. Benzene exposure: An overview of monitoring methods and their findings. *Chem. Biol. Interact.* **2010**, *184*, 58–66. [CrossRef] [PubMed]

354. Kim, S.R.; Halden, R.U.; Buckley, T.J. Volatile organic compounds in human milk: Methods and measurements. *Environ. Sci. Technol.* **2007**, *41*, 1662–1667. [CrossRef]

355. Medeiros Vinci, R.; Jacxsens, L.; Van Loco, J.; Matsiko, E.; Lachat, C.; de Schaetzen, T.; Canfyn, M.; Van Overmeire, I.; Kolsteren, P.; De Meulenaer, B. Assessment of human exposure to benzene through foods from the Belgian market. *Chemosphere* **2012**, *88*, 1001–1007. [CrossRef] [PubMed]

356. Cheasley, R.; Keller, C.P.; Setton, E. Lifetime excess cancer risk due to carcinogens in food and beverages: Urban versus rural differences in Canada. *Can. J. Public Health* **2017**, *108*, e288–e295. [CrossRef]

357. Galbraith, D.; Gross, S.A.; Paustenbach, D. Benzene and human health: A historical review and appraisal of associations with various diseases. *Crit. Rev. Toxicol.* **2010**, *40* (Suppl. S2), 1–46. [CrossRef]

358. JECFA, Joint FAO/WHO Expert Committee on Food Additives. *Evaluation of Certain Food Additives Prepared by the Twenty-third Meeting of the Joint FAO/WHO Expert Committee on Food Additives (JECFA)*; WHO: Geneva, Switzerland, 1980; Volume 648.

359. Lachenmeier, D.W.; Reusch, H.; Sproll, C.; Schoeberl, K.; Kuballa, T. Occurrence of benzene as a heat-induced contaminant of carrot juice for babies in a general survey of beverages. *Food Addit. Contam. Part A Chem. Anal. Control Expo. Risk Assess.* **2008**, *25*, 1216–1224. [CrossRef]

360. Hamlet, C.G.; Sadd, P.A.; Crews, C.; Velíšek, J.; Baxter, D.E. Occurrence of 3-chloro-propane-1,2-diol (3-MCPD) and related compounds in foods: A review. *Food Addit. Contam.* **2002**, *19*, 619–631. [CrossRef] [PubMed]

361. Andres, S.; Appel, K.E.; Lampen, A. Toxicology, occurrence and risk characterisation of the chloropropanols in food: 2-monochloro-1,3-propanediol, 1,3-dichloro-2-propanol and 2,3-dichloro-1-propanol. *Food Chem. Toxicol.* **2013**, *58*, 467–478. [CrossRef] [PubMed]

362. EFSA CONTAM Panel, European Food Safety Authority, Panel on Contaminants in the Food Chain. Update of the risk assessment on 3-monochloropropanediol and its fatty acid esters. *EFSA J.* **2018**, *16*, 5083. [CrossRef]

363. Crews, C.; Hasnip, S.; Chapman, S.; Hough, P.; Potter, N.; Todd, J.; Brereton, P.; Matthews, W. Survey of chloropropanols in soy sauces and related products purchased in the UK in 2000 and 2002. *Food Addit. Contam.* **2003**, *20*, 916–922. [CrossRef] [PubMed]

364. Nyman, P.J.; Diachenko, G.W.; Perfetti, G.A. Survey of chloropropanols in soy sauces and related products. *Food Addit. Contam.* **2003**, *20*, 909–915. [CrossRef] [PubMed]

365. JECFA, Joint FAO/WHO Expert Committee on Food Additives. *Safety Evaluation of Certain Food Additives and Contaminants*; Prepared by the Fifty-Seven Meeting of the Joint FAO/WHO Expert Committee on Food Additives (JECFA); WHO: Geneva, Switzerland, 2002.

366. Korte, R.; Schulz, S.; Brauer, B. Chloropropanols (3-MCPD, 1,3-DCP) from food contact materials: GC-MS method improvement, market survey and investigations on the effect of hot water extraction. *Food Addit. Contam. Part A Chem. Anal. Control Expo. Risk Assess.* **2021**, *38*, 904–913. [CrossRef] [PubMed]

367. Becalski, A.; Zhao, T.; Breton, F.; Kuhlmann, J. 2- and 3-Monochloropropanediols in paper products and their transfer to foods. *Food Addit. Contam. Part A Chem. Anal. Control Expo. Risk Assess.* **2016**, *33*, 1499–1508. [CrossRef]

368. NTP, National Toxicology Program. *1,3-Dichloro-2-Propanol [CAS No. 96-23-1]: Review of Toxicological Literature*; National Institute of Environmental Health Sciences, National Institutes of Health and U.S. Department of Health and Human Services: Research Triangle Park, NC, USA, 2005.

369. JECFA, Joint FAO/WHO Expert Committee on Food Additives. 1,3-Dichloro-2-propanol (addendum). In *Safety Evaluation of Certain Food Additives and Contaminants*; Prepared by the Fifty-Seven Meeting of the Joint FAO/WHO Expert Committee on Food Additives (JECFA); WHO: Geneva, Switzerland, 2007; pp. 209–238.

370. El Ramy, R.; Ould Elhkim, M.; Lezmi, S.; Poul, J.M. Evaluation of the genotoxic potential of 3-monochloropropane-1,2-diol (3-MCPD) and its metabolites, glycidol and β-chlorolactic acid, using the single cell gel/comet assay. *Food Chem. Toxicol.* **2007**, *45*, 41–48. [CrossRef]

371. Aasa, J.; Törnqvist, M.; Abramsson-Zetterberg, L. Measurement of micronuclei and internal dose in mice demonstrates that 3-monochloropropane-1,2-diol (3-MCPD) has no genotoxic potency in vivo. *Food Chem. Toxicol.* **2017**, *109*, 414–420. [CrossRef]

372. Lynch, B.S.; Bryant, D.W.; Hook, G.J.; Nestmann, E.R.; Munro, I.C. Carcinogenicity of monochloro-1,2-propanediol (α-chlorohydrin, 3-MCPD). *Int. J. Toxicol.* **1998**, *17*, 47–76. [CrossRef]

373. Buhrke, T.; Voss, L.; Briese, A.; Stephanowitz, H.; Krause, E.; Braeuning, A.; Lampen, A. Oxidative inactivation of the endogenous antioxidant protein DJ-1 by the food contaminants 3-MCPD and 2-MCPD. *Arch. Toxicol.* **2018**, *92*, 289–299. [CrossRef]

374. Abt, E.; Incorvati, V.; Robin, L.P.; Redan, B.W. Occurrence of ethyl carbamate in foods and beverages: Review of the formation mechanisms, advances in analytical methods, and mitigation strategies. *J. Food Prot.* **2021**, *84*, 2195–2212. [CrossRef] [PubMed]

375. Ough, C.S.; Crowell, E.A.; Gutlove, B.R. Carbamyl compound reactions with ethanol. *Am. J. Enol. Vitic.* **1988**, *39*, 239–242.

376. Gowd, V.; Su, H.; Karlovsky, P.; Chen, W. Ethyl carbamate: An emerging food and environmental toxicant. *Food Chem.* **2018**, *248*, 312–321. [CrossRef]

377. IARC, International Agency for Research on Cancer. Alcohol Consumption and Ethyl Carbamate. In *IARC Monographs on the Evaluation of Carcinogenic Risks to Humans*; International Agency for Research on Cancer: Lyon, France, 2010; Volume 96.

378. JECFA, Joint FAO/WHO Expert Committee on Food Additives. *Safety Evaluation of Certain Food Contaminants*; Prepared by the Sixty-Fourth Meeting of the Joint FAO/WHO Expert Committee on Food Additives (JECFA); FAO Food and Nutrition Paper; World Health Organization: Geneva, Switzerland, 2006; Volume 82, pp. 1–778.

379. EFSA CONTAM Panel, European Food Safety Authority, Panel on Contaminants in the Food Chain. Opinion of the Scientific Panel on Contaminants in the Food chain on a request from the European Commission on ethyl carbamate and hydrocyanic acid in food and beverages. *EFSA J.* **2007**, *551*, 1–44.

380. Choi, B.; Koh, E. Effect of fruit thermal processing on ethyl carbamate content in maesil (Prunus mume) liqueur. *Food Sci. Biotechnol.* **2021**, *30*, 1427–1434. [CrossRef]

381. NTP, National Toxicology Program. *Toxicology and Carcinogenesis Studies of Urethane, Ethanol and Urethane/Ethanol (Urethane, CAS No. 51-79-6; Ethanol CAS No. 64-17-5) in B6C3F1 Mice (Drinking Water Studies)*; National Toxicology Program: Research Triangle, NC, USA, 2004.

382. Chan, P.C. *NTP Technical Report on Toxicity Studies of Urethane in Drinking Water and Urethane in 5% Ethanol Administered to F344/N Rats and B6C3F1 Mice*; Toxicity Report Series; United States Department of Health and Human Services Public Health Service National Institutes of Health: Research Triangle Park, NC, USA, 1996; pp. 1–91.

383. Hübner, P.; Groux, P.M.; Weibel, B.; Sengstag, C.; Horlbeck, J.; Leong-Morgenthaler, P.M.; Lüthy, J. Genotoxicity of ethyl carbamate (urethane) in Salmonella, yeast and human lymphoblastoid cells. *Mutat. Res.* **1997**, *390*, 11–19. [CrossRef]

384. Allen, J.W.; Stoner, G.D.; Pereira, M.A.; Backer, L.C.; Sharief, Y.; Hatch, G.G.; Campbell, J.A.; Stead, A.G.; Nesnow, S. Tumorigenesis and genotoxicity of ethyl carbamate and vinyl carbamate in rodent cells. *Cancer Res.* **1986**, *46*, 4911–4915.

385. Sotomayor, R.E.; Washington, M.C. Formation of etheno and oxoethyl adducts in liver DNA from rats exposed subchronically to urethane in drinking water and ethanol. *Cancer Lett.* **1996**, *100*, 155–161. [CrossRef]

386. Fernando, R.C.; Nair, J.; Barbin, A.; Miller, J.A.; Bartsch, H. Detection of 1,N6-ethenodeoxyadenosine and 3,N4-ethenodeoxycytidine by immunoaffinity/32P-postlabelling in liver and lung DNA of mice treated with ethyl carbamate (urethane) or its metabolites. *Carcinogenesis* **1996**, *17*, 1711–1718. [CrossRef] [PubMed]

387. Barbin, A. Formation of DNA etheno adducts in rodents and humans and their role in carcinogenesis. *Acta Biochim. Pol.* **1998**, *45*, 145–161. [CrossRef]

388. Forkert, P.G. Mechanisms of lung tumorigenesis by ethyl carbamate and vinyl carbamate. *Drug Metab. Rev.* **2010**, *42*, 355–378. [CrossRef] [PubMed]

389. Park, K.K.; Liem, A.; Stewart, B.C.; Miller, J.A. Vinyl carbamate epoxide, a major strong electrophilic, mutagenic and carcinogenic metabolite of vinyl carbamate and ethyl carbamate (urethane). *Carcinogenesis* **1993**, *14*, 441–450. [CrossRef] [PubMed]

390. Hoffler, U.; El-Masri, H.A.; Ghanayem, B.I. Cytochrome P450 2E1 (CYP2E1) is the principal enzyme responsible for urethane metabolism: Comparative studies using CYP2E1-null and wild-type mice. *J. Pharm. Exp.* **2003**, *305*, 557–564. [CrossRef]

391. Guengerich, F.P.; Kim, D.H. Enzymatic oxidation of ethyl carbamate to vinyl carbamate and its role as an intermediate in the formation of 1,N6-ethenoadenosine. *Chem. Res. Toxicol.* **1991**, *4*, 413–421. [CrossRef]

392. Forkert, P.G.; Lee, R.P.; Reid, K. Involvement of CYP2E1 and carboxylesterase enzymes in vinyl carbamate metabolism in human lung microsomes. *Drug Metab. Dispos.* **2001**, *29*, 258–263.

393. Sozio, F.; Schioppa, T.; Sozzani, S.; Del Prete, A. Urethane-induced lung carcinogenesis. *Methods Cell Biol.* **2021**, *163*, 45–57. [CrossRef]

394. Zimmerli, B.; Schlatter, J. Ethyl carbamate: Analytical methodology, occurrence, formation, biological activity and risk assessment. *Mutat. Res.* **1991**, *259*, 325–350. [CrossRef]

395. Schlatter, J.; Lutz, W.K. The carcinogenic potential of ethyl carbamate (urethane): Risk assessment at human dietary exposure levels. *Food Chem. Toxicol.* **1990**, *28*, 205–211. [CrossRef]

396. Jägerstad, M.; Skog, K. Genotoxicity of heat-processed foods. *Mutat. Res./Fundam. Mol. Mech. Mutagen.* **2005**, *574*, 156–172. [CrossRef]

397. Mottram, D.S.; Wedzicha, B.L.; Dodson, A.T. Acrylamide is formed in the Maillard reaction. *Nature* **2002**, *419*, 448–449. [CrossRef]

398. Stadler, R.H.; Blank, I.; Varga, N.; Robert, F.; Hau, J.; Guy, P.A.; Robert, M.-C.; Riediker, S. Acrylamide from Maillard reaction products. *Nature* **2002**, *419*, 449–450. [CrossRef]

399. Friedman, M. Chemistry, biochemistry, and safety of acrylamide. A review. *J. Agric. Food Chem.* **2003**, *51*, 4504–4526. [CrossRef]

400. Tritscher, A. Human health risk assessment of processing-related compounds in food. *Toxicol. Lett.* **2004**, *149*, 177–186. [CrossRef] [PubMed]

401. Dybing, E.; Farmer, P.B.; Andersen, M.; Fennell, T.R.; Lalljie, S.P.D.; Müller, D.J.G.; Olin, S.; Petersen, B.J.; Schlatter, J.; Scholz, G.; et al. Human exposure and internal dose assessments of acrylamide in food. *Food Chem. Toxicol.* **2005**, *43*, 365–410. [CrossRef] [PubMed]

402. IARC, International Agency for Research on Cancer. Some Industrial Chemicals. In *IARC Monographs on the Evaluation of Carcinogenic Risks to Humans*; International Agency for Research on Cancer: Lyon, France, 1994; Volume 60.

403. JECFA, Joint FAO/WHO Expert Committee on Food Additives. *Safety Evaluation of Certain Contaminants in Food*; Prepared by the Seventy-Second Meeting of the Joint FAO/WHO Expert Committee on Food Additives (JECFA); FAO JECFA Monographs 8; WHO: Geneva, Switzerland, 2011.

404. Commission Regulation (EU). *Commission Regulation (EU) 2017/2158 of 20 November 2017 Establishing Mitigation Measures and Benchmark Levels for the Reduction of the Presence of Acrylamide in Food*; Commission Regulation (EU): Brussels, Belgium, 2017; pp. 24–44.

405. Xu, F.; Oruna-Concha, M.J.; Elmore, J.S. The use of asparaginase to reduce acrylamide levels in cooked food. *Food Chem.* **2016**, *210*, 163–171. [CrossRef] [PubMed]

406. Rice, J.M. The carcinogenicity of acrylamide. *Mutat. Res./Genet. Toxicol. Environ. Mutagen.* **2005**, *580*, 3–20. [CrossRef] [PubMed]

407. Klaunig, J.E. Acrylamide carcinogenicity. *J. Agric. Food Chem.* **2008**, *56*, 5984–5988. [CrossRef]

408. NTP, National Toxicology Program. *Toxicology and Carcinogenesis Studies of Acrylamide (CAS No. 79-06-1) in F344/N Rats and B6C3F1 Mice (Feed and Drinking Water Studies)*; 1551-8272; National Toxicology Program, National Institutes of Health: Research Triangle Park, NC, USA, 2011; pp. 25–28.

409. Besaratinia, A.; Pfeifer, G.P. A review of mechanisms of acrylamide carcinogenicity. *Carcinogenesis* **2007**, *28*, 519–528. [CrossRef]
410. Watzek, N.; Böhm, N.; Feld, J.; Scherbl, D.; Berger, F.; Merz, K.H.; Lampen, A.; Reemtsma, T.; Tannenbaum, S.R.; Skipper, P.L.; et al. N7-glycidamide-guanine DNA adduct formation by orally ingested acrylamide in rats: A dose-response study encompassing human diet-related exposure levels. *Chem. Res. Toxicol.* **2012**, *25*, 381–390. [CrossRef]
411. Klaunig, J.E.; Kamendulis, L.M. Mechanisms of acrylamide induced rodent carcinogenesis. *Adv. Exp. Med. Biol.* **2005**, *561*, 49–62. [CrossRef]
412. Hagio, S.; Tsuji, N.; Furukawa, S.; Takeuchi, K.; Hayashi, S.; Kuroda, Y.; Honma, M.; Masumura, K. Effect of sampling time on somatic and germ cell mutations induced by acrylamide in gpt delta mice. *Genes Environ.* **2021**, *43*, 4. [CrossRef]
413. Shimamura, Y.; Iio, M.; Urahira, T.; Masuda, S. Inhibitory effects of Japanese horseradish (Wasabia japonica) on the formation and genotoxicity of a potent carcinogen, acrylamide. *J. Sci. Food Agric.* **2017**, *97*, 2419–2425. [CrossRef] [PubMed]
414. Katen, A.L.; Stanger, S.J.; Anderson, A.L.; Nixon, B.; Roman, S.D. Chronic acrylamide exposure in male mice induces DNA damage to spermatozoa; Potential for amelioration by resveratrol. *Reprod. Toxicol.* **2016**, *63*, 1–12. [CrossRef] [PubMed]
415. Dobrovolsky, V.N.; Pacheco-Martinez, M.M.; McDaniel, L.P.; Pearce, M.G.; Ding, W. In vivo genotoxity assessment of acrylamide and glycidyl methacrylate. *Food Chem. Toxicol.* **2016**, *87*, 120–127. [CrossRef] [PubMed]
416. Calleman, C.J. The metabolism and pharmacokinetics of acrylamide: Implications for mechanisms of toxicity and human risk estimation. *Drug Metab. Rev.* **1996**, *28*, 527–590. [CrossRef]
417. Exon, J.H. A review of the toxicology of acrylamide. *J. Toxicol. Environ. Health Part B* **2006**, *9*, 397–412. [CrossRef]
418. Gamboa da Costa, G.; Churchwell, M.I.; Hamilton, L.P.; Von Tungeln, L.S.; Beland, F.A.; Marques, M.M.; Doerge, D.R. DNA adduct formation from acrylamide via conversion To glycidamide in adult and neonatal mice. *Chem. Res. Toxicol.* **2003**, *16*, 1328–1337. [CrossRef]
419. Dearfield, K.L.; Douglas, G.R.; Ehling, U.H.; Moore, M.M.; Sega, G.A.; Brusick, D.J. Acrylamide: A review of its genotoxicity and an assessment of heritable genetic risk. *Mutat. Res./Fundam. Mol. Mech. Mutagen.* **1995**, *330*, 71–99. [CrossRef]
420. Besaratinia, A.; Pfeifer, G.P. Genotoxicity of acrylamide and glycidamide. *J. Natl. Cancer Inst.* **2004**, *96*, 1023–1029. [CrossRef]
421. Jones, D.J.L.; Singh, R.; Emms, V.; Farmer, P.B.; Grant, D.; Quinn, P.; Maxwell, C.; Mina, A.; Ng, L.L.; Schumacher, S.; et al. Determination of N7-glycidamide guanine adducts in human blood DNA following exposure to dietary acrylamide using liquid chromatography/tandem mass spectrometry. *Rapid Commun. Mass Spectrom.* **2022**, *36*, e9245. [CrossRef]
422. Fennell, T.R.; Friedman, M.A. Comparison of acrylamide metabolism in humans and rodents. *Adv. Exp. Med. Biol.* **2005**, *561*, 109–116. [CrossRef] [PubMed]
423. Nowak, A.; Zakłos-Szyda, M.; Żyżelewicz, D.; Koszucka, A.; Motyl, I. Acrylamide decreases cell viability, and provides oxidative stress, DNA damage, and apoptosis in human colon adenocarcinoma cell line Caco-2. *Molecules* **2020**, *25*, 368. [CrossRef] [PubMed]
424. De Conti, A.; Tryndyak, V.; VonTungeln, L.S.; Churchwell, M.I.; Beland, F.A.; Antunes, A.M.M.; Pogribny, I.P. Genotoxic and epigenotoxic alterations in the lung and liver of mice induced by acrylamide: A 28 day drinking water study. *Chem. Res. Toxicol.* **2019**, *32*, 869–877. [CrossRef] [PubMed]
425. EFSA CONTAM Panel, European Food Safety Authority, Panel on Contaminants in the Food Chain. Scientific Opinion on acrylamide in food. *EFSA J.* **2015**, *13*, 4104.
426. Mucci, L.A.; Wilson, K.M. Acrylamide intake through diet and human cancer risk. *J. Agric. Food Chem.* **2008**, *56*, 6013–6019. [CrossRef]
427. Pelucchi, C.; Bosetti, C.; Galeone, C.; La Vecchia, C. Dietary acrylamide and cancer risk: An updated meta-analysis. *Int. J. Cancer* **2015**, *136*, 2912–2922. [CrossRef]
428. Duale, N.; Bjellaas, T.; Alexander, J.; Becher, G.; Haugen, M.; Paulsen, J.E.; Frandsen, H.; Olesen, P.T.; Brunborg, G. Biomarkers of human exposure to acrylamide and relation to polymorphisms in metabolizing genes. *Toxicol. Sci.* **2009**, *108*, 90–99. [CrossRef]
429. Hogervorst, J.G.; Schouten, L.J.; Konings, E.J.; Goldbohm, R.A.; van den Brandt, P.A. A prospective study of dietary acrylamide intake and the risk of endometrial, ovarian, and breast cancer. *Cancer Epidemiol. Biomark. Prev.* **2007**, *16*, 2304–2313. [CrossRef]
430. Bonneck, S. Acrylamide Risk Governance in Germany. In *Global Risk Governance*; Springer: Dordrecht, The Netherlands, 2008; pp. 231–274.
431. Baum, M.; Fauth, E.; Fritzen, S.; Herrmann, A.; Mertes, P.; Merz, K.; Rudolphi, M.; Zankl, H.; Eisenbrand, G. Acrylamide and glycidamide: Genotoxic effects in V79-cells and human blood. *Mutat. Res.* **2005**, *580*, 61–69. [CrossRef]
432. Sugimura, T.; Wakabayashi, K.; Nakagama, H.; Nagao, M. Heterocyclic amines: Mutagens/carcinogens produced during cooking of meat and fish. *Cancer Sci.* **2004**, *95*, 290–299. [CrossRef]
433. Nagao, M.; Ushijima, T.; Watanabe, N.; Okochi, E.; Ochiai, M.; Nakagama, H.; Sugimura, T. Studies on mammary carcinogenesis induced by a heterocyclic amine, 2-amino-1-methyl-6-phenylimidazo[4,5-b]pyridine, in mice and rats. *Environ. Mol. Mutagen.* **2002**, *39*, 158–164. [CrossRef] [PubMed]
434. Weisburger, J.H.; Rivenson, A.; Kingston, D.G.; Wilkins, T.D.; Van Tassell, R.L.; Nagao, M.; Sugimura, T.; Hara, Y. Dietary modulation of the carcinogenicity of the heterocyclic amines. *Princess Takamatsu Symp.* **1995**, *23*, 240–250. [PubMed]
435. Sasaki, Y.F.; Saga, A.; Akasaka, M.; Nishidate, E.; Watanabe-Akanuma, M.; Ohta, T.; Matsusaka, N.; Tsuda, S. In vivo genotoxicity of heterocyclic amines detected by a modified alkaline single cell gel electrophoresis assay in a multiple organ study in the mouse. *Mutat. Res.* **1997**, *395*, 57–73. [CrossRef]

436. Bellamri, M.; Walmsley, S.J.; Turesky, R.J. Metabolism and biomarkers of heterocyclic aromatic amines in humans. *Genes Environ.* **2021**, *43*, 29. [CrossRef] [PubMed]
437. Fuccelli, R.; Rosignoli, P.; Servili, M.; Veneziani, G.; Taticchi, A.; Fabiani, R. Genotoxicity of heterocyclic amines (HCAs) on freshly isolated human peripheral blood mononuclear cells (PBMC) and prevention by phenolic extracts derived from olive, olive oil and olive leaves. *Food Chem. Toxicol.* **2018**, *122*, 234–241. [CrossRef]
438. Bellamri, M.; Le Hegarat, L.; Vernhet, L.; Baffet, G.; Turesky, R.J.; Langouët, S. Human T lymphocytes bioactivate heterocyclic aromatic amines by forming DNA adducts. *Environ. Mol. Mutagen.* **2016**, *57*, 656–667. [CrossRef]
439. Dingley, K.H.; Curtis, K.D.; Nowell, S.; Felton, J.S.; Lang, N.P.; Turteltaub, K.W. DNA and protein adduct formation in the colon and blood of humans after exposure to a dietary-relevant dose of 2-amino-1-methyl-6-phenylimidazo[4,5-b]pyridine. *Cancer Epidemiol. Biomark. Prev.* **1999**, *8*, 507–512.
440. Reistad, R.; Rossland, O.J.; Latva-Kala, K.J.; Rasmussen, T.; Vikse, R.; Becher, G.; Alexander, J. Heterocyclic aromatic amines in human urine following a fried meat meal. *Food Chem. Toxicol.* **1997**, *35*, 945–955. [CrossRef]
441. Friesen, M.D.; Kaderlik, K.; Lin, D.; Garren, L.; Bartsch, H.; Lang, N.P.; Kadlubar, F.F. Analysis of DNA adducts of 2-amino-1-methyl-6-phenylimidazo[4,5- b]pyridine in rat and human tissues by alkaline hydrolysis and gas chromatography/electron capture mass spectrometry: Validation by comparison with 32P-postlabeling. *Chem. Res. Toxicol.* **1994**, *7*, 733–739. [CrossRef]
442. Totsuka, Y.; Fukutome, K.; Takahashi, M.; Takahashi, S.; Tada, A.; Sugimura, T.; Wakabayashi, K. Presence of N2-(deoxyguanosin-8-yl)-2-amino-3,8-dimethylimidazo[4,5-f]quinoxaline (dG-C8-MeIQx) in human tissues. *Carcinogenesis* **1996**, *17*, 1029–1034. [CrossRef]
443. Pathak, K.V.; Chiu, T.L.; Amin, E.A.; Turesky, R.J. Methemoglobin formation and characterization of hemoglobin adducts of carcinogenic aromatic amines and heterocyclic aromatic amines. *Chem. Res. Toxicol.* **2016**, *29*, 255–269. [CrossRef] [PubMed]
444. Kim, D.; Guengerich, F.P. Cytochrome P450 activation of arylamines and heterocyclic amines. *Annu. Rev. Pharm. Toxicol.* **2005**, *45*, 27–49. [CrossRef] [PubMed]
445. Chen, J.; Stampfer, M.J.; Hough, H.L.; Garcia-Closas, M.; Willett, W.C.; Hennekens, C.H.; Kelsey, K.T.; Hunter, D.J. A prospective study of N-acetyltransferase genotype, red meat intake, and risk of colorectal cancer. *Cancer Res.* **1998**, *58*, 3307–3311. [PubMed]
446. Gooderham, N.J.; Murray, S.; Lynch, A.M.; Yadollahi-Farsani, M.; Zhao, K.; Boobis, A.R.; Davies, D.S. Food-derived heterocyclic amine mutagens: Variable metabolism and significance to humans. *Drug Metab. Dispos.* **2001**, *29*, 529–534. [PubMed]
447. McQueen, C.A.; Maslansky, C.J.; Williams, G.M. Role of the acetylation polymorphism in determining susceptibility of cultured rabbit hepatocytes to DNA damage by aromatic amines. *Cancer Res.* **1983**, *43*, 3120–3123. [PubMed]
448. Layton, D.W.; Bogen, K.T.; Knize, M.G.; Hatch, F.T.; Johnson, V.M.; Felton, J.S. Cancer risk of heterocyclic amines in cooked foods: An analysis and implications for research. *Carcinogenesis* **1995**, *16*, 39–52. [CrossRef]
449. Augustsson, K.; Skog, K.; Jägerstad, M.; Dickman, P.W.; Steineck, G. Dietary heterocyclic amines and cancer of the colon, rectum, bladder, and kidney: A population-based study. *Lancet* **1999**, *353*, 703–707. [CrossRef]
450. Bogen, K.T.; Keating, G.A. U.S. dietary exposures to heterocyclic amines. *J. Expo. Anal. Environ. Epidemiol.* **2001**, *11*, 155–168. [CrossRef]
451. Snyderwine, E.G. Some perspectives on the nutritional aspects of breast cancer research. Food-derived heterocyclic amines as etiologic agents in human mammary cancer. *Cancer* **1994**, *74*, 1070–1077. [CrossRef]
452. Ward, M.H.; Sinha, R.; Heineman, E.F.; Rothman, N.; Markin, R.; Weisenburger, D.D.; Correa, P.; Zahm, S.H. Risk of adenocarcinoma of the stomach and esophagus with meat cooking method and doneness preference. *Int. J. Cancer* **1997**, *71*, 14–19. [CrossRef]
453. Sinha, R.; Kulldorff, M.; Swanson, C.A.; Curtin, J.; Brownson, R.C.; Alavanja, M.C. Dietary heterocyclic amines and the risk of lung cancer among Missouri women. *Cancer Res.* **2000**, *60*, 3753–3756. [PubMed]
454. Zhu, J.; Chang, P.; Bondy, M.L.; Sahin, A.A.; Singletary, S.E.; Takahashi, S.; Shirai, T.; Li, D. Detection of 2-amino-1-methyl-6-phenylimidazo[4,5-b]-pyridine-DNA adducts in normal breast tissues and risk of breast cancer. *Cancer Epidemiol. Biomark. Prev.* **2003**, *12*, 830–837.
455. Tang, D.; Kryvenko, O.N.; Wang, Y.; Trudeau, S.; Rundle, A.; Takahashi, S.; Shirai, T.; Rybicki, B.A. 2-Amino-1-methyl-6-phenylimidazo[4,5-b]pyridine (PhIP)-DNA adducts in benign prostate and subsequent risk for prostate cancer. *Int. J. Cancer* **2013**, *133*, 961–971. [CrossRef] [PubMed]
456. Jakszyn, P.; Agudo, A.; Ibáñez, R.; García-Closas, R.; Pera, G.; Amiano, P.; González, C.A. Development of a food database of nitrosamines, heterocyclic amines, and polycyclic aromatic hydrocarbons. *J. Nutr.* **2004**, *134*, 2011–2014. [CrossRef]
457. Zelinkova, Z.; Wenzl, T. The occurrence of 16 EPA PAHs in food—A review. *Polycycl. Aromat. Compd.* **2015**, *35*, 248–284. [CrossRef]
458. Lijinsky, W. The formation and occurrence of polynuclear aromatic hydrocarbons associated with food. *Mutat. Res.* **1991**, *259*, 251–261. [CrossRef]
459. Park, K.C.; Pyo, H.; Kim, W.; Yoon, K.S. Effects of cooking methods and tea marinades on the formation of benzo[a]pyrene in grilled pork belly (Samgyeopsal). *Meat Sci.* **2017**, *129*, 1–8. [CrossRef]
460. Kazerouni, N.; Sinha, R.; Hsu, C.H.; Greenberg, A.; Rothman, N. Analysis of 200 food items for benzo[a]pyrene and estimation of its intake in an epidemiologic study. *Food Chem. Toxicol.* **2001**, *39*, 423–436. [CrossRef]
461. IARC, International Agency for Research on Cancer. Some non-heterocyclic polycyclic aromatic hydrocarbons and some related exposures. In *IARC Monographs on the Evaluation of Carcinogenic Risks to Humans*; International Agency for Research on Cancer: Lyon, France, 2010; Volume 92.

462. Kroese, E.D.; Muller, J.J.A.; Mohn, G.R.; Dortant, P.M.; Wester, P.W. *Tumorigenic Effects in Wistar Rats Orally Administered benzo[a]pyrene for Two Years (Gavage Studies): Implications for Human Cancer Risks Associated with Oral Exposure to Polycyclic Aromatic Hydrocarbons*; 658603 010; National Institute for Public Health and the Environment (RIVM): Bilthoven, The Netherlands, 2001.

463. IARC, International Agency for Research on Cancer. Bitumens and Bitumen Emissions, and some N- and S-Heterocyclic Polycyclic Aromatic Hydrocarbons. In *IARC Monographs on the Evaluation of Carcinogenic Risks to Humans*; International Agency for Research on Cancer: Lyon, France, 2013; Volume 103.

464. Moorthy, B.; Chu, C.; Carlin, D.J. Polycyclic aromatic hydrocarbons: From metabolism to lung cancer. *Toxicol. Sci.* **2015**, *145*, 5–15. [CrossRef] [PubMed]

465. Gelboin, H.V. Benzo[alpha]pyrene metabolism, activation and carcinogenesis: Role and regulation of mixed-function oxidases and related enzymes. *Physiol. Rev.* **1980**, *60*, 1107–1166. [CrossRef]

466. Jerina, D.M.; Sayer, J.M.; Agarwal, S.K.; Yagi, H.; Levin, W.; Wood, A.W.; Conney, A.H.; Pruess-Schwartz, D.; Baird, W.M.; Pigott, M.A.; et al. Reactivity and tumorigenicity of bay-region diol epoxides derived from polycyclic aromatic hydrocarbons. *Adv. Exp. Med. Biol.* **1986**, *197*, 11–30. [CrossRef] [PubMed]

467. Kim, K.B.; Lee, B.M. Oxidative stress to DNA, protein, and antioxidant enzymes (superoxide dismutase and catalase) in rats treated with benzo(a)pyrene. *Cancer Lett.* **1997**, *113*, 205–212. [CrossRef]

468. Kawabata, T.T.; White, K.L., Jr. Suppression of the vitro humoral immune response of mouse splenocytes by benzo(a)pyrene metabolites and inhibition of benzo(a)pyrene-induced immunosuppression by alpha-naphthoflavone. *Cancer Res.* **1987**, *47*, 2317–2322. [PubMed]

469. Myers, J.N.; Harris, K.L.; Rekhadevi, P.V.; Pratap, S.; Ramesh, A. Benzo(a)pyrene-induced cytotoxicity, cell proliferation, DNA damage, and altered gene expression profiles in HT-29 human colon cancer cells. *Cell Biol. Toxicol.* **2021**, *37*, 891–913. [CrossRef] [PubMed]

470. Damiani, L.A.; Yingling, C.M.; Leng, S.; Romo, P.E.; Nakamura, J.; Belinsky, S.A. Carcinogen-induced gene promoter hypermethylation is mediated by DNMT1 and causal for transformation of immortalized bronchial epithelial cells. *Cancer Res.* **2008**, *68*, 9005–9014. [CrossRef]

471. Commission Regulation (EU). *Commission Regulation (EU) No 835/2011 of 19 August 2011 Amending Regulation (EC) No 1881/2006 as Regards Maximum Levels for Polycyclic Aromatic Hydrocarbons in Foodstuffs*; Commission Regulation (EU): Brussels, Belgium, 2011; pp. 4–8.

472. Commission Regulation (EU). *Commission Regulation (EU) No 2020/1255 of 7 September 2020 Amending Regulation (EC) No 1881/2006 as Regards Maximum Levels of Polycyclic Aromatic Hydrocarbons (PAHs) in Traditionally Smoked Meat and Smoked Meat Products and Traditionally Smoked Fish and Smoked Fishery Products and Establishing a Maximum Level of PAHs in Powders of Food of Plant Origin Used for the Preparation of Beverages*; Commission Regulation (EU): Brussels, Belgium, 2020.

473. Sinha, R.; Kulldorff, M.; Gunter, M.J.; Strickland, P.; Rothman, N. Dietary benzo[a]pyrene intake and risk of colorectal adenoma. *Cancer Epidemiol. Biomark. Prev.* **2005**, *14*, 2030–2034. [CrossRef]

474. Yoon, E.; Park, K.; Lee, H.; Yang, J.-H.; Lee, C. Estimation of excess cancer risk on time-weighted lifetime average daily intake of PAHs from food ingestion. *Hum. Ecol. Risk Assess. Int. J.* **2007**, *13*, 669–680. [CrossRef]

475. Bononi, M.; Quaglia, G.; Tateo, F. Identification of ethylene oxide in herbs, spices and other dried vegetables imported into Italy. *Food Addit. Contam. Part A Chem. Anal. Control Expo. Risk Assess.* **2014**, *31*, 271–275. [CrossRef]

476. Heuser, S.G.; Scudamore, K.A. Fumigant residues in wheat and flour: Solvent extraction and gas-chromatographic determination of free methyl bromide and ethylene oxide. *Analyst* **1968**, *93*, 252–258. [CrossRef]

477. Jensen, K.G. Determination of ethylene oxide residues in processed food products by gas-liquid chromatography after derivatization. *Z. Lebensm. Unters.* **1988**, *187*, 535–540. [CrossRef]

478. Tthe European Parliament and the Council of the European Union. *Regulation (EC) No 1107/2009 of the European Parliament and of the Council of 21 October 2009 Concerning the Placing of Plant Protection Products on the Market and Repealing Council Directives 79/117/EEC and 91/414/EEC*; The European Parliament: Brussels, Belgium, 2009.

479. The European Parliament. *Regulation (EC) No 396/2005 of the European Parliament and of the Council of 23 February 2005 on Maximum Residue Levels of Pesticides in or on Food and Feed of Plant and Animal Origin and Amending Council Directive 91/414/EEC*; The European Parliament: Brussels, Belgium, 2005.

480. Kowalska, A.; Manning, L. Food Safety Governance and Guardianship: The role of the private sector in addressing the EU ehylene oxide incident. *Foods* **2022**, *11*, 204. [CrossRef] [PubMed]

481. Bessaire, T.; Stroheker, T.; Eriksen, B.; Mujahid, C.; Hammel, Y.A.; Varela, J.; Delatour, T.; Panchaud, A.; Mottier, P.; Stadler, R.H. Analysis of ethylene oxide in ice creams manufactured with contaminated carob bean gum (E410). *Food Addit. Contam. Part A Chem. Anal. Control Expo. Risk Assess.* **2021**, *38*, 2116–2127. [CrossRef] [PubMed]

482. Kirman, C.R.; Li, A.A.; Sheehan, P.J.; Bus, J.S.; Lewis, R.C.; Hays, S.M. Ethylene oxide review: Characterization of total exposure via endogenous and exogenous pathways and their implications to risk assessment and risk management. *J. Toxicol. Environ. Health B Crit. Rev.* **2021**, *24*, 1–29. [CrossRef] [PubMed]

483. Bolt, H.M. Quantification of endogenous carcinogens. *The ethylene oxide paradox. Biochem. Pharm.* **1996**, *52*, 1–5. [CrossRef]

484. Törnqvist, M.; Gustafsson, B.; Kautiainen, A.; Harms-Ringdahl, M.; Granath, F.; Ehrenberg, L. Unsaturated lipids and intestinal bacteria as sources of endogenous production of ethene and ethylene oxide. *Carcinogenesis* **1989**, *10*, 39–41. [CrossRef]

485. Kirman, C.R.; Hays, S.M. Derivation of endogenous equivalent values to support risk assessment and risk management decisions for an endogenous carcinogen: Ethylene oxide. *Regul. Toxicol. Pharm.* **2017**, *91*, 165–172. [CrossRef]
486. NTP, National Toxicology Program. Toxicology and carcinogenesis studies of ethylene oxide (CAS No. 75-21-8) in B6C3F1 mice (inhalation studies). *Natl. Toxicol. Program. Tech. Rep. Ser.* **1987**, *326*, 1–114.
487. Bolt, H.M. Carcinogenicity and genotoxicity of ethylene oxide: New aspects and recent advances. *Crit. Rev. Toxicol.* **2000**, *30*, 595–608. [CrossRef]
488. ATSDR, Agency for Toxic Substances and Disease Registry. *Toxicological Profile for Ethylene Oxide*; Draft for Public Comment; Agency for Toxic Substances and Disease Registry: Atlanta, GA, USA, 2020.
489. Lynch, H.; Kozal, J.S.; Russell, A.J.; Thompson, W.J.; Divis, H.R.; Freid, R.D.; Calabrese, E.J.; Mundt, K.A. Systematic review of the scientific evidence on ethylene oxide as a human carcinogen. *Chem. Biol. Interact.* **2022**, *364*, 110031. [CrossRef]
490. Dunkelberg, H. Carcinogenicity of ethylene oxide and 1,2-propylene oxide upon intragastric administration to rats. *Br. J. Cancer* **1982**, *46*, 924–933. [CrossRef]
491. Natarajan, A.T.; Preston, R.J.; Dellarco, V.; Ehrenberg, L.; Generoso, W.; Lewis, S.; Tates, A.D. Ethylene oxide: Evaluation of genotoxicity data and an exploratory assessment of genetic risk. *Mutat. Res.* **1995**, *330*, 55–70. [CrossRef]
492. Farooqi, Z.; Törnqvist, M.; Ehrenberg, L.; Natarajan, A.T. Genotoxic effects of ethylene oxide and propylene oxide in mouse bone marrow cells. *Mutat. Res.* **1993**, *288*, 223–228. [CrossRef]
493. Ghosh, M.; Godderis, L. Genotoxicity of ethylene oxide: A review of micronucleus assay results in human population. *Mutat. Res. Rev. Mutat. Res.* **2016**, *770*, 84–91. [CrossRef] [PubMed]
494. Kolman, A.; Chovanec, M.; Osterman-Golkar, S. Genotoxic effects of ethylene oxide, propylene oxide and epichlorohydrin in humans: Update review (1990–2001). *Mutat. Res.* **2002**, *512*, 173–194. [CrossRef]
495. Recio, L.; Donner, M.; Abernethy, D.; Pluta, L.; Steen, A.M.; Wong, B.A.; James, A.; Preston, R.J. In vivo mutagenicity and mutation spectrum in the bone marrow and testes of B6C3F1 lacI transgenic mice following inhalation exposure to ethylene oxide. *Mutagenesis* **2004**, *19*, 215–222. [CrossRef] [PubMed]
496. Bolt, H.M.; Peter, H.; Föst, U. Analysis of macromolecular ethylene oxide adducts. *Int. Arch. Occup. Environ. Health* **1988**, *60*, 141–144. [CrossRef]
497. Rusyn, I.; Asakura, S.; Li, Y.; Kosyk, O.; Koc, H.; Nakamura, J.; Upton, P.B.; Swenberg, J.A. Effects of ethylene oxide and ethylene inhalation on DNA adducts, apurinic/apyrimidinic sites and expression of base excision DNA repair genes in rat brain, spleen, and liver. *DNA Repair.* **2005**, *4*, 1099–1110. [CrossRef]
498. Marsden, D.A.; Jones, D.J.; Lamb, J.H.; Tompkins, E.M.; Farmer, P.B.; Brown, K. Determination of endogenous and exogenously derived N7-(2-hydroxyethyl)guanine adducts in ethylene oxide-treated rats. *Chem. Res. Toxicol.* **2007**, *20*, 290–299. [CrossRef]
499. Walker, V.E.; Fennell, T.R.; Upton, P.B.; Skopek, T.R.; Prevost, V.; Shuker, D.E.; Swenberg, J.A. Molecular dosimetry of ethylene oxide: Formation and persistence of 7-(2-hydroxyethyl)guanine in DNA following repeated exposures of rats and mice. *Cancer Res.* **1992**, *52*, 4328–4334.
500. Fennell, T.R.; Brown, C.D. A physiologically based pharmacokinetic model for ethylene oxide in mouse, rat, and human. *Toxicol. Appl. Pharm.* **2001**, *173*, 161–175. [CrossRef]
501. Müller, M.; Krämer, A.; Angerer, J.; Hallier, E. Ethylene oxide-protein adduct formation in humans: Influence of glutathione-S-transferase polymorphisms. *Int. Arch. Occup. Environ. Health* **1998**, *71*, 499–502. [CrossRef] [PubMed]
502. Yong, L.C.; Schulte, P.A.; Wiencke, J.K.; Boeniger, M.F.; Connally, L.B.; Walker, J.T.; Whelan, E.A.; Ward, E.M. Hemoglobin adducts and sister chromatid exchanges in hospital workers exposed to ethylene oxide: Effects of glutathione S-transferase T1 and M1 genotypes. *Cancer Epidemiol. Biomark. Prev.* **2001**, *10*, 539–550.
503. Fennell, T.R.; MacNeela, J.P.; Morris, R.W.; Watson, M.; Thompson, C.L.; Bell, D.A. Hemoglobin adducts from acrylonitrile and ethylene oxide in cigarette smokers: Effects of glutathione S-transferase T1-null and M1-null genotypes. *Cancer Epidemiol. Biomark. Prev.* **2000**, *9*, 705–712.
504. Haufroid, V.; Merz, B.; Hofmann, A.; Tschopp, A.; Lison, D.; Hotz, P. Exposure to ethylene oxide in hospitals: Biological monitoring and influence of glutathione S-transferase and epoxide hydrolase polymorphisms. *Cancer Epidemiol. Biomark. Prev.* **2007**, *16*, 796–802. [CrossRef]
505. Philippin, G.; Cadet, J.; Gasparutto, D.; Mazon, G.; Fuchs, R.P. Ethylene oxide and propylene oxide derived N7-alkylguanine adducts are bypassed accurately in vivo. *DNA Repairr.* **2014**, *22*, 133–136. [CrossRef]
506. Tompkins, E.M.; McLuckie, K.I.; Jones, D.J.; Farmer, P.B.; Brown, K. Mutagenicity of DNA adducts derived from ethylene oxide exposure in the pSP189 shuttle vector replicated in human Ad293 cells. *Mutat. Res.* **2009**, *678*, 129–137. [CrossRef]
507. Pottenger, L.H.; Boysen, G.; Brown, K.; Cadet, J.; Fuchs, R.P.; Johnson, G.E.; Swenberg, J.A. Understanding the importance of low-molecular weight (ethylene oxide- and propylene oxide-induced) DNA adducts and mutations in risk assessment: Insights from 15 years of research and collaborative discussions. *Environ. Mol. Mutagen.* **2019**, *60*, 100–121. [CrossRef]
508. Gollapudi, B.B.; Su, S.; Li, A.A.; Johnson, G.E.; Reiss, R.; Albertini, R.J. Genotoxicity as a toxicologically relevant endpoint to inform risk assessment: A case study with ethylene oxide. *Environ. Mol. Mutagen.* **2020**, *61*, 852–871. [CrossRef]
509. van Sittert, N.J.; Boogaard, P.J.; Natarajan, A.T.; Tates, A.D.; Ehrenberg, L.G.; Törnqvist, M.A. Formation of DNA adducts and induction of mutagenic effects in rats following 4 weeks inhalation exposure to ethylene oxide as a basis for cancer risk assessment. *Mutat. Res.* **2000**, *447*, 27–48. [CrossRef]

510. SCF, Scientific Committee on Food. *Opinion of the Scientific Committee on Food on Impurities of Ethylene Oxide in Food Additives*; SCF/CS/ADD/EMU/186 Final; Scientific Committee on Food: Brussels, Belgium, 2002.
511. Fowles, J.; Mitchell, J.; McGrath, H. Assessment of cancer risk from ethylene oxide residues in spices imported into New Zealand. *Food Chem. Toxicol.* **2001**, *39*, 1055–1062. [CrossRef]
512. Bisanti, L.; Maggini, M.; Raschetti, R.; Alegiani, S.S.; Ippolito, F.M.; Caffari, B.; Segnan, N.; Ponti, A. Cancer mortality in ethylene oxide workers. *Br. J. Ind. Med.* **1993**, *50*, 317–324. [CrossRef]
513. Steenland, K.; Whelan, E.; Deddens, J.; Stayner, L.; Ward, E. Ethylene oxide and breast cancer incidence in a cohort study of 7576 women (United States). *Cancer Causes Control* **2003**, *14*, 531–539. [CrossRef]
514. Jinot, J.; Fritz, J.M.; Vulimiri, S.V.; Keshava, N. Carcinogenicity of ethylene oxide: Key findings and scientific issues. *Toxicol. Mech. Methods* **2018**, *28*, 386–396. [CrossRef]
515. Marsh, G.M.; Keeton, K.A.; Riordan, A.S.; Best, E.A.; Benson, S.M. Ethylene oxide and risk of lympho-hematopoietic cancer and breast cancer: A systematic literature review and meta-analysis. *Int. Arch. Occup. Environ. Health* **2019**, *92*, 919–939. [CrossRef] [PubMed]
516. Vincent, M.J.; Kozal, J.S.; Thompson, W.J.; Maier, A.; Dotson, G.S.; Best, E.A.; Mundt, K.A. Ethylene oxide: Cancer evidence integration and dose-response implications. *Dose Response* **2019**, *17*, 1559325819888317. [CrossRef]
517. EPA, US Environmental Protection Agency. Evaluation of the Inhalation Carcinogenicity of Ethylene Oxide (CASRN 75-21-8). In *Support of Summary Information on the Integrated Risk Information System (IRIS)*; EPA/635/R-16/350Fa; US Environmental Protection Agency: Washington, DC, USA, 2016.
518. German Federal Institute for Risk Assessment, Bundesinstitut für Risikobewertung (BfR). *Updated BfR Opinion on Health Risk Assessment of Ethylene Oxide Residues in Sesame Seeds. Opinion no 024/2021 Issued 1 September 2021*; German Federal Institute for Risk Assessment: Berlin, Germany, 2021; p. 9.
519. Lijinsky, W. N-Nitroso compounds in the diet. *Mutat. Res.* **1999**, *443*, 129–138. [CrossRef]
520. Loeppky, R.N.; Yu, H. Amidine nitrosation. *J. Org. Chem.* **2004**, *69*, 3015–3024. [CrossRef] [PubMed]
521. EMA, European Medicines Agency. *Assessment Report. Procedure under Article 5(3) of Regulation EC (No) 726/2004. Nitrosamine Impurities in Human Medicinal Products. Procedure Number: EMEA/H/A-5(3)/1490*; EMA/369136/2020; European Medicines Agency: Amsterdam, The Netherlands, 2020.
522. Leaf, C.D.; Wishnok, J.S.; Tannenbaum, S.R. Mechanisms of endogenous nitrosation. *Cancer Surv.* **1989**, *8*, 323–334.
523. Bartsch, H. N-nitroso compounds and human cancer: Where do we stand? *IARC Sci. Publ.* **1991**, *105*, 1–10.
524. Bartsch, H.; Ohshima, H.; Pignatelli, B.; Calmels, S. Endogenously formed N-nitroso compounds and nitrosating agents in human cancer etiology. *Pharmacogenetics* **1992**, *2*, 272–277. [CrossRef] [PubMed]
525. Griesenbeck, J.S.; Steck, M.D.; Huber, J.C., Jr.; Sharkey, J.R.; Rene, A.A.; Brender, J.D. Development of estimates of dietary nitrates, nitrites, and nitrosamines for use with the Short Willet Food Frequency Questionnaire. *Nutr. J.* **2009**, *8*, 16. [CrossRef] [PubMed]
526. Haorah, J.; Zhou, L.; Wang, X.; Xu, G.; Mirvish, S.S. Determination of total N-nitroso compounds and their precursors in frankfurters, fresh meat, dried salted fish, sauces, tobacco, and tobacco smoke particulates. *J. Agric. Food Chem.* **2001**, *49*, 6068–6078. [CrossRef] [PubMed]
527. Li, L.; Wang, P.; Xu, X.; Zhou, G. Influence of various cooking methods on the concentrations of volatile N-nitrosamines and biogenic amines in dry-cured sausages. *J. Food Sci.* **2012**, *77*, C560–C565. [CrossRef] [PubMed]
528. Havery, D.C.; Kline, D.A.; Miletta, E.M.; Joe, F.L., Jr.; Fazio, T. Survey of food products for volatile N-nitrosamines. *J. Assoc. Off. Anal. Chem.* **1976**, *59*, 540–546. [CrossRef]
529. IARC, International Agency for Research on Cancer. Some industrial chemicals. In *IARC Monographs on the Evaluation of Carcinogenic Risks to Humans*; International Agency for Research on Cancer: Lyon, France, 2000; Volume 77.
530. IARC, International Agency for Research on Cancer. Ingested nitrate and nitrite, and cyanobacterial peptide toxins. In *IARC Monographs on the Evaluation of Carcinogenic Risks to Humans*; International Agency for Research on Cancer: Lyon, France, 2010; Volume 94.
531. Jeffrey, A.M.; Iatropoulos, M.J.; Williams, G.M. Nasal cytotoxic and carcinogenic activities of systemically distributed organic chemicals. *Toxicol. Pathol.* **2006**, *34*, 827–852. [CrossRef]
532. Pour, P.; Gingell, R.; Langenbach, R.; Nagel, D.; Grandjean, C.; Lawson, T.; Salmasi, S. Carcinogenicity of N-nitrosomethyl(2-oxopropyl)amine in Syrian hamsters. *Cancer Res.* **1980**, *40*, 3585–3590. [PubMed]
533. Thresher, A.; Foster, R.; Ponting, D.J.; Stalford, S.A.; Tennant, R.E.; Thomas, R. Are all nitrosamines concerning? A review of mutagenicity and carcinogenicity data. *Regul. Toxicol. Pharm.* **2020**, *116*, 104749. [CrossRef]
534. Lijinsky, W. Mutagenesis, carcinogenesis and alkylating properties of nitrosamines and related compounds. *Prog. Clin. Biol. Res.* **1986**, *209*, 141–148.
535. Tsuda, S.; Matsusaka, N.; Madarame, H.; Miyamae, Y.; Ishida, K.; Satoh, M.; Sekihashi, K.; Sasaki, Y.F. The alkaline single cell electrophoresis assay with eight mouse organs: Results with 22 mono-functional alkylating agents (including 9 dialkyl N-nitrosoamines) and 10 DNA crosslinkers. *Mutat. Res.* **2000**, *467*, 83–98. [CrossRef]
536. Verna, L. N-Nitrosodiethylamine mechanistic data and risk assessment: Bioactivation, DNA-adduct formation, mutagenicity, and tumor initiation. *Pharmacol. Ther.* **1996**, *71*, 57–81. [CrossRef]

537. Arimoto-Kobayashi, S.; Kaji, K.; Sweetman, G.M.; Hayatsu, H. Mutation and formation of methyl- and hydroxylguanine adducts in DNA caused by N-nitrosodimethylamine and N-nitrosodiethylamine with UVA irradiation. *Carcinogenesis* **1997**, *18*, 2429–2433. [CrossRef] [PubMed]

538. Otteneder, M.; Lutz, W.K. Correlation of DNA adduct levels with tumor incidence: Carcinogenic potency of DNA adducts. *Mutat. Res.* **1999**, *424*, 237–247. [CrossRef]

539. Loeppky, R.N.; Ye, Q.; Goelzer, P.; Chen, Y. DNA adducts from N-nitrosodiethanolamine and related beta-oxidized nitrosamines in vivo: (32)P-postlabeling methods for glyoxal- and O(6)-hydroxyethyldeoxyguanosine adducts. *Chem. Res. Toxicol.* **2002**, *15*, 470–482. [CrossRef] [PubMed]

540. Rajewsky, M.F.; Engelbergs, J.; Thomale, J.; Schweer, T. Relevance of DNA repair to carcinogenesis and cancer therapy. *Recent Results Cancer Res.* **1998**, *154*, 127–146.

541. Kamataki, T.; Fujita, K.-i.; Nakayama, K.; Yamazaki, Y.; Miyamoto, M.; Ariyoshi, N. Role of human cytochrome P450 (CYP) in the metabolic activation of nitrosamine derivatives: Application of genetically engineered Salmonella expressing human CYP. *Drug Metab. Rev.* **2002**, *34*, 667–676. [CrossRef]

542. Yang, C.S.; Smith, T.; Ishizaki, H.; Hong, J.Y. Enzyme mechanisms in the metabolism of nitrosamines. *IARC Sci. Publ.* **1991**, *105*, 265–274.

543. Johnson, G.E.; Dobo, K.; Gollapudi, B.; Harvey, J.; Kenny, J.; Kenyon, M.; Lynch, A.; Minocherhomji, S.; Nicolette, J.; Thybaud, V.; et al. Permitted daily exposure limits for noteworthy N-nitrosamines. *Environ. Mol. Mutagen.* **2021**, *62*, 293–305. [CrossRef]

544. Gushgari, A.J.; Halden, R.U. Critical review of major sources of human exposure to N-nitrosamines. *Chemosphere* **2018**, *210*, 1124–1136. [CrossRef]

545. Poirier, S.; Hubert, A.; de-Thé, G.; Ohshima, H.; Bourgade, M.C.; Bartsch, H. Occurrence of volatile nitrosamines in food samples collected in three high-risk areas for nasopharyngeal carcinoma. *IARC Sci. Publ.* **1987**, *84*, 415–419.

546. O'Brien, P.J.; Siraki, A.G.; Shangari, N. Aldehyde sources, metabolism, molecular toxicity mechanisms, and possible effects on human health. *Crit. Rev. Toxicol.* **2005**, *35*, 609–662. [CrossRef]

547. Sousa, B.C.; Pitt, A.R.; Spickett, C.M. Chemistry and analysis of HNE and other prominent carbonyl-containing lipid oxidation compounds. *Free Radic. Biol. Med.* **2017**, *111*, 294–308. [CrossRef] [PubMed]

548. Adams, T.B.; Gavin, C.L.; Taylor, S.V.; Waddell, W.J.; Cohen, S.M.; Feron, V.J.; Goodman, J.; Rietjens, I.M.; Marnett, L.J.; Portoghese, P.S.; et al. The FEMA GRAS assessment of alpha,beta-unsaturated aldehydes and related substances used as flavor ingredients. *Food Chem. Toxicol.* **2008**, *46*, 2935–2967. [CrossRef] [PubMed]

549. JECFA, Joint FAO/WHO Expert Committee on Food Additives. *Evaluation of Certain Food Additives and Contaminants*; Prepared by the Sixty-First Meeting of the Joint FAO/WHO Expert Committee on Food Additives (JECFA); WHO: Geneva, Switzerland, 2004.

550. Gasc, N.; Taché, S.; Rathahao, E.; Bertrand-Michel, J.; Roques, V.; Guéraud, F. 4-Hydroxynonenal in foodstuffs: Heme concentration, fatty acid composition and freeze-drying are determining factors. *Redox Rep.* **2007**, *12*, 40–44. [CrossRef] [PubMed]

551. Feron, V.J.; Til, H.P.; de Vrijer, F.; Woutersen, R.A.; Cassee, F.R.; van Bladeren, P.J. Aldehydes: Occurrence, carcinogenic potential, mechanism of action and risk assessment. *Mutat. Res./Genet. Toxicol.* **1991**, *259*, 363–385. [CrossRef]

552. IARC, International Agency for Research on Cancer. Acrolein, crotonaldehyde, and arecoline. In *IARC Monographs on the Evaluation of Carcinogenic Risks to Humans*; International Agency for Research on Cancer: Lyon, France, 2021; Volume 128.

553. EFSA CONTAM Panel, European Food Safety Authority, Panel on Contaminants in the Food Chain. Scientific Opinion on Flavouring Group Evaluation 200, Revision 1 (FGE.200 Rev 1): 74 a,b-unsaturated aliphatic aldehydes and precursors from chemical subgroup 1.1.1 of FGE.19. *EFSA J.* **2018**, *16*, 5422. [CrossRef]

554. Chung, F.L.; Tanaka, T.; Hecht, S.S. Induction of liver tumors in F344 rats by crotonaldehyde. *Cancer Res.* **1986**, *46*, 1285–1289.

555. NTP, National Toxicology Program. *Toxicology and Carcinogenesis Studies of Malonaldehyde, Sodium Salt (3-hydroxy-2-propenal, Sodium Salt) (CAS No. 24382-04-5) in F344/N Rats and B6C3F1 Mice (Gavage Studies)*; 0888-8051 (Print); National Toxicology Program: Research Triangle, NC, USA, 1988; pp. 1–182.

556. NTP, National Toxicology Program. *Toxicology and Carcinogensis Studies of 2,4-Hexadienal (89% trans,trans Isomer, CAS No. 142-83-6; 11% cis, trans Isomer) (Gavage Studies)*; 0888-8051 (Print); National Toxicology Program: Research Triangle, NC, USA, 2003; pp. 1–290.

557. EFSA CONTAM Panel, European Food Safety Authority, Panel on Contaminants in the Food Chain. Scientific Opinion on Flavouring Group Evaluation 203, Revision 2 (FGE.203Rev2): a,b-unsaturated aliphatic aldehydes and precursors from chemical subgroup 1.1.4 of FGE.19 with two or more conjugated double-bonds and with or without additional non-conjugated double-bonds. *EFSA J.* **2018**, *16*, 5322. [CrossRef]

558. IARC, International Agency for Research on Cancer. Re-evaluation of some organic chemicals, hydrazine and hydrogen Peroxide. In *IARC Monographs on the Evaluation of Carcinogenic Risks to Humans*; International Agency for Research on Cancer: Lyon, France, 1999; Volume 71.

559. Von Tungeln, L.S.; Yi, P.; Bucci, T.J.; Samokyszyn, V.M.; Chou, M.W.; Kadlubar, F.F.; Fu, P.P. Tumorigenicity of chloral hydrate, trichloroacetic acid, trichloroethanol, malondialdehyde, 4-hydroxy-2-nonenal, crotonaldehyde, and acrolein in the B6C3F(1) neonatal mouse. *Cancer Lett.* **2002**, *185*, 13–19. [CrossRef]

560. Eder, E.; Deininger, C.; Neudecker, T.; Deininger, D. Mutagenicity of β-alkyl substituted acrolein congeners in the Salmonella typhimurium strain TA100 and genotoxicity testing in the SOS chromotest. *Environ. Mol. Mutagen.* **1992**, *19*, 338–345. [CrossRef]

561. Eder, E.; Scheckenbach, S.; Deininger, C.; Hoffman, C. The possible role of alpha, beta-unsaturated carbonyl compounds in mutagenesis and carcinogenesis. *Toxicol Lett.* **1993**, *67*, 87–103. [CrossRef]

562. Brambilla, G.; Sciabà, L.; Faggin, P.; Maura, A.; Marinari, U.M.; Ferro, M.; Esterbauer, H. Cytotoxicity, DNA fragmentation and sister-chromatid exchange in Chinese hamster ovary cells exposed to the lipid peroxidation product 4-hydroxynonenal and homologous aldehydes. *Mutat. Res.* **1986**, *171*, 169–176. [CrossRef]

563. Esterbauer, H. Cytotoxicity and genotoxicity of lipid-oxidation products. *Am. J. Clin. Nutr.* **1993**, *57*, 779S–785S, discussion 785S–786S. [CrossRef] [PubMed]

564. Zhang, S.; Villalta, P.W.; Wang, M.; Hecht, S.S. Analysis of crotonaldehyde- and acetaldehyde-derived 1,n(2)-propanodeoxyguanosine adducts in DNA from human tissues using liquid chromatography electrospray ionization tandem mass spectrometry. *Chem. Res. Toxicol.* **2006**, *19*, 1386–1392. [CrossRef] [PubMed]

565. Guéraud, F. 4-Hydroxynonenal metabolites and adducts in pre-carcinogenic conditions and cancer. *Free Radic. Biol. Med.* **2017**, *111*, 196–208. [CrossRef]

566. Hu, W.; Feng, Z.; Eveleigh, J.; Iyer, G.; Pan, J.; Amin, S.; Chung, F.L.; Tang, M.S. The major lipid peroxidation product, trans-4-hydroxy-2-nonenal, preferentially forms DNA adducts at codon 249 of human p53 gene, a unique mutational hotspot in hepatocellular carcinoma. *Carcinogenesis* **2002**, *23*, 1781–1789. [CrossRef]

567. Singh, S.P.; Chen, T.; Chen, L.; Mei, N.; McLain, E.; Samokyszyn, V.; Thaden, J.J.; Moore, M.M.; Zimniak, P. Mutagenic effects of 4-hydroxynonenal triacetate, a chemically protected form of the lipid peroxidation product 4-hydroxynonenal, as assayed in L5178Y/Tk+/− mouse lymphoma cells. *J. Pharmacol. Exp. Ther.* **2005**, *313*, 855–861. [CrossRef]

568. Gutteridge, J.M.C. Aspects to consider when detecting and measuring lipid peroxidation. *Free Radic. Res. Commun.* **1986**, *1*, 173–184. [CrossRef]

569. Tsikas, D. Assessment of lipid peroxidation by measuring malondialdehyde (MDA) and relatives in biological samples: Analytical and biological challenges. *Anal. Biochem.* **2017**, *524*, 13–30. [CrossRef]

570. Nielsen, N.S.; Timm-Heinrich, M.; Jacobsen, C. Comparison of wet-chemical methods for determination of lipid hydroperoxides. *J. Food Lipids* **2003**, *10*, 35–50. [CrossRef]

571. Calingasan, N.Y.; Uchida, K.; Gibson, G.E. Protein-bound acrolein. *J. Neurochem.* **1999**, *72*, 751–756. [CrossRef]

572. Yuan, J.M.; Gao, Y.T.; Wang, R.; Chen, M.; Carmella, S.G.; Hecht, S.S. Urinary levels of volatile organic carcinogen and toxicant biomarkers in relation to lung cancer development in smokers. *Carcinogenesis* **2012**, *33*, 804–809. [CrossRef] [PubMed]

573. Yuan, J.M.; Butler, L.M.; Gao, Y.T.; Murphy, S.E.; Carmella, S.G.; Wang, R.; Nelson, H.H.; Hecht, S.S. Urinary metabolites of a polycyclic aromatic hydrocarbon and volatile organic compounds in relation to lung cancer development in lifelong never smokers in the Shanghai Cohort Study. *Carcinogenesis* **2014**, *35*, 339–345. [CrossRef] [PubMed]

574. Grigoryan, H.; Schiffman, C.; Gunter, M.J.; Naccarati, A.; Polidoro, S.; Dagnino, S.; Dudoit, S.; Vineis, P.; Rappaport, S.M. Cys34 adductomics links colorectal cancer with the gut microbiota and redox biology. *Cancer Res.* **2019**, *79*, 6024–6031. [CrossRef] [PubMed]

575. Eder, E.; Schuler, D.; Budiawan. Cancer risk assessment for crotonaldehyde and 2-hexenal: An approach. *IARC Sci. Publ.* **1999**, *150*, 219–232.

576. Kunishima, M.; Yamauchi, Y.; Mizutani, M.; Kuse, M.; Takikawa, H.; Sugimoto, Y. Identification of (Z)-3:(E)-2-hexenal isomerases essential to the production of the leaf aldehyde in plants. *J. Biol Chem.* **2016**, *291*, 14023–14033. [CrossRef]

577. Jordán, M.J.; Tandon, K.; Shaw, P.E.; Goodner, K.L. Aromatic profile of aqueous banana essence and banana fruit by Gas Chromatography–Mass Spectrometry (GC-MS) and Gas Chromatography–Olfactometry (GC-O). *J. Agric. Food Chem.* **2001**, *49*, 4813–4817. [CrossRef]

578. EFSA CONTAM Panel, European Food Safety Authority, Panel on Contaminants in the Food Chain. Scientific Opinion on Flavouring Group Evaluation 71 Revision 1 (FGE.71Rev1): Consideration of aliphatic, linear, a,b-unsaturated alcohols, aldehydes, carboxylic acids, and related esters evaluated by JECFA (63rd and 69th meeting) structurally related to flavouring substances evaluated in FGE.05Rev3. *EFSA J.* **2020**, *18*, 5924. [CrossRef]

579. Nádasi, E.; Varjas, T.; Pajor, L.; Ember, I. Carcinogenic potential of trans-2-hexenal is based on epigenetic effect. *In Vivo* **2005**, *19*, 559–562.

580. Dittberner, U.; Eisenbrand, G.; Zankl, H. Genotoxic effects of the α, β-unsaturated aldehydes 2-trans-butenal,2-trans-hexenal and 2-trans, 6-cis-rmnonadienal. *Mutat. Res./Environ. Mutagen. Relat. Subj.* **1995**, *335*, 259–265. [CrossRef]

581. Dittberner, U.; Schmetzer, B.; Gölzer, P.; Eisenbrand, G.; Zankl, H. Genotoxic effects of 2-trans-hexenal in human buccal mucosa cells in vivo. *Mutat. Res.* **1997**, *390*, 161–165. [CrossRef]

582. Eder, E.; Schuler, D. An approach to cancer risk assessment for the food constituent 2-hexenal on the basis of 1,N2-propanodeoxyguanosine adducts of 2-hexenal in vivo. *Arch. Toxicol.* **2000**, *74*, 642–648. [CrossRef] [PubMed]

583. Schuler, D.; Eder, E. Detection of 1,N2-propanodeoxyguanosine adducts of 2-hexenal in organs of Fischer 344 rats by a 32P-post-labeling technique. *Carcinogenesis* **1999**, *20*, 1345–1350. [CrossRef]

584. Gölzer, P.; Janzowski, C.; Pool-Zobel, B.L.; Eisenbrand, G. (E)-2-hexenal-induced DNA damage and formation of cyclic 1,N2-(1,3-propano)-2'-deoxyguanosine adducts in mammalian cells. *Chem. Res. Toxicol.* **1996**, *9*, 1207–1213. [CrossRef]

585. Stout, M.D.; Jeong, Y.C.; Boysen, G.; Li, Y.; Sangaiah, R.; Ball, L.M.; Gold, A.; Swenberg, J.A. LC/MS/MS method for the quantitation of trans-2-hexenal-derived exocyclic 1,N(2)-propanodeoxyguanosine in DNA. *Chem. Res. Toxicol.* **2006**, *19*, 563–570. [CrossRef] [PubMed]

586. Stout, M.D.; Bodes, E.; Schoonhoven, R.; Upton, P.B.; Travlos, G.S.; Swenberg, J.A. Toxicity, DNA binding, and cell proliferation in male F344 rats following short-term gavage exposures to trans-2-hexenal. *Toxicol. Pathol.* **2008**, *36*, 232–246. [CrossRef] [PubMed]
587. Kiwamoto, R.; Rietjens, I.M.; Punt, A. A physiologically based in silico model for trans-2-hexenal detoxification and DNA adduct formation in rat. *Chem. Res. Toxicol.* **2012**, *25*, 2630–2641. [CrossRef]
588. Janzowski, C.; Glaab, V.; Mueller, C.; Straesser, U.; Kamp, H.G.; Eisenbrand, G. Alpha,beta-unsaturated carbonyl compounds: Induction of oxidative DNA damage in mammalian cells. *Mutagenesis* **2003**, *18*, 465–470. [CrossRef]
589. Palmer, S. Diet, nutrition, and cancer. *Prog. Food Nutr. Sci.* **1985**, *9*, 283–341.
590. Weisburger, J.H. Carcinogenesis in our food and cancer prevention. *Adv. Exp. Med. Biol.* **1991**, *289*, 137–151. [CrossRef]
591. Yoo, J.Y.; Cho, H.J.; Moon, S.; Choi, J.; Lee, S.; Ahn, C.; Yoo, K.Y.; Kim, I.; Ko, K.P.; Lee, J.E.; et al. Pickled vegetable and salted fish intake and the risk of gastric cancer: Two prospective cohort studies and a meta-analysis. *Cancers* **2020**, *12*, 996. [CrossRef]
592. Gallicchio, L.; Matanoski, G.; Tao, X.; Chen, L.; Lam, T.K.; Boyd, K.; Robinson, K.A.; Balick, L.; Mickelson, S.; Caulfield, L.E.; et al. Adulthood consumption of preserved and nonpreserved vegetables and the risk of nasopharyngeal carcinoma: A systematic review. *Int. J. Cancer* **2006**, *119*, 1125–1135. [CrossRef] [PubMed]
593. IARC, International Agency for Research on Cancer. Red meat and processed meat. In *IARC Monographs on the Evaluation of Carcinogenic Risks to Humans*; International Agency for Research on Cancer: Lyon, France, 2015; Volume 114.
594. Farvid, M.S.; Sidahmed, E.; Spence, N.D.; Mante Angua, K.; Rosner, B.A.; Barnett, J.B. Consumption of red meat and processed meat and cancer incidence: A systematic review and meta-analysis of prospective studies. *Eur. J. Epidemiol.* **2021**, *36*, 937–951. [CrossRef] [PubMed]
595. Turner, N.D.; Lloyd, S.K. Association between red meat consumption and colon cancer: A systematic review of experimental results. *Exp. Biol. Med.* **2017**, *242*, 813–839. [CrossRef] [PubMed]
596. Anderson, J.J.; Darwis, N.D.M.; Mackay, D.F.; Celis-Morales, C.A.; Lyall, D.M.; Sattar, N.; Gill, J.M.R.; Pell, J.P. Red and processed meat consumption and breast cancer: UK Biobank cohort study and meta-analysis. *Eur. J. Cancer* **2018**, *90*, 73–82. [CrossRef] [PubMed]
597. Turesky, R.J. Mechanistic evidence for red meat and processed meat intake and cancer risk: A follow-up on the International Agency for Research on Cancer evaluation of 2015. *Chimia* **2018**, *72*, 718–724. [CrossRef]
598. Tannenbaum, S.R.; Bishop, W.; Yu, M.C.; Henderson, B.E. Attempts to isolate N-nitroso compounds from Chinese-style salted fish. *Natl. Cancer Inst. Monogr.* **1985**, *69*, 209–211.
599. IARC, International Agency for Research on Cancer. Review of human carcinogens. Personal habits and indoor combustions. In *IARC Monographs on the Evaluation of Carcinogenic Risks to Humans*; International Agency for Research on Cancer: Lyon, France, 2012; Volume 100E.
600. Qiu, Y.; Chen, J.H.; Yu, W.; Wang, P.; Rong, M.; Deng, H. Contamination of Chinese salted fish with volatile N-nitrosamines as determined by QuEChERS and gas chromatography-tandem mass spectrometry. *Food Chem.* **2017**, *232*, 763–769. [CrossRef]
601. Poirier, S.; Bouvier, G.; Malaveille, C.; Ohshima, H.; Shao, Y.M.; Hubert, A.; Zeng, Y.; de Thé, G.; Bartsch, H. Volatile nitrosamine levels and genotoxicity of food samples from high-risk areas for nasopharyngeal carcinoma before and after nitrosation. *Int. J. Cancer* **1989**, *44*, 1088–1094. [CrossRef]
602. Zheng, X.; Luo, Y.; Christensson, B.; Drettner, B. Induction of nasal and nasopharyngeal tumours in Sprague-Dawley rats fed with Chinese salted fish. *Acta Otolaryngol.* **1994**, *114*, 98–104. [CrossRef]
603. Yu, M.C.; Nichols, P.W.; Zou, X.N.; Estes, J.; Henderson, B.E. Induction of malignant nasal cavity tumours in Wistar rats fed Chinese salted fish. *Br. J. Cancer* **1989**, *60*, 198–201. [CrossRef]
604. Huang, D.P.; Ho, J.H.; Saw, D.; Teoh, T.B. Carcinoma of the nasal and paranasal regions in rats fed Cantonese salted marine fish. *IARC Sci. Publ.* **1978**, *20*, 315–328.
605. Weisburger, J.H.; Marquardt, H.; Hirota, N.; Mori, H.; Williams, G.M. Induction of cancer of the glandular stomach in rats by an extract of nitrite-treated fish. *J. Natl. Cancer Inst.* **1980**, *64*, 163–167. [PubMed]
606. Widlak, P.; Zheng, X.; Osterdahl, B.G.; Drettner, B.; Christensson, B.; Kumar, R.; Hemminki, K. N-nitrosodimethylamine and 7-methylguanine DNA adducts in tissues of rats fed Chinese salted fish. *Cancer Lett.* **1995**, *94*, 85–90. [CrossRef]
607. Yuan, J.M.; Wang, X.L.; Xiang, Y.B.; Gao, Y.T.; Ross, R.K.; Yu, M.C. Preserved foods in relation to risk of nasopharyngeal carcinoma in Shanghai, China. *Int. J. Cancer* **2000**, *85*, 358–363. [CrossRef]
608. Lian, M. Salted fish and processed foods intake and nasopharyngeal carcinoma risk: A dose-response meta-analysis of observational studies. *Eur. Arch. Otorhinolaryngol.* **2022**, *279*, 2501–2509. [CrossRef]
609. Hara, N.; Sakata, K.; Nagai, M.; Fujita, Y.; Hashimoto, T.; Yanagawa, H. Statistical analyses on the pattern of food consumption and digestive-tract cancers in Japan. *Nutr. Cancer* **1984**, *6*, 220–228. [CrossRef]
610. FDA, Food and Drug Administration. *Food Additive Regulations. Synthetic Flavoring Agents and Adjuvants. A Rule by the Food and Drug Administration on 10/09/2018 83 FR 50490. 21 CFR 172. 21 CFR 177. Document Number: 2018-21807. Final Rule; Notif. Part Denial Petition*; Food and Drug Administration: Rome, Italy, 2018; pp. 50490–50503.
611. IARC, International Agency for Research on Cancer. Some chemicals that cause tumours of the urinary tract in rodents. In *IARC Monographs on the Evaluation of Carcinogenic Risks to Humans*; International Agency for Research on Cancer: Lyon, France, 2019; Volume 119.
612. Felter, S.P.; Llewelyn, C.; Navarro, L.; Zhang, X. How the 62-year old Delaney Clause continues to thwart science: Case study of the flavor substance β-myrcene. *Regul. Toxicol. Pharm.* **2020**, *115*, 104708. [CrossRef]

613. Api, A.M.; Belmonte, F.; Belsito, D.; Biserta, S.; Botelho, D.; Bruze, M.; Burton, G.A., Jr.; Buschmann, J.; Cancellieri, M.A.; Dagli, M.L.; et al. RIFM fragrance ingredient safety assessment, myrcene, CAS Registry Number 123-35-3. *Food Chem. Toxicol.* **2020**, *144* (Suppl. S1), 111631. [CrossRef]

614. NTP, National Toxicology Program. Technical report on the toxicology and carcinogenesis studies of beta-myrcene (CAS No. 123-35-3) in F344/N rats and B6C3F1 mice (gavage studies). *Natl. Toxicol. Program Tech. Rep. Ser.* **2010**, *557*, 1–163.

615. JECFA, Joint FAO/WHO Expert Committee on Food Additives. *Safety Evaluation of Certain Food Additives*; Prepared for the Seventy-Ninth Meeting of the Joint FAO/WHO Expert Committee on Food Additives; WHO: Geneva, Switzerland, 2015.

616. EFSA CEF Panel, European Food Safety Authority, Panel on Food Contact Materials, Enzymes, Flavourings and Processing Aids. Scientific Opinion on Flavouring Group Evaluation 78, Revision 2 (FGE.78Rev2): Consideration of aliphatic and alicyclic and aromatic hydrocarbons evaluated by JECFA (63rd meeting) structurally related to aliphatic hydrocarbonsevaluated by EFSA in FGE.25Rev3. *EFSA J.* **2015**, *13*, 4067. [CrossRef]

617. Kauderer, B.; Zamith, H.; Paumgartten, F.J.; Speit, G. Evaluation of the mutagenicity of beta-myrcene in mammalian cells in vitro. *Environ. Mol. Mutagen.* **1991**, *18*, 28–34. [CrossRef] [PubMed]

618. Zamith, H.P.; Vidal, M.N.; Speit, G.; Paumgartten, F.J. Absence of genotoxic activity of beta-myrcene in the in vivo cytogenetic bone marrow assay. *Braz. J. Med. Biol. Res.* **1993**, *26*, 93–98. [PubMed]

619. Cohen, S.M.; Eisenbrand, G.; Fukushima, S.; Gooderham, N.J.; Guengerich, F.P.; Hecht, S.S.; Rietjens, I.; Bastaki, M.; Davidsen, J.M.; Harman, C.L.; et al. FEMA GRAS assessment of natural flavor complexes: Citrus-derived flavoring ingredients. *Food Chem. Toxicol.* **2019**, *124*, 192–218. [CrossRef]

620. Madyastha, K.M.; Srivatsan, V. Metabolism of beta-myrcene in vivo and in vitro: Its effects on rat-liver microsomal enzymes. *Xenobiotica* **1987**, *17*, 539–549. [CrossRef] [PubMed]

621. Ishida, T.; Asakawa, Y.; Takemoto, T.; Aratani, T. Terpenoids biotransformation in mammals III: Biotransformation of alpha-pinene, beta-pinene, pinane, 3-carene, carane, myrcene, and p-cymene in rabbits. *J. Pharm. Sci.* **1981**, *70*, 406–415. [CrossRef]

622. Adams, T.B.; Gavin, C.L.; McGowen, M.M.; Waddell, W.J.; Cohen, S.M.; Feron, V.J.; Marnett, L.J.; Munro, I.C.; Portoghese, P.S.; Rietjens, I.M.; et al. The FEMA GRAS assessment of aliphatic and aromatic terpene hydrocarbons used as flavor ingredients. *Food Chem. Toxicol.* **2011**, *49*, 2471–2494. [CrossRef]

623. De-Oliveira, A.C.; Ribeiro-Pinto, L.F.; Paumgartten, J.R. In vitro inhibition of CYP2B1 monooxygenase by beta-myrcene and other monoterpenoid compounds. *Toxicol. Lett.* **1997**, *92*, 39–46. [CrossRef]

624. De-Oliveira, A.C.; Ribeiro-Pinto, L.F.; Otto, S.S.; Gonçalves, A.; Paumgartten, F.J. Induction of liver monooxygenases by beta-myrcene. *Toxicology* **1997**, *124*, 135–140. [CrossRef]

625. Cesta, M.F.; Hard, G.C.; Boyce, J.T.; Ryan, M.J.; Chan, P.C.; Sills, R.C. Complex histopathologic response in rat kidney to oral β-myrcene: An unusual dose-related nephrosis and low-dose alpha2u-globulin nephropathy. *Toxicol. Pathol.* **2013**, *41*, 1068–1077. [CrossRef]

626. JECFA, Joint FAO/WHO Expert Committee on Food Additives. *Evaluation of Certain Food Additives*; Prepared by the Sixty-third meeting of the Joint FAO/WHO Expert Committee on Food Additives (JECFA); WHO: Geneva, Switzerland, 2005.

627. Mog, S.R.; Zang, Y.J. Safety assessment of food additives: Case example with myrcene, a synthetic flavoring agent. *Toxicol. Pathol.* **2019**, *47*, 1035–1037. [CrossRef]

628. IARC, International Agency for Research on Cancer. Some drugs and herbal products. In *IARC Monographs on the Evaluation of Carcinogenic Risks to Humans*; International Agency for Research on Cancer: Lyon, France, 2016; Volume 108.

629. JECFA, Joint FAO/WHO Expert Committee on Food Additives. *Evaluation of Certain Food Additives and Contaminants*; Prepared by the Fifty-Fifth Meeting of the Joint FAO/WHO Expert Committee on Food Additives (JECFA); WHO: Geneva, Switzerland, 2001.

630. EFSA AFC Panel, European Food Safety Authority, Panel on Food Additives, Flavourings, Processing Aids and Materials in contact with Food. Opinion of the Scientific Panel on Food Additives, Flavourings, Processing Aids and Materials in contact with Foods on a request from the Commission on Pulegone and Menthofuran in flavourings and other food ingredients with flavouring properties. Question number EFSA-Q-2003-119. *EFSA J.* **2005**, *298*, 1–32. [CrossRef]

631. EMA, European Medicines Agency. *Public Statement on the Use of Herbal Medicinal Products Containing Pulegone and Menthofuran*; EMA/HMPC/138386/2005 Rev. 1; European Medicines Agency: Amsterdam, The Netherlands, 2016.

632. NTP, National Toxicology Program. Toxicology and carcinogenesis studies of pulegone (CAS No. 89-82-7) in F344/N rats and B6C3F1 mice (gavage studies). *Natl. Toxicol. Program. Tech. Rep. Ser.* **2011**, *563*, 1–201.

633. Api, A.M.; Belsito, D.; Biserta, S.; Botelho, D.; Bruze, M.; Burton, G.A., Jr.; Buschmann, J.; Cancellieri, M.A.; Dagli, M.L.; Date, M.; et al. RIFM fragrance ingredient safety assessment, pulegone, CAS Registry Number 89-82-7. *Food Chem. Toxicol.* **2021**, *149* (Suppl. S1), 112092. [CrossRef] [PubMed]

634. Khojasteh, S.C.; Hartley, D.P.; Ford, K.A.; Uppal, H.; Oishi, S.; Nelson, S.D. Characterization of rat liver proteins adducted by reactive metabolites of menthofuran. *Chem. Res. Toxicol.* **2012**, *25*, 2301–2309. [CrossRef] [PubMed]

635. Chen, L.J.; Lebetkin, E.H.; Burka, L.T. Metabolism of (R)-(+)-pulegone in F344 rats. *Drug Metab. Dispos.* **2001**, *29*, 1567–1577.

636. Chen, L.J.; Lebetkin, E.H.; Burka, L.T. Comparative disposition of (R)-(+)-pulegone in B6C3F1 mice and F344 rats. *Drug Metab. Dispos.* **2003**, *31*, 892–899. [CrossRef]

637. Khojasteh-Bakht, S.C.; Chen, W.; Koenigs, L.L.; Peter, R.M.; Nelson, S.D. Metabolism of (R)-(+)-pulegone and (R)-(+)-menthofuran by human liver cytochrome P-450s: Evidence for formation of a furan epoxide. *Drug Metab. Dispos.* **1999**, *27*, 574–580.

638. Gordon, W.P.; Huitric, A.C.; Seth, C.L.; McClanahan, R.H.; Nelson, S.D. The metabolism of the abortifacient terpene, (R)-(+)-pulegone, to a proximate toxin, menthofuran. *Drug Metab. Dispos.* **1987**, *15*, 589–594.

639. Nelson, S.D.; McClanahan, R.H.; Thomassen, D.; Gordon, W.P.; Knebel, N. Investigations of mechanisms of reactive metabolite formation from (R)-(+)-pulegone. *Xenobiotica* **1992**, *22*, 1157–1164. [CrossRef]

640. Engel, W. In vivo studies on the metabolism of the monoterpene pulegone in humans using the metabolism of ingestion-correlated amounts (MICA) approach: Explanation for the toxicity differences between (S)-(-)- and (R)-(+)-pulegone. *J. Agric. Food Chem.* **2003**, *51*, 6589–6597. [CrossRef]

641. McClanahan, R.H.; Thomassen, D.; Slattery, J.T.; Nelson, S.D. Metabolic activation of (R)-(+)-pulegone to a reactive enonal that covalently binds to mouse liver proteins. *Chem. Res. Toxicol.* **1989**, *2*, 349–355. [CrossRef] [PubMed]

642. Da Rocha, M.S.; Dodmane, P.R.; Arnold, L.L.; Pennington, K.L.; Anwar, M.M.; Adams, B.R.; Taylor, S.V.; Wermes, C.; Adams, T.B.; Cohen, S.M. Mode of action of pulegone on the urinary bladder of F344 rats. *Toxicol. Sci.* **2012**, *128*, 1–8. [CrossRef] [PubMed]

643. Alshannaq, A.; Yu, J.H. Occurrence, toxicity, and analysis of major mycotoxins in food. *Int. J. Environ. Res. Public Health* **2017**, *14*, 632. [CrossRef]

644. Riley, R.T.; Merrill, A.H., Jr. Ceramide synthase inhibition by fumonisins: A perfect storm of perturbed sphingolipid metabolism, signaling, and disease. *J. Lipid Res.* **2019**, *60*, 1183–1189. [CrossRef]

645. Schroeder, J.J.; Crane, H.M.; Xia, J.; Liotta, D.C.; Merrill, A.H., Jr. Disruption of sphingolipid metabolism and stimulation of DNA synthesis by fumonisin B1. A molecular mechanism for carcinogenesis associated with Fusarium moniliforme. *J. Biol. Chem.* **1994**, *269*, 3475–3481. [CrossRef]

646. NTP, National Toxicology Program. Toxicology and carcinogenesis studies of fumonisin B1 (cas no. 116355-83-0) in F344/N rats and B6C3F1 mice (feed studies). *Natl. Toxicol. Program. Tech. Rep. Ser.* **2001**, *496*, 1–352.

647. JECFA, Joint FAO/WHO Expert Committee on Food Additives. *Safety Evaluation of Certain Mycotoxins in Food*; Prepared by the Fifty-Sixth Meeting of the Joint FAO/WHO Expert Committee on Food Additives (JECFA); WHO: Geneva, Switzerland, 2001.

648. Gelderblom, W.C.A.; Marasas, W.F.O.; Lebepe-Mazur, S.; Swanevelder, S.; Abel, S. Cancer initiating properties of fumonisin B1 in a short-term rat liver carcinogenesis assay. *Toxicology* **2008**, *250*, 89–95. [CrossRef]

649. Gelderblom, W.C.A.; Snyman, S.D.; Lebepe-Mazur, S.; van der Westhuizen, L.; Kriek, N.P.J.; Marasas, W.F.O. The cancer-promoting potential of fumonisin B1 in rat liver using diethylnitrosamine as a cancer initiator. *Cancer Lett.* **1996**, *109*, 101–108. [CrossRef]

650. Sakai, A.; Suzuki, C.; Masui, Y.; Kuramashi, A.; Takatori, K.; Tanaka, N. The activities of mycotoxins derived from Fusarium and related substances in a short-term transformation assay using v-Ha-ras-transfected BALB/3T3 cells (Bhas 42 cells). *Mutat. Res.* **2007**, *630*, 103–111. [CrossRef]

651. Carlson, D.B.; Williams, D.E.; Spitsbergen, J.M.; Ross, P.F.; Bacon, C.W.; Meredith, F.I.; Riley, R.T. Fumonisin B1 promotes aflatoxin B1 and N-methyl-N'-nitro-nitrosoguanidine-initiated liver tumors in rainbow trout. *Toxicol. Appl. Pharm.* **2001**, *172*, 29–36. [CrossRef]

652. Mobio, T.A.; Tavan, E.; Baudrimont, I.; Anane, R.; Carratú, M.R.; Sanni, A.; Gbeassor, M.F.; Shier, T.W.; Narbonne, J.F.; Creppy, E.E. Comparative study of the toxic effects of fumonisin B1 in rat C6 glioma cells and p53-null mouse embryo fibroblasts. *Toxicology* **2003**, *183*, 65–75. [CrossRef]

653. Galvano, F.; Russo, A.; Cardile, V.; Galvano, G.; Vanella, A.; Renis, M. DNA damage in human fibroblasts exposed to fumonisin B(1). *Food Chem. Toxicol.* **2002**, *40*, 25–31. [CrossRef]

654. Domijan, A.M.; Zeljezić, D.; Milić, M.; Peraica, M. Fumonisin B(1): Oxidative status and DNA damage in rats. *Toxicology* **2007**, *232*, 163–169. [CrossRef] [PubMed]

655. Hassan, A.M.; Abdel-Aziem, S.H.; El-Nekeety, A.A.; Abdel-Wahhab, M.A. Panax ginseng extract modulates oxidative stress, DNA fragmentation and up-regulate gene expression in rats sub chronically treated with aflatoxin B1 and fumonisin B 1. *Cytotechnology* **2015**, *67*, 861–871. [CrossRef] [PubMed]

656. Ehrlich, V.; Darroudi, F.; Uhl, M.; Steinkellner, H.; Zsivkovits, M.; Knasmueller, S. Fumonisin B(1) is genotoxic in human derived hepatoma (HepG2) cells. *Mutagenesis* **2002**, *17*, 257–260. [CrossRef] [PubMed]

657. Aranda, M.; Pérez-Alzola, L.P.; Ellahueñe, M.F.; Sepúlveda, C. Assessment of in vitro mutagenicity in Salmonella and in vivo genotoxicity in mice of the mycotoxin fumonisin B(1). *Mutagenesis* **2000**, *15*, 469–471. [CrossRef]

658. Mary, V.S.; Valdehita, A.; Navas, J.M.; Rubinstein, H.R.; Fernández-Cruz, M.L. Effects of aflatoxin B_1, fumonisin B_1 and their mixture on the aryl hydrocarbon receptor and cytochrome P450 1A induction. *Food Chem. Toxicol.* **2015**, *75*, 104–111. [CrossRef] [PubMed]

659. Spotti, M.; Maas, R.F.; de Nijs, C.M.; Fink-Gremmels, J. Effect of fumonisin B(1) on rat hepatic P450 system. *Environ. Toxicol. Pharm.* **2000**, *8*, 197–204. [CrossRef]

660. Müller, S.; Dekant, W.; Mally, A. Fumonisin B1 and the kidney: Modes of action for renal tumor formation by fumonisin B1 in rodents. *Food Chem. Toxicol.* **2012**, *50*, 3833–3846. [CrossRef]

661. Harrer, H.; Humpf, H.U.; Voss, K.A. In vivo formation of N-acyl-fumonisin B1. *Mycotoxin Res.* **2015**, *31*, 33–40. [CrossRef]

662. Wang, E.; Norred, W.P.; Bacon, C.W.; Riley, R.T.; Merrill, A.H., Jr. Inhibition of sphingolipid biosynthesis by fumonisins. Implications for diseases associated with Fusarium moniliforme. *J. Biol. Chem.* **1991**, *266*, 14486–14490. [CrossRef]

663. Voss, K.A.; Howard, P.C.; Riley, R.T.; Sharma, R.P.; Bucci, T.J.; Lorentzen, R.J. Carcinogenicity and mechanism of action of fumonisin B1: A mycotoxin produced by Fusarium moniliforme (F. verticillioides). *Cancer Detect. Prev.* **2002**, *26*, 1–9. [CrossRef]

664. Riley, R.T.; Enongene, E.; Voss, K.A.; Norred, W.P.; Meredith, F.I.; Sharma, R.P.; Spitsbergen, J.; Williams, D.E.; Carlson, D.B.; Merrill, A.H., Jr. Sphingolipid perturbations as mechanisms for fumonisin carcinogenesis. *Environ. Health Perspect.* **2001**, *109* (Suppl. S2), 301–308. [CrossRef] [PubMed]

665. Merrill, A.H., Jr.; Sullards, M.C.; Wang, E.; Voss, K.A.; Riley, R.T. Sphingolipid metabolism: Roles in signal transduction and disruption by fumonisins. *Environ. Health Perspect.* **2001**, *109* (Suppl. S2), 283–289. [CrossRef]

666. Gelderblom, W.C.; Abel, S.; Smuts, C.M.; Marnewick, J.; Marasas, W.F.; Lemmer, E.R.; Ramljak, D. Fumonisin-induced hepatocarcinogenesis: Mechanisms related to cancer initiation and promotion. *Environ. Health Perspect.* **2001**, *109* (Suppl. S2), 291–300. [CrossRef]

667. Rumora, L.; Domijan, A.M.; Grubisić, T.Z.; Peraica, M. Mycotoxin fumonisin B1 alters cellular redox balance and signalling pathways in rat liver and kidney. *Toxicology* **2007**, *242*, 31–38. [CrossRef]

668. Yin, J.J.; Smith, M.J.; Eppley, R.M.; Page, S.W.; Sphon, J.A. Effects of fumonisin B1 on lipid peroxidation in membranes. *Biochim. Biophys. Acta* **1998**, *1371*, 134–142. [CrossRef]

669. Sun, G.; Wang, S.; Hu, X.; Su, J.; Huang, T.; Yu, J.; Tang, L.; Gao, W.; Wang, J.S. Fumonisin B1 contamination of home-grown corn in high-risk areas for esophageal and liver cancer in China. *Food Addit. Contam.* **2007**, *24*, 181–185. [CrossRef]

670. Wang, H.; Wei, H.; Ma, J.; Luo, X. The fumonisin B1 content in corn from North China, a high-risk area of esophageal cancer. *J. Environ. Pathol. Toxicol. Oncol.* **2000**, *19*, 139–141.

671. Isaacson, C. The change of the staple diet of black South Africans from sorghum to maize (corn) is the cause of the epidemic of squamous carcinoma of the oesophagus. *Med. Hypotheses* **2005**, *64*, 658–660. [CrossRef]

672. van der Westhuizen, L.; Shephard, G.S.; Scussel, V.M.; Costa, L.L.; Vismer, H.F.; Rheeder, J.P.; Marasas, W.F. Fumonisin contamination and fusarium incidence in corn from Santa Catarina, Brazil. *J. Agric. Food Chem.* **2003**, *51*, 5574–5578. [CrossRef] [PubMed]

673. Claeys, L.; Romano, C.; De Ruyck, K.; Wilson, H.; Fervers, B.; Korenjak, M.; Zavadil, J.; Gunter, M.J.; De Saeger, S.; De Boevre, M.; et al. Mycotoxin exposure and human cancer risk: A systematic review of epidemiological studies. *Compr. Rev. Food Sci. Food Saf.* **2020**, *19*, 1449–1464. [CrossRef] [PubMed]

674. Persson, E.C.; Sewram, V.; Evans, A.A.; London, W.T.; Volkwyn, Y.; Shen, Y.J.; Van Zyl, J.A.; Chen, G.; Lin, W.; Shephard, G.S.; et al. Fumonisin B1 and risk of hepatocellular carcinoma in two Chinese cohorts. *Food Chem. Toxicol.* **2012**, *50*, 679–683. [CrossRef] [PubMed]

675. Cantalejo, M.J.; Torondel, P.; Amate, L.; Carrasco, J.M.; Hernández, E. Detection of fusarin C and trichothecenes in Fusarium strains from Spain. *J. Basic Microbiol.* **1999**, *39*, 143–153. [CrossRef]

676. Kleigrewe, K.; Söhnel, A.C.; Humpf, H.U. A new high-performance liquid chromatography-tandem mass spectrometry method based on dispersive solid phase extraction for the determination of the mycotoxin fusarin C in corn ears and processed corn samples. *J. Agric. Food Chem.* **2011**, *59*, 10470–10476. [CrossRef]

677. Han, Z.; Tangni, E.K.; Huybrechts, B.; Munaut, F.; Scauflaire, J.; Wu, A.; Callebaut, A. Screening survey of co-production of fusaric acid, fusarin C, and fumonisins B_1, B_2 and B_3 by Fusarium strains grown in maize grains. *Mycotoxin Res.* **2014**, *30*, 231–240. [CrossRef]

678. Zhu, B.; Jeffrey, A.M. Stability of Fusarin C: Effects of the normal cooking procedure used in china and ph. *Nutr. Cancer* **1992**, *18*, 53–58. [CrossRef]

679. Gelderblom, W.C.; Thiel, P.G.; Jaskiewicz, K.; Marasas, W.F. Investigations on the carcinogenicity of fusarin C–a mutagenic metabolite of Fusarium moniliforme. *Carcinogenesis* **1986**, *7*, 1899–1901. [CrossRef]

680. Cheng, S.J.; Jiang, Y.Z.; Li, M.H.; Lo, H.Z. A mutagenic metabolite produced by Fusarium moniliforme isolated from Linxian county, China. *Carcinogenesis* **1985**, *6*, 903–905. [CrossRef]

681. Gelderblom, W.C.; Snyman, S.D. Mutagenicity of potentially carcinogenic mycotoxins produced by Fusarium moniliforme. *Mycotoxin Res.* **1991**, *7*, 46–52. [CrossRef]

682. Norred, W.P.; Plattner, R.D.; Vesonder, R.F.; Bacon, C.W.; Voss, K.A. Effects of selected secondary metabolites of Fusarium moniliforme on unscheduled synthesis of DNA by rat primary hepatocytes. *Food Chem. Toxicol.* **1992**, *30*, 233–237. [CrossRef]

683. Bever, R.J., Jr.; Couch, L.H.; Sutherland, J.B.; Williams, A.J.; Beger, R.D.; Churchwell, M.I.; Doerge, D.R.; Howard, P.C. DNA adduct formation by Fusarium culture extracts: Lack of role of fusarin C. *Chem. Biol. Interact.* **2000**, *128*, 141–157. [CrossRef]

684. Lu, S.J.; Ronai, Z.A.; Li, M.H.; Jeffrey, A.M. Fusarium moniliforme metabolites: Genotoxicity of culture extracts. *Carcinogenesis* **1988**, *9*, 1523–1527. [CrossRef] [PubMed]

685. Gelderblom, W.C.; Swart, P.; Kramer, P.S. Investigations on the spectral interactions of fusarin C with rat liver microsomal cytochrome P-450. *Xenobiotica* **1988**, *18*, 1005–1014. [CrossRef] [PubMed]

686. Lu, S.J.; Li, M.H.; Park, S.S.; Gelboin, H.V.; Jeffrey, A.M. Metabolism of fusarin C by rat liver microsomes. Role of esterase and cytochrome P-450 enzymes with respect to the mutagenicity of fusarin C in Salmonella typhimurium. *Biochem. Pharm.* **1989**, *38*, 3811–3817. [CrossRef] [PubMed]

687. Zhu, B.; Jeffrey, A.M. Fusarin C: Isolation and identification of two microsomal metabolites. *Chem. Res. Toxicol.* **1993**, *6*, 97–101. [CrossRef]

688. Sondergaard, T.E.; Hansen, F.T.; Purup, S.; Nielsen, A.K.; Bonefeld-Jørgensen, E.C.; Giese, H.; Sørensen, J.L. Fusarin C acts like an estrogenic agonist and stimulates breast cancer cells in vitro. *Toxicol. Lett.* **2011**, *205*, 116–121. [CrossRef]

689. Marasas, W.F.; Jaskiewicz, K.; Venter, F.S.; Van Schalkwyk, D.J. Fusarium moniliforme contamination of maize in oesophageal cancer areas in Transkei. *S. Afr. Med. J.* **1988**, *74*, 110–114.
690. IARC, International Agency for Research on Cancer. Polychlorinated biphenyls and polybrominted biphenyls. In *IARC Monographs on the Evaluation of Carcinogenic Risks to Humans*; International Agency for Research on Cancer: Lyon, France, 2016; Volume 107.
691. IARC, International Agency for Research on Cancer. DDT, lindane, and 2,4-D. In *IARC Monographs on the Evaluation of Carcinogenic Risks to Humans*; International Agency for Research on Cancer: Lyon, France, 2018; Volume 113.
692. Walker, K. Cost-comparison of DDT and alternative insecticides for malaria control. *Med. Vet. Entomol.* **2000**, *14*, 345–354. [CrossRef]
693. Pedercini, M.; Movilla Blanco, S.; Kopainsky, B. Application of the malaria management model to the analysis of costs and benefits of DDT versus non-DDT malaria control. *PLoS ONE* **2011**, *6*, e27771. [CrossRef]
694. ATSDR, Agency for Toxic Substances and Disease Registry. *Toxicological Profile for DDT, DDE, and DDD*; Agency for Toxic Substances and Disease Registry: Atlanta, GA, USA, 2022.
695. WHO, World Health Organization. *The Use of DDT in Malaria Vector Control*; WHO Position Statement; World Health Organization: Geneva, Switzerland, 2011.
696. Smith, D. Worldwide trends in DDT levels in human breast milk. *Int. J. Epidemiol.* **1999**, *28*, 179–188. [CrossRef]
697. van den Berg, M.; Kypke, K.; Kotz, A.; Tritscher, A.; Lee, S.Y.; Magulova, K.; Fiedler, H.; Malisch, R. WHO/UNEP global surveys of PCDDs, PCDFs, PCBs and DDTs in human milk and benefit-risk evaluation of breastfeeding. *Arch. Toxicol.* **2017**, *91*, 83–96. [CrossRef] [PubMed]
698. NCI, National Cancer Institute. *Bioassay of DDT, TDE, and p,p'-DDE for Possible Carcinogenicity*; National Institutes of Health: Bethesda, MD, USA, 1978.
699. IARC, International Agency for Research on Cancer. DDT and associated compounds. Occupational exposures in insecticide application, and some pesticides. In *IARC Monographs on the Evaluation of Carcinogenic Risks to Humans*; International Agency for Research on Cancer: Lyon, France, 1991; Volume 53.
700. EFSA CONTAM Panel, European Food Safety Authority, Panel on Contaminants in the Food Chain. Opinion of the Scientific Panel on contaminants in the food chain [CONTAM] related to DDT as an undesirable substance in animal feed. *EFSA J.* **2006**, *433*, 1–69. [CrossRef]
701. Tebourbi, O.; Driss, M.R.; Sakly, M.; Rhouma, K.B. Metabolism of DDT in different tissues of young rats. *J. Environ. Sci. Health B* **2006**, *41*, 167–176. [CrossRef] [PubMed]
702. Morgan, D.P.; Roan, C.C. Absorption, storage, and metabolic conversion of ingested DDT and DDT metabolites in man. *Arch. Environ. Health* **1971**, *22*, 301–308. [CrossRef] [PubMed]
703. Nims, R.W.; Lubet, R.A.; Fox, S.D.; Jones, C.R.; Thomas, P.E.; Reddy, A.B.; Kocarek, T.A. Comparative pharmacodynamics of CYP2B induction by DDT, DDE, and DDD in male rat liver and cultured rat hepatocytes. *J. Toxicol. Environ. Health A* **1998**, *53*, 455–477. [CrossRef] [PubMed]
704. Lund, B.O.; Bergman, A.; Brandt, I. Metabolic activation and toxicity of a DDT-metabolite, 3-methylsulphonyl-DDE, in the adrenal zona fasciculata in mice. *Chem. Biol. Interact.* **1988**, *65*, 25–40. [CrossRef]
705. Williams, G.M.; Numoto, S. Promotion of mouse liver neoplasms by the organochlorine pesticides chlordane and heptachlor in comparison to dichlorodiphenyltrichlorpethane. *Carcinogenesis* **1984**, *5*, 1689–1696. [CrossRef]
706. Williams, G.M.; Telang, S.; Tong, C. Inhibition of intercellular communication between liver cells by the liver tumor promoter 1,1,1-trichloro-2,2-bis(p-chlorophenyl)ethane. *Cancer Lett.* **1981**, *11*, 339–344. [CrossRef]
707. López-Cervantes, M.; Torres-Sánchez, L.; Tobías, A.; López-Carrillo, L. Dichlorodiphenyldichloroethane burden and breast cancer risk: A meta-analysis of the epidemiologic evidence. *Environ. Health Perspect.* **2003**, *112*, 207–214. [CrossRef]
708. Gatto, N.M.; Longnecker, M.P.; Press, M.F.; Sullivan-Halley, J.; McKean-Cowdin, R.; Bernstein, L. Serum organochlorines and breast cancer: A case–control study among African-American women. *Cancer Causes Control* **2007**, *18*, 29–39. [CrossRef]
709. Mouly, T.A.; Toms, L.L. Breast cancer and persistent organic pollutants (excluding DDT): A systematic literature review. *Environ. Sci. Pollut. Res. Int.* **2016**, *23*, 22385–22407. [CrossRef] [PubMed]
710. VoPham, T.; Bertrand, K.A.; Hart, J.E.; Laden, F.; Brooks, M.M.; Yuan, J.M.; Talbott, E.O.; Ruddell, D.; Chang, C.H.; Weissfeld, J.L. Pesticide exposure and liver cancer: A review. *Cancer Causes Control* **2017**, *28*, 177–190. [CrossRef] [PubMed]
711. Baris, D.; Zahm, S.H.; Cantor, K.P.; Blair, A. Agricultural use of DDT and risk of non-Hodgkin's lymphoma: Pooled analysis of three case-control studies in the United States. *Occup. Environ. Med.* **1998**, *55*, 522–527. [CrossRef] [PubMed]
712. Kulkarni, P.S.; Crespo, J.G.; Afonso, C.A. Dioxins sources and current remediation technologies—A review. *Environ. Int.* **2008**, *34*, 139–153. [CrossRef] [PubMed]
713. Institute of Medicine, (US) Committee on the Implications of Dioxin in the Food Supply. *Dioxins and Dioxin-like Compounds in the Food Supply: Strategies to Decrease Exposure*; National Academies Press (US): Washington, DC, USA, 2003.
714. Van den Berg, M.; Birnbaum, L.; Bosveld, A.T.; Brunström, B.; Cook, P.; Feeley, M.; Giesy, J.P.; Hanberg, A.; Hasegawa, R.; Kennedy, S.W.; et al. Toxic equivalency factors (TEFs) for PCBs, PCDDs, PCDFs for humans and wildlife. *Environ. Health Perspect.* **1998**, *106*, 775–792. [CrossRef]

715. Van den Berg, M.; Birnbaum, L.S.; Denison, M.; De Vito, M.; Farland, W.; Feeley, M.; Fiedler, H.; Hakansson, H.; Hanberg, A.; Haws, L.; et al. The 2005 World Health Organization reevaluation of human and Mammalian toxic equivalency factors for dioxins and dioxin-like compounds. *Toxicol. Sci.* **2006**, *93*, 223–241. [CrossRef]

716. Schecter, A.; Birnbaum, L.; Ryan, J.J.; Constable, J.D. Dioxins: An overview. *Environ. Res.* **2006**, *101*, 419–428. [CrossRef]

717. Charnley, G.; Kimbrough, R.D. Overview of exposure, toxicity, and risks to children from current levels of 2,3,7,8-tetrachlorodibenzo-p-dioxin and related compounds in the USA. *Food Chem. Toxicol.* **2006**, *44*, 601–615. [CrossRef]

718. Scialli, A.R.; Watkins, D.K.; Ginevan, M.E. Agent Orange eposure and 2,3,7,8-tetrachlorodibenzo-p-dioxin (TCDD) in human milk. *Birth Defects Res. B Dev. Reprod. Toxicol.* **2015**, *104*, 129–139. [CrossRef]

719. Ulaszewska, M.M.; Zuccato, E.; Davoli, E. PCDD/Fs and dioxin-like PCBs in human milk and estimation of infants' daily intake: A review. *Chemosphere* **2011**, *83*, 774–782. [CrossRef] [PubMed]

720. Arisawa, K. Recent decreasing trends of exposure to PCDDs/PCDFs/dioxin-like PCBs in general populations, and associations with diabetes, metabolic syndrome, and gout/hyperuricemia. *J. Med. Investig.* **2018**, *65*, 151–161. [CrossRef] [PubMed]

721. NTP, National Toxicology Program. Toxicology and carcinogenesis studies of a mixture of 2,3,7,8-tetrachlorodibenzo-p-dioxin (TCDD) (Cas No. 1746-01-6), 2,3,4,7,8-pentachlorodibenzofuran (PeCDF) (Cas No. 57117-31-4), and 3,3',4,4',5-pentachlorobiphenyl (PCB 126) (Cas No. 57465-28-8) in female Harlan Sprague-Dawley rats (gavage studies). *Natl. Toxicol. Program. Tech. Rep. Ser.* **2006**, *526*, 1–180.

722. NTP, National Toxicology Program. Toxicology and carcinogenesis studies of 3,3',4,4',5-pentachlorobiphenyl (PCB 126) (CAS No. 57465-28-8) in female Harlan Sprague-Dawley rats (Gavage Studies). *Natl. Toxicol. Program. Tech. Rep. Ser.* **2006**, *520*, 4–246.

723. NTP, National Toxicology Program. Technical report on the toxicology and carcinogenesis studies of 2,3,7,8-tetrachlorodibenzo-p-dioxin (TCDD) (CAS No. 1746-01-6) in female Harlan Sprague-Dawley rats (Gavage Studies). *Natl. Toxicol. Program. Tech. Rep. Ser.* **2006**, *521*, 4–232.

724. Knerr, S.; Schrenk, D. Carcinogenicity of "non-dioxinlike" polychlorinated biphenyls. *Crit. Rev. Toxicol.* **2006**, *36*, 663–694. [CrossRef]

725. Huff, J.E.; Salmon, A.G.; Hooper, N.K.; Zeise, L. Long-term carcinogenesis studies on 2,3,7,8-tetrachlorodibenzo-p-dioxin and hexachlorodibenzo-p-dioxins. *Cell Biol. Toxicol.* **1991**, *7*, 67–94. [CrossRef] [PubMed]

726. Turteltaub, K.W.; Felton, J.S.; Gledhill, B.L.; Vogel, J.S.; Southon, J.R.; Caffee, M.W.; Finkel, R.C.; Nelson, D.E.; Proctor, I.D.; Davis, J.C. Accelerator mass spectrometry in biomedical dosimetry: Relationship between low-level exposure and covalent binding of heterocyclic amine carcinogens to DNA. *Proc. Natl. Acad. Sci. USA* **1990**, *87*, 5288–5292. [CrossRef]

727. Randerath, K.; Putman, K.L.; Randerath, E.; Mason, G.; Kelley, M.; Safe, S. Organ-specific effects of long term feeding of 2,3,7,8-tetrachlorodibenzo-p-dioxin and 1,2,3,7,8-pentachlorodibenzo-p-dioxin on I-compounds in hepatic and renal DNA of female Sprague-Dawley rats. *Carcinogenesis* **1988**, *9*, 2285–2289. [CrossRef]

728. Dragan, Y.P.; Schrenk, D. Animal studies addressing the carcinogenicity of TCDD (or related compounds) with an emphasis on tumour promotion. *Food Addit. Contam.* **2000**, *17*, 289–302. [CrossRef]

729. Whysner, J.; Montandon, F.; McClain, R.M.; Downing, J.; Verna, L.K.; Steward, R.E.; Williams, G.M. Absence of DNA adduct formation by phenobarbital, polychlorinated biphenyls, and chlordane in mouse liver using the 32P-postlabeling assay. *Toxicol. Appl. Pharmacol.* **1998**, *148*, 14–23. [CrossRef] [PubMed]

730. Schilderman, P.A.; Maas, L.M.; Pachen, D.M.; de Kok, T.M.; Kleinjans, J.C.; van Schooten, F.J. Induction of DNA adducts by several polychlorinated biphenyls. *Environ. Mol. Mutagen.* **2000**, *36*, 79–86. [CrossRef]

731. Pohjanvirta, R.; Vartiainen, T.; Uusi-Rauva, A.; Mönkkönen, J.; Tuomisto, J. Tissue distribution, metabolism, and excretion of 14C-TCDD in a TCDD-susceptible and a TCDD-resistant rat strain. *Pharm. Toxicol.* **1990**, *66*, 93–100. [CrossRef] [PubMed]

732. Olson, J.R. Metabolism and disposition of 2,3,7,8-tetrachlorodibenzo-p-dioxin in guinea pigs. *Toxicol. Appl. Pharm.* **1986**, *85*, 263–273. [CrossRef]

733. Kahn, P.C.; Gochfeld, M.; Nygren, M.; Hansson, M.; Rappe, C.; Velez, H.; Ghent-Guenther, T.; Wilson, W.P. Dioxins and dibenzofurans in blood and adipose tissue of Agent Orange-exposed Vietnam veterans and matched controls. *JAMA* **1988**, *259*, 1661–1667. [CrossRef] [PubMed]

734. Diliberto, J.J.; DeVito, M.J.; Ross, D.G.; Birnbaum, L.S. Subchronic exposure of [3H]- 2,3,7,8-tetrachlorodibenzo-p-dioxin (TCDD) in female B6C3F1 mice: Relationship of steady-state levels to disposition and metabolism. *Toxicol. Sci.* **2001**, *61*, 241–255. [CrossRef]

735. Gasiewicz, T.A.; Geiger, L.E.; Rucci, G.; Neal, R.A. Distribution, excretion, and metabolism of 2,3,7,8-tetrachlorodibenzo-p-dioxin in C57BL/6J, DBA/2J, and B6D2F1/J mice. *Drug Metab. Dispos.* **1983**, *11*, 397–403.

736. Inui, H.; Itoh, T.; Yamamoto, K.; Ikushiro, S.; Sakaki, T. Mammalian cytochrome P450-dependent metabolism of polychlorinated dibenzo-p-dioxins and coplanar polychlorinated biphenyls. *Int. J. Mol. Sci.* **2014**, *15*, 14044–14057. [CrossRef]

737. Wroblewski, V.J.; Olson, J.R. Hepatic metabolism of 2,3,7,8-tetrachlorodibenzo-p-dioxin (TCDD) in the rat and guinea pig. *Toxicol. Appl. Pharm.* **1985**, *81*, 231–240. [CrossRef]

738. Whysner, J.; Williams, G.M. 2,3,7,8-Tetrachlorodibenzo-p-dioxin mechanistic data and risk assessment: Gene regulation, cytotoxicity, enhanced cell proliferation, and tumor promotion. *Pharmacol. Ther.* **1996**, *71*, 193–223. [CrossRef]

739. Patrizi, B.; Siciliani de Cumis, M. TCDD toxicity mediated by epigenetic mechanisms. *Int. J. Mol. Sci.* **2018**, *19*, 4101. [CrossRef]

740. Knerr, S.; Schrenk, D. Carcinogenicity of 2,3,7,8-tetrachlorodibenzo-p-dioxin in experimental models. *Mol. Nutr. Food Res.* **2006**, *50*, 897–907. [CrossRef] [PubMed]

741. Lucier, G.; Clark, G.; Hiermath, C.; Tritscher, A.; Sewall, C.; Huff, J. Carcinogenicity of TCDD in laboratory animals: Implications for risk assessment. *Toxicol. Ind. Health* **1993**, *9*, 631–668. [CrossRef]

742. Schecter, A.; Startin, J.; Wright, C.; Kelly, M.; Päpke, O.; Lis, A.; Ball, M.; Olson, J. Dioxins in U.S. food and estimated daily intake. *Chemosphere* **1994**, *29*, 2261–2265. [CrossRef]

743. González, N.; Domingo, J.L. Polychlorinated dibenzo-p-dioxins and dibenzofurans (PCDD/Fs) in food and human dietary intake: An update of the scientific literature. *Food Chem. Toxicol.* **2021**, *157*, 112585. [CrossRef] [PubMed]

744. Danjou, A.M.; Fervers, B.; Boutron-Ruault, M.C.; Philip, T.; Clavel-Chapelon, F.; Dossus, L. Estimated dietary dioxin exposure and breast cancer risk among women from the French E3N prospective cohort. *Breast Cancer Res.* **2015**, *17*, 39. [CrossRef]

745. Boffetta, P.; Mundt, K.A.; Adami, H.O.; Cole, P.; Mandel, J.S. TCDD and cancer: A critical review of epidemiologic studies. *Crit Rev. Toxicol.* **2011**, *41*, 622–636. [CrossRef]

746. Cole, P.; Trichopoulos, D.; Pastides, H.; Starr, T.; Mandel, J.S. Dioxin and cancer: A critical review. *Regul. Toxicol. Pharm.* **2003**, *38*, 378–388. [CrossRef]

747. WHO, World Health Organization. *Assessment of the Health Risk of Dioxins: Re-Evaluation of the Tolerable Daily Intake (TDI)*; Executive Summary; WHO Regional Office for Europe: Copenhagen, Denmark, 1998.

748. van Leeuwen, F.X.; Feeley, M.; Schrenk, D.; Larsen, J.C.; Farland, W.; Younes, M. Dioxins: WHO's tolerable daily intake (TDI) revisited. *Chemosphere* **2000**, *40*, 1095–1101. [CrossRef]

749. SCF, Scientific Committee on Food. *Opinion of the Scientific Committee on Food on the Risk Assessment of Dioxins and Dioxin-like PCBs in Food*; Adopted on 30 May 2001; Scientific Committee on Food: Brussels, Belgium, 2001.

750. EPA, US Environmental Protection Agency. *Dioxin Reassessment—An SAB Review of the Office of Research and Development's Reassessment of Dioxin: Review of the Revised Sections (Dose Response Modeling, Integrated Summary, Risk Characterization, and Toxicity Equivalence Factors) of the EPA's Reassessment of Dioxin by the Dioxin Reassessment Review Subcommittee of the EPA Science Advisory Board (SAB)*; EPA-SAB-EC-01–006; US Environmental Protection Agency: Washington, DC, USA, 2001.

751. Starr, T.B. Significant shortcomings of the U.S. Environmental Protection Agency's latest draft risk characterization for dioxin-like compounds. *Toxicol. Sci.* **2001**, *64*, 7–13. [CrossRef]

752. Paustenbach, D.J. The U.S. EPA Science Advisory Board Evaluation (2001) of the EPA dioxin reassessment. *Regul. Toxicol. Pharm.* **2002**, *36*, 211–219. [CrossRef] [PubMed]

753. Popp, J.A.; Crouch, E.; McConnell, E.E. A Weight-of-evidence analysis of the cancer dose-response characteristics of 2,3,7,8-tetrachlorodibenzodioxin (TCDD). *Toxicol. Sci.* **2006**, *89*, 361–369. [CrossRef] [PubMed]

754. NRC, National Research Council. *Health Risks from Dioxin and Related Compounds*; Evaluation of EPA Reassessment; The National Academies Press: Washington, DC, USA, 2006. [CrossRef]

755. EFSA CEP Panel, European Food Safety Authority, Panel on Food Contact Materials, Enzymes and Processing Aids. Scientific Opinion on safety of benzophenone to be used as flavouring. *EFSA J.* **2017**, *15*, 5013.

756. Adams, T.B.; McGowen, M.M.; Williams, M.C.; Cohen, S.M.; Feron, V.J.; Goodman, J.I.; Marnett, L.J.; Munro, I.C.; Portoghese, P.S.; Smith, R.L.; et al. The FEMA GRAS assessment of aromatic substituted secondary alcohols, ketones, and related esters used as flavor ingredients. *Food Chem. Toxicol.* **2007**, *45*, 171–201. [CrossRef] [PubMed]

757. Hu, L.; Tian, M.; Feng, W.; He, H.; Wang, Y.; Yang, L. Sensitive detection of benzophenone-type ultraviolet filters in plastic food packaging materials by sheathless capillary electrophoresis-electrospray ionization-tandem mass spectrometry. *J. Chromatogr. A* **2019**, *1604*, 460469. [CrossRef] [PubMed]

758. Anderson, W.A.; Castle, L. Benzophenone in cartonboard packaging materials and the factors that influence its migration into food. *Food Addit. Contam.* **2003**, *20*, 607–618. [CrossRef] [PubMed]

759. Castle, L.; Damant, A.P.; Honeybone, C.A.; Johns, S.M.; Jickells, S.M.; Sharman, M.; Gilbert, J. Migration studies from paper and board food packaging materials. Part 2. Survey for residues of dialkylamino benzophenone UV-cure ink photoinitiators. *Food Addit. Contam.* **1997**, *14*, 45–52. [CrossRef] [PubMed]

760. Jung, T.; Simat, T.J.; Altkofer, W.; Fügel, D. Survey on the occurrence of photo-initiators and amine synergists in cartonboard packaging on the German market and their migration into the packaged foodstuffs. *Food Addit. Contam. Part A Chem. Anal. Control Expo. Risk Assess.* **2013**, *30*, 1993–2016. [CrossRef]

761. Rhodes, M.C.; Bucher, J.R.; Peckham, J.C.; Kissling, G.E.; Hejtmancik, M.R.; Chhabra, R.S. Carcinogenesis studies of benzophenone in rats and mice. *Food Chem. Toxicol.* **2007**, *45*, 843–851. [CrossRef]

762. NTP, National Toxicology Program. *Technical Report on the Toxicity Studies of Benzophenone (CAS No. 119-61-9) Administered in Feed to F344/N Rats and B6C3F1 Mice 0888-8051 (Print)*; National Toxicology Program: Research Triangle, NC, USA, 2000; pp. 1–412.

763. NTP, National Toxicology Program. Toxicology and carcinogenesis studies of benzophenone (CAS No. 119-61-9) in F344/N rats and B6C3F1 mice (feed studies). *Natl. Toxicol. Program. Tech. Rep. Ser.* **2006**, *533*, 1–264.

764. JECFA, Joint FAO/WHO Expert Committee on Food Additives. *Safety Evaluation of Certain Food Additives and Contaminants*; Prepared by the Seventy-Third Meeting of the Joint FAO/WHO Expert Committee on Food Additives (JECFA); WHO: Geneva, Switzerland, 2011.

765. Abramsson-Zetterberg, L.; Svensson, K. 4-Methylbenzophenone and benzophenone are inactive in the micronucleus assay. *Toxicol. Lett.* **2011**, *201*, 235–239. [CrossRef] [PubMed]

766. Takemoto, K.; Yamazaki, H.; Nakajima, M.; Yokoi, T. Genotoxic activation of benzophenone and its two metabolites by human cytochrome P450s in SOS/umu assay. *Mutat. Res.* **2002**, *519*, 199–204. [CrossRef]

767. Cuquerella, M.C.; Lhiaubet-Vallet, V.; Cadet, J.; Miranda, M.A. Benzophenone photosensitized DNA damage. *Acc. Chem. Res.* **2012**, *45*, 1558–1570. [CrossRef] [PubMed]

768. Dumont, E.; Wibowo, M.; Roca-Sanjuán, D.; Garavelli, M.; Assfeld, X.; Monari, A. Resolving the benzophenone DNA-photosensitization mechanism at QM/MM level. *J. Phys. Chem. Lett.* **2015**, *6*, 576–580. [CrossRef]

769. Nakagawa, Y.; Suzuki, T.; Tayama, S. Metabolism and toxicity of benzophenone in isolated rat hepatocytes and estrogenic activity of its metabolites in MCF-7 cells. *Toxicology* **2000**, *156*, 27–36. [CrossRef]

770. Jeon, H.K.; Sarma, S.N.; Kim, Y.J.; Ryu, J.C. Toxicokinetics and metabolisms of benzophenone-type UV filters in rats. *Toxicology* **2008**, *248*, 89–95. [CrossRef]

771. Nakagawa, Y.; Tayama, K. Benzophenone-induced estrogenic potency in ovariectomized rats. *Arch. Toxicol.* **2002**, *76*, 727–731. [CrossRef]

772. Nakagawa, Y.; Tayama, K. Estrogenic potency of benzophenone and its metabolites in juvenile female rats. *Arch. Toxicol.* **2001**, *75*, 74–79. [CrossRef]

773. Suzuki, T.; Kitamura, S.; Khota, R.; Sugihara, K.; Fujimoto, N.; Ohta, S. Estrogenic and antiandrogenic activities of 17 benzophenone derivatives used as UV stabilizers and sunscreens. *Toxicol. Appl. Pharm.* **2005**, *203*, 9–17. [CrossRef]

774. Chen, M.L.; Chen, C.H.; Huang, Y.F.; Chen, H.C.; Chang, J.W. Cumulative dietary risk assessment of benzophenone-type photoinitiators from packaged foodstuffs. *Foods* **2022**, *11*, 152. [CrossRef]

775. Erythropel, H.C.; Maric, M.; Nicell, J.A.; Leask, R.L.; Yargeau, V. Leaching of the plasticizer di(2-ethylhexyl)phthalate (DEHP) from plastic containers and the question of human exposure. *Appl. Microbiol. Biotechnol.* **2014**, *98*, 9967–9981. [CrossRef] [PubMed]

776. Serrano, S.E.; Braun, J.; Trasande, L.; Dills, R.; Sathyanarayana, S. Phthalates and diet: A review of the food monitoring and epidemiology data. *Environ. Health* **2014**, *13*, 43. [CrossRef] [PubMed]

777. ATSDR, Agency for Toxic Substances and Disease Registry. *Toxicological Profile for di(2-Ethylhexyl)Phthalate (DEHP)*; Agency for Toxic Substances and Disease Registry: Atlanta, GA, USA, 2022.

778. Fierens, T.; Servaes, K.; Van Holderbeke, M.; Geerts, L.; De Henauw, S.; Sioen, I.; Vanermen, G. Analysis of phthalates in food products and packaging materials sold on the Belgian market. *Food Chem. Toxicol.* **2012**, *50*, 2575–2583. [CrossRef] [PubMed]

779. Sharman, M.; Read, W.A.; Castle, L.; Gilbert, J. Levels of di-(2-ethylhexyl)phthalate and total phthalate esters in milk, cream, butter and cheese. *Food Addit. Contam.* **1994**, *11*, 375–385. [CrossRef] [PubMed]

780. EFSA CEP Panel, European Food Safety Authority, Panel on Food Contact Materials, Enzymes and Processing Aids. Scientific Opinion on the update of the risk assessment of di-butylphthalate(DBP), butyl-benzyl-phthalate (BBP), bis(2-ethylhexyl)phthalate (DEHP), di-isononylphthalate (DINP) anddi-isodecylphthalate (DIDP) for use in food contact materials. *EFSA J.* **2019**, *17*, 5838.

781. Bosgra, S.; Bos, P.M.; Vermeire, T.G.; Luit, R.J.; Slob, W. Probabilistic risk characterization: An example with di(2-ethylhexyl) phthalate. *Regul. Toxicol. Pharm.* **2005**, *43*, 104–113. [CrossRef] [PubMed]

782. Khedr, A. Optimized extraction method for LC-MS determination of bisphenol A, melamine and di(2-ethylhexyl) phthalate in selected soft drinks, syringes, and milk powder. *J. Chromatogr. B Anal. Technol. Biomed. Life Sci.* **2013**, *930*, 98–103. [CrossRef]

783. NTP, National Toxicology Program. Carcinogenesis bioassay of di(2-ethylhexyl)phthalate (CAS No. 117-81-7) in F344 rats and B6C3F1 mice (feed studies). *Natl. Toxicol. Program. Tech. Rep. Ser.* **1982**, *217*, 1–127.

784. NTP, National Toxicology Program. Toxicology and carcinogenesis studies of di(2-ethylhexyl) phthalate administered in feed to Sprague Dawley (Hsd:Sprague Dawley SD) rats. *Natl. Toxicol. Program. Tech. Rep. Ser.* **2021**, *601*, NTP-TR-601. [CrossRef]

785. Voss, C.; Zerban, H.; Bannasch, P.; Berger, M.R. Lifelong exposure to di-(2-ethylhexyl)-phthalate induces tumors in liver and testes of Sprague-Dawley rats. *Toxicology* **2005**, *206*, 359–371. [CrossRef]

786. Ito, Y.; Yamanoshita, O.; Asaeda, N.; Tagawa, Y.; Lee, C.H.; Aoyama, T.; Ichihara, G.; Furuhashi, K.; Kamijima, M.; Gonzalez, F.J.; et al. Di(2-ethylhexyl)phthalate induces hepatic tumorigenesis through a peroxisome proliferator-activated receptor alpha-independent pathway. *J. Occup. Health* **2007**, *49*, 172–182. [CrossRef] [PubMed]

787. Doull, J.; Cattley, R.; Elcombe, C.; Lake, B.G.; Swenberg, J.; Wilkinson, C.; Williams, G.; van Gemert, M. A cancer risk assessment of di(2-ethylhexyl)phthalate: Application of the new U.S. EPA Risk Assessment Guidelines. *Regul. Toxicol. Pharm.* **1999**, *29*, 327–357. [CrossRef] [PubMed]

788. Ward, J.M.; Ohshima, M.; Lynch, P.; Riggs, C. Di(2-ethylhexyl)phthalate but not phenobarbital promotes N-nitrosodiethylamine-initiated hepatocellular proliferative lesions after short-term exposure in male B6C3F1 mice. *Cancer Lett.* **1984**, *24*, 49–55. [CrossRef]

789. Diwan, B.A.; Ward, J.M.; Rice, J.M.; Colburn, N.H.; Spangler, E.F. Tumor-promoting effects of di(2-ethylhexyl)phthalate in JB6 mouse epidermal cells and mouse skin. *Carcinogenesis* **1985**, *6*, 343–347. [CrossRef] [PubMed]

790. Butterworth, B.E.; Bermudez, E.; Smith-Oliver, T.; Earle, L.; Cattley, R.; Martin, J.; Popp, J.A.; Strom, S.; Jirtle, R.; Michalopoulos, G. Lack of genotoxic activity of di(2-ethylhexyl)phthalate (DEHP) in rat and human hepatocytes. *Carcinogenesis* **1984**, *5*, 1329–1335. [CrossRef] [PubMed]

791. Caldwell, J.C. DEHP: Genotoxicity and potential carcinogenic mechanisms-a review. *Mutat. Res.* **2012**, *751*, 82–157. [CrossRef]

792. Erkekoglu, P.; Kocer-Gumusel, B. Genotoxicity of phthalates. *Toxicol. Mech. Methods* **2014**, *24*, 616–626. [CrossRef] [PubMed]

793. Karabulut, G.; Barlas, N. Genotoxic, histologic, immunohistochemical, morphometric and hormonal effects of di-(2-ethylhexyl)-phthalate (DEHP) on reproductive systems in pre-pubertal male rats. *Toxicol. Res.* **2018**, *7*, 859–873. [CrossRef]

794. Wang, X.; Jiang, L.; Ge, L.; Chen, M.; Yang, G.; Ji, F.; Zhong, L.; Guan, Y.; Liu, X. Oxidative DNA damage induced by di-(2-ethylhexyl) phthalate in HEK-293 cell line. *Environ. Toxicol. Pharm.* **2015**, *39*, 1099–1106. [CrossRef]

795. Erkekoglu, P.; Rachidi, W.; Yuzugullu, O.G.; Giray, B.; Favier, A.; Ozturk, M.; Hincal, F. Evaluation of cytotoxicity and oxidative DNA damaging effects of di(2-ethylhexyl)-phthalate (DEHP) and mono(2-ethylhexyl)-phthalate (MEHP) on MA-10 Leydig cells and protection by selenium. *Toxicol. Appl. Pharm.* **2010**, *248*, 52–62. [CrossRef]
796. She, Y.; Jiang, L.; Zheng, L.; Zuo, H.; Chen, M.; Sun, X.; Li, Q.; Geng, C.; Yang, G.; Jiang, L.; et al. The role of oxidative stress in DNA damage in pancreatic β cells induced by di-(2-ethylhexyl) phthalate. *Chem. Biol. Interact.* **2017**, *265*, 8–15. [CrossRef] [PubMed]
797. Takagi, A.; Sai, K.; Umemura, T.; Hasegawa, R.; Kurokawa, Y. Significant increase of 8-hydroxydeoxyguanosine in liver DNA of rats following short-term exposure to the peroxisome proliferators di(2-ethylhexyl)phthalate and di(2-ethylhexyl)adipate. *Jpn. J. Cancer Res.* **1990**, *81*, 213–215. [CrossRef] [PubMed]
798. Takagi, A.; Sai, K.; Umemura, T.; Hasegawa, R.; Kurokawa, Y. Relationship between hepatic peroxisome proliferation and 8-hydroxydeoxyguanosine formation in liver DNA of rats following long-term exposure to three peroxisome proliferators; di(2-ethylhexyl) phthalate, aluminium clofibrate and simfibrate. *Cancer Lett.* **1990**, *53*, 33–38. [CrossRef]
799. von Däniken, A.; Lutz, W.K.; Jäckh, R.; Schlatter, C. Investigation of the potential for binding of Di(2-ethylhexyl) phthalate (DEHP) and Di(2-ethylhexyl) adipate (DEHA) to liver DNA in vivo. *Toxicol. Appl. Pharm.* **1984**, *73*, 373–387. [CrossRef]
800. Lutz, W.K. Investigation of the potential for binding of di(2-ethylhexyl) phthalate (DEHP) to rat liver DNA in vivo. *Environ. Health Perspect.* **1986**, *65*, 267–269. [CrossRef]
801. Gupta, R.C.; Goel, S.K.; Earley, K.; Singh, B.; Reddy, J.K. 32P-postlabeling analysis of peroxisome proliferator-DNA adduct formation in rat liver in vivo and hepatocytes in vitro. *Carcinogenesis* **1985**, *6*, 933–936. [CrossRef]
802. Koch, H.M.; Preuss, R.; Angerer, J. Di(2-ethylhexyl)phthalate (DEHP): Human metabolism and internal exposure—An update and latest results. *Int. J.* **2006**, *29*, 155–165, discussion 181-155. [CrossRef]
803. Silva, M.J.; Samandar, E.; Preau, J.L., Jr.; Needham, L.L.; Calafat, A.M. Urinary oxidative metabolites of di(2-ethylhexyl) phthalate in humans. *Toxicology* **2006**, *219*, 22–32. [CrossRef]
804. Frederiksen, H.; Skakkebaek, N.E.; Andersson, A.M. Metabolism of phthalates in humans. *Mol. Nutr. Food Res.* **2007**, *51*, 899–911. [CrossRef]
805. Ito, Y.; Yokota, H.; Wang, R.; Yamanoshita, O.; Ichihara, G.; Wang, H.; Kurata, Y.; Takagi, K.; Nakajima, T. Species differences in the metabolism of di(2-ethylhexyl) phthalate (DEHP) in several organs of mice, rats, and marmosets. *Arch. Toxicol.* **2005**, *79*, 147–154. [CrossRef]
806. Ito, Y.; Kamijima, M.; Hasegawa, C.; Tagawa, M.; Kawai, T.; Miyake, M.; Hayashi, Y.; Naito, H.; Nakajima, T. Species and inter-individual differences in metabolic capacity of di(2-ethylhexyl)phthalate (DEHP) between human and mouse livers. *Environ. Health Prev. Med.* **2014**, *19*, 117–125. [CrossRef] [PubMed]
807. Ito, Y.; Kamijima, M.; Nakajima, T. Di(2-ethylhexyl) phthalate-induced toxicity and peroxisome proliferator-activated receptor alpha: A review. *Environ. Health Prev. Med.* **2019**, *24*, 47. [CrossRef] [PubMed]
808. Melnick, R.L. Is peroxisome proliferation an obligatory precursor step in the carcinogenicity of di(2-ethylhexyl)phthalate (DEHP)? *Environ. Health Perspect.* **2001**, *109*, 437–442. [CrossRef]
809. Corton, J.C.; Peters, J.M.; Klaunig, J.E. The PPARα-dependent rodent liver tumor response is not relevant to humans: Addressing misconceptions. *Arch. Toxicol.* **2018**, *92*, 83–119. [CrossRef]
810. Rusyn, I.; Corton, J.C. Mechanistic considerations for human relevance of cancer hazard of di(2-ethylhexyl) phthalate. *Mutat. Res.* **2012**, *750*, 141–158. [CrossRef]
811. Shelby, M.D. NTP-CERHR monograph on the potential human reproductive and developmental effects of di (2-ethylhexyl) phthalate (DEHP). *Ntp. Cerhr. Mon.* **2006**, *5*, 1–13.
812. Schecter, A.; Lorber, M.; Guo, Y.; Wu, Q.; Yun, S.H.; Kannan, K.; Hommel, M.; Imran, N.; Hynan, L.S.; Cheng, D.; et al. Phthalate concentrations and dietary exposure from food purchased in New York State. *Environ. Health Perspect.* **2013**, *121*, 473–494. [CrossRef]
813. Qu, J.; Xia, W.; Qian, X.; Wu, Y.; Li, J.; Wen, S.; Xu, S. Geographic distribution and time trend of human exposure of Di(2-ethylhexyl) phthalate among different age groups based on global biomonitoring data. *Chemosphere* **2022**, *287*, 132115. [CrossRef] [PubMed]
814. Morgan, M.; Deoraj, A.; Felty, Q.; Roy, D. Environmental estrogen-like endocrine disrupting chemicals and breast cancer. *Mol. Cell Endocrinol.* **2017**, *457*, 89–102. [CrossRef] [PubMed]
815. Stickney, J.A.; Sager, S.L.; Clarkson, J.R.; Smith, L.A.; Locey, B.J.; Bock, M.J.; Hartung, R.; Olp, S.F. An updated evaluation of the carcinogenic potential of 1,4-dioxane. *Regul. Toxicol. Pharm.* **2003**, *38*, 183–195. [CrossRef]
816. ATSDR, Agency for Toxic Substances and Disease Registry. *Toxicological Profile for 1,4-Dioxane*; Agency for Toxic Substances and Disease Registry: Atlanta, GA, USA, 2012.
817. SCF, Scientific Committee on Food. *Opinion of the Scientific Committee on Food on Impurities of 1,4-Dioxane, 2-Chloroethanol and Mono- and Diethylene Glycol in Currently Permitted Food Additives and in Proposed Use of Ethyl Hydroxyethyl Cellulose in Gluten-Free Bread*; SCF/CS/ADD/EMU/198 Final; Scientific Committee on Food: Brussels, Belgium, 2002.
818. Nishimura, T.; Iizuka, S.; Kibune, N.; Ando, M. Study of 1,4-dioxane intake in the total diet using the market-basket method. *J. Health Sci.* **2004**, *50*, 101–107. [CrossRef]
819. NCI, National Cancer Institute. Bioassay of 1,4-dioxane for possible carcinogenicity. *Natl. Cancer Inst. Carcinog. Tech. Rep. Ser.* **1978**, *80*, 1–123.

820. Kano, H.; Umeda, Y.; Kasai, T.; Sasaki, T.; Matsumoto, M.; Yamazaki, K.; Nagano, K.; Arito, H.; Fukushima, S. Carcinogenicity studies of 1,4-dioxane administered in drinking-water to rats and mice for 2 years. *Food Chem. Toxicol.* **2009**, *47*, 2776–2784. [CrossRef]

821. Kano, H.; Umeda, Y.; Saito, M.; Senoh, H.; Ohbayashi, H.; Aiso, S.; Yamazaki, K.; Nagano, K.; Fukushima, S. Thirteen-week oral toxicity of 1,4-dioxane in rats and mice. *J. Toxicol. Sci.* **2008**, *33*, 141–153. [CrossRef]

822. Lundberg, I.; Högberg, J.; Kronevi, T.; Holmberg, B. Three industrial solvents investigated for tumor promoting activity in the rat liver. *Cancer Lett.* **1987**, *36*, 29–33. [CrossRef]

823. Rosenkranz, H.S.; Klopman, G. 1,4-Dioxane: Prediction of in vivo clastogenicity. *Mutat. Res.* **1992**, *280*, 245–251. [CrossRef]

824. Mirkova, E.T. Activity of the rodent carcinogen 1,4-dioxane in the mouse bone marrow micronucleus assay. *Mutat. Res.* **1994**, *322*, 142–144. [CrossRef]

825. Kitchin, K.T.; Brown, J.L. Is 1,4-dioxane a genotoxic carcinogen? *Cancer Lett.* **1990**, *53*, 67–71. [CrossRef]

826. Morita, T.; Hayashi, M. 1,4-Dioxane is not mutagenic in five in vitro assays and mouse peripheral blood micronucleus assay, but is in mouse liver micronucleus assay. *Environ. Mol. Mutagen.* **1998**, *32*, 269–280. [CrossRef]

827. Itoh, S.; Hattori, C. In vivo genotoxicity of 1,4-dioxane evaluated by liver and bone marrow micronucleus tests and Pig-a assay in rats. *Mutat. Res. Genet. Toxicol. Environ. Mutagen.* **2019**, *837*, 8–14. [CrossRef]

828. Gi, M.; Fujioka, M.; Kakehashi, A.; Okuno, T.; Masumura, K.; Nohmi, T.; Matsumoto, M.; Omori, M.; Wanibuchi, H.; Fukushima, S. In vivo positive mutagenicity of 1,4-dioxane and quantitative analysis of its mutagenicity and carcinogenicity in rats. *Arch. Toxicol.* **2018**, *92*, 3207–3221. [CrossRef]

829. Charkoftaki, G.; Golla, J.P.; Santos-Neto, A.; Orlicky, D.J.; Garcia-Milian, R.; Chen, Y.; Rattray, N.J.W.; Cai, Y.; Wang, Y.; Shearn, C.T.; et al. Identification of dose-dependent DNA damage and repair responses from subchronic exposure to 1,4-dioxane in mice using a systems analysis approach. *Toxicol. Sci.* **2021**, *183*, 338–351. [CrossRef] [PubMed]

830. Roy, S.K.; Thilagar, A.K.; Eastmond, D.A. Chromosome breakage is primarily responsible for the micronuclei induced by 1,4-dioxane in the bone marrow and liver of young CD-1 mice. *Mutat. Res.* **2005**, *586*, 28–37. [CrossRef] [PubMed]

831. Totsuka, Y.; Maesako, Y.; Ono, H.; Nagai, M.; Kato, M.; Gi, M.; Wanibuchi, H.; Fukushima, S.; Shiizaki, K.; Nakagama, H. Comprehensive analysis of DNA adducts (DNA adductome analysis) in the liver of rats treated with 1,4-dioxane. *Proc. Jpn. Acad. Ser. B* **2020**, *96*, 180–187. [CrossRef] [PubMed]

832. Göen, T.; von Helden, F.; Eckert, E.; Knecht, U.; Drexler, H.; Walter, D. Metabolism and toxicokinetics of 1,4-dioxane in humans after inhalational exposure at rest and under physical stress. *Arch. Toxicol.* **2016**, *90*, 1315–1324. [CrossRef]

833. Nannelli, A.; De Rubertis, A.; Longo, V.; Gervasi, P.G. Effects of dioxane on cytochrome P450 enzymes in liver, kidney, lung and nasal mucosa of rat. *Arch. Toxicol.* **2005**, *79*, 74–82. [CrossRef]

834. Lafranconi, M.; Budinsky, R.; Corey, L.; Klapacz, J.; Crissman, J.; LeBaron, M.; Golden, R.; Pleus, R. A 90-day drinking water study in mice to characterize early events in the cancer mode of action of 1,4-dioxane. *Regul. Toxicol. Pharm.* **2021**, *119*, 104819. [CrossRef]

835. Dourson, M.L.; Higginbotham, J.; Crum, J.; Burleigh-Flayer, H.; Nance, P.; Forsberg, N.D.; Lafranconi, M.; Reichard, J. Update: Mode of action (MOA) for liver tumors induced by oral exposure to 1,4-dioxane. *Regul Toxicol Pharm.* **2017**, *88*, 45–55. [CrossRef]

836. Dourson, M.; Reichard, J.; Nance, P.; Burleigh-Flayer, H.; Parker, A.; Vincent, M.; McConnell, E.E. Mode of action analysis for liver tumors from oral 1,4-dioxane exposures and evidence-based dose response assessment. *Regul. Toxicol. Pharm.* **2014**, *68*, 387–401. [CrossRef]

837. Chappell, G.A.; Heintz, M.M.; Haws, L.C. Transcriptomic analyses of livers from mice exposed to 1,4-dioxane for up to 90 days to assess potential mode(s) of action underlying liver tumor development. *Curr. Res. Toxicol.* **2021**, *2*, 30–41. [CrossRef]

838. Chen, Y.; Wang, Y.; Charkoftaki, G.; Orlicky, D.J.; Davidson, E.; Wan, F.; Ginsberg, G.; Thompson, D.C.; Vasiliou, V. Oxidative stress and genotoxicity in 1,4-dioxane liver toxicity as evidenced in a mouse model of glutathione deficiency. *Sci. Total Environ.* **2022**, *806*, 150703. [CrossRef]

839. Goldsworthy, T.L.; Monticello, T.M.; Morgan, K.T.; Bermudez, E.; Wilson, D.M.; Jäckh, R.; Butterworth, B.E. Examination of potential mechanisms of carcinogenicity of 1,4-dioxane in rat nasal epithelial cells and hepatocytes. *Arch. Toxicol.* **1991**, *65*, 1–9. [CrossRef]

840. FDA, Food and Drug Administration. *Indirect Food Additives: Adhesives and Components of Coatings*; Food and Drug Administration: Silver Spring, MD, USA, 1998; pp. 350–351.

841. Nishimura, T.; Iizuka, S.; Kibune, N.; Ando, M.; Magara, Y. Study of 1,4-dioxane intake in the total diet. *J. Health Sci.* **2005**, *51*, 514–517. [CrossRef]

842. EFSA CONTAM Panel, European Food Safety Authority, Panel on Contaminants in the Food Chain. Scientific Opinion on the evaluation of the substances currently on the list in the annex to Commission Directive 96/3/EC as acceptable previous cargoes for edible fats and oils—Part II of III. *EFSA J.* **2012**, *10*, 2703. [CrossRef]

843. Johnson, W., Jr. Safety assessment of MIBK (methyl isobutyl ketone). *Int. J. Toxicol.* **2004**, *23* (Suppl. S1), 29–57. [CrossRef] [PubMed]

844. Api, A.M.; Belsito, D.; Botelho, D.; Bruze, M.; Burton, G.A., Jr.; Buschmann, J.; Dagli, M.L.; Date, M.; Dekant, W.; Deodhar, C.; et al. RIFM fragrance ingredient safety assessment, 4-methyl-2-pentanone, CAS Registry Number 108-10-1. *Food Chem. Toxicol.* **2019**, *130* (Suppl. S1), 110587. [CrossRef]

845. Stout, M.D.; Herbert, R.A.; Kissling, G.E.; Suarez, F.; Roycroft, J.H.; Chhabra, R.S.; Bucher, J.R. Toxicity and carcinogenicity of methyl isobutyl ketone in F344N rats and B6C3F1 mice following 2-year inhalation exposure. *Toxicology* **2008**, *244*, 209–219. [CrossRef]

846. NTP, National Toxicology Program. Toxicology and carcinogenesis studies of methyl isobutyl ketone (Cas No. 108-10-1) in F344/N rats and B6C3F1 mice (inhalation studies). *Natl. Toxicol. Program. Tech. Rep. Ser.* **2007**, *538*, 1–236.

847. Brooks, T.M.; Meyer, A.L.; Hutson, D.H. The genetic toxicology of some hydrocarbon and oxygenated solvents. *Mutagenesis* **1988**, *3*, 227–232. [CrossRef]

848. O'Donoghue, J.L.; Haworth, S.R.; Curren, R.D.; Kirby, P.E.; Lawlor, T.; Moran, E.J.; Phillips, R.D.; Putnam, D.L.; Rogers-Back, A.M.; Slesinski, R.S.; et al. Mutagenicity studies on ketone solvents: Methyl ethyl ketone, methyl isobutyl ketone, and isophorone. *Mutat. Res.* **1988**, *206*, 149–161. [CrossRef]

849. JECFA, Joint FAO/WHO Expert Committee on Food Additives. *Safety Evaluation of Certain Food Additives*; Prepared by the Fifty-First Meeting of the Joint FAO/WHO Expert Committee on Food Additives (JECFA); WHO: Geneva, Switzerland, 1999.

850. Granvil, C.P.; Sharkawi, M.; Plaa, G.L. Metabolic fate of methyl n-butyl ketone, methyl isobutyl ketone and their metabolites in mice. *Toxicol. Lett.* **1994**, *70*, 263–267. [CrossRef]

851. Duguay, A.B.; Plaa, G.L. Plasma concentrations in methyl isobutyl ketone-potentiated experimental cholestasis after inhalation or oral administration. *Fundam. Appl. Toxicol.* **1993**, *21*, 222–227. [CrossRef] [PubMed]

852. Duguay, A.B.; Plaa, G.L. Tissue concentrations of methyl isobutyl ketone, methyl n-butyl ketone and their metabolites after oral or inhalation exposure. *Toxicol. Lett.* **1995**, *75*, 51–58. [CrossRef]

853. Vézina, M.; Kobusch, A.B.; du Souich, P.; Greselin, E.; Plaa, G.L. Potentiation of chloroform-induced hepatotoxicity by methyl isobutyl ketone and two metabolites. *Can. J. Physiol. Pharm.* **1990**, *68*, 1055–1061. [CrossRef]

854. Gingell, R.; Régnier, J.F.; Wilson, D.M.; Guillaumat, P.O.; Appelqvist, T. Comparative metabolism of methyl isobutyl carbinol and methyl isobutyl ketone in male rats. *Toxicol. Lett.* **2003**, *136*, 199–204. [CrossRef]

855. Raymond, P.; Plaa, G.L. Ketone potentiation of haloalkane-induced hepato- and nephrotoxicity. I. Dose-response relationships. *J. Toxicol. Environ. Health* **1995**, *45*, 465–480. [CrossRef]

856. Raymond, P.; Plaa, G.L. Ketone potentiation of haloalkane-induced hepato- and nephrotoxicity. II. Implication of monooxygenases. *J. Toxicol. Environ. Health* **1995**, *46*, 317–328. [CrossRef]

857. Borghoff, S.J.; Poet, T.S.; Green, S.; Davis, J.; Hughes, B.; Mensing, T.; Sarang, S.S.; Lynch, A.M.; Hard, G.C. Methyl isobutyl ketone exposure-related increases in specific measures of $\alpha 2u$-globulin ($\alpha 2u$) nephropathy in male rats along with in vitro evidence of reversible protein binding. *Toxicology* **2015**, *333*, 1–13. [CrossRef]

858. Borghoff, S.J.; Hard, G.C.; Berdasco, N.M.; Gingell, R.; Green, S.M.; Gulledge, W. Methyl isobutyl ketone (MIBK) induction of alpha2u-globulin nephropathy in male, but not female rats. *Toxicology* **2009**, *258*, 131–138. [CrossRef]

859. Vézina, M.; Plaa, G.L. Potentiation by methyl isobutyl ketone of the cholestasis induced in rats by a manganese-bilirubin combination or manganese alone. *Toxicol. Appl. Pharm.* **1987**, *91*, 477–483. [CrossRef]

860. Joseph, L.D.; Yousef, I.M.; Plaa, G.L.; Sharkawi, M. Potentiation of lithocholic-acid-induced cholestasis by methyl isobutyl ketone. *Toxicol. Lett.* **1992**, *61*, 39–47. [CrossRef]

861. Hughes, B.J.; Thomas, J.; Lynch, A.M.; Borghoff, S.J.; Green, S.; Mensing, T.; Sarang, S.S.; LeBaron, M.J. Methyl isobutyl ketone-induced hepatocellular carcinogenesis in B6C3F(1) mice: A constitutive androstane receptor (CAR)-mediated mode of action. *Regul. Toxicol. Pharm.* **2016**, *81*, 421–429. [CrossRef] [PubMed]

862. Grober, E.; Schaumburg, H.H. Occupational exposure to methyl isobutyl ketone causes lasting impairment in working memory. *Neurology* **2000**, *54*, 1853–1855. [CrossRef] [PubMed]

863. Wong, J.M.; Bernhard, R.A. Effect of nitrogen source on pyrazine formation. *J. Agric. Food Chem.* **1988**, *36*, 123–129. [CrossRef]

864. Wu, X.; Huang, M.; Kong, F.; Yu, S. Short communication: Study on the formation of 2-methylimidazole and 4-methylimidazole in the Maillard reaction. *J. Dairy Sci.* **2015**, *98*, 8565–8571. [CrossRef]

865. Chan, P.C. NTP technical report on the toxicity studies of 2- and 4-Methylimidazole (CAS No. 693-98-1 and 822-36-6) administered in feed to F344/N rats and B6C3F1 mice. *Toxic Rep. Ser.* **2004**, *67*, 1-g12.

866. Schlee, C.; Markova, M.; Schrank, J.; Laplagne, F.; Schneider, R.; Lachenmeier, D.W. Determination of 2-methylimidazole, 4-methylimidazole and 2-acetyl-4-(1,2,3,4-tetrahydroxybutyl)imidazole in caramel colours and cola using LC/MS/MS. *J. Chromatogr. B Anal. Technol. Biomed. Life Sci.* **2013**, *927*, 223–226. [CrossRef]

867. Jacobs, G.; Voorspoels, S.; Vloemans, P.; Fierens, T.; Van Holderbeke, M.; Cornelis, C.; Sioen, I.; De Maeyer, M.; Vinkx, C.; Vanermen, G. Caramel colour and process by-products in foods and beverages: Part I—Development of a UPLC-MS/MS isotope dilution method for determination of 2-acetyl-4-(1,2,3,4-tetrahydroxybutyl)imidazole (THI), 4-methylimidazole (4-MEI) and 2-methylimidazol (2-MEI). *Food Chem.* **2018**, *255*, 348–356. [CrossRef]

868. Choi, S.J.; Jung, M.Y. Simple and fast sample preparation followed by Gas Chromatography-Tandem Mass Spectrometry (GC-MS/MS) for the analysis of 2- and 4-methylimidazole in cola and dark beer. *J. Food Sci.* **2017**, *82*, 1044–1052. [CrossRef] [PubMed]

869. JECFA, Joint FAO/WHO Expert Committee on Food Additives. *Toxicological Evaluation of Certain Food Additives and Contaminants*; Prepared by the Twenty-Ninth Meeting of the Joint FAO/WHO Expert Committee on Food Additives (JECFA); WHO: Geneva, Switzerland, 1985.

870. EFSA ANS Panel, European Food Safety Authority, Panel on Food Additives and Nutrient Sources added to Food. Scientific Opinion on the re-evaluation of caramel colours (E 150a,b,c,d) as food additives. *EFSA J.* **2011**, *9*, 2004. [CrossRef]

871. Vollmuth, T.A. Caramel color safety—An update. *Food Chem. Toxicol.* **2018**, *111*, 578–596. [CrossRef]

872. Morgan, S.E.; Edwards, W.C. Pilot studies in cattle and mice to determine the presence of 4-methylimidazole in milk after oral ingestion. *Vet. Hum. Toxicol.* **1986**, *28*, 240–242.

873. Müller, L.; Sivertsen, T.; Langseth, W. Ammoniated forage poisoning: Concentrations of alkylimidazoles in ammoniated forage and in milk, plasma and urine in sheep and cow. *Acta Vet. Scand.* **1998**, *39*, 511–514. [CrossRef]

874. Mottier, P.; Mujahid, C.; Tarres, A.; Bessaire, T.; Stadler, R.H. Process-induced formation of imidazoles in selected foods. *Food Chem.* **2017**, *228*, 381–387. [CrossRef] [PubMed]

875. NTP, National Toxicology Program. Toxicology and carcinogenesis studies of 4-methylimidazole (Cas No. 822-36-6) in F344/N rats and B6C3F1 mice (feed studies). *Natl. Toxicol. Program. Tech. Rep. Ser.* **2007**, *535*, 1–274.

876. Chan, P.C.; Sills, R.C.; Kissling, G.E.; Nyska, A.; Richter, W. Induction of thyroid and liver tumors by chronic exposure to 2-methylimidazole in F344/N rats and B6C3F1 mice. *Arch. Toxicol.* **2008**, *82*, 399–412. [CrossRef] [PubMed]

877. Chan, P.C.; Hill, G.D.; Kissling, G.E.; Nyska, A. Toxicity and carcinogenicity studies of 4-methylimidazole in F344/N rats and B6C3F1 mice. *Arch. Toxicol.* **2008**, *82*, 45–53. [CrossRef]

878. NTP, National Toxicology Program. Toxicology and carcinogensis studies of 2-methylimidazole (Cas No. 693-98-1) in B6C3F1 mice (feed studies). *Natl. Toxicol. Program. Tech. Rep. Ser.* **2004**, *516*, 1–292.

879. Chan, P.; Mahler, J.; Travlos, G.; Nyska, A.; Wenk, M. Induction of thyroid lesions in 14-week toxicity studies of 2 and 4-methylimidazole in Fischer 344/N rats and B6C3F1 mice. *Arch. Toxicol.* **2006**, *80*, 169–180. [CrossRef]

880. Beevers, C.; Adamson, R.H. Evaluation of 4-methylimidazole, in the Ames/Salmonella test using induced rodent liver and lung S9. *Environ. Mol. Mutagen.* **2016**, *57*, 51–57. [CrossRef]

881. Morita, T.; Uneyama, C. Genotoxicity assessment of 4-methylimidazole: Regulatory perspectives. *Genes Environ.* **2016**, *38*, 20. [CrossRef]

882. Celik, R.; Topaktas, M. Genotoxic effects of 4-methylimidazole on human peripheral lymphocytes in vitro. *Drug Chem. Toxicol.* **2018**, *41*, 27–32. [CrossRef] [PubMed]

883. Johnson, J.D.; Reichelderfer, D.; Zutshi, A.; Graves, S.; Walters, D.; Smith, C. Toxicokinetics of 2-methylimidazole in male and female F344 rats. *J. Toxicol. Environ. Health Part A* **2002**, *65*, 869–879. [CrossRef] [PubMed]

884. Fennell, T.R.; Watson, S.L.; Dhungana, S.; Snyder, R.W. Metabolism of 4-methylimidazole in Fischer 344 rats and B6C3F1 mice. *Food Chem. Toxicol* **2019**, *123*, 181–194. [CrossRef] [PubMed]

885. Sanders, J.M.; Griffin, R.J.; Burka, L.T.; Matthews, H.B. Disposition of 2-methylimidazole in rats. *J. Toxicol. Environ. Health A* **1998**, *54*, 121–132. [CrossRef]

886. Dalvie, D.K.; Kalgutkar, A.S.; Khojasteh-Bakht, S.C.; Obach, R.S.; O'Donnell, J.P. Biotransformation reactions of five-membered aromatic heterocyclic rings. *Chem. Res. Toxicol.* **2002**, *15*, 269–299. [CrossRef] [PubMed]

887. Ohta, K.; Fukasawa, Y.; Yamaguchi, J.; Kohno, Y.; Fukushima, K.; Suwa, T.; Awazu, S. Retention mechanism of imidazoles in connective tissue. IV. Identification of a nucleophilic imidazolone metabolite in rats. *Biol. Pharm. Bull.* **1998**, *21*, 1334–1337. [CrossRef] [PubMed]

888. Capen, C.C. Mechanisms of chemical injury of thyroid gland. *Prog. Clin. Biol Res.* **1994**, *387*, 173–191.

889. Williams, G.M.; Iatropoulos, M.J. Alteration of liver cell function and proliferation: Differentiation between adaptation and toxicity. *Toxicol. Pathol.* **2002**, *30*, 41–53. [CrossRef]

890. Borghoff, S.J.; Fitch, S.E.; Black, M.B.; McMullen, P.D.; Andersen, M.E.; Chappell, G.A. A systematic approach to evaluate plausible modes of actions for mouse lung tumors in mice exposed to 4-methylimidozole. *Regul. Toxicol. Pharm.* **2021**, *124*, 104977. [CrossRef]

891. Brusick, D.; Aardema, M.J.; Allaben, W.T.; Kirkland, D.J.; Williams, G.; Llewellyn, G.C.; Parker, J.M.; Rihner, M.O. A weight of evidence assessment of the genotoxic potential of 4-methylimidazole as a possible mode of action for the formation of lung tumors in exposed mice. *Food Chem. Toxicol.* **2020**, *145*, 111652. [CrossRef] [PubMed]

892. Folmer, D.E.; Doell, D.L.; Lee, H.S.; Noonan, G.O.; Carberry, S.E. A U.S. population dietary exposure assessment for 4-methylimidazole (4-MEI) from foods containing caramel colour and from formation of 4-MEI through the thermal treatment of food. *Food Addit. Contam. Part A Chem. Anal. Control Expo. Risk Assess.* **2018**, *35*, 1890–1910. [CrossRef] [PubMed]

893. Fierens, T.; Van Holderbeke, M.; Cornelis, C.; Jacobs, G.; Sioen, I.; De Maeyer, M.; Vinkx, C.; Vanermen, G. Caramel colour and process contaminants in foods and beverages: Part II—Occurrence data and exposure assessment of 2-acetyl-4-(1,2,3,4-tetrahydroxybutyl)imidazole (THI) and 4-methylimidazole (4-MEI) in Belgium. *Food Chem.* **2018**, *255*, 372–379. [CrossRef] [PubMed]

894. JECFA, Joint FAO/WHO Expert Committee on Food Additives. *Combined Compendium of Food Additive Specification*; Caramel Colours; Monograph 11; WHO: Geneva, Switzerland, 2011.

895. Yaylayan, V.A. Precursors, formation and determination of furan in food. *J. Verbrauch. Lebensm.* **2006**, *1*, 5–9. [CrossRef]

896. EFSA CONTAM Panel, European Food Safety Authority, Panel on Contaminants in the Food Chain. Scientific opinion on the risks for publichealth related to the presence of furan and methylfurans in food. *EFSA J.* **2017**, *15*, 5005. [CrossRef]

897. Moro, S.; Chipman, J.K.; Wegener, J.W.; Hamberger, C.; Dekant, W.; Mally, A. Furan in heat-treated foods: Formation, exposure, toxicity, and aspects of risk assessment. *Mol. Nutr. Food Res.* **2012**, *56*, 1197–1211. [CrossRef]

898. FDA, Food and Drug Administration. *Exploratory Data on Furan in Food: Individual Food Products*; Food and Drug Administration: Silver Spring, MD, USA, 2009.

899. Van Lancker, F.; Adams, A.; Owczarek, A.; De Meulenaer, B.; De Kimpe, N. Impact of various food ingredients on the retention of furan in foods. *Mol. Nutr. Food Res.* **2009**, *53*, 1505–1511. [CrossRef]

900. NTP, National Toxicology Program. *Toxicology and Carcinogenesis Studies of Furan (CAS No. 110-00-9) in F344 Rats and B6C3F1 Mice(Gavage Studies)*; 0888-8051 (Print); National Toxicology Program: Research Triangle, NC, USA, 1993; pp. 1–286.

901. IARC, International Agency for Research on Cancer. Dry cleaning, some chlorinated solvents and other industrial chemicals. In *IARC Monographs on the Evaluation of Carcinogenic Risk to Humans*; International Agency for Research on Cancer: Lyon, France, 1995; Volume 63.

902. Von Tungeln, L.S.; Walker, N.J.; Olson, G.R.; Mendoza, M.C.; Felton, R.P.; Thorn, B.T.; Marques, M.M.; Pogribny, I.P.; Doerge, D.R.; Beland, F.A. Low dose assessment of the carcinogenicity of furan in male F344/N Nctr rats in a 2-year gavage study. *Food Chem. Toxicol.* **2017**, *99*, 170–181. [CrossRef]

903. Moser, G.J.; Foley, J.; Burnett, M.; Goldsworthy, T.L.; Maronpot, R. Furan-induced dose-response relationships for liver cytotoxicity, cell proliferation, and tumorigenicity (furan-induced liver tumorigenicity). *Exp. Toxicol. Pathol.* **2009**, *61*, 101–111. [CrossRef]

904. Durling, L.; Svensson, K.; Abramssonzetterberg, L. Furan is not genotoxic in the micronucleus assay in vivo or in vitro. *Toxicol. Lett.* **2007**, *169*, 43–50. [CrossRef]

905. Wilson, D.M.; Goldsworthy, T.L.; Popp, J.A.; Butterworth, B.E. Evaluation of genotoxicity, pathological lesions, and cell proliferation in livers of rats and mice treated with furan. *Environ. Mol. Mutagen.* **1992**, *19*, 209–222. [CrossRef] [PubMed]

906. McDaniel, L.P.; Ding, W.; Dobrovolsky, V.N.; Shaddock, J.G., Jr.; Mittelstaedt, R.A.; Doerge, D.R.; Heflich, R.H. Genotoxicity of furan in Big Blue rats. *Mutat. Res.* **2012**, *742*, 72–78. [CrossRef] [PubMed]

907. Ding, W.; Petibone, D.M.; Latendresse, J.R.; Pearce, M.G.; Muskhelishvili, L.; White, G.A.; Chang, C.W.; Mittelstaedt, R.A.; Shaddock, J.G.; McDaniel, L.P.; et al. In vivo genotoxicity of furan in F344 rats at cancer bioassay doses. *Toxicol. Appl. Pharm.* **2012**, *261*, 164–171. [CrossRef]

908. Jeffrey, A.M.; Brunnemann, K.D.; Duan, J.D.; Schlatter, J.; Williams, G.M. Furan induction of DNA cross-linking and strand breaks in turkey fetal liver in comparison to 1,3-propanediol. *Food Chem. Toxicol.* **2012**, *50*, 675–678. [CrossRef] [PubMed]

909. Peterson, L.A.; Naruko, K.C.; Predecki, D.P. A reactive metabolite of furan, cis-2-butene-1,4-dial, is mutagenic in the Ames assay. *Chem. Res. Toxicol.* **2000**, *13*, 531–534. [CrossRef] [PubMed]

910. Byrns, M.C.; Vu, C.C.; Neidigh, J.W.; Abad, J.-L.; Jones, R.A.; Peterson, L.A. Detection of DNA adducts derived from the reactive metabolite of furan, cis-2-butene-1,4-dial. *Chem. Res. Toxicol.* **2006**, *19*, 414–420. [CrossRef]

911. Churchwell, M.I.; Scheri, R.C.; Von Tungeln, L.S.; Gamboa da Costa, G.; Beland, F.A.; Doerge, D.R. Evaluation of serum and liver toxicokinetics for furan and liver DNA adduct formation in male Fischer 344 rats. *Food Chem. Toxicol.* **2015**, *86*, 1–8. [CrossRef]

912. Neuwirth, C.; Mosesso, P.; Pepe, G.; Fiore, M.; Malfatti, M.; Turteltaub, K.; Dekant, W.; Mally, A. Furan carcinogenicity: DNA binding and genotoxicity of furan in rats in vivo. *Mol. Nutr. Food Res.* **2012**, *56*, 1363–1374. [CrossRef]

913. Kedderis, G.L.; Carfagna, M.A.; Held, S.D.; Batra, R.; Murphy, J.E.; Gargas, M.L. Kinetic analysis of furan biotransformation by F-344 rats in vivo and in vitro. *Toxicol. Appl. Pharm.* **1993**, *123*, 274–282. [CrossRef]

914. Chen, L.J.; Hecht, S.S.; Peterson, L.A. Identification of cis-2-butene-1,4-dial as a microsomal metabolite of furan. *Chem. Res. Toxicol.* **1995**, *8*, 903–906. [CrossRef]

915. Byrns, M.C.; Predecki, D.P.; Peterson, L.A. Characterization of nucleoside adducts of cis-2-butene-1,4-dial, a reactive metabolite of furan. *Chem. Res. Toxicol.* **2002**, *15*, 373–379. [CrossRef] [PubMed]

916. Russo, M.T.; De Luca, G.; Palma, N.; Leopardi, P.; Degan, P.; Cinelli, S.; Pepe, G.; Mosesso, P.; Di Carlo, E.; Sorrentino, C.; et al. Oxidative stress, mutations and chromosomal aberrations induced by in vitro and in vivo exposure to furan. *Int. J. Mol. Sci.* **2021**, *22*, 9687. [CrossRef] [PubMed]

917. De Conti, A.; Kobets, T.; Escudero-Lourdes, C.; Montgomery, B.; Tryndyak, V.; Beland, F.A.; Doerge, D.R.; Pogribny, I.P. Dose- and time-dependent epigenetic changes in the livers of Fisher 344 rats exposed to furan. *Toxicol. Sci.* **2014**, *139*, 371–380. [CrossRef] [PubMed]

918. de Conti, A.; Kobets, T.; Tryndyak, V.; Burnett, S.D.; Han, T.; Fuscoe, J.C.; Beland, F.A.; Doerge, D.R.; Pogribny, I.P. Persistence of furan-induced epigenetic aberrations in the livers of F344 rats. *Toxicol. Sci.* **2015**, *144*, 217–226. [CrossRef]

919. FDA, Food and Drug Administration. *An Updated Exposure Assessment for Furan from the Consumption of Adult and Baby Foods*; 18 April 2007; Food and Drug Administration: Silver Spring, MD, USA, 2007.

920. Lachenmeier, D.W.; Maser, E.; Kuballa, T.; Reusch, H.; Kersting, M.; Alexy, U. Detailed exposure assessment of dietary furan for infants consuming commercially jarred complementary food based on data from the DONALD study. *Matern Child. Nutr.* **2012**, *8*, 390–403. [CrossRef] [PubMed]

921. IFIC, International Food Information Council; FDA, US Food and Drug Administration. *Overview of Food Ingredients, Additives & Colors*; US Food and Drug Administration: Washington, DC, USA, 2010.

922. Krishan, M.; Navarro, L.; Beck, B.; Carvajal, R.; Dourson, M. A regulatory relic: After 60 years of research on cancer risk, the Delaney Clause continues to keep us in the past. *Toxicol. Appl. Pharm.* **2021**, *433*, 115779. [CrossRef]

923. Williams, G.M.; Karbe, E.; Fenner-Crisp, P.; Iatropoulos, M.J.; Weisburger, J.H. Risk assessment of carcinogens in food with special consideration of non-genotoxic carcinogens. *Exp. Toxicol. Pathol.* **1996**, *48*, 209–215. [CrossRef]

924. Felter, S.P.; Zhang, X.; Thompson, C. Butylated hydroxyanisole: Carcinogenic food additive to be avoided or harmless antioxidant important to protect food supply? *Regul. Toxicol. Pharm.* **2021**, *121*, 104887. [CrossRef]

925. Xu, X.; Liu, A.; Hu, S.; Ares, I.; Martínez-Larrañaga, M.R.; Wang, X.; Martínez, M.; Anadón, A.; Martínez, M.A. Synthetic phenolic antioxidants: Metabolism, hazards and mechanism of action. *Food Chem.* **2021**, *353*, 129488. [CrossRef]

926. EFSA ANS Panel, European Food Safety Authority, Panel on Food Additives and Nutrient Sources added to Food. Scientific Opinion on the reevaluation of butylated hydroxyanisole-BHA (E 320) as a food additive. *EFSA J.* **2011**, *9*, 2392. [CrossRef]

927. EFSA ANS Panel, European Food Safety Authority, Panel on Food Additives and Nutrient Sources added to Food. Scientific Opinion on the reevaluation of butylated hydroxytoluene BHT (E 321) as a food additive. *EFSA J.* **2012**, *10*, 2588.

928. EFSA ANS Panel, European Food Safety Authority, Panel on Food Additives and Nutrient Sources added to Food. Statement on the refined exposure assessment of tertiary-butyl hydroquinone (E 319). *EFSA J.* **2016**, *14*, 4363.

929. Deisinger, P.J. Human exposure to naturally occurring hydroquinone. *J. Toxicol. Environ. Health* **1996**, *47*, 31–46. [CrossRef] [PubMed]

930. Whysner, J. Butylated hydroxyanisole mechanistic data and risk assessment: Conditional species-specific cytotoxicity, enhanced cell proliferation, and tumor promotion. *Pharmacol. Ther.* **1996**, *71*, 137–151. [CrossRef]

931. Williams, G.M.; Iatropoulos, M.J.; Whysner, J. Safety assessment of butylated hydroxyanisole and butylated hydroxytoluene as antioxidant food additives. *Food Chem. Toxicol.* **1999**, *37*, 1027–1038. [CrossRef]

932. Hirose, M.; Fukushima, S.; Sakata, T.; Inui, M.; Ito, N. Effect of quercetin on two-stage carcinogenesis of the rat urinary bladder. *Cancer Lett.* **1983**, *21*, 23–27. [CrossRef]

933. Gharavi, N.; Haggarty, S.; El-Kadi, A.S. Chemoprotective and carcinogenic effects of tert-butylhydroquinone and its metabolites. *Curr. Drug Metab.* **2007**, *8*, 1–7. [CrossRef]

934. NTP, National Toxicology Program. Toxicology and carcinogenesis studies of hydroquinone (CAS No. 123-31-9) in F344/N rats and B6C3F1 mice (gavage studies). *Natl. Toxicol. Program. Tech. Rep. Ser.* **1989**, *366*, 1–248.

935. Lanigan, R.S.; Yamarik, T.A. Final report on the safety assessment of BHT(1). *Int. J. Toxicol.* **2002**, *21* (Suppl. S2), 19–94. [CrossRef]

936. Williams, G.M.; McQueen, C.A.; Tong, C. Toxicity studies of butylated hydroxyanisole and butylated hydroxytoluene. I. Genetic and cellular effects. *Food Chem. Toxicol.* **1990**, *28*, 793–798. [CrossRef]

937. van Esch, G.J. Toxicology of tert-butylhydroquinone (TBHQ). *Food Chem. Toxicol.* **1986**, *24*, 1063–1065. [CrossRef]

938. Eskandani, M.; Hamishehkar, H.; Ezzati Nazhad Dolatabadi, J. Cytotoxicity and DNA damage properties of tert-butylhydroquinone (TBHQ) food additive. *Food Chem.* **2014**, *153*, 315–320. [CrossRef] [PubMed]

939. Saito, K.; Nakagawa, S.; Yoshitake, A.; Miyamoto, J.; Hirose, M.; Ito, N. DNA-adduct formation in the forestomach of rats treated with 3-tert-butyl-4-hydroxyanisole and its metabolites as assessed by an enzymatic 32P-postlabeling method. *Cancer Lett.* **1989**, *48*, 189–195. [CrossRef]

940. Jagetia, G.C.; Menon, K.S.L.; Jain, V. Genotoxic effect of hydroquinone on the cultured mouse spleenocytes. *Toxicol. Lett.* **2001**, *121*, 15–20. [CrossRef]

941. Peters, M.M.; Lau, S.S.; Dulik, D.; Murphy, D.; van Ommen, B.; van Bladeren, P.J.; Monks, T.J. Metabolism of tert-butylhydroquinone to S-substituted conjugates in the male Fischer 344 rat. *Chem. Res. Toxicol.* **1996**, *9*, 133–139. [CrossRef] [PubMed]

942. Malkinson, A.; Radcliffe, R.; Bauer, A.; Dwyer-Nield, L.; Kleeberger, S. Quantitative trait loci that regulate susceptibility to both butylated hydroxytoluene-induced pulmonary inflammation and lung tumor promotion in CXB recombinant inbred mice. *Chest* **2002**, *121*, 82s. [CrossRef]

943. Whysner, J.; Wang, C.X.; Zang, E.; Iatropoulos, M.J.; Williams, G.M. Dose response of promotion by butylated hydroxyanisole in chemically initiated tumours of the rat forestomach. *Food Chem. Toxicol.* **1994**, *32*, 215–222. [CrossRef]

944. Bauer, A.K.; Dwyer-Nield, L.D.; Keil, K.; Koski, K.; Malkinson, A.M. Butylated hydroxytoluene (BHT) induction of pulmonary inflammation: A role in tumor promotion. *Exp. Lung Res.* **2001**, *27*, 197–216. [CrossRef]

945. Malkinson, A.M. Role of inflammation in mouse lung tumorigenesis: A review. *Exp. Lung Res.* **2005**, *31*, 57–82. [CrossRef]

946. English, J.C.; Perry, L.G.; Vlaovic, M.; Moyer, C.; O'Donoghue, J.L. Measurement of cell proliferation in the kidneys of Fischer 344 and Sprague-Dawley rats after gavage administration of hydroquinone. *Fundam. Appl. Toxicol.* **1994**, *23*, 397–406. [CrossRef]

947. Whysner, J.; Verna, L.; English, J.C.; Williams, G.M. Analysis of studies related to tumorigenicity induced by hydroquinone. *Regul. Toxicol. Pharmacol.* **1995**, *21*, 158–176. [CrossRef] [PubMed]

948. Peters, M. Cytotoxicity and cell-proliferation induced by the nephrocarcinogen hydroquinone and its nephrotoxic metabolite 2,3,5-(tris-glutathion-S- yl)hydroquinone. *Carcinogenesis* **1997**, *18*, 2393–2401. [CrossRef] [PubMed]

949. Hard, G.C.; Whysner, J.; English, J.C.; Zang, E.; Williams, G.M. Relationship of hydroquinone-associated rat renal tumors with spontaneous chronic progressive nephropathy. *Toxicol. Pathol.* **1997**, *25*, 132–143. [CrossRef] [PubMed]

950. Botterweck, A.A.; Verhagen, H.; Goldbohm, R.A.; Kleinjans, J.; van den Brandt, P.A. Intake of butylated hydroxyanisole and butylated hydroxytoluene and stomach cancer risk: Results from analyses in the Netherlands Cohort Study. *Food Chem. Toxicol.* **2000**, *38*, 599–605. [CrossRef]

951. IARC, International Agency for Research on Cancer. *Cruciferous Vegetables, Isothiocyanates and Indoles*; IARC Handbooks of Cancer Prevention; International Agency for Research on Cancer: Lyon, France, 2004; Volume 9.

952. Williams, G.M.; Iatropoulos, M.J.; Jeffrey, A.M. Anticarcinogenicity of monocyclic phenolic compounds. *Eur. J. Cancer Prev.* **2002**, *11* (Suppl. S2), S101–S107.

953. O'Donoghue, J.L. Hydroquinone and its analogues in dermatology—A risk-benefit viewpoint. *J. Cosmet. Dermatol.* **2006**, *5*, 196–203. [CrossRef]

954. Suh, H.J.; Chung, M.S.; Cho, Y.H.; Kim, J.W.; Kim, D.H.; Han, K.W.; Kim, C.J. Estimated daily intakes of butylated hydroxyanisole (BHA), butylated hydroxytoluene (BHT) andtert-butyl hydroquinone (TBHQ) antioxidants in Korea. *Food Addit. Contam.* **2005**, *22*, 1176–1188. [CrossRef]

955. Haque, A.; Brazeau, D.; Amin, A.R. Perspectives on natural compounds in chemoprevention and treatment of cancer: An update with new promising compounds. *Eur. J. Cancer* **2021**, *149*, 165–183. [CrossRef]

956. Langner, E.; Rzeski, W. Dietary derived compounds in cancer chemoprevention. *Contemp. Oncol.* **2012**, *16*, 394–400. [CrossRef]

957. Murakami, A.; Ohigashi, H.; Koshimizu, K. Anti-tumor promotion with food phytochemicals: A strategy for cancer chemoprevention. *Biosci. Biotechnol. Biochem.* **1996**, *60*, 1–8. [CrossRef]

958. Ranjan, A.; Ramachandran, S.; Gupta, N.; Kaushik, I.; Wright, S.; Srivastava, S.; Das, H.; Srivastava, S.; Prasad, S.; Srivastava, S.K. Role of phytochemicals in cancer prevention. *Int. J. Mol. Sci.* **2019**, *20*, 4981. [CrossRef]

959. Wattenberg, L.W. Inhibition of carcinogenesis by naturally-occurring and synthetic compounds. *Basic Life Sci.* **1990**, *52*, 155–166. [CrossRef] [PubMed]

960. Newman, D.J.; Cragg, G.M. Natural products as sources of new drugs over the nearly four decades from 01/1981 to 09/2019. *J. Nat. Prod.* **2020**, *83*, 770–803. [CrossRef] [PubMed]

961. Gescher, A.J.; Sharma, R.A.; Steward, W.P. Cancer chemoprevention by dietary constituents: A tale of failure and promise. *Lancet Oncol.* **2001**, *2*, 371–379. [CrossRef]

962. Potter, J.D. The failure of cancer chemoprevention. *Carcinogenesis* **2014**, *35*, 974–982. [CrossRef] [PubMed]

963. Steward, W.P.; Brown, K. Cancer chemoprevention: A rapidly evolving field. *Br. J. Cancer* **2013**, *109*, 1–7. [CrossRef]

964. Patterson, S.L.; Colbert Maresso, K.; Hawk, E. Cancer chemoprevention: Successes and failures. *Clin. Chem.* **2013**, *59*, 94–101. [CrossRef]

965. Johnson, I.T. Phytochemicals and cancer. *Proc. Nutr. Soc.* **2007**, *66*, 207–215. [CrossRef]

966. Kleiner, H.E. Oral administration of naturally occurring coumarins leads to altered phase I and II enzyme activities and reduced DNA adduct formation by polycyclic aromatic hydrocarbons in various tissues of SENCAR mice. *Carcinogenesis* **2001**, *22*, 73–82. [CrossRef]

967. Prince, M.; Campbell, C.T.; Robertson, T.A.; Wells, A.J.; Kleiner, H.E. Naturally occurring coumarins inhibit 7,12-dimethylbenz[a]anthracene DNA adduct formation in mouse mammary gland. *Carcinogenesis* **2005**, *27*, 1204–1213. [CrossRef]

968. Huber, W.W.; McDaniel, L.P.; Kaderlik, K.R.; Teitel, C.H.; Lang, N.P.; Kadlubar, F.F. Chemoprotection against the formation of colon DNA adducts from the food-borne carcinogen 2-amino-1-methyl-6-phenylimidazo[4,5-b]pyridine (PhIP) in the rat. *Mutat. Res./Fundam. Mol. Mech. Mutagen.* **1997**, *376*, 115–122. [CrossRef]

969. Cavin, C. The coffee-specific diterpenes cafestol and kahweol protect against aflatoxin B1-induced genotoxicity through a dual mechanism. *Carcinogenesis* **1998**, *19*, 1369–1375. [CrossRef] [PubMed]

970. Miller, E.G.; McWhorter, K.; Rivera-Hidalgo, F.; Wright, J.M.; Hirsbrunner, P.; Sunahara, G.I. Kahweol and cafestol: Inhibitors of hamster buccal pouch carcinogenesis. *Nutr. Cancer* **1991**, *15*, 41–46. [CrossRef]

971. Alhusainy, W.; Williams, G.M.; Jeffrey, A.M.; Iatropoulos, M.J.; Taylor, S.; Adams, T.B.; Rietjens, I.M.C.M. The natural basil flavonoid nevadensin protects against a methyleugenol-induced marker of hepatocarcinogenicity in male F344 rat. *Food Chem. Toxicol.* **2014**, *74*, 28–34. [CrossRef]

972. Azuine, M.A.; Bhide, S.V. Chemopreventive effect of turmeric against stomach and skin tumors induced by chemical carcinogens in Swiss mice. *Nutr. Cancer* **1992**, *17*, 77–83. [CrossRef] [PubMed]

973. Ikezaki, S.; Nishikawa, A.; Furukawa, F.; Kudo, K.; Nakamura, H.; Tamura, K.; Mori, H. Chemopreventive effects of curcumin on glandular stomach carcinogenesis induced by N-methyl-N'-nitro-N-nitrosoguanidine and sodium chloride in rats. *Anticancer Res.* **2001**, *21*, 3407–3411. [PubMed]

974. Singletary, K.W.; Nelshoppen, J.M.; Scardefield, S.; Wallig, M. Inhibition by butylated hydroxytoluene and its oxidative metabolites of DMBA-induced mammary tumorigenesis and of mammary DMBA-DNA adduct formation in vivo in the female rat. *Food Chem. Toxicol.* **1992**, *30*, 455–465. [CrossRef]

975. Williams, G.M.; Tanaka, T.; Maeura, Y. Dose-related inhibition of aflatoxin B1 induced hepatocarcinogenesis by the phenolic antioxidants, butylated hydroxyanisole and butylated hydroxytoluene. *Carcinogenesis* **1986**, *7*, 1043–1050. [CrossRef]

976. Williams, G.M.; Iatropoulos, M.J. Inhibition of the hepatocarcinogenicity of aflatoxin B1 in rats by low levels of the phenolic antioxidants butylated hydroxyanisole and butylated hydroxytoluene. *Cancer Lett.* **1996**, *104*, 49–53. [CrossRef]

977. Simonich, M.T.; Egner, P.A.; Roebuck, B.D.; Orner, G.A.; Jubert, C.; Pereira, C.; Groopman, J.D.; Kensler, T.W.; Dashwood, R.H.; Williams, D.E.; et al. Natural chlorophyll inhibits aflatoxin B1-induced multi-organ carcinogenesis in the rat. *Carcinogenesis* **2007**, *28*, 1294–1302. [CrossRef]

978. Kensler, T.W.; Egner, P.A.; Dolan, P.M.; Groopman, J.D.; Roebuck, B.D. Mechanism of protection against aflatoxin tumorigenicity in rats fed 5-(2-pyrazinyl)-4-methyl-1,2-dithiol-3-thione (oltipraz) and related 1,2-dithiol-3-thiones and 1,2-dithiol-3-ones. *Cancer Res.* **1987**, *47*, 4271–4277. [PubMed]

979. Wattenberg, L.W.; Loub, W.D. Inhibition of polycyclic aromatic hydrocarbon-induced neoplasia by naturally occurring indoles. *Cancer Res.* **1978**, *38*, 1410–1413. [PubMed]

980. Rauscher, R.; Edenharder, R.; Platt, K.L. In vitro antimutagenic and in vivo anticlastogenic effects of carotenoids and solvent extracts from fruits and vegetables rich in carotenoids. *Mutat. Res./Genet. Toxicol. Environ. Mutagen.* **1998**, *413*, 129–142. [CrossRef]

981. Tanaka, T.; Shnimizu, M.; Moriwaki, H. Cancer chemoprevention by carotenoids. *Molecules* **2012**, *17*, 3202–3242. [CrossRef]

982. Hecht, S.S. Chemoprevention of cancer by isothiocyanates, modifiers of carcinogen metabolism. *J. Nutr.* **1999**, *129*, 768S–774S. [CrossRef]

983. Wattenberg, L.W. Inhibition of carcinogen-induced neoplasia by sodium cyanate, tert-butyl isocyanate, and benzyl isothiocyanate administered subsequent to carcinogen exposure. *Cancer Res.* **1981**, *41*, 2991–2994.

984. Zhang, Y.; Kensler, T.W.; Cho, C.G.; Posner, G.H.; Talalay, P. Anticarcinogenic activities of sulforaphane and structurally related synthetic norbornyl isothiocyanates. *Proc. Natl. Acad. Sci. USA* **1994**, *91*, 3147–3150. [CrossRef]

985. Walters, D.G. Cruciferous vegetable consumption alters the metabolism of the dietary carcinogen 2-amino-1-methyl-6-phenylimidazo[4,5-b]pyridine (PhIP) in humans. *Carcinogenesis* **2004**, *25*, 1659–1669. [CrossRef]

986. Bayat Mokhtari, R.; Baluch, N.; Homayouni, T.S.; Morgatskaya, E.; Kumar, S.; Kazemi, P.; Yeger, H. The role of Sulforaphane in cancer chemoprevention and health benefits: A mini-review. *J. Cell Commun. Signal.* **2018**, *12*, 91–101. [CrossRef]

987. Cömert, E.D.; Gökmen, V. Evolution of food antioxidants as a core topic of food science for a century. *Food Res. Int.* **2018**, *105*, 76–93. [CrossRef]

988. Liu, R.H. Potential synergy of phytochemicals in cancer prevention: Mechanism of action. *J. Nutr.* **2004**, *134*, 3479S–3485S. [CrossRef] [PubMed]

989. Racchi, M.L. Antioxidant defenses in plants with attention to Prunus and *Citrus spp.*. *Antioxidants* **2013**, *2*, 340–369. [CrossRef] [PubMed]

990. Bishayee, A. Cancer prevention and treatment with resveratrol: From rodent studies to clinical trials. *Cancer Prev. Res.* **2009**, *2*, 409–418. [CrossRef] [PubMed]

991. Heidor, R.; de Conti, A.; Ortega, J.F.; Furtado, K.S.; Silva, R.C.; Tavares, P.E.L.M.; Purgatto, E.; Ract, J.N.R.; de Paiva, S.A.R.; Gioielli, L.A.; et al. The chemopreventive activity of butyrate-containing structured lipids in experimental rat hepatocarcinogenesis. *Mol. Nutr. Food Res.* **2015**, *60*, 420–429. [CrossRef]

992. Guariento, A.H.; Furtado, K.S.; de Conti, A.; Campos, A.; Purgatto, E.; Carrilho, J.; Shinohara, E.M.G.; Tryndyak, V.; Han, T.; Fuscoe, J.C.; et al. Transcriptomic responses provide a new mechanistic basis for the chemopreventive effects of folic acid and tributyrin in rat liver carcinogenesis. *Int. J. Cancer* **2014**, *135*, 7–18. [CrossRef]

993. Ulrich, S.; Wolter, F.; Stein, J.M. Molecular mechanisms of the chemopreventive effects of resveratrol and its analogs in carcinogenesis. *Mol. Nutr. Food Res.* **2005**, *49*, 452–461. [CrossRef]

994. Dashwood, R.H. Indole-3-carbinol: Anticarcinogen or tumor promoter in brassica vegetables? *Chem. Biol. Interact.* **1998**, *110*, 1–5. [CrossRef]

995. Lake, B.G. Coumarin metabolism, toxicity and carcinogenicity: Relevance for human risk assessment. *Food Chem. Toxicol.* **1999**, *37*, 423–453. [CrossRef]

Review

Alkenylbenzenes in Foods: Aspects Impeding the Evaluation of Adverse Health Effects

Andreas Eisenreich [1,*], **Mario E. Götz** [1], **Benjamin Sachse** [1], **Bernhard H. Monien** [1], **Kristin Herrmann** [2] **and Bernd Schäfer** [1]

1. Department of Food Safety, German Federal Institute for Risk Assessment (BfR), Max-Dohrn-Str. 8-10, 10589 Berlin, Germany; mario.goetz@bfr.bund.de (M.E.G.); benjamin.sachse@bfr.bund.de (B.S.); bernhard.monien@bfr.bund.de (B.H.M.); bernd.schaefer@bfr.bund.de (B.S.)
2. Department of Pesticides Safety, German Federal Institute for Risk Assessment (BfR), Max-Dohrn-Str. 8-10, 10589 Berlin, Germany; kristin.herrmann@bfr.bund.de
* Correspondence: andreas.eisenreich@bfr.bund.de; Tel./Fax: +49-30-18412-25202

Abstract: Alkenylbenzenes are naturally occurring secondary plant metabolites, primarily present in different herbs and spices, such as basil or fennel seeds. Thus, alkenylbenzenes, such as safrole, methyleugenol, and estragole, can be found in different foods, whenever these herbs and spices (or extracts thereof) are used for food production. In particular, essential oils or other food products derived from the aforementioned herbs and spices, such as basil-containing pesto or plant food supplements, are often characterized by a high content of alkenylbenzenes. While safrole or methyleugenol are known to be genotoxic and carcinogenic, the toxicological relevance of other alkenylbenzenes (e.g., apiol) regarding human health remains widely unclear. In this review, we will briefly summarize and discuss the current knowledge and the uncertainties impeding a conclusive evaluation of adverse effects to human health possibly resulting from consumption of foods containing alkenylbenzenes, especially focusing on the genotoxic compounds, safrole, methyleugenol, and estragole.

Keywords: alkenylbenzenes; food; consumption; regulation; mixtures

Citation: Eisenreich, A.; Götz, M.E.; Sachse, B.; Monien, B.H.; Herrmann, K.; Schäfer, B. Alkenylbenzenes in Foods: Aspects Impeding the Evaluation of Adverse Health Effects. *Foods* **2021**, *10*, 2139. https://doi.org/10.3390/foods10092139

Academic Editor: Rinaldo Botondi

Received: 9 August 2021
Accepted: 7 September 2021
Published: 10 September 2021

Publisher's Note: MDPI stays neutral with regard to jurisdictional claims in published maps and institutional affiliations.

1. Introduction

Alkenylbenzenes primarily occur as secondary plant metabolites in various herbs and spices (e.g., basil, fennel, and parsley) but are also present—albeit at lower levels in agricultural crops, e.g., in tomatoes and apples [1,2]. Alkenylbenzenes are components of essential oils. Therefore, high concentrations can be found in food products made from aromatic parts of the abovementioned herbs and spices (e.g., fennel tea, basil-containing pesto, and plant food supplements) [3–6]. Since alkenylbenzenes have strong aromatic properties, they are also used as flavoring substances in foods and as fragrances in cosmetics [1]. Several alkenylbenzenes, such as safrole, methyleugenol, and estragole, are known to be toxic, and the most relevant toxicological endpoints include genotoxicity and carcinogenicity, whereby the toxicity is not caused by the parent compounds themselves but by their highly reactive metabolites [1].

The toxicity of alkenylbenzenes—especially their genotoxic and carcinogenic potential—is a controversially debated issue. Results of various toxicological studies demonstrated that single alkenylbenzenes, such as safrole, methyleugenol, and estragole, cause—amongst other things—genotoxic and carcinogenic effects in animal studies [1,7,8]. However, some other alkenylbenzenes, such as elemicin and apiol, have not yet been sufficiently assessed regarding their genotoxic and carcinogenic properties.

Beside the toxicity of single compounds, it has to be kept in mind that different foods may contain more than one alkenylbenzene, such as basil, which contains methyleugenol, estragole, and other compounds [9]. This is of particular importance for substances exhibiting a similar mode of action, since it may result in additive toxicity [9]. On the other hand, it was shown in different studies that the genotoxic potential of alkenylbenzenes may be reduced by other plant components, such as the sulfotransferase (SULT) inhibitor nevadensin [10,11]. This was called the matrix-derived combination effect [10]. However, the relevance of this effect in different food matrices is still an intensively discussed issue [12,13].

Occurrence data of alkenylbenzenes in different foods are necessary to assess human exposure. However, there are significant variations in occurrence levels, depending on, e.g., the analyzed samples (parts of plants, time of harvesting, region of origin), methodology (i.e., not standardized sample preparation and analytical methods), etc. Due to these differences, it is often not possible to assess occurrence data of different origins in a comparative manner, which complicates conducting a reliable exposure assessment.

Moreover, structural differences in alkenylbenzenes, such as in estragole vs. *trans*-anethole (see Table 1) influence toxicokinetics of these compounds. This, in turn, also affects the toxic (especially the genotoxic) potential of different alkenylbenzenes, which has to be taken into account for an assessment of the risks possibly resulting from exposure to these substances.

Table 1. Occurrence of safrole, methyleugenol, estragole, *trans*-anethole, and myristicin found in essential oils (EO) from herbs and spices.

	Safrole	Methyleugenol	Estragole (=Methylchavicol)	*trans*-Anethole	Myristicin
CAS N°	94-59-7	93-15-2	140-67-0	4180-23-8	607-91-0
Structural formula					
IUPAC name	5-prop-2-enyl-1,3-benzodioxole	1,2-dimethoxy-4-prop-2-enylphenol	1-methoxy-4-prop-2-enylbenzene	1-methoxy-4-(E)-prop-1-enyl]benzene	4-methoxy-6-prop-2-enyl-1,3-benzodioxole
Synonyms (select.)	5-Allyl-1,3-benzodioxole Shikimole Safrene Sassafras Rhyuno oil	4-Allyl-1,2-dimethoxybenzene 4-Allylveratrole Eugenol methyl ether Eugenyl methyl ether	4-Allylanisole 1-Allyl-4-methoxybenzene p-Allylanisole Chavicol methyl ether Tarragon	(E)-Anethole p-Propenylanisole 4-Propenylanisole Anise camphor (E)-1-Methoxy-4-(prop-1-en-1-yl)benzene	6-Allyl-4-methoxy-1,3-benzodioxole 5-Allyl-1-methoxy-2,3-(methylenedioxy)benzene Asaricin
Occurrence in essential oils (%) + (reference)					
Allspice berries		62.7 [14]			
Allspice berries		4–9 [15]			
Allspice berries		8.8 [16]			
Anise seeds		0.1–0.2 [17]	1.44–7.08 [17]	79.49–89.99 [17]	
Anise seeds			0.5–2.3 [18]	76.9–93.7 [18]	
Anise seeds				>90 [18]	
Chinese Star anise seeds			0.5–5.5 [19,20]	88.5–92 [19,21,22]	
Japanese Star Anise seeds	6.6 [23]	9.8 [23]		1.2 [23]	3.5 [23]
Sweet Fennel aerial parts			2–3 [24]	9.7–54.7 [24]	
Sweet Fennel roots					2.5–10 [24]
Basil oil *Ocimum basilicum* leaves		9.27–87.04 [25–28]	0–81 [25–28]		
Western tarragon		0.51–28.87 [29,30]	17.26–75 [30,31]		
Eastern tarragon	no Safrole (but 21.45–38.90 Elemicin) [32]	9.59–28.40 [32]	0.29–0.31 [32]		
Nutmeg kernel Eastern Indonesia	1.6 (and 1.7 Elemicin) [33]	16.7 (and 16.8. Methyl-iso-eugenol) [33]			2.3 [33]
Nutmeg kernel Sri Lanka	1.4 (and 2.1 Elemicin) [34]	0.6 [34]			4.9 [34]

In the following parts, we will briefly summarize and discuss the current knowledge and the uncertainties impeding a reliable evaluation of the health risks resulting from alkenylbenzene exposure, especially focusing on the genotoxic compounds, safrole, methyleugenol and estragole. Moreover, we will shed some more light on ongoing discussions (e.g., the toxic potential of single compounds and mixtures) and some strengths as well as weaknesses of current experimental and analytical strategies regarding the risks possibly resulting from exposure to alkenylbenzenes in general.

2. Current Knowledge

2.1. Occurrence of Alkenylbenzenes

Consumers need to know which of their food consumption habits might result in high intake levels of genotoxic alkenylbenzenes in order to become able to draw informed decisions, whether to choose or not to choose a certain alkenylbenzene-containing aromatized or natural food.

2.1.1. Alkenylbenzenes in Herbs and Spices

In the following part, some examples are described to shed more light on the complexity of alkenylbenzene composition in different herbs and spices.

Fennel

The herb fennel (*Foeniculum vulgare* Mill., Umbelliferae or Apiaceae) is cultivated in many countries all over the world. Essential oils can be obtained by steam distillation of the dried ripe fruits or other parts of the plant such as leaves, stems, or roots, as described by Trenkle [24]. The wild common fennel is bitter (var. *vulgare*), and the cultivated one is rather sweet (var. *dulce*). Essential oil yields can be 2–6%, the major constituent of which is usually *trans*-anethole (60–90%) [35,36]. Depending on the extraction methods used, estragole contents vary between 3.3–5.3% in the aerial parts of the plant [37]. Trenkle, in 1972 in the aerial parts of the sweet fennel, found (stems, leaves, and seeds) *trans*-anethole (9.7–54.7%), *cis*-anethole (0.1–0.8%), and estragole (2.0–3.0%) but no myristicin [24]. However, the fennel roots contained neither anetholes nor estragole but contained instead dill-apiol (45.6–62.7%), myristicin (2.5–10%), and parsley apiol (0.2%). The oil from sweet fennel fruits is used as a flavor component in many products. Hydrodistillation of fennel fruits may yield up to 88% estragole [38]. Very common in, e.g., Europe is the consumption of fennel tea infusions. The determination of estragole in infusions from different widely used commercial herbal teas based on *Foeniculum vulgare* seeds by an optimized headspace solid-phase microextraction followed by gas chromatography–mass spectrometry (GC–MS) analysis revealed levels of estragole to range within 50–250 µg/L [39] or even reach levels from 241–2058 µg/L in teas from teabags [40]. In preparations of tea extracts from herbal tea mixtures (n = 16) of the fennel–anise–caraway type, estragole contents ranged from 4.0–76.7 µg/L, whilst *trans*-anethole concentrations ranged from 83.2–7266.4 µg/L [41]. Interestingly, one hour following ingestion of fennel–anise–caraway tea by breastfeeding women, approximately 1% (i.e., 0.13 µg/L milk) of the consumed estragole via tea ingestion and up to 5% (i.e., 4.23 µg/L milk) of *trans*-anethole consumed via tea ingestion was found in the human milk of lactating mothers [41]. An earlier study could identify in breast milk from breastfeeding women, at the time point of two hours following ingestion of a 100 mg *trans*-anethole containing capsule, a mean concentration of 9.9 µg *trans*-anethole per liter of human breast milk. Peak concentrations of *trans*-anethole were 23.2 µg/L milk [42]. These results indicate that some alkenylbenzenes may even, as parent compounds, escape maternal hepatic metabolism and can be transferred into breast milk, albeit at very low concentrations. To our knowledge, other systematic studies that investigated metabolites of estragole and *trans*-anethole and other alkenylbenzenes in human breast milk are missing.

Basil

Sweet basil herb (*Ocimum basilicum* L., Labiatae or Lamiaceae) is, nowadays, cultivated in many countries around the world, originating probably from Africa and tropical Asia. The essential oil is generated from dried leaves and stems (aerial parts of the plant) by steam distillation. In a systematic study of essential oils obtained from the aerial parts of seven varieties of *Ocimum basilicum*, it was found that basil oils may contain, relative to other identified components, high amounts of methyleugenol (9.27–87.04%) and estragole (0–48.28%; only in the varieties "Lettuce Leaf" and "Dark Green"). The alkenylbenzene content depends on the basil variety, season, and the environmental conditions, as well as the maturation state at harvest, such as growth height. Another alkenylbenzene found in nearly all sweet basil oils investigated is eugenol (0–33.5%). All the studied varieties of *Ocimum basilicum*, except "Lettuce Leaf" (lowest contents of methyleugenol 9.24–15.45%), were very rich in methyleugenol (up to 87.04%) with dependence on solar irradiance, temperature and relative humidity as determining factors [25]. Earlier studies on the chemical components of *Ocimum basilicum* plants focused on the age and the leaf position at the stem [26], as well as differentiated the essential oil analysis derived from the flowers, leaves, and stems [27]. Eugenol levels were slightly higher in younger leaves, and methyleugenol levels predominated in older leaves, but appears to be more affected by leaf position. The flowers of basil collected in Turkey contained 58.26% estragole, 0.23% *trans*-anethole, and only 0.03% methyleugenol. The respective leaves contained 52.60% estragole, 0.55% *trans*-anethole, and 0.18% methyleugenol. Interestingly, the basil stems contained less estragole (15.91%), *trans*-anethole (0.10%) and methyleugenol (0.06%), but in addition, and exclusively found in stems, were dill-apiol (50.07%), apiol (9.48%), elemicin (0.30%), and low amounts of eugenol (0.12%). There still appears to be no full clarity on the biosynthetic pathways of alkenylbenzenes in basil species and the environmental factors influencing the expression of biosynthetic enzymes. As discussed by Vani and colleagues, chavicol *O*-methyltransferase identified in crude protein extracts of sweet basil may be responsible for the conversion of chavicol to estragole [28]. Eugenol may be transformed into methyleugenol by eugenol *O*-methyltransferase, both enzymes most likely use *S*-adenosylmethionine (SAM) as the methyl donor. However, formation of estragole and methyleugenol is strongly dependent on season and on solar irradiance. Estragole contents may even reach 81% if leaves of *Ocimum basilicum* are extracted with n-hexane before analysis with GC–MS [28]. Using the same techniques, Vani et al. identified high contents of methyleugenol (36–76%) in n-hexane extracts of another basil species *Ocimum tenuiflorum* (also named *Ocimum sanctum*), mainly grown in India.

Exposure to estragole and methyleugenol might be low at common use levels of fresh basil, but there are only a few systematic investigations of alkenylbenzene contents in food preparations of various recipes. Moreover, with consumption of an essential oil merchandised as a food supplement or the plants being part of dishes in which basil is prepared together with other culinary oils, consumer exposure to alkenylbenzenes may increase considerably. An example is given by Bousova and colleagues who found estragole at a high concentration of 101 mg/kg pesto product [43]. This traditional dish from Genova, Italy, mainly consists of olive oil, hard cheese, pine nuts, garlic, salt, and basil leaves. Very varying levels of estragole in pesto preparations have been reported (0.05–19.30 mg/kg versus fresh basil containing 10.21–16.05 mg/kg [44]. Another study reported levels of estragole in "Pesto Genovese" (3.2–34.1 mg/kg estragole) [6]. The same study additionally reported levels of methyleugenol (22.9–56.4 mg/kg) and even myristicin (13.2–15.8 mg/kg), and in one sample apiol (3.4 mg/kg).

Recently, Sestili et al. concluded that maximum level should be precautionarily defined for alkenylbenzenes from different basil species and thus different chemotypes that contain high amounts of methyleugenol and estragole in essential oils intended for consumption with food [45]. Currently, no precise data regarding the consumption of basil or the realistic levels of different alkenylbenzenes in this herb is available.

Further occurrence data of alkenylbenzenes in other herbs and spices, such as allspice, anise, and tarragon, are summarized in Table 1.

2.1.2. Alkenylbenzenes in Aromatized and Fortified Food Products

Many essential oils contain alkenylbenzenes. The most prominent examples of essential oils used in food and beverages are oils produced from basil, fennel, tarragon, parsley, anise, star anise, nutmeg, and mace [46]. Such oils are mostly obtained from plant components by hydrodistillation, steam distillation, solvent extraction, supercritical fluid extraction, ultrasound- or microwave-assisted extractions, or a combination of diverse techniques [47].

When consumed with food products, essential oils can contribute significantly to the overall exposure to potentially genotoxic and carcinogenic compounds. These oils usually contain between 30–90 weight% of the critical ingredient. Depending on the amount of essential oils added to processed foods for reasons of flavoring or food supplements, unknown amounts of alkenylbenzenes exist as undefined mixtures in finished food products. Although these oils are generally meant to be used in very small volumes to refine culinary products, it is, however, difficult to calculate people's overall exposure, also because of the individual food intake habits. As a special case, plant food supplements may contain high amounts of essential oils.

Essential Oils Used as Food Flavorings

Depending on the origin of the plants, basil oils and tarragon oils contain variable but very high amounts of estragole (methylchavicol). Whilst basil oils may be widely used by consumers, tarragon oils are mainly used for food aroma compositions [32]. Parsley seed oils are used for seasonings for meat and sauces. They contain apiol, myristicin, and 2,3,4,5-tetramethoxy-allylbenzene. Pimento oils from berries or leaves of that tree predominantly contain eugenol and can also be used for food aroma compositions. In most essential oils containing anethole, the *trans*-anethole isomer by far predominates the *cis* isomer. *trans*-Anethole contents are high in fennel, anise, and star anise oils [46].

Nutmeg oils and mace oils are mainly used for cola-flavored soft drinks and may contain myristicin and other alkenylbenzenes. Thus, it is expected that all the ingredients of nutmeg are part of cola-flavored soft drinks to various extents. Major compounds of nutmeg and mace oils are sabinene, *alpha*- and *beta*-pinene, myrcene, limonene, and at least five different alkenylbenzenes. Myristicin, safrole, and elemicin determine the flavor of these oils to a great extent. Myristicin, safrole, elemicin, methyleugenol, and eugenol could be quantified in cola-flavored soft drinks [48]. However, an at least 30-fold variation in the levels of safrole and myristicin, for example, has been reported in different nutmeg oils of specific geographical origins, ranging from 0.1–3.2% and from 0.5–13.5%, respectively [48]. Consequently, the amounts of safrole and myristicin were quantified in cola-flavored soft drinks of different brands and following different processing procedures, including various storage conditions. Variation in the contents of safrole and myristicin in different cola-flavored soft drinks were identified to be approximately two to three orders of magnitude [49]. Minimum contents of safrole and myristicin were 0.6–0.4 µg/L, and maximum levels ranged from 43.9–325.6 µg/L for safrole and myristicin, respectively. Other alkenylbenzenes than safrole and myristicin were not evaluated in those cola-flavored soft drinks, so that the total content of alkenylbenzenes in cola-flavored soft drinks remains to be elucidated.

2.2. Toxicity of Alkenylbenzenes

2.2.1. Toxicokinetic Impact on Toxic Properties of Alkenylbenzenes

Following oral exposure, alkenyl benzenes are rapidly absorbed from the gastrointestinal tract. The low systemic bioavailability of the ingested parent compounds, however, points to a pronounced first pass metabolism [50–58]. Different metabolic routes have been observed for alkenylbenzenes, resulting either in bioactivation (toxification) or in detoxification of the parent compounds. The extent of the different pathways depends on species and dose [59,60]. Important metabolic steps of estragole as an example for the alkoxyallylbenzenes are shown in Figure 1.

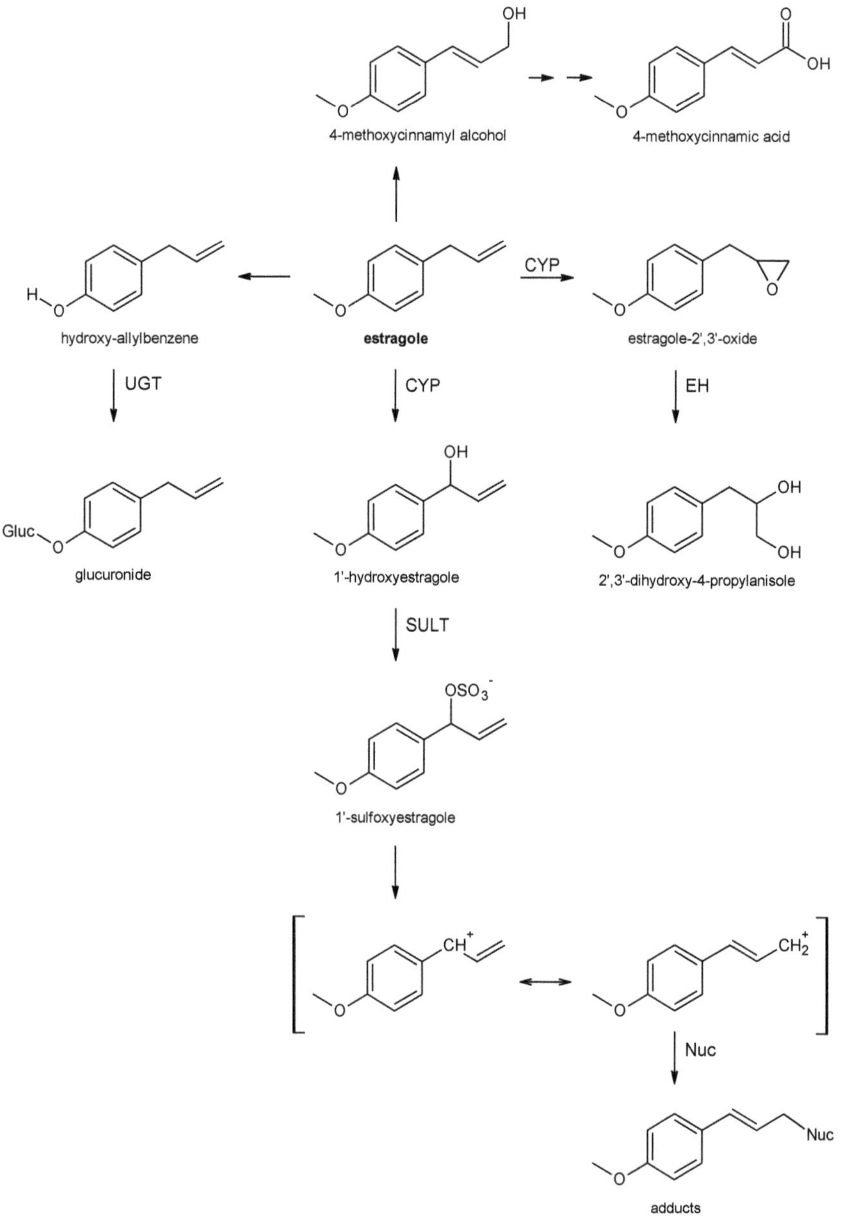

Figure 1. Important metabolic steps of estragole as an example for the allylalkoxybenzenes. CYP, cytochrome P450-monooxygenase; UGT, uridine 5′-diphospho-glucuronosyltransferase; EH, epoxide hydrolase; SULT, sulfotransferase; Nuc, nucleophile (e.g., DNA, protein).

For alkoxyallylbenzenes, such as safrole, methyleugenol, and estragole, metabolic pathways include *O*-dealkylation of the alkoxy substituents at the aromatic ring, epoxidation at the double bond of the allylic side chain, and $1'$-hydroxylation of the allylic side chain [59,60]. *O*-dealkylation of an aromatic alkoxy group (or demethylenation) leads to the formation of the corresponding phenolic (catecholic) derivatives [50,61–67]. The resulting phenol group can be further metabolized via phase II enzymes to stable glucuronides or sulfate conjugates that are rapidly excreted in the urine [54,59]. Therefore, this metabolic route can be considered a detoxifying pathway.

Epoxide formation at the double bond of the allylic side chain represents another metabolic route. Following epoxidation, the epoxide ring can be cleaved by epoxide hydrolases to form diols. The occurrence of $2',3'$-dihydrodiols (and sometimes the epoxides) in urine of rodents treated with different alkenylbenzenes points to the formation of these metabolites in vivo [64,67,68]. Additionally, detoxification of epoxides by glutathione-*S*-transferases was observed [69,70]. If detoxification does not occur fast enough and/or to a critical extent, epoxides may be attacked by nucleophilic structures of the cell. Experiments of Guenthner and Luo have demonstrated that the epoxides are capable forming covalent adducts with proteins and DNA in vitro, suggesting a potential for genotoxicity [69,71]. However, the toxicological relevance of that pathway is generally considered low, since the epoxide is rapidly detoxified by epoxide hydrolases or via glutathione conjugation, with humans generally having a higher epoxide hydrolase activity than rats [59,60,69].

The first step of the third pathway is the cytochrome P450 (CYP)-mediated $1'$-hydroxylation of the allylic side chain [59,60]. $1'$-hyroxy derivatives were detected as metabolites in the urine of rodents and humans following oral exposure to alkenylbenzenes [50,57,58,64,68]. On one hand, $1'$-hydroxy derivatives can be further metabolized by glucuronidation, leading to detoxification, as demonstrated for $1'$-hydroxyestragole [72]. Another option for detoxification, especially in humans, is the oxidation to the corresponding oxo derivative, which may be conjugated with glutathione [73–75]. However, bioactivation is also possible as the $1'$-hydroxy alkenylbenzenes can subsequently be sulfoconjugated by SULTs. The resulting allylic sulfate esters are instable and may react with cellular nucleophiles, such as proteins or DNA [76,77]. This metabolic pathway is considered primarily responsible for the tumorigenic activity of some allylalkoxybenzenes, such as safrole, estragole, and methyleugenol [8,78,79]. The toxicological relevance of this pathway is also underlined by the finding that co-administration of the SULT-inhibitor pentachlorophenol (PCP) drastically reduced the carcinogenic activity of safrole in rodents [80]. Apart from that, the reactive sulfate esters may by detoxified by glutathione conjugation, yielding mercapturic acid derivatives. Indeed, the occurrence of *N*-acetyl-*S*-[$3'$-(4-methoxyphenyl)allyl]-L-cysteine—the mercapturic acid formed from $1'$-sulfoxy estragole—has been detected in the urine of human volunteers after drinking fennel tea containing approximately 2 mg estragole [52]. Of note, although the highly reactive metabolites are formed via $1'$-hydroxylation followed by sulfoconjugation at $1'$-position, the final adducts are formed at the sterically less hindered $3'$-position [52,77,81].

In addition, modifications at the side chain, yielding the $3'$-hydroxylated isomers with the double bond in $1',2'$-position, is also possible. Such metabolites, as well as the parent compounds, may undergo further conversion to $3'$-hydroxy and $3'$-oxo derivatives via different chemical reactions [61,81].

Whereas the detoxifying *O*-dealkylation appears to be predominant at relatively low dose levels in rodents, the fraction of $1'$-hydroxylation at the allylic side chain—leading to the proximate carcinogenic metabolite—seems to increase at higher doses in rodent studies [68,82,83]. However, the formation of $1'$-hydroxy metabolites is also possible at relevant dose levels in humans, as the $1'$-hydroxy metabolite of estragole has already been detected in the urine of human volunteers after drinking fennel tea [58]. Likewise, the occurrence of *N*-acetyl-*S*-[$3'$-(4-methoxyphenyl)allyl]-L-cysteine—the mercapturic acid formed from $1'$-sulfoxy estragole—in the urine of human volunteers after drinking fennel tea [52], as well as the detection of DNA adducts of 1′-sulfoxymethyleugenol in human liver samples [84,85], underlines the formation of reactive cations in humans. Of note, interindividual human variations, such as polymorphisms and lifestyle factors influencing the activity of certain enzymes involved in the metabolism of allylalkoxybenzenes, may also influence the level of bioactivation of these compounds [85–88]. To illustrate this, Tremmel et al. have shown that the number of methyleugenol-derived DNA adducts in human liver samples is associated with the SULT1A1 copy number polymorphism [85].

In contrast to safrole, methyleugenol, and estragole, no carcinogenic effects have yet been observed for other members of the alkoxyallylbenzenes, such as elemicin and apiol [79]. Of note, the available studies generally do not meet today's standards for carcinogenicity studies, e.g., study duration was often too short. Results from physiologically based biokinetic modeling studies, however, suggest that the extent of bioactivation to the ultimate carcinogenic 1′-sulfoxy metabolites is in the same order of magnitude for safrole, methyleugenol, estragole, elemicin, and myristicin, also pointing to a toxicological relevance for the latter two compounds [74,89–91].

Alkoxyprop-1-enylbenzenes, such as *trans*-anethole, are generally considered less toxic compared to the alkoxyallylbenzenes [60], although *trans*-anethole also acts as a liver carcinogen at high dose levels [92]. This class undergoes similar metabolic changes, such as the alkoxyallylbenzenes [57,65–67,93,94]. However, formation of the 1′,2′-epoxide is assumed to be primarily responsible for the hepatotoxic effects of *trans*-anethole observed in rodent studies at higher dose levels [95]. The efficient detoxification by epoxide hydrolases and glutathione—as described for the alkoxyallylbenzenes—may limit the toxicological relevance at low exposure levels. Another major metabolic pathway for this class is hydroxylation at the 3′-position at the propenyl side chain. Interestingly, in contrast to the 1′-hydroxy metabolites formed from the alkoxyallylbenzenes, the 3′-metabolites of the alkoxyprop-1-enylbenzenes are not efficiently metabolized by SULTs but mainly undergo oxidative side chain modification, yielding alkoxy cinnamoyl derivatives and alkoxy benzoic acid derivatives that are further conjugated with glycine [51,57,62,66]. Nevertheless, it has recently been demonstrated that both *trans*-anethole and estragole may principally lead to the formation of the same DNA adducts and hemoglobin adducts at the 3′-position, although adduct formation resulting from *trans*-anethole is much lower. This adduct formation by *trans*-anethole observed in hepatic S9-mix was efficiently blocked by PCP, indicating that some of the primarily formed 3′-hydroxyanethole is also converted by SULTs into the reactive 3′-sulfoxyanethole [76].

No carcinogenicity has been observed for eugenol, a hydroxyallylbenzene [79,96]. For this structural class, the free phenolic hydroxyl group enables a rapid phase II conjugation, leading to hydrophilic and non-toxic metabolites that are subject to fast renal elimination [54]. This difference may explain the lower toxicity of hydroxyallylbenzenes compared to alkoxyallylbenzenes. Further metabolic routes exist for hydoxyallylbenzenes, e.g., isomerization of the double bond and quinone methide formation [59,60]. However, these pathways shall not be described here, as hydroxyallylbenzenes are not in the primary focus of this review.

For the hydroxyprop-1-enylbenzene derivative isoeugenol, evidence of carcinogenicity was observed in a two year study [97]. However, the relevance of these findings is not yet fully clear [60]. Generally, this class of compounds may undergo similar metabolic changes, similar to those of the other alkenylbenzenes. However, the combination of the free phenol group and the double bond in 1′,2′-position facilitates rapid detoxifying metabolism via phase II conjugation at the phenolic hydroxyl group and hydroxylation at 3′-positions [53].

2.2.2. Aspects Regarding Genotoxic and Carcinogenic Effects of Alkenylbenzenes

Safrole

Regarding the toxicity of safrole, animal studies showed that administration led to the induction of tumors, e.g., in mice and rats [98]. In the early 1960s, the first data were published indicating that safrole causes carcinogenic effects in rat liver [99]. In the following years, results of different animal studies (e.g., in rat and mice) confirmed that safrole is a carcinogen in the liver and other tissues, such as the lung [100,101]. Moreover, it was demonstrated that this carcinogenic effect was—at least in parts—mediated via active metabolites, such as 1′-hydroxysafrole or, rather, 1′-sulfoxysafrole [80,98,101–103]. The mutagenic effect of safrole and its metabolites was also verified in vitro and in vivo [7,104]. The toxicological relevance of the genotoxic 1′-sulfoxysafrole is underlined by the finding that co-administration of the SULT-inhibitor PCP drastically reduced the carcinogenic activity of safrole in rodents [80]. Therefore, the Scientific Committee on Food (SCF) of the European Commission (EC) considered safrole as a genotoxic carcinogen in 2002 [98]. In line with this, International Agency for Research on Cancer (IARC) also classified safrole as "possibly carcinogenic to humans" (Group 2B) [105].

Estragole

Results of different in vivo studies indicated that treatment of mice with estragole or its metabolite 1′-hydroxyestragole led to the induction of hepatic tumors [8,79,106]. Results of further studies conducted in bacteria and in cell culture indicated that mutagenic effects were more pronounced following treatment with the metabolite 1′-hydroxyestragole than with the parent compound estragole [7,107]. Therefore, induction of liver tumors seems to depend on the formation of 1′-hydroxymetabolites [8,31] that are further activated to highly reactive 1′-sulfoxy metabolites [108]. Based on the available data, the SCF of the EC concluded that estragole is genotoxic and carcinogenic [109]. Therefore, it was not possible to establish a safe exposure limit, and usage restrictions were recommended [109].

Methyleugenol

Long-term studies have revealed that methyleugenol induces liver and neuroendocrine tumors in rodents [79,110]. In this context, the National Toxicology Program (NTP) stated that there was clear evidence for carcinogenic activity in rats and in mice [110]. Methyleugenol was considered to be a multisite and multispecies carcinogen [111]. Different in vitro studies provided inconclusive results regarding mutagenicity of methyleugenol. In bacterial test systems, no mutagenic activity of methyleugenol was found without metabolic activation, whereas, e.g., in mammalian cell culture, a genotoxic activity was observed [110,112–114]. Moreover, the 1′-hydroxy- and 2′,3′-epoxy-metabolites were also found to be mutagenic in vitro [112,114]. In 2000, de Vincenzi et al. concluded from these findings that methyleugenol is a naturally occurring genotoxic carcinogen, exhibiting a DNA-binding potency similar to that of safrole. In line with this, the SCF of the EC also stated that methyleugenol has been demonstrated to be genotoxic and carcinogenic and recommended reduction in exposure and restrictions in use levels for this substance [111]. Substantiating this, IARC classified methyleugenol in 2013 as "possibly carcinogenic to humans" (Group 2B) [115].

Other Alkenylbenzenes

Even if some alkenylbenzenes are structurally closely related, such as estragole and *trans*-anethole or eugenol and methyleugenol (see Table 1), they often exhibit a different genotoxic and carcinogenic potential.

For eugenol, long-term studies revealed no mutagenic potential and no carcinogenic effects in rodents [79,96,116]. Based on the available data and a corresponding quality assessment, different authorities considered eugenol as not genotoxic or carcinogenic [117,118].

In contrast to eugenol, isoeugenol was found to exhibit carcinogenic activity in rodents (e.g., in the liver of mice [97,119]). The relevance of this finding, however, is not fully clear, yet [60]. Results of various in vitro and in vivo studies showed no mutagenic activity [113,120], whereas some few studies indicated a potential genotoxic activity at higher concentration in vitro [121]. However, based on the available data, isoeugenol was considered to be a non-genotoxic carcinogen by Joint Food and Agriculture Organization/World Health Organization (FAO/WHO) Expert Committee on Food Additives (JECFA) and European Food Safety Authority (EFSA). Moreover, it was concluded that isoeugenol would not rise a safety concern at the estimated intake levels arising from use as a flavoring substance [122–124].

In chronic rodent studies, *trans*-anethole did not increase the tumor incidence [79,92,95]. Moreover, most studies performed on the mutagenicity of *trans*-anethole failed to show a mutagenic activity, whereas only a few studies—often offering a limited reliability or reproducibility—indicated a mutagenic activity [7,95,125,126]. Based on the available data, JECFA concluded that *trans*-anethole is not genotoxic [126]. Due to the hepatotoxic effects observed in rats (considered as secondary to its cytotoxic properties and possibly mediated via the anethole epoxide), safety concerns were formulated by JECFA regarding the use of *trans*-anethole as flavoring agent [126]. In this context, recently published results have to be mentioned, showing that both *trans*-anethole as well as the structurally related estragole are able to form DNA and hemoglobin adducts, even if the adduct formation resulting from *trans*-anethole is much lower [76].

2.2.3. Toxicity of Alkenylbenzenes from Complex Food Matrices

In 2019, EFSA published a guideline document regarding the genotoxicity assessment of chemical mixtures [127]. In this guideline, EFSA recommends the application of a component-based approach to chemical mixtures, in which the genotoxic potential of all components are assessed individually. Consequently, this means that, if a mixture contains one or more chemical substances that are individually assessed to be genotoxic (in vivo via a relevant route of administration), the whole mixture raises concern for genotoxicity [127]. In line with this, the toxicity of alkenylbenzenes is, in most cases, tested in vivo only with pure substances [128].

In 2008, Rietjens and colleagues mentioned that exposure of consumers to alkenylbenzenes under everyday conditions most often occurs in presence of a "normal" food matrix, such as herbs (e.g., basil) or in processed food products (e.g., pesto sauces or beverages), respectively [4,128]. In this context, it has to be kept in mind that—due to the presence of several alkenylbenzenes in most foods/matrices, such as anise or basil—additive effects must be assumed regarding the genotoxicity of mixtures. In line with this, Dusemund et al. stated that specific foods may contain more than one alkenylbenzene [9]. This, in turn, could possibly lead to additive or combined toxicity effects of different alkenylbenzenes taken up via the same food. Moreover, other researchers have also concluded that the consumption of food containing different alkenylbenzenes may contribute to combined toxic effects [3,6,129,130]. This may also apply to other compounds present in a distinct food matrix (e.g., contaminants), which can affect similar endpoints like alkenylbenzenes, such as genotoxicity by a comparable or different mode of action [131–133].

On the other hand, Rietjens et al. stated that certain food matrices may also reduce the genotoxic potential of alkenylbenzenes via interaction on a metabolic level [128]. Bioactivation of alkenylbenzenes, such as safrole and estragole, plays an important role for mediating their genotoxic effects via the generation of proximate (1′-hydroxy metabolites) and ultimate (1′-sulfoxy metabolites) carcinogenic intermediates [89,102,128].

In 2008, Jeurissen et al. found that methanolic basil extract reduces the genotoxic effect of 1′-hydroxyestragole in vitro via inhibition of SULT-mediated bioactivation to the 1′-sulfoxy metabolite [134]. In contrast to that, Müller et al. failed to show any protective matrix-derived effect related to other matrix compounds in an in vitro study characterizing the genotoxic effects of estragole vs. estragole-containing basil oil [135]. The EFSA Scientific Co-operation (ESCO) working group discussed this discrepancy in 2009. Comparing the available in vitro data for these specific basil-based preparations, they concluded that the occurrence of potentially protective matrix effects have to be shown in vivo at relevant intake levels for each botanical or botanical preparation of interest [12].

Several years later, Alhusainy et al. identified nevadensin to be the compound in basil extract responsible for reducing the generation of the ultimate carcinogenic metabolites of estragole and methyleugenol [11,136–138]. Additionally, van den Berg et al. found apigenin, a less potent SULT-inhibitor, to be present in powdered basil material, too [10]. It was speculated that bioactivation of estragole may be reduced by matrix-derived combination effect of SULT-inhibitors, such as nevadensin in basil-containing foods [10,11,136]. However, regarding realistic human intake of foods containing high levels of estragole, such as basil-containing plant food supplements, this matrix effect was predicted to be of limited relevance [10]. Therefore, van den Berg and colleagues critically stated that the matrix-derived combination effect for those basil-containing foods should be judged on a case-by-case basis [10].

Furthermore, the presence of those SULT-inhibitors in botanical matrices was only shown in some distinct botanical preparations, such as methanolic basil extract and basil-containing plant food supplements [10,136]. Therefore, the existence, validity, and potential efficacy of those protective effects has been unknown for most other botanicals or botanical preparations, until now. Moreover, the relevance of those potential matrix-derived effects seems to be rather low in the context of human-relevant exposure level [139].

2.2.4. Genotoxicity and Carcinogenicity of Alkenylbenzenes Required Restrictions for Their Use in Foods

Due to their genotoxic properties, the use of safrole as a flavoring substance for human food has been prohibited in the USA since 1960 [140]. Moreover, the use of methyleugenol as a flavoring substance in food was also forbidden in the USA by the United States (U.S.) Food and Drug Administration (FDA) in 2018 [141]. By contrast, the use of estragole in foods is not restricted in the USA [9,31]. The U.S. FDA approved *trans*-anethole as a food additive [95]. Isoeugenol is also approved as a flavoring substance in food in the USA [142]. The same applies for eugenol [116,143]. Moreover, the use of eugenol is also permitted in other regions, including Australia, Indonesia, and the European Union (EU) [116]. Besides this, eugenol and isoeugenol are also approved as fish anesthetic, e.g., in Australia, New Zealand and Finland, but not in the EU or the USA [144,145]. In this context, it is not surprising that residues of both substances were also found in the fillet tissue of freshwater fish previously exposed to this compound [146]. Finally, the use of myristicin in food products is not regulated in the USA [147]. Moreover, there are currently no specific guidelines or laws concerning the production or sale of synthetic myristicin or myristicin isolated from natural sources [147]. This also applies to the EU and other regions/countries in the world. The aforementioned information shows that the use of different alkenylbenzenes—some of which have genotoxic potential, e.g., estragole or *trans*-anethole—is not adequately regulated in the USA. Moreover, the use of most alkenylbenzenes is currently not regulated in most other regions of the world, including Asia, Africa, and South America.

In 2001 and 2002, respectively, the Scientific Committee on Food (SCF) of the EC evaluated safrole, methyleugenol, and estragole and concluded that these compounds are genotoxic carcinogens and suggested restrictions for their use in foods [98,109,111]. Based on the SCF's recommendations, the EC prohibited the addition of pure safrole, methyleugenol, and estragole as a flavoring substance to food and established maximum levels for these substances—when naturally present in corresponding ingredients—in certain compound foodstuffs, such as soups and sauces or non-alcoholic beverages [148]. Thus, in the EU, estragole, methyleugenol, safrole, and *beta*-asarone shall not be added as such to food (see Annex III Part A of Regulation (EC) No 1334/2006). Further EU restrictions apply to these alkenylbenzenes (in Annex III Part B of Regulation (EC) No 1334/2006). The maximum levels of estragole, methyleugenol, and safrole, naturally present in flavorings and food ingredients with flavoring properties or in certain food compounds to which flavorings and/or food ingredients with flavoring properties have been added, have been defined by the EU Parliament and the Council. Accordingly, estragole may not be present in amounts greater than 50 mg/kg food in dairy products, processed fruits, vegetables (including mushrooms, fungi, roots, tubers, pulses, and legumes), nuts and seeds, and fish products. Non-alcoholic beverages may not contain more than 10 mg estragole per kg. As for methyleugenol, soups and sauces may not contain more than 60 mg/kg;

dairy products and ready to eat savories, no more than 20 mg/kg; meat preparations and meat products, including poultry and game, no more than 15 mg/kg; fish preparations and fish products, no more than 10 mg/kg; and non-alcoholic beverages, no more than 1 mg methyleugenol/kg. In addition, even up to 25 mg/kg safrole may be present in soups and sauces; 15 mg safrole/kg in meat preparations and meat products, including poultry and game; safrole is still permitted in fish preparations and fish products. In non-alcoholic beverages, 1 mg safrole/kg shall not be exceeded. Furthermore, the content of beta-asarone, a major constituent of *Calamus* oils, is legally restricted in Europe for alcoholic beverages to a maximum of 1 mg/kg (see Annex II Part B of Regulation (EC) No 1334/2006). The tetraploid form *of Acorus calamus* L. shall not be used as a source for the production of flavorings and food ingredients with flavoring properties (see Annex IV Part A of Regulation (EC) No 1334/2006). In addition, according to the abovementioned regulation, "the maximum levels shall not apply where a compound food contains no added flavorings and the only food ingredients with flavoring properties which have been added are fresh, dried or frozen herbs and spices". However, the usage of other structurally related (see Table 1) and potentially toxic alkenylbenzenes, such as elemicin or apiol, is, so far, not regulated in the EU, whereas some derivatives, such as eugenol, isoeugenol, and *trans*-anethole are listed as authorized flavoring compounds in Regulation (EU) No 872/2012.

3. Aspects Impeding the Evaluation of Adverse Health Effects of Alkenylbenzenes

3.1. Uncertainties Regarding the Occurrence of Alkenylbenzenes

Adequate and comparable occurrence data are of high importance in order to estimate oral exposure of humans to certain alkenylbenzenes via consumption of foods containing these substances. Currently, there are several issues, which make it difficult to perform such a reliable oral exposure assessment of consumers for alkenylbenzenes. These aspects will be discussed in the following parts of the text.

3.1.1. Conclusions Regarding Aromatized Foods and Their Potential Alkenylbenzene Contents

Even if the amount of a specific alkenylbenzene appears to be low in a certain food category, there is a risk of dose addition depending on the dietary habits of consumers if many alkenylbenzene-containing food products are frequently ingested in a short period of time. Some alkenylbenzenes, such as elemicin and apiol, have not yet been fully assessed for their hepatotoxic and genotoxic potential and have not been sufficiently monitored in the potentially relevant food products. Since, for all existing food matrices, specific extraction, separation, and detection procedures for each of the alkenylbenzenes would have to be elaborated, standardized, and validated, this appears to be a difficult endeavor. However, given the high hepatotoxic potential of other alkenylbenzenes, there is an undoubted necessity to analytically determine all the possible alkenylbenzenes in raw and prepared food products and food preparations. Analytical techniques have advanced significantly in specificity and sensitivity in recent years, and the first promising approaches have been made to quantitate some alkenylbenzenes in foods and beverages [6,49,149–152]. Further ambitious experimental monitoring activities could build on those approaches.

3.1.2. Issues Regarding Currently Available Occurrence Data for Alkenylbenzenes

The herbs and spices that we discussed above contain variable amounts of alkenylbenzenes. Depending on the family, the genus, the species and their varieties, the geographical origin (e.g., soil, humidity, solar irradiance, etc.), and the plant parts analyzed (fruits, seeds, flowers, leaves, stems, roots, in different maturation states at harvest, etc.) very different contents of alkenylbenzenes and their metabolites may prevail in foods. In addition, depending on the procedures of post-harvest treatment, sample preparation (extraction methods and duration, solvents used, etc.), and analytical methods utilized (e.g., GC, LC, etc., which are described in detail elsewhere [153]), the reliability of quantitative data may vary considerably. These circumstances call for the elaboration of standardized analytical procedures to enable reliable quantification of alkenylbenzenes in crude spice extracts, essential oils and their oleoresins, and finished aromatized foods. Ideally, such methods would become internationally harmonized. These efforts have to be complemented with efforts to measure all known alkenylbenzenes in a representative set of well-defined finished food products that belong to food categories naturally containing alkenylbenzenes, and in those that intentionally become aromatized.

3.2. Consumption of Alkenylbenzene-Containing Foodstuffs

Besides information on occurrence, data on consumption also play an important role for the exposure assessment. In the following section, we will discuss some aspects leading to uncertainties regarding the currently available consumption data on alkenylbenzene-containing foods

3.2.1. Limited Availability of Data Regarding Consumption

Besides data regarding some few individual foods (e.g., fennel-based teas or plant food supplements) [5,12,154], the availability of consumption data—especially of current information—in the context of alkenylbenzenes is rather limited [98,109,111]. This leads to uncertainties regarding the current exposure of humans to alkenylbenzenes via the consumption of food.

Among other things, the availability of current consumption data on naturally alkenylbenzene-containing foods, such as herbs, spices, and flavored foods (e.g., baked goods or beverages) is mandatory to perform a reliable risk assessment for these substances.

Currently available consumption data regarding alkenylbenzene-containing food are not up to date, since they were largely captured and evaluated approximately 20 years ago [98,109,111]. In this context, it is important to note that the consumer behavior may have changed over the last two decades in different regions of the world [155,156]. Moreover, consumption habits may also vary between different countries in the EU as well as worldwide [157]. This may lead to differences regarding exposure of consumers to alkenylbenzenes via distinct locally favored foods or mainly regionally consumed food products, such as basil-containing pesto sauce, herbal Indonesian beverages, or herb-based Chinese medicinal teas [6,116,130].

3.2.2. Lack of Biomarker Prevents Exposure Estimation

Until now, no biomarker has been identified that reliably reflects the external exposure of humans to alkenylbenzenes via food consumption under realistic conditions. However, the analytical quantification of a biomarker is a possible alternative for the estimation of the external exposure. Two types of biomarkers are commonly used for other compounds in food, i.e., mercapturic acids (MAs) in urine samples [158] or protein adducts (usually determined in hemoglobin) [159]. Amounts of a urinary MA excreted within 24 h may be used for the estimation of the daily exposure to the parent compound, if the compound ratio is known, which is converted into the MA (reverse dosimetry) [160]. Recently, *N*-acetyl-*S*-[3'-(4-methoxyphenyl)allyl]-L-cysteine was described as the main MA, which is formed from estragole and its structural congener *trans*-anethole [52]. A single controlled exposure to fennel tea (n = 12) resulted in the excretion of *N*-acetyl-*S*-[3'-(4-methoxyphenyl)allyl]-L-cysteine in the urine samples within 24 h. The interindividual variation of total *N*-acetyl-*S*-[3'-(4-methoxyphenyl)allyl]-L-cysteine excreted (93–1076 ng) reflected the complexity of estragole/*trans*-anethole metabolism involving different enzyme families, i.e., CYP, SULT and alcohol dehydrogenases (ADH). It hinders an accurate estimation of the external exposure for individuals from the *N*-acetyl-*S*-[3'-(4-methoxyphenyl)allyl]-L-cysteine determined in 24 h urine samples. In addition, the biomarker is not specific; one cannot distinguish between *N*-acetyl-*S*-[3'-(4-methoxyphenyl)allyl]-L-cysteine formed from estragole or *trans*-anethole. These considerations also hold true for another biomarker of exposure to these compounds, the hemoglobin adduct *N*-(isoestragole-3-yl)-valine (IES-Val) [76]. Hemoglobin adducts are considered as biomarkers of medium-term exposure, because hemoglobin can accumulate adducts in its lifetime of ~120 d. IES-Val was shown to increase steadily when fennel tea was consumed over 28 d. However, the complexity of metabolism and the missing specificity for one compound also hinders the exposure estimation from IES-Val.

In summary, the conjugate *N*-acetyl-*S*-[3′-(4-methoxyphenyl)allyl]-L-cysteine (24 h urine) and the adduct IES-Val (hemoglobin) can only be used as biomarkers for the internal exposure to the ultimate carcinogen 1′-sulfoxyestragole or to 3′-sulfoxyisoestragole, the reactive sulfate ester metabolites of estragole and *trans*-anethole, respectively [52,76]. Similar studies have not yet been published for the other alkenylbenzenes, such as safrole or methyleugenol. The biomarkers equivalent to those described for estragole/*trans*-anethole may offer a higher specificity. However, the metabolism of safrole and methyleugenol may be as complex as that of estragole, which renders an exposure assessment for individuals at least difficult.

Together, the aforementioned data show that more research is needed regarding the exposure of humans to alkenylbenzenes via the consumption of food, especially in the context of real-life influences. In this context, the development of specific biomarkers and reliable measurement strategies is also of high importance.

3.3. Issues Regarding the Toxicity of Alkenylbenzenes

3.3.1. The Genotoxic and Carcinogenic Potential of Alkenylbenzenes

Toxicity data regarding estragole, safrole, and methyleugenol show that the genotoxic and carcinogenic potential of these compounds is complex and may differ—at least in part—from that of other structurally related alkenylbenzenes, such as *trans*-anethole and eugenol. This may be based—amongst other things—on toxicokinetic differences of alkenylbenzenes, albeit having only slight structural differences, such as estragole vs. *trans*-anethole or methyleugenol vs. eugenol. In many cases, there are no adequate studies regarding carcinogenicity of different alkenylbenzenes, such as elemicin and apiol. In addition, studies investigating the genotoxic potential are often missing or the study design is not adequate to reliably address the genotoxic potential of that class of compounds. Therefore, additional studies are needed, especially those designed according to international guidelines and taking into account the alkenylbenzene specific bioactivation via SULTs to allow a comparative analyses and assessment of the (geno-)toxic potential of alkenylbenzenes in a conclusive manner.

3.3.2. Weaknesses of Standard Genotoxicity Tests and Implications for Hazard Assessment

In order to identify the possible genotoxic activity of a given substance, genotoxicity studies are conducted and evaluated in several legal sectors. As a general rule, at least one mutagenicity test with bacteria and one cytogenicity test with mammalian cells is required [161–165]. Depending on the legal area, in vivo genotoxicity studies are either generally requested or may be subsequently required based on the findings of the in vitro tests.

There are many new or revised Organisation for Economic Co-operation and Development (OECD) test guidelines [166] for several genotoxic endpoints available. These test guidelines comprise the requirements for a reliable study design as well as an acceptable presentation of the study results.

Nevertheless, the tests described by OECD should be regarded as standard tests, which are generally suitable for identifying a possible genotoxic activity of a test substance, but need to be adapted to the individual case. As a prerequisite, information on metabolism as well as mechanistic understanding of the test compound is needed before the genotoxicity test is carried out in a modified way.

The weaknesses of the in vitro and in vivo standard test systems can manifest themselves in either false-positive or false-negative results. False test results are to be avoided in the regulatory process in order to prevent unnecessary animal studies, but also to enable protection of human health. Some pitfalls of standard genotoxicity studies—with special focus on the situation for alkenylbenzenes—will be discussed in the following section.

False-Negative Results

Depending on the existence of particular functional chemical groups in the molecular structure of the test substance, it might be possible that the test substance reacts with components of the metabolic activation system (i.e., proteins of the S9 mix) or the solvent (i.e., DMSO). As described by Nestmann et al. in 1985, DMSO can undergo chemical reactions with alkyl halides [167]. Consequently, the lowered effective concentration of the test substance reduces the sensitivity of the test system and can provoke false-negative results.

Another possibility for an artificial negative result in the Ames test can be extreme experimental conditions, such as drastic changes in pH leading to cytotoxicity. This could, ultimately, mask the quantitative formation and detection of revertant colonies.

Moreover, the solubility as well as the stability of the test substance plays an important role in the sensitivity of the test system. Substances with a short half-life could decay before passing through the bacterial cell wall. Thus, contact with the genetic material in the bacterium would be prevented. This circumstance is particularly critical as the test substance is per se reactive due to its low stability. Thus, nucleophilic reactions of the DNA with the test substance are considered likely.

For some substances, it has also been shown that they can be detected better in the Ames test with the pre-incubation method than with the standard plate incorporation assay. Among these are substances with special structural characteristics, such as short chain aliphatic nitrosamines, divalent metals, aldehydes, azo-dyes and diazo compounds, pyrollizidine alkaloids, allyl compounds, and nitro compounds [168]. In order to avoid false-negative results, the appropriate modification of the Ames test should be favored.

For other genotoxic endpoints, such as clastogenicity, false-negative results are also described in the literature. Substances producing crosslinks with DNA should be handled with care in Comet assays in which DNA strand breaks are detected. To DNA/DNA-intra-strand and DNA/DNA-inter-strand-crosslinkers belong cisplatin and mytomycin C, respectively [169]. Both compounds impair the sensitivity of the Comet assay due to their potential to reduce DNA fragmentation. Consequently, the formation of comets is not adequately captured.

Many substances are not inherently genotoxic but require critical enzymatic steps to form reactive intermediates. The bioactivation process is often mediated by so-called phase I enzymes. On the contrary, phase II enzymes serve the purpose of making substances more inert, water-soluble and thus easier to excrete. Metabolic competence of a test system depends on various factors (e.g., cell type, S9 mix, co-factors). For instance, different cell types express different enzymes. To illustrate this, the metabolic competence of liver cells is greater than that of human lung fibroblasts (V79 cells). However, the latter are often used in common genotoxicity studies, and even the addition of S9 mix—as recommended by OECD test guidelines—cannot compensate for many phase II enzymes or their co-factors. The individual enzyme capacity and activity level within a given test system depends on many further factors. For instance, S9 mix can be obtained from the liver, but also from other organs, such as the lungs or kidney. Of note, the enzyme composition and level are also species-dependent and can be affected by the use of chemical inductors. Although S9 mix is often obtained from livers of arochlor-treated rats, the use of hamster S9 mix or even human S9 mix is advantageous in specific cases. S9 mix largely represents phase I enzymes (CYP). The addition of the appropriate co-factor (nicotinamide adenine dinucleotide phosphate (NADPH)-generating system) thus primarily boosts oxidative conversions. Other enzymes, such as the epoxide hydrolases, require no co-factors other than water, and are thus also addressed by using S9 mix. The situation is different for most phase II enzymes. These are either not expressed by the bacteria or mammalian cells, are not components of the standard S9 mix, or are less active due to a lacking co-factor. Even if a reactive metabolite is formed extracellularly by a metabolic activation system, such as S9 mix, it is questionable whether this metabolite can permeate the cellular barrier and reach the genetic material. All these aforementioned factors regarding biotransformation can contribute to false-negative results.

False-negative results also play an important role in animal experiments. The test for micronuclei formation in bone marrow or peripheral blood is generally only considered reliable and valid if the test substance (or its metabolites) is systemically available. To ensure this, the corresponding OECD test guideline 474 [170] recommends using a reduction in the polychromatic erythrocytes/normochromatic erythrocytes (PCE/NCE) ratio as a surrogate for bioavailability. If the ratio is not decreased compared to the PCE/NCE ratio of the control animals, systemic availability of the test substance is difficult to prove without further information (e.g., clinical signs of animals, detection of the test substance in blood plasma). However, study evaluators are often faced with the problem that many test substances are not cytotoxic in the bone marrow. Consequently, they do not modify the PCE/NCE ratio. In this respect, the test substance could be systemically available without a change in the PCE/NCE ratio.

False-Positive Results

False positive results can often be explained by the presence of extreme conditions in the culture medium. For example, strong fluctuations in pH or osmolality can lead to cytotoxicity and eventually to artificial positive test results.

Another example for false-positive results might be the enzyme equipment of the test system. The classical Salmonella strains applied in the Ames test express nitroreductases. These enzymes allow for azo- and nitroreduction. However, these enzymes are not present in mammals. Owing to this uncertainty, an extrapolation to the human situations might be difficult. Theoretically, metabolic activation of nitro compounds can be mediated by intestinal bacteria in humans. To clarify this, an absorption test should be carried out in these cases. If the test substance (or its metabolites) is completely absorbed, the risk of intestinal bioactivation is low, as direct contact with the intestinal bacteria is unlikely [169].

Possible Ways to Optimize Standard Genotoxicity Tests

One option to mimic possible bioactivation and detoxification steps—taking into account phase I and also phase II enzymes—is the use of bacteria, mammalian cells, and animals that have been genetically modified to artificially express certain enzymes. In this way, bacteria and cells with murine enzymes could serve the purpose of studying possible biotransformation processes in mice. Toxicological endpoints could be the bacterial reverse mutation assay, but also the hypoxanthine-guanine phosphoribosyltransferase (HPRT) assay or the mouse lymphoma assay. Apart from that, indicator assays such as DNA adduct formation could also give indications about the genotoxic activity of a test compound. Furthermore, genotoxicity studies applying bacteria, mammalian cells, or even animals expressing the corresponding murine enzymes might be of particular importance if, for example, tumor formation was observed in mice and the underlying mechanism has to be clarified. Finally, experiments utilizing human enzymes could help to better extrapolate findings from animal studies to the human situation.

In order to select a suitable genotoxicity model, information on the species-dependent metabolism as well as a hypothesis for the underlying genotoxic mode of action is indispensable. A number of in silico tools have been established for the identification of possible structural alerts for genotoxicity. Many of them are well-trained and generate reliable predictions about metabolism. These prediction programs can, therefore, help to identify critical metabolites that might mediate genotoxic events.

Of note, in silico programs should be selected with care. If a chemical structure is too dissimilar for what the in silico model is trained for, a reliable prediction for metabolism or genotoxicity can be challenging. For this reason, a prerequisite for conducting valid and robust predictions is the availability of appropriate representative training data within the in silico model [171].

Optimization of Standard Genotoxicity Tests Using the Example of the Alkenylbenzene Methyleugenol

The dominant metabolic pathway relevant for genotoxicity is conversion of methyleugenol to 1′-hydroxymethyleugenol via CYP enzymes [61]. After sulfo-conjugation of the allylic hydroxyl group by SULTs, electrophilic esters are formed, which can be attacked by nucleophilic structures in the cell (e.g., DNA or proteins). This critical bioactivation step has been described not only for methyleugenol, but also for other alkenylbenzenes, such as safrole and estragole [79,80,102,120,172–177]. If DNA adducts are not error-free repaired by the cell's repair system, they can manifest as mutations. This is particularly concerning if proto-oncogenes or tumor suppressor genes are affected, as cancer development might be triggered.

The genotoxic and mutagenic activity of methyleugenol has been tested in numerous standard in vitro tests. In the bacterial reverse mutation test with conventional bacterial strains, methyleugenol was not mutagenic [113,178,179]. This finding did not change with the addition of an exogenous activation system (S9 mix). The main reason for this observation is that SULTs are not considered in conventional genotoxicity studies. Whereas the use of an S9 mix can increase the metabolic competence for phase I enzymes (if appropriate co-factors are added) many phase II enzymes—such as SULTs—remain unconsidered.

In contrast, mutagenic findings have been observed in bacteria being genetically modified for the expression of murine and human SULTs [77]. Likewise, DNA adduct formation was higher in bacteria, cells, and animals expressing murine or human SULTs in comparison to the wild-type [77,180]. This illustrates that standard genotoxicity tests should be optimized and adapted to the relevant question. However, when quantitatively comparing DNA adduct levels and the mutagenic potential of alkenylbenzenes between conventional, murine, and humanized test systems, attention should be paid to how much SULT (and which form) is expressed in which bacterial strain and animal model [77,180]. Furthermore, the SULT status in mice may also vary between tissues [181].

Outlook for Future Studies and Testing

While, for the alkenylbenzene methyleugenol, the bioactivation via SULTs could be described very well, corresponding experiments with SULT proficient test systems for other relevant alkenylbenzenes, such as estragole, safrole, or elemicin, are still widely lacking. Such experiments should be made up in order to better understand the influence of SULTs, thereby enabling a more realistic extrapolation to the human situation.

3.3.3. Toxicity of Mixtures Is Still a Controversially Debated Issue

The Scientific Panel on food additives, flavorings, processing aids, and materials in contact with food (AFC) of EFSA commented regarding the genotoxic potential of estragole and tarragon that the modification of inherent toxicity of a naturally occurring substance by the matrix in which it is present (e.g., the herb) can be considered plausible [13]. However, the panel further stated that, besides a reduction in toxicity, effects related to additional compounds present in the corresponding matrix could also lead to unchanged as well as to increased toxicity, depending on the mode of action. Moreover, it was mentioned that research on individual substance/matrix interactions cannot be used to draw general conclusions about herbs and spices under all conditions of use, ingestion, and metabolism [13].

Discrepancies between experimental settings and real-life human consumption scenarios raise another important issue in this context. In animal studies, alkenylbenzenes, such as methyleugenol, were administered to test subjects as pure substances. However, this does not reflect the eating habits of the consumer, who mainly consumes methyleugenol via herbs and spices [182,183].

The group of Rietjens could demonstrate that flavonoids such as nevadensin—which, in addition to methyleugenol, is also present in certain matrices, such as herbs and spices—exhibit SULT-inhibiting effects. In animal studies, it was shown that methyleugenol-derived

DNA adduct levels were lowered by simultaneous administration of methyleugenol and nevadensin in the liver of rats [137]. These experimental findings may indicate that such matrix-derived effects should be considered in the evaluation of genotoxicity studies to reliably assess the risk of adduct formation in humans. For methyleugenol, however, adducts could also be detected in human lung and liver samples [84,85,184] raising the question whether possible matrix effects would be sufficient to protect humans against methyleugenol derived mutagenicity. This should be taken seriously, especially since a copy number variation in humans exists for SULT1A1—the SULT form with the highest activity towards 1′-hydroxymethyleugenol [77]. An association of methyleugenol-mediated DNA adducts in human livers with this copy number variation and their expression levels has already been shown [85].

Together, these data and arguments show that the toxicity of mixtures is still a controversially debated issue. Therefore, further studies are needed to shed more light on this controversially debated issue.

3.3.4. Transferability of Findings in Animal Studies to Human

Experimental animal models (e.g., mice or rats) are typically used to study toxicological effects of substances occurring in food, such as alkenylbenzenes, to assess their potential impact on human health through the oral intake of food [96,147,185]. This is in line with current recommendations of international scientific bodies, such as EFSA or OECD, published in corresponding testing guidelines [186,187]. However, the utilization of experimental animals and the transfer or extrapolation of the obtained results to a human setting poses different problems, leading to uncertainties regarding data interpretation and assignability.

Results of rodent studies regarding safrole showed that genotoxic effects were mediated via its active metabolites, such as the proximate carcinogen 1′-hydroxysafrole or, rather, 1′-sulfoxysafrole [80,98,102]. However, findings of a comparative study performed in rats and humans showed that the safrole metabolite 1′-hydroxysafrole was only found in rat but not in human urine [50]. These findings suggest species-specific differences regarding the metabolisms of rodents vs. humans, and—as discussed by Bode and Dong in 2015—this discrepancy raises the question of whether the genotoxic effects observed in experimental animals are also expectable in humans [101]. In line with this interpretation, absence of carcinogenic alkenylbenzene metabolites in human urine was used by Smith et al. as argument against a potential cancer risk to humans through the consumption of food containing methyleugenol and estragole [107]. However, such results might be influenced by the study design, e.g., by the dose administered. Indeed, the formation of 1′-hydroxy metabolites is also possible at relevant dose levels in humans, as the 1′-hydroxy metabolite of estragole has already been detected in the urine of human volunteers after drinking fennel tea [58]. Moreover, *N*-acetyl-*S*-[3′-(4-methoxyphenyl)allyl]-L-cysteine—the mercapturic acid formed from 1′-sulfoxy estragole—was also found in the urine of human volunteers after drinking fennel tea [52], and DNA adducts of 1′-sulfoxymethyleugenol were detected in human liver samples [84,85]. This indicates that the intake of relatively low doses of alkenylbenzenes via food in humans may lead to the generation of instable cations. In addition, it should be noted in this context that a lack of certain metabolites in urine gives no information regarding the presence of metabolites, especially of phase I intermediates, in liver or other metabolizing tissues. In addition, metabolites with certain structural characteristics, such as some sulfates or even carbo cations, may have only a short half-life due to their reactivity and are therefore difficult to detect analytically.

Differences between rodent and human metabolism influencing potential genotoxic effects of alkenylbenzenes were also indicated by others. Sulfoconjugation-mediating SULTs are known to play an important role in the generation of ultimate carcinogenic metabolites of different compounds, including alkenylbenzenes (e.g., metabolites of safrole and methyleugenol) or heat-induced food contaminants, such as furfuryl alcohol [77,89,188]. Under the conditions of the test system used, human SULT1A1 and murine Sult1a1 activated the test compounds at lower concentrations than other members of the SULT family did [77,188]. In this context, the efficacy of human SULT1A1, regarding the activation of methyleugenol, was demonstrated in DNA adduct studies to be higher than that of its murine orthologue in vitro and in vivo [77,180]. Of note, quantitative comparisons should be handled with caution, as the level of adduct formation depends on the level of SULT enzymes. SULT expression varies in a tissue-specific manner and depends on the selected species (e.g., transgenic humanized versus wild-type mice). Nevertheless, these data indicate that SULTs may influence the genotoxic effects of alkenylbenzenes and other genotoxic compounds in a species-dependent manner. Further substantiating this, Al-Malahmeh and colleagues also described, in 2017, species-specific differences between rat and human regarding the metabolism of the alkenylbenzene myristicin and generation of its genotoxic 1′-sulfoxy metabolite. However, physiologically based kinetic modelling indicated that these differences were within a default factor of four [89].

Therefore, it has not yet been fully clarified to what extent carcinogenicity data from animal studies regarding alkenylbenzenes is transferable to humans.

4. Conclusions

In this review, we summarized several aspects regarding the occurrence, toxicokinetics, and toxicity of alkenylbenzenes.

The currently available information summarized in this article clearly show that a number of different alkenylbenzenes, such as safrole, methyleugenol, and estragole, have genotoxic and carcinogenic properties. Although the toxicological relevance for these well-investigated derivatives is still under discussion when these substances are taken up in low amounts via herbs and spices, it seems very clear, from a toxicological point of view, that high intake levels—as may result from specific plant food supplements, for example—should be avoided.

However, there are still several uncertainties impeding a reliable evaluation of the health risks that may result from the intake of different alkenylbenzenes via food. These uncertainties are based on the following data gaps, which need to be closed by appropriate research:

- valid occurrence data reflecting the occurrence of all toxicologically relevant alkenyl-benzenes in different food products
- comprehensive consumption data for such alkenylbenzene-containing products, which should be collected via appropriate consumption surveys
- determination of toxicological properties of yet insufficiently investigated derivatives, such as elemicin and apiol, via adequate studies designed according to international guidelines and taking into account the alkenylbenzene-specific bioactivation (e.g., via SULTs)

The aforementioned uncertainties and associated discussions underline that it is currently not possible to perform a conclusive evaluation of possible adverse effects to human health related to the consumption of alkenylbenzene-containing foods.

Author Contributions: Conceptualization, A.E. and B.S. (Bernd Schäfer); writing—original draft preparation, A.E., M.E.G., B.S. (Benjamin Sachse), B.H.M., K.H. and B.S. (Bernd Schäfer); writing—review and editing, A.E., M.E.G., B.S. (Benjamin Sachse), B.H.M., K.H., and B.S. (Bernd Schäfer). All authors have read and agreed to the published version of the manuscript.

Funding: This research received no external funding.

Institutional Review Board Statement: Not applicable.

Informed Consent Statement: Not applicable.

Data Availability Statement: Not applicable.

Conflicts of Interest: The authors declare no conflict of interest.

References

1. Kristanc, L.; Kreft, S. European medicinal and edible plants associated with subacute and chronic toxicity part I: Plants with carcinogenic, teratogenic and endocrine-disrupting effects. *Food Chem. Toxicol.* **2016**, *92*, 150–164. [CrossRef]
2. Atkinson, R.G. Phenylpropenes: Occurrence, Distribution, and Biosynthesis in Fruit. *J. Agric. Food Chem.* **2016**, *66*, 2259–2272. [CrossRef] [PubMed]
3. Alajlouni, A.M.; Al-Malahmeh, A.J.; Isnaeni, F.N.; Wesseling, S.; Vervoort, J.; Rietjens, I.M. Level of Alkenylbenzenes in Parsley and Dill Based Teas and Associated Risk Assessment Using the Margin of Exposure Approach. *J. Agric. Food Chem.* **2016**, *64*, 8640–8646. [CrossRef] [PubMed]
4. van den Berg, S.J.; Alhusainy, W.; Restani, P.; Rietjens, I.M. Chemical analysis of estragole in fennel based teas and associated safety assessment using the Margin of Exposure (MOE) approach. *Food Chem. Toxicol.* **2014**, *65*, 147–154. [CrossRef] [PubMed]
5. Uusitalo, L.; Salmenhaara, M.; Isoniemi, M.; Garcia-Alvarez, A.; Serra-Majem, L.; Ribas-Barba, L.; Finglas, P.; Plumb, J.; Tuominen, P.; Savela, K. Intake of selected bioactive compounds from plant food supplements containing fennel (*Foeniculum vulgare*) among Finnish consumers. *Food Chem.* **2016**, *194*, 619–625. [CrossRef] [PubMed]
6. Al-Malahmeh, A.J.; Al-Ajlouni, A.M.; Wesseling, S.; Vervoort, J.; Rietjens, I.M. Determination and risk assessment of naturally occurring genotoxic and carcinogenic alkenylbenzenes in basil-containing sauce of pesto. *Toxicol. Rep.* **2016**, *4*, 1–8. [CrossRef]
7. Swanson, A.B.; Chambliss, D.D.; Blomquist, J.C.; Miller, E.C.; Miller, J.A. The mutagenicities of safrole, estragole, eugenol, trans-anethole, and some of their known or possible metabolites for *Salmonella typhimurium* mutants. *Mutat. Res. Mol. Mech. Mutagen.* **1979**, *60*, 143–153. [CrossRef]
8. Drinkwater, N.R.; Miller, E.C.; Miller, J.A.; Pitot, H.C. Hepatocarcinogenicity of Estragole (1-Allyl-4-methoxybenzene) and 1'-Hydroxyestragole in the Mouse and Mutagenicity of 1'-Acetoxyestragole in Bacteria 2. *J. Natl. Cancer Inst.* **1976**, *57*, 1323–1331. [CrossRef]
9. Dusemund, B.; Rietjens, I.M.; Abraham, K.; Cartus, A.; Schrenk, D. Chapter 16—Undesired Plant-Derived Components in Food. In *Chemical Contaminants and Residues in Food*, 2nd ed.; Woodhead Publishing: Southston, UK, 2017; pp. 379–424.
10. van den Berg, S.J.; Klaus, V.; Alhusainy, W.; Rietjens, I.M. Matrix-derived combination effect and risk assessment for estragole from basil-containing plant food supplements (PFS). *Food Chem. Toxicol.* **2013**, *62*, 32–40. [CrossRef]
11. Alhusainy, W.; Paini, A.; Berg, J.H.J.V.D.; Punt, A.; Scholz, G.; Schilter, B.; Van Bladeren, P.J.; Taylor, S.; Adams, T.B.; Rietjens, I.M. In vivo validation and physiologically based biokinetic modeling of the inhibition of SULT-mediated estragole DNA adduct formation in the liver of male Sprague-Dawley rats by the basil flavonoid nevadensin. *Mol. Nutr. Food Res.* **2013**, *57*, 1969–1978. [CrossRef]
12. EFSA-ESCO. ESCO report: Advice on the EFSA guidance document for the safety assessment of botanicals and botanical preparations intended for use as food supplements, based on real case studies. In *EFSA Supporting Publications*; Wiley: Hoboken, NJ, USA, 2009; Volume 6, p. 280.
13. EFSA. *Minutes of the 23rd Plenary Meeting of the Scientific Panel on Food Additives, Flavourings, Processing Aids and Materials in Contact with Food (AFC)*; EFSA: Parma, Italy, 2007.
14. Martinez-Velazquez, M.; Castillo-Herrera, G.A.; Rosario-Cruz, R.; Flores-Fernandez, J.M.; Lopez-Ramirez, J.; Hernandez-Gutierrez, R.; Lugo-Cervantes, E.D.C.; Rosario-Cruz, R. Acaricidal effect and chemical composition of essential oils extracted from *Cuminum cyminum*, *Pimenta dioica* and *Ocimum basilicum* against the cattle tick Rhipicephalus (Boophilus) microplus (Acari: Ixodidae). *Parasitol. Res.* **2010**, *108*, 481–487. [CrossRef] [PubMed]
15. Padmakumari, K.; Sasidharan, I.; Sreekumar, M. Composition and antioxidant activity of essential oil of pimento (*Pimenta dioica* (L.) Merr.) from Jamaica. *Nat. Prod. Res.* **2011**, *25*, 152–160. [CrossRef]
16. Nabney, J.; Robinson, F.V. Constituents of pimento berry oil (*Pimenta dioica*). *Flavour Ind.* **1972**, *3*, 50–51.
17. Anastasopoulou, E.; Graikou, K.; Ganos, C.; Calapai, G.; Chinou, I. *Pimpinella anisum* seeds essential oil from Lesvos Island: Effect of hydrodistillation time, comparison of its aromatic profile with other samples of the Greek market. Safe use. *Food Chem. Toxicol.* **2019**, *135*, 110875. [CrossRef] [PubMed]
18. Orav, A.; Raal, A.; Arak, E. Essential oil composition of *Pimpinella anisum* L. fruits from various European countries. *Nat. Prod. Res.* **2008**, *22*, 227–232. [CrossRef] [PubMed]
19. Bricout, J. Sur la constitution de l'huile essentielle de badiane. *Bull. Société Chim. Fr.* **1974**, *384*, 1901–1903.
20. De Maack, F.; Brunet, D.; Malnati, J.-C.; Estienne, J. Étude des constituants mineurs de l'anéthole obtenu à partir d'huile essentielle de badiane—Contribution à la recherche de l'origine d'un anéthole par l'identification des sesquiterpènes. *Ann. Falsif. L'expertise Chim. Toxicol.* **1982**, *810*, 357–367.
21. Tuan, D.Q.; Ilangantileket, S.G. Liquid CO_2 extraction of essential oil from Star anise fruits (*Illicium verum* H.). *J. Food Eng.* **1997**, *31*, 47–57. [CrossRef]
22. Huang, Y.; Zhao, J.; Zhou, L.; Wang, J.; Gong, Y.; Chen, X.; Guo, Z.; Wang, Q.; Jiang, W. Antifungal Activity of the Essential Oil of *Illicium verum* Fruit and Its Main Component trans-Anethole. *Molecules* **2010**, *15*, 7558–7569. [CrossRef] [PubMed]
23. Cook, W.B.; Howard, A.S. The essential oil of illicium anisatum linn. *Can. J. Chem.* **1966**, *44*, 2461–2464. [CrossRef]
24. Trenkle, K. Recent studies on fennel (*Foeniculum vulgare* M.) 2. The volatile oil of the fruit, herbs and roots of fruit-bearing plants. *Die Pharm.* **1972**, *27*, 319–324.
25. Muráriková, A.; Ťažký, A.; Neugebauerová, J.; Planková, A.; Jampílek, J.; Mučaji, P.; Mikuš, P. Characterization of essential oil composition in different basil species and pot cultures by a GC-MS method. *Molecules* **2017**, *22*, 1221. [CrossRef] [PubMed]

26. Fischer, R.; Nitzan, N.; Chaimovitsh, D.; Rubin, B.; Dudai, N. Variation in Essential Oil Composition within Individual Leaves of Sweet Basil (*Ocimum basilicum* L.) Is More Affected by Leaf Position than by Leaf Age. *J. Agric. Food Chem.* **2011**, *59*, 4913–4922. [CrossRef]

27. Chalchat, J.-C.; Özcan, M.M. Comparative essential oil composition of flowers, leavesand stems of basil (*Ocimum basilicum* L.) used as herb. *Food Chem.* **2008**, *110*, 501–503. [CrossRef]

28. Vani, S.R.; Cheng, S.; Chuah, C. Comparative study of volatile compounds from genus Ocimum. *Am. J. Appl. Sci.* **2009**, *6*, 523. [CrossRef]

29. Tateo, F.; Santamaria, L.; Bianchi, L.; Bianchi, A. Basil Oil and Tarragon Oil: Composition and Genotoxicity Evaluation. *J. Essent. Oil Res.* **1989**, *1*, 111–118. [CrossRef]

30. Vostrowsky, O.; Michaelis, K.; Ihm, H.; Zintl, R.; Knobloch, K. On the essential oil components from tarragon (*Artemisia dracunculus* L.). *Z. Lebensm. Unters. Forschung* **1981**, *173*, 365–367. [CrossRef]

31. De Vincenzi, M.; Silano, M.; Maialetti, F.; Scazzocchio, B. Constituents of aromatic plants: II. Estragole. *Fitoterapia* **2000**, *71*, 725–729. [CrossRef]

32. Lawrence, B.M. Progress in essential oils/Progrès dans le domaine des huiles essentielles. *Perfum. Flavorist* **1988**, *13*, 44–50.

33. Du, S.-S.; Yang, K.; Wang, C.-F.; You, C.X.; Geng, Z.-F.; Guo, S.-S.; Deng, Z.-W.; Liu, Z.-L. Chemical constituents and activities of the essential oil from myristica fragrans against cigarette beetle lasioderma serricorne. *Chem. Biodivers.* **2014**, *11*, 1449–1456. [CrossRef] [PubMed]

34. Committee, A.M. Application of gas-liquid chromatography to the analysis of essential oils. Part XIV. Monographs for five essential oils. *Analyst* **1988**, *113*, 1125–1136.

35. Embong, M.B.; Hadziyev, D.; Molnar, S. Essential oils from spices grown in Alberta. Anise oil (*Pimpinella anisum*). *Can. J. Plant Sci.* **1977**, *57*, 681–688. [CrossRef]

36. Piccaglia, R.; Marotti, M. Characterization of several aromatic plants grown in northern Italy. *Flavour Fragr. J.* **1993**, *8*, 115–122. [CrossRef]

37. Simándi, B.; Deák, A.; Rónyai, E.; Yanxiang, G.; Veress, T.; Lemberkovics, E.; Then, M.; Sass-Kiss, A.; Vámos-Falusi, Z. Supercritical carbon dioxide extraction and fractionation of fennel oil. *J. Agric. Food Chem.* **1999**, *47*, 1635–1640. [CrossRef]

38. Miguel, M.G.; Cruz, C.; Faleiro, L.; Simões, M.T.F.; Figueiredo, A.C.; Barroso, J.; Pedro, L. *Foeniculum vulgare* Essential Oils: Chemical Composition, Antioxidant and Antimicrobial Activities. *Nat. Prod. Commun.* **2010**, *5*, 319–328. [CrossRef]

39. Basaglia, G.; Fiori, J.; Leoni, A.; Gotti, R. Determination of Estragole in Fennel Herbal Teas by HS-SPME and GC–MS. *Anal. Lett.* **2013**, *47*, 268–279. [CrossRef]

40. Raffo, A.; Nicoli, S.; Leclercq, C. Quantification of estragole in fennel herbal teas: Implications on the assessment of dietary exposure to estragole. *Food Chem. Toxicol.* **2010**, *49*, 370–375. [CrossRef]

41. Denzer, M.Y.; Kirsch, F.; Buettner, A. Are odorant constituents of herbal tea transferred into human milk? *J. Agric. Food Chem.* **2015**, *63*, 104–111. [CrossRef]

42. Hausner, H.; Bredie, W.L.; Mølgaard, C.; Petersen, M.A.; Møller, P. Differential transfer of dietary flavour compounds into human breast milk. *Physiol. Behav.* **2008**, *95*, 118–124. [CrossRef] [PubMed]

43. Bousova, K.; Mittendorf, K.; Senyuva, H. A solid-phase microextraction GC/MS/MS method for rapid quantitative analysis of food and beverages for the presence of legally restricted biologically active flavorings. *J. Assoc. Off. Anal. Chem.* **2011**, *94*, 1189–1199. [CrossRef]

44. Siano, F.; Ghizzoni, C.; Gionfriddo, F.; Colombo, E.; Servillo, L.; Castaldo, D. Determination of estragole, safrole and eugenol methyl ether in food products. *Food Chem.* **2003**, *81*, 469–475. [CrossRef]

45. Sestili, P.; Ismail, T.; Calcabrini, C.; Guescini, M.; Catanzaro, E.; Turrini, E.; Layla, A.; Akhtar, S.; Fimognari, C. The potential effects of *Ocimum basilicum* on health: A review of pharmacological and toxicological studies. *Expert Opin. Drug Metab. Toxicol.* **2018**, *14*, 679–692. [CrossRef]

46. Surburg, H.; Panten, J. *Common Fragrance and Flavor Materials: Preparation, Properties and Uses*; Wiley: Hoboken, NJ, USA, 2006.

47. Roohinejad, S.; Koubaa, M.; Barba, F.J.; Leong, S.Y.; Khelfa, A.; Greiner, R.; Chemat, F. Extraction Methods of Essential Oils From Herbs and Spices. In *Essential Oils in Food Processing*; John Wiley & Sons: Chichester, UK, 2017; pp. 21–55. [CrossRef]

48. Krishnamoorty, B.; Rema, J. Nutmeg and mace. In *Handbook of Herbs and Spices*; Peter, K.V., Ed.; Woodhead Publishing Ltd.: Cambridge, UK, 2001; pp. 238–248.

49. Raffo, A.; D'Aloise, A.; Magrì, A.L.; Leclercq, C. Quantitation of tr-cinnamaldehyde, safrole and myristicin in cola-flavoured soft drinks to improve the assessment of their dietary exposure. *Food Chem. Toxicol.* **2013**, *59*, 626–635. [CrossRef]

50. Benedetti, M.S.; Malnoe, A.; Broillet, A.L. Absorption, metabolism and excretion of safrole in the rat and man. *Toxicology* **1977**, *7*, 69–83. [CrossRef]

51. Caldwell, J.; Sutton, J. Influence of dose size on the disposition of trans-[methoxy-14C]anethole in human volunteers. *Food Chem. Toxicol.* **1988**, *26*, 87–91. [CrossRef]

52. Monien, B.H.; Sachse, B.; Niederwieser, B.; Abraham, K. Detection of N-Acetyl-S-[3′-(4-methoxyphenyl)allyl]-l-Cys (AMPAC) in Human Urine Samples after Controlled Exposure to Fennel Tea: A New Metabolite of Estragole and trans-Anethole. *Chem. Res. Toxicol.* **2019**, *32*, 2260–2267. [CrossRef]

53. Badger, D.; Smith, R.; Bao, J.; Kuester, R.; Sipes, I. Disposition and metabolism of isoeugenol in the male Fischer 344 rat. *Food Chem. Toxicol.* **2002**, *40*, 1757–1765. [CrossRef]

54. Fischer, I.U.; Von Unruh, G.E.; Dengler, H.J. The metabolism of eugenol in man. *Xenobiotica* **1990**, *20*, 209–222. [CrossRef] [PubMed]

55. Hong, S.P.; Fuciarelli, A.F.; Johnson, J.D.; Graves, S.W.; Bates, D.J.; Smith, C.S.; Waidyanatha, S. Toxicokinetics of Isoeugenol in F344 rats and B6C3F 1 mice. *Xenobiotica* **2013**, *43*, 1010–1017. [CrossRef] [PubMed]

56. Hong, S.P.; Fuciarelli, A.F.; Johnson, J.D.; Graves, S.W.; Bates, D.J.; Waidyanatha, S.; Smith, C.S. Toxicokinetics of methyleugenol in F344 rats and B6C3F 1 mice. *Xenobiotica* **2012**, *43*, 293–302. [CrossRef]

57. Sangster, S.A.; Caldwell, J.; Hutt, A.J.; Anthony, A.; Smith, R.L. The metabolic disposition of [methoxy-14c]-labelled trans-anethole, estragole and p-propylanisole in human volunteers. *Xenobiotica* **1987**, *17*, 1223–1232. [CrossRef]

58. Zeller, A.; Horst, K.; Rychlik, M. Study of the Metabolism of Estragole in Humans Consuming Fennel Tea. *Chem. Res. Toxicol.* **2009**, *22*, 1929–1937. [CrossRef]

59. Martins, C.; Rueff, J.; Rodrigues, A.S. Genotoxic alkenylbenzene flavourings, a contribution to risk assessment. *Food Chem. Toxicol.* **2018**, *118*, 861–879. [CrossRef] [PubMed]

60. Rietjens, I.M.; Cohen, S.M.; Fukushima, S.; Gooderham, N.; Hecht, S.; Marnett, L.J.; Smith, R.L.; Adams, T.B.; Bastaki, M.; Harman, C.G.; et al. Impact of Structural and Metabolic Variations on the Toxicity and Carcinogenicity of Hydroxy- and Alkoxy-Substituted Allyl- and Propenylbenzenes. *Chem. Res. Toxicol.* **2014**, *27*, 1092–1103. [CrossRef]

61. Cartus, A.; Herrmann, K.; Weishaupt, L.W.; Merz, K.-H.; Engst, W.; Glatt, H.; Schrenk, D. Metabolism of Methyleugenol in Liver Microsomes and Primary Hepatocytes: Pattern of Metabolites, Cytotoxicity, and DNA-Adduct Formation. *Toxicol. Sci.* **2012**, *129*, 21–34. [CrossRef]

62. Cartus, A.T.; Merz, K.-H.; Schrenk, D. Metabolism of Methylisoeugenol in Liver Microsomes of Human, Rat, and Bovine Origin. *Drug Metab. Dispos.* **2011**, *39*, 1727–1733. [CrossRef]

63. Kamienski, F.X.; Casida, J.E. Importance of demethylenation in the metabolism in vivo and in vitro of methylenedioxyphenyl synergists and related compounds in mammals. *Biochem. Pharmacol.* **1970**, *19*, 91–112. [CrossRef]

64. Klungsøyr, J.; Scheline, R.R. Metabolism of safrole in the rat. *Acta Pharmacol. Toxicol.* **1983**, *52*, 211–216. [CrossRef] [PubMed]

65. Solheim, E.; Scheline, R.R. Metabolism of alkenebenzene derivatives in the rat I. P-methoxyallylbenzene (estragole) and p-methoxypropenylbenzene (anethole). *Xenobiotica* **1973**, *3*, 493–510. [CrossRef]

66. Solheim, E.; Scheline, R.R. Metabolism of Alkenebenzene Derivatives in the Rat. II. Eugenol and Isoeugenol Methyl Ethers. *Xenobiotica* **1976**, *6*, 137–150. [CrossRef] [PubMed]

67. Solheim, E.; Scheline, R.R. Metabolism of alkenebenzene derivatives in the rat III. Elemicin and isoelemicin. *Xenobiotica* **1980**, *10*, 371–380. [CrossRef]

68. Anthony, A.; Caldwell, J.; Hutt, A.; Smith, R. Metabolism of estragole in rat and mouse and influence of dose size on excretion of the proximate carcinogen 1′-hydroxyestragole. *Food Chem. Toxicol.* **1987**, *25*, 799–806. [CrossRef]

69. Guenthner, T.M.; Luo, G. Investigation of the role of the 2′,3′-epoxidation pathway in the bioactivation and genotoxicity of dietary allylbenzene analogs. *Toxicology* **2001**, *160*, 47–58. [CrossRef]

70. Luo, G.; Guenthner, T.M. Detoxication of the 2′,3′-epoxide metabolites of allylbenzene and estragole: Conjugation with glutathione. *Drug Metab. Dispos.* **1994**, *22*, 731–737.

71. Luo, G.; Guenthner, T.M. Covalent binding to DNA in vitro of 2′,3′-oxides derived from allylbenzene analogs. *Drug Metab. Dispos.* **1996**, *24*, 1020–1027.

72. Iyer, L.V.; Ho, M.N.; Shinn, W.M.; Bradford, W.W.; Tanga, M.J.; Nath, S.S.; Green, C.E. Glucuronidation of 1′-Hydroxyestragole (1′-HE) by Human UDP-Glucuronosyltransferases UGT2B7 and UGT1A9. *Toxicol. Sci.* **2003**, *73*, 36–43. [CrossRef] [PubMed]

73. Al-Subeihi, A.A.; Spenkelink, B.; Punt, A.; Boersma, M.G.; van Bladeren, P.J.; Rietjens, I.M. Physiologically based kinetic modeling of bioactivation and detoxification of the alkenylbenzene methyleugenol in human as compared with rat. *Toxicol. Appl. Pharmacol.* **2012**, *260*, 271–284. [CrossRef]

74. Martati, E.; Boersma, M.G.; Spenkelink, A.; Khadka, D.B.; van Bladeren, P.J.; Rietjens, I.M.; Punt, A. Physiologically Based Biokinetic (PBBK) Modeling of Safrole Bioactivation and Detoxification in Humans as Compared With Rats. *Toxicol. Sci.* **2012**, *128*, 301–316. [CrossRef]

75. Punt, A.; Paini, A.; Boersma, M.G.; Freidig, A.P.; Delatour, T.; Scholz, G.; Schilter, B.; van Bladeren, P.J.; Rietjens, I.M. Use of Physiologically Based Biokinetic (PBBK) Modeling to Study Estragole Bioactivation and Detoxification in Humans as Compared with Male Rats. *Toxicol. Sci.* **2009**, *110*, 255–269. [CrossRef]

76. Bergau, N.; Herfurth, U.M.; Sachse, B.; Abraham, K.; Monien, B.H. Bioactivation of estragole and anethole leads to common adducts in DNA and hemoglobin. *Food Chem. Toxicol.* **2021**, *153*, 112253. [CrossRef] [PubMed]

77. Herrmann, K.; Engst, W.; Appel, K.E.; Monien, B.H.; Glatt, H. Identification of human and murine sulfotransferases able to activate hydroxylated metabolites of methyleugenol to mutagens in Salmonella typhimurium and detection of associated DNA adducts using UPLC-MS/MS methods. *Mutagenesis* **2012**, *27*, 453–462. [CrossRef]

78. NTP USNTP. *NTP Technical Report on the Toxicology and Carcinogenesis Studies of Methyleugenol (CAS NO. 93-15-2) in F344/N Rats and B6C3F1 Mice (Gavage Studies)*; NTP TR 491; National Toxicology Program Technical Report Series (NTPTR); Environmental Health Information Service (EHS): Springfield, IL, USA, 2000; pp. 1–412.

79. Miller, E.C.; Swanson, A.B.; Phillips, D.H.; Fletcher, T.L.; Liem, A.; Miller, J.A. Structure-activity studies of the carcinogenicities in the mouse and rat of some naturally occurring and synthetic alkenylbenzene derivatives related to safrole and estragole. *Cancer Res.* **1983**, *43*, 1124–1134.

80. Boberg, E.W.; Miller, E.C.; Miller, J.A.; Poland, A.; Liem, A. Strong Evidence from Studies with Brachymorphic Mice and Pentachlorophenol That 1′-Sulfoöxysafrole Is the Major Ultimate Electrophilic and Carcinogenic Metabolite of 1′-Hydroxysafrole in Mouse Liver. *Cancer Res.* **1983**, *43*, 5163–5173. [PubMed]

81. Punt, A.; Delatour, T.; Scholz, G.; Schilter, B.; van Bladeren, P.J.; Rietjens, I. Tandem Mass Spectrometry Analysis of N2-(trans-Isoestragol-3′-yl)-2′-deoxyguanosine as a Strategy to Study Species Differences in Sulfotransferase Conversion of the Proximate Carcinogen 1′-Hydroxyestragole. *Chem. Res. Toxicol.* **2007**, *20*, 991–998. [CrossRef]

82. Al-Subeihi, A.A.; Spenkelink, B.; Rachmawati, N.; Boersma, M.G.; Punt, A.; Vervoort, J.; van Bladeren, P.J.; Rietjens, I. Physiologically based biokinetic model of bioactivation and detoxification of the alkenylbenzene methyleugenol in rat. *Toxicol. Vitro.* **2011**, *25*, 267–285. [CrossRef] [PubMed]

83. Punt, A.; Freidig, A.P.; Delatour, T.; Scholz, G.; Boersma, M.G.; Schilter, B.; Van Bladeren, P.J.; Rietjens, I.M. A physiologically based biokinetic (PBBK) model for estragole bioactivation and detoxification in rat. *Toxicol. Appl. Pharmacol.* **2008**, *231*, 248–259. [CrossRef] [PubMed]

84. Herrmann, K.; Schumacher, F.; Engst, W.; Appel, K.E.; Klein, K.; Zanger, U.M.; Glatt, H. Abundance of DNA adducts of methyleugenol, a rodent hepatocarcinogen, in human liver samples. *Carcinogenesis* **2013**, *34*, 1025–1030. [CrossRef] [PubMed]

85. Tremmel, R.; Herrmann, K.; Engst, W.; Meinl, W.; Klein, K.; Glatt, H.; Zanger, U.M. Methyleugenol DNA adducts in human liver are associated with SULT1A1 copy number variations and expression levels. *Arch. Toxicol.* **2017**, *91*, 3329–3339. [CrossRef]

86. Al-Subeihi, A.A.; Alhusainy, W.; Kiwamoto, R.; Spenkelink, B.; van Bladeren, P.J.; Rietjens, I.M.; Punt, A. Evaluation of the interindividual human variation in bioactivation of methyleugenol using physiologically based kinetic modeling and Monte Carlo simulations. *Toxicol. Appl. Pharmacol.* **2015**, *283*, 117–126. [CrossRef]

87. Ning, J.; Louisse, J.; Spenkelink, B.; Wesseling, S.; Rietjens, I.M. Study on inter-ethnic human differences in bioactivation and detoxification of estragole using physiologically based kinetic modeling. *Arch. Toxicol.* **2017**, *91*, 3093–3108. [CrossRef]

88. Punt, A.; Paini, A.; Spenkelink, A.; Scholz, G.; Schilter, B.; Van Bladeren, P.J.; Rietjens, I.M. Evaluation of Interindividual Human Variation in Bioactivation and DNA Adduct Formation of Estragole in Liver Predicted by Physiologically Based Kinetic/Dynamic and Monte Carlo Modeling. *Chem. Res. Toxicol.* **2016**, *29*, 659–668. [CrossRef]

89. Al-Malahmeh, A.J.; Al-Ajlouni, A.; Wesseling, S.; Soffers, A.E.M.F.; Al-Subeihi, A.; Kiwamoto, R.; Vervoort, J.; Rietjens, I.M. Physiologically based kinetic modeling of the bioactivation of myristicin. *Arch. Toxicol.* **2016**, *91*, 713–734. [CrossRef]

90. Martati, E.; Boersma, M.G.; Spenkelink, A.; Khadka, D.B.; Punt, A.; Vervoort, J.; Van Bladeren, P.J.; Rietjens, I.M. Physiologically Based Biokinetic (PBBK) Model for Safrole Bioactivation and Detoxification in Rats. *Chem. Res. Toxicol.* **2011**, *24*, 818–834. [CrossRef]

91. van den Berg, S.J.P.L.; Punt, A.; Soffers, A.E.; Vervoort, J.; Ngeleja, S.; Spenkelink, B.; Rietjens, I.M. Physiologically based kinetic models for the alkenylbenzene elemicin in rat and human and possible implications for risk assessment. *Chem. Res. Toxicol.* **2012**, *25*, 2352–2367. [CrossRef]

92. Truhaut, R.; Le Bourhis, B.; Attia, M.; Glomot, R.; Newman, J.; Caldwell, J. Chronic toxicity/carcinogenicity study of trans-anethole in rats. *Food Chem. Toxicol.* **1989**, *27*, 11–20. [CrossRef]

93. Sangster, S.; Caldwell, J.; Smith, R. Metabolism of anethole. II. Influence of dose size on the route of metabolism of trans-anethole in the rat and mouse. *Food Chem. Toxicol.* **1984**, *22*, 707–713. [CrossRef]

94. Sangster, S.; Caldwell, J.; Smith, R.; Farmer, P. Metabolism of anethole. I. Pathways of metabolism in the rat and mouse. *Food Chem. Toxicol.* **1984**, *22*, 695–706. [CrossRef]

95. Newberne, P.; Smith, R.; Doull, J.; Goodman, J.; Munro, I.; Portoghese, P.; Wagner, B.; Weil, C.; Woods, L.; Adams, T.; et al. The FEMA GRAS Assessment of trans-Anethole Used as a Flavouring Substance. *Food Chem. Toxicol.* **1999**, *37*, 789–811. [CrossRef]

96. NTP. *Carcinogenesis Studies of Eugenol (CAS No. 93-15-2) in F344/N Rats and B6C3F1 Mice (Feed Studies)*; NTP-80-068 No. 223 (=NIH Publication No. 84-1779); National Toxicology Program Technical Report Series (NTPTR); European Commission: Brussel, Belgium, 1983; pp. 1–159.

97. NTP USNTP. *NTP Technical Report on the Toxicology and Carcinogenesis Studies of Isoeugenol (CAS No. 97-54-1) in F344/N Rats and B6C3F1 Mice (Gavage Studies)*; NTP TR 551 National Toxicology Program Technical Report Series (NTPTR); European Commission: Brussel, Belgium, 2010; pp. 1–178.

98. SCF ECSCoF. *Opinion of the Scientific Committee on Food on the Safety of the Presence of Safrole (1-allyl-3,4-methylene Dioxy Benzene) in Flavourings and other Food Ingredients with Flavouring Properties*; European Commission: Brussel, Belgium, 2002; pp. 1–15.

99. Homburger, F.; Kelley, J.T.; Friedler, G.; Russfield, A.; Homburger, H. Toxic and Possible Carcinogenic Effects of 4-Allyl-1,2-Methylenedioxybenzene (Safrole) in Rats on Deficient Diets. *Pharmacology* **1961**, *4*, 1–11. [CrossRef] [PubMed]

100. Innes, J.R.; Ulland, B.M.; Valerio, M.G.; Petrucelli, L.; Fishbein, L.; Hart, E.R.; Pallotta, A.J.; Bates, R.R.; Falk, H.L.; Gart, J.J.; et al. Bioassay of pesticides and industrial chemicals for tumorigenicity in mice: A preliminary note. *J. Natl. Cancer Inst.* **1969**, *42*, 1101–1114.

101. Bode, A.M.; Dong, Z. Toxic Phytochemicals and Their Potential Risks for Human Cancer. *Cancer Prev. Res.* **2014**, *8*, 1–8. [CrossRef]

102. Borchert, P.; Miller, J.A.; Miller, E.C.; Shires, T.K. 1′-Hydroxysafrole, a proximate carcinogenic metabolite of safrole in the rat and mouse. *Cancer Res.* **1973**, *33*, 590–600.

103. Jeurissen, S.M.F.; Bogaards, J.J.P.; Awad, H.M.; Boersma, M.G.; Brand, W.; Fiamegos, Y.C.; Van Beek, T.A.; Alink, G.M.; Sudhölter, E.J.R.; Cnubben, N.H.P.; et al. Human Cytochrome P450 Enzyme Specificity for Bioactivation of Safrole to the Proximate Carcinogen 1′-Hydroxysafrole. *Chem. Res. Toxicol.* **2004**, *17*, 1245–1250. [CrossRef] [PubMed]

104. Munerato, M.C.; Sinigaglia, M.; Reguly, M.L.; de Andrade, H.H.R. Genotoxic effects of eugenol, isoeugenol and safrole in the wing spot test of Drosophila melanogaster. *Mutat. Res. Toxicol. Environ. Mutagen.* **2005**, *582*, 87–94. [CrossRef]
105. IARC. Overall evaluations of carcinogenicity: An updating of IARC Monographs Volumes 1 to 42—Eugenol. In *IARC Monographs on the Evaluation of Carcinogenic Risk of Chemicals to Humans*; IARC: Lyon, France, 1987; Volume 36, pp. 1–63.
106. Suzuki, Y.; Umemura, T.; Hibi, D.; Inoue, T.; Jin, M.; Ishii, Y.; Sakai, H.; Nohmi, T.; Yanai, T.; Nishikawa, A.; et al. Possible involvement of genotoxic mechanisms in estragole-induced hepatocarcinogenesis in rats. *Arch. Toxicol.* **2012**, *86*, 1593–1601. [CrossRef]
107. Smith, R.; Adams, T.; Doull, J.; Feron, V.; Goodman, J.; Marnett, L.; Portoghese, P.; Waddell, W.; Wagner, B.; Rogers, A.; et al. Safety assessment of allylalkoxybenzene derivatives used as flavouring substances—methyl eugenol and estragole. *Food Chem. Toxicol.* **2002**, *40*, 851–870. [CrossRef]
108. Probert, P.M.; Palmer, J.M.; Alhusainy, W.; Amer, A.; Rietjens, I.M.; White, S.A.; Jones, D.E.; Wright, M.C. Progenitor-derived hepatocyte-like (B-13/H) cells metabolise 1′-hydroxyestragole to a genotoxic species via a SULT2B1-dependent mechanism. *Toxicol. Lett.* **2015**, *243*, 98–110. [CrossRef] [PubMed]
109. SCF ECSCoF. *Opinion of the Scientific Committee on Food on Estragole (1-Allyl-4-methoxybenzene)*; Environmental Health Information Service (EHS): Springfield, IL, USA, 2001; pp. 1–10.
110. NTP. *Report on Carcinogens—Background Document for Methyleugenol*; National Toxicology Program Technical Report Series (NTPTR); Environmental Health Information Service (EHS): Springfield, IL, USA, 2000; pp. 1–47.
111. SCF ECSCoF. *Opinion of the Scientific Committee on Food on Methyleugenol (4-Allyl-1,2-dimethoxybenzene)*; SCF/CS/FLAV/FLAVOUR/4 ADD1 FINAL; Environmental Health Information Service (EHS): Springfield, IL, USA, 2001.
112. De Vincenzi, M.; Silano, M.; Stacchini, P.; Scazzocchio, B. Constituents of aromatic plants: I. Methyleugenol. *Fitoterapia* **2000**, *71*, 216–221. [CrossRef]
113. Sekizawa, J.; Shibamoto, T. Genotoxicity of safrole-related chemicals in microbial test systems. *Mutat. Res. Toxicol.* **1982**, *101*, 127–140. [CrossRef]
114. Chan, V.; Caldwell, J. Comparative induction of unscheduled DNA synthesis in cultured rat hepatocytes by allylbenzenes and their 1′-hydroxy metabolites. *Food Chem. Toxicol.* **1992**, *30*, 831–836. [CrossRef]
115. IARC. *Methyleugenol. IARC Monographs Working Group on the Evaluation of Carcinogenic Risks to Humans, Lyon, France, 2011 IARC Monographs on the Evaluation of Carcinogenic Risk of Chemicals to Man*; IARC: Geneva, Switzerland, 2013; Volume 101, pp. 407–433.
116. Suparmi, S.; Ginting, A.J.; Mariyam, S.; Wesseling, S.; Rietjens, I.M. Levels of methyleugenol and eugenol in instant herbal beverages available on the Indonesian market and related risk assessment. *Food Chem. Toxicol.* **2019**, *125*, 467–478. [CrossRef] [PubMed]
117. JECFA JFWECoFA. Sixty-fifth meeting of the Joint FAO/WHO Expert Committee on Food Additives. *WHO Food Addit. Series* **2006**, *56*, 155–200.
118. EFSA. Consideration of eugenol and related hydroxyallylbenzene derivatives evaluated by JECFA (65th meeting) structurally related to ring-substituted phenolic substances evaluated by EFSA in FGE.22 (2006). *EFSA J.* **2009**, 1–53. [CrossRef]
119. NTP USNTP. *Isoeugenol: Reproductive Assessment by Continuous Breeding when Administered to Sprague-Dawley Rats by Gavage*; National Toxicology Program Technical Report Series (NTPTR); Environmental Health Information Service (EHS): Springfield, IL, USA, 2002; pp. 1–167.
120. Burkey, J.L.; Sauer, J.-M.; McQueen, C.A.; Sipes, I.G. Cytotoxicity and genotoxicity of methyleugenol and related congeners—a mechanism of activation for methyleugenol. *Mutat. Res. Mol. Mech. Mutagen.* **2000**, *453*, 25–33. [CrossRef]
121. Jansson, T.; Curvall, M.; Hedin, A.; Enzell, C.R. In vitro studies of biological effects of cigarette smoke condensate: II. Induction of sister-chromatid exchanges in human lymphocytes by weakly acidic, semivolatile constituents. *Mutat. Res. Toxicol.* **1986**, *169*, 129–139. [CrossRef]
122. EFSA. Enzymes, Flavourings and Processing Aids (CEF). Scientific Opinion on Flavouring Group Evaluation 81 (FGE.81): Consideration of hydroxypropenylbenzenes evaluated by JECFA (61st meeting) structurally related to 2-methoxy-4-(prop-1-enyl)phenyl 3-methylbutyrate from chemical group 17 evaluated by EFSA in FGE.30. *EFSA J.* **2010**, *8*, 1899.
123. EFSA. Scientific Opinion on Flavouring Group Evaluation 30, Revision 1 (FGE.30Rev1): 4-Prop-1-enylphenol and 2-methoxy-4-(prop-1-enyl)phenyl 3-methylbutyrate from chemical group 17. *EFSA J.* **2011**, *9*, 1991. [CrossRef]
124. JECFA JFWECoFA. Sixty-first meeting of the Joint FAO/WHO Expert Committee on Food Addtives. *WHO Tech. Rep. Ser.* **2004**, *922*, 96.
125. Gorelick, N. Genotoxicity of trans-anethole in vitro. *Mutat. Res. Mol. Mech. Mutagen.* **1995**, *326*, 199–209. [CrossRef]
126. JECFA JFWECoFA. Evaluation of certain food additives. Fifty-first report of the Joint FAO/WHO Expert Committee on Food Additives. *WHO Tech. Rep. Ser.* **2000**, *891*, 1–168.
127. EFSA Scientific Committee; More, S.; Bampidis, V.; Benford, D.; Boesten, J.; Bragard, C.; Halldorsson, T.; Hernandez-Jerez, A.; Hougaard-Bennekou, S.; Koutsoumanis, K.; et al. Genotoxicity assessment of chemical mixtures. *EFSA J.* **2019**, *17*, e05519. [CrossRef]
128. Rietjens, I.M.; Slob, W.; Galli, C.; Silano, V. Risk assessment of botanicals and botanical preparations intended for use in food and food supplements: Emerging issues. *Toxicol. Lett.* **2008**, *180*, 131–136. [CrossRef]

129. Al-Malahmeh, A.J.; Alajlouni, A.M.; Ning, J.; Wesseling, S.; Vervoort, J.; Rietjens, I.M. Determination and risk assessment of naturally occurring genotoxic and carcinogenic alkenylbenzenes in nutmeg-based plant food supplements. *J. Appl. Toxicol.* **2017**, *37*, 1254–1264. [CrossRef]
130. Ning, J.; Cui, X.; Kong, X.; Tang, Y.; Wulandari, R.; Chen, L.; Wesseling, S.; Rietjens, I.M. Risk assessment of genotoxic and carcinogenic alkenylbenzenes in botanical containing products present on the Chinese market. *Food Chem. Toxicol.* **2018**, *115*, 344–357. [CrossRef] [PubMed]
131. Yang, R.S.H. *Toxicology of Chemical Mixtures: Case Studies, Mechanisms, and Novel Approaches*; Elsevier Science: Amsterdam, The Netherlands, 2016.
132. Könemann, W.; Pieters, M. Confusion of concepts in mixture toxicology. *Food Chem. Toxicol.* **1996**, *34*, 1025–1031. [CrossRef]
133. EFSA. Guidance on harmonised methodologies for human health, animal health and ecological risk assessment of combined exposure to multiple chemicals. *EFSA J.* **2019**, *17*, e05634.
134. Jeurissen, S.M.; Punt, A.; Delatour, T.; Rietjens, I.M. Basil extract inhibits the sulfotransferase mediated formation of DNA adducts of the procarcinogen 1′-hydroxyestragole by rat and human liver S9 homogenates and in HepG2 human hepatoma cells. *Food Chem. Toxicol.* **2008**, *46*, 2296–2302. [CrossRef] [PubMed]
135. Müller, L.; Kasper, P.; Müller-Tegethoff, K.; Petr, T. The genotoxic potential in vitro and in vivo of the allyl benzene etheric oils estragole, basil oil and trans-anethole. *Mutat. Res. Lett.* **1994**, *325*, 129–136. [CrossRef]
136. Alhusainy, W.; Paini, A.; Punt, A.; Louisse, J.; Spenkelink, A.; Vervoort, J.; Delatour, T.; Scholz, G.; Schilter, B.; Adams, T.; et al. Identification of nevadensin as an important herb-based constituent inhibiting estragole bioactivation and physiology-based biokinetic modeling of its possible in vivo effect. *Toxicol. Appl. Pharmacol.* **2010**, *245*, 179–190. [CrossRef]
137. Alhusainy, W.; Williams, G.M.; Jeffrey, A.M.; Iatropoulos, M.J.; Taylor, S.; Adams, T.B.; Rietjens, I.M. The natural basil flavonoid nevadensin protects against a methyleugenol-induced marker of hepatocarcinogenicity in male F344 rat. *Food Chem. Toxicol.* **2014**, *74*, 28–34. [CrossRef]
138. Al-Subeihi, A.A.; Alhusainy, W.; Paini, A.; Punt, A.; Vervoort, J.; van Bladeren, P.J.; Rietjens, I.M. Inhibition of methyleugenol bioactivation by the herb-based constituent nevadensin and prediction of possible in vivo consequences using physiologically based kinetic modeling. *Food Chem. Toxicol.* **2013**, *59*, 564–571. [CrossRef] [PubMed]
139. Rietjens, I.M.; Tyrakowska, B.; Berg, S.J.P.L.V.D.; Soffers, A.E.M.F.; Punt, A. Matrix-derived combination effects influencing absorption, distribution, metabolism and excretion (ADME) of food-borne toxic compounds: Implications for risk assessment. *Toxicol. Res.* **2014**, *4*, 23–35. [CrossRef]
140. IARC. Safrole. In *IARC Monographs on the Evaluation of Carcinogenic Risk of Chemicals to Man*; IARC: Geneva, Switzerland, 1972; Volume 10, pp. 169–174.
141. Cohen, S.M.; Eisenbrand, G.; Fukushima, S.; Gooderham, N.J.; Guengerich, F.P.; Hecht, S.S.; Rietjens, I.M.C.M.; Harman, C.; Taylor, S.V. GRAS flavoring substances. *Food Technol.* **2020**, *3*, 45–65.
142. Zhao, D.-H.; Ke, C.-L.; Liu-Dong, L.; Wang, X.-F.; Wang, Q.; Chang-Liang, K. Elimination kinetics of eugenol in grass carp in a simulated transportation setting. *BMC Vet. Res.* **2017**, *13*, 346. [CrossRef]
143. Kamatou, G.P.; Vermaak, I.; Viljoen, A.M. Eugenol—From the Remote Maluku Islands to the International Market Place: A Review of a Remarkable and Versatile Molecule. *Molecules* **2012**, *17*, 6953–6981. [CrossRef]
144. Priborsky, J.; Velisek, J. A Review of Three Commonly Used Fish Anesthetics. *Rev. Fish. Sci. Aquac.* **2018**, *26*, 417–442. [CrossRef]
145. Tuckey, N.P.; Forster, M.E.; Gieseg, S.P. Effects of rested harvesting on muscle metabolite concentrations and K-values in Chinook salmon (*Oncorhynchus tshawytscha*) fillets during storage at 15 °C. *J. Food Sci.* **2010**, *75*, C459–C464. [CrossRef]
146. Meinertz, J.R.; Schreier, T.M.; Porcher, S.T.; Smerud, J.R. Evaluation of a Method for Quantifying Eugenol Concentrations in the Fillet Tissue from Freshwater Fish Species. *J. AOAC Int.* **2016**, *99*, 558–564. [CrossRef]
147. NTP USNTP. NTP Technical Report on the Toxicity Studies of Myristicin (CASRN 607-91-0) Administered by Gavage to F344/NTac Rats and B6C3F1/N Mice. *Natl. Toxicol. Program Tech. Rep. Ser.* **2019**, *95*, 1–65.
148. The European Parliament and of the Council. Regulation (EC) No 1334/2008 of the European Parliament and of the Council of 16 December 2008 on flavourings and certain food ingredients with flavouring properties for use in and on foods and amending Council Regulation (EEC) No 1601/91, Regulations (EC) No 2232/96 and (EC) No 110/2008 and Directive 2000/13/EC. *Off. J. Eur. Union* **2008**, *L354*, 1–207.
149. Cao, X.; Xie, Q.; Zhang, S.; Xu, H.; Su, J.; Zhang, J.; Deng, C.; Song, G. Fabrication of functionalized magnetic microspheres based on monodispersed polystyrene for quantitation of allyl-benzodioxoles coupled with gas chromatography and mass spectrometry. *J. Chromatogr. A* **2019**, *1607*, 460402. [CrossRef] [PubMed]
150. Ávila, M.; Zougagh, M.; Escarpa, A.; Ríos, A. Determination of alkenylbenzenes and related flavour compounds in food samples by on-column preconcentration-capillary liquid chromatography. *J. Chromatogr. A* **2009**, *1216*, 7179–7185. [CrossRef] [PubMed]
151. López, V.; Gerique, J.; Langa, E.; Berzosa, C.; Valero, M.S.; Gómez-Rincón, C. Antihelmintic effects of nutmeg (*Myristica fragans*) on Anisakis simplex L3 larvae obtained from *Micromesistius potassou*. *Res. Vet. Sci.* **2015**, *100*, 148–152. [CrossRef] [PubMed]
152. Chiu, S.; Wang, T.; Belski, M.; Abourashed, E.A. HPLC-Guided Isolation, Purification and Characterization of Phenylpropanoid and Phenolic Constituents of Nutmeg Kernel (*Myristica fragrans*). *Nat. Prod. Commun.* **2016**, *11*. [CrossRef]
153. Dang, H.; Quirino, J. Analytical Separation of Carcinogenic and Genotoxic Alkenylbenzenes in Foods and Related Products (2010–2020). *Toxins* **2021**, *13*, 387. [CrossRef]

154. Mihats, D.; Pilsbacher, L.; Gabernig, R.; Routil, M.; Gutternigg, M.; Laenger, R. Levels of estragole in fennel teas marketed in Austria and assessment of dietary exposure. *Int. J. Food Sci. Nutr.* **2016**, *68*, 569–576. [CrossRef]

155. Rehm, C.; Penalvo, J.L.; Afshin, A.; Mozaffarian, D. Dietary Intake among US Adults, 1999-2012. *JAMA* **2016**, *315*, 2542–2553. [CrossRef]

156. Somi, M.H.; Mousavi, S.M.; Naghashi, S.; Faramarzi, E.; Jafarabadi, M.A.; Ghojazade, M.; Majdi, A.; Alavi, S.A.N. Is there any relationship between food habits in the last two decades and gastric cancer in North-Western Iran? *Asian Pac. J. Cancer Prev.* **2015**, *16*, 283–290. [CrossRef]

157. Rosi, A.; Paolella, G.; Biasini, B.; Scazzina, F.; Sinu Working Group on Nutritional Surveillance in Adolescents. Dietary habits of adolescents living in North America, Europe or Oceania: A review on fruit, vegetable and legume consumption, sodium intake, and adherence to the Mediterranean Diet. *Nutr. Metab. Cardiovasc Dis.* **2019**, *29*, 544–560. [CrossRef]

158. Haufroid, V.; Lison, D. Mercapturic acids revisited as biomarkers of exposure to reactive chemicals in occupational toxicology: A minireview. *Int. Arch. Occup. Environ. Health* **2005**, *78*, 343–354. [CrossRef]

159. Hartwig, A.; Arand, M.; Epe, B.; Guth, S.; Jahnke, G.; Lampen, A.; Martus, H.J.; Monien, B.; Rietjens, I.M.; Schmitz-Spanke, S.; et al. Mode of action-based risk assessment of genotoxic carcinogens. *Arch Toxicol.* **2020**, *94*, 1787–1877. [CrossRef]

160. Clewell, H.J.; Tan, Y.M.; Campbell, J.L.; Andersen, M. Quantitative Interpretation of Human Biomonitoring Data. *Toxicol. Appl. Pharmacol.* **2008**, *231*, 122–133. [CrossRef] [PubMed]

161. EFSA EFSASC. Scientific opinion on genotoxicity testing strategies applicable to food and feed safety assessment. *EFSA J.* **2011**, *9*, 2379.

162. ICH. *ICH S2 (R1) Guidance on Genotoxicity Testing and Data Interpretation for Pharmaceuticals Intended for Human Use. Step 4. ICH Harmonised Tripartite Guideline*; Bibra Toxicology Advice & Consulting: Wallington, UK, 2011.

163. The European Parliament and of the Council of the European Union. Regulation (EU) No 528/2012 of the European Parliament and of the Council of 22 May 2012 concerning the making available on the market and use of biocidal products. *Off. J. Eur. Union* **2012**, *L167*, 1–123.

164. The European Parliament and of the Council of the European Union. Regulation (EC) No 1107/2009 of the European Parliament and of the Council of 21 October 2009 concerning the placing of plant protection products on the market and repealing Council Directives 79/117/EEC and 91/414/EEC. *Off. J. Eur. Union* **2009**, *L309*, 1–50.

165. EC. Commission Communication in the framework of the implementation of Commission Regulation (EU) No 283/2013 of 1 March 2013 setting out the data requirements for active substances, in accordance with Regulation (EC) No 1107/2009 of the European Parliament and of the Council concerning the placing of plant protection products on the market. *Off. J. Eur. Union* **2013**, *C95*, 1–20.

166. OECD. OECD Test Guidelines for Chemicals. Available online: https://www.oecd.org/chemicalsafety/testing/oecdguidelinesforthetestingofchemicals.htm (accessed on 8 September 2021).

167. Nestmann, E.R.; Douglas, G.R.; Kowbel, D.J.; Harrington, T.R. Solvent interactions with test compounds and recommendations for testing to avoid artifacts. *Environ. Mutagen.* **1985**, *7*, 163–170. [CrossRef] [PubMed]

168. Gatehouse, D.; Haworth, S.; Cebula, T.; Gocke, E.; Kier, L.; Matsushima, T.; Melcion, C.; Nohmi, T.; Ohta, T.; Venitt, S.; et al. Recommendations for the performance of bacterial mutation assays. *Mutat. Res. Mutagen. Relat. Subj.* **1994**, *312*, 217–233. [CrossRef]

169. Nesslany, F.; Zennouche, N.; Simar-Meintières, S.; Talahari, I.; Nkili-Mboui, E.-N.; Marzin, D. In vivo Comet assay on isolated kidney cells to distinguish genotoxic carcinogens from epigenetic carcinogens or cytotoxic compounds. *Mutat. Res. Toxicol. Environ. Mutagen.* **2007**, *630*, 28–41. [CrossRef] [PubMed]

170. OECD. *Test No. 474: Mammalian Erythrocyte Micronucleus Test*; OECD Guidelines for the Testing of Chemicals, Section 4; OECD Publishing: Paris, France, 2016. [CrossRef]

171. Herrmann, K.; Holzwarth, A.; Rime, S.; Fischer, B.C.; Kneuer, C. (Q)SAR tools for the prediction of mutagenic properties: Are they ready for application in pesticide regulation? *Pest Manag. Sci.* **2020**, *76*, 3316–3325. [CrossRef]

172. Wislocki, P.G.; Miller, E.C.; Miller, J.A.; McCoy, E.C.; Rosenkranz, H.S. Carcinogenic and mutagenic activities of safrole, 1′-hydroxysafrole, and some known or possible metabolites. *Cancer Res.* **1977**, *37*, 1883–1891. [PubMed]

173. Boberg, E.W.; Liem, A.; Miller, E.C.; Miller, J.A. Inhibition by pentachlorophenol of the initiating and promoting activities of 1′-hydroxysafrole for the formation of enzyme-altered foci and tumors in rat liver. *Carcinogenesis* **1987**, *8*, 531–539. [CrossRef]

174. Groh, I.A.M.; Cartus, A.T.; Vallicotti, S.; Kajzar, J.; Merz, K.-H.; Schrenk, D.; Esselen, M. Genotoxic potential of methyleugenol and selected methyleugenol metabolites in cultured Chinese hamster V79 cells. *Food Funct.* **2012**, *3*, 428. [CrossRef] [PubMed]

175. Randerath, K.; Haglund, R.E.; Phillips, D.H.; Reddy, M.V. 32P-post-labelling analysis of DNA adducts formed in the livers of animals treated with safrole, estragole and other naturally-occurring alkenylbenzenes. I. adult female CD-1 mice. *Carcinogenesis* **1984**, *5*, 1613–1622. [CrossRef]

176. Daimon, H.; Sawada, S.; Asakura, S.; Sagami, F. Inhibition of sulfotransferase affecting in vivo genotoxicity and DNA adducts induced by safrole in rat liver. *Teratog. Carcinog. Mutagen.* **1997**, *17*, 327–337. [CrossRef]

177. Borchert, P.; Wislocki, P.G.; Miller, J.A.; Miller, E.C. The metabolism of the naturally occurring hepatocarcinogen safrole to 1′-hydroxysafrole and the electrophilic reactivity of 1′-acetoxysafrole. *Cancer Res.* **1973**, *33*, 575–589. [PubMed]

178. Schiestl, R.H.; Chan, W.-S.; Gietz, R.D.; Mehta, R.D.; Hastings, P. Safrole, eugenol and methyleugenol induce intrachromosomal recombination in yeast. *Mutat. Res. Toxicol.* **1989**, *224*, 427–436. [CrossRef]

179. Mortelmans, K.; Haworth, S.; Lawlor, T.; Speck, W.; Tainer, B.; Zeiger, E. Salmonella mutagenicity tests: II. Results from the testing of 270 chemicals. *Environ. Mutagen.* **1986**, *8*, 1–119. [CrossRef]
180. Herrmann, K.; Engst, W.; Meinl, W.; Florian, S.; Cartus, A.T.; Schrenk, D.; Appel, K.E.; Nolden, T.; Himmelbauer, H.; Glatt, H. Formation of hepatic DNA adducts by methyleugenol in mouse models: Drastic decrease by Sult1a1 knockout and strong increase by transgenic human SULT1A1/2. *Carcinogenesis* **2014**, *35*, 935–941. [CrossRef]
181. Herrmann, K.; Engst, W.; Florian, S.; Lampen, A.; Meinl, W.; Glatt, H.R. The influence of the SULT1A status-wild-type, knockout or humanized-on the DNA adduct formation by methyleugenol in extrahepatic tissues of mice. *Toxicol. Res.* **2016**, *5*, 808–815. [CrossRef]
182. Tan, K.H.; Nishida, R. Methyl Eugenol: Its Occurrence, Distribution, and Role in Nature, Especially in Relation to Insect Behavior and Pollination. *J. Insect Sci.* **2012**, *12*, 1–60. [CrossRef]
183. Smith, B.; Cadby, P.; Leblanc, J.-C.; Setzer, R.W. Application of the margin of exposure (MoE) approach to substances in food that are genotoxic and carcinogenic: Example: Methyleugenol, CASRN: 93-15-2. *Food Chem. Toxicol.* **2010**, *48*, S89–S97. [CrossRef] [PubMed]
184. Monien, B.H.; Schumacher, F.; Herrmann, K.; Glatt, H.; Turesky, R.J.; Chesné, C. Simultaneous Detection of Multiple DNA Adducts in Human Lung Samples by Isotope-Dilution UPLC-MS/MS. *Anal. Chem.* **2014**, *87*, 641–648. [CrossRef] [PubMed]
185. Abdo, K.; Cunningham, M.; Snell, M.; Herbert, R.; Travlos, G.; Eldridge, S.; Bucher, J. 14-Week toxicity and cell proliferation of methyleugenol administered by gavage to F344 rats and B6C3F1 mice. *Food Chem. Toxicol.* **2001**, *39*, 303–316. [CrossRef]
186. EFSA. Guidance on conducting repeated-dose 90-day oral toxicity study in rodents on whole food/feed. *EFSA J.* **2011**, *9*, 2438. [CrossRef]
187. OECD. Test No. 407: Repeated Dose 28-day Oral Toxicity Study in Rodents. In *OECD Guidelines for the Testing of Chemicals; Section 4*; OECD Publishing: Paris, France, 2008.
188. Sachse, B.; Meinl, W.; Sommer, Y.; Glatt, H.; Seidel, A.; Monien, B.H. Bioactivation of food genotoxicants 5-hydroxymethylfurfural and furfuryl alcohol by sulfotransferases from human, mouse and rat: A comparative study. *Arch. Toxicol.* **2014**, *90*, 137–148. [CrossRef] [PubMed]

Article

Phase II Metabolism of Asarone Isomers In Vitro and in Humans Using HPLC-MS/MS and HPLC-qToF/MS

Lena Hermes, Janis Römermann, Benedikt Cramer and Melanie Esselen *

Institute of Food Chemistry, University of Muenster, Corrensstraße 45, 48149 Muenster, Germany;
lena.hermes@uni-muenster.de (L.H.); ja.roemermann@gmail.com (J.R.); cramerb@uni-muenster.de (B.C.)
* Correspondence: esselen@uni-muenster.de; Tel.: +49-251-8333874

Abstract: (1) Background: Metabolism data of asarone isomers, in particular phase II, in vitro and in humans is limited so far. For the first time, phase II metabolites of asarone isomers were characterized and human kinetic as well as excretion data after oral intake of asarone-containing tea infusion was determined. (2) Methods: A high pressure liquid chromatography coupled with quadrupole time-of-flight mass spectrometry (HPLC-qTOF-MS) approach was used to identify phase II metabolites using liver microsomes of different species and in human urine samples. For quantitation of the respective glucuronides, a beta-glucuronidase treatment was performed prior to analysis via high pressure liquid chromatography coupled with tandem mass spectrometry (HPLC-MS/MS). (3) Results: Ingested beta-asarone and *erythro* and *threo*-asarone diols were excreted as diols and respective diol glucuronide conjugates within 24 h. An excretion rate about 42% was estimated. *O*-Demethylation of beta-asarone was also indicated as a human metabolic pathway because a corresponding glucuronic acid conjugate was suggested. (4) Conclusions: Already reported *O*-demethylation and epoxide-derived diols formation in phase I metabolism of beta-asarone in vitro was verified in humans and glucuronidation was characterized as main conjugation reaction. The excretion rate of 42% as *erythro* and *threo*-asarone diols and respective asarone diol glucuronides suggests that epoxide formation is a key step in beta-asarone metabolism, but further, as yet unknown metabolites should also be taken into consideration.

Keywords: asarone isomers; human study; metabolites; asarone glucuronides

Citation: Hermes, L.; Römermann, J.;
Cramer, B.; Esselen, M. Phase II
Metabolism of Asarone Isomers In
Vitro and in Humans Using
HPLC-MS/MS and HPLC-qToF/MS.
Foods **2021**, *10*, 2032. https://doi.org/
10.3390/foods10092032

Academic Editors: Andreas
Eisenreich and Bernd Schaefer

Received: 14 July 2021
Accepted: 26 August 2021
Published: 29 August 2021

Publisher's Note: MDPI stays neutral
with regard to jurisdictional claims in
published maps and institutional affiliations.

1. Introduction

Alpha-asarone (aA) and beta-asarone (bA) are phenylpropanoids, whose chemical structures only differ in the conformation of the double bound of the respective C3 side chain. These compounds mainly occur in essential oils of *Acorus* species, which are widely disseminated in Europe, North America and East Asia [1,2]. The predominant class is *Acorus (A.) calamus* L. Nevertheless, asarone contents strongly vary depending on several factors such as country of origin, polyploidy and plant materials e.g., roots or leaves [3]. In Indian tetraploid varieties, bA is present at high amounts of up to 95%, whereas aA amounts are on average about 15% [4].

Besides flavoring properties and bioactivity, essential oils as well as dried rhizomes or leaves of calamus are mainly used in herbal teas, frozen desserts, yoghurts, alcoholic and non-alcoholic beverages or food supplements [5]. Traditional phytomedicine and cosmetics also represent significant human exposure routes [2,4,5]. Moreover, asarone isomers and calamus-derived preparations are considered to show positive effects on human health such as antioxidant, anti-inflammatory, antidepressant, anti-microbial, neuro-, chemo- and radioprotective properties that facilitate their use as phytopharmaceutical summarized in Das et al. [6] and Chellian et al. [7]. Besides all positive effects, these substances are also of toxicological concern. Among acute and chronic toxicity, hepatotoxic properties in rodents as well as cytotoxic, genotoxic and mutagenic effects in vitro are reported for parent compounds and several oxidative phase I metabolites [7–11]. Furthermore, DNA repair mechanisms are also activated in response to asarone-mediated genotoxic effects in cells [10].

Phase I metabolism of aA and bA is elucidated in liver microsomes of different species. The main metabolite of aA is reported as 3′-oxoasarone, which arises out of (*E*)-3′-hydroxyasarone (3′OH) via further oxidation steps, whereas bA is mainly metabolized via an epoxide intermediate to *erythro*- and *threo*-1′,2′-dihydroxyasarone (*erythro*- and *threo*-asarone diols) and 2,4,5 trimethoxy-phenylacetone (asarone ketone) [12–14]. Even though in vitro phase I metabolism has been deeply characterised in the literature, little is known so far about potential phase II conjugation of these compounds. Nonetheless, the intake of *A. calamus* oil preparations in a human intervention trial proposes a renal hydroxylated bA metabolite after glucuronidase treatment, emphasizing glucuronic acid conjugation as possible phase II pathway [15]. This metabolite is also observed in urine of rats after an intraperitoneal application of aA [16].

Currently, there is no human data available regarding toxicokinetic parameters. In studies with rodents, for aA, bioavailability of 34% up to 78% is reported, which is further enhanced by inhalation [7]. Plasma half-life times of 29 min for aA and 13 min for bA are found after intravenous treatment of rats [17]. Depending on application form, longer half-life periods for oral and inhalative aA application, with values of 65 min up to 95 min, are reported [17,18]. In blood serum of rats, the half-life value of approximately 54 min indicate that bA is rapidly excreted [19]. Furthermore, marginal oral bioavailability has been considered, due the hydrophilic character of the asarone isomers, summarized in Chellian et al. [7]. Nevertheless, both isomers are shown to cross the blood brain barrier [18,20].

So far, in the European Union a maximum level of 1 mg/kg for bA is defined for alcoholic beverages, and for food a recommended maximum of 0.1 mg/kg is given [5,21]. For calamus preparations, it is recommended to use diploid varieties, which contain almost no or little bA amounts. For herbal medicine products, an intake of 2 µg/kg body weight/day is temporarily acceptable [5,22]. aA is not further regulated by law in the European Union, while bA is classified as a genotoxic carcinogen, thus the margin of exposure approach is used for risk assessment and a value below 10,000 is considered a high priority [23]. Toxicokinetic and toxicodynamic data of the genotoxin bA have to be urgently improved for adequate risk evaluation.

In consideration of the limited metabolism data of asarone isomers, in this study microsomal phase II metabolites were characterized using the respective phase I metabolites. Moreover, a human study was performed to identify main renal phase II metabolites. Ten participants consumed a commercially available calamus-containing infusion and gave urine samples over a period of 48 h. Furthermore, the study provided new insights into the excretion kinetic over the observation period and the overall rate of excretion, which was calculated considering the ingested asarone amounts. A high-pressure liquid chromatography coupled with quadrupole time-of-flight mass spectrometry (HPLC-qTOF-MS) approach was used to identify new metabolites in microsome samples and in human urine. Kinetic and excretion data were generated using validated high-pressure liquid chromatography with tandem mass spectrometric (HPLC-MS/MS) detection for quantitation.

2. Materials and Methods

2.1. Chemicals and Reagents

All solvents used for sample dilution or chromatography were of LC-MS-grade and purchased from Carl Roth (Karlsruhe, Germany), Fisher Scientific (Schwerte, Germany) or Sigma-Aldrich (Steinheim, Germany), if not stated otherwise. Water was purified using a Purelab Flex 2 system (Veolia Water Technologies, Celle, Germany). Formic acid (FA), and magnesium chloride ($MgCl_2$) were ordered from Merck (Darmstadt, Germany), tris(hydroxymethyl)aminomethane (TRIS) from VWR (Darmstadt, Germany) and ammonium hydrogen carbonate from Carl Roth (Karlsruhe, Germany). Uridine-5′-diphosphate glucuronic acid (UDPGA), glucose-6-phosphate-dehydrogenase (G6P-DH) and NADP+

were purchased from Sigma-Aldrich and *E. coli* beta-glucuronidase from Romer Labs GmbH (Butzbach, Germany). G6P was from AppliChem (Darmstadt, Germany).

Microsomes from the species pig and horse were isolated according to the protocol of LAKE 1987 and stored at −80 °C in storage buffer [24]. Human microsomes (Corning® UltraPool™, Corning, NY, USA) were purchased from Corning Inc. (Durham, NC, USA).

bA, 3′OH and (Z)-asarone-1′,2′-epoxide (bAE) were kindly provided by our project partner from the University of Kaiserslautern (Germany). *Threo-* and *erythro*-asarone diols were isolated from a decomposed epoxide solution. Further information can be found in the literature [10]. Additionally, 7′-hydroxycoumarin and 4-methylumbelliferyl-β-D-glucuronide were purchased from Sigma-Aldrich (Steinheim, Germany). All analyte solutions were prepared in acetonitrile and stored at −20 °C.

2.2. In Vitro Experiments

Phase II metabolites were generated by incubation of liver microsomes (rat, pig, human) with the respective phase I metabolites 3′OH and bAE. Subsequently characterization was carried out by a HPLC-qTOF-MS approach. Experimental conditions were set for glucuronic and sulfuric acid conjugation and experimental settings are described below. As sulfuric acid conjugation was not successful, further sample preparation focuses on glucuronidation. Details for sulfuric acid conjugation are described in Supporting Information Table S1.

Sample Preparation

The method to simulate glucuronic acid conjugation by microsomes using phase I metabolites was conducted in accordance with WU et al., 2007 [25]. In each reaction mixture protein concentration was normalized to 1 mg/mL, independent of the microsome species. The reaction mixture consisted of 0.3 mM UDPGA, 0.4 mM $MgCl_2$ and 0.1 mM 3′OH or bAE, respectively. The volume of each reaction mixture was filled with 83.4 mM Tris buffer to 200 µL. Microsomes were directly added before the reaction was started. Each tube was gently vortexed and incubated at 37 °C for 4 h while gently shaking. To stop the reaction, 400 µL acetonitrile were added to each tube and the samples were centrifugated at 4 °C for 5 min and 14,000× g. 150 µL of supernatant were diluted with 850 µL of water achieving a final concentration of acetonitrile of 12%, noting the starting conditions of the following HPLC-qTOF-MS method. A blank sample without analyte was used to distinguish analyte signals from matrix signals. Moreover, a second blank sample without liver microsomes was implemented to distinguish enzymatic from chemical reactions. Efficiency of the used microsomal systems was verified with the control 7′-hydroxycoumarin (100 µM).

2.3. Human Study

2.3.1. Study Conditions and Subjects

Ten healthy enrolled participants (five females and five males, age 25.8 ± 4.0, BMI 23.8 ± 1.9) were informed about the aim and scope of the study and gave their written consent to the study conditions prior to their commencement. Samples and food diaries were equipped with a six-digit number and assigned to each participant randomly. The study was approved by the research ethical committee of the University Hospital Münster, Germany (File reference: 2020-002-f-S).

2.3.2. Study Design

Study participants were not allowed to consume *A. calamus*-derived preparations or herbal products three days before intake of the tea infusion (wash-out phase), and wrote a food diary during the complete study progress (Figure 1).

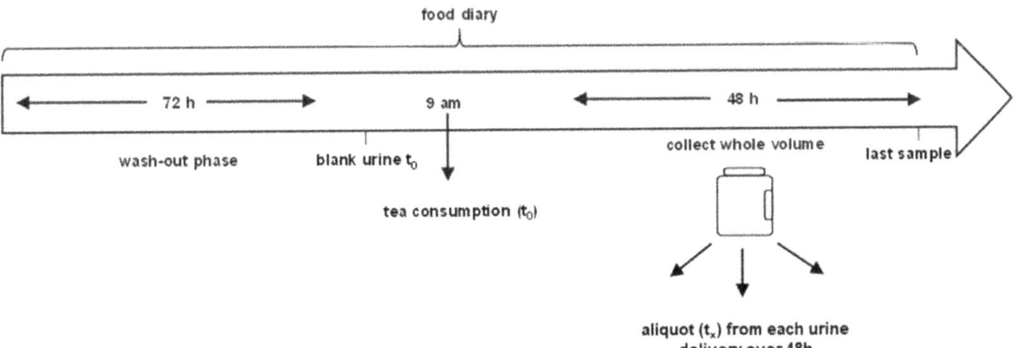

Figure 1. Study design including wash-out phase, diary keeping and urine collection after intake of 300 mL calamus tea infusion. Urine samples were collected for 48 h and total urine volume was determined. The participants collected a spot urine sample every time they urinated.

On day one, participants passed a morning urine sample as blank before calamus tea consumption. Tea preparation is described in (Section 2.3.3 "Calamus Tea selection"). All participants consumed at the same time 300 mL of the prepared tea infusion to ensure a consistent intake of asarone isomers. The tea infusion was consumed within half an hour and urine was collected for 48 h. The total urine volume over 48 h was determined by each participant by summarizing the volumes of each spot urine sample. Thereafter, an aliquot of 2 mL of each urine sample was collected for every time point (Figure 1). Urine samples were stored at 20 °C prior to sample preparation.

The total urine volume and the concentrations of *erythro*- and *threo*-asarone diols after beta-glucuronidase treatment was used to determine the excretion rate. Excretion values over 48 h were totalized and were compared with the amounts of *erythro*- and *threo*-asarone diols and bA in 300 mL of the consumed tea infusion. The average of all values for each participant was used as overall excretion rate (%).

The kinetic curve was determined using point in time of urination (aliquot of each urine delivery of each participant), given urine volume and concentrations of *erythro*- and *threo*-asarone diols in the urine samples over the period of 48 h. Time points were classified in two-hour blocks, except for the night hours (14–20 h) and the last 24 h, because in this period only a small number of samples was above the limit of quantitation (LOQ). The results were represented in a box-plot-whisker diagram. The average 50% of the data is located within the box, whose dimensions are defined by the lower and upper quartiles. The lowest and highest data points are shown by the whiskers if the values fall within the 1.5 interquartile range otherwise they are outliers and demonstrated as dots below and above the whiskers.

2.3.3. Calamus Tea Selection

Calamus tea consisted of organically grown calamus roots, which have been dried and chopped prior to disposal. The used calamus infusion classified as food was analyzed within a previous product screening study [26]. Tea was prepared by weighing 48.5 g of dried calamus roots, infusing it with 3300 mL of boiling water and steeping for 15 min. For analysis of the asarone amount, an aliquot was filtrated using a 0.45 µM regenerated cellulose (RC) membrane (Phenomenex, Aschaffenburg, Germany), diluted with acetonitrile/0.1% formic acid in water (12/88, *v/v*) and analyzed by HPLC-MS/MS as reported previously [26]. 300 mL of the infusion contained 0.76 mg bA, 0.65 mg *erythro*-asarone diols and 1.38 mg *threo*-asarone diols. The amounts are in the mean of commercially available calamus infusions [26]. The fresh calamus tea was prepared by a separation of the rhizome from the *A. calamus* plant, which was bought in a local garden center (Vechta, Germany)

The outer root layer was cleaned, and 3 g of calamus roots were crushed. Thereafter the roots were infused with 200 mL boiling water (100 °C). Steeping time, filtration, dilution and analysis were carried out as described above.

2.3.4. Urine Sample Preparation

To 100 µL of each urine sample of each collection point, 100 µL of ammonium hydrogen carbonate (NH_4HCO_3) buffer (pH 6.6) were added which contained 6000 U/mL of beta-glucuronidase. The samples were incubated for 16 h at 37 °C by gently shaking. A volume of 200 µL acetonitrile were added to stop the enzyme reaction and after homogenization samples were centrifuged at 14,000× *g* for 5 min at 5 °C. Afterwards, the supernatant was diluted 1:10 with acetonitrile/0.1% formic acid in water (12/88, *v/v*) prior to HPLC-MS/MS or qTOF-MS analysis. Workability of beta-glucuronidase was verified by the 4-methylumbelliferyl-β-D-glucuronide converted to the highly fluorescent 4-methylumbelliferon and glucuronic acid. Fluorescence was measured at 365/440 nm.

2.4. Method Validation

The developed HPLC-MS/MS method for quantitation of *erythro-* and *threo-*asarone diols was evaluated with regard to the following parameters: linearity, limit of detection (LOD), LOQ, recovery as well as intraday and interday repeatability [27].

For quantitation of *erythro-* and *threo-*asarone diols in the urine samples, a matrix-matched calibration in blank urine of different volunteers was prepared. To that end, urine was processed as described in Section 2.3.4. and fortified with the analytes at concentrations of 0.25, 0.5, 1, 2.5, 5, 10, 25, 50 ng/mL, respectively. Linearity across the whole working range was verified by the Mandel's fitting test and a coefficient of determination (R^2) $\geq$ 0.995.

LOD and LOQ were determined using a matrix-matched approach in blank urine of different volunteers. Analytes were spiked in the following concentrations 0.05, 0.1, 0.25, 0.5, 1.0, 2.5, 5.0 and 10 ng/mL. Procedure was performed in triplicate. LOD and LOQ were determined using a Signal to Noise (S/N) approach receiving a S/N ratio of three for LOD and ten for LOQ.

For determination of recovery rates, blank urine was spiked with distinct analyte concentrations (0.25, 0.5, 1, 2.5, 5, 10, 25, 50 ng/mL) prior to sample preparation (Section 2.3.4). For calculation, the matrix-matched calibration was measured along with the matrix calibration and the slope of both calibration curves was compared. Each calibration point was prepared in triplicate.

The precision of the method is described by interday and intraday repeatability. Intraday repeatability was evaluated by preparing and analyzing one randomly chosen urine sample of one test person ten times. For interday repeatability one sample was prepared three times and repeatedly injected throughout the measurement of all urine samples of the study.

2.5. HPLC-MS/MS and HPLC-qTOF-MS Settings

Chromatographic separation for MS/MS analysis was achieved using a 1260 Infinity LC system (Agilent Technologies, Waldbronn, Germany). MS/MS analysis was performed using a QTrap® 5500 mass spectrometer equipped with a Turbo V ion source and operated with Analyst software 1.6.2 (Sciex, Darmstadt, Germany). The obtained values for the MS parameters declustering potential (DP), collision energy (CE) and collision cell exit potential (CXP) were individually determined by infusing standard solutions into the MS system. MS parameters are as follows: Q_1 (*m/z*), 225; Q_3 (*m/z*) Q_N (quantifier transition)/Q_L (qualifier transition), 193/167; declustering potential (DP), 104; collision energy (CE) Q_N/Q_L, 18/23; CXP (V), 11. Retention time (RT): 5.39 min for *erythro-*asarone diols and 5.69 min for *threo-*asarone diols. Further HPLC-MS/MS setup details are presented in Supporting Information Table S2.

Chromatographic separation for qTOF-MS analysis was achieved using a Bruker Elute system (Bruker, Bremen, Germany). Mass spectrometric analysis was carried out on a Bruker impact II qTOF system equipped with an ESI Apollo II ion source operated in positive and negative ionization mode, depending on the analyte of interest (Bruker, Bremen, Germany). For identification of phase II metabolites, a full scan mode within a mass range of m/z 50 to 1300 as well as Auto MS/MS scan modes, were used. Further HPLC-qTOF-MS setup details for the analysis of liver microsome samples, as well as urine samples are given in the Supporting Information Tables S3 and S4.

3. Results

3.1. Microsome Experiments

Incubation of the selected phase I metabolites 3′OH and bAE with pig liver microsomes resulted in formation of different glucuronic acid conjugates. For bAE, it is re-ported that this compound is not stable and hydrolyzes in aqueous solution, with a half-life between 2.4 min and 4 min to *erythro-* and *threo-*asarone diols and asarone ketone [13,28].

Consequently, incubation of bAE with microsomes resulted in diol-derived glucuronic acid conjugates. The extracted ion chromatograms (XICs) with m/z 399.1297 for 3′OH glucuronide (Figure 2a) and m/z 417.1402 for *erythro-* and threo-asarone diols-derived glucuronic acid conjugates (Figure 2b) allowed the detection of two peaks with mass differences (Δm) of 0.5 ppm and 0.8 ppm to the calculated masses of $[M–H]^-$. Figure 2c shows the qTOF-MS spectrum of the 3′OH-glucuronide. The fragment with m/z 223.0984 can be assigned to the loss of the glucuronic acid moiety and corresponds to the $[M–H]^-$ of 3′OH (Figure 2c). Due to its low concentration, the spectrum of the *erythro-* and threo-asarone diol-glucuronides did not provide significant fragmentation data. Liver microsomes of human and horse were also used to investigate the phase II metabolism of both phase I metabolites (3′OH, bAE). The respective glucuronic acid conjugates were formed by all species but with slightly different turnover rates (data not shown). Detailed information about species-specific phase II-Metabolism has to be considered in subsequent analyses and are not in the scope of the presented investigations. Sulfuric acid conjugation was not observed at all, indicating that glucuronidation can be considered as the main metabolic phase II pathway in microsomes from all species.

3.2. Method Validation

Method validation of the used HPLC-MS/MS method was performed prior to analysis of the urine samples from the human study. As *erythro-* and *threo* asarone diols were found to be the dominant metabolites in urine after beta-glucuronidase treatment, quantitation of these compounds with a matrix-matched calibration in blank urine was performed. *Erythro-* and *threo-*asarone diols are diastereomers, which represent a pair of enantiomers, respectively (Figure 3a).

Accordingly, with the used HPLC-MS/MS method, for the diastereomers *erythro-* and *threo-*asarone diols could be chromatographically separated, while the enantiomers coeluted. Figure 3b shows the analysis of one selected urine sample spiked with erythro and threo-asarone diols at a concentration of 5 ng/mL

From matrix-matched calibration, the validation parameters LOD and LOQ as well as linearity were determined. Linearity across the applied concentration range was confirmed by means of the Mandel's fitting test as well as a $R^2 \geq 0.995$ for the analytes. The analytical precision via interday and intraday repeatability reached values of between 3% and 12% and recovery rates of 83% or 103% were determined. All values fall into an acceptable range considering the respective US Food and Drug Administration regulations [27]. The validation parameters are illustrated in Table 1.

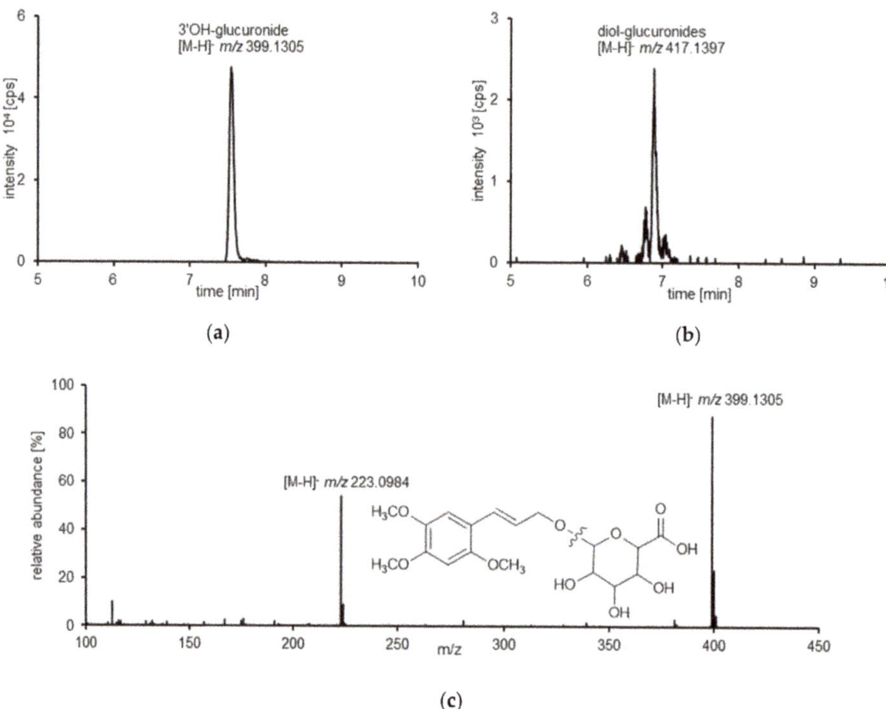

Figure 2. HPLC-qTOF-MS chromatograms after incubation of (**a**) 3′OH and (**b**) bAE in pig liver microsomes. Presented are the extracted-ion chromatogram (XICs) with the calculated mass of (**a**) m/z 399.1305 ± 0.01 for the 3′OH glucuronide and (**b**) m/z 417.1397 ± 0.01 for *erythro-* and *threo*-asarone diols-derived glucuronic acid conjugates. (**c**) HPLC-qTOF-MS spectrum of 3′OH glucuronide (m/z 399.1305 ± 0.01) with the respective structural formula and the suggested cleavage of the glucuronic acid majority to m/z 223.0984.

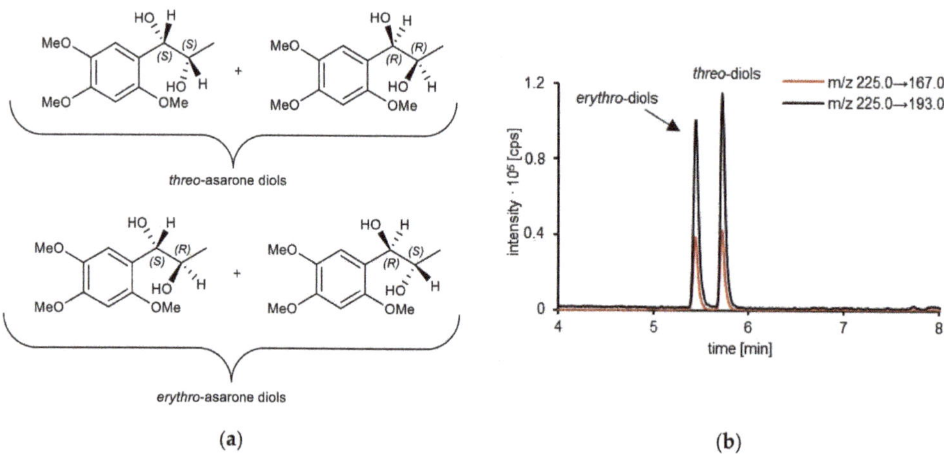

Figure 3. (**a**) Structural illustration of *erythro-* and *threo*-asarone diols and their stereochemistry. (**b**) HPLC-MS/MS chromatogram of a 1:10 diluted urine sample spiked with 5 ng/mL of erythro- and threo-asarone diols. Presented are the quantifier (m/z 225→193) and qualifier (m/z 225→167) SRM transition.

Table 1. Method performance characteristics of the LC-MS/MS method used for quantitation of *erythro-* and *threo*-asarone diols in urine samples.

Substance	Linear Range [ng/mL]	LOQ [ng/mL]	LOQ [ng/mL]	Interday Repeatability [%]	Intraday Repeatability [%]	Recovery [%]
erythro-asarone diols	0.25–50	0.09	0.30	12.3	3.4	103
threo-asarone diols	0.25–50	0.06	0.25	8.5	8.3	83

3.3. Human Study

3.3.1. Analysis of the Consumed Tea Infusion

The amounts of bA (0.76 mg) as well as *erythro-* (0.65 mg) and *threo*-diols (1.38 mg) in 300 mL of the consumed tea were used in total (2.79 mg) for calculation of the excretion rates.

3.3.2. HPLC-MS/MS and qTOF-MS Analysis of Urine Samples

Figure 4 shows HPLC-MS/MS chromatograms of an exemplary urine sample from one randomly selected participant before (a) and after beta-glucuronidase treatment (b), recorded in MRM-mode. The subsequently mentioned metabolism was observed in the urine of all participants with marginal differences in individual metabolite concentrations and excretion rates. The two peaks (5.39 and 5.69 min) represent the *erythro-* and *threo*-asarone diols, respectively, whereas the peak with a retention time of 5.80 min showing the same MRM transition could not be identified with the available standards (Figure 4a). No signal corresponding to 3′OH or asarone ketone was detected in all analyzed urine samples. Furthermore, no hints for a 3′OH glucuronide were found. However, after beta-glucuronidase treatment, the signal at 5.80 min disappeared, while the *erythro*-asarone diols peak (5.39 min) slightly and the *threo*-asarone diols peak (5.69 min) strongly increased (Figure 4b). These results suggest that the peak eluting at 5.80 min represents glucuronidated metabolites of the consumed asarone derivatives.

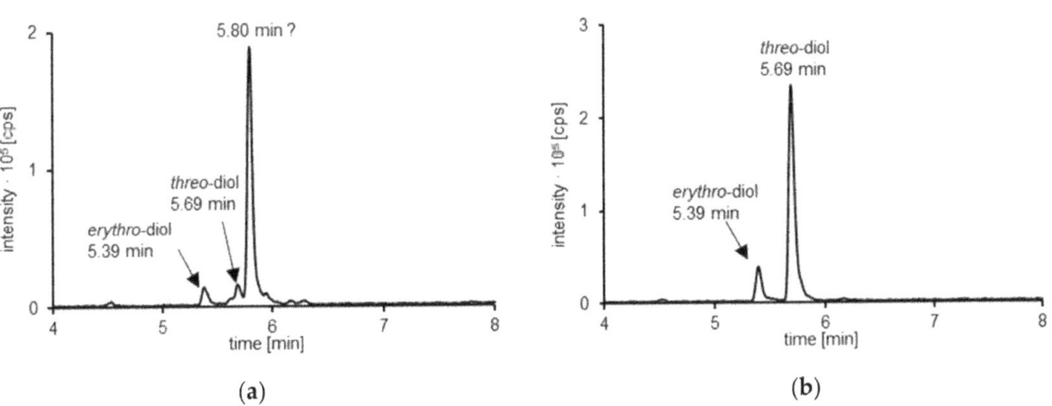

Figure 4. HPLC-MS/MS chromatogram of a randomly selected urine sample, which was given after consumption of a calamus tea infusion, (**a**) before; (**b**) after treatment with beta-glucuronidase.

To verify these findings and further to identify further new phase II metabolites, an untargeted HPLC-qTOF-MS approach was applied to human urine samples before beta-glucuronidase treatment. For the main peak, a mass of m/z 417.1404 ($[C_{18}H_{26}O_{11}-H]^-$, Δm: 0.2 ppm) supports the suggestion that *erythro-* and *threo*-asarone diol-glucuronides are potential phase II metabolites in humans (Figure 5a). Moreover, an unknown metabolite with an exact mass of m/z 403.1256 was detected in human urine. Based on a calculated m/z of 403.1256 for $[C_{17}H_{24}O_{11}-H]^-$, a mass difference of 1 ppm to the calculated mass suggested that also demethylated *erythro*- and *threo*-asarone diols-derived glucuronides were formed (Figure 5b). The recorded qTOF-MS spectrum supports our suggestions. The detected fragment ions of m/z 227.0923 are reported to arise due to the loss of the glucuronic acid moiety, and m/z 212.0685 with a further loss of a methyl group (Figure 5c).

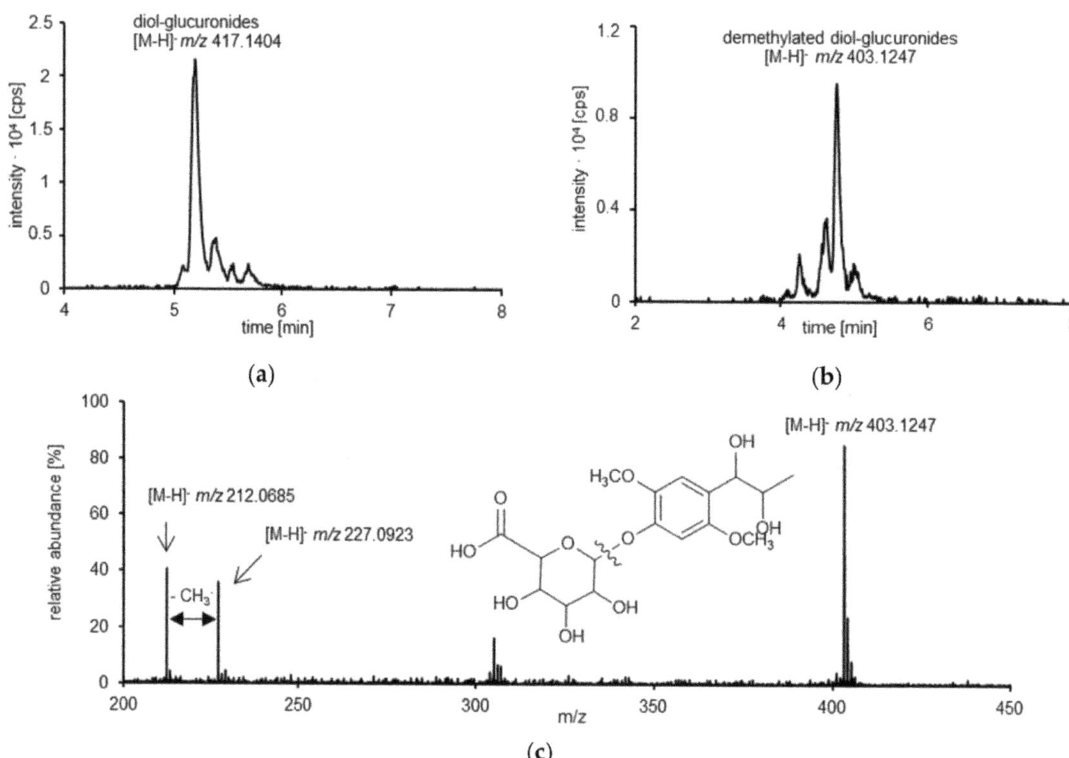

Figure 5. Exemplary HPLC-qTOF-MS chromatograms of a randomly selected urine sample before enzyme treatment. Presented are the extracted-ion chromatogram (XICs) with the calculated mass for (**a**) *erythro*- and *threo*-asarone diol glucuronides (diol-glucuronides, m/z 417.1404 ± 0.02) and (**b**) demethylated *erythro*- and *threo*-asarone diol glucuronides (demethylated diol-glucuronides, m/z 403.1247 ± 0.01). (**c**) HPLC-qTOF-MS spectrum of the *O*-demethylated *erythro*- and *threo*-asarone diol glucuronides (m/z 403.1247 ± 0.01) with the respective structural formula. The fragment m/z 227.0923 ± 0.02 corresponds to the *O*-demethylated metabolites after glucuronic acid cleavage. The loss of a further methyl group is shown by the exact mass of m/z 212.0685 ± 0.01.

As mentioned before, the calamus infusion used for the human study contains bA as well as *erythro*- and *threo*-asarone diols, thus the potential metabolization of bA to the identified phase II metabolites is not possible. Therefore, in a second proof of concept experiment, an infusion of fresh calamus roots was consumed by three participants and urine samples were collected as described above. It is reported that this tea infusion contains only bA (20 mg/kg) [26]. The analysis of these three urine samples showed that all characterized phase II metabolites are also excreted after single bA intake. Corresponding chromatograms are shown in Supporting Information Figure S1.

3.3.3. Kinetic Studies and Excretion Rate Determination

Kinetic data and excretion rates were determined based on the analysis of the *erythro*- and *threo*-asarone diol peaks formed after beta-glucuronidase treatment, because the respective glucuronides were characterized as main human metabolites. The *O*-demethylated *erythro*- and *threo*-asarone diols-derived glucuronides were not included, because no corresponding reference compounds were available. Based on the above-mentioned finding, that an oral intake of bA also results in a renal excretion of *erythro*- and *threo*-asarone diols glucuronides, the overall amount of bA and *erythro*- and *threo*-asarone-diols was used to calculate total excretion rates. In sum, a total excretion of 42 ± 6% of all participants was determined. Kinetic data over a period of 48 h is shown in Figure 6. The respective phase II metabolites were rapidly excreted, with a maximum excretion between one and six hours. After 24 h only marginal amounts of *erythro*- and *threo*-asarone diols were further detected, thus the period between 24 h and 48 h was summarized in one bar.

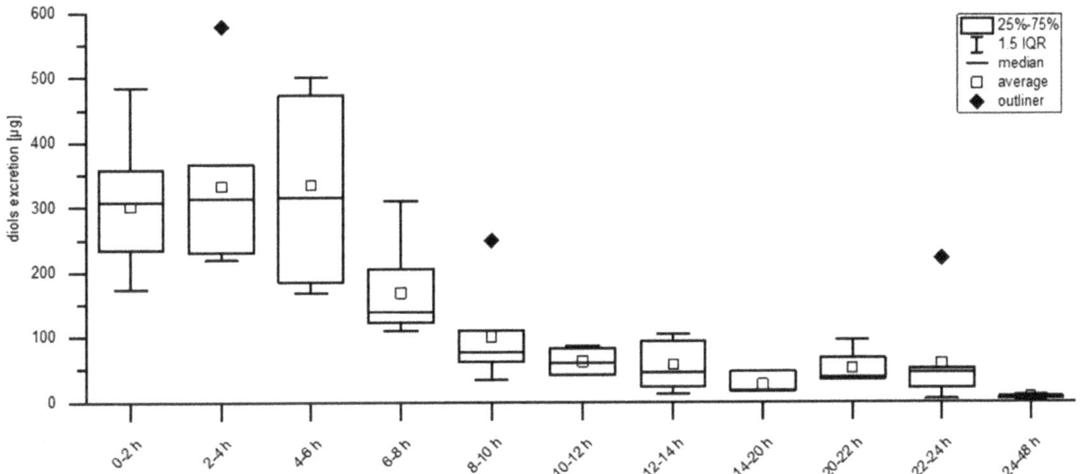

Figure 6. *Erythro*- and *threo*-asarone diols (diols) excretion kinetic [µg] of ten participants over a period of 48 h. Excretion is classified in two-hour blocks, except for the night hours (14–20 h) and the last 24 h, because the concentrations of the metabolites were mostly under the Limit of Quantification (LOQ).

4. Discussion

The scope of this work was the characterization of so far unknown asarone isomers-derived phase II metabolites in vitro and in humans, as well as a quantitative assessment of their excretion profile and kinetic in humans. In microsome experiments using the selected phase I metabolites bAE and 3′OH, new glucuronic acid conjugates, *erythro*- and *threo*-asarone diol glucuronides and 3′OH glucuronide, were identified based on their exact mass and mass spectrometric fragmentation pattern. After the intake of a bA- and *erythro*- and *threo*-asarone diols-containing tea infusion, signals for *erythro*- and *threo*-asarone diol glucuronides were detected, while no signals for further reported phase I metabolites e.g., 3′OH or asarone ketone were observed. However, in the presented human study, signals indicating *O*-demethylation reactions and respective glucuronic acid conjugation were found and suggest a further metabolic pathway.

In literature, a human renal glucuronic acid conjugate of hydroxylated aA is postulated after consumption of *A. calamus* oil without any further characterization [15]. The results of the conducted human study exclude the formation of a 3′OH-derived glucuronide. Side chain hydroxylation is hypothesized as main phase I metabolic pathway of aA [14]. Furthermore, the results suggest that side chain hydroxylation at position 3′ of bA does not occur in humans, even though it is a published pathway using liver microsomes of different species [13].

It is known that bAE rapidly decomposes to *erythro*- and *threo*-asarone diols and asarone ketone in aqueous solutions [13,28,29], hence glucuronic acid conjugation of *ery*-

thro- and *threo-*asarone diols in liver microsomes after an incubation with bAE was detected. The consumed commercially available tea infusion contained besides bA also the *erythro-* and *threo-*asarone diols, thus raising the question as to whether the corresponding glucuronides are solely formed from the *erythro-* and *threo-*asarone diols present in the beverage. To answer this question, a second proof of the concept study with a calamus infusion of fresh non-dried roots, containing only the parent compound bA, was performed. However, the *erythro-* and *threo-*asarone diol-derived glucuronides were also found in a comparable pattern in human urine after consumption of this infusion. For the first time, these results explicitly emphasize that the epoxide-diol-pathway, which is identified as the main toxification pathway using liver microsomes [13], is also of special relevance in humans. In addition, these results are of high toxicological concern because the bAE is postulated as ultimate carcinogen [13]. Moreover, in mammalian cell systems bA-derived DNA adducts are identified and genotoxic effects of bAE are reported [9,28]. Considering the observed excretion rate of 42%, it is quite reasonable that the highly reactive epoxide intermediate is formed to a significant extent, which promotes its binding to macromolecules such as DNA or proteins. However, a fast repair of epoxide-derived genotoxic DNA-damage in liver tumor HepG2 cells and also a time-dependent decrease of DNA adducts in rat hepatocytes are reported [10,28]. Epoxide hydrolases catalyze the formation of less reactive dihydro-diol derivatives and they are suggested to play a major role in the detoxification of epoxides in vivo [30]. Nevertheless, missing data on the stability of bAE in vivo makes it difficult to assess potential further risks.

O-Demethylation was identified as a second metabolism pathway in humans because glucuronic acid conjugates of demethylated *erythro-* and *threo-*asarone diols were detected in human urine after intake of tea infusion of fresh or dried calamus roots, respectively. It is reported that O-demethylation, besides hydroxylation and epoxide formation plays a crucial role in microsomal metabolization of bA [13].

For the phenylpropenes elemicin, myristicin, and safrole several metabolites are characterized in human urine after nutmeg abuse or in urine samples of rats after drug administration. *O*-Demethylation and side chain hydroxylation are identified as main phase I reactions, whereas in phase II the functionalized metabolites are found to be partly conjugated to glucuronic acid or sulfuric acid [31]. Urinary recovery rates of the phenylpropenes estragole and eugenol of between 65–70% and 95% are reported and 50% of recovered eugenol is also excreted as glucuronic or sulfuric acid conjugates [32,33]. In contrast asarone derivatives were exclusively excreted as glucuronic acid conjugates, but with recovery rates below 50%. Regarding the excretion rate, it has to be mentioned that demethylated reference compounds are not available so far. Consequently, the detected demethylated asarone derivatives cannot be quantified after beta-glucuronidase treatment. Furthermore, no sulfonated metabolites were determined in human urine. These findings are in line with further in vitro investigations showing that the incubation of 3'OH and bAE with cytosolic fractions did not lead to sulfuric acid conjugates. It is also reported that sulfonation of 1'-hydroxy-estragole is less dominant in humans than in rodent species using respective liver S9-fractions [34]. Another work postulates 2,4,6-trimethoxycinnamic acid as a further metabolite of aA and bA in rat hepatocytes via LC-MS [35], which was not considered in this investigation.

The results show that, after calamus tea consumption by ten healthy participants, bA and *erythro-* and *threo-*asarone diols were quickly excreted within 24 h as respective glucuronides and, to a smaller extent, as non-conjugated *erythro-* and *threo-*asarone diols, reaching their maximum levels between 1 h and 6 h. A fast excretion and also a rapid glucuronidation is reported for the structure-related compound estragole after the intake of a fennel tea or the oral administration eugenol [34,36].

5. Conclusions

In sum, the results point out that uridine 5′-diphospho-glucuronosyltransferase-catalyzed conjugation reactions play a crucial role in phase II metabolism of asarone derivatives. This study confirms *O*-demethylation as an important metabolism pathway in humans for the first time. A fast renal excretion within 24 h is observed after calamus tea consumption. The recovery rate of only 42% emphasize that further work is still needed to characterize yet unknown human bA-related metabolites. A metabolization of bA via the epoxide-diol-pathway is suggested and should be considered for human risk assessment.

Supplementary Materials: The following are available online at https://www.mdpi.com/article/10.3390/foods10092032/s1, Figure S1: HPLC-qTOF-MS chromatogram of a selected urine sample after intake of fresh prepared bA-containing calamus tea, Table S1: Substances and concentrations used for phase II sulfonation experiments, Table S2: HPLC-MS/MS setup for the quantitation of *threo-* and *erythro-*asarone diols in urine samples, Table S3: HPLC-qTOF-MS setup for the screening of liver microsome samples for phase-II-metabolites originating from beta-asarone epoxide (bAE) and 3′hydroxyasarone (3′OH), Table S4: Characterization of unknown metabolites: Differences to HPLC-qTOF-MS setup presented in Table S2.

Author Contributions: Conceptualization, M.E.; methodology, L.H. and B.C.; investigation, L.H. and J.R.; data curation, L.H.; writing—original draft preparation, L.H.; writing—review and editing, B.C. and M.E.; visualization, B.C. and M.E.; supervision, M.E.; funding acquisition, M.E. All authors have read and agreed to the published version of the manuscript.

Funding: This research was funded by German Federal Ministry of Education and Research, grant number 01FP13061F.

Institutional Review Board Statement: The study was conducted according to the guidelines of the Declaration of Helsinki, and approved by the Ethics Committee of University Hospital Münster, Germany (protocol code 2020-002-f-S, date of approval: 17 February 2020).

Informed Consent Statement: Informed consent was obtained from all subjects involved in the study.

Data Availability Statement: Data is contained within the article or Supplementary Materials.

Acknowledgments: We would like to thank Alexander Cartus and Simone Stegmüller for isolating and providing beta-asarone and beat-asarone epoxide.

Conflicts of Interest: The authors declare no conflict of interest.

References

1. Zuo, H.L.; Yang, F.Q.; Zhang, X.M.; Xia, Z.N. Separation of cis- and trans-asarone from *Acorus tatarinowii* by preparative gas chromatography. *J. Anal. Methods Chem.* **2012**, *2012*, 402081. [CrossRef]
2. Rajput, S.B.; Tonge, M.B.; Karuppayil, S.M. An overview on traditional uses and pharmacological profile of *Acorus* calamus Linn. (Sweet flag) and other *Acorus* species. *Phytomedicine* **2014**, *21*, 268–276. [CrossRef]
3. Varshney, V.K.; Song, B.H.; Ginwal, H.S.; Mittal, N. High Levels of diversity in the phytochemistry, ploidy and genetics of the medicinal plant *Acorus calamus* L. *J. Med. Aromat. Plants* **2015**, *2015*, 1–9. [CrossRef]
4. Rana, T.S.; Mahar, K.S.; Pandey, M.M.; Srivastava, S.K.; Rawat, A.K.S. Molecular and chemical profiling of 'sweet flag' (*Acorus calamus* L.) germplasm from India. *Physiol. Mol. Biol. Plants* **2013**, *19*, 231–237. [CrossRef] [PubMed]
5. Scientific Committee on Food; European Commission. Opinion of the Scientific Committee on Food on the Presence of Beta-asarone in Flavourings and Other Food Ingredients with Flavouring Properties (SCF/CS/FLAV/FLAVOUR/9 ADD1 Final). 2002. Available online: https://ec.europa.eu/food/system/files/2016-10/fs_food-improvement-agents_flavourings-out111.pdf (accessed on 27 August 2021).
6. Das, B.K.; Swamy, A.V.; Koti, B.C.; Gadad, P.C. Experimental evidence for use of *Acorus* calamus (asarone) for cancer chemoprevention. *Heliyon* **2019**, *5*, e01585. [CrossRef] [PubMed]
7. Chellian, R.; Pandy, V.; Mohamed, Z. Pharmacology and toxicology of α- and β-asarone: A review of preclinical evidence. *Phytomedicine* **2017**, *32*, 41–58. [CrossRef]
8. Haupenthal, S.; Berg, K.; Gründken, M.; Vallicotti, S.; Hemgesberg, M.; Sak, K.; Schrenk, D.; Esselen, M. In vitro genotoxicity of carcinogenic asarone isomers. *Food Func.* **2017**, *8*, 1227–1234. [CrossRef]

9.   Berg, K.; Bischoff, R.; Stegmüller, S.; Cartus, A.; Schrenk, D. Comparative investigation of the mutagenicity of propenylic and allylic asarone isomers in the Ames fluctuation assay. *Mutagenesis* **2016**, *31*, 443–451. [CrossRef]
10. Hermes, L.; Haupenthal, S.; Uebel, T.; Esselen, M. DNA double strand break repair as cellular response to genotoxic asarone isomers considering phase I metabolism. *Food Chem. Toxicol.* **2020**, *142*, 111484. [CrossRef]
11. Wiseman, R.W.; Miller, E.C.; Miller, J.A.; Liem, A. Structure-activity studies of the hepatocarcinogenicities of alkenylbenzene derivatives related to estragole and safrole on administration to preweanling male C57BL/6J x C3H/HeJ F1 mice. *Cancer Res.* **1987**, *47*, 2275–2283. [CrossRef]
12. Cartus, A.T.; Schrenk, D. Metabolism of carcinogenic alpha-asarone by human cytochrome P450 enzymes. *Naunyn-Schmiedebergs Arch. Pharmacol.* **2020**, *393*, 213–223. [CrossRef]
13. Cartus, A.T.; Stegmüller, S.; Simson, N.; Wahl, A.; Neef, S.; Kelm, H.; Schrenk, D. Hepatic metabolism of carcinogenic β-asarone. *Chem. Res. Toxicol.* **2015**, *28*, 1760–1773. [CrossRef]
14. Cartus, A.T.; Schrenk, D. Metabolism of the carcinogen alpha-asarone in liver microsomes. *Food Chem. Toxicol.* **2016**, *87*, 103–112. [CrossRef]
15. Björnstad, K.; Helander, A.; Hultén, P.; Beck, O. Bioanalytical investigation of asarone in connection with *Acorus* calamus oil intoxications. *J. Anal. Toxicol.* **2009**, *33*, 604–609. [CrossRef]
16. Morales-Ramírez, P.; Madrigal-Bujaidar, E.; Mercader-Martínez, J.; Cassani, M.; González, G.; Chamorro-Cevallos, G.; Salazar-Jacobo, M. Sister-chromatid exchange in-duction produced by in vivo and in vitro exposure to alpha-asarone. *Mutat. Res. Genet. Toxicol. Environ. Mutagen* **1992**, *279*, 269–273. [CrossRef]
17. Yang, Q.; Deng, Z.; Zhang, F.; Sun, P.; Li, J.; Zheng, W. Development of an LC–MS/MS method for quantification of two isomeric phenylpropenes and the application to pharmacokinetic studies in rats. *Biomed. Chromatogr.* **2018**, *32*, e4115. [CrossRef]
18. Lu, J.; Fu, T.; Qian, Y.; Zhang, Q.; Zhu, H.; Pan, L.; Guo, L.; Zhang, M. Distribution of α-asarone in brain following three different routes of administration in rats. *Eur. J. Pharm. Sci.* **2014**, *63*, 63–70. [CrossRef]
19. Meng, X.; Zhao, X.; Wang, S.; Jia, P.; Bai, Y.; Liao, S.; Zheng, X. Simultaneous determination of volatile constituents from *Acorus tatarinowii* Schott in rat plasma by gas chromatography-mass spectrometry with selective ion monitoring and application in pharmacokinetic study. *J. Anal. Methods Chem.* **2013**, *2013*, 949830. [CrossRef]
20. Fang, Y.Q.; Shi, C.; Liu, L.; Fang, R.M. Pharmacokinetics of beta-asarone in rabbit blood, hippocampus, cortex, brain stem, thalamus and cerebellum. *Pharmazie* **2012**, *67*, 120–123. [CrossRef]
21. Regulation (EC) No 1334/2008 of the European Parliament and of the Council of 16 December 2008 on Flavourings and Certain food Ingredients with Flavouring Properties for use in and on Foods and Amending Council Regulation (EEC) No 1601/91, Regulations (EC) No 2232/96 and (EC) No 110/2008 and Directive 2000/13/EC. 2008. Available online: https://eur-lex.europa.eu/legal-content/EN/TXT/PDF/?uri=CELEX:32008R1334&from=en (accessed on 27 August 2021).
22. European Medicines Agency, Commitee on Herbal Medicinal Products. Public Statement on the Use of Herbal Medicinal Products Containing Asarone. 2005. Available online: https://www.ema.europa.eu/en/documents/scientific-guideline/public-statement-use-herbal-medicinal-products-containing-asarone_en.pdf (accessed on 27 August 2021).
23. Van den Berg, S.J.P.L.; Restani, P.; Boersma, M.G.; Delmulle, L.; Rietjens, I.M.C.M. Levels of genotoxic and carcinogenic ingredients in plant food supplements and associated risk assessment. *Food Nutr. Sci.* **2011**, *2*, 989–1010. [CrossRef]
24. Lake, B.G. Preparation and characterisation of microsomal fractions for studies of xenobiotic metabolism. In *Biochemical Toxicology—A Practical Approach*; Snell, K., Mullock, B., Eds.; IRL Press: Oxford, UK, 1987; pp. 183–215.
25. Wu, X.; Murphy, P.; Cunnick, J.; Hendrich, S. Synthesis and characterization of deoxynivalenol glucuronide: Its comparative immunotoxicity with deoxynivalenol. *Food Chem. Toxicol.* **2007**, *45*, 1846–1855. [CrossRef] [PubMed]
26. Hermes, L.; Römermann, J.; Cramer, B.; Esselen, M. Quantitative analysis of β-asarone derivatives in Acorus calamus and herbal food products by HPLC-MS/MS. *J. Agric. Food Chem.* **2021**, *69*, 776–782. [CrossRef]
27. US Food & Drug Administration Office of Foods and Veterinary Medicine. *Guidelines for the Validation of Chemical Methods for the FDA FVM Program*, 3rd ed.; Regulatory Science Steering Committee (RSSC): Silver Spring, MD, USA, 2019. Available online: https://www.fda.gov/media/81810/download (accessed on 27 August 2021).
28. Stegmüller, S.; Schrenk, D.; Cartus, A.T. Formation and fate of DNA adducts of alpha- and beta-asarone in rat hepatocytes. *Food Chem. Toxicol.* **2018**, *116*, 138–146. [CrossRef]
29. Kim, S.G.; Liem, A.; Stewart, B.C.; Miller, J.A. New studies on trans-anethole oxide and trans-asarone oxide. *Carcinogenesis* **1999**, *20*, 1303–1307. [CrossRef]
30. Luo, G.; Guenthner, T.M. Metabolism of allylbenzene 2′,3′-oxide and estragole 2′,3′-oxide in the isolated perfused rat liver. *J. Pharm. Exp. Ther.* **1995**, *272*, 588–596.
31. Beyer, J.; Ehlers, D.; Maurer, H.H. Abuse of nutmeg (*Myristica* fragrans Houtt.): Studies on the metabolism and the toxicologic detection of its ingredients elemicin, myristicin, and safrole in rat and human urine using gas chromatography/mass spectrometry. *Ther. Drug. Monit.* **2006**, *28*, 568–575. [CrossRef] [PubMed]
32. Fischer, I.U.; von Unruh, G.E.; Dengler, H.J. The metabolism of eugenol in man. *Xenobiotica* **1990**, *20*, 209–222. [CrossRef]
33. Sangster, S.A.; Caldwell, J.; Hutt, A.J.; Anthony, A.; Smith, R.L. The metabolic disposition of methoxy-14C-labelled trans-anethole, estragole and p-propylanisole in human volunteers. *Xenobiotica* **1987**, *17*, 1223–1232. [CrossRef]

34. Punt, A.; Delatour, T.; Scholz, G.; Schilter, B.; van Bladeren, P.J.; Rietjens, I.M.C.M. Tandem mass spectrometry analysis of *N*2-(trans-Isoestragol-3′-yl)-2′-deoxyguanosine as a strategy to study species differences in sulfotransferase conversion of the proximate carcinogen 1′-hydroxyestragole. *Chem. Res. Toxicol.* **2007**, *20*, 991–998. [CrossRef]
35. Hasheminejad, G.; Caldwell, J. Genotoxicity of the alkenylbenzenes alpha- and beta-asarone, myristicin and elimicin as determined by the UDS assay in cultured rat hepatocytes. *Food Chem. Toxicol.* **1994**, *32*, 223–231. [CrossRef]
36. Zeller, A.; Horst, K.; Rychlik, M. Study of the metabolism of estragole in humans consuming fennel tea. *Chem. Res. Toxicol.* **2009**, *22*, 1929–1937. [CrossRef] [PubMed]

Review

Myristicin and Elemicin: Potentially Toxic Alkenylbenzenes in Food

Mario E. Götz, Benjamin Sachse, Bernd Schäfer and Andreas Eisenreich *

Department of Food Safety, German Federal Institute for Risk Assessment (BfR), Max-Dohrn-Str. 8-10, 10589 Berlin, Germany; mario.goetz@bfr.bund.de (M.E.G.); benjamin.sachse@bfr.bund.de (B.S.); bernd.schaefer@bfr.bund.de (B.S.)
* Correspondence: andreas.eisenreich@bfr.bund.de; Tel.: +49-30-18412-25202; Fax: +49-30-18412-625202

Abstract: Alkenylbenzenes represent a group of naturally occurring substances that are synthesized as secondary metabolites in various plants, including nutmeg and basil. Many of the alkenylbenzene-containing plants are common spice plants and preparations thereof are used for flavoring purposes. However, many alkenylbenzenes are known toxicants. For example, safrole and methyleugenol were classified as genotoxic carcinogens based on extensive toxicological evidence. In contrast, reliable toxicological data, in particular regarding genotoxicity, carcinogenicity, and reproductive toxicity is missing for several other structurally closely related alkenylbenzenes, such as myristicin and elemicin. Moreover, existing data on the occurrence of these substances in various foods suffer from several limitations. Together, the existing data gaps regarding exposure and toxicity cause difficulty in evaluating health risks for humans. This review gives an overview on available occurrence data of myristicin, elemicin, and other selected alkenylbenzenes in certain foods. Moreover, the current knowledge on the toxicity of myristicin and elemicin in comparison to their structurally related and well-characterized derivatives safrole and methyleugenol, especially with respect to their genotoxic and carcinogenic potential, is discussed. Finally, this article focuses on existing data gaps regarding exposure and toxicity currently impeding the evaluation of adverse health effects potentially caused by myristicin and elemicin.

Keywords: alkenylbenzenes; myristicin; elemicin; safrole; methyleugenol; flavoring

Citation: Götz, M.E.; Sachse, B.; Schäfer, B.; Eisenreich, A. Myristicin and Elemicin: Potentially Toxic Alkenylbenzenes in Food. *Foods* **2022**, *11*, 1988. https://doi.org/10.3390/foods11131988

Academic Editor: Isabel Sierra Alonso

Received: 31 May 2022
Accepted: 1 July 2022
Published: 5 July 2022

Publisher's Note: MDPI stays neutral with regard to jurisdictional claims in published maps and institutional affiliations.

1. Introduction

Myristicin (CAS N°: 607-91-0; IUPAC name: 4-methoxy-6-prop-2-enyl-1,3-benzodioxole), elemicin (CAS N°: 487-11-6; IUPAC name: 1,2,3-trimethoxy-5-prop-2-enylbenzene), safrole (CAS N°: 94-59-7; IUPAC name: 5-prop-2-enyl-1,3-benzodioxole), and methyleugenol (CAS N°: 93-15-2; IUPAC name: 1,2-dimethoxy-4-prop-2-enylbenzene) are secondary plant metabolites belonging to the group of alkenylbenzenes (Figure 1). Such derivatives may particularly be found in Umbelliferae (anise, star anise, fennel, sweet fennel, and parsley), Myristicaceae (nutmeg and mace), Labiatae (sweet and exotic basil), and Compositae (tarragon).

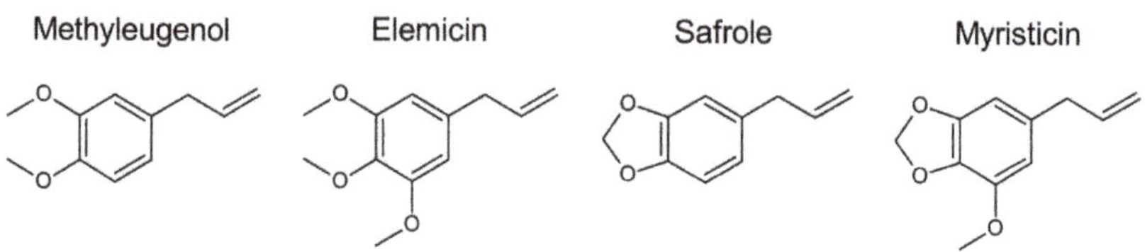

Figure 1. Structural formulas of methyleugenol, elemicin, safrole, and myristicin.

Several alkenylbenzene derivatives are known toxicants. For example, safrole and methyleugenol were classified by the International Agency for Research on Cancer (IARC)

as "possibly carcinogenic to humans" (Group 2B) [1,2]. Moreover, the Scientific Committee on Food (SCF) of the European Commission (EC) considered safrole and methyleugenol as genotoxic carcinogens and suggested restrictions for their use in foods [3,4]. In consequence, the EC prohibited the addition of safrole and methyleugenol as pure flavoring substances to food and established maximum levels for these substances–when naturally present–for certain flavored foodstuffs, such as soups and sauces or non-alcoholic beverages (Regulation (EC) No 1334/2008). In contrast to safrole and methyleugenol, the structurally closely related alkenylbenzenes myristicin and elemicin were not assessed by international expert bodies in a comparable manner until now (Figure 1). This is probably due to the lack of reliable data regarding genotoxicity, carcinogenicity, and other toxic effects.

It is assumed that the toxicity of safrole and methyleugenol is mainly caused by metabolic activation at the allylic side chain, namely 1′-hydroxylation and subsequent sulfonation to the resulting allylic sulfate esters. These intermediates are instable and may subsequently react with cellular nucleophiles including DNA and proteins [5]. The relevance of further metabolites, e.g., quinone methides or other so far unknown reactive intermediates possibly involved in hepatotoxic effects are far less clear.

Paying attention to the chemical structure, it may be noted that the less intensively studied derivatives, myristicin and elemicin, each bear just one additional methoxy substituent at the benzene core as compared to safrole and methyleugenol, respectively (Figure 1). The allylic side chain is a common feature of all the four alkenylbenzenes discussed here. Metabolic differences with respect to the allylic side chain, the suspected site of metabolic activation leading to toxicity, are expected to be marginal. However, oxidative demethylenation of myristicin and oxidative demethylation of elemicin at the benzene core are not entirely comparable to similar reactions on safrole and methyleugenol. Consequently, the resulting patterns of urinary benzyl metabolites that could be detected in rats [6,7] and in humans [8] following elemicin and nutmeg ingestion, respectively, are different from those formed from safrole and methyleugenol in rodents.

Nevertheless, for all the four compounds, the toxic principle that is assumed to lead to toxification is the 1′-hydroxylation and subsequent sulfonation to the resulting allylic sulfate esters [5].

Intake of myristicin, elemicin, safrole, and methyleugenol is presumed to occur mainly from the consumption of spices and essential oils made thereof [9,10]. High levels may also be present in plant-based food supplements (PFS) as well as in various processed flavored foods such as sauces, baked goods, and beverages, such as cola-flavored softdrinks [5,10–13].

Taken together, the toxicological relevance of alkenylbenzene occurrence, especially for less intensively investigated members such as myristicin and elemicin, is still under discussion.

In this review, we summarize and discuss the current knowledge regarding occurrence, toxicokinetics, and the toxicity of myristicin vs. safrole, as well as elemicin vs. methyleugenol. Moreover, we highlight data gaps currently impeding the assessment of the adverse health effects of these substances.

2. Occurrence of Myristicin and Elemicin

The occurrence of the alkenylbenzenes safrole and methyleugenol in plants used as foods was described in detail elsewhere [5]. Beside this, aromatic plants, as well as powders, extracts, or essential oils made thereof can also serve as sources for myristicin and elemicin, which are described here.

Of note, when interpreting occurrence data, it should always be kept in mind that the chemical constituents of culinary spices, including alkenylbenzenes, are widely dependent not only on species but also on environmental factors such as geographic location, seasonal variation, and harvest time. Furthermore, the analytical method–especially the extraction procedure–may also affect the obtained occurrence data [14–16]. In the following part, as well as in Table 1, we will present relevant examples of (culinary) plants and their essential oils containing myristicin and elemicin.

2.1. Myristicin and Elemicin in Nutmeg and Mace

The name "myristicin" originally referred to the solids that crystallize from nutmeg oil while in prolonged storage. These, however, are today known to be myristic acid [17]. Elemicin was first identified as a component of the myristicin fraction from nutmeg oil [18].

The nutmeg tree is a tropical tree indigenous to the Maluku Islands of Indonesia (Myristica fragrans Houtt., family Myristicaceae). Its seeds consist of a kernel and a covering aril surrounding the kernel. Whereas mace designates the red lacy aril, the dried kernels of the ripe seeds are named nutmeg. When grounded material or powders are hydrodistilled, about 2.4% crude oil can be obtained [19]. These oils are rich in alkenylbenzenes, such as eugenol (19.9%), methyleugenol (16.7%), methyl iso-eugenol (16.8%), myristicin (2.3%), safrole (1.6%), and elemicin (1.7%). In contrast, a slightly different composition of oils from the dried kernels of Myristica fragrans originating from Sri Lanka was reported, with almost no eugenol (0.2%), or methyleugenol (0.6%), nearly equal amounts of safrole (1.4%) and elemicin (2.1%), but higher levels of myristicin (4.9%) [20]. Numerous reports on the composition of nutmeg oils are published, reporting varying levels of alkenylbenzenes in nutmeg seeds. It is, on the one hand, the storage of ground powders [21], but on the other hand, very much the geographical origin that determines the volatile composition of nutmeg extracts as recognized by Baldry and colleagues [22]. They showed high variabilities in alkenylbenzene contents, with myristicin ranging from 0.5% to 12.4%, safrole from 0.1% to 3.2%, elemicin from 0.3% to 4.6%, methyleugenol from 0.1% to 1.2%, and eugenol from 0.1% to 0.7% in different nutmeg oils from West India and South East Asia. Mace powders and mace oils contain similar constituents as nutmeg powders and nutmeg oils. For example, in 10 powdered genuine Indonesian nutmeg seeds extracted with boiling methanol, myristicin accounted for up to 2.9% and safrole accounted for up to 0.39%. Nutmeg oil from Indonesian nutmegs contained 9.73% myristicin and 2.16% safrole [23].

Nutmeg and mace are used as domestic spices and as flavoring ingredients in many food products, such as in gelatins, puddings, sweet sauces, baked goods, meats, fish, pickles (processed vegetables), candy, ice cream, and non-alcoholic beverages [24,25]. In addition, several globally available PFS contain nutmeg seed powders or nutmeg oils to very varying extents [11].

Further relevant examples for plants and their essential oils containing myristicin and elemicin, as well as other selected alkenylbenzenes, are listed in Table 1.

Table 1. Occurrence of safrole, myristicin, methyleugenol, and elemicin found in essential oils (EO) from culinary plants.

Source	Safrole	Myristicin	Methyleugenol	Elemicin
Nutmeg	0.1–3.2% [19,20,22,23]	0.5–12.4% [19,20,22,23] (16.9 ± 0.6 mg [26])	0.1–16.7% [19,20,22]	0.3–4.6% [19,20,22]
Parsley		20.3–94.1% (seed) [27,28]; 3.1–91.9% (leaf) [27]; 6.6–30.1% (root) [27] (1435 ppm [29]; 3.6–526 ppm (leaf) [30])		
Sweet fennel		2.5–10% (root) [31,32]		
Dill		0.21% (seed) [33]; 4.38% (root) [32]		
Parsnip		18.3–66.2% (root) [34,35] (200 ppm (root) [36])		
Sweet basil		9.24–87.04% [15]; 0.03% (flower) [37,38]; 0.06% (stem) [37,38]; 0.18–76% (leaf) [37–39]		0.30% (stem) [37,38]

Table 1. *Cont.*

Source	Safrole	Myristicin	Methyleugenol	Elemicin
Carrot		34.4% (leaf) [40]; 43.9% (fruit) [40]; 0.4–29.7% (root) [41,42]; (0.5–15 ppm (root) [43])	2.51% (fruit) [44]	1.4–35.3% [45]; 32.89% (fruit) [44]
Pepper (Piper)	0.2–3.0 mg/kg (fruit) [46]; <3% (fruit) [47]; 4.81% (fruit) [48]; 24% [49]; 49% [50]; 64–98% [51]	<1% (fruit) [47]; 16.55% (fruit) [48]; 0.3–7.6% (leaf) [52]	<3% (fruit) [47]; 1.53% (fruit) [48]	<1% (fruit) [47]; 3.91% (fruit) [48]; 0.2–1.6% (leaf) [52]
Japanese star anise	6.6% [53]	3.5% [53]	9.8% [53]	
Tarragon			9.59–28.40% (seeds) [54]	21.45–38.90% (seeds) [54]
Sweet bay			3.1% (flower) [55]; 4.7% (bark) [55]; 16.0% (stem) [55]; 11.8–21.3% (leaf) [55,56]	0.8% (stem) [55]; 5% (leaf) [56]

2.2. Myristicin and Elemicin in Food Flavorings

Due to the intentional use of essential oils and the dried powder of nutmeg or mace for flavoring reasons, certain types of soft drinks, pastries, and some types of crisps contain high levels of myristicin and elemicin.

Cola-flavored soft drinks may contain nutmeg oil and/or mace oil, which consist of different major compounds, such as sabinenes and myrcene, as well as at least five different alkenylbenzenes. Myristicin, safrole, and elemicin mainly determine the flavor of these oils. Accordingly, myristicin, safrole, elemicin, methyleugenol, and eugenol were detected in cola-flavored soft drinks [57]. In 2013, Raffo et al. published quantitative data on the amounts of safrole and myristicin in the cola-flavored soft drinks of different brands following different processing procedures, including various storage conditions. Levels of safrole and myristicin varied approximately 2–3 orders of magnitude. In flavored soft drinks, average concentrations of safrole and myristicin were 23.0 and 168.3 μg/L, with minimum contents of 0.6 and 0.4 μg/L and maximum levels of 43.9 and 325.6 μg/L, respectively [12]. These variations might be due to variable levels of alkenylbenzenes in the added essential oils. For example, measurements of alkenylbenzene concentrations in different nutmeg oils of specific geographical origins revealed an at least 30-fold variation, e.g., in the levels of safrole (ranging from 0.1 to 3.2%) and myristicin (0.5 to 13.5%), respectively [57]. In the study of Raffo et al., only the levels of myristicin and safrole were measured in cola-flavored soft drinks, but not those of other alkenylbenzenes. Therefore, the total amount of alkenylbenzenes in cola-flavored soft drinks remains unknown so far.

Another important example for processed foods containing alkenylbenzenes is *"Pesto"*. This traditional dish from Genova, Italy, mainly consists of olive oil, hard cheese, pine nuts, garlic, salt, and basil leaves. Different alkenylbenzenes were detected at varying levels in basil-containing *"Pesto"*, including methyleugenol (22.9–56.4 mg/kg), myristicin (13.2–15.8 mg/kg), estragole (3.2–34.1 mg/kg), and apiol (3.4 mg/kg) [58].

Parsley and dill teas can be purchased without restriction. Recently levels of alkenylbenzenes in such teas were investigated. Myristicin, methyleugenol, apiol, and estragole are detected to varying extents in dry tea samples or in hot water herbal extracts containing parsley, dill leaves, or seeds, or being in a mixture with other herbs. The total amount of alkenylbenzenes in the dry tea samples ranged from 18 to 1269 μg/g dry preparation [59]. In 2017, Alajlouni and colleagues also found relevant levels of the alkenylbenzenes myristicin, apiol, and estragole (17–6487 μg/g) in parsley and dill-based PFS [60].

Beside this, baked goods, meat products, condiments, relishes, soft candy, gelatin, pudding, soups, alcoholic beverages, and gravies may also contain myristicin and elemicin to various and often unknown amounts if refined with oils from parsley, nutmeg or mace [24,61]. This also applies to other alkenylbenzenes [62,63]. Therefore, monitoring of myristicin, elemicin, and other alkenylbenzenes in many food commodities appears justified in order to gain a reliable database for future exposure assessments [25].

2.3. Myristicin and Elemicin in Foods

Analytical methods are already in place to monitor myristicin and elemicin in complex food matrices [25]. An early study reported 16.9 mg myristicin per gram dried nutmeg powder following 12 h methanol extraction at 50 °C [26]. Other methods for analyzing ground nutmeg, wine and beer spices, and many food commodities utilize ultrasonic assisted extractions, followed by solid phase extraction and gas chromatography (GC)–mass spectrometry (MS) [64,65]. Other methods for analyzing myristicin from ground nutmeg (502 µg/g), from wine and beer spices (11.87 µg/g), from some food commodities (2.46–15.22 µg/g), and even from human serum (17.60–33.25 µg/g from human volunteers who incorporated 100 mg myristicin 1 h before blood sampling) utilize ultrasonic assisted extractions, followed by solid phase extraction and gas chromatography (GC)–mass spectrometry (MS) [64,65]. Other methods use functionalized magnetic microspheres for isolation of allyl-benzodioxoles, followed by gas chromatography–mass spectrometry, such as myristicin (264.2–599.6 µg/L) and safrole (14.0–40.35 µg/L) from cola drinks [66]. However, in these methods, varying and often not fully validated analytical procedures were used, hampering the comparability of the analytical results. In addition, for many food categories no data are available at all. Since there is no legal mandate for monitoring all potentially toxic alkenylbenzenes in all relevant food categories, the availability of comprehensive and reliable occurrence data is currently rather limited. Taken together, the actual occurrence levels of myristicin and elemicin, as well as of many other alkenylbenzenes, are still widely unknown for many foods.

3. Toxicity of Myristicin and Elemicin: Lessons Learned from Safrole and Methyleugenol

In the following, the current knowledge on the toxicity of myristicin and elemicin in comparison to the structurally related and well-investigated alkenylbenzenes, safrole and methyleugenol, is summarized.

3.1. Metabolism of Myristicin and Elemicin vs. Safrole and Methyleugenol
3.1.1. Common Structural Features

Initial steps of the hepatic activation of methylenedioxy- and methoxy-substituted allylic alkenylbenzenes include epoxidation of the exocyclic double bond followed by its cleavage by microsomal or cytosolic epoxide hydrolases or spontaneous hydration to generate $2',3'$-dihydrodiols [67]. Such metabolites are detected in the urine of animals treated with allylbenzenes [6,68–70]. Another pathway may be the hydroxylation of the $1'$-carbon atom adjacent to this $2',3'$-double bond [71]. Side chain reactions of alkenylbenzenes are catalyzed by various cytochrome P450 monooxygenases (CYPs). Epoxides and dihydrodiols may be derived not only from the allylbenzene compounds but also from some of their metabolites, which still possess an intact allyl group, such as the allylcatechols [72]. However, phenolic and catecholic compounds typically undergo rapid phase II conjugation, which might be a predominant pathway for such metabolites as also shown for the alkenylbenzene eugenol containing a free phenolic group [73]. Thus, in contrast to alkenylbenzenes that bear only methoxy or methylenedioxy substituents, the high first-pass conjugation and rapid elimination may explain why eugenol is deemed to be less toxic as compared to the well-known hepatocarcinogens methyleugenol and safrole.

Following hydroxylation at the $1'$-position (Figure 2), the alcoholic metabolite can be sulfonated. Subsequent heterolytic cleavage of the formed sulfate moiety would generate an electrophilic carbenium ion intermediate, which is highly reactive towards nucleophilic sites [74,75], and that may, for example, generate glutathione (GSH) conjugates, as well as adducts with proteins, RNA, or DNA [76]. Since the carbenium ionic charge is delocalized, adducts can be formed at the 1'- or 3'-position, with the 3'-position being the preferred site [77,78].

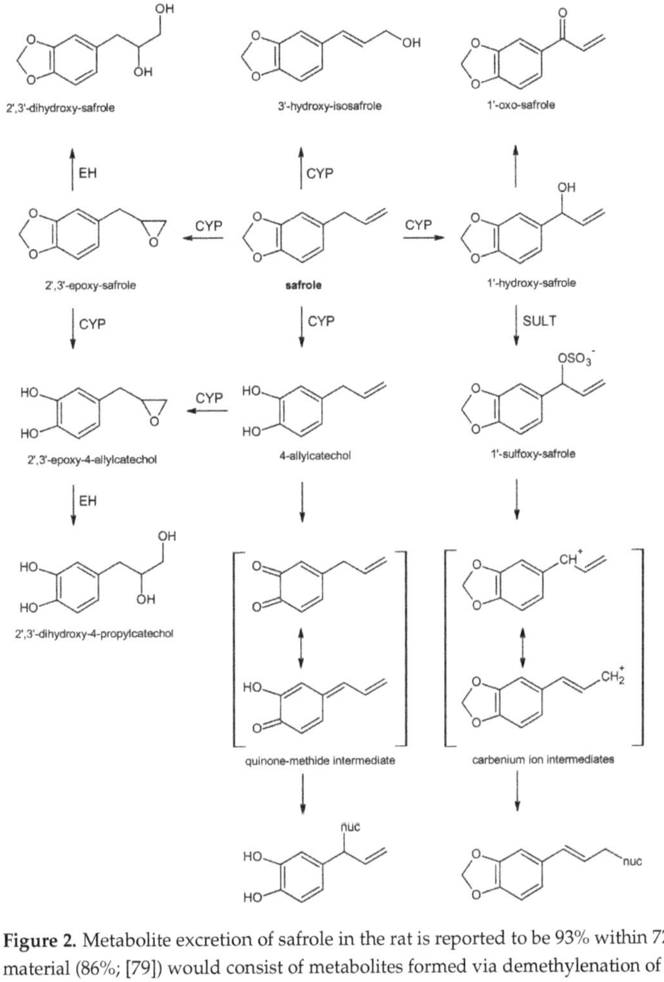

Figure 2. Metabolite excretion of safrole in the rat is reported to be 93% within 72 h, and most of this material (86%; [79]) would consist of metabolites formed via demethylenation of the methylenedioxy moiety to yield carbon monoxide or formate and the dihydroxy-benzene moiety [80]. The other metabolic routes observed were allylic hydroxylation and the epoxide-diol pathway [70,79]. Oxidations of the allylic side chain of safrole may proceed (i) via an epoxide resulting in side chain propane diols during different stages of the metabolic steps [72], or (ii) via 1'-hydroxylation followed by sulfonation that might lead to a reactive carbocation intermediate [5]. Other possible steps of metabolic ways of safrole are (iii) the subsequent oxidation of the 1'-hydroxysafrole to the 1'-oxo-safrole [81], (iv) oxidation at the 3'-position to yield 3'-hydroxy-isosafrole, and (v) the demethylenation of safrole to 4-allylcatechol that may isomerize to its quinone-methide [82–84]. The occurrence of glutathione conjugates at the 1'-position may be indicative of the intermediate formation of *para*-quinone methide tautomers [82], whereas glutathione conjugates at the benzene ring point to reactions with *ortho*-quinone intermediates [82]. CYP: cytochrome P450 monooxygenases; SULT: sulfotransferases; EH: epoxide hydrolases; nuc: nucleophilic structures such as DNA or proteins.

However, the metabolic pathway to the carbenium ions is only one selected pathway, already often discussed with respect to the cyto- and genotoxic activity of alkenylbenzenes. This metabolic pathway presumes the presence of sulfotransferases (SULT) and cofactors such as 3′-phosphoadenosine-5′-phosphosulfate (PAPS) [5].

On the other hand, alkenylbenzenes and their metabolites that bear ortho- and/or para-phenolic groups may form quinone methide intermediates (Figure 2) [82,85] that are also prone to be conjugated by GSH or react directly with other nucleophiles in the cell. The transient formation of a quinone methide of eugenol appears plausible [85] since an eugenol GSH conjugate was detected utilizing rat liver or rat lung microsomes [86]. The cytotoxic effects of eugenol recognized in rat hepatocytes are reasoned to be due to the formation of a reactive quinone methide intermediate [87]. In 1990, Fischer et al. tentatively identified metabolites including thiophenol metabolites (11%) following eugenol ingestion in the urine of humans, presumably formed by GSH conjugation at an aromatic ring position [73]. Thus, methoxylated non-phenolic substances (e.g., methyleugenol and elemicin) may as well undergo CYP enzyme-mediated O-demethylation and subsequent quinone methide formation followed by GSH conjugation. Similarly, there can be oxidative demethylation of methoxy groups in elemicin by CYP1B1 [88], creating the possibility to yield also catechols or other phenols and conjugates, as was shown from benzodioxole-substituted alkenylbenzenes myristicin and safrole in rat and human urine using GC-MS [8].

A number of CYP isoenzymes are capable of catalyzing the 1′-hydroxylation of akenylbenzenes [88–93]. An overview of CYPs that are demonstrated to be involved in the oxidation of methyleugenol, elemicin, safrole, and myristicin is shown in Table 2.

Table 2. Human cytochrome P450 isoenzymes mediating the 1′-hydroxylation of alkenylbenzenes.

Substance	Cytochrome P450 Subtype	Reference
Methyleugenol	CYP1A2 (CYP2C9, 2C19)	[90,92]
Elemicin	CYP1A1, CYP1A2, CYP3A4	[88]
Safrole	CYP2A6 (CYP1A2, CYP2C19, CYP2E1)	[89,91]
Myristicin	CYP3A4 (CYP1A1)	[93]

Main human CYPs and, in brackets, contributing human CYPs involved in the metabolism of methyleugenol, elemicin, safrole, and myristicin.

Apart from the epoxide, the carbenium ion, and the quinone methide metabolic pathways of the alkenylbenzenes already discussed, another metabolic pathway that may occur after rearrangement of the double bond from 2′,3′-position to 1′,2′-position is the oxidation of 3′-hydroxy metabolites of alkenylbenzenes leading to cinnamic acids and propionic acids [6,68,69]. In principle, 3′-hydroxy-1′,2′-propenylbenzenes may be equivalent to 1′-hydroxy-allylbenzenes as substrates for hepatic SULTs. On the other hand, due to steric reasons further side chain oxidation of the 3′-hydroxy-propenylbenzenes yielding cinnamaldehydes and cinnamic acids, which can be conjugated with GSH or glycine, appear to dominate. Further oxidation, probably via the fatty acid β-oxidation cycle, would lead to side chain cleavage and the formation of benzoic acids and its glycine conjugates [5].

Seemingly small, but relevant structural molecular differences in benzene ring substituents of the parent alkenylbenzenes call for a closer look at the potential metabolic pathways of elemicin, myristicin, methyleugenol, and safrole. In an attempt to identify similarities and possible differences of elemicin and myristicin, we compared their metabolic features to the closely related derivatives methyleugenol and safrole. Those two compounds bearing only methoxy groups at the benzene ring without methylene bridge are methyleugenol and elemicin. The two compounds with a methylenedioxy moiety are safrole and myristicin, which are categorized as benzodioxoles.

3.1.2. Metabolism of Methyleugenol

Results of ADME experiments performed in 2000 within a study of the National Toxicology Program (NTP) led to the conclusion that absorption of orally ingested methyleugenol in rats and mice is rapid and complete, and that the distribution of methyleugenol to tissues is fast. In rodents, methyleugenol is extensively metabolized in the liver and more than 70% of the dose administered is found in the urine of rats and mice as hydroxylated, sulfated, or glucuronidated metabolites [92].

With view on the toxicity of methyleugenol, it is generally assumed that bioactivation is mainly mediated via 1'-hydroxylation at the allylic side chain followed by sulfo conjugation, yielding a highly reactive sulfate ester [10,94].

In the NTP study, it was shown that repeated ingestion of methyleugenol may saturate metabolic enzymes [92], leading to greater tissue accumulation and thus higher probability for genotoxicity, mutations, and malignant cell transformations. Saturability of metabolism is of special concern in cases when 1'-hydoxylation of the allylic side chain becomes more prominent over other pathways. This may enhance hepatocarcinogenesis in rodents at higher dose levels [95].

In rat bile, methyleugenol could be found in the form of GSH conjugates. These conjugates detected by Yao and colleagues potentially resulted from reactions with methyleugenol-derived epoxide metabolites, α,β-unsaturated aldehydes, carbenium ions, and quinone methides [96]. These conjugates were further metabolized, yielding the cysteine conjugates found in rat urine. In GSH-fortified microsomal preparations that lack SULT and PAPS, it was generally not expected that carbenium intermediates would be formed. However, Yao et al. found 1'-bound GSH and related cysteine conjugates in such incubations [96]. Thus, it is hypothesized that 1'-hydroxy metabolites or other metabolites than the sulfate esters may directly react with GSH under certain conditions.

Beside 1'-hydroxylation, the metabolites observed in rats and mice suggest that methyleugenol can also undergo demethylation, ring, and/or further side chain oxidations [92].

The NTP authors further concluded that the risk to humans ingesting methyleugenol is expected to be subject to marked inter-individual metabolic variability. Indeed, hydroxylation of methyleugenol investigated in human liver microsomes varied considerably (37-fold), with the highest hydroxylation rate being similar to that observed with liver microsomes from rats [97]. Moreover, one study by Tremmel et al. demonstrated that methyleugenol-induced DNA adduct levels in human liver samples were dependent on the SULT1A1 copy number [94].

3.1.3. Metabolism of Elemicin

Elemicin is the natural continuation of methyleugenol, bearing two *meta-* and one *para-*methoxy group relative to the allyl side chain. For this compound, the *O*-demethylation pathway becomes more prominent, which leads to some divergent metabolites, compared to methyleugenol. In 1980, Solheim and Scheline revealed that the two major metabolic pathways of elemicin in rats follow the cinnamoyl pathway and the epoxide-diol pathway [6]. The former route gives 3,4,5-trimethoxyphenyl-propionic acid and its glycine conjugate as major urinary metabolites, whereas 2',3'-dihydroxy-elemicin is the most prominent metabolite of the latter route. In addition, elemicin can also be 1'-hydroxylated at the allylic side chain. When comparing the kinetic constants for conversion of elemicin and 1'-hydroxy-elemicin by male rat liver and mixed gender pooled human liver fractions, van den Berg et al. concluded that glucuronidation of 1'-hydroxy-elemicin, representing a detoxification pathway, is the most important pathway in rats and in humans. In contrast, bioactivation of 1'-hydroxy-elemicin by sulfonation was suggested to be only a minor pathway in both rat and human liver [76].

In 2019, Wang et al. confirmed and extended these studies. They found a total of 22 metabolites for elemicin in mice, e.g., in urine, feces, and plasma [88]. In vivo, elemicin and most of its metabolites were mainly excreted in urine collected from 0 to 24 h post-procedure in metabolic cages of male C57BL/6 mice that were orally administered

100 mg/kg elemicin. The obtained results indicate that phase I metabolic reactions of elemicin included demethylation, hydroxylation, hydration, allyl rearrangement, reduction, hydroformylation, and carboxylation. Phase II metabolism of elemicin yielded several conjugates, e.g., with cysteine, *N*-acetyl cysteine, glucuronic acid, glycine, or taurine [88]. In addition, the 4-demethoxylated forms of elemicin and of $2',3'$-dihydroxy-elemicin could be detected in human urine after nutmeg abuse [8].

3.1.4. Metabolism of Safrole

From a toxicological point of view, safrole bioactivation by sequential $1'$-hydroxylation and sulfonation, resulting in reactive sulfate esters capable of forming adducts with cellular nucleophiles such as DNA, is of high relevance [71,89]. In 1983, Boberg et al. identified $1'$-sulfoxy-safrole as an ultimate electrophilic metabolite of safrole and as an initiator of hepatic carcinogenicity in vivo. The toxicological relevance of this pathway was demonstrated in mice co-treated with the hepatic SULT inhibitor pentachlorophenol (0.05% added to the diet of mice) in vivo and in mice being genetically defective with respect to the hepatic synthesis of PAPS [75].

However, work in rats, mice, and guinea pigs elucidated multiple metabolic pathways of safrole far beyond $1'$-hydroxylation and sulfo conjugation. Upon intraperitoneal (i.p.) injection safrole is metabolized in rat and guinea pig by the epoxide-diol pathway and by cleavage of the methylenedioxy ring to form a catechol [98,99]. Since an allylic double bond is still present in the catechol and $1'$-hydroxy-safrole, both metabolites can be further metabolized via epoxides to the corresponding dihydrodiols. A small amount of a triol $1',2',3'$-trihydroxy-safrole was found in rat urine by Stillwell and colleagues [70]. Interestingly, $2',3'$-epoxy-safrole apparently has sufficient stability in vivo to be absorbed from the peritoneal cavity to the circulatory system, and to persist even in urine. The major urinary metabolites identified by GC-MS were 4-allylcatechol, $1'$-hydroxy-safrole, $2',3'$-dihydroxy-safrole, $2',3'$-dihydroxy-4-propyl-catechol, $2'$-hydroxy-$1'$- (3,4-methylenedioxy-phenyl)-propanoic acid, and 3,4-methylenedioxy-benzoyl glycine [70] (Figure 2).

Urinary metabolites of safrole in the rat were also identified via GC-MS in a further study performed in 1982. Metabolite excretion was 93% within 72 h, and most of this material (86%) consisted of metabolites formed via demethylenation of the methylenedioxy moiety. The other metabolic routes observed were allylic hydroxylation and the epoxide-diol pathway [79].

3.1.5. Metabolism of Myristicin

Myristicin is well absorbed following oral exposure and is metabolized extensively.

Metabolism of the volatile alkenylbenzene myristicin results in the formation of less volatile metabolites, predominantly remaining in the aqueous phase on extraction with ether [99].

Early experiments highlighted the cleavage of the methylenedioxyphenyl moiety concomitant with CO_2 release from myristicin as an important metabolic pathway. Within 48 h after oral administration of radiolabeled myristicin to male albino mice, 73% of the radiocarbon was set free as $^{14}CO_2$ [98], which was potentially formed from the hydroxylation of the methylene group of myristicin and subsequent release and degradation of formate-^{14}C. This demethylenation reaction was found to be catalyzed by microsomal CYPs and would yield the corresponding catechol derivative.

Later analytical studies in rat and human urine indeed revealed further water-soluble metabolites of myristicin, including the catecholic derivatives. In male Wistar rats that were administered myristicin once by oral gavage (100 mg/kg), different metabolites were identified in urine, including $1'$-hydroxy-myristicin, 5-allyl-2,3-dihydroxy-1-methoxy-benzene, 5-allyl-2-hydroxy-1,3-methoxy-benzene, 5-allyl-1-hydroxy-2,3-methylendioxy-benzene, 5-($2',3'$-dihydroxypropyl)-1-hydroxy-2,3-methylendioxy-benzene [8].

Incubation of myristicin in rat liver microsomes formed two major metabolites, 1′-hydroxy-myristicin and 5-allyl-1-methoxy-2,3-dihydroxy-benzene, bearing a catechol moiety [7]. Those metabolites were also identified in the above mentioned study by Beyer et al. in 2006 [8].

Isolation of metabolites from male Sprague–Dawley rat urine after a single oral administration of 100 mg/kg myristicin, and comparison before and after glucuronidase treatment, suggests that the catecholic metabolites 5-allyl-1-methoxy-2,3-dihydroxy-benzene and 1′-hydroxy-myristicin are also excreted in their respective conjugated forms [7].

Currently, no comprehensive studies with respect to quantitative metabolism and excretion of myristicin in humans are available. However, one study examined metabolites present in the urine of a patient who ingested five nutmeg seeds, resulting in an intoxication [8].

3.2. Genotoxicity

As described above, different metabolic pathways may lead to the formation of reactive intermediates capable of binding DNA, thereby causing genotoxicity. For many alkenylbenzenes, it is widely accepted that the 1′-hydroxylation at the allylic side chain, followed by SULT-mediated sulfo conjugation yielding a highly electrophilic sulfate ester, might be the most relevant pathway leading to toxicity [10]. The sulfate ester may form inter alia DNA adducts as demonstrated by ^{32}P-postlabeling techniques and mass spectrometry [78,100–105]. Structures of four DNA adducts formed in mouse liver after administration of the proximate hepatocarcinogen 1′-hydroxy-estragole were initially described by Phillips et al. in 1981 [106,107]. Similar kinds of studies, as well as studies on other genotoxicity endpoints and mutagenicity, were performed for many alkenylbenzenes, as systematically reviewed in detail elsewhere [5,10,108]. In the following part, the most relevant studies on genotoxicity of methyleugenol, elemicin, safrole, and myristicin are exemplarily described in brief.

3.2.1. Genotoxicity of Methyleugenol vs. Elemicin

Methyleugenol was found to induce sister chromatid exchange (SCE) in Chinese hamster ovary (CHO) cells after metabolic activation, as well as intrachromosomal recombination in yeast with and without metabolic activation [92]. Some years later, Groh and colleagues further characterized the impact of methyleugenol and its metabolites on DNA damage induction in vitro. It was observed that 1′-hydroxy-methyleugenol and 2′,3′-epoxy-methyleugenol had a higher DNA strand breaking activity than the parent compound methyleugenol in Chinese hamster lung fibroblast (V79) cells, demonstrating the marked relevance of these metabolites. However, in the same study, only 3′-oxomethylisoeugenol and 2′,3′-epoxy-methyleugenol induced the formation of micronucleated V79 cells [109]. Furthermore, methyleugenol and the oxidative metabolites concentration dependently increased the amount of DNA strand breaks, as measured using the in vitro alkaline comet assay in human colon carcinoma HT29 cells [110,111].

In 1992, Chan and Caldwell found that methyleugenol, 1′-hydroxy-methyleugenol and 2′,3′-epoxy-methyleugenol caused unscheduled DNA synthesis (UDS) in rat hepatocytes, and that the inducing potency of the 1′-hydroxy metabolite was higher than that of the parent substance in vitro [112]. In 2006, methyleugenol was also shown to form DNA adducts after hydroxylation and sulfonation. DNA adducts of methyleugenol were detected using ^{32}P-postlabeling techniques in the livers of F344 rats (*n* = 4 out of 8) exposed orally to 5 mg/kg/day for 28 days. No adducts were found after exposure to 1 mg/kg/day [113].

In 2013, Herrmann et al. detected methyleugenol-induced DNA adducts also in human liver samples [114]. Twenty-nine human liver samples unambiguously contained the N^2- (*trans*-methylisoeugenol-3′-yl)-2′-deoxyguanosine adduct (N^2-MIE-dG). A second adduct, N^6- (*trans*-methylisoeugenol-3′-yl)-2′-deoxyadenosine (N^6-MIE-dA), was also found in most samples, but at much lower levels. The median methyleugenol DNA adduct level detected in human non-tumorous liver samples was $13/10^8$ nucleotides for

N^2-MIE-dG and N^6-MIE-dA combined, corresponding to 1700 adducts per diploid genome (6.6×10^9 base pairs). As further elegantly reported, hepatic DNA adduct formation by methyleugenol in mice is strongly affected by their SULT1A content [115,116], proving the toxicological relevance of this metabolic pathway. Indeed, also in human liver samples, an association between the SULT1A1 copy number and the adduct level was demonstrated [94]. Moreover, it is shown in vitro for the structural derivative estragole that the resulting DNA adducts are inefficiently repaired [117], which might contribute to the accumulation of substantial levels of DNA adducts upon prolonged dietary exposure.

Beside this, Yang et al. recently showed that reactive metabolites of methyleugenol were also able to form RNA adducts [118]. However, the biological consequences of these RNA adductions are so far unclear, as also mentioned by the authors.

As shown for methyleugenol [112], also elemicin was found positive in a DNA binding assay and in UDS assays [24,76,119,120].

Despite the well-recognized DNA damages, methyleugenol is reported to be only weakly or non-mutagenic in different bacterial test systems with or without metabolic activation [3,92,121,122]. In another study done by Groh et al. in 2012, it was shown that methyleugenol did not cause mutations at the *hprt* locus in cultured V79 cells after 1 h of incubation. After extended treatment (24 h) only 2′,3′-epoxy-methyleugenol exhibited slight mutagenic activity with a mutation frequency being 4–5 times higher than the spontaneous mutation frequency of the solvent control [109]. A possible explanation for the lack of mutagenicity, especially in bacterial systems, might be the lack of metabolic competence, especially in view on SULTs or the cofactor PAPS [5,122].

The mutagenic potential of methyleugenol was also studied in vivo [3]. Data published by NTP indicates that oral administration of methyleugenol via gavage (10–1000 mg/kg bw; 5 days/week) to B6C3F1 mice does not cause micronucleus formation in peripheral blood erythrocytes [92]. Likewise, it was unable to induce chromosomal aberrations in CHO cells or micronucleus formation in peripheral blood erythrocytes of mice in other studies [92,122]. In contrast, Devereux and colleagues observed a higher frequency of β-catenin gene mutations (20/29; 69%) in hepatocellular carcinomas of mice exposed to methyleugenol (37–150 mg/kg) than in spontaneous liver tumors (2/22; 9%) from unexposed mice [123]. Since deregulation of Wnt/β-catenin signaling is considered an early event in chemically induced hepatocarcinogenesis, this observation represents an indication of the genotoxic potential of methyleugenol [3,122].

Beside this, mutagenicity of methyleugenol was recently verified in vivo, utilizing a xanthine-guanine phosphoribosyltransferase (*gpt*) delta rodent gene mutation assay [124]. For this in vivo mutation assay, transgenic *gpt* delta rats ($n = 10$/group, both sexes) were treated for 13 weeks with different doses of methyleugenol via gavage (0, 10, 30, and 100 mg/kg). A significant increase in mutagenicity assessed, via *gpt* and *Spi⁻* mutant frequencies, was observed in rat hepatocytes of the highest dose group. Mutant frequencies were further associated with pro-carcinogenic processes. From these data, the authors concluded that genotoxic mechanisms might be involved in methyleugenol-induced hepatocarcinogenesis [124].

In contrast to methyleugenol, there is currently no literature available regarding the mutagenic potential of elemicin. However, the structural features and the few data on genotoxicity suggest such an activity also for elemicin.

3.2.2. Genotoxicity of Safrole vs. Myristicin

The genotoxic activity of safrole is known. For example, it was demonstrated that safrole is capable of inducing sister chromatid exchanges, chromosomal aberrations, replicative DNA synthesis, and DNA adducts in rat liver in vivo [125]. It appears that these effects result from the 1′-hydroxylation followed by sulfo conjugation yielding reactive sulfate esters. This is because the concomitant application of the SULT-inhibitor PCP or the use of brachymorphic mice, being deficient in the SULT cofactor PAPS, strongly reduced the genotoxic effects [126].

Already in 1986, Reddy and Randerath reported that two DNA adducts were detected by ^{32}P-postlabeling techniques in the liver of adult female CD1 mice treated with safrole [104]. These DNA adducts were identified as N^2- (*trans*-isosafrol–3′-yl)-2′-deoxyguanosine and N^2- (safrol-1′-yl)-2′-deoxyguanosine. In 1998, using the same ^{32}P-postlabeling assay, Daimon et al. studied DNA adduct formation in hepatocytes isolated from male F344 rats exposed to safrole [127]. The sum of the two above mentioned major DNA adducts was 898 DNA adducts/10^8 nucleotides. In this study, hepatocytes were isolated 24 h after a single dose of safrole or five repeated doses (once a day) by gavage and allowed to proliferate in Williams' medium E supplemented with an epidermal growth factor. This enabled a certain percentage of DNA repair in situ. Beside this, safrole was shown to cause UDS in cultured rat hepatocytes, but not in HeLa cells [128,129].

Randerath et al. investigated the DNA adduct formation of a series of alkenylbenzenes in the liver of adult female CD-1 mice by ^{32}P-postlabeling 24 h after i.p. administration of non-radioactive test compounds (2 or 10 mg/mouse). The known hepatocarcinogens, safrole, estragole, and methyleugenol, exhibited the strongest binding to mouse liver DNA. However, the formation of DNA adducts in the liver were demonstrated also for myristicin in male B6C3F1 mice and female CD-1 mice. In comparison to safrole, estragole, and methyleugenol, substitution at the 3-, 4-, and 5-positions of the benzene ring of allylbenzenes (elemicin, myristicin) results in compounds with intermediate DNA binding capability [100]. In 2007, Zhou et al. further proved that myristicin forms DNA adducts comparable to those of safrole and methyleugenol in cultured human hepatocytes as well as in adult mouse liver, as analyzed via ^{32}P-postlabeling [130]. With the exception of methyleugenol, DNA adduction was dose-dependent in these experiments, decreasing in the order, methyleugenol > safrole ~ myristicin.

In other experiments, female mice were exposed to soft drinks. Covalent liver DNA adducts detected by ^{32}P-postlabeling were identical to those detectable with the single compounds myristicin and safrole. Liver adduct levels increased with exposure duration [101].

DNA adduct formation and DNA damage by myristicin were also assessed using an avian egg model [131–133]. Medium white turkey eggs with 22- to 24-day-old fetuses received three injections of nine alkenylbenzenes: safrole (1, 2 mg/egg), methyleugenol (2, 4 mg/egg), estragole (20, 40 mg/egg), myristicin (25, 50 mg/egg), elemicin (20, 50 mg/egg), anethole (5, 10 mg/egg), methyl isoeugenol (40, 80 mg/egg), eugenol (1, 2.5 mg/egg), and isoeugenol (1, 4 mg/egg). Fetal livers were harvested 3 h after the last injection. Measurements of DNA strand breaks were executed using the comet assay and DNA adduct formation, and were analyzed via ^{32}P-postlabeling. At the highest doses tested, estragole, myristicin, elemicin, safrole, methyleugenol, and anethole induced DNA adduct formation. Estragole, myristicin, and elemicin also induced DNA strand breaks as measured with the comet assay.

In freshly isolated hepatocytes from male F344 rats, myristicin induced a dose-dependent but slight increase in UDS, an indicator of DNA excision repair activity [120]. However, the authors concluded from the obtained data that myristicin was negative in that assay [120]. Decreased DNA damage repair might be an important indirect genotoxic mode of action, as highlighted by Martins et al. in 2014 and 2018 [10,134]. They showed in vitro that exposure of human leukemia cells (K562) for 6 h with 100 µM myristicin led to reduced expression of various DNA damage response genes including *OGG1* (base excision repair), *ERCC1* (nucleotide excision repair), *RAD50* (double strand break repair), *ATM* (DNA damage signaling), and *GADD45G* (stress response). As summarized by Célia Maria da Silva Martins in 2016 in her dissertation, myristicin appears to activate apoptotic mechanisms and downregulate DNA damage response genes involved in nucleotide excision repair, double strand break repair, DNA damage signaling, and stress response [135]. In 2011, Martins et al. studied the mutagenic potential of myristicin in vitro in mammalian cells [136]. In this experimental setting, myristicin tested without metabolic activation was negative in a comet assay used to evaluate DNA breaks, as well as in a γH2AX assay (sometimes recognized as an indicator for DNA double strand breaks) performed in CHO cells.

The DNA damaging activity may lead to the manifestation of heritable mutations. The mutagenic potential of safrole and its metabolites was studied in different experimental settings [4]. In the bacterial reverse mutation assay (Ames test), safrole was generally negative, or at most, weakly positive [137–139]. In contrast to the parent compound safrole, 1′-hydroxy-safrole, as well as other metabolites (2′,3′-epoxy-safrole, 1′-acetoxysafrole and 1′-oxo-safrole), were demonstrated to be directly mutagenic in the Ames test [139,140]. In addition, safrole was shown to be mutagenic in other experimental settings (bacteria and yeast) and to induce cell transformation in vitro [141,142]. The mutagenic potential of safrole, including the induction of gene mutations, chromosomal aberrations, DNA single-strand breaks, and SCEs was also demonstrated in mammalian cells [143–145].

Safrole's mutagenic potential was also studied in vivo [4]. In 1972, Epstein and colleagues obtained negative results for safrole in a mouse dominant lethal assay [146]. In line with this, testing of safrole in a bone marrow micronucleus assay and in a rat liver UDS assay also led to negative outcomes [147,148].

However, other studies clearly indicated the mutagenic potential of safrole in vivo. The first studies performed by Green and Savage in 1978 showed that safrole was positive in an in vivo i.p. host-mediated assay with *Salmonella typhimurium* [138]. Similar findings were published by Poirier and de Serres in 1979, utilizing the same assay with *S. typhimurium* or *Saccharomyces cerevisiae* [141]. Some years later, Daimon and colleagues showed that repeated-dose treatment of F344 rats with 125 or 250 mg safrole/kg bw dose-dependently induced chromosome aberrations in rat liver cells [127]. Moreover, singe-dose treatment of rats with 10–500 mg safrole/kg bw caused SCEs in rat livers in a dose-dependent manner, too. These effects were associated with the generation of DNA adducts in the hepatocytes of these rats [127].

The aforementioned indication of safrole's mutagenic activity was substantiated by the findings of Jin et al., who observed an increased *gpt* mutant frequency in transgenic *gpt* delta rats after a 13-week exposure to safrole via diet at the highest dose group tested (0, 0.1%, 0.5%; $n = 10$/group, both sexes). The authors concluded that these data clearly demonstrated the mutagenicity of safrole in vivo [149]. These findings were confirmed by results of another study performed in 2013, utilizing a similar in vivo transgenic rodent model [150]. In this study, male F344/NSlc-Tg (*gpt* delta) rats ($n = 15$ per dose) were fed with 0.5% safrole via diet for 4 weeks. This dose was identified as a carcinogenic dose from an earlier study [140]. In this experimental setting, safrole caused a significantly increased *gpt* mutant frequency, which was associated with a tumor-promoting activity, as suggested by an elevated number and area of GST-P-positive foci in rat livers, compared to controls. The authors stated that these data confirm that safrole is a genotoxic carcinogen [150].

In contrast to safrole, data on myristicin's mutagenic potential is sparse. A study published by Damhoeri et al. in 1985, studied the mutagenic activity of oleoresins prepared from myristicin-containing nutmeg fruits without metabolic activation in an in vitro mutagenicity assay in *S. typhimurium*. The authors reported that the tested oleoresins were mutagenic. Moreover, pure myristicin was also positive in the mutagenicity test [151]. Based on these data, it was suggested by Hallstrom and Thuvander in 1997 that both nutmeg and myristicin may be weakly mutagenic, but additional studies were required to finally conclude on the mutagenic potential [24].

Just recently in 2019, NTP characterized the mutagenic potential of myristicin. Myristicin was not mutagenic in *S. typhimurium* with or without metabolic activation. In addition, a micronucleus test was integrated in the subchronic toxicity study in which myristicin (0, 10, 30, 100, 300, or 600 mg/kg bw; 5 days/week) was administered via gavage to F344/NTac rats and B6C3F1/N mice (10 male and 10 female/group) for 13 weeks. There was a significant dose-dependent decrease in the percentage of polychromatic erythrocytes (PCEs) in the peripheral blood of male and female mice, illustrating toxicity to the bone marrow in mice and suggesting that the test compound reached the target tissue. In mice, however, no significant effect of myristicin on micronucleated red blood cells was observed. A significant increase in micronucleated immature erythrocytes in the peripheral blood was observed in male and female rats of the highest dose group (600 mg/kg bw). This was accompanied by significantly elevated amounts of circulating PCEs. Therefore, the authors suggested that myristicin might have stimulated erythropoiesis in rats. It was concluded that studies performed by the NTP provide limited evidence for the genotoxicity of myristicin [152]. However, findings from others indicated that myristicin, similar to other genotoxic alkenylbenzenes, e.g., safrole and methyleugenol, forms DNA adducts in vivo [100,130,132]. However, NTP authors stated that the consequence of these adducts is unknown, as myristicin was not tested for mutation induction in vivo [152]. Therefore, further and more adequate studies are needed to allow for a conclusive evaluation of the mutagenic potential of myristicin. Ideally, those studies should be designed as comparative studies (e.g., testing of myristicin vs. other alkenylbenzenes, such as safrole, in a similar experimental setting, e.g., as proposed by Nohmi and colleagues [150,153]), to allow a ranking regarding the genotoxic potential of these substances. As demonstrated with other alkenylbenzenes, in vivo assays capable of detecting gene mutations, i.e., transgenic rodent assays like the *gpt* delta assay, might be appropriate test systems for detecting potential alkenylbenzene-induced mutations.

3.2.3. Genotoxic Effects in Pregnant Mice and in Offsprings

Since the altered hormone constitution in pregnancy may profoundly affect the activity of maternal xenobiotic metabolizing enzymes [154], a period of heightened susceptibility to chemical carcinogenesis may exist not only for the developing conceptus, but also for the dam [155,156]. For example, the effects of pregnancy on the covalent binding of several carcinogens to DNA were investigated in mice. Non-pregnant or timed-pregnant (18th day of gestation) mice of similar age were treated with safrole or 1′-hydroxy-safrole per os. Tissue DNA adduct levels at 24 h after treatment were analyzed via ^{32}P-postlabeling. Binding of safrole and its proximate carcinogen, 1′-hydroxy-safrole, to maternal liver and kidney DNA was increased by a factor of 2.3–3.5 during pregnancy in mice [157]. In 1993, Randerath et al. observed a similar effect in the liver of pregnant mice exposed to myristicin (48,000 adducts/10^9 nucleotides in liver DNA from dams vs. 17,000 adducts/10^9 nucleotides in liver DNA from non-pregnant mice) [101]. This indicates that exposure to genotoxic compounds may be more hazardous for the maternal body during pregnancy than for non-pregnant adult females. In addition, safrole and myristicin may not quantitatively react in a first pass manner in mouse maternal liver alone. Part of the amount of safrole administered maternally and some reactive metabolites may reach the fetus transplacentally. Indeed, DNA adduct formation was observed by Randerat et al. in fetal liver after exposure to myristicin in pregnant mice [101]. The ability to form DNA adducts of myristicin transplacentally is of concern with respect to rapid cell divisions occurring in fetal liver cells, thus increasing the possibility of fixing potential mutagenic lesions which may further lead to carcinogenesis. In this context, administration of safrole to pregnant mice during the second half of gestation also led to the development of epithelial kidney tumors in female offsprings, demonstrating transplacental carcinogenesis [155]. In this study, a strong age- and sex-dependent difference ($p < 0.01$) in offspring renal carcinogenesis by safrole was observed. For comparison, in the case of the direct alkylating carcinogen ethylnitrosourea, no significant sex-dependent differences were observed [158], and preweaning as well as adult mice were equally sensitive to renal carcinogenesis by ethylnitrosourea [159].

3.3. Carcinogenicity of Safrole, Methyleugenol, Myristicin and Elemicin

Mutagenicity may lead to the development of cancer. For example, mutations in tumor suppressor genes or proto-oncogenes can cause uncontrolled cell division [160,161].

Safrole and methyleugenol are known hepatocarcinogens in experimental animals [126,162–167]. This was demonstrated by several rodent studies described and discussed in detail elsewhere [3,4,122,168].

In contrast to safrole and methyleugenol [92,122,168,169], data on the carcinogenicity of myristicin and elemicin are sparse. However, some limited experimental information is available suggesting the possible carcinogenic activity of these compounds [24,152,170].

Although results from an early experimental study using a preweaning mouse model suggest that myristicin is not hepatocarcinogenic [166], the reliability of this study must be questioned. Based on an in silico analysis, Auerbach et al., in 2010, reported that myristicin might potentially act as a weak carcinogen [170]. They predicted that administration of myristicin at 2 mmol/kg/day for 2 years would lead to a weak, albeit significant, increase in hepatic tumor burden in male rats. However, it should be noted that the informative value of in silico testing, with respect to the endpoint carcinogenicity, is rather limited [171].

For elemicin, first indications of tumorigenicity were reported by Wiseman and colleagues, who administered male B6C3F1 mice i.p 1'-hydroxy-elemicin or 1'-acetoxyelemicin in 4 doses during the first 21 days postnatally [126]. In this study, an average of 0.8 hepatoma/mouse relative to 0.1 hepatoma/mouse for the solvent-treated controls was observed after 13 months. An earlier and similar assay with 1'-hydroxy-elemicin, but using only 50% of the doses used by Wiseman et al., however, provided no evidence for its hepatocarcinogenicity when administered to preweaning male mice [166]. Data from two-year combined toxicity and carcinogenicity studies do not exist so far, neither for myristicin nor for elemicin. Those studies are crucial for a conclusive evaluation of the carcinogenic potential of myristicin and elemicin, as also stated by others [10,24,61]. Thus, the possible carcinogenic potential (including the underlying mode of action) of myristicin and elemicin merit further attention.

3.4. Other Toxicological Endpoints

In the following part of the manuscript, further toxicologically relevant effects of methyleugenol, elemicin, safrole, and myristicin will be described in a comparative manner. This includes acute, as well as subchronic toxicity, studied in vivo.

3.4.1. Acute Toxicity of Methyleugenol vs. Elemicin and Safrole vs. Myristicin

In 2000, results of a short-term animal study done by NTP showed that methyleugenol is moderately toxic following a single oral dose. The median lethal oral dose (LD_{50}) was 810 to 1560 mg/kg body weight (bw) for rats and 540 mg/kg bw for mice [92]. The undiluted chemical (98% purity) was found to be neither an eye irritant nor a skin irritant to rats and mice [92,172]. In contrast to methyleugenol, there is currently no literature available regarding the acute toxicity of elemicin.

Safrole was shown to be moderately toxic [173]. Its LD_{50} following oral administration was 1950 mg/kg bw and 2350 mg/kg bw in rats and mice, respectively [174,175]. Moreover, for safrole, acute neurological effects were described, including depression, ataxia in rats, as well as psychoactive and hallucinogenic effects in humans, which were considered as being similar to those reported for other methylendioxybenzene compounds, including myristicin [173,176,177]. The availability of literature regarding acute toxicity of myristicin is limited. In 1961, Truit and colleagues performed an acute toxicity study in rats treated i.p. with myristicin (200–1000 mg/kg bw) [178]. In the highest dose group, myristicin induced hyperexcitability followed by central nervous depression in rats. From these data, authors derived an LD_{50} > 1000 mg myristicin/kg in rats following i.p. application [178]. Although the database on myristicin is rather limited, its acute toxicity after oral administration was considered to be low [24]. Taken together, acute toxicity of myristicin seems to be comparable to that of safrole, especially regarding neurological effects.

3.4.2. Subchronic Toxic Effects of Methyleugenol vs. Elemicin and Safrole vs. Myristicin

In 2000, NTP published the results of 14-week rat and mouse studies, in which subchronic toxicity of the oral administration of methyleugenol (0, 10, 30, 100, 300 or 1000 mg/kg bw via gavage; 5 days/week) to male and female F344/N rats and B6C3F1 mice was investigated [92]. Regarding the experiments done with rats, all animals survived until the end of the study. However, exposure to methyleugenol reduced body weight gain and caused cholestasis, hepatic dysfunction with hypoproteinemia and hypoalbuminemia, as well as atrophic gastritis. Moreover, this led to increased liver and testis weight and adrenal gland hypertrophy [92]. A no observed effect level (NOEL) of 30 mg/kg bw per day was identified [3,92]. In the mouse study, 9 out of 10 males and all females of the highest dose group died before the end of the study [92]. Methyleugenol exposure was associated with reduced body weight gain, elevated liver weight in mice, and increased incidences of cytological alteration, necrosis, bile duct hyperplasia, and subacute inflammation in livers. Furthermore, there were increased incidences for atrophy, necrosis, oedema, mitotic alteration, and cystic glands of the fundic region of the glandular stomach in mice of both sexes [92]. A NOEL of 10 mg methyleugenol/kg bw and day was identified for mice [3,92]. In sum, the available subchronic studies indicated that methyleugenol is moderately toxic, which includes different adverse effects, primarily in liver and stomach [3,92,179].

In contrast to methyleugenol, there is currently no literature available regarding elemicin's subchronic toxicity.

In 1965, Hagan et al. performed a subchronic toxicity study, in which safrole (250, 500 and 750 mg/kg bw per day) was administered via gavage to Osborne–Mendel rats of both sexes for 105 days [180]. In the two highest dose groups, several rats died before the scheduled end of the study. In the lowest dose group, all rats survived until the end of the study. Several organotoxic effects were observed in this rat study, including liver hypertrophy, focal necrosis with slight fibrosis, steatosis, bile duct proliferation, and adrenal enlargement with fatty infiltration [4,180].

Comparable findings were obtained by Jin and colleagues in a rat study performed in 2011 [149]. In this study, safrole was administered to rats via diet (doses: 0, 0.1, and 0.5%; $n = 10$/group; both sexes) for 13 weeks. The main findings of this study were significantly reduced final body weights in male and female rats of all dose groups and hepatotoxic effects, including increased relative liver weights and significantly increased incidences of centrilobular hypertrophy, centrilobular vacuolar degeneration, and single cell necrosis of hepatocytes. Moreover, the authors found that the relative kidney weights of male and female rats were significantly increased after 13 weeks. Accompanying this, different nephrotoxic effects were observed in male rats of the highest dose group, such as significantly increased incidences of tubular hyaline droplets, granular cast, pelvic calcification, and interstitial cell infiltration in the kidney [149]. Taken together, the liver and kidney appeared to be the target organs with the most severe effects.

Regarding myristicin's subchronic toxicity, NTP published in 2019 the results of 90-day toxicity studies performed in F344/NTac rats and B6C3F1/N mice [152]. In these studies, different doses of myristicin (0, 10, 30, 100, 300, or 600 mg/kg bw) were administered via gavage 5 days/week for 13 weeks to rats and mice of both sexes ($n = 10$). In the rat study, all males survived until the end, whereas, three female rats of the highest dose group died within 4 days of the study [152]. Exposure of rats to myristicin led to various treatment-related effects, including reduced mean body weight, enlarged livers, increased relative liver and kidney weights, as well as increased triglycerides and alanine aminotransferase activity regarding clinical pathology. Accompanying this, several treatment-related lesions were identified in rats, such as centrilobular hepatocyte hypertrophy and necrosis in the liver; epithelium atrophy and hyperplasia as well as necrosis in the glandular stomach; and renal tubule hyaline droplet accumulation as well as a slightly increased severity of nephropathy [152]. Moreover, myristicin also affected the reproductive system of male rats, which included decreased absolute left cauda and left epididymis weights, as well as a lowered number of sperm per cauda epididymis, germinal epithelium degeneration, elongated spermatid retention in seminiferous tubules of the testis, and exfoliated germ cells in epididymal duct lumina. Therefore, the authors concluded that oral myristicin exposure exhibited the potential to induce reproductive toxicity in male F344/NTac rats [152]. In the mouse study, all animals survived until the end. In mice exposed to myristicin, mean body weights were reduced, livers were enlarged, absolute and relative liver weights were elevated and hematology parameters were affected, which included increased leukocyte counts and segmented neutrophil number. Moreover, various treatment-related lesions were observed in mice, such as oval cell hyperplasia, centrilobular hepatocyte hypertrophy, and necrosis of the liver, epithelial and nerve atrophy, glands hyperplasia, hyaline droplet accumulation, and cytoplasmic vacuolization of the respiratory epithelium in the nose. Beside this, there was a significantly increased incidence of atrophy and hyperplasia in the epithelium of the glandular stomach as well as of chronic and epithelial suppurative inflammation in the forestomach [152]. From these findings, authors concluded that the major targets after oral myristicin administration in rats and mice were the liver and glandular stomach. Additional targets were salivary glands, the nose, kidney, testis, epididymis, and the forestomach. Study authors identified a lowest observed effect level (LOEL) of 30 mg/kg bw (increased relative liver weight) for male rats, 10 mg/kg bw (clinical chemistry) for female rats, 100 mg/kg bw (increased liver weights) for male mice, and 10 mg/kg bw (increased liver weights) for female mice. Moreover, a NOEL of 10 mg/kg bw for male rats and of 30 mg/kg bw for male mice was identified, but not for female rats or mice [152]. Together, the aforementioned data clearly indicate that the spectrum of toxic effects following subchronic myristicin exposure is at least in part, and especially regarding the hepatic and renal effects, comparable to that of the structurally similar compound safrole.

4. Conclusions

The limited toxicological data and the lack of occurrence and consumption data preclude a comprehensive evaluation of adverse health effects potentially associated with myristicin, elemicin, and other alkenylbenzenes.

Therefore, additional occurrence data is needed for all toxicologically relevant alkenylbenzenes in different food products, especially those containing high levels of alkenylbenzenes (e.g., essential oils, basil-containing pesto, or PFS) [5,11,58]. Alkenylbenzenes can be separated either via GC or high-performance liquid chromatography techniques (HPLC) followed by MS [12,25,181–184]. However, to increase the specificity and accuracy of methods used for sample preparation, extraction, as well as substance separation constant standardization efforts are needed. Furthermore, data on the consumption of alkenylbenzene-containing foods is required. This data should be collected via appropriate consumption surveys.

The alkenylbenzenes safrole and myristicin as well as methyleugenol and elemicin are structurally closely related (Figure 1). This in turn suggests that the hazard potential of those compounds could exhibit similarities. In this regard, it appears reasonable to identify potential hazards of the toxicologically widely unexplored alkenylbenzenes myristicin and elemicin in comparison to those of the known genotoxic carcinogens safrole and methyleugenol. The available toxicological data, e.g., data on toxicokinetics and genotoxicity, already suggest that both myristicin and elemicin might form reactive metabolites being similar to those being formed from safrole and methyleugenol. However, the sparse data also indicate that there might be quantitative differences that may result in an altered toxicity profile. This in turn, cannot be finally evaluated at present. Indeed, their genotoxic and carcinogenic potential is widely unknown, so far. In this context, two-year combined oral toxicity and carcinogenicity studies are mandatory for the evaluation of the long-term effects, as well as of the carcinogenic potential of myristicin and elemicin, as also recommended by others [10,24,61]. Moreover, the underlying modes of action of these compounds merit further attention, too. In this context, an appropriate experimental setting should be designed taking into account the alkenylbenzene-specific bioactivation (e.g., via SULTs) discussed in detail before [5].

It is important to note that the conventional bacterial reverse mutation test (Organization for Economic Co-Operation and Development (OECD) Test Guideline (TG) 471; Ames test [185]) lacks the metabolic competence to yield the ultimate carcinogenic sulf-oxy intermediates from alkenylbenzenes [186]. However, genetic modifications of the bacteria enabling SULT expression may lead to a more adequate in vitro setting for the mutagenicity testing of compounds metabolically activated via this pathway, such as methyleugenol, myristicin, and elemicin [5,186]. Substantiating this, Monien et al. demonstrated in 2011 that furfuryl alcohol was negative in the standard Ames test, whereas it was mutagenic in a modified setting utilizing *S. typhimurium* TA100 engineered for the expression of human SULT1 [187]. In line with this, in 2016, Honda and colleagues found methyleugenol, which is not mutagenic in standard Ames test [92], to be mutagenic in a modified Ames test using a human SULT1-expressing *S. typhimurium* TA100 strain [186]. Although scientific approaches exist that augment bacteria with human sulfotransferases, these systems are not yet internationally standardized and validated for regulatory purposes.

An alternative approach is the hypoxanthine guanine phosphoribosyltransferase (HPRT) assay (OECD TG 476), which is an in vitro mammalian cell gene mutation test using the *hprt* and *xprt* genes for gene mutation measurement in mammalian cells [188]. The method is described in detail elsewhere [189]. Modification of the HPRT assay via the use of replication competent cells (e.g., human liver cells) expressing human SULT1A1 could also offer an appropriate setting for in vitro mutagenicity testing of compounds bioactivated in a SULT-dependent manner, such as safrole, methyleugenol, myristicin, and elemicin.

From a toxicological point of view, and for the sake of animal welfare, initial mutagenicity testing of alkenylbenzenes with unknown modes of action, such as elemicin and myristicin, should be done in vitro. This might be sufficient, if initial testing of mutagenicity is conducted using appropriate test systems, enabling the intracellular activation to reactive sulfate esters by SULT-proficient bacterial or mammalian cells. For regulatory purposes, it appears however reasonable to recommend transgenic rodent (TGR) models (OECD TG 488 [190]) as ultimate confirmatory assays to decide on mutagenic potencies of alkenylbenzenes in vivo following a positive in vitro finding [189].

A promising candidate among those TGR models appears to be the *gpt* delta rodent gene mutation assay, developed by Nohmi et al. [153,191–193]. Since its development, it was already successfully used in various studies in the context of food safety research [153,193]. Regarding alkenylbenzenes, the *gpt* delta TGR model was demonstrated to reliably identify safrole, methyleugenol, and estragole as mutagens [124,149,194], as also concluded by others [3,4,195]. One additional benefit of such test systems is the option to evaluate mutagenicity in any tissue of interest [189]. This is of particular interest when mutagenicity would have to be tested in distinct organs, such as in the liver, e.g., for testing of suspected hepatocarcinogens, such as methyleugenol and elemicin [189]. Moreover, such an approach might pave the way for simultaneous testing of mutagenicity in different tissues at the same time. Moreover, such in vivo assays are needed to distinguish between genotoxic (e.g., aflatoxin B1) and non-genotoxic carcinogens (e.g., 3-chloro-1,2-propanediol) [153].

Together, the aforementioned approaches would shed more light on the existing, and currently still serious, data gaps, and could help to reduce considerable uncertainties currently impeding the evaluation of adverse health effects potentially associated with the consumption of foods containing alkenylbenzenes.

Author Contributions: Conceptualization, M.E.G. and A.E.; writing—original draft preparation M.E.G., A.E., B.S. (Benjamin Sachse) and B.S. (Bernd Schäfer); writing—review and editing, M.E.G., A.E., B.S. (Benjamin Sachse) and B.S. (Bernd Schäfer). All authors have read and agreed to the published version of the manuscript.

Funding: This research received no external funding.

Institutional Review Board Statement: Not applicable.

Informed Consent Statement: Not applicable.

Data Availability Statement: Not applicable.

Conflicts of Interest: The authors declare no conflict of interest.

References

1. IARC. Overall evaluations of carcinogenicity: An updating of IARC Monographs Volumes 1 to 42—Eugenol. *IARC Monogr. Eval. Carcinog. Risk Chem. Hum.* **1987**, *36*, 1–63.
2. IARC. Some chemicals present in industrial and consumer products, food and drinking-water/Methyleugenol. *IARC Monogr. Eval. Carcinog. Risks Hum.* **2013**, *101*, 407–433.
3. SCF. Opinion of the Scientific Committee on Food on Methyleugenol (4-Allyl-1,2-dimethoxybenzene). In *SCF/CS/FLAV/FLAVOUR/4 ADD1 FINAL*; European Commission: Brussels, Belgium, 2001; pp. 1–10.
4. SCF. Opinion of the Scientific Committee on Food on the safety of the presence of safrole (1-allyl-3,4-methylene dioxy benzene) in flavourings and other food ingredients with flavouring properties. In *SCF/CS/FLAV/FLAVOUR/6 ADD3 FINAL*; European Commission: Brussels, Belgium, 2002; pp. 1–15.
5. Eisenreich, A.; Götz, M.E.; Sachse, B.; Monien, B.H.; Herrmann, K.; Schäfer, B. Alkenylbenzenes in Foods: Aspects Impeding the Evaluation of Adverse Health Effects. *Foods* **2021**, *10*, 2139. [CrossRef] [PubMed]
6. Solheim, E.; Scheline, R.R. Metabolism of alkenebenzene derivatives in the rat III. Elemicin and isoelemicin. *Xenobiotica* **1980**, *10*, 371–380. [CrossRef]
7. Lee, H.S.; Jeong, T.C.; Kim, J.H. In vitro and in vivo metabolism of myristicin in the rat. *J. Chromatogr. B Biomed. Sci. Appl.* **1998**, *705*, 367–372. [CrossRef]
8. Beyer, J.; Ehlers, D.; Maurer, H.H. Abuse of nutmeg (*Myristica fragrans* Houtt.): Studies on the metabolism and the toxicologic detection of its ingredients elemicin, myristicin, and safrole in rat and human urine using gas chromatography/mass spectrometry. *Ther. Drug Monit.* **2006**, *28*, 568–575. [CrossRef]
9. Kristanc, L.; Kreft, S. European medicinal and edible plants associated with subacute and chronic toxicity part I: Plants with carcinogenic, teratogenic and endocrine-disrupting effects. *Food Chem. Toxicol.* **2016**, *92*, 150–164. [CrossRef]
10. Martins, C.; Rueff, J.; Rodrigues, A.S. Genotoxic alkenylbenzene flavourings, a contribution to risk assessment. *Food Chem. Toxicol.* **2018**, *118*, 861–879. [CrossRef]
11. Al-Malahmeh, A.J.; Alajlouni, A.M.; Ning, J.; Wesseling, S.; Vervoort, J.; Rietjens, I. Determination and risk assessment of naturally occurring genotoxic and carcinogenic alkenylbenzenes in nutmeg-based plant food supplements. *J. Appl. Toxicol.* **2017**, *37*, 1254–1264. [CrossRef]
12. Raffo, A.; D'Aloise, A.; Magri, A.L.; Leclercq, C. Quantitation of tr-cinnamaldehyde, safrole and myristicin in cola-flavoured soft drinks to improve the assessment of their dietary exposure. *Food Chem. Toxicol.* **2013**, *59*, 626–635. [CrossRef]
13. Siano, F.; Ghizzoni, C.; Gionfriddo, F.; Colombo, E.; Servillo, L.; Castaldo, D. Determination of estragole, safrole and eugenol methyl ether in food products. *Food Chem.* **2003**, *81*, 469–475. [CrossRef]
14. Ghimire, B.K.; Yoo, J.H.; Yu, C.Y.; Chung, I.M. GC-MS analysis of volatile compounds of Perilla frutescens Britton var. Japonica accessions: Morphological and seasonal variability. *Asian Pac. J. Trop. Med.* **2017**, *10*, 643–651. [CrossRef]
15. Muráriková, A.; Ťažký, A.; Neugebauerová, J.; Planková, A.; Jampílek, J.; Mučaji, P.; Mikuš, P. Characterization of essential oil composition in different basil species and pot cultures by a GC-MS method. *Molecules* **2017**, *22*, 1221. [CrossRef]
16. Simándi, B.; Deák, A.; Rónyai, E.; Yanxiang, G.; Veress, T.; Lemberkovics, É.; Then, M.; Sass-Kiss, Á.; Vámos-Falusi, Z. Supercritical carbon dioxide extraction and fractionation of fennel oil. *J. Agric. Food Chem.* **1999**, *47*, 1635–1640. [CrossRef]
17. Shulgin, A.T. Possible implication of myristicin as a psychotropic substance. *Nature* **1966**, *210*, 380–384. [CrossRef]
18. Shulgin, A.T. Composition of the myristicin fraction from oil of nutmeg. *Nature* **1963**, *197*, 379. [CrossRef]
19. Du, S.S.; Yang, K.; Wang, C.F.; You, C.X.; Geng, Z.F.; Guo, S.S.; Deng, Z.W.; Liu, Z.L. Chemical constituents and activities of the essential oil from myristica fragrans against cigarette beetle lasioderma serricorne. *Chem. Biodivers.* **2014**, *11*, 1449–1456. [CrossRef]
20. Analytical Methods Committee. Application of gas-liquid chromatography to the analysis of essential oils. Part XIV. Monographs for five essential oils. *Analyst* **1988**, *113*, 1125–1136. [CrossRef]
21. Sanford, K.J.; Heinz, D. Effects of storage on the volatile composition of nutmeg. *Phytochemistry* **1971**, *10*, 1245–1250. [CrossRef]
22. Baldry, J.; Dougan, J.; Matthews, W.S.; Nabney, J.; Pickering, G.R.; Robinson, F.V. Composition and Flavour of Nutmeg oils. *Int. Flavours Food Addit.* **1976**, *7*, 28–30.
23. Archer, A.W. Determination of safrole and myristicin in nutmeg and mace by high-performance liquid chromatography. *J. Chromatogr. A* **1988**, *438*, 117–121. [CrossRef]
24. Hallstrom, H.; Thuvander, A. Toxicological evaluation of myristicin. *Nat. Toxins* **1997**, *5*, 186–192. [CrossRef]
25. Lopez, P.; van Sisseren, M.; De Marco, S.; Jekel, A.; de Nijs, M.; Mol, H.G. A straightforward method to determine flavouring substances in food by GC-MS. *Food Chem.* **2015**, *174*, 407–416. [CrossRef] [PubMed]
26. Ávila, M.; Zougagh, M.; Escarpa, A.; Rios, A. Determination of alkenylbenzenes and related flavour compounds in food samples by on-column preconcentration-capillary liquid chromatography. *J. Chromatogr. A* **2009**, *1216*, 7179–7185. [CrossRef] [PubMed]
27. Franz, C.; Glasl, H. Zur Kenntnis der Ätherischen Öle von Petersilie II. Vergleichende Untersuchung des Frucht-, Blatt- und Wurzelöles einiger Petersiliensorten. *Qual. Plant.* **1976**, *25*, 253–262. [CrossRef]
28. Simon, J.E.; Quinn, J. Characterization of essential oil of parsley. *J. Agric. Food Chem.* **1988**, *36*, 467–472. [CrossRef]
29. Macleod, A.J.; Snyder, C.H.; Subramanian, G. Volatile aroma constituents of parsley leaves. *Phytochemistry* **1985**, *24*, 2623–2627. [CrossRef]
30. Masanetz, C.; Grosch, W. Key odorants of parsley leaves (*Petroselinum crispum* [Mill.] Nym. ssp. crispum) by Odour–activity values. *Flavour. Frag. J.* **1998**, *13*, 115–124. [CrossRef]

31. Trenkle, K. Recent studies on fennel (*Foeniculum vulgare* M.) 2. The volatile oil of the fruit, herbs and roots of fruit-bearing plants. *Die Pharm.* **1972**, *27*, 319–324.
32. Hao, Y.; Kang, J.; Guo, X.; Yang, R.; Chen, Y.; Li, J.; Shi, L. Comparison of Nutritional Compositions and Essential Oil Profiles of Different Parts of a Dill and Two Fennel Cultivars. *Foods* **2021**, *10*, 1784. [CrossRef]
33. Al-Sheddi, E.S.; Al-Zaid, N.A.; Al-Oqail, M.M.; Al-Massarani, S.M.; El-Gamal, A.A.; Farshori, N.N. Evaluation of cytotoxicity, cell cycle arrest and apoptosis induced by *Anethum graveolens* L. essential oil in human hepatocellular carcinoma cell line. *Saudi Pharm. J.* **2019**, *27*, 1053–1060. [CrossRef]
34. Kubeczka, K.H.; Stahl, E. [On the essential oils from the Apiaceae (Umbelliferae). I. The oil of roots from Pastinaca sativa (author's transl)]. *Planta Med.* **1975**, *27*, 235–241. [CrossRef]
35. Stahl, E. Variation of Myristicin Content in Cultivated Parsnip Roots (*Pastinaca sativa* ssp. sativa var. hortensis). *J. Agric. Food Chem.* **1981**, *29*, 890–892. [CrossRef]
36. Lichtenstein, E.P.; Casida, J.E. Naturally occurring insecticides: Myristicin, an Insecticide and Synergist Occurring Naturally in the Edible Parts of Parsnips. *J. Agric. Food Chem.* **1963**, *11*, 410–415. [CrossRef]
37. Fischer, R.; Nitzan, N.; Chaimovitsh, D.; Rubin, B.; Dudai, N. Variation in essential oil composition within individual leaves of sweet basil (*Ocimum basilicum* L.) is more affected by leaf position than by leaf age. *J. Agric. Food Chem.* **2011**, *59*, 4913–4922. [CrossRef]
38. Chalchat, J.-C.; Özcan, M.M. Comparative essential oil composition of flowers, leaves and stems of basil (*Ocimum basilicum* L.) used as herb. *Food Chem.* **2008**, *110*, 501–503. [CrossRef]
39. Vani, S.R.; Cheng, S.; Chuah, C. Comparative study of volatile compounds from genus Ocimum. *Am. J. Appl. Sci.* **2009**, *6*, 523. [CrossRef]
40. Smaili, T.; Zellagui, A.; Cioni, P.L.; Flamini, G. A myristicin-rich essential oil from Daucus sahariensis growing in Algeria. *Nat. Prod. Commun.* **2011**, *6*, 883–886. [CrossRef]
41. Jabrane, A.; Jannet, H.B.; Harzallah-Skhiri, F.; Mastouri, M.; Casanova, J.; Mighri, Z. Flower and root oils of the tunisian *Daucus carota* L. ssp.maritimus (*Apiaceae*): Integrated analyses by GC, GC/MS, and 13C-NMR spectroscopy, and in vitro antibacterial activity. *Chem. Biodivers.* **2009**, *6*, 881–889. [CrossRef]
42. Buttery, R.G.; Seifert, R.M.; Guadagni, D.G.; Black, D.R.; Ling, L. Characterization of some volatile constituents of carrots. *J. Agric. Food Chem.* **1968**, *16*, 1009–1015. [CrossRef]
43. Wulf, L.W.; Nagel, C.W.; Branen, A.L. Analysis of Myristicin and Falcarinol in Carrots by High-Pressure Liquid Chromatography. *J. Agric. Food Chem.* **1978**, *26*, 1390–1393. [CrossRef]
44. Mansour, E.S.S.; Maatooq, G.T.; Khalil, A.T.; Marwan, E.S.M.; Sallam, A.A. Essential oil of Daucus glaber Forssk. *Z. Naturforsch. Sect. C J. Biosci.* **2004**, *59*, 373–378. [CrossRef]
45. Marzouki, H.; Khaldi, A.; Falconieri, D.; Piras, A.; Marongiu, B.; Molicotti, P.; Zanetti, S. Essential oils of Daucus carota subsp. carota of Tunisia obtained by supercritical carbon dioxide extraction. *Nat. Prod. Commun.* **2010**, *5*, 1955–1958. [CrossRef]
46. Rivera-Pérez, A.; López-Ruiz, R.; Romero-González, R.; Frenich, A.G. A new strategy based on gas chromatography–high resolution mass spectrometry (GC–HRMS-Q-Orbitrap) for the determination of alkenylbenzenes in pepper and its varieties. *Food Chem.* **2020**, *321*, 126727. [CrossRef]
47. Jirovetz, L.; Buchbauer, G.; Ngassoum, M.B.; Geissler, M. Aroma compound analysis of Piper nigrum and Piper guineense essential oils from Cameroon using solid-phase microextraction–gas chromatography, solid-phase microextraction–gas chromatography–mass spectrometry and olfactometry. *J. Chromatogr. A* **2002**, *976*, 265–275. [CrossRef]
48. Ekundayo, O.; Laakso, I.; Hiltunen, R.; Adegbola, R.M.; Oguntimein, B.; Sofowora, A. Essential oil constituents of ashanti pepper (piper guineense) fruits (Berries). *J. Agric. Food Chem.* **1988**, *36*, 880–882. [CrossRef]
49. Bessiere, J.; Menut, C.; Lamaty, G.; Joseph, H. Variations in the volatile constituents of peperomia rotundifolia schlecht. & cham. grown on different host-trees in guadeloupe. *Flavour. Frag. J.* **1994**, *9*, 131–133.
50. De Diaz, A.M.; Diaz, P.P.; Cardoso, H. Volatile Constituents of Peperomia subespatulata. *Planta Med.* **1988**, *54*, 92–93. [CrossRef]
51. Andrade, E.H.A.; Carreira, L.M.M.; da Silva, M.H.L.; da Silva, J.D.; Bastos, C.N.; Sousa, P.J.C.; Guimarães, E.F.; Maia, J.G.S. Variability in Essential-Oil Composition of Piper marginatum sensu lato. *Chem. Biodivers.* **2008**, *5*, 197–208. [CrossRef]
52. De Lira, P.N.B.; Da Silva, J.K.R.; Andrade, E.H.A.; Sousa, P.J.C.; Silva, N.N.S.; Maia, J.G.S. Essential oil composition of three Peperomia species from the Amazon, Brazil. *Nat. Pro. Comm.* **2009**, *4*, 427–430. [CrossRef]
53. Cook, W.; Howard, A. The essential oil of Illicium anisatum Linn. *Can. J. Chem.* **1966**, *44*, 2461–2464. [CrossRef]
54. Lawrence, B.M. Progress in essential oils/Progrès dans le domaine des huiles essentielles. *Perfum. Flavor.* **1988**, *13*, 44–50.
55. Fiorini, C.; Fouraste, I.; David, B.; Bessiere, J. Composition of the Flower, Leaf and Stem Essential Oils from Laurus nobilis. *Flavour. Frag. J.* **1997**, *12*, 91–93. [CrossRef]
56. Conforti, F.; Statti, G.; Uzunov, D.; Menichini, F. Comparative chemical composition and antioxidant activities of wild and cultivated *Laurus nobilis* L. leaves and *Foeniculum vulgare* subsp. piperitum (Ucria) coutinho seeds. *Biol. Pharm. Bull.* **2006**, *29*, 2056–2064. [CrossRef]
57. Krishnamoorty, B.; Rema, J. Nutmeg and mace. In *Handbook of Herbs and Spices*; Peter, K.V., Ed.; CRC Press-Woodhead Publishing Ltd.: Boca Raton, FL, USA, 2001; Volume 1, pp. 238–248.
58. Al-Malahmeh, A.J.; Alajlouni, A.M.; Wesseling, S.; Vervoort, J.; Rietjens, I.M.C.M. Determination and risk assessment of naturally occurring genotoxic and carcinogenic alkenylbenzenes in basil-containing sauce of pesto. *Toxicol. Rep.* **2017**, *4*, 1–8. [CrossRef]

59. Alajlouni, A.M.; Al-Malahmeh, A.J.; Isnaeni, F.N.; Wesseling, S.; Vervoort, J.; Rietjens, I.M.C.M. Level of Alkenylbenzenes in Parsley and Dill Based Teas and Associated Risk Assessment Using the Margin of Exposure Approach. *J. Agric. Food Chem.* **2016**, *64*, 8640–8646. [CrossRef]

60. Alajlouni, A.M.; Al-Malahmeh, A.J.; Wesseling, S.; Kalli, M.; Vervoort, J.; Rietjens, I.M.C.M. Risk assessment of combined exposure to alkenylbenzenes through consumption of plant food supplements containing parsley and dill. *Food Addit. Contam. Part A Chem. Anal. Control Expo. Risk Assess.* **2017**, *34*, 2201–2211. [CrossRef]

61. De Vincenzi, M.; De Vincenzi, A.; Silano, M. Constituents of aromatic plants: Elemicin. *Fitoterapia* **2004**, *75*, 615–618. [CrossRef]

62. De Vincenzi, M.; Silano, M.; Maialetti, F.; Scazzocchio, B. Constituents of aromatic plants: II. Estragole. *Fitoterapia* **2000**, *71*, 725–729. [CrossRef]

63. De Vincenzi, M.; Silano, M.; Stacchini, P.; Scazzocchio, B. Constituents of aromatic plants: I. Methyleugenol. *Fitoterapia* **2000**, *71*, 216–221. [CrossRef]

64. Dawidowicz, A.L.; Dybowski, M.P. Determination of myristicin in commonly spices applying SPE/GC. *Food Chem. Toxicol.* **2012**, *50*, 2362–2367. [CrossRef]

65. Dawidowicz, A.L.; Dybowski, M.P. Simple and rapid determination of myristicin in human serum. *Forensic Toxicol.* **2013**, *31*, 119–123. [CrossRef]

66. Cao, X.; Xie, Q.; Zhang, S.; Xu, H.; Su, J.; Zhang, J.; Deng, C.; Song, G. Fabrication of functionalized magnetic microspheres based on monodispersed polystyrene for quantitation of allyl-benzodioxoles coupled with gas chromatography and mass spectrometry. *J. Chromatogr. A* **2019**, *1607*, 460402. [CrossRef]

67. Guenthner, T.M.; Luo, G. Investigation of the role of the 2′,3′-epoxidation pathway in the bioactivation and genotoxicity of dietary allylbenzene analogs. *Toxicology* **2001**, *160*, 47–58. [CrossRef]

68. Solheim, E.; Scheline, R.R. Metabolism of alkenebenzene derivatives in the rat I. P-methoxyallylbenzene (estragole) and p-methoxypropenylbenzene (anethole). *Xenobiotica* **1973**, *3*, 493–510. [CrossRef]

69. Solheim, E.; Scheline, R.R. Metabolism of alkenebenzene derivatives in the rat. II. Eugenol and isoeugenol methyl ethers. *Xenobiotica* **1976**, *6*, 137–150. [CrossRef]

70. Stillwell, W.; Carman, J.K.; Bell, L.; Horning, M.G. The metabolism of safrole and 2′,3′-epoxysafrole in the rat and guinea pig. *Drug Metab. Dispos.* **1974**, *2*, 489–498.

71. Wislocki, P.; Borchert, P.; Miller, J.; Miller, E. The metabolic activation of the carcinogen 1′-hydroxysafrole in vivo and in vitro and the electrophilic reactivities of possible ultimate carcinogens. *Cancer Res.* **1976**, *36*, 1686–1695.

72. Delaforge, M.; Janiaud, P.; Levi, P.; Morizot, J.P. Biotransformation of allylbenzene analogues in vivo and in vitro through the epoxide-diol pathway. *Xenobiotica* **1980**, *10*, 737–744. [CrossRef]

73. Fischer, I.U.; Von Unruh, G.E.; Dengler, H.J. The metabolism of eugenol in man. *Xenobiotica* **1990**, *20*, 209–222. [CrossRef]

74. Boberg, E.W.; Miller, E.C.; Miller, J.A. The metabolic sulfonation and side-chain oxidation of 3′-hydroxyisosafrole in the mouse and its inactivity as a hepatocarcinogen relative to 1′-hydroxysafrole. *Chem. Biol. Interact.* **1986**, *59*, 73–97. [CrossRef]

75. Boberg, E.W.; Miller, E.C.; Miller, J.A.; Poland, A.; Liem, A. Strong evidence from studies with brachymorphic mice and pentachlorophenol that 1′-sulfoöxysafrole is the major ultimate electrophilic and carcinogenic metabolite of 1′-hydroxysafrole in mouse liver. *Cancer Res.* **1983**, *43*, 5163–5173.

76. Van den Berg, S.J.P.L.; Punt, A.; Soffers, A.E.; Vervoort, J.; Ngeleja, S.; Spenkelink, B.; Rietjens, I.M.C.M. Physiologically based kinetic models for the alkenylbenzene elemicin in rat and human and possible implications for risk assessment. *Chem. Res. Toxicol.* **2012**, *25*, 2352–2367. [CrossRef]

77. Monien, B.H.; Sachse, B.; Niederwieser, B.; Abraham, K. Detection of N-Acetyl-S-[3′-(4-methoxyphenyl)allyl]-l-Cys (AMPAC) in Human Urine Samples after Controlled Exposure to Fennel Tea: A New Metabolite of Estragole and trans-Anethole. *Chem. Res. Toxicol.* **2019**, *32*, 2260–2267. [CrossRef]

78. Herrmann, K.; Engst, W.; Appel, K.E.; Monien, B.H.; Glatt, H. Identification of human and murine sulfotransferases able to activate hydroxylated metabolites of methyleugenol to mutagens in Salmonella typhimurium and detection of associated DNA adducts using UPLC-MS/MS methods. *Mutagenesis* **2012**, *27*, 453–462. [CrossRef]

79. Klungsøyr, J.; Scheline, R.R. Metabolism of Safrole in the Rat. *Acta Pharmacol. Toxicol.* **1983**, *52*, 211–216. [CrossRef]

80. Anders, M.W.; Sunram, J.M.; Wilkinson, C.F. Mechanism of the metabolism of 1,3-benzodioxoles to carbon monoxide. *Biochem. Pharm.* **1984**, *33*, 577–580. [CrossRef]

81. Martati, E.; Boersma, M.G.; Spenkelink, A.; Khadka, D.B.; van Bladeren, P.J.; Rietjens, I.M.; Punt, A. Physiologically based biokinetic (PBBK) modeling of safrole bioactivation and detoxification in humans as compared with rats. *Toxicol. Sci.* **2012**, *128*, 301–316. [CrossRef]

82. Bolton, J.L.; Acay, N.M.; Vukomanovic, V. Evidence That 4-Allyl-o-quinones Spontaneously Rearrange to Their More Electrophilic Quinone Methides: Potential Bioactivation Mechanism for the Hepatocarcinogen Safrole. *Chem. Res. Toxicol.* **1994**, *7*, 443–450. [CrossRef]

83. Dietz, B.M.; Bolton, J.L. Biological reactive intermediates (BRIs) formed from botanical dietary supplements. *Chem. Biol. Interact.* **2011**, *192*, 72–80. [CrossRef]

84. Bolton, J.L.; Dunlap, T.L.; Dietz, B.M. Formation and biological targets of botanical o-quinones. *Food Chem. Toxicol.* **2018**, *120*, 700–707. [CrossRef] [PubMed]

85. Tsai, R.S.; Carrupt, P.A.; Testa, B.; Caldwell, J. Structure-genotoxicity relationships of allylbenzenes and propenylbenzenes: A quantum chemical study. *Chem. Res. Toxicol.* **1994**, *7*, 73–76. [CrossRef] [PubMed]

86. Thompson, D.; Constantin-Teodosiu, D.; Egestad, B.; Mickos, H.; Moldeus, P. Formation of glutathione conjugates during oxidation of eugenol by microsomal fractions of rat liver and lung. *Biochem. Pharm.* **1990**, *39*, 1587–1595. [CrossRef]

87. Thompson, D.C.; Constantin-Teodosiu, D.; Moldeus, P. Metabolism and cytotoxicity of eugenol in isolated rat hepatocytes. *Chem. Biol. Interact.* **1991**, *77*, 137–147. [CrossRef]

88. Wang, Y.K.; Yang, X.N.; Zhu, X.; Xiao, X.R.; Yang, X.W.; Qin, H.B.; Gonzalez, F.J.; Li, F. Role of Metabolic Activation in Elemicin-Induced Cellular Toxicity. *J. Agric. Food Chem.* **2019**, *67*, 8243–8252. [CrossRef] [PubMed]

89. Jeurissen, S.M.F.; Bogaards, J.J.P.; Awad, H.M.; Boersma, M.G.; Brand, W.; Fiamegos, Y.C.; Van Beek, T.A.; Alink, G.M.; Sudhölter, E.J.R.; Cnubben, N.H.P.; et al. Human cytochrome P450 enzyme specificity for bioactivation of safrole to the proximate carcinogen 1′-hydroxysafrole. *Chem. Res. Toxicol.* **2004**, *17*, 1245–1250. [CrossRef] [PubMed]

90. Jeurissen, S.M.F.; Bogaards, J.J.P.; Boersma, M.G.; Ter Horst, J.P.F.; Awad, H.M.; Fiamegos, Y.C.; Van Beek, T.A.; Alink, G.M.; Sudhölter, E.J.R.; Cnubben, N.H.P.; et al. Human cytochrome P450 enzymes of importance for the bioactivation of methyleugenol to the proximate carcinogen 1′-hydroxymethyleugenol. *Chem. Res. Toxicol.* **2006**, *19*, 111–116. [CrossRef] [PubMed]

91. Jeurissen, S.M.F.; Punt, A.; Boersma, M.G.; Bogaards, J.J.P.; Fiamegos, Y.C.; Schilter, B.; Van Bladeren, P.J.; Cnubben, N.H.P.; Rietjens, I.M.C.M. Human cytochrome P450 enzyme specificity for the bioactivation of estragole and related alkenylbenzenes. *Chem. Res. Toxicol.* **2007**, *20*, 798–806. [CrossRef]

92. NTP. NTP Technical Report on the Toxicology and Carcinogenesis Studies of Methyleugenol (CAS NO. 93-15-2) in F344/N Rats and B6C3F1 Mice (Gavage Studies). *Natl. Toxicol. Program Tech. Rep. Ser.* **2000**, *491*, 1–412.

93. Yun, C.H.; Lee, H.S.; Lee, H.Y.; Yim, S.K.; Kim, K.H.; Kim, E.; Yea, S.S.; Guengerich, F.P. Roles of human liver cytochrome P450 3A4 and 1A2 enzymes in the oxidation of myristicin. *Toxicol. Lett.* **2003**, *137*, 143–150. [CrossRef]

94. Tremmel, R.; Herrmann, K.; Engst, W.; Meinl, W.; Klein, K.; Glatt, H.; Zanger, U.M. Methyleugenol DNA adducts in human liver are associated with SULT1A1 copy number variations and expression levels. *Arch. Toxicol.* **2017**, *91*, 3329–3339. [CrossRef]

95. Zangouras, A.; Caldwell, J.; Hutt, A.J.; Smith, R.L. Dose dependent conversion of estragole in the rat and mouse to the carcinogenic metabolite, 1 -hydroxyestragole. *Biochem. Pharmacol.* **1981**, *30*, 1383–1386. [CrossRef]

96. Yao, H.; Peng, Y.; Zheng, J. Identification of glutathione and related cysteine conjugates derived from reactive metabolites of methyleugenol in rats. *Chem. Biol. Interact.* **2016**, *253*, 143–152. [CrossRef]

97. Gardner, I.; Wakazono, H.; Bergin, P.; De Waziers, I.; Beaune, P.; Kenna, J.G.; Caldwell, J. Cytochrome P450 mediated bioactivation of methyleugenol to 1′-hydroxymethyleugenol in Fischer 344 rat and human liver microsomes. *Carcinogenesis* **1997**, *18*, 1775–1783. [CrossRef]

98. Casida, J.E.; Engel, J.L.; Essac, E.G.; Kamienski, F.X.; Kuwatsuka, S. Methylene-C14-dioxyphenyl compounds: Metabolism in relation to their synergistic action. *Science* **1966**, *153*, 1130–1133. [CrossRef]

99. Kamienski, F.X.; Casida, J.E. Importance of demethylenation in the metabolism in vivo and in vitro of methylenedioxyphenyl synergists and related compounds in mammals. *Biochem. Pharmacol.* **1970**, *19*, 91–112. [CrossRef]

100. Randerath, K.; Haglund, R.E.; Phillips, D.H.; Reddy, M.V. 32P-post-labelling analysis of DNA adducts formed in the livers of animals treated with safrole, estragole and other naturally-occurring alkenylbenzenes. I. adult female CD-1 mice. *Carcinogenesis* **1984**, *5*, 1613–1622. [CrossRef]

101. Randerath, K.; Putman, K.L., Randerath, E. Flavor constituents in cola drinks induce hepatic DNA adducts in adult and fetal mice. *Biochem. Biophys. Res. Commun.* **1993**, *192*, 61–68. [CrossRef]

102. Randerath, K.; Randerath, E.; Danna, T.F.; van Golen, L.; Putman, K.L. A new sensitive 32P-postlabeling assay based on the specific enzymatic conversion of bulky DNA lesions to radiolabeled dinucleotides and nucleoside 5′-monophosphates. *Carcinogenesis* **1989**, *10*, 1231–1239. [CrossRef]

103. Randerath, K.; Reddy, M.V.; Gupta, R.C. 32P-labeling test for DNA damage. *Proc. Natl. Acad. Sci. USA* **1981**, *78*, 6126–6129. [CrossRef]

104. Reddy, M.V.; Randerath, K. Nuclease p1-mediated enhancement of sensitivity of 32P-postlabeling test for structurally diverse DNA adducts. *Carcinogenesis* **1986**, *7*, 1543–1551. [CrossRef]

105. Bergau, N.; Herfurth, U.M.; Sachse, B.; Abraham, K.; Monien, B.H. Bioactivation of estragole and anethole leads to common adducts in DNA and hemoglobin. *Food Chem. Toxicol.* **2021**, *153*, 112253. [CrossRef]

106. Phillips, D.H.; Miller, J.A.; Miller, E.C.; Adams, B. Structures of the DNA adducts formed in mouse liver after administration of the proximate hepatocarcinogen 1′-hydroxyestragole. *Cancer Res.* **1981**, *41*, 176–186.

107. Phillips, D.H.; Miller, J.A.; Miller, E.C.; Adams, B. N2 atom of guanine and N6 atom of adenine residues as sites for covalent binding of metabolically activated 1′-hydroxysafrole to mouse liver DNA in vivo. *Cancer Res.* **1981**, *41*, 2664–2671.

108. Rietjens, I.M.; Cohen, S.M.; Fukushima, S.; Gooderham, N.J.; Hecht, S.; Marnett, L.J.; Smith, R.L.; Adams, T.B.; Bastaki, M.; Harman, C.G.; et al. Impact of structural and metabolic variations on the toxicity and carcinogenicity of hydroxy- and alkoxy-substituted allyl- and propenylbenzenes. *Chem. Res. Toxicol.* **2014**, *27*, 1092–1103. [CrossRef]

109. Groh, I.A.M.; Cartus, A.T.; Vallicotti, S.; Kajzar, J.; Merz, K.H.; Schrenk, D.; Esselen, M. Genotoxic potential of methyleugenol and selected methyleugenol metabolites in cultured Chinese hamster V79 cells. *Food Funct.* **2012**, *3*, 428–436. [CrossRef]

110. Groh, I.A.M.; Rudakovski, O.; Gründken, M.; Schroeter, A.; Marko, D.; Esselen, M. Methyleugenol and oxidative metabolites induce DNA damage and interact with human topoisomerases. *Arch. Toxicol.* **2016**, *90*, 2809–2823. [CrossRef]

111. Groh, I.A.M.; Esselen, M. Methyleugenol and selected oxidative metabolites affect DNA-Damage signalling pathways and induce apoptosis in human colon tumour HT29 cells. *Food Chem. Toxicol.* **2017**, *108*, 267–275. [CrossRef]

112. Chan, V.S.; Caldwell, J. Comparative induction of unscheduled DNA synthesis in cultured rat hepatocytes by allylbenzenes and their 1′-hydroxy metabolites. *Food Chem. Toxicol.* **1992**, *30*, 831–836. [CrossRef]

113. Ellis, J.K.; Carmichael, P.L.; Gooderham, N.J. DNA adduct levels in the liver of the F344 rat treated with the natural flavour methyleugenol. *Toxicology* **2006**, *1*, 73–74. [CrossRef]

114. Herrmann, K.; Schumacher, F.; Engst, W.; Appel, K.E.; Klein, K.; Zanger, U.M.; Glatt, H. Abundance of DNA adducts of methyleugenol, a rodent hepatocarcinogen, in human liver samples. *Carcinogenesis* **2013**, *34*, 1025–1030. [CrossRef] [PubMed]

115. Herrmann, K.; Engst, W.; Florian, S.; Lampen, A.; Meinl, W.; Glatt, H.R. The influence of the SULT1A status-wild-type, knockout or humanized-on the DNA adduct formation by methyleugenol in extrahepatic tissues of mice. *Toxicol. Res.* **2016**, *5*, 808–815. [CrossRef] [PubMed]

116. Herrmann, K.; Engst, W.; Meinl, W.; Florian, S.; Cartus, A.T.; Schrenk, D.; Appel, K.E.; Nolden, T.; Himmelbauer, H.; Glatt, H. Formation of hepatic DNA adducts by methyleugenol in mouse models: Drastic decrease by Sult1a1 knockout and strong increase by transgenic human SULT1A1/2. *Carcinogenesis* **2014**, *35*, 935–941. [CrossRef] [PubMed]

117. Yang, S.; Wesseling, S.; Rietjens, I. Estragole DNA adduct accumulation in human liver HepaRG cells upon repeated in vitro exposure. *Toxicol. Lett.* **2021**, *337*, 1–6. [CrossRef] [PubMed]

118. Yang, X.; Feng, Y.; Zhang, Z.; Wang, H.; Li, W.; Wang, D.O.; Peng, Y.; Zheng, J. In Vitro and In Vivo Evidence for RNA Adduction Resulting from Metabolic Activation of Methyleugenol. *J. Agric. Food Chem.* **2020**, *68*, 15134–15141. [CrossRef]

119. Van den Berg, S.; Restani, P.; Boersma, M.; Delmulle, L.; Rietjens, I. Levels of Genotoxic and Carcinogenic Compounds in Plant Food Supplements and Associated Risk Assessment. *Food Nutr. Sci.* **2011**, *2*, 989–1010. [CrossRef]

120. Hasheminejad, G.; Caldwell, J. Genotoxicity of the alkenylbenzenes α- and β-asarone, myristicin and elemicin as determined by the UDS assay in cultured rat hepatocytes. *Food Chem. Toxicol.* **1994**, *32*, 223–231. [CrossRef]

121. Sekizawa, J.; Shibamoto, T. Genotoxicity of safrole-related chemicals in microbial test systems. *Mutat. Res.* **1982**, *101*, 127–140. [CrossRef]

122. NTP. Methyleugenol CAS No. 93-15-2. Report on Carcinogens. 2016; pp. 1–2. Available online: https://ntp.niehs.nih.gov/ntp/roc/content/profiles/methyleugenol.pdf (accessed on 13 May 2022).

123. Devereux, T.R.; Anna, C.H.; Foley, J.F.; White, C.M.; Sills, R.C.; Barrett, J.C. Mutation of beta-catenin is an early event in chemically induced mouse hepatocellular carcinogenesis. *Oncogene* **1999**, *18*, 4726–4733. [CrossRef]

124. Jin, M.; Kijima, A.; Hibi, D.; Ishii, Y.; Takasu, S.; Matsushita, K.; Kuroda, K.; Nohmi, T.; Nishikawa, A.; Umemura, T. In vivo genotoxicity of methyleugenol in gpt delta transgenic rats following medium-term exposure. *Toxicol. Sci.* **2013**, *131*, 387–394. [CrossRef]

125. Daimon, H.; Sawada, S.; Asakura, S.; Sagami, F. Inhibition of sulfotransferase affecting in vivo genotoxicity and DNA adducts induced by safrole in rat liver. *Teratog. Carcinog. Mutag.* **1997**, *17*, 327–337. [CrossRef]

126. Wiseman, R.W.; Miller, E.C.; Miller, J.A.; Liem, A. Structure-Activity Studies of the Hepatocarcinogenicities of Alkenylbenzene Derivatives Related to Estragole and Safrole on Administration to Preweanling Male C57BL/6JXC3H/HeJ F1Mice. *Cancer Res.* **1987**, *47*, 2275–2283. [CrossRef]

127. Daimon, H.; Sawada, S.; Asakura, S.; Sagami, F. In vivo genotoxicity and DNA adduct levels in the liver of rats treated with safrole. *Carcinogenesis* **1998**, *19*, 141–146. [CrossRef]

128. Howes, A.J.; Chan, V.S.; Caldwell, J. Structure-specificity of the genotoxicity of some naturally occurring alkenylbenzenes determined by the unscheduled DNA synthesis assay in rat hepatocytes. *Food Chem. Toxicol.* **1990**, *28*, 537–542. [CrossRef]

129. Martin, C.N.; McDermid, A.C.; Garner, R.C. Testing of known carcinogens and noncarcinogens for their ability to induce unscheduled DNA synthesis in HeLa cells. *Cancer Res.* **1978**, *38*, 2621–2627.

130. Zhou, G.D.; Moorthy, B.; Bi, J.; Donnelly, K.C.; Randerath, K. DNA adducts from alkoxyallylbenzene herb and spice constituents in cultured human (HepG2) cells. *Environ. Mol. Mutagen.* **2007**, *48*, 715–721. [CrossRef]

131. Kobets, T.; Cartus, A.T.; Fuhlbrueck, J.A.; Brengel, A.; Stegmuller, S.; Duan, J.D.; Brunnemann, K.D.; Williams, G.M. Assessment and characterization of DNA adducts produced by alkenylbenzenes in fetal turkey and chicken livers. *Food Chem. Toxicol.* **2019**, *129*, 424–433. [CrossRef]

132. Kobets, T.; Duan, J.D.; Brunnemann, K.D.; Etter, S.; Smith, B.; Williams, G.M. Structure-activity relationships for DNA damage by alkenylbenzenes in Turkey egg fetal liver. *Toxicol. Sci.* **2016**, *150*, 301–311. [CrossRef]

133. Kobets, T.; Iatropoulos, M.J.; Williams, G.M. Mechanisms of DNA-reactive and epigenetic chemical carcinogens: Applications to carcinogenicity testing and risk assessment. *Toxicol. Res.* **2019**, *8*, 123–145. [CrossRef]

134. Martins, C.; Doran, C.; Silva, I.C.; Miranda, C.; Rueff, J.; Rodrigues, A.S. Myristicin from nutmeg induces apoptosis via the mitochondrial pathway and down regulates genes of the DNA damage response pathways in human leukaemia K562 cells. *Chem. Biol. Interact.* **2014**, *218*, 1–9. [CrossRef]

135. Martins, C.M.d.S. *Study of the Natural Alkenylbenzenes Compounds: Mechanisms of DNA Lesions and Implications for Humana Health*; Universidade Nova de Lisboa: Lisboa, Portugal, 2016; pp. 1–247.

136. Martins, C.; Doran, C.; Laires, A.; Rueff, J.; Rodrigues, A.S. Genotoxic and apoptotic activities of the food flavourings myristicin and eugenol in AA8 and XRCC1 deficient EM9 cells. *Food Chem. Toxicol.* **2011**, *49*, 385–392. [CrossRef]

137. Baker, R.S.U.; Bonin, A.M. Tests with the Salmonella plate-incorporation assay. *Prog. Mutat. Res.* **1985**, *5*, 177–180.

138. Green, N.R.; Savage, J.R. Screening of safrole, eugenol, their ninhydrin positive metabolites and selected secondary amines for potential mutagenicity. *Mutat. Res.* **1978**, *57*, 115–121. [CrossRef]
139. Swanson, A.B.; Chambliss, D.D.; Blomquist, J.C.; Miller, E.C.; Miller, J.A. The mutagenicities of safrole, estragole, eugenol, trans-anethole, and some of their known or possible metabolites for Salmonella typhimurium mutants. *Mutat. Res. Fundam. Mol. Mech. Mutagen.* **1979**, *60*, 143–153. [CrossRef]
140. Wislocki, P.G.; Miller, E.C.; Miller, J.A.; Mccoy, E.C.; Rosenkranz, H.S. Carcinogenic and mutagenic activities of safrole, 1′-hydroxysafrole, and some known or possible metabolites. *Cancer Res.* **1977**, *37*, 1883–1891. [PubMed]
141. Poirier, L.A.; de Serres, F.J. Initial National Cancer Institute studies on mutagenesis as a prescreen for chemical carcinogens: An appraisal. *J. Natl. Cancer Inst.* **1979**, *62*, 919–926. [PubMed]
142. Purchase, I.F.; Longstaff, E.; Ashby, J.; Styles, J.A.; Anderson, D.; Lefevre, P.A.; Westwood, F.R. An evaluation of 6 short-term tests for detecting organic chemical carcinogens. *Br. J. Cancer* **1978**, *37*, 873–903. [CrossRef] [PubMed]
143. Ishidate, M, Jr. The in vitro chromosomal aberration test using Chinese hamster lung (CHL) fibroblast cells in culture. *Prog. Mutat. Res.* **1985**, *5*, 427–432.
144. Myhr, B.; Bowers, L.; Caspary, W. Assays for the induction of gene mutations at the thymidine kinase locus in L5178Y mouse lymphoma cells in culture. *Prog. Mutat. Res.* **1985**, *5*, 555–568.
145. Bradley, M. Measurement of DNA single-strand breaks by alkaline elution in rat hepatocytes. *Prog. Mutat. Res.* **1985**, *5*, 353–357.
146. Epstein, S.S.; Arnold, E.; Andrea, J.; Bass, W.; Bishop, Y. Detection of chemical mutagens by the dominant lethal assay in the mouse. *Toxicol Appl Pharm.* **1972**, *23*, 288–325. [CrossRef]
147. Gocke, E.; King, M.T.; Eckhardt, K.; Wild, D. Mutagenicity of cosmetics ingredients licensed by the European Communities. *Mutat. Res.* **1981**, *90*, 91–109. [CrossRef]
148. Mirsalis, J.C.; Tyson, C.K.; Butterworth, B.E. Detection of genotoxic carcinogens in the in vivo-in vitro hepatocyte DNA repair assay. *Env. Mutagen* **1982**, *4*, 553–562. [CrossRef]
149. Jin, M.; Kijima, A.; Suzuki, Y.; Hibi, D.; Inoue, T.; Ishii, Y.; Nohmi, T.; Nishikawa, A.; Ogawa, K.; Umemura, T. Comprehensive toxicity study of safrole using a medium-term animal model with gpt delta rats. *Toxicology* **2011**, *290*, 312–321. [CrossRef]
150. Matsushita, K.; Kijima, A.; Ishii, Y.; Takasu, S.; Jin, M.; Kuroda, K.; Kawaguchi, H.; Miyoshi, N.; Nohmi, T.; Ogawa, K.; et al. Development of a medium-term animal model using gpt delta rats to evaluate chemical carcinogenicity and genotoxicity. *J. Toxicol. Pathol.* **2013**, *26*, 19–27. [CrossRef]
151. Damhoeri, A.; Hosono, A.; Itoh, T.; Matsuyama, A. In vitro mutagenicity tests on capsicum pepper, shallot and nutmeg oleoresins. *Agric. Biol. Chem.* **1985**, *49*, 1519–1520. [CrossRef]
152. NTP. NTP Technical Report on the Toxicity Studies of Myristicin (CASRN 607-91-0) Administered by Gavage to F344/NTac Rats and B6C3F1/N Mice. *Natl. Toxicol. Program Tech. Rep. Ser.* **2019**, *95*, 1–65.
153. Nohmi, T. Past, Present and Future Directions of gpt delta Rodent Gene Mutation Assays. *Food Saf.* **2016**, *4*, 1–13. [CrossRef]
154. Feuer, G. Action of pregnancy and various progesterones on hepatic microsomal activites. *Drug Metab. Rev.* **1979**, *9*, 147–169. [CrossRef]
155. Vesselinovitch, S.D.; Rao, K.V.; Mihailovich, N. Transplacental and Lactational Carcinogenesis by Safrole. *Cancer Res.* **1979**, *39*, 4378–4380.
156. Rice, J.M. Exposure to chemical carcinogens during pregnancy: Consequences for mother and conceptus. *Adv. Exp. Med. Biol.* **1984**, *176*, 13–49. [CrossRef]
157. Lu, L.J.W.; Disher, R.M.; Randerath, K. Differences in the covalent binding of benzo[a]pyrene, safrole, 1′-hydroxysafrole, and 4-aminobiphenyl to DNA of pregnant and non-pregnant mice. *Cancer Lett.* **1986**, *31*, 43–52. [CrossRef]
158. Vesselinovitch, S.D.; Koka, M.; Rao, K.V.; Mihailovich, N.; Rice, J.M. Prenatal multicarcinogenesis by ethylnitrosourea in mice. *Cancer Res.* **1977**, *37*, 1822–1828.
159. Vesselinovitch, S.D.; Rao, K.V.; Mihailovich, N.; Rice, J.M.; Lombard, L.S. Development of broad spectrum of tumors by ethylnitrosourea in mice and the modifying role of age, sex, and strain. *Cancer Res.* **1974**, *34*, 2530–2538.
160. Giglia-Mari, G.; Zotter, A.; Vermeulen, W. DNA damage response. *Cold Spring Harb. Perspect. Biol.* **2011**, *3*, a000745. [CrossRef]
161. Rodrigues, A.S.; Gomes, B.C.; Martins, C.; Gromicho, M.; Oliveira, N.G.; Guerreiro, P.S.; Rueff, J. DNA repair and resistance to cancer therapy. In *New Research Directions in DNA Repair*; IntechOpen: London, UK, 2013. [CrossRef]
162. Long, E.; Nelson, A.; Fitzhugh, O.; Hansen, W. Liver tumours produced in rats by feeding safrole. *Arch. Pathol.* **1963**, *75*, 595–604.
163. Borchert, P.; Miller, J.A.; Miller, E.C.; Shires, T.K. 1′-Hydroxysafrole, a Proximate Carcinogenic Metabolite of Safrole in the Rat and Mouse. *Cancer Res.* **1973**, *33*, 590–600.
164. Borchert, P.; Wislocki, P.G.; Miller, J.A.; Miller, E.C. The Metabolism of the Naturally Occurring Hepatocarcinogen Safrole to 1′-Hydroxysafrole and the Electrophilic Reactivity of 1′-Acetoxysafrole. *Cancer Res.* **1973**, *33*, 575–589. [PubMed]
165. Drinkwater, N.R.; Miller, E.C.; Miller, J.A.; Pitot, H.C. Hepatocarcinogenicity of estragole (1-allyl-4-methoxybenzene) and 1′-hydroxyestragole in the mouse and mutagenicity of 1′-acetoxyestragole in bacteria. *J. Natl. Cancer Inst.* **1976**, *57*, 1323–1331. [CrossRef] [PubMed]
166. Miller, E.C.; Swanson, A.B.; Phillips, D.H.; Liem, A.; Miller, J.A.; Fletcher, T.L. Structure-activity studies of the carcinogenicities in the mouse and rat of some naturally occurring and synthetic alkenylbenzene derivatives related to safrole and estragole. *Cancer Res.* **1983**, *43*, 1124–1134. [PubMed]

167. Johnson, J.D.; Ryan, M.J.; Toft Ii, J.D.; Graves, S.W.; Hejtmancik, M.R.; Cunningham, M.L.; Herbert, R.; Abdo, K.M. Two-year toxicity and carcinogenicity study of methyleugenol in F344/N rats and B6C3F1 mice. *J. Agric. Food Chem.* **2000**, *48*, 3620–3632. [CrossRef]

168. NTP. Safrole-CAS No. 94-59-7. Report on Carcinogens; 2016; pp. 1–2. Available online: https://ntp.niehs.nih.gov/ntp/roc/content/profiles/safrole.pdf (accessed on 13 May 2022).

169. Miller, J.A.; Miller, E.C. The metabolic activation and nucleic acid adducts of naturally-occurring carcinogens: Recent results with ethyl carbamate and the spice flavors safrole and estragole. *Br. J. Cancer* **1983**, *48*, 1–15. [CrossRef]

170. Auerbach, S.S.; Shah, R.R.; Mav, D.; Smith, C.S.; Walker, N.J.; Vallant, M.K.; Boorman, G.A.; Irwin, R.D. Predicting the hepatocarcinogenic potential of alkenylbenzene flavoring agents using toxicogenomics and machine learning. *Toxicol. Appl. Pharm.* **2010**, *243*, 300–314. [CrossRef]

171. Glück, J.; Buhrke, T.; Frenzel, F.; Braeuning, A.; Lampen, A. In silico genotoxity and carcinogenicity prediction for food-relevant secondary plant metabolites. *Food Chem. Toxicol.* **2018**, *116*, 298–306. [CrossRef]

172. Beroza, M.; Inscoe, M.N.; Schwartz Jr, P.H.; Keplinger, M.L.; Mastri, C.W. Acute toxicity studies with insect attractants. *Toxicol. Appl. Pharm.* **1975**, *31*, 421–429. [CrossRef]

173. Singleton, V.L.; Kratzer, F.H. Toxicity and related physiological activity of phenolic substances of plant origin. *J. Agric. Food Chem.* **1969**, *17*, 497–512. [CrossRef]

174. Liu, G.X.; Xu, F.; Shang, M.Y.; Wang, X.; Cai, S.Q. The Relative Content and Distribution of Absorbed Volatile Organic Compounds in Rats Administered Asari Radix et Rhizoma Are Different between Powder- and Decoction-Treated Groups. *Molecules* **2020**, *25*, 4441. [CrossRef]

175. Taylor, J.; Jones, W.; Hagan, E.; Gross, M.; Davis, D.; Cook, E. Toxicity of oil of calamus (Jammu variety). *Toxicol. Appl. Pharm.* **1967**, *10*, 405.

176. Buchanan, R.L. Toxicity of spices containing methylenedioxybenzene derivatives: A review. *J. Food Saf.* **1978**, *1*, 275–293. [CrossRef]

177. Von Oettingen, W.F. *Phenol and Its Derivatives: The Relation between Their Chemical Constitution and Their Effect on the Organism*; US Government Printing Office: Washington, DC, USA, 1949.

178. Truitt, E.B., Jr.; Callaway, E., III; Braude, M.C.; Krantz, J.C., Jr. The pharmacology of myristicin. A contribution to the psychopharmacology of nutmeg. *J. Neuropsychiatr.* **1961**, *2*, 205–210.

179. Abdo, K.M.; Cunningham, M.L.; Snell, M.L.; Herbert, R.A.; Travlos, G.S.; Eldridge, S.R.; Bucher, J.R. 14-Week toxicity and cell proliferation of methyleugenol administered by gavage to F344 rats and B6C3F1 mice. *Food Chem. Toxicol.* **2001**, *39*, 303–316. [CrossRef]

180. Hagan, E.C.; Jenner, P.M.; Jones, W.I.; Fitzhugh, O.G.; Long, E.L.; Brouwer, J.G.; Webb, W.K. Toxic Properties of Compounds Related to Safrole. *Toxicol. Appl. Pharm.* **1965**, *7*, 18–24. [CrossRef]

181. Chiu, S.; Wang, T.; Belski, M.; Abourashed, E.A. HPLC-guided isolation, purification and characterization of phenylpropanoid and phenolic constituents of nutmeg kernel (*Myristica fragrans*). *Nat. Pro. Comm.* **2016**, *11*, 483–488. [CrossRef]

182. Dang, H.N.P.; Quirino, J.P. High Performance Liquid Chromatography versus Stacking-Micellar Electrokinetic Chromatography for the Determination of Potentially Toxic Alkenylbenzenes in Food Flavouring Ingredients. *Molecules* **2021**, *27*, 13. [CrossRef]

183. Dang, H.N.P.; Quirino, J.P. Analytical Separation of Carcinogenic and Genotoxic Alkenylbenzenes in Foods and Related Products (2010–2020). *Toxins* **2021**, *13*, 387. [CrossRef]

184. Rivera-Pérez, A.; Romero-González, R.; Garrido Frenich, A. Determination and Occurrence of Alkenylbenzenes, Pyrrolizidine and Tropane Alkaloids in Spices, Herbs, Teas, and Other Plant-derived Food Products Using Chromatographic Methods: Review from 2010–2020. *Food Rev. Int.* **2021**, 1–27. [CrossRef]

185. OECD. *Test No. 471: Bacterial Reverse Mutation Test*; OECD Publishing: Paris, France, 2020. [CrossRef]

186. Honda, H.; Minegawa, K.; Fujita, Y.; Yamaguchi, N.; Oguma, Y.; Glatt, H.; Nishiyama, N.; Kasamatsu, T. Modified Ames test using a strain expressing human sulfotransferase 1C2 to assess the mutagenicity of methyleugenol. *Genes Env.* **2016**, *38*, 1. [CrossRef]

187. Monien, B.H.; Herrmann, K.; Florian, S.; Glatt, H. Metabolic activation of furfuryl alcohol: Formation of 2-methylfuranyl DNA adducts in Salmonella typhimurium strains expressing human sulfotransferase 1A1 and in FVB/N mice. *Carcinogenesis* **2011**, *32*, 1533–1539. [CrossRef] [PubMed]

188. OECD. *Test No. 476: In Vitro Mammalian Cell Gene Mutation Tests Using the Hprt and Xprt Genes*; OECD Publishing: Paris, France, 2016. [CrossRef]

189. Dobrovolsky, V.N.; Heflich, R.H. Chapter Eleven—Detecting Mutations Vivo. In *Mutagenicity: Assays and Applications*; Kumar, A., Dobrovolsky, V.N., Dhawan, A., Shanker, R., Eds.; Academic Press: Cambridge, MA, USA, 2018; pp. 229–249. [CrossRef]

190. OECD. *Test No. 488: Transgenic Rodent Somatic and Germ Cell Gene Mutation Assays*; OECD Publishing: Paris, France, 2011.

191. Lambert, I.B.; Singer, T.M.; Boucher, S.E.; Douglas, G.R. Detailed review of transgenic rodent mutation assays. *Mutat. Res.* **2005**, *590*, 1–280. [CrossRef] [PubMed]

192. Nohmi, T.; Katoh, M.; Suzuki, H.; Matsui, M.; Yamada, M.; Watanabe, M.; Suzuki, M.; Horiya, N.; Ueda, O.; Shibuya, T.; et al. A new transgenic mouse mutagenesis test system using Spi- and 6-thioguanine selections. *Env. Mol. Mutagen.* **1996**, *28*, 465–470. [CrossRef]

193. Nohmi, T.; Masumura, K.; Toyoda-Hokaiwado, N. Transgenic rat models for mutagenesis and carcinogenesis. *Genes Env.* **2017**, *39*, 11. [CrossRef]

194. Suzuki, Y.; Umemura, T.; Hibi, D.; Inoue, T.; Jin, M.; Ishii, Y.; Sakai, H.; Nohmi, T.; Yanai, T.; Nishikawa, A.; et al. Possible involvement of genotoxic mechanisms in estragole-induced hepatocarcinogenesis in rats. *Arch. Toxicol.* **2012**, *86*, 1593–1601. [CrossRef]
195. SCF. Opinion of the Scientific Committee on Food on Estragole (1-Allyl-4-methoxybenzene). In *SCF/CS/FLAV/FLAVOUR/6 ADD2 FINAL*; European Commission: Brussels, Belgium, 2001; pp. 1–10.

 *foods*

Review

Safety Aspects of the Use of Isolated Piperine Ingested as a Bolus

Rainer Ziegenhagen, Katharina Heimberg, Alfonso Lampen and Karen Ildico Hirsch-Ernst *

German Federal Institute for Risk Assessment (BfR), Max-Dohrn-Str. 8-10, 10589 Berlin, Germany; rainer.ziegenhagen@bfr.bund.de (R.Z.); Katharina.Heimberg@bfr.bund.de (K.H.); Alfonso.Lampen@bfr.bund.de (A.L.)
* Correspondence: Karen-Ildico.Hirsch-Ernst@bfr.bund.de

Abstract: Piperine is a natural ingredient of *Piper nigrum* (black pepper) and some other *Piper* species. Compared to the use of pepper for food seasoning, piperine is used in food supplements in an isolated, concentrated form and ingested as a bolus. The present review focuses on the assessment of the possible critical health effects regarding the use of isolated piperine as a single ingredient in food supplements. In human and animal studies with single or short-term bolus application of isolated piperine, interactions with several drugs, in most cases resulting in increased drug bioavailability, were observed. Depending on the drug and extent of the interaction, such interactions may carry the risk of unintended deleteriously increased or adverse drug effects. Animal studies with higher daily piperine bolus doses than in human interaction studies provide indications of disturbance of spermatogenesis and of maternal reproductive and embryotoxic effects. Although the available human studies rarely reported effects that were regarded as being adverse, their suitability for detailed risk assessment is limited due to an insufficient focus on safety parameters apart from drug interactions, as well as due to the lack of investigation of the potentially adverse effects observed in animal studies and/or combined administration of piperine with other substances. Taken together, it appears advisable to consider the potential health risks related to intake of isolated piperine in bolus form, e.g., when using certain food supplements.

Keywords: piperine; food safety; drug interaction; reproductive toxicity; bolus administration

Citation: Ziegenhagen, R.; Heimberg, K.; Lampen, A.; Hirsch-Ernst, K.I. Safety Aspects of the Use of Isolated Piperine Ingested as a Bolus. *Foods* **2021**, *10*, 2121. https://doi.org/10.3390/foods10092121

Academic Editor: Dirk W. Lachenmeier

Received: 13 August 2021
Accepted: 6 September 2021
Published: 8 September 2021

Publisher's Note: MDPI stays neutral with regard to jurisdictional claims in published maps and institutional affiliations.

1. Introduction

The alkaloid piperine ((E,E)-piperine; IUPAC-name: (2E,4E)-5-(2H-1,3-benzodioxol-5-yl)-1-(piperidin-1-yl)penta-2,4-dien-1-one; CAS-No.: 94-62-2; FEMA-No: 2909; molecular formula: $C_{17}H_{19}NO_3$; molecular weight: 285.34 g/mol) is a natural ingredient of *Piper nigrum*, Piper longum and some other *Piper* species, as well as of *Aframomum melegueta* K. Schum. (Grains of Paradise) [1–7]. The alkaloid is the main compound imparting the pungent flavour to fruits of *Piper nigrum* and *Piper longum*. *Piper nigrum* fruits are used to produce black, white and green pepper. Black peppercorns are produced from whole, dried, full-grown, not yet fully ripe fruits, while white peppercorns are produced from dried, ripe fruits after removal of the outer layer [8,9]. Green peppercorns are obtained from unripe fruits subjected to processing methods by which the green colour is maintained.

The occurrence of piperine in the European/Western diet primarily results from use of pepper for food seasoning, but also from the use of the substance in isolated form for spicing/flavouring purposes, e.g., in beverages and spirits [2]. The substance can occur in four stereoisomeric forms: (E,E)-piperine (= piperine), (Z,E)-piperine (= isopiperine), (E,Z)-piperine (= isochavicine) and (Z,Z)-piperine (= chavicine). In black and white pepper, (E,E)-piperine constitutes by far the main and most pungent isomer. The other three isomers seem to be formed primarily via light-induced or enzymatic isomerization [10].

Primarily in in vitro and in animal studies, as well as in some human studies, piperine has been shown to be a biologically versatile compound that can interact with a variety of chemically and functionally diverse biomolecular targets, such as enzymes, membrane transporters, receptors or other biomolecules. For example, piperine may provide protection against forms of oxidative damage and improve the activities of compromised anti-oxidative defence mechanisms (e.g., related to superoxide dismutase or catalase), but depending on the study settings, may also decrease the anti-oxidative defence mechanisms and among others, piperine further displays the potential to influence the activity of drug-metabolizing enzymes, including enzymes involved in phase I (cytochrome P-450-enzymes) and phase II metabolism (e.g., UDP-glucuronosyltransferases), to interact with cellular drug transporters (e.g., P-glycoprotein) or to modulate the cellular targets (monoamine oxidase) associated with neurodegenerative diseases [1,11–18].

Currently, attention is largely focused on the potential of piperine to influence the bioavailability of certain drugs via interaction with drug-metabolizing enzymes and/or inhibition of drug transporters or efflux pumps, thereby in many cases increasing the drug bioavailability and efficacy. In addition, inclusion of piperine into drug-loaded nanoparticles or lipospheres is being investigated as a means of increasing the effectiveness of advanced drug delivery systems [19,20].

In food supplements, piperine (primarily in the form of highly piperine-enriched pepper extracts, frequently with a piperine content in the range of $\geq 95\%$) is often used and promoted, among others, as a bio-enhancer to increase the bioavailability of other ingredients contained in these food supplements. Based on the use of piperine in various food supplements, its multi-facetted biological activities, and on the differences regarding the pattern of piperine intake when comparing the use of food supplements to the use of pepper for food seasoning, a closer look into the safety aspects of the use of piperine as an ingredient of food supplements appears to be warranted.

This review focuses on the possible critical health effects regarding the use of isolated piperine as a single ingredient in food supplements (i.e., without the addition of other bioactive substances). The use of isolated piperine as a flavouring agent is outside the scope of this review. In the context of the present review, the focus was laid on adult persons; thus, children and adolescents were not considered. To this end, a literature search was performed in the scientific databases Pubmed and Embase, with the last update performed in February 2021. To initially retrieve a broad spectrum of references, the search term "piperine" was used, without combination with other search terms. To further identify the relevant scientific publications that are within the scope of the present review, the abstracts of the retrieved references were screened to facilitate the selection of a subset of publications that were subsequently subjected to further scrutiny of the full texts. In addition, reference lists of the identified relevant publications as well as websites of acknowledged scientific bodies or national authorities were checked.

2. Occurrence and Exposure

2.1. Occurrence

Piper nigrum is the main source of piperine in European/Western cuisine. Other potential sources are foods flavoured with piperine in isolated form or foods flavoured with other *Piper* species (e.g., *Piper longum* and *Piper retrofractum* Vahl)or with the spice "grains of paradise" (*Aframomum melegueta*) [2]. Regarding the piperine content of black pepper, ranges of 2–7% [21], 2–9% [13] or 4–6% with contents up to 10% [2] have been reported. For *Piper longum*, piperine contents of 1.2–5% [13,22–24], and for *Piper retrofractum* Vahl contents of 3.1–4.5% have been indicated [13,23]. In an investigation of four commercial brands of pure ground black pepper with high piperine contents (10–11%), E,E-piperine was the most abundant ($\geq$99% of detected piperine isomers) and Z,Z-piperine (= chavicine) the least abundant ($\leq$0.07%) piperine isomere [10].

During storage of ground black, white and green pepper at 4 °C for 6 months, a decrease in piperine content of about 12–30% was observed [25]. Different findings were made regarding heat treatment of pepper, ranging from mild piperine losses of 4–12.5% during cooking in an open pan (30 min) or pressure cooking (20 min) [26] to losses of approximately 28% during cooking (20 min) or about 34% during pressure cooking (10 min) [27].

2.2. Exposure

In 2007, by extrapolation from limited consumption data, the Australian Therapeutic Goods Administration (TGA) estimated that the piperine intake in New Zealand was about 25 mg per person per day, in the USA approximately 60 mg and in India approximately 120 mg per person per day [28]. According to another source, which was based on annual US import data of black pepper, with an estimated average per capita intake of approximately 0.7 g pepper/day, a corresponding per capita piperine intake of 14–54 mg/day was calculated for the US population [29].

In an exposure estimation performed by the German Federal Institute for Risk Assessment (BfR) in 2018, which was based on food consumption data from the National Consumption Survey II of the Max Rubner-Institute (2008) [30], including approximately 20,000 individuals, a mean per capita pepper intake of the male German population (14–80 years) of 0.6 g/day was estimated, with an estimated per capita intake at the 95th intake percentile of 1.6 g pepper/day. Assuming an average piperine content in pepper of 4–6%, this would correspond to an estimated mean per capita intake by the male population of 24–36 mg piperine/day and an estimated intake at the 95th percentile of 64–96 mg piperine/day. Regarding this estimation, it is noted on the one hand that information on the consumption of herbs/spices is generally subject to greater uncertainty as their consumption is often not documented, and an underestimation of the amount consumed can therefore be assumed. On the other hand, it should be borne in mind that this intake estimation does not take into account possible piperine losses caused by storage or food preparation.

In India, a consumption survey conducted from December 2006 to July 2008 in three regions recorded the median monthly per capita intakes of black pepper of 3–18.5 g (0.1–0.62 g/day) and in the 90th percentile of 16.7–41.7 g (0.56–1.39 g/day) [31]

Taken together, the available estimations of daily piperine intake resulting from the use of pepper in food preparation are afflicted with considerable scientific uncertainty. It should also be kept in mind that when using pepper for food seasoning, the piperine intake occurs in conjunction with all other pepper constituents and with different degrees of comminution of the peppercorn, potentially bringing about matrix effects influencing the bioavailability or pharmacodynamic effects of piperine ingested in this way, which may differ from the intake of piperine as an isolated substance.

In its assessment of isolated piperine and several aliphatic and arylalkyl amines and amides as flavouring agents, the European Food Safety Authority EFSA (2015) reported an estimated European per capita intake of 6.2 µg piperine/day for the use of isolated piperine as a flavouring substance based on the EU Maximised Survey-derived Daily Intake (MSDI) method (see also below). However, EFSA noted in this assessment that the use levels were needed for some of the abovementioned flavouring substances, including piperine, to calculate the Modified Theoretical Added Maximum Daily Intakes (mTAMDIs) in order to identify those flavouring substances that required a more refined exposure assessment and to finalise the evaluation [32].

In food supplements, piperine is usually used in combination with other ingredients to increase their bioavailability, and commonly its addition occurs via highly piperine-enriched black pepper extracts (piperine content frequently in the range of ≥ 95%). Therefore, black pepper extract is often mentioned on the ingredient list of food supplements and the piperine content is only indicated in second place. The piperine content of food supplements is frequently in the range of 5–30 mg per daily dose, with single products reaching dosages of 40 or up to about 50–100 mg per daily dose, but the market may be subject to change.

The piperine content of the highly piperine-enriched black pepper extracts (frequently in the range of ≥95%) is very similar to the piperine content of chemically defined piperine used as an isolated flavouring substance (piperine content ≥97% [33]) or to the piperine content of the substance used in scientific investigations, for which it was procured as a chemical from chemical companies (usually ≥97%).

3. Kinetics and Metabolism

In animal studies conducted by Bhat and Chandrasekhara [34,35] and Suresh and Srinivasan [36], with rats receiving an oral dose of 170 mg piperine/kg body weight (bw), only about 3–4% of the dose was detected in faeces in unchanged form over a period of 4 or 5 days, respectively, and it was concluded that 96–97% of the administered piperine dosage was absorbed [34,36]. In an accompanying investigation with everted sacs of rat intestines, only piperine was detected in serosal fluid and intestinal tissue, which led to the conclusion that piperine did not undergo any metabolic change during absorption [34]. However, in both the abovementioned animal studies, only small portions of the administered oral dose could be detected in serum and investigated tissues. In the more recent study of Suresh and Srinivasan (2010), maximum levels were reached 6 h after oral administration of a piperine dose of 170 mg/kg bw, with approximately 38.8 μmol piperine/L in serum and 0.39% of the administered piperine dose in liver, 0.37% in kidney and about 9.7% in the flushed intestine [36]. In both studies, no piperine was detectable in urine [34,36], but Bhat and Chandrasekhara detected piperine metabolites, i.e., piperonylic acid, piperonyl alcohol, piperonal and vanillic acid, and their conjugates, in urine, which in their free forms represented about 15.5% of the administered dose (measured within 96 h after piperine administration) [35]. The latter authors assumed that most of the administered piperine was absorbed and that it was not transformed during intestinal absorption but was probably later metabolized rapidly by other tissues [34].

In a more recent study in rats, the bioavailability of an oral dose of 3.5 mg piperine/kg bw was calculated to be about 25% by comparing plasma AUC values following oral and i.v. administration [37]. Regarding piperine metabolites, Gao et al. (2017) identified 12 metabolites in rat plasma, bile, urine and faeces, with 10 piperine metabolites occurring both in plasma and urine. The metabolites were grouped into metabolites resulting from methylenedioxycyclic ring-opening, from methylenedioxycyclic ring-oxidation and from piperidine ring-cleavage [38]. Shang et al. (2017) even detected and tentatively characterized 148 piperine metabolites in rat plasma, urine and faeces after oral administration of 250 mg piperine/kg bw. Piperine mainly underwent hydrogenation, dehydrogenation, hydroxylation, glucuronide conjugation, sulphate conjugation, ring cleavage and their composite reactions. However, information on plasma or urine levels of the detected piperine metabolites is not available from this study [39]. In laying hens receiving piperine–enriched feed (80 mg/kg feed), significant proportions of piperine isomers were observed in egg yolks (3.0 μg piperine, 0.7 μg chavicine, 2.9 μg isopiperine and 5.3 μg isochavicine per g egg yolk), indicating that piperine metabolism can also comprise substance isomerization [40].

Information on piperine serum or plasma levels observed in rats after oral administration is not uniform. With oral doses of 3.5, 20, 35 or 250 mg piperine/kg bw, corresponding plasma C_{max} values of approximately 0.45, 3.4, 5.4–6.0 or 12.7 μmol piperine/L were observed in different studies [37,41–43]. However, other studies observed higher C_{max} values with approximately 9.9 μmol/L after an oral dose of 20 mg piperine/kg bw [44] or levels of approximately 28–39 μmol/L after a dose of 170 mg piperine/kg bw [34,36]. Plasma protein binding was about 98% in rats receiving an oral dose of 35 mg piperine/kg bw [42]. Furthermore, piperine was shown to efficiently penetrate and homogeneously distribute into the brain of rats after oral piperine doses (35 mg/kg bw), leading to comparable $AUC_{(0-\infty)}$-values in brain and plasma with a brain–plasma AUC ratio of 0.95. However, based on the $AUC_{(0-\infty)}$ values, the piperine level in cerebrospinal fluid was around 50 times lower than in brain or plasma [42].

In humans, information on kinetics and metabolism of oral piperine doses are sparse. In an investigation with two individuals receiving a single oral dose of 50 mg piperine (approximately 0.71–0.83 mg/kg bw, assuming a body weight of 60–70 kg), the plasma peak concentrations reached 2.7–3.3 μmol/L (T_{max} = 1–3 h) [45].

In human urine, the piperine metabolites 5-(3-4-dihydroxphenyl)valeric acid piperidide (which was excreted as sulphate) and its derivate hydroxylated in position 4 of the piperidine ring, 5-(3-4-dihydroxphenyl)valeric acid-4-hydroxypiperidide, were observed one or two days after oral administration of piperine (25 mg) or a high dose of pepper, respectively. Interestingly, these two urine metabolites could not be detected in 2 out of 14 investigated individuals who instead excreted 5-(3-4-dihydroxphenyl)-2-4-pentadienoic acid piperidide, providing first indications for individual differences in human piperine metabolism [46]. In rat urine, all three metabolites could be detected [38]. In an in vitro study comparing the hepatic piperine metabolism in mouse, rat, dog and human hepatocytes, the predominant metabolic pathways included formation of a catechol derivate for all species; however, the metabolic pathways displayed species-specific differences in terms of types and quantities of metabolites [47].

4. Safety Aspects

4.1. Information Based on Evaluations by Scientific Bodies and National Authorities

The European Food Safety Authority (EFSA) has evaluated the use of piperine as a flavouring substance. In its evaluations, EFSA (2008; 2011; 2015) disagreed with a No Observed Effect Level (NOEL) of 20 mg piperine/kg bw/day that had previously been identified by the Joint FAO/WHO Expert Committee on Food Additives (JECFA) (2006), due to the shortcomings of the underlying animal study (lack of histopathology, study duration) and in 2015 identified a No Observed Adverse Effect Level (NOAEL) of 5 mg piperine/kg bw/day, based on a newly available 90-day rat feeding study performed according to OECD guideline 408 (endpoint: dose-dependent increase in cholesterol level in male animals) [29,32,48–50] (see (2) in Section 4.2.2). In its final conclusion, EFSA (2015) agreed with the JECFA (2006) conclusion "no safety concern at estimated levels of intake as flavouring substance" based on the MSDI approach (estimated European per capita intake by the MSDI approach: 6.2 μg piperine/day) [32,50]. Currently, isolated piperine is approved as a flavouring agent in the European Union with no restrictions on use or maximum levels set in regulation (EC) No. 1334/2008. However, its use level may be self-limiting due to the pungent taste of piperine.

In 2007, the Australian Complementary Medicines Evaluation Committee (CMEC) evaluated the use of piperine as a component in herbal preparations for use in listed medicines. Due to the possible effects on the bioavailability of medicinal products (leading to increased bioavailability in most cases, see Section 4.2.4) and the risk of inadvertent interactions with medicinal products, the committee recommended a maximum daily dose limit of 10 mg/day for piperine (based on a person's body weight of 50 kg) when present as a component in herbal preparations for use in listed medicines [28].

The Canadian authority Health Canada (2019) has elaborated a monograph on the use of *Piper nigrum* (black pepper) as an ingredient in *Natural Health Products*, which also includes piperine isolated from the fruits of *Piper nigrum*. For adults (≥18 years), a daily dose of 250–420 mg for the unextracted powder of *Piper nigrum* fruits and a daily maximum dose of 14 mg for the use of piperine as an isolated substance in these products were established. For these products, a label statement is required that persons taking other medicines or natural health products should consult a healthcare practitioner/provider/professional or physician before use, as black pepper/piperine may alter their effectiveness. The same applies to pregnant or breastfeeding women. The monograph does not list any contraindications or known adverse reactions [51].

None of the evaluations described above mentioned the paternal reproductive toxicological effects observed in some animal studies (see (3) in Section 4.2.2).

In 2016, on request of the Norwegian Food Safety Authority, the Norwegian Scientific Committee for Food Safety (VKM) carried out a risk assessment of a daily dose of 1.5 mg piperine in food supplements. The panel applied the Margin of Exposure (MOE) approach in its assessment and used the NOAEL of 5 mg/kg bw/day identified by EFSA (2015), which was based on an animal study (endpoint: dose-dependent increase in cholesterol level in male animals) as the starting point for the MOE calculation. As the margin of exposure for all age groups considered was greater than 100, the panel concluded that this intake was unlikely to produce adverse effects in individuals aged 10 years or older [2].

4.2. Potential Hazards

4.2.1. Genotoxicity

In its assessment of piperine as a flavouring agent, JECFA (2006) concluded regarding genotoxicity that piperine belongs to a group of aliphatic and aromatic amine and amide derivates for which negative results were reported in bacterial assays for reverse mutation and that piperine consistently gave negative results in a variety of in vivo studies [50,52–54]. EFSA agreed in its assessment with JECFA that the available studies on genotoxicity did not preclude the evaluation of piperine (and some other aliphatic and arylalkyl amines and amides) as a flavouring agent [32,48,49].

In a more recent study, piperine displayed negative results in an in vitro micronucleus test with Chinese hamster ovary cells in the presence or absence of metabolic activation and caused no increase in the numbers of micronucleated polychromatic erythrocytes in an in vivo micronucleus test in mice with all the tested doses (highest tested dose 574 mg/kg bw), leading to the conclusion that in this study piperine was not genotoxic [55].

4.2.2. Animal Studies

(1) Acute and Subacute Toxicity

Piyachaturawat et al. investigated the acute toxicity of piperine (dissolved in equal volumes of DMSO and 95% ethanol) following a single administration via intragastric (i.g.) gavage to mice and rats [56]. For adult male mice, the resulting calculated LD_{50} value was 330 mg/kg bw, compared to an LD_{50} of 15.1 mg/kg bw based on i.v. administration. In the same study, an LD_{50} value of 514 mg/kg bw was derived for single i.g. administration of piperine to adult female rats, while a higher LD_{50} value was calculated for young female weanling rats (LD_{50} > 585 mg/kg bw). Animals receiving lethal doses experienced convulsion and died of respiratory paralysis [56].

In another study with mice, all animals survived oral doses of 143 to 574 mg/kg/day given on two consecutive days, but displayed lethargy ranging from a slight degree in the low-dose group to a severe one in the high-dose group [55].

In a subacute oral toxicity study with female rats receiving 0, 100, 250, 350 or 500 mg piperine/kg bw/day for seven days, the body weight gain of animals receiving 100 mg/kg bw was comparable to that of the control group. Daily doses of 250 mg/kg bw led to reduced body weight gain and caused haemorrhage in the stomach of 3 out of 8 animals, whereas the higher doses of 350 and 500 mg/kg bw caused the death of 2 and 5 out of 8 animals, respectively. Animals receiving 500 mg piperine/kg bw/day displayed histopathological changes of different types and degrees in the stomach, urinary bladder, adrenal glands and small intestine. In addition, luteal cells in the central portion of the corpora lutea were degenerated in this dosage group [56].

(2) Subchronic Toxicity Studies

In a study with groups of young male rats that received 100 mg piperine/kg feed for 56 days or 110, 220 or 440 mg pepper oleoresin/kg feed (equivalent to 50–200 mg piperine/kg feed), or 2 g pepper/kg feed, no adverse effects on growth, food efficacy, organ weights, blood count and investigated clinical chemistry parameters were observed compared to the control group [50,57]. In its evaluation of piperine as a flavouring agent, JECFA (2006) based its NOEL for piperine of 20 mg/kg bw/day on this animal study, but

EFSA (2008) considered this study inadequate for NOAEL identification due to study-design limitations [48,50]. It should be mentioned that this animal study reported increased haemoglobin values for the piperine group (242 g/L) compared to the control and to the pepper/pepper oleoresin groups (~140 g/L); however, the red blood cell count was similar to the other groups, making a typing error highly possible [57].

The subchronic toxicity study (90-day study) in rats that was finalised in 2013 and used by EFSA (2015) [32] for the assessment of piperine as a flavouring agent has been published meanwhile (Bastaki et al. (2018) [29]). In this study, 0, 5, 15 or 50 mg piperine/kg bw/day were administered via feed for 90 days. According to the EFSA assessment, the reduced weight gain observed in the highest male dose group was due to reduced feed intake (possibly related to food palatability). There were no mortalities, no gross and microscopic changes nor clinical pathology or organ weight changes attributable to piperine. Some statistically significant changes in haematology, coagulation or clinical chemistry parameters were considered by EFSA as not dose-dependent, small in magnitude and within the range of historical values. However, statistically significant dose-dependent increases in cholesterol levels were observed in male animals receiving 15 and 50 mg piperine/kg bw/day (approximately by 30 and 55%, respectively), which was used by EFSA for NOAEL identification.

According to EFSA, reduced relative epididymides weights were observed in male animals administered 5 and 50 mg/kg bw/day, respectively, but these changes were considered small and not dose-dependent and therefore of limited toxicological relevance [32]. Regarding this finding, it is noted on the one hand that reduced relative (to-brain weight) epididymis weights (17–23%) were observed in all three piperine dosage groups without displaying dose dependence, with reductions being statistically significant only in the 5 and 50 mg/kg bw dosage groups. On the other hand, it must be mentioned that these changes were without histopathological findings. From this study, EFSA identified a NOAEL of 5 mg piperine/kg bw/day due to the dose-dependent elevated cholesterol plasma levels in male animals at the mid and high dose (15 and 50 mg/kg bw/day) [32].

Contrary to the EFSA (2015) assessment, Bastaki et al. (2018) ascribed no toxicological relevance to the observed increases in cholesterol levels described above because, in their opinion, the cholesterol levels were within the historical control range for male animals and because of corroborating evidence from other studies showing an absence of a cholesterol increase. Rather, they identified a NOAEL of 50 mg piperine/kg bw/day (the highest dose tested) from this study [29].

(3)   Paternal Reproductive Toxicological Effects

In 1999, as a consequence of administration of 0, 5 or 10 mg piperine/kg bw/day (suspended in 0.9% saline) via a gastric catheter to young adult male rats for 30 days (n = 10 per group), Malini et al. [58,59] reported in the high-dose group statistically significantly ($p < 0.05$) reduced sperm concentrations in caput and cauda epididymides, statistically significantly reduced relative weight of testis (relative to body weight) and reduced absolute weight of the cauda epididymides, vas deferens, seminal vesicle and ventral prostate. The relative organ weights were also reduced, according to the authors' own calculations based on the mean body and organ weights stated in one of the two publications by Malini et al. [59]; however, no information on statistical significance is available for these calculations. Furthermore, histopathological changes in the testis, increased serum gonadotropins (FSH, LH) and reduced the intra-testicular testosterone concentration, as well as reduced the testicular lipid content, changes in the testicular lipid profile and reduced activity of some testicular lipogenic enzymes were observed (for details, see Table 1). Histopathological changes in the testes were also observed at 5 mg piperine/kg bw/day, but to a lesser extent than in the high-dose group. Within this low-dose group, the other parameters mentioned above showed changes in the same direction as seen in the high-dose group, but these changes were also of smaller magnitude and, in the majority of cases, no longer statistically significant. The reduced testicular weight was attributed to disturbed spermatogenesis and the reduced total lipid content of the testes caused by piperine administration [58,59].

Table 1. Paternal reproductive effects of piperine observed in animal studies.

Authors	Species	Application Mode	Duration	Dosage	Effects
(1) Studies for which bolus application of piperine can be assumed					
(1.1) Studies with male rats with the age of $\geq$90 days at study start (young adult rats)					
Malini et al., 1999 [58,59]	male albino rats (3–4 months old)	suspended in 0.9% saline administered via gastric catheter	30 days	5 mg/kg bw/day	• histopathology: partial degeneration of germ cell types. • serum LH stat. significantly [1] increased. • absolute organ weight of cauda epididymides and vas deferens stat. significantly reduced. Effects on other parameters mentioned below with the administration of 10 mg piperine/kg bw/day (sperm concentration, relative weight of testes, absolute weight of seminal vesicle and ventral prostate, serum FSH and intratesticular testosterone level) were in the same direction as seen with the dose of 10 mg/kg bw/day, but the changes were milder in nature and did not reach statistical significance.
				10 mg/kg bw/day	• sperm concentrations in caput and cauda epididymides stat. significantly reduced. • histopathology: severe damage to seminiferous tubule, decreased seminiferous tubular and Leydig cell nuclear diameter, desquamation of spermatocytes and spermatids. • serum FSH and LH stat. significantly increased; intratesticular testosterone stat. significantly decreased. • relative organ weight of testes stat. significantly reduced; absolute organ weight of cauda epididymides, vas deferens, seminal vesicle and ventral prostate stat. significantly reduced. • total lipids in testes stat. significantly reduced; changes in testicular lipid composition. • enzyme activities in testes of NAD+-dependent malate dehydrogenase and NADP+-isocitrate dehydrogenase stat. significantly decreased.

167

Table 1. *Cont.*

Authors	Species	Application Mode	Duration	Dosage	Effects
D'Cruz and Mathur, 2005; and D'Cruz et al., 2008 [60,61]	male Wistar rats (90 days old)	dissolved in vehicle (10% DMSO in ethanol and groundnut oil ration of 1:1) administered via micropipette	30 day	1 mg/kg bw/day	• no relevant effects on sperm parameters, weight of testes and accessory sex organs or investigated enzyme activities.
				10 mg/kg bw/day	• stat. significantly decreased epididymal sperm count and sperm motility. • stat. significantly reduced absolute organ weight of testes and caput, corpus and cauda epididymides. • stat. significantly reduced levels of sialic acid in testes and caput epididymides. • reduced activity of antioxidant enzymes in testes and corpus and cauda epididymides (in most cases statistically significant). • stat. significantly increased hydrogen peroxide generation in testes and corpus and cauda epididymides.
				100 mg/kg bw and day	• stat. significantly decreased epididymal sperm count, sperm motility and sperm viability. • significantly reduced relative organ weight of testes (as indicated in text, data not shown); • stat. significantly reduced absolute organ weight of caput, corpus, cauda epididymides, seminal vesicle and ventral prostate. • stat. significantly reduced levels of sialic acid in testes and caput, corpus and cauda epididymides, • stat. significantly reduced activity of antioxidant enzymes in testes and epididymides. • stat. significantly increased hydrogen peroxide generation in testes and caput, corpus and cauda epididymides.

Table 1. *Cont.*

Authors	Species	Application Mode	Duration	Dosage	Effects
Chinta and Periyasamy 2016; and Chinta et al., 2017 [16,62]	male Wistar rats (90 days old)	suspended in 0.5% carboxymethyl cellulose (no further information on application mode)	60 days of piperine administration followed by recovery period of 60 days without piperine administration	10 mg/kg bw administered every day	Effects at the end of **the piperine administration period** [2] • stat. significantly decreased epididymal sperm count, sperm motility and sperm viability. • histopathology: desquamated spermatozoa and decreased thickness of germ layer in seminiferous tubules; histopathological changes in epididymides and seminal vesicle. • serum FSH, LH and sex hormone-binding globulin stat. significantly increased; testicular testosterone stat. significantly decreased. • stat. significantly reduced level of sialic acid in epididymides and fructose level in seminal vesicles, • stat. significantly reduced activity of antioxidant enzymes (super oxide dismutase, catalase) in testes and epididymides. • stat. significantly increased lipid peroxidation in testes and epidydimides. • in Leydig cells stat significantly decreased activity of enzymes involved in testosterone synthesis • absolute liver weight was reduced by approximately 36% (stat. significant) **After a recovery period** of 60 days without piperine administration, the observed adverse effects on reproductive organs were reversible Absolute liver weight was still reduced by approximately 29% (stat. not significant).
				10 mg/kg bw administered every 4th day	Effects at the end of **the piperine administration period** [2] • With this administration regime, several of the adverse effects seen with the daily piperine administration were also observed but these changes were smaller, however, in some cases still statistically significant: (i.e., stat significantly reduced sperm motility and viability; stat significantly reduced testicular testosterone; stat significantly reduced levels of sialic acid in epididymides and fructose in seminal vesicle; stat significantly reduced activity of antioxidant enzymes in epididymides). Some other changes did not reach statistical significance. • Absolute liver weight was reduced by approximately 35% (stat significant). **After a recovery period** of 60 days without piperine administration the observed adverse effects on reproductive organs were reversible with no significant deviations from control group. Absolute liver weight was still reduced
				10 mg/kg bw administered every 7th day	Effects at the end of the **piperine administration period** • Sperm count or testicular testosterone were somewhat reduced but without reaching statistical significance. • Absolute liver weight was reduced by approximately 39% (stat. significant). **After a recovery period** of 60 days without piperine administration, no significant deviations from control group

Table 1. *Cont.*

Authors	Species	Application Mode	Duration	Dosage	Effects
(1.2)	Studies with male rats with age of 35 days at study start (juvenile rats)				
Chen et al., 2018 [63,64]	male Sprague-Dawley rats (35 days old)	suspended in normal saline administration via gavage	30 days	5 mg/kg bw/day	• Dose-dependent increase in serum testosterone; • decrease in serum FSH (similar serum levels in both piperine groups). • increase in Leydig cell size and slight increase in Leydig cell number (similar increases in both piperine groups). • authors state that sperm count in epididymides was reduced (only pictures of histological organ sections shown but no data on sperm count provided) [64].
				10 mg/kg bw/day	• dose-dependent increase in serum testosterone; • decrease in serum FSH (similar serum levels in both piperine groups). • Increase in Leydig cell size and slight increase in Leydig cell number (similar increases in both piperine groups). • authors state that sperm count in testis and epididymides was reduced (only pictures of histological organ sections shown but no data on sperm count provided).
(2) Studies with piperine application spread over the course of the day					
Bastaki et al., 2018 [29]	female and male Sprague-Dawley rats (48–57 days old)	added to feed	90 days	5 mg/kg bw/day	• statistically significant decrease of relative (to-brain ratio) epididymides weight (by about 23% compared to control) without histologic correlates [3].
				15 mg/kg bw/day	• statistically not significant decrease of relative (to-brain ratio) epididymides weight (by about 17% compared to control) without histologic correlates [3].
				50 mg/kg bw/day	• statistically significant decrease of relative (to-brain ratio) epididymides weight (by about 21% compared to control) without histologic correlates [3].

[1] stat. significantly = statistically significantly; $p < 0.05$; [2] Data on male reproductive organ weights were also recorded in this study, but in one of these two publications inconsistencies were noticed between the testes weights and calculated indexes or a coefficient based on testes and body weights [62]. For this reason, no further detailed information on the weights of the male reproductive organs is provided here. [3] Only potentially reprotoxicologically relevant effects reported. For other effects observed in this study, see (2) in Section 4.2.2.

In another study [60], D'Cruz and co-workers administered 0, 1, 10 and 100 mg piperine/kg bw/day dissolved in a vehicle (10% DMSO in ethanol and groundnut oil at the ratio of 1:1) to young adult male rats via a micropipette for 30 days and observed in the 10 mg/kg bw group statistically significant ($p < 0.05$) reductions in the cauda epididymal sperm count and sperm motility, statistically significantly reduced absolute weight of the testes and caput, corpus and cauda epididymides, and reduced activity of the antioxidant enzymes in the testes and corpus and cauda epididymides, accompanied by increased hydrogen peroxide generation and lipid peroxidation. In this study, the observed effects in the 10 mg/kg bw group on epididymal sperm count and testicular and epididymal weight were markedly smaller than in the study by Malini and co-workers. In the high-dose group (100 mg/kg bw), the effects were more pronounced and additionally included statistically significantly reduced sperm viability and significantly reduced relative testes weight (relative to body weight) (for details, see Table 1). Further immunofluorescence studies by the same group [61] revealed a dose-dependent increase in caspase 3 and FAS protein in testicular germ cells that was related to piperine administration and was accompanied by dose-dependent changes in the testicular antioxidant system (reduced activity of antioxidant enzymes, increases in hydrogen peroxide generation and in lipid peroxidation). In the low-dose group (1 mg/kg bw), no relevant reproductive toxicological effects were observed in either of the studies by D'Cruz et al. [60,61]. These studies used small animal groups with only four animals each, which reduces the scientific significance of the study findings.

In young adult male rats that orally received 10 mg piperine/kg bw/day (suspended in normal saline containing 0.5% carboxymethyl cellulose) for 60 days, Chinta and co-workers observed a statistically significantly ($p < 0.05$) reduced epididymal sperm count, sperm motility and sperm viability compared to the control group (the latter receiving the vehicle only), accompanied by histopathological changes in the testes and epididymides; statistically significantly increased serum gonadotropins (FSH, LH), with a reduced intra-testicular testosterone concentration; and statistically significantly reduced activity of anti-oxidant enzymes or other enzymes in the testes and epididymides (for details, see Table 1) [16,62]. Data on reproductive organ weights were also recorded in this study, but in one of the two publications dealing with this study, inconsistencies were noticed between the testes weights and calculated indexes or a coefficient based on testes and body weights [62]. For this reason, no further detailed information on the weights of the reproductive organs is provided here. After a recovery period of 60 days without piperine administration, the observed changes were reversible.

Animals receiving the piperine dose (10 mg/kg bw) every 4th day in the same study displayed some of the abovementioned adverse effects that were also observed with the daily piperine administration, e.g., reduced sperm viability and mobility, which were less pronounced but in some cases still statistically significant (for details, see Table 1). With piperine administration (10 mg/kg bw) every 7th day, the individual parameters that were mentioned in relation to the daily piperine administration showed changes in the same direction but the changes were considerably milder and no longer statistically significant (e.g., sperm count, testicular testosterone). Chinta and co-workers concluded from their data that piperine might be a good lead molecule for the development of a reversible oral male contraceptive [16,62]. Animals receiving piperine every day displayed decreased body weight of about 10% after 60 days of piperine administration, which was not statistically significant. All animal groups receiving piperine (n = 6 per study group) showed statistically significantly reduced absolute liver weights (approximately 35–39% reduction compared to the control group) at the end of the piperine administration period. The authors did not comment on those data. These reductions were still existent in the group that had received piperine every day (about −9%) and the group that had received piperine every 4th day (about −41%) after the recovery period, without reaching statistical significance ($p < 0.05$). Due to the administration of piperine suspended with carboxymethyl cellulose, it is assumed that piperine was given as a bolus once per day.

In a fourth study with male rats (juvenile animals, 35 days old at baseline; n = 6 per group) that involved administration of 0, 5 or 10 mg piperine/kg bw/day for 30 days by gavage, stimulation of pubertal Leydig cell development (increased Leydig cell number and promoted maturation) and an inhibited spermatogenesis were observed. Regarding the latter effect, the authors only cite histological findings of the testes and epididymides and indicated that already the dose of 5 mg piperine/kg bw/day reduced the epididymal sperm count, but no concrete figures on sperm counts were provided. This lack of data presentation reduces the scientific weight of the evidence provided in this study. Serum testosterone levels were elevated and the FSH levels were lowered in both piperine groups [63,64]. In this regard, the findings in juvenile rats differ from those observed in older rats [59,62].

Overall, the findings from these four studies [16,58–64] largely point in the same direction, with some differences of observed adverse reproductive effects between young adult and juvenile male rats. Concordantly, in young adult male rats, reproductive toxicological effects, i.e., disturbed spermatogenesis (and accompanying effects on testes, epididymides and accessory male reproductive organs of different nature and degrees) were observed with intakes of 10 mg/kg bw/day [16,58–62]. The less pronounced effects, which were only partly statistically significant, were observed in adult male rats already at 5 mg/kg bw/day. From these studies, a LOAEL of 5 mg piperine/kg bw/day [58,59] and a NOAEL of 1 mg/kg bw/day [60,61] can be identified for the endpoint male reproductive toxicity (disturbed spermatogenesis). However, the study of D'Cruz and co-workers used a wide spacing between the tested piperine doses (factor of 10) [60,61]. In one study, adverse paternal reproductive effects observed with repeated daily piperine (bolus) doses of 10 mg/kg bw were reversible after piperine discontinuation for several weeks [16,62]. Based on information of the four studies on piperine administration, it can be assumed that piperine administration in these studies was carried out via bolus administration. In three of these four studies, piperine administration was via gavage, micropipette or gastric catheter, suggesting bolus administration. The piperine dosage form of the fourth study, i.e., piperine in carboxymethyl cellulose, suggests bolus application as well.

It is noted that these studies are afflicted with certain limitations (statistical analysis of organ weights mainly comprising data on the absolute organ weights and the data on relative organ weights not being available in most cases; small animal group sizes in some studies; in the study of Chen et al., only histological findings were cited but no concrete data on sperm counts were provided; and reduced absolute liver weights not having been reported in other animal studies at this daily dose and contradictory information on the parameters related to testes weights in the study of Chinta and co-workers). However, taken as a whole, the aggregated study findings all point in the same direction and the paternal toxicological reproductive effects, i.e., disturbed spermatogenesis, are corroborated by findings at different levels, such as histopathology, sperm parameters, hormonal changes and changes at the level of enzyme activities, as well as changes in absolute organ weights. The limited scientific significance of absolute organ weights is acknowledged; however, a statistically significant change in the relative testes weights was seen at least in one study [58] with daily doses of 10 mg/kg bw. The observed differences in hormone levels between the studies of Malini et al. and Chinta et al. on the one hand compared to Chen et al. on the other hand may be related to the different life stages of the investigated male animals (juvenile versus young adult rats) [16,58,59,62–64].

The question of whether the mode of piperine administration in the study by D'Cruz and co-workers (piperine dissolved in 10% DMSO, ethanol and groundnut oil) [60,61] or in the study by Chinta and co-workers (together with carboxymethyl cellulose) [16,62] could possibly affect the bioavailability of piperine, leading to increased adverse effects, remains elusive. Adequate data to compare the influence of these modes of administration with the influences of the currently available piperine-containing dietary supplements and the food additives or galenic technics used in their manufacturing on the bioavailability of piperine, are currently not available.

In contrast to the largely consistent findings from the four studies cited above, different results are available from the 90-day toxicity study with rats used by EFSA (2015) for the evaluation of piperine as a flavouring agent, and which has already been described in (2) in Section 4.2.2 [29,32]. In this study, intakes of 0, 5, 15 or 50 mg piperine/kg bw/day were administered via feed. In male animals of the 5 and 50 mg/kg bw-groups, statistically significantly reduced relative epididymis weights (relative to brain weight) were observed. EFSA attributed only limited toxicological relevance to these findings due to the small changes and the non-existent dose dependence, and these changes were without histopathological findings (see also (2) in Section 4.2.2 and Table 1). This study did not include any specific examinations of sperm parameters or LH and FSH blood levels, as these types of investigations are not common in 90-day toxicity studies.

In its assessment, EFSA (2015) did not address the findings of Malini et al. [58,59] and D'Cruz and co-workers [60,61], which were available at that time (since a review of the available scientific literature was not foreseen at this time as part of this assessment procedure).

A major difference between the four animal studies cited first [16,58–64] and the 90-day toxicity study used by EFSA seems to be that in the 90-day toxicity study, piperine was administered via feed, resulting in multiple intakes of small quantities spread throughout the day, whereas in the four first-cited animals studies, it can be assumed that piperine was administered as a bolus dose, possibly resulting in higher maximum blood or tissue levels or otherwise increased bioavailability. The bolus administration of piperine in the first four animal studies more closely resembles the usual human use of food supplements, which often bear recommendations relating to 1–3 doses per day.

In this context, it is noted that Daware et al. observed no increased numbers of abnormal sperm cells in a sperm shape abnormality test performed with male mice receiving daily doses of 35–75 mg piperine/kg bw for 5 days [54]. However, this test is primarily performed regarding a genotoxicity assessment.

In in vitro studies, reduced viability and motility of goat sperm cells were seen with high doses of piperine (40–100 µmol/L) added to the sperm culture media [65], as well as impaired fertilization ability of hamster sperms directly exposed to high piperine doses (180–1005 µmol/L) in the capacitation medium [66]. However, the scientific relevance of these in vitro findings remains elusive due to the high piperine concentrations used and the direct exposure of the sperm cells to piperine via culture media, which differs from the exposure of sperm cells resulting from oral piperine intakes.

The mode of action of the bolus doses of piperine on spermatogenesis and the accompanying effects on male reproductive organs remains elusive. With young adult male rats, it has been hypothesized that induced oxidative stress due to depletion of antioxidant enzymes and increased generation of reactive oxygen species (ROS) in epididymis and testis, and activation of the Fas-mediated pathway in testicular germ cells, may contribute to the observed antifertility effects. However, inhibition of the cytochrome P-450 enzymes or other enzymes involved in the synthesis of testicular steroid hormones, interaction of piperine with the active site of the androgen binding protein, induction of hormonal imbalances (effects on serum levels of FSH, LH, sex hormone-binding globulin and testicular testosterone) or other effects on the functional integrity of the testis and the male reproductive organs are also being discussed, and appear possible [16,61,62].

It is noted that even with high bolus doses of fine *Piper nigrum* fruit powder (25 or 100 mg/kg bw/day) administered for 20 or 90 days to male mice, negative effects on the sperm count in the cauda epididymis, sperm motility, viability and number of morphologically abnormal spermatozoa were observed (viability not affected with 25 mg/kg bw dose administered for 20 days), which increased with escalating daily dose and duration of application from 20 to 90 days. After 90 days of pepper powder administration, statistically significantly reduced relative weights (relative to body weight) of the testis, epididymis and seminal vesicle were observed in both dosage groups. No male animal receiving 100 mg/kg bw/day for 90 days (other animals were not examined) was fertile in mating trials with untreated female mice 24 h and 14 days after the termination of pepper administration, respectively. The fertility of the treated male animals improved after an 8-week recovery period, but was still (statistically not significantly) reduced at this time point [67]. The fine fruit powder (suspended in water containing milk powder) was administered by a feeding needle; therefore, it can be assumed that the administration occurred as a bolus.

(4)   Maternal Reproductive Toxicological and Embryotoxic Effects

Depending on the time point of piperine administration before or during pregnancy, different reproductive toxicological or embryotoxic effects were observed in female animals.

In the study by Daware et al. (2000), young female mice (n = 6 per group) receiving 0, 10 or 20 mg piperine/kg bw/day for 14 days until the day of mating with untreated male animals displayed a statistically significantly reduced mating rate in the high-dose group (mating performance: 50% versus 83% in control group) and in both piperine dose groups a statistically significantly reduced fertility index (fewer mated animals became pregnant; fertility index: 60% and 66%, respectively). With this piperine administration protocol, the litter size of the pregnant animals and the growth of the pups were not affected. When the piperine doses (10 or 20 mg/kg bw/day) were administered to female mice from Day 1 through to Day 5 of gestation, significantly reduced implantation rates were observed in both dose groups, with implantations in 1 out of 6 mated animals each in the low- or high-dose group, respectively, versus 6 out of 6 mated animals in the control group. The post-implantation survival was not affected. In this study, piperine was given suspended in a formulation containing 1% carboyxmethyl cellulose, which was most likely done by bolus administration [54].

Significant implantation-inhibiting effects were also observed in another study by Piyachaturawat et al. (1982) with mice receiving oral bolus doses of 2 × 12.5 or 2 × 50 mg piperine/kg bw/day (dissolved in equal volume of DMSO and 95% ethanol; n = 19–21) from Day 2 through to Day 5 of gestation, with the implantation rates reduced by 71% and 90%, respectively, compared to the control group receiving the vehicle only (other oral doses not investigated). In addition, significant abortive effects were observed in mice given the same piperine bolus doses (2 × 12.5 or 2 × 50 mg/kg bw/day; n = 17–21) from Day 8 through to Day 12 of gestation (interrupted pregnancies in 58.8% and 71.4% of pregnant animals, respectively; other oral doses not investigated). Bolus doses of 25 mg piperine/kg bw/day given from gestation Day 15 onwards (n = 8) resulted in delayed labour and significantly increased number of dead foetuses (6.1 dead foetuses/litter vs. 0.3 dead foetuses/litter in the control group; other oral doses not investigated) [68].

In a study with female rats (n = 6), a reduced implantation rate (33% reduction) was observed with an oral piperine dose of 100 mg piperine/kg bw/day administered from gestation Day 1 through to Day 7 (other doses not investigated) [69].

In female hamsters, which received intra-gastric daily doses of 50 or 100 mg piperine/kg bw/day from Day 1 through to Day 4 of the oestrus cycle, followed by hormonally induced superovulation and artificial insemination with spermatozoa of untreated male animals, increased fertilization of eggs in the early phase of fertilization 9 or 24 h after insemination were observed compared to the control animals [70]. However, due to the applied methodology (hormonally induced superovulation which might interfere with piperine effects on female reproduction, artificial insemination, no information on pregnancy outcome), the scientific significance of these findings remains elusive regarding the effects of piperine on maternal reproduction or embryonic development.

In conclusion, from these studies, a LOAEL of 10 mg/kg bw/day can be identified with regard to adverse maternal reproductive and embryotoxic effects in mice [54]. This was the lowest daily dose investigated in these studies.

(5)   Interactions with Drugs

Animal studies on the interactions of piperine with drugs are discussed together with the corresponding human studies in Section 4.2.4.

4.2.3. Human Studies

(1) Intervention Studies

In human single-dose studies, piperine doses of 50 or 500 mg were applied either alone (50 mg piperine) or in combination (500 mg piperine) with curcumin. However, neither of both studies was designed to address piperine safety issues or provided data on the safety parameters (occurrence/absence of adverse events, haematological or clinical chemistry lab parameters) [45,71].

Human studies with repeated piperine administrations comprise only a small number of studies using piperine without concomitant administration of other substances (drug interaction studies with piperine-only run-in phases of 3–10 days and piperine doses of 15–20 mg/day) [72–79]. In addition, there are a number of other studies in which piperine (in several cases in form of highly piperine-enriched pepper extracts) was given in combination with other substances, e.g., curcumin, resveratrol, *Camellia sinensis* extract, herbal extracts or others, in order to increase the bioavailability or effectiveness of these substances [80–116]. Some of these piperine combination studies included control groups receiving piperine without concomitant administration of the other substances mentioned above, but these studies lacked control groups receiving no piperine, and were conducted in individuals suffering from different diseases and using various drugs during the course of the study, or no information on product tolerance was provided [88,89,99,116]. The combination studies frequently used piperine doses in the range of 4–15 mg/day with study durations of 4–17 weeks, but also other studies are available, in which 60 mg piperine/day administered for 4 days [90], doses of 10 mg/day for 6 months [96,97] or doses of 40 mg/day for 3 or 6 month, respectively [86,111], were used. However, available human studies with piperine administration were primarily conducted to evaluate the efficacy of piperine or the efficacy of the accompanying substances, and in most cases safety issues were only marginally addressed or reported. Most of the studies provided no information or only inadequate information on the occurrence or absence of adverse events and/or no data of the relevant safety lab parameters (a situation that is often found with substances used as ingredients in food supplements [117]). In this regard, it should be noted that the fact that no information on the absence or occurrence of adverse events was provided in several studies cannot be taken as a proof that actually no adverse events occurred [117]. In some of the available studies in which piperine was administered concomitantly with other substances, it is stated that no serious adverse events or severe undesirable effects were reported, leaving open questions regarding the less severe effects. A few studies with combined administration of piperine (4–10 mg/day or unspecified doses) with other substances (iron preparation, resveratrol, curcumin, multi-ingredients or others) reported that no adverse events occurred [82,83,101,106,111–113,118]. Individual studies with combined administration of piperine with other substances (a multi-ingredient product, curcumin) conducted in patients with non-alcoholic fatty liver disease or COPD reported the occurrence of adverse events, comprising gastrointestinal adverse effects (abdominal discomfort or diarrhoea in two studies with 3/40 individuals or 4/45 individuals, respectively; none in placebo groups with n = 40 or 12, respectively) or rash (in one study with 1/45 individuals; none in placebo group with n = 12) [84,103].

None of the identified human studies included investigations regarding the potential effects of piperine on male reproductive capacity (i.e., sperm parameters).

Taken together, due to insufficient data on safety parameters, the lack of investigations into the effects of piperine on human spermatogenesis and/or the combined administration of piperine with other substances conducted in most studies, the available human studies involving piperine administration provide no adequate scientific basis for the assessment of the possible health risks of oral intake of isolated piperine used as single ingredient and ingested in bolus form.

(2) Studies on Reproductive Toxicological Effects

No published human intervention study could be identified, which included investigations into the effects of bolus intakes of isolated piperine on male reproductive organs, reproductive capacity or sperm parameters. The same applies regarding the reproductive effects in women (pregnant women or women who intend to become pregnant).

In an epidemiological study, a statistically significant inverse association was observed between plasma testosterone concentrations and, among others, plasma piperine levels in healthy middle-aged men (median: 50 years) [119]. However, as already mentioned by the study authors, a statistical association does not imply causality and therefore the scientific significance of these data from this particular study alone regarding male reproductive effects remains elusive (i.e., piperine plasma levels could just be a lifestyle marker).

4.2.4. Interactions of Piperine with Medicinal Products and Other Substances

The pharmacokinetic interactions of piperine with various chemically and pharmacologically diverse drugs have been observed in both human and animal studies. In most instances, interactions of piperine with drugs resulted in a better bioavailability of the investigated drugs, exemplified by increases in the maximum plasma/serum concentrations (C_{max}) and/or increased AUC values (AUC = area under the curve) for the respective drugs. These effects are in line with the purported "bio-enhancing" activity of piperine.

In human studies, oral administration of 20 mg piperine/day ($\sim$ 0,29 mg piperine/kg bw based on a body weight of 70 kg) for one or several days resulted in improved bioavailability (elevated serum/plasma concentrations and/or elevated AUC values) of the following drugs: propranolol (antihypertensive drug), theophylline (bronchodilatory drug), phenytoin, carbamazepine (antiepileptic drugs), nevirapine (HIV-1 reverse transcriptase inhibitor), chlorzoxazone (muscle relaxant), diclofenac (non-steroidal anti-inflammatory drug) and fexofenadine (antihistaminic drug). In the cases involving administration of 20 mg of piperine/day, the increases in the drug C_{max} and AUC values were approximately 1.07- to 2.2-fold (C_{max}) and 1.09- to 2.7-fold (AUC values), depending on the drug and drug dosage. Regarding rifampicin (antibiotic drug), such interactions were reported in the context of concomitant administration of 50 mg piperine/day (for details, see Table 2) [72–77,79,120–122].

For midazolam (sedative), prolonged and increased pharmacological effects of the drug (prolonged duration of sedation, increased number of individuals with amnesia) were observed with piperine administration of 15 mg/day for 3 days and subsequent midazolam administration on the 4th day [78].

For the substances β-carotene and coenzyme Q_{10}, increased bioavailability was already observed in the case of combined administration of 5 mg piperine/day with 15 mg β-carotene/day for 14 days or with 120 mg coenzyme Q_{10}/day, respectively, for 21 days [123,124]. Regarding curcumin, the C_{max} or AUC values were approx. 30 or 20 times higher, respectively, when administered together with 20 mg piperine/day [125].

Table 2. Effects of piperine on the bioavailability of drugs in human studies.

Drug	Therapeutic Use	Drug Dose and Duration	Piperine Dose and Duration	Drug Bioavailability	Reference
propranolol	antihypertensive drug	40 mg/day as single dose, preceded by piperine administration for 7 days	20 mg/days for 7 days	increased bioavailability (AUC: 2.03 fold, C_{max}: 2.04 fold)	Bano et al., 1991 [72]
theophylline	bronchodilatory drug	150 mg as single dose, preceded by piperine administration for 7 days	20 mg/days for 7 days	increased bioavailability (AUC: 1.96 fold, C_{max} 1.62 fold)	
rifampicine	antibiotic drug	450 mg as single dose	50 mg as single dose	increased bioavailability (AUC: 1.71 fold, C_{max}: 1.25 fold)	Zutshi et al., 1985 [122]
phenytoin	antiepileptic drug	300 mg as single dose preceded by piperine administration for 7 days, given to healthy subjects	20 mg/day for 7 days	increased bioavailability (AUC: 1.50 fold, C_{max}: 1.27 fold) C_{max} increase was statistically not significant	Bano et al., 1987 [79]
		150 mg given to patients with epilepsy who were on phenytoin therapy with 2×150 mg/day for at least 2 months	20 mg as single dose	slightly increased bioavailability (AUC: 1.09 fold, C_{max}: 1.10 fold)	Pattanaik et al., 2006 [120]
		200 mg given to patients with epilepsy who were on phenytoin therapy with 2×200 mg/days for at least 2 months	20 mg as single dose	slightly increased bioavailability (AUC: 1.17 fold, C_{max}: 1.22 fold)	
carbamazepine	antiepileptic drug	200 mg as single dose given to healthy subjects on day 11 (applied 1 day after last piperine dose)	20 mg/day for 10 days	increased bioavailability (AUC_{last}: 1.48 fold, C_{max}: 1.68 fold)	Bedada et al., 2017 [73]
		300 mg given to patients with epilepsy who were on carbamazepine therapy with 2×300 mg/day for at least 2 months	20 mg as single dose	slightly increased bioavailability (AUC: 1.10 fold, C_{max}: 1.07 fold; C_{max} increase was statistically not significant	Pattanaik et al., 2009 [121]
		500 mg given to patients with epilepsy who were on carbamazepine therapy with 2×500 mg/day for at least 2 months	20 mg as single dose	slightly increased bioavailability (AUC: 1.13 fold, C_{max}: 1.10 fold)	
nevirapine	non-nucleoside HIV-1 reverse transcriptase inhibitor	200 mg as single dose on day 7 of piperine administration	20 mg/day for 7 days	increased bioavailability (AUC: 2.67 fold, C_{max}: 2.20 fold)	Kasibhatta and Naidu 2007 [77]
chlorzoxazone	centrally acting muscle relaxant	250 mg as single dose on day 11 (applied 1 day after last piperine dose)	20 mg/day for 10 days	increased bioavailability (AUC_∞: 1.69 fold, C_{max}: 1.58 fold)	Bedada and Boga 2017 [74]
fexofenadine	antihistaminic drug	120 mg as single dose on day 11 (applied 1 day after last piperine dose)	20 mg/day for 10 days	increased bioavailability (AUC: 1.68 fold, C_{max}: 1.88 fold)	Bedada and Boga 2017 [75]
diclofenac	nonsteroidal anti-inflammatory drug	100 mg as single dose on day 11 (applied 1 day after last piperine dose)	20 mg/day for 10 days	increased bioavailability (AUC: 1.66 fold, C_{max}: 1.64 fold)	Bedada et al., 2017 [76]
midazolam	sedative	10 mg as single dose on day 4 (applied 1 day after last piperine dose)	15 mg/day for 3 days	increased clinical effects - increased duration of sedation ≈ 1.89 fold; - increased number of individuals with drug-induced amnesia	Rezaee et al., 2014 [78]

In these human studies referred to above, piperine was administered once daily as a bolus and the piperine and/or drug administration was performed for one or more days according to different administration schedules. The effects of piperine were influenced by the dose of the administered drug and/or the duration of the piperine or the drug administration and/or possibly by the investigated population group, i.e., healthy individuals or individuals suffering from certain diseases (for details, see Table 2 and below). With the exception of the studies on β-carotene and coenzyme Q_{10}, currently no published human study could be identified with simultaneous longer-term piperine and drug/substance administration to achieve steady-state levels of piperine and the drug/substance.

In animal studies (rodents or rabbits), higher piperine doses were generally used (2.1–30 mg/kg bw/day), as compared to the doses employed in human studies in which interactions were investigated. The interactions described for the animal studies in most cases also resulted in increased bioavailability (increased serum/plasma C_{max} and/or increased AUC values) of the investigated drugs. However, interactions in animal studies were observed with several additional drugs, such as ibuprofen, nimesulide, oxyphenylbutazone (non-steroidal anti-inflammatory drugs); amoxycillin, ampicillin, norfloxacin, marbofloxacin, metronidazole (antibiotic drugs); glimepiride, nateglinide (antidiabetic drugs); simvastatin, rosuvastatin (lipid-lowering drugs); verapamil, diltiazeme (calcium channel blockers); sodium valproate (antiepileptic drug); darunavir ethanolate (HIV protease inhibitor); losartan (angiotensin II receptor type 1 antagonist); domperidone (antiemetic drug); almotriptan (anti-migraine drug); and fexofenadine (antihistaminic drug) [126–144].

Increased bioavailability in animal studies has also been observed for a number of other substances (for example for puerarin, resveratrol, emodin, linarin or cannabidiol).

In the case of the drugs nimesulide, oxyphenylbutazone, ibuprofen or nateglinide, the piperine-mediated elevated bioavailability was reported to also be associated with increased drug efficacy. Moreover, increased pharmacological effects, without concomitant measurement of drug levels, have been observed for pentobarbitone (short-acting barbiturate), pentazocine (opioid), sertraline (selective serotonin reuptake inhibitor), midazolam or diazepam [126,134,140,141,145,146].

By contrast, animal studies were also identified in which no improved drug bioavailability or a decreased bioavailability was observed for the combined administration of piperine with drugs, i.e., with cefadroxil, carbamazepine, warfarin or diltiazem [127,147–149].

For diltiazem, a reduced drug bioavailability was observed with administration of 10 or 20 mg of piperine/kg bw/day for 14 days and diltiazem administration on Day 15 (i.e., one day after termination of piperine administration). The repeated dosing of piperine in this study led to induced gene expression of the multidrug transporter P-glycoprotein, which may have played a role in limiting the bioavailability of diltiazem [147].

As a further example, co-administration of warfarin (2 mg/kg bw) with piperine (10 mg/kg bw) to rats resulted in a decrease in the warfarin C_{max} and AUC values by 32 and 20%, respectively, combined with a reduced clinical anti-coagulant effect of warfarin [149].

The influence of the study conditions (investigated species, investigated human population, duration of drug treatment and/or schedule of piperine application and piperine dose) on the results of available interaction studies is illustrated by data obtained with carbamazepine. In healthy humans receiving a single dose of carbamazepine (200 mg) on Day 11 after 10 days of piperine administration (20 mg/day), significant increases in C_{max} (1.7 fold) and AUC (1.5 fold) values were observed compared to the control group receiving carbamazepine only, which were attributed to piperine-mediated inhibition of the CYP3A4 enzyme [73]. In epilepsy patients, with long-time carbamazepine monotherapy and steady-state carbamazepine plasma levels, small increases in C_{max} (1.1 fold) and AUC (1.1) values were observed after administration of a single dose of 20 mg piperine combined with 500 mg carbamazepine [121]. In an animal study with rats receiving a single combined dose of carbamazepine with 3.5 or 35 mg piperine/kg bw, no significant changes in plasma levels (C_{max} or AUC-values) of carbamazepine and the major carbamazepine metabolite, carbamazepine-10,11-epoxide, were seen compared to the control group receiving carbamazepine only. Rats concomitantly receiving the same carbamazepine and piperine doses for 14 days showed significantly decreased plasma levels of carbamazepine (plasma C_{max} 25–39% reduced; plasma AUC 29–37% reduced). The plasma levels of carbamazepine-10,11-epoxide and brain levels of carbamazepine were also reduced in both piperine-treated groups, but reached statistical significance only in the high-dose group. Carbamazepine-10,11-epoxide brain levels were also reduced but reached no statistical significance. A decreased plasma concentration of carbamazepine was observed in the high-dose piperine group, despite decreased CYP3A2 protein expression in rat liver. The reduced carbamazepine drug and metabolite levels were mainly attributed to a reduced carbamazepine absorption (animals displayed increased defecation and wet faeces with piperine) and a decreased brain penetration of carbamazepine caused by piperine [148].

In addition to the abovementioned study conditions, it can be assumed that the concomitant administration of other bioactive substances together with piperine might also influence the effects of piperine on the bioavailability of the investigated drugs.

Concerning the improved bioavailability seen with several drugs, different underlying mechanisms have been discussed, such as improved drug absorption and/or inhibition of degradation or elimination. In this context, depending on the drug or investigated substance, different mechanisms for influencing absorption and different molecular targets relating to the metabolism of xenobiotics are at the center of discussion, viz. the unspecific gastrointestinal effects (increased splanchnic blood flow), altered membrane dynamics, inhibition of cytochrome P-450 enzymes (CYP3A4, CYP2E1, CYP2C9, etc.), inhibition of multispecific efflux transporters, such as P-glycoprotein, or other mechanisms [1,15,73–78,120,124,150].

It is noted that for the reduced bioavailability of diltiazem observed in an animal study, induction of P-glycoprotein has been considered as an underlying principle [147].

In summary, single or short-term administration of piperine bolus doses combined with several chemically and pharmacologically diverse drugs resulted in interactions, which in most cases led to increased drug bioavailability. However, it can be assumed that such interactions with piperine may also occur with other drugs that have not be tested in this respect so far. Such interactions may vary over time and depend on the used drug, drug dosage and piperine dosage, and may also possibly vary depending on concomitant administration of further bioactive substances. Appropriate investigations in humans are required with the drugs in question to further clarify this issue. Interactions (increased bioavailability of drugs) in humans were observed with several drugs at bolus doses of 20 mg piperine/day. For one drug (midazolam), increased clinical efficacy was reported with bolus administration of 15 mg piperine/day. The findings with β-carotene and coenzyme Q_{10} indicate that already bolus doses of 5 mg piperine/day may possibly cause interactions with certain substances or drugs.

5. Discussion

This review focuses on the evaluation of the safety of isolated piperine used as a single ingredient in food supplements, i.e., in bolus form, in adult individuals. Children and adolescents were not included in the consideration since adult persons constitute the prime target population for food supplements containing piperine.

Piperine is a natural ingredient of *Piper nigrum* (black pepper) and some other *Piper* species, e.g., *Piper longum*. Due to the use of peppercorns as a spice, it is a common component of the human diet. Apart from this source, piperine may be used in isolated form as a flavouring agent in food production.

When pepper is used for food seasoning purposes, piperine is added to food in combination with all other components of the peppercorns (which may influence observed biological effects) and in varying degrees of comminution (which may influence its bioavailability). In this context, it is usually consumed together with large quantities of food and in several portions throughout the day. By contrast, when piperine is used in food supplements, the piperine supply differs greatly in that piperine is ingested in isolated or highly concentrated form and as a bolus (usually in 1–3 portions per day), without any substantial amounts of other pepper components. These differences may influence the bioavailability and biological effects of piperine ingested as a bolus and in isolated form compared to its usual intake via food seasoning.

Currently available human studies with oral piperine administrations provide no adequate scientific basis for the final assessment of the potential health risks associated with intakes of isolated piperine used as a single ingredient and ingested in bolus form. This is a situation also frequently found with other substances used as ingredients of dietary supplements. Although hardly any adverse effects (apart from potentially undesirable interactions of piperine with various drugs) were reported in available human studies, it is noted that in most cases safety issues were only marginally addressed or reported. In addition, is to be considered that in many studies piperine was administered in combination with other substances and that the combined administration complicates data interpretation.

However, some currently available human and animal studies already provided indications of potential health risks of bolus doses of isolated piperine.

In human and animal studies with single or short-time application, piperine was shown to interact with various drugs. In most cases, these interactions led to an improved bioavailability of the investigated drugs and are in line with the purported "bio-enhancing" activity of piperine. It can be assumed that such interactions may also occur with other drugs, and can vary, depending on the individual drug, drug dosage, piperine dosage, duration of intake and time span between piperine and drug intake. Additionally, interactions may also possibly vary depending on the concomitant administration of further bioactive substances. For clarification, appropriate investigations in humans are required with the drugs in questions, involving in particular studies with repeated drug and piperine application to reach steady-state levels.

Improved bioavailability can offer advantages in drug therapy if this is performed under medical supervision. Without adequate medical supervision, however, depending on the drug, piperine-based drug interactions may carry the risk of unintended and/or deleteriously increased medicinal drug effects or of the occurrence of adverse drug effects, especially in the case of drugs with a narrow therapeutic range. Increased bioavailability of certain drugs has been observed in several human studies with bolus doses of 20 mg/day (= LOEL), and findings with β-carotene and coenzyme Q_{10} suggest that in some cases bolus doses of 5 mg piperine/day might also cause such interactions. Taken together, against the background of these data, it seems advisable that individuals taking medicinal products, especially drugs with known piperine interactions or drugs for which no interaction data are available, should consult a physician prior to the use of isolated piperine as a food supplement.

Based on animal data, further potential health risks or potential risk groups, respectively, of bolus doses of isolated piperine can be identified. In four animal studies with juvenile and young adult rats, largely consistent paternal reproductive toxic effects (see Table 1) were observed at piperine bolus doses of 10 mg/kg bw/day [16,58–64].

In young adult rats, male reproductive toxic effects, which were significantly weaker and only partly statistically significant, were observed by Malini et al., already at bolus doses of 5 mg/kg bw/day [58,59]. On the other hand, a NOAEL of 1 mg piperin/kg bw/day (as bolus) can be identified from the study by D'Cruz and co-workers, who used a large spacing between the tested piperine doses (factor 10) [60,61]. Consistently clearer adverse paternal reproductive effects, i.e., disturbed spermatogenesis and accompanying adverse male reproductive effects, however, were observed with doses of 10 mg/kg bw/day.

It is to be acknowledged that these studies are afflicted with certain limitations; however, taken as a whole, the aggregated study findings point in the same direction and paternal toxicological reproductive effects, i.e., disturbed spermatogenesis, are corroborated by findings at different levels (i.e., histopathology, sperm parameters, hormonal changes and other parameters).

The reason for the different findings in terms of paternal reproductive toxic effects of these four studies compared to a sub-chronic 90-day animal toxicity study [29] remains elusive. The major difference seems to lie in the method of piperine administration: in the sub-chronic 90-day animal study, piperine was administered via feed, resulting in multiple intakes of small piperine quantities spread throughout the day. In the case of the other four abovementioned animal studies, piperine administration as a bolus can be assumed, possibly resulting in higher peak blood or tissue levels or otherwise increased bioavailability.

Regarding the reversibility of the disturbed spermatogenesis observed by Chinta and colleagues (see above) [16,62], it remains to be investigated to what extent such reversibility may still be present with piperine administrations lasting longer than 60 days. In addition, it remains to be clarified to what extent these findings may be extrapolated to humans. Even taking into account these findings of Chinta and co-workers with respect to the potential reversibility of the perturbation of spermatogenesis under certain conditions, the consistent paternal reproductive effects observed with bolus doses of 10 mg/kg bw/day should be classified as being toxicologically relevant and of potential concern.

Human studies that included adequate investigations to clarify the possible effects of bolus doses of isolated piperine on the male reproductive system could not be identified in the course of preparation of the current review. Therefore, as long as this knowledge gap has not been resolved and against the background of the adverse male reproductive effects observed in animal studies with high bolus intakes of isolated piperine, it seems advisable to maintain an adequate margin of exposure between those bolus doses of isolated piperine for which adverse male reproductive effects in animal studies were reported and the daily amounts of isolated piperine used in food supplements.

By using a dose of 10 mg piperine/kg bw/day for which adverse male reproductive effects were reported in several animal studies as a point of departure, the application of an uncertainty factor of 3 seems warranted for extrapolation to a NOAEL from this intake level. In addition, for deriving health-based guidance values from animal data, EFSA recommends to use an overall default assessment factor of 100 to account for inter-species and intra-human variability (10 for inter-species variability x 10 for inter-human variability) in the absence of chemical-specific data on the kinetics and/or dynamics [151]. Based on an assumed body weight of 70 kg, this approach would lead to a health-based guidance value of 2.3 mg/day (10 mg/kg bw/day $\times$ 70 kg bw/(3 $\times$ 100)) for piperine when used in isolated form via bolus administration, i.e., as a food supplement. This intake seems to be low compared to estimates of daily piperine intakes resulting from the usual culinary use of pepper in normal human diet. However, it should be taken into account that the intake of piperine in isolated form may not be directly comparable to its intake in conjunction with all other pepper ingredients and various degrees of comminution of the peppercorn. On the other hand, the effect of pepper consumption (in particular of high bolus doses with a high degree of comminution) on human male reproductive capacity remains to be clarified.

The health-based guidance value of 2.3 mg/day, as calculated above, which would currently be expected to provide an adequate level of protection with respect to potential male reproductive toxicity, would also be below the piperine bolus doses for which interactions with concomitantly administered drugs/substances have been observed.

In animal studies, maternal reproductive toxicity and embryotoxic effects were observed with piperine bolus doses of 10–25 mg/kg bw/day, which varied, depending on the time of piperine administration with regard to the day of mating, the duration of time covered during gestation and the piperine dose (10 mg/kg bw/day: reduced fertility index, reduced implantation rates; 25 mg kg bw/day: reduced implantation rates, abortive effects, delayed labour and increased foetal mortality) [54,68]. It should be noted that the observed adverse effects occurred in these studies at the lowest piperine bolus doses tested and that certain adverse effects were only investigated at 25 mg/kg bw/day. From the available animal studies, a LOAEL of 10 mg/kg bw/day, but no NOAEL regarding maternal reproductive toxicity and embryotoxic effects could be identified. It remains to be clarified whether the use of DMSO in the oral piperine application by Piyachaturawat et al. (1982) might have had an influence on the bioavailability or observed effects of piperine. However, adverse maternal reproductive and embryotoxic effects were also observed in a second study [54].

With regard to the potential health risks for which there are indications when using isolated piperine in bolus doses (drug interactions, disturbed spermatogenesis, adverse maternal reproductive and embryotoxic effects), it remains to be clarified whether these risks may be influenced by other bio-active, additional ingredients in food supplements ingested together with piperine. In principal, the same considerations apply to highly piperine-enriched pepper extracts, although the difference between some highly piperine-enriched extracts and the piperine preparations supplied from chemical companies (purity: $\geq 97\%$), used for instance in the cited studies of Chinta et al. [16,62] or D'Cruz et al. [60,61], does not seem to be substantial. However, in cases involving either other bio-active ingredients and/or highly piperine-enriched pepper extracts, any claimed mitigation or elimination of the adverse effects described above should be supported by adequate scientific investigations and data.

In conclusion, human and animal studies with single or short-term application of isolated piperine used in bolus form revealed interactions of the substance with several drugs, which can give rise to potential health risks, and based on which individuals taking medications can be identified as a potential risk group. For individuals taking medicinal products (especially drugs with known piperine interactions or drugs for which no interaction data are available), it seems advisable to consult a physician prior to the use of isolated piperine as a food supplement.

Animal studies with higher daily piperine bolus doses provide indications for further potential health risks (disturbed spermatogenesis; adverse maternal reproductive and embryotoxic effects), for which no adequate human data are currently available.

Considering that a distinct NOAEL for maternal reproductive and embryotoxic effects could currently not be identified from the available animal studies, it seems advisable for pregnant women to abstain from the use of food supplements containing isolated piperine. Moreover, the reduced implantation rates that were reported in animal studies may also be of relevance for women who wish to become pregnant.

Regarding the observed adverse paternal reproduction effects in animal studies and the lack of information on the effects of bolus doses of isolated piperine on the human male reproductive system, it appears prudent to maintain an adequate margin of exposure between those bolus doses that produced adverse paternal reproductive effects in animal studies and the maximum daily amounts of isolated piperine in food supplements.

Considering the uncertainties outlined above, the importance of addressing the existing knowledge gaps regarding effects of bolus doses of isolated piperine on human male reproductive capacity in future human intervention studies, by including specific investigations into this endpoint in the study design, is emphasised.

Author Contributions: Conceptualization, R.Z.; writing—original draft preparation, R.Z.; writing—review & editing, R.Z., K.H., A.L. and K.I.H.-E. All authors have read and agreed the published version of the manuscript.

Funding: This research received no external funding.

Institutional Review Board Statement: Not applicable.

Informed Consent Statement: Not applicable.

Data Availability Statement: Not applicable.

Acknowledgments: The authors would like to thank the Dietary Exposure and Aggregated Exposure Unit of the German Federal Institute for Risk Assessment's (BfR), in particular Christine Sommerfeld, for conducting the estimation of the daily pepper intake of the German population (14–80 years) and Britta Nagl and Carolin Schopf for technical assistance in the preparation of the manuscript.

Conflicts of Interest: The authors declare no conflict of interest.

Abbreviations

AUC	area under the curve
bw	body weight
C_{max}	maximum (or peak) serum/plasma concentration of a drug/substance
CMEC	(Australian) Complementary Medicines Evaluation Committee
COPD	chronic obstructive pulmonary disease
DMSO	dimethyl sulfoxide
EFSA	European Food Safety Authority
FAO	Food and Agriculture Organization
FSH	follicle-stimulating hormone
i.g.	intragastric
i.v.	intravenous
JECFA	Joint FAO/WHO Expert Committee on Food Additives
LH	luteinizing hormone
LOAEL	Lowest observed adverse effect level
LOEL	Lowest observed effect level
MOE	Margin of Exposure
MSDI	Maximised Survey-derived Daily Intake
NOAEL	No observed adverse effect level
NOEL	No observed effect level
OECD	Organization for Economic Co-operation and Development
mTAMDIs	Modified Theoretical Added Maximum Daily Intakes
TGA	(Australien) Therapuetic Goods Administration
T_{max}	time to reach maximum serum/plasma concentration after administration of a drug/substance
WHO	World Health Organization

References

1. Lee, S.H.; Kim, H.Y.; Back, S.Y.; Han, H.K. Piperine-mediated drug interactions and formulation strategy for piperine: Recent advances and future perspectives. *Expert Opin. Drug Metab. Toxicol.* **2018**, *14*, 43–57. [CrossRef]
2. VKM. Norwegian Scientific Committee for Food Safety. Risk assessment of "other substances"—Piperine. Opinion of the Panel Food Additives, Flavourings, Processing Aids, Materials in Contact with Food and Cosmetics of the Norwegian Scientific Committee for Food Safety. *VKM Rep.* **2016**, *31*, 1–43. Available online: https://vkm.no/download/18.645b840415d03a2fe8f25ff2/1502802968337/08fcbfacc1.pdf (accessed on 2 June 2021).
3. EFSA. European Food Safety Authority. Compendium of botanicals reported to contain naturally occuring substances of possible concern for human health when used in food and food supplements. *EFSA J.* **2012**, *10*, 2663. [CrossRef]
4. ECHA. European Chemicals Agency. Piperine. Substance Infocard. Available online: https://echa.europa.eu/de/substance-information/-/substanceinfo/100.002.135 (accessed on 2 June 2021).
5. ChemIDplus. Substance name: Piperine [USP]. In *National Institutes of Health/U.S. National Library of Medicine.* Available online: https://chem.nlm.nih.gov/chemidplus/name/piperine%20%5Busp%5D (accessed on 2 June 2021).
6. Stohr, J.R.; Xiao, P.G.; Bauer, R. Constituents of Chinese Piper species and their inhibitory activity on prostaglandin and leukotriene biosynthesis in vitro. *J. Ethnopharmacol.* **2001**, *75*, 133–139. [CrossRef]
7. Haq, I.U.; Imran, M.; Nadeem, M.; Tufail, T.; Gondal, T.A.; Mubarak, M.S. Piperine: A review of its biological effects. *Phytother. Res. PTR* **2021**, *35*, 680–700. [CrossRef] [PubMed]
8. Vasavirama, K.; Upender, M. Piperine: A valuable alkaloid from piper species. *Int. J. Pharm. Pharm. Sci.* **2014**, *6*, 34–38.
9. Orav, A.; Stulova, I.; Kailas, T.; Muurisepp, M. Effect of storage on the essential oil composition of *Piper nigrum* L. fruits of different ripening states. *J. Agric. Food Chem.* **2004**, *52*, 2582–2586. [CrossRef]
10. Kozukue, N.; Park, M.S.; Choi, S.H.; Lee, S.U.; Ohnishi-Kameyama, M.; Levin, C.E.; Friedman, M. Kinetics of light-induced cis-trans isomerization of four piperines and their levels in ground black peppers as determined by HPLC and LC/MS. *J. Agric. Food Chem.* **2007**, *55*, 7131–7139. [CrossRef]
11. Chinta, G.; Syed, S.B.; Coumar, M.S.; Periyasamy, L. Piperine: A comprehensive review of pre-clinical and clinical investigations. *Curr. Bioact. Compd.* **2015**, *11*, 156–169. [CrossRef]
12. Chavarria, D.; Silva, T.; Magalhaes e Silva, D.; Remiao, F.; Borges, F. Lessons from black pepper: Piperine and derivatives thereof. *Expert Opin. Ther. Pat.* **2016**, *26*, 245–264. [CrossRef] [PubMed]
13. Gorgani, L.; Mohammadi, M.; Najafpour, G.D.; Nikzad, M. Piperine—the bioactive compound of black pepper: From isolation to medicinal formulations. *Compr. Rev. Food Sci. Food Saf.* **2017**, *16*, 124–140. [CrossRef]
14. Singh, A.; Duggal, S. Piperine—Review of advances in pharmacology. *Int. J. Pharm. Sci. Nanotechnol.* **2009**, *2*, 615–620. [CrossRef]
15. Srinivasan, K. Black pepper and its pungent principle-piperine: A review of diverse physiological effects. *Crit. Rev. Food Sci. Nutr.* **2007**, *47*, 735–748. [CrossRef]

16. Chinta, G.; Periyasamy, L. Reversible anti-spermatogenic effect of piperine on epididymis and seminal vesicles of albino rats. *Drug Res.* **2016**, *66*, 420–426. [CrossRef]
17. Vijayakumar, R.S.; Surya, D.; Nalini, N. Antioxidant efficacy of black pepper (*Piper nigrum* L.) and piperine in rats with high fat diet induced oxidative stress. *Redox Rep. Commun. Free. Radic. Res.* **2004**, *9*, 105–110. [CrossRef]
18. Bakshi, H.; Nagpal, M.; Singh, M.; Dhingra, G.A.; Aggarwal, G. Propitious profile of peppery piperine. *Curr. Mol. Pharmacol.* **2020**. E-pub ahead of print. [CrossRef]
19. Raza, K.; Kumar, D.; Kiran, C.; Kumar, M.; Guru, S.K.; Kumar, P.; Arora, S.; Sharma, G.; Bhushan, S.; Katare, O.P. Conjugation of Docetaxel with multiwalled carbon nanotubes and codelivery with piperine: Implications on pharmacokinetic profile and anticancer activity. *Mol. Pharm.* **2016**, *13*, 2423–2432. [CrossRef]
20. Cherniakov, I.; Izgelov, D.; Barasch, D.; Davidson, E.; Domb, A.J.; Hoffman, A. Piperine-pro-nanolipospheres as a novel oral delivery system of cannabinoids: Pharmacokinetic evaluation in healthy volunteers in comparison to buccal spray administration. *J. Control. Release* **2017**, *266*, 1–7. [CrossRef]
21. Zachariah, T.J.; Parthasarathy, V.A. Black pepper. In *Chemistry of Spices*; Parthasarathy, V.A., Chempakam, B., Zachariah, T.J., Eds.; CAB International: Wallingford, UK, 2008; pp. 21–40.
22. Liu, H.L.; Luo, R.; Chen, X.Q.; Ba, Y.Y.; Zheng, L.; Guo, W.W.; Wu, X. Identification and simultaneous quantification of five alkaloids in Piper longum L. by HPLC-ESI-MS(n) and UFLC-ESI-MS/MS and their application to *Piper nigrum* L. *Food Chem.* **2015**, *177*, 191–196. [CrossRef]
23. Bao, N.; Ochir, S.; Sun, Z.; Borjihan, G.; Yamagishi, T. Occurrence of piperidine alkaloids in Piper species collected in different areas. *J. Nat. Med.* **2014**, *68*, 211–214. [CrossRef]
24. Zaveri, M.; Khandhar, A.; Patel, S.; Patel, A. Chemistry and pharmacology of *Piper longum* L. *Int. J. Pharm. Sci. Rev. Res.* **2010**, *5*, 67–76.
25. Liu, H.; Zheng, J.; Liu, P.; Zeng, F. Pulverizing processes affect the chemical quality and thermal property of black, white, and green pepper (*Piper nigrum* L.). *J. Food Sci. Technol.* **2018**, *55*, 2130–2142. [CrossRef]
26. Nisha, P.; Singhal, R.S.; Pandit, A.B. The degradation kinetics of flavor in black pepper (*Piper nigrum* L.). *J. Food Eng.* **2009**, *92*, 44–49. [CrossRef]
27. Suresh, D.; Manjunatha, H.; Srinivasan, K. Effect of heat processing of spices on the concentrations of their bioactive principles: Turmeric (*Curcuma longa*), red pepper (*Capsicum annuum*) and black pepper (*Piper nigrum*). *J. Food Compost. Anal.* **2007**, *20*, 346–351. [CrossRef]
28. Therapeutic Goods Administration. CMEC 64 Complementary Medicines Evaluation Committee, Extracted Ratified Minutes. Sixty-Fourth Meeting, 14 December 2007. Available online: https://www.tga.gov.au/sites/default/files/cmec-minutes-64.pdf (accessed on 2 June 2021).
29. Bastaki, M.; Aubanel, M.; Bauter, M.; Cachet, T.; Demyttenaere, J.; Diop, M.M.; Harman, C.L.; Hayashi, S.M.; Krammer, G.; Li, X.; et al. Absence of adverse effects following administration of piperine in the diet of Sprague-Dawley rats for 90 days. *Food Chem. Toxicol.* **2018**, *120*, 213–221. [CrossRef]
30. MRI. *Nationale Verzehrsstudie II*; Ergebnisbericht, Teil 2; Max Rubner-Institut: Karlsruhe, Germany, 2008.
31. Ferrucci, L.M.; Daniel, C.R.; Kapur, K.; Chadha, P.; Shetty, H.; Graubard, B.I.; George, P.S.; Osborne, W.; Yurgalevitch, S.; Devasenapathy, N.; et al. Measurement of spices and seasonings in India: Opportunities for cancer epidemiology and prevention. *Asian Pac. J. Cancer Prev. APJCP* **2010**, *11*, 1621–1629. [PubMed]
32. EFSA. European Food Safety Authority. Scientific Opinion on Flavouring Group Evaluation 86, Revision 2 (FGE.86Rev2): Consideration of aliphatic and arylalkyl amines and amides evaluated by JECFA (65th meeting). *EFSA J.* **2015**, *13*, 3998. [CrossRef]
33. Joint FAO/WHO Expert Committee on Food Additives. Compendium of food additive specifications Addendum 13. In Proceedings of the Joint FAO/WHO Expert Committee on Food Additives (JECFA) 65th Meeting, Geneva, Switzerland, 7–15 June 2005.
34. Bhat, B.G.; Chandrasekhara, N. Studies on the metabolism of piperine: Absorption, tissue distribution and excretion of urinary conjugates in rats. *Toxicology* **1986**, *40*, 83–92. [CrossRef]
35. Bhat, B.G.; Chandrasekhara, N. Metabolic disposition of piperine in the rat. *Toxicology* **1987**, *44*, 99–106. [CrossRef]
36. Suresh, D.; Srinivasan, K. Tissue distribution & elimination of capsaicin, piperine & curcumin following oral intake in rats. *Indian J. Med. Res.* **2010**, *131*, 682–691. [PubMed]
37. Li, C.; Wang, Q.; Ren, T.; Zhang, Y.; Lam, C.W.K.; Chow, M.S.S.; Zuo, Z. Non-linear pharmacokinetics of piperine and its herb-drug interactions with docetaxel in Sprague-Dawley rats. *J. Pharm. Biomed. Anal.* **2016**, *128*, 286–293. [CrossRef]
38. Gao, T.; Xue, H.; Lu, L.; Zhang, T.; Han, H. Characterization of piperine metabolites in rats by ultra-high-performance liquid chromatography with electrospray ionization quadruple time-of-flight tandem mass spectrometry. *Rapid Commun. Mass Spectrom. RCM* **2017**, *31*, 901–910. [CrossRef] [PubMed]
39. Shang, Z.; Cai, W.; Cao, Y.; Wang, F.; Wang, Z.; Lu, J.; Zhang, J. An integrated strategy for rapid discovery and identification of the sequential piperine metabolites in rats using ultra high-performance liquid chromatography/high resolution mass spectrometery. *J. Pharm. Biomed. Anal.* **2017**, *146*, 387–401. [CrossRef]
40. Ternes, W.; Krause, E.L. Characterization and determination of piperine and piperine isomers in eggs. *Anal. Bioanal. Chem.* **2002**, *374*, 155–160. [CrossRef] [PubMed]

41. Sahu, P.K.; Sharma, A.; Rayees, S.; Kour, G.; Singh, A.; Khullar, M.; Magotra, A.; Paswan, S.K.; Gupta, M.; Ahmed, I.; et al. Pharmacokinetic study of piperine in wistar rats after oral and intravenous administration. *Int. J. Drug Deliv.* **2014**, *6*, 82–87.
42. Ren, T.; Wang, Q.; Li, C.; Yang, M.; Zuo, Z. Efficient brain uptake of piperine and its pharmacokinetics characterization after oral administration. *Xenobiotica* **2018**, *48*, 1249–1257. [CrossRef]
43. Shao, B.; Cui, C.; Ji, H.; Tang, J.; Wang, Z.; Liu, H.; Qin, M.; Li, X.; Wu, L. Enhanced oral bioavailability of piperine by self-emulsifying drug delivery systems: In vitro, in vivo and in situ intestinal permeability studies. *Drug Deliv.* **2015**, *22*, 740–747. [CrossRef]
44. Bajad, S.; Singla, A.K.; Bedi, K.L. Liquid chromatographic method for determination of piperine in rat plasma: Application to pharmacokinetics. *J. Chromatogr. B Anal. Technol. Biomed. Life Sci.* **2002**, *776*, 245–249. [CrossRef]
45. Kakarala, M.; Dubey, S.K.; Tarnowski, M.; Cheng, C.; Liyanage, S.; Strawder, T.; Tazi, K.; Sen, A.; Djuric, Z.; Brenner, D.E. Ultra-low flow liquid chromatography assay with ultraviolet (UV) detection for piperine quantitation in human plasma. *J. Agric. Food Chem.* **2010**, *58*, 6594–6599. [CrossRef]
46. Hölzel, C.; Spiteller, G. Piperin—ein Beispiel für individuell unterschiedliche (polymorphe) Metabolisierung einer allgegenwärtigen Nahrungskomponente. *Liebigs Ann. Chem.* **1984**, *1984*, 1319–1331. [CrossRef]
47. Li, Y.; Li, M.; Wang, Z.; Wen, M.; Tang, J. Identification of the metabolites of piperine via hepatocyte incubation and liquid chromatography combined with diode-array detection and high-resolution mass spectrometry. *Rapid Commun. Mass Spectrom. RCM* **2020**, *34*, e8947. [CrossRef]
48. EFSA. European Food Safety Authority. Flavouring Group Evaluation 86, (FGE.86)—Consideration of aliphatic and aromatic amines and amides evaluated by JECFA (65th meeting)—Scientific Opinion of the Panel on Food Additives—Flavourings, Processing Aids and Materials in Contact with Food. *EFSA J.* **2008**, *745*, 1–46. [CrossRef]
49. EFSA. European Food Safety Authority. Scientific Opinion on Flavouring Group Evaluation 86, Revision 1 (FGE.86Rev1): Consideration of aliphatic and aromatic amines and amides evaluated by JECFA (65th meeting). *EFSA J.* **2011**, *9*, 1926. [CrossRef]
50. JECFA. Joint FAO/WHO Expert Committee on Food Additives. Safety evaluation of certain food additives. *WHO Food Addit. Ser.* **2006**, *56*, 327–403.
51. Health Canada. Monograph black pepper—Piper nigrum (25 March 2019). Available online: http://webprod.hc-sc.gc.ca/nhpid-bdipsn/atReq.do?atid=blackpepper.poivrenoir&lang=eng (accessed on 2 June 2021).
52. Muralidhara; Narasimhamurthy, K. Lack of genotoxic effects of piperine, (the active principle of black pepper) in albino mice. *J. Food Saf.* **1990**, *11*, 39–48. [CrossRef]
53. Karekar, V.R.; Mujumdar, A.M.; Joshi, S.S.; Dhuley, J.; Shinde, S.L.; Ghaskadbi, S. Assessment of genotoxic effect of piperine using Salmonella typhimurium and somatic and somatic and germ cells of Swiss albino mice. *Arzneimittel-Forschung* **1996**, *46*, 972–975.
54. Daware, M.B.; Mujumdar, A.M.; Ghaskadbi, S. Reproductive toxicity of piperine in Swiss albino mice. *Planta Med.* **2000**, *66*, 231–236. [CrossRef]
55. Thiel, A.; Buskens, C.; Woehrle, T.; Etheve, S.; Schoenmakers, A.; Fehr, M.; Beilstein, P. Black pepper constituent piperine: Genotoxicity studies in vitro and in vivo. *Food Chem. Toxicol.* **2014**, *66*, 350–357. [CrossRef]
56. Piyachaturawat, P.; Glinsukon, T.; Toskulkao, C. Acute and subacute toxicity of piperine in mice, rats and hamsters. *Toxicol. Lett.* **1983**, *16*, 351–359. [CrossRef]
57. Bhat, B.G.; Chandrasekhara, N. Lack of adverse influence of black pepper, its oleoresin and piperine in the weanling rat. *J. Food Saf.* **1986**, *7*, 215–223. [CrossRef]
58. Malini, T.; Arunakaran, J.; Aruldhas, M.M.; Govindarajulu, P. Effects of piperine on the lipid composition and enzymes of the pyruvate-malate cycle in the testis of the rat in vivo. *Biochem. Mol. Biol. Int.* **1999**, *47*, 537–545. [CrossRef]
59. Malini, T.; Manimaran, R.R.; Arunakaran, J.; Aruldhas, M.M.; Govindarajulu, P. Effects of piperine on testis of albino rats. *J. Ethnopharmacol.* **1999**, *64*, 219–225. [CrossRef]
60. D'Cruz, S.C.; Mathur, P.P. Effect of piperine on the epididymis of adult male rats. *Asian J. Androl.* **2005**, *7*, 363–368. [CrossRef]
61. D'Cruz, S.C.; Vaithinathan, S.; Saradha, B.; Mathur, P.P. Piperine activates testicular apoptosis in adult rats. *J. Biochem. Mol. Toxicol.* **2008**, *22*, 382–388. [CrossRef]
62. Chinta, G.; Coumar, M.S.; Periyasamy, L. Reversible testicular toxicity of piperine on male albino rats. *Pharmacogn. Mag.* **2017**, *13*, S525–S532. [CrossRef]
63. Chen, X.; Ge, F.; Lian, Q.; Ge, F. Piperine promotes pubertal leydig cell development but inhibits spermatogenesis in rats. *Andrology* **2018**, *6*, 60. [CrossRef]
64. Chen, X.; Ge, F.; Liu, J.; Bao, S.; Chen, Y.; Li, D.; Li, Y.; Huang, T.; Chen, X.; Zhu, Q.; et al. Diverged effects of piperine on testicular development: Stimulating leydig cell development but inhibiting spermatogenesis in rats. *Front. Pharmacol.* **2018**, *9*, 244. [CrossRef]
65. Janarthanan, R.; Chinta, G.; Jesthadi, D.; Shanmuganathan, B.; Periyasamy, L. Effect of piperine on goat epididymal spermatozoa: An in vitro study. *Asian J. Pharm. Clin. Res.* **2014**, *7*, 57–61.
66. Piyachaturawat, P.; Sriwattana, W.; Damrongphol, P.; Pholpramool, C. Effects of piperine on hamster sperm capacitation and fertilization in vitro. *Int. J. Androl.* **1991**, *14*, 283–290. [CrossRef]
67. Mishra, R.K.; Singh, S.K. Antispermatogenic and antifertility effects of fruits of *Piper nigrum* L. in mice. *Indian J. Exp. Biol.* **2009**, *47*, 706–714. [PubMed]

68. Piyachaturawat, P.; Glinsukon, T.; Peugvicha, P. Postcoital antifertility effect of piperine. *Contraception* **1982**, *26*, 625–633. [CrossRef]

69. Chandhoke, N.; Gupta, S.; Dhar, S. Interceptive activity of various species of Piper, their natural amides and semi-synthetic analogs. *Indian J. Pharm. Sci.* **1978**, *40*, 113–116.

70. Piyachaturawat, P.; Pholpramool, C. Enhancement of fertilization by piperine in hamsters. *Cell Biol. Int.* **1997**, *21*, 405–409. [CrossRef]

71. Sethi, P.; Dua, V.K.; Mohanty, S.; Mishra, S.K.; Jain, R.; Edwards, G. Development and validation of a reversed phase hplc method for simultaneous determination of curcumin and piperine in human plasma for application in clinical pharmacological studies. *J. Liq. Chromatogr. Relat. Technol.* **2009**, *32*, 2961–2974. [CrossRef]

72. Bano, G.; Raina, R.K.; Zutshi, U.; Bedi, K.L.; Johri, R.K.; Sharma, S.C. Effect of piperine on bioavailability and pharmacokinetics of propranolol and theophylline in healthy volunteers. *Eur. J. Clin. Pharmacol.* **1991**, *41*, 615–617. [CrossRef]

73. Bedada, S.K.; Appani, R.; Boga, P.K. Effect of piperine on the metabolism and pharmacokinetics of carbamazepine in healthy volunteers. *Drug Res.* **2017**, *67*, 46–51. [CrossRef]

74. Bedada, S.K.; Boga, P.K. Effect of piperine on CYP2E1 enzyme activity of chlorzoxazone in healthy volunteers. *Xenobiotica* **2017**, *47*, 1035–1041. [CrossRef] [PubMed]

75. Bedada, S.K.; Boga, P.K. The influence of piperine on the pharmacokinetics of fexofenadine, a P-glycoprotein substrate, in healthy volunteers. *Eur. J. Clin. Pharmacol.* **2017**, *73*, 343–349. [CrossRef]

76. Bedada, S.K.; Boga, P.K.; Kotakonda, H.K. Study on influence of piperine treatment on the pharmacokinetics of diclofenac in healthy volunteers. *Xenobiotica* **2017**, *47*, 127–132. [CrossRef]

77. Kasibhatta, R.; Naidu, M.U. Influence of piperine on the pharmacokinetics of nevirapine under fasting conditions: A randomised, crossover, placebo-controlled study. *Drugs R&D* **2007**, *8*, 383–391. [CrossRef]

78. Rezaee, M.M.; Kazemi, S.; Kazemi, M.T.; Gharooee, S.; Yazdani, E.; Gharooee, H.; Shiran, M.R.; Moghadamnia, A.A. The effect of piperine on midazolam plasma concentration in healthy volunteers, a research on the CYP3A-involving metabolism. *DARU J. Pharm. Sci.* **2014**, *22*, 8. [CrossRef] [PubMed]

79. Bano, G.; Amla, V.; Raina, R.K.; Zutshi, U.; Chopra, C.L. The effect of piperine on pharmacokinetics of phenytoin in healthy volunteers. *Planta Med.* **1987**, *53*, 568–569. [CrossRef]

80. Rondanelli, M.; Opizzi, A.; Perna, S.; Faliva, M.; Solerte, S.B.; Fioravanti, M.; Klersy, C.; Cava, E.; Paolini, M.; Scavone, L.; et al. Improvement in insulin resistance and favourable changes in plasma inflammatory adipokines after weight loss associated with two months' consumption of a combination of bioactive food ingredients in overweight subjects. *Endocrine* **2013**, *44*, 391–401. [CrossRef] [PubMed]

81. Rondanelli, M.; Opizzi, A.; Perna, S.; Faliva, M.; Solerte, S.B.; Fioravanti, M.; Klersy, C.; Edda, C.; Maddalena, P.; Luciano, S.; et al. Acute effect on satiety, resting energy expenditure, respiratory quotient, glucagon-like peptide-1, free fatty acids, and glycerol following consumption of a combination of bioactive food ingredients in overweight subjects. *J. Am. Coll. Nutr.* **2013**, *32*, 41–49. [CrossRef]

82. Peterson, C.T.; Vaughn, A.R.; Sharma, V.; Chopra, D.; Mills, P.J.; Peterson, S.N.; Sivamani, R.K. Effects of turmeric and curcumin dietary supplementation on human gut microbiota: A double-blind, randomized, placebo-controlled pilot study. *J. Evid.-Based Integr. Med.* **2018**, *23*, 1–8. [CrossRef]

83. Polley, K.R.; Jenkins, N.; O'Connor, P.; McCully, K. Influence of exercise training with resveratrol supplementation on skeletal muscle mitochondrial capacity. *Appl. Physiol. Nutr. Metab.* **2016**, *41*, 26–32. [CrossRef]

84. Cicero, A.F.G.; Sahebkar, A.; Fogacci, F.; Bove, M.; Giovannini, M.; Borghi, C. Effects of phytosomal curcumin on anthropometric parameters, insulin resistance, cortisolemia and non-alcoholic fatty liver disease indices: A double-blind, placebo-controlled clinical trial. *Eur. J. Nutr.* **2020**, *59*, 477–483. [CrossRef] [PubMed]

85. Di Pierro, F.; Settembre, R. Safety and efficacy of an add-on therapy with curcumin phytosome and piperine and/or lipoic acid in subjects with a diagnosis of peripheral neuropathy treated with dexibuprofen. *J. Pain Res.* **2013**, *6*, 497–503. [CrossRef]

86. Haghpanah, S.; Zarei, T.; Eshghi, P.; Zekavat, O.; Bordbar, M.; Hoormand, M.; Karimi, M. Efficacy and safety of resveratrol, an oral hemoglobin F-augmenting agent, in patients with beta-thalassemia intermedia. *Ann. Hematol.* **2018**, *97*, 1919–1924. [CrossRef]

87. Panahi, Y.; Hosseini, M.S.; Khalili, N.; Naimi, E.; Majeed, M.; Sahebkar, A. Antioxidant and anti-inflammatory effects of curcuminoid-piperine combination in subjects with metabolic syndrome: A randomized controlled trial and an updated meta-analysis. *Clin. Nutr.* **2015**, *34*, 1101–1108. [CrossRef]

88. Panahi, Y.; Khalili, N.; Hosseini, M.S.; Abbasinazari, M.; Sahebkar, A. Lipid-modifying effects of adjunctive therapy with curcuminoids-piperine combination in patients with metabolic syndrome: Results of a randomized controlled trial. *Complementary Ther. Med.* **2014**, *22*, 851–857. [CrossRef]

89. Panahi, Y.; Khalili, N.; Sahebi, E.; Namazi, S.; Simental-Mendia, L.E.; Majeed, M.; Sahebkar, A. Effects of curcuminoids plus piperine on glycemic, hepatic and inflammatory biomarkers in patients with type 2 diabetes mellitus: A randomized double-blind placebo-controlled trial. *Drug Res.* **2018**, *68*, 403–409. [CrossRef]

90. Delecroix, B.; Abaidia, A.E.; Leduc, C.; Dawson, B.; Dupont, G. Curcumin and piperine supplementation and recovery following exercise induced muscle damage: A randomized controlled trial. *J. Sports Sci. Med.* **2017**, *16*, 147–153. [PubMed]

91. Gilardini, L.; Pasqualinotto, L.; Di Pierro, F.; Risso, P.; Invitti, C. Effects of Greenselect Phytosome(R) on weight maintenance after weight loss in obese women: A randomized placebo-controlled study. *BMC Complementary Altern. Med.* **2016**, *16*, 233. [CrossRef]

92. Panahi, Y.; Khalili, N.; Sahebi, E.; Namazi, S.; Atkin, S.L.; Majeed, M.; Sahebkar, A. Curcuminoids plus piperine modulate adipokines in type 2 diabetes mellitus. *Curr. Clin. Pharmacol.* **2017**, *12*, 253–258. [CrossRef] [PubMed]
93. Panahi, Y.; Khalili, N.; Sahebi, E.; Namazi, S.; Karimian, M.S.; Majeed, M.; Sahebkar, A. Antioxidant effects of curcuminoids in patients with type 2 diabetes mellitus: A randomized controlled trial. *Inflammopharmacology* **2017**, *25*, 25–31. [CrossRef] [PubMed]
94. Panahi, Y.; Khalili, N.; Sahebi, E.; Namazi, S.; Reiner, Z.; Majeed, M.; Sahebkar, A. Curcuminoids modify lipid profile in type 2 diabetes mellitus: A randomized controlled trial. *Complementary Ther. Med.* **2017**, *33*, 1–5. [CrossRef]
95. Rahimnia, A.R.; Panahi, Y.; Alishiri, G.; Sharafi, M.; Sahebkar, A. Impact of supplementation with curcuminoids on systemic inflammation in patients with knee osteoarthritis: Findings from a randomized double-blind placebo-controlled trial. *Drug Res.* **2015**, *65*, 521–525. [CrossRef] [PubMed]
96. Nageswari, A.D.; Rajanandh, M.G.; Uday, M.K.R.A.; Nasreen, R.J.; Pujitha, R.R.; Prathiksha, G. Effect of rifampin with bio-enhancer in the treatment of newly diagnosed sputum positive pulmonary tuberculosis patients: A double-center study. *J. Clin. Tuberc. Other. Mycobact. Dis.* **2018**, *12*, 73–77. [CrossRef]
97. Patel, N.; Jagannath, K.; Vora, A.; Patel, M.; Patel, A. A randomized, controlled, phase III clinical trial to evaluate the efficacy and tolerability of risorine with conventional rifampicin in the treatment of newly diagnosed pulmonary tuberculosis patients. *J. Assoc. Physicians India* **2017**, *65*, 48–54.
98. Esmaily, H.; Sahebkar, A.; Iranshahi, M.; Ganjali, S.; Mohammadi, A.; Ferns, G.; Ghayour-Mobarhan, M. An investigation of the effects of curcumin on anxiety and depression in obese individuals: A randomized controlled trial. *Chin. J. Integr. Med.* **2015**, *21*, 332–338. [CrossRef]
99. Khonche, A.; Biglarian, O.; Panahi, Y.; Valizadegan, G.; Soflaei, S.S.; Ghamarchehreh, M.E.; Majeed, M.; Sahebkar, A. Adjunctive therapy with curcumin for peptic ulcer: A randomized controlled trial. *Drug Res.* **2016**, *66*, 444–448. [CrossRef] [PubMed]
100. Lieberman, S.; Spahrs, R.; Stanton, A.; Martinez, L.; Grinder, M. Weight loss, body measurements, and compliance: A 12 week total lifestyle intervention pilot study. *Altern. Complement. Ther.* **2005**, *11*, 307–313. [CrossRef]
101. Majeed, M.; Vaidyanathan, P.; Kiradi, P.; Lad, P.S.; Vuppala, K.K. A clinical study on iron deficiency anaemia with bioiron. *Int. J. Ayurveda Pharma Res.* **2016**, *4*, 18–24.
102. Panahi, Y.; Alishiri, G.H.; Parvin, S.; Sahebkar, A. Mitigation of systemic oxidative stress by curcuminoids in osteoarthritis: Results of a randomized controlled trial. *J. Diet. Suppl.* **2016**, *13*, 209–220. [CrossRef]
103. Sharafkhaneh, A.; Lee, J.J.; Liu, D.; Katz, R.; Caraway, N.; Acosta, C.; Wistuba, I.I.; Aggarwal, B.; Dickey, B.; Moghaddam, S.J.; et al. A pilot double-blind, randomized, placebo-controlled trial of curcumin/bioperine for lung cancer chemoprevention in patients with chronic obstructive pulmonary disease. *Adv. Lung Cancer* **2013**, *2*, 62–69. [CrossRef]
104. Mirhafez, S.R.; Farimani, A.R.; Gholami, A.; Hooshmand, E.; Tavallaie, S.; Nobakht, M.G.B.F. The effect of curcumin with piperine supplementation on pro-oxidant and antioxidant balance in patients with non-alcoholic fatty liver disease: A randomized, double-blind, placebo-controlled trial. *Drug Metab. Pers. Ther.* **2019**, *34*. [CrossRef]
105. Panahi, Y.; Valizadegan, G.; Ahamdi, N.; Ganjali, S.; Majeed, M.; Sahebkar, A. Curcuminoids plus piperine improve nonalcoholic fatty liver disease: A clinical trial. *J. Cell. Biochem.* **2019**, *120*, 15989–15996. [CrossRef]
106. Galluccio, F.; Barskova, T.; Cerinic, M.M. Short-term effect of the combination of hyaluronic acid, chondroitin sulfate, and keratin matrix on early symptomatic knee osteoarthritis. *Eur. J. Rheumatol.* **2015**, *2*, 106–108. [CrossRef]
107. Hatab, H.M.; Abdel Hamid, F.F.; Soliman, A.F.; Al-Shafie, T.A.; Ismail, Y.M.; El-Houseini, M.E. A combined treatment of curcumin, piperine, and taurine alters the circulating levels of IL-10 and miR-21 in hepatocellular carcinoma patients: A pilot study. *J. Gastrointest. Oncol.* **2019**, *10*, 766–776. [CrossRef]
108. Directo, D.; Wong, M.W.H.; Elam, M.L.; Falcone, P.; Osmond, A.; Jo, E. The Effects of a Multi-Ingredient Performance Supplement Combined with Resistance Training on Exercise Volume, Muscular Strength, and Body Composition. *Sports* **2019**, *7*, 152. [CrossRef] [PubMed]
109. Saberi-Karimian, M.; Keshvari, M.; Ghayour-Mobarhan, M.; Salehizadeh, L.; Rahmani, S.; Behnam, B.; Jamialahmadi, T.; Asgary, S.; Sahebkar, A. Effects of curcuminoids on inflammatory status in patients with non-alcoholic fatty liver disease: A randomized controlled trial. *Complementary Ther. Med.* **2020**, *49*, 102322. [CrossRef]
110. Heidari-Beni, M.; Moravejolahkami, A.R.; Gorgian, P.; Askari, G.; Tarrahi, M.J.; Bahreini-Esfahani, N. Herbal formulation "turmeric extract, black pepper, and ginger" versus Naproxen for chronic knee osteoarthritis: A randomized, double-blind, controlled clinical trial. *Phytother. Res.* **2020**, *34*, 2067–2073. [CrossRef] [PubMed]
111. Majeed, M.; Majeed, S.; Nagabhushanam, K. An open-label pilot study on Macumax supplementation for dry-type age-related macular degeneration. *J. Med. Food* **2021**, *24*, 551–557. [CrossRef] [PubMed]
112. Pastor, R.F.; Repetto, M.G.; Lairion, F.; Lazarowski, A.; Merelli, A.; Manfredi Carabetti, Z.; Pastor, I.; Pastor, E.; Iermoli, L.V.; Bavasso, C.A.; et al. Supplementation with resveratrol, piperine and alpha-tocopherol decreases chronic inflammation in a cluster of older adults with metabolic syndrome. *Nutrients* **2020**, *12*, 3149. [CrossRef]
113. Sousa, D.F.; Araújo, M.F.M.; de Mello, V.D.; Damasceno, M.M.C.; Freitas, R. Cost-effectiveness of passion fruit albedo versus turmeric in the glycemic and lipaemic control of people with type 2 diabetes: Randomized clinical trial. *J. Am. Coll. Nutr.* **2020**, *10*, 1–10. [CrossRef]
114. Shadnoush, M.; Zahedi, H.; Norouzy, A.; Sahebkar, A.; Sadeghi, O.; Najafi, A.; Hosseini, S.; Qorbani, M.; Ahmadi, A.; Ardehali, S.H.; et al. Effects of supplementation with curcuminoids on serum adipokines in critically ill patients: A randomized double-blind placebo-controlled trial. *Phytother. Res. PTR* **2020**, *34*, 3180–3188. [CrossRef]

115. Ablon, G.; Kogan, S. A six-month, randomized, double-blind, placebo-controlled study evaluating the safety and efficacy of a nutraceutical supplement for promoting hair growth in women with self-perceived thinning hair. *J. Drugs Dermatol.* **2018**, *17*, 558–565.

116. Mohajer, A.; Ghayour-Mobarhan, M.; Parizadeh, S.M.R.; Tavallaie, S.; Rajabian, M.; Sahebkar, A. Effects of supplementation with curcuminoids on serum copper and zinc concentrations and superoxide dismutase enzyme activity in obese subjects. *Trace Elem. Electrolytes* **2014**, *32*, 16–21. [CrossRef]

117. Andres, S.; Pevny, S.; Ziegenhagen, R.; Bakhiya, N.; Schafer, B.; Hirsch-Ernst, K.I.; Lampen, A. Safety aspects of the use of quercetin as a dietary supplement. *Mol. Nutr. Food Res.* **2018**, *62*, 1700447. [CrossRef]

118. Di Pierro, F.; Bressan, A.; Ranaldi, D.; Rapacioli, G.; Giacomelli, L.; Bertuccioli, A. Potential role of bioavailable curcumin in weight loss and omental adipose tissue decrease: Preliminary data of a randomized, controlled trial in overweight people with metabolic syndrome. Preliminary study. *Eur. Rev. Med. Pharmacol. Sci.* **2015**, *19*, 4195–4202.

119. Piontek, U.; Wallaschofski, H.; Kastenmuller, G.; Suhre, K.; Volzke, H.; Do, K.T.; Artati, A.; Nauck, M.; Adamski, J.; Friedrich, N.; et al. Sex-specific metabolic profiles of androgens and its main binding protein SHBG in a middle aged population without diabetes. *Sci. Rep.* **2017**, *7*, 2235. [CrossRef]

120. Pattanaik, S.; Hota, D.; Prabhakar, S.; Kharbanda, P.; Pandhi, P. Effect of piperine on the steady-state pharmacokinetics of phenytoin in patients with epilepsy. *Phytother. Res. PTR* **2006**, *20*, 683–686. [CrossRef]

121. Pattanaik, S.; Hota, D.; Prabhakar, S.; Kharbanda, P.; Pandhi, P. Pharmacokinetic interaction of single dose of piperine with steady-state carbamazepine in epilepsy patients. *Phytother. Res. PTR* **2009**, *23*, 1281–1286. [CrossRef]

122. Zutshi, R.K.; Singh, R.; Zutshi, U.; Johri, R.K.; Atal, C.K. Influence of piperine on rifampicin blood levels in patients of pulmonary tuberculosis. *J. Assoc. Physicians India* **1985**, *33*, 223–224.

123. Badmaev, V.; Majeed, M.; Norkus, E.P. Piperine, an alkaloid derived from black pepper increases serum response of beta-carotene during 14-days of oral beta-carotene supplementation. *Nutr. Res.* **1999**, *19*, 381–388. [CrossRef]

124. Badmaev, V.; Majeed, M.; Prakash, L. Piperine derived from black pepper increases the plasma levels of coenzyme Q10 following oral supplementation. *J. Nutr. Biochem.* **2000**, *11*, 109–113. [CrossRef]

125. Shoba, G.; Joy, D.; Joseph, T.; Majeed, M.; Rajendran, R.; Srinivas, P.S. Influence of piperine on the pharmacokinetics of curcumin in animals and human volunteers. *Planta Med.* **1998**, *64*, 353–356. [CrossRef]

126. Gupta, S.K.; Velpandian, T.; Sengupta, S.; Mathur, P.; Sapra, P. Influence of piperine on nimesulide induced antinociception. *Phytother. Res. PTR* **1998**, *12*, 266–269. [CrossRef]

127. Hiwale, A.R.; Dhuley, J.N.; Naik, S.R. Effect of co-administration of piperine on pharmacokinetics of beta-lactam antibiotics in rats. *Indian J. Exp. Biol.* **2002**, *40*, 277–281.

128. Janakiraman, K.; Manavalan, R. Studies on effect of piperine on oral bioavailability of ampicillin and norfloxacin. *Afr. J. Tradit. Complementary Altern. Med. AJTCAM* **2008**, *5*, 257–262. [CrossRef]

129. Jin, M.J.; Han, H.K. Effect of piperine, a major component of black pepper, on the intestinal absorption of fexofenadine and its implication on food-drug interaction. *J. Food Sci.* **2010**, *75*, H93–H96. [CrossRef]

130. Alhumayyd, M.S.; Bukhari, I.A.; Almotrefi, A.A. Effect of piperine, a major component of black pepper, on the pharmacokinetics of domperidone in rats. *J. Physiol. Pharmacol.* **2014**, *65*, 785–789.

131. Veeresham, C.; Sujatha, S.; Rani, T.S. Effect of piperine on the pharmacokinetics and pharmacodynamics of glimepiride in normal and streptozotocin-induced diabetic rats. *Nat. Prod. Commun.* **2012**, *7*, 1283–1286. [CrossRef]

132. Auti, P.; Choudhary, A.; Gabhe, S.; Mahadik, K. Bioanalytical method development, validation and its application in pharmacokinetic studies of verapamil in the presence of piperine in rats. *Int. J. Pharm. Res.* **2018**, *10*, 118–123.

133. Basu, S.; Jana, S.; Patel, V.B.; Patel, H. Effects of piperine, cinnamic acid and gallic acid on rosuvastatin pharmacokinetics in rats. *Phytother. Res. PTR* **2013**, *27*, 1548–1556. [CrossRef]

134. Mujumdar, A.M.; Dhuley, J.N.; Deshmukh, V.K.; Naik, S.R. Effect of piperine on bioavailability of oxyphenylbutazone in rats. *Indian Drugs* **1999**, *36*, 123–126.

135. Parveen, B.; Pillai, K.K.; Tamboli, E.T.; Ahmad, S. Effect of piperine on pharmacokinetics of sodium valproate in plasma samples of rats using gas chromatography-mass spectrometry method. *J. Pharm. Bioallied Sci.* **2015**, *7*, 317–320. [CrossRef] [PubMed]

136. Suvarna, V.M.; Sangave, P.C. HPLC Estimation, Ex vivo Everted Sac Permeability and In Vivo Pharmacokinetic Studies of Darunavir. *J. Chromatogr. Sci.* **2018**, *56*, 307–316. [CrossRef]

137. Auti, P.; Gabhe, S.; Mahadik, K. Bioanalytical method development and its application to pharmacokinetics studies on Simvastatin in the presence of piperine and two of its synthetic derivatives. *Drug Dev. Ind. Pharm.* **2019**, *45*, 664–668. [CrossRef] [PubMed]

138. Rewanthwar, S.; Lakshmi, P.K. Effect of piperine, quercetin, polysorbate 80 on the oral bioavailability of losartan in Male Wistar rats. *Pharm. Nanotechnol.* **2014**, *2*, 49–55. [CrossRef]

139. Singh, A.; Pawar, V.K.; Jakhmola, V.; Parabia, M.H.; Awasthi, R.; Sharma, G. In vivo assessment of enhanced bioavailability of metronidazole with piperine in rabbits. *Res. J. Pharm. Biol. Chem. Sci.* **2010**, *1*, 273–278.

140. Sama, V.; Nadipelli, M.; Yenumula, P.; Bommineni, M.R.; Mullangi, R. Effect of piperine on antihyperglycemic activity and pharmacokinetic profile of nateglinide. *Arzneimittel-Forschung* **2012**, *62*, 384–388. [CrossRef] [PubMed]

141. Venkatesh, S.; Durga, K.D.; Padmavathi, Y.; Reddy, B.M.; Mullangi, R. Influence of piperine on ibuprofen induced antinociception and its pharmacokinetics. *Arzneimittel-Forschung* **2011**, *61*, 506–509. [CrossRef]

142. Chauhan, V.B.; Modi, C.M.; Patel, U.D.; Patel, H.B. The effect of piperine pre-conditioning on the pharmacokinetics of orally administered marbofloxacin in rats. *Vet. Arh.* **2020**, *90*, 69–75. [CrossRef]
143. Kolakota, R.; Bothsa, A.; Mugada, V. Impact of N-acyl piperidine (Piperine) from *Piper nigrum* on the pharmacokinetics of CYP3A substrate almotriptan in rats. *Trop. J. Nat. Prod. Res.* **2020**, *4*, 378–384. [CrossRef]
144. Tiwari, A.; Gabhe, S.Y.; Mahadik, K.R. Comparative study on the pharmacokinetics of ibuprofen alone or in combination with piperine and its synthetic derivatives as a potential bioenhancer. *Int. J. Pharm. Sci. Res.* **2021**, *12*, 363–371. [CrossRef]
145. Mishra, P.S.; Phadnis, P.; Vyas, S.; Nyati, P. Study of cns activities of piperine perse and its bio-enhancing effect on various drugs in experimental animal models. *Int. J. Pharm. Pharm. Sci.* **2016**, *8*, 135–140.
146. Atal, S.; Atal, S.; Vyas, S.; Phadnis, P. Bio-enhancing effect of piperine with metformin on lowering blood glucose level in alloxan induced diabetic mice. *Pharmacogn. Res.* **2016**, *8*, 56–60. [CrossRef]
147. Qiang, F.; Kang, K.W.; Han, H.K. Repeated dosing of piperine induced gene expression of P-glycoprotein via stimulated pregnane-X-receptor activity and altered pharmacokinetics of diltiazem in rats. *Biopharm. Drug Dispos.* **2012**, *33*, 446–454. [CrossRef]
148. Ren, T.; Xiao, M.; Yang, M.; Zhao, J.; Zhang, Y.; Hu, M.; Cheng, Y.; Xu, H.; Zhang, C.; Yan, X.; et al. Reduced systemic and brain exposure with inhibited liver metabolism of carbamazepine after its long-term combination treatment with piperine for epilepsy control in rats. *AAPS J.* **2019**, *21*, 90. [CrossRef] [PubMed]
149. Zayed, A.; Babaresh, W.M.; Darweesh, R.S.; El-Elimat, T.; Hawamdeh, S.S. Piperine alters the pharmacokinetics and anticoagulation of warfarin in rats. *J. Exp. Pharmacol.* **2020**, *12*, 169–179. [CrossRef]
150. Han, H.K. The effects of black pepper on the intestinal absorption and hepatic metabolism of drugs. *Expert Opin. Drug Metab. Toxicol.* **2011**, *7*, 721–729. [CrossRef] [PubMed]
151. EFSA. European Food Safety Authority. Guidance on selected default values to be used by the EFSA Scientific Committee, Scientific Panels and Units in the absence of actual measured data. *EFSA J.* **2012**, *10*, 2579.

 *foods*

Article

Health Benefits and Risks of Consuming Spices on the Example of Black Pepper and Cinnamon

Joanna Newerli-Guz * and Maria Śmiechowska 

Department of Quality Management, Gdynia Maritime University, Morska 83, 81-225 Gdynia, Poland
* Correspondence: j.newerli-guz@wznj.umg.edu.pl

Abstract: The aim of this study is to present the benefits and risks associated with the consumption of black pepper and cinnamon, which are very popular spices in Poland. The article presents the current state of knowledge about health properties and possible dangers, such as liver damage, associated with their consumption. The experimental part presents the results of the research on the antioxidant properties against the DPPH radical, which was 80.85 ± 3.84–$85.42 \pm 2.34\%$ for black pepper, and 55.52 ± 7.56–$91.87 \pm 2.93\%$ for cinnamon. The total content of polyphenols in black pepper was 10.67 ± 1.30–32.13 ± 0.24 mg GAE/g, and in cinnamon 52.34 ± 0.96–94.71 ± 3.34 mg GAE/g. In addition, the content of piperine and pepper oil in black pepper was determined, as well as the content of coumarin in cinnamon. The content of piperine in the black pepper samples was in the range of 3.92 ± 0.35–$9.23 \pm 0.05\%$. The tested black pepper samples contained 0.89 ± 0.08–2.19 ± 0.15 mL/100 g d.m. of essential oil. The coumarin content in the cinnamon samples remained in the range of 1027.67 ± 50.36–4012.00 ± 79.57 mg/kg. Taking into account the content of coumarin in the tested cinnamon samples, it should be assumed that the majority of cinnamon available in Polish retail is *Cinnamomum cassia* (L.) J. Presl.

Keywords: black pepper; cinnamon; health benefits and risks; antioxidant properties

Citation: Newerli-Guz, J.; Śmiechowska, M. Health Benefits and Risks of Consuming Spices on the Example of Black Pepper and Cinnamon. *Foods* **2022**, *11*, 2746. https://doi.org/10.3390/foods11182746

Academic Editors: Andreas Eisenreich and Bernd Schaefer

Received: 5 August 2022
Accepted: 2 September 2022
Published: 7 September 2022

Publisher's Note: MDPI stays neutral with regard to jurisdictional claims in published maps and institutional affiliations.

1. Introduction

Herbs and spices are very important in food technology, gastronomy and home cooking. Spices are most often defined as products of plant origin [1]. Initially their use was mainly combined with flavoring and improving the appearance of food. Then, it was pointed out that spices have preservative, antioxidant and antimicrobial properties, which significantly affect the shelf life and value of food [2]. Although the preservative properties of spices have been known to mankind since ancient times, nowadays, thanks to highly advanced analytical methods, we are learning about the compounds that are responsible for this action [3]. Health-protective properties are another benefit of the use of spices. Herbs and spices are carriers of numerous chemical compounds with health-improving properties that can provide potential protection against cardiovascular diseases, neurodegenerative diseases, diabetes type 2, and cancer [4]. However, we must remember the differences in terminology and we want to pay special attention to the difference between medicinal herbs, which are classified as medicinal products, and herbs, which are in turn food products.

Black pepper [*Piper nigrum* L.] and cinnamon [*Cinnamomum* spp.] are very popular spices in Poland. In 2020, imports of black pepper to Europe amounted to 80,000 tons, of which 6% was imported to Poland, and it was the 6th largest importer of black pepper in Europe. Over 75% of Polish pepper imports are whole grains. Part of the imported pepper is re-exported, and the estimated domestic consumption in 2020 was 5400 tons [5]. Among European countries, Poland ranks 6th when it comes to cinnamon imports (other than cinnamon Cinnamomum zeylanicum Blume neither crushed nor ground) and it accounts for 0.74% of total world imports. When it comes to cinnamon Cinnamomum zeylanicum B. neither crushed nor ground, Poland ranks 8th among the European countries with imports amounting to 0.32% [6]. Those two spices are one of the most consumed spices in Europe and Poland with a wide range of uses [7,8]. The aim of this work was to make a commodity characterization of these spices in the context of their potential importance for the consumer. In the above-mentioned spices, the main bioactive substances were determined and compared with the results of other authors' research. Specifically, the antioxidant properties of the spices were determined as these are most often mentioned in the discussion of health-promoting properties of spices. Additionally, piperine and coumarine were chosen as representative biomarkers of the quality of black pepper and cinnamon, respectively. Piperine is the most important bioactive compound of black pepper, while coumarin due to its hepatotoxic activity should be closely monitored in any products with the addition of cinnamon.

An attempt was also made to determine the significance of the addition of a given spice to food on its health-promoting impact.

2. Characteristics of Selected Spices and Its Relevance to the Consumer

2.1. Black Pepper [Piper nigrum L.]—Origin, Types and Properties

Black pepper [*Piper nigrum* L.] is one of the most popular and one of the oldest spices [9]. It comes from the Malabar Coast of south-western India from where its journey around the world began. It is grown in many countries, including India, China, Indonesia, Malaysia, Brazil, Sri Lanka, and Vietnam. The great career of pepper began in the fifteenth and sixteenth centuries with the era of geographical discoveries. The spice is the dried seeds of this tropical evergreen plant. It occurs in several stages of harvesting as white, green and black pepper and differs in the degree of maturity and method of treatment.

Black pepper is obtained from unripe fruits by drying, most often in the sun, until the skin acquires a very dark color, almost black. Many factors influence the quality of black pepper (Figure 1). One of the factors affecting the quality of this spice is the variety of the plant. The Indian Spice Research Institute in India maintains the world's largest collection of over 2000 varieties [9]. Other factors shaping the quality of black pepper are details of production including, inter alia, conditions of growing, harvesting, threshing, drying, cleaning, storage and standardization. The quality of black pepper depends largely on the size of the corns, color, the content of light, damaged peppercorns, humidity, the presence of foreign bodies such as animal excrement, and the presence of insects. These factors are fundamentally determined by the practices of harvesting, processing and the treatment of peppercorns on plantations and the classification and storage procedures adopted by exporters. Another aspect of quality is the level of microbial contamination, which should not exceed the permissible limits [10]. The commercial quality of black pepper is determined by pre-treatment and treatment of grains after harvest. Most often, peppercorn is blanched and intensively dried, as a result of which it usually contains less than 10% water.

However, pepper drying is not always carried out in mechanical dryers. In India, a very popular method is the solarization of pepper, which involves drying pepper in the sun after blanching in hot water [11]. Pepper is grown in tropical conditions characterized by high temperature, high precipitation and humidity. Such conditions are favorable to the development of fungi, which leads to an increased occurrence of mycotoxins, especially in the absence of Good Agricultural Practices (GAP), Good Manufacturing Practices (GMP) and Good Hygiene Practices (GHP) [12]. For additional protection of black pepper against mycotoxins, as well as for decontamination, gamma irradiation is used [13]. Increasingly, however, consumers are looking for organic or bio-based products. Such alternative cultivation methods include, for example, the combined use of *Rhizosphere* bacteria with endophytic bacteria, which inhibits root diseases and increases the productivity of black pepper. Organic cultivation, on the other hand, provides pepper without pesticide and herbicide residues [14–16]. Additional factors affecting the quality of pepper are climate change and economic and social factors. Research conducted by Karamawati et al. [17] shows that Indonesian farmers are switching from pepper to oil palms farming for this reason, which brings higher profits and requires less work. The growing conditions and commercial quality of black pepper affect the sensory characteristics, which gives satisfaction to consumers. Unfortunately, the consumers are often disappointed with the quality of the spice when they discern the lack of information about the origin of the spice or the content does not correspond to the name of the product. To identify and study the authenticity of black pepper, very advanced research methods such as infrared spectroscopy with Fourier-transform using chemometry or metabolomic studies using ultra-high-performance liquid chromatography coupled with high-resolution mass spectrometry must be used [18,19]. These methods are very expensive, time-consuming and require a huge commitment of forces and resources and very well-equipped research laboratories. Therefore, the study of authenticity and traceability seeks fast, non-destructive methods based on the analysis of various data using chemometry, which will shorten and accelerate the research at least at the initial stage of identification [20,21]. The sensory and health properties of black pepper arise from its composition. Black pepper (*Piper nigrum* L.) contains compounds that can be classified into three groups. The first group consists of compounds that determine the sharpness of black pepper, the second includes substances that determine its aroma, and the third group consists of other compounds such as: fiber, starch, polyphenols, mineral salts, lipids [22]. The compound, which determines the pungent aroma of black pepper is piperine, together with its analogues.

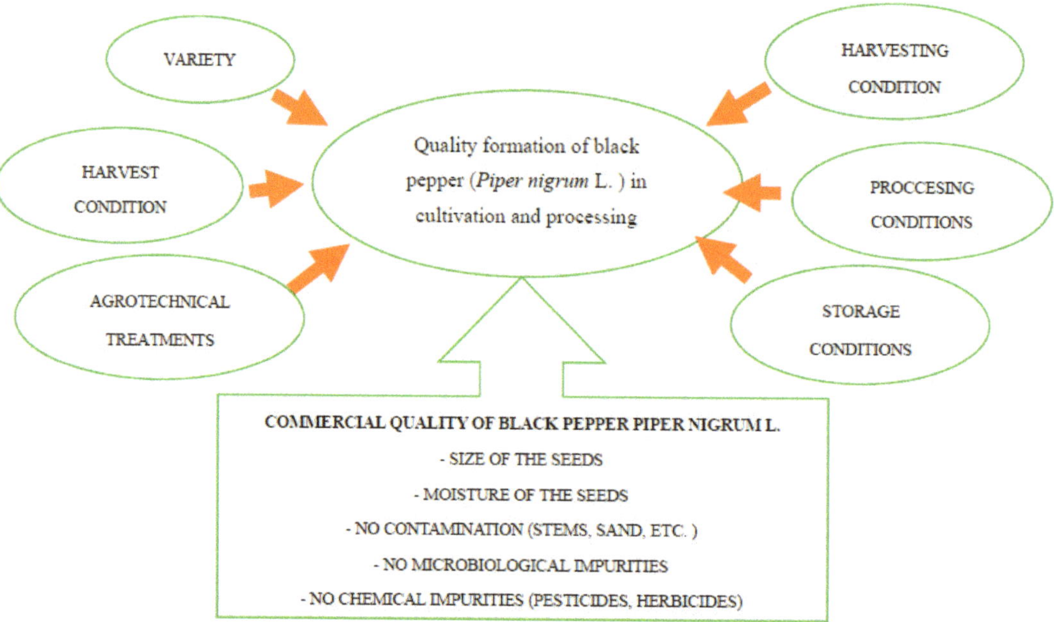

Figure 1. Factors influencing the black pepper (*Piper nigrum* L.) quality. *Source: own study.*

From its detection by Hans Christian Ørsted in 1819, 55 piperine analogues were identified in black pepper [23]. The essential oil of black pepper also has very rich composition. It contains terpene and sesquiterpene hydrocarbons and their oxidized forms in the amount of 1–3%. Its composition is influenced by geographical origin and respective production process. A very important ingredient of black pepper is oleoresin, derived by extraction in the amount of 6–13%. It consists of 15–20% of essential oil and 35–55% piperine [24–26]. For the consumer, the quality of spice is determined by its seasoning properties (the smell and taste giving and improving in different dishes). The health properties of black pepper and piperine are presented on Figure 2 and in Table 1.

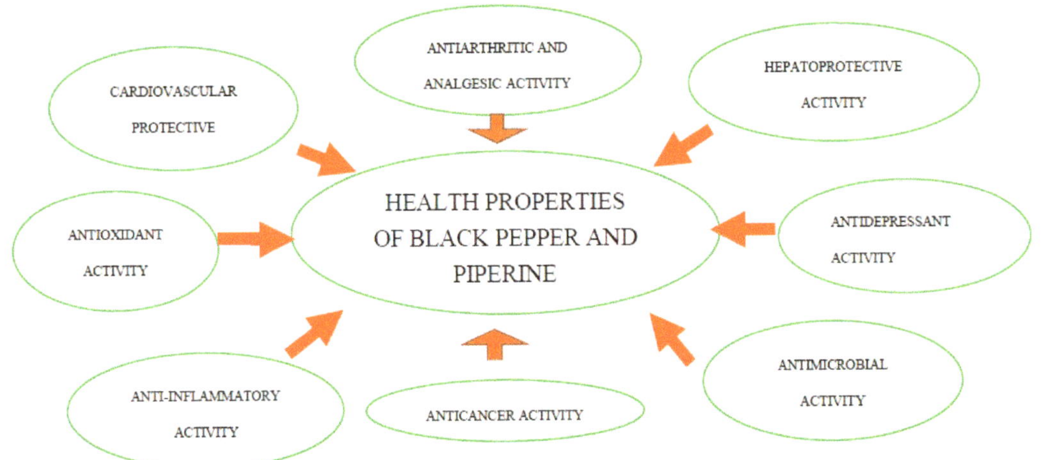

Figure 2. Health properties of black pepper and piperine. *Source: own study.*

Table 1. Health properties of black pepper and piperine based on other authors' studies.

Health Effects	Source
Antiallergic effect	Aswar et al., 2015; Kim & Lee 2009 [27,28]
Analgetic activity	Bukhari et al., 2013; Jeena et al., 2014; Tasleem et al., 2014 [29–31]
Antiarthritic activity	Bang et al., 2009; Umar et al., 2013 [32,33]
Anticancer activity	Banerjee et al., 2021; Greenshields et al., 2015; Morsy & Abd El-Salam 2017; Prashant et al., 2017 [34–37]
Antidiabetic activity	Atal et al., 2016; Kharbanda et al., 2016; Oboh et al., 2013; Sarfraz et al., 2021 [38–41]
Antidepressant activity	Emon et al., 2020; Ghosh et al., 2021; Hritcu et al., 2015; Mao et al., 2014 [42–45]
Antihypertensive activity	Hlavačková et al., 2011; Lee et al., 2015; Taqvi et al., 2008; [46–48]
Antiinflammatory activity	Bang et al., 2009; Jeena et al., 2014; Tasleem et al., 2014; Yu et al., 2020 [30–32,49]
Antimicrobial activity	Bawazeer et al., 2022; Chen et al., 2019; Daigham & Mahfouz 2020; Hien & Dao 2022; Martinelli et al., 2017; Morsy & Abd El-Salam 2017; Zhang et al., 2017; Zou et al., 2015 [36,50–56]
Antineurodegenerative activity	Chonpathompikunlert 2010; Elnaggar et al., 2015; Etman et al., 2018 [57–59]
Antiobesity activity	Du et al., 2020; Lailiyah et al., 2021; Meriga et al., 2017; Shah et al., 2011 [60–63]
Antioxidant activity	Jeena et al., 2014; Srinivasan 2007; Vijayakumar & Nalini 2006 [30,64,65]
Cardiovascular protection	Dutta et al., 2014; Taqvi et al., 2008; Wang et al., 2021 [48,66,67]
Gastrointestinal and antidiarrheal activity	Mehmood & Gilani 2010; Srinivasan 2007; Shamkuwar et al., 2012; Shamkuwar 2013 [64,68–70]
Hepatoprotective and pancreatitis activity	Bae et al., 2011; Christina et al., 2006; Gurumurthy et al., 2012; Matsuda et al., 2008; Nirwane & Bapat 2011 [71–75]

Source: own study.

2.2. Cinnamon (Cinnamomum spp.)—Characteristics, Properties, Use

Cinnamon, next to pepper, is one of the most widely used spices in the world. Cinnamon (*Cinnamomum* Scheffer) belongs to the *Lauracae* family, which includes more than 200 species. Cinnamon trees are evergreen trees growing in tropical climates in countries such as India, Indonesia, Philippines, Sri Lanka, Myanmar, Vietnam and China. However, of these, only a few have a unique significance in medicine, cosmetology and as a spice (Table 2).

Table 2. Main varieties of cinnamon for medicinal and seasoning purposes.

Systematic Term	Synonym	Common Term	Occurrence	Source
Cinnamomum verum J. Presl	*Cinnamomum zeylanicum* Blume	Ceylon cinnamon True cinnamon	Sri Lanca	Weerasekera et al., 2021 [76]
Cinnamomum cassia (L.) J.Presl	*Cinnamomum aromaticum* Ness	Chinese cinnamon	China, North-east asia	Zhang et al., 2017 [55]
Cinnamomum burmannii (Nees & T.Nees) Blume	—	Indonesian cinnamon Java cinnamon	Indonesia, Wietnam, Filipines	Al-Dhubiab 2012 [77]
Cinnamomum tamala (Buch.-Ham.) T.Ness & C.H. Eberm.	*Cinnamomum albiflorum* Ness *Cinnamomum lindleyi* Lukman *Laurus tamala* Buch.-Ham.	Indian cassia Indian bay leaf Tejpatta	India, Nepal, Bhutan, China	Tiwari & Talreja 2020 [78]
Cinnamomum loureiroi Ness	—	Saigon cinnamon Vietnamese cinnamon	Wietnam	Kumar et al., 2019 [79]

Source: own study.

Species of cinnamon trees differ in the morphological features of the leaves, such as shape, color, size and veining. Other distinguishing features of cinnamon trees are the morphological features of flowers, fruits and bark. The properties of cinnamon are mainly due to flowers, fruits, leaves and bark composition. The most valued part of cinnamon due to its unique healing properties is the bark of *C. verum* J. Presl (synonym of *C. zeylanicum* Blume) [76]. The bark of *C. verum* J. Presl is thin, softer than other cinnamon trees bark, paper and curls inwards on both sides, while the bark of *C. cassia* L. is hard, thicker and curls only on one side, see Figure 3. Individual species of cinnamon bark also have a different color, aroma, tenderness and taste.

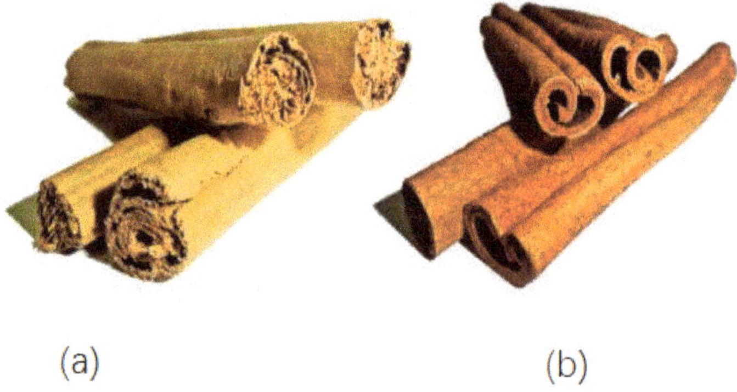

Figure 3. Cinnamon bark (**a**) *C. verum* J. Presl and (**b**) *C. cassia* L. *Source: [80].*

The medicinal and spicy properties of *C. verum* J. Presl cinnamon bark result mainly from its composition. In addition, Ceylon cinnamon is more expensive due to its chemical composition, high quality, health benefits, and trace amounts of coumarin, which is found in higher concentrations in Cassia cinnamon. Table 3 shows the bark composition of two popular cinnamon trees.

Table 3. Basic composition of the bark of *Cinnamomum verum* J. Presl and *Cinnamomum cassia* L.

Cinnamon	Composition
Cinnamomum verum	cinnamaldehyde 1.99% (1.49–3.20%), cinnamylacetate, cinnamon alcohol 0.043% (nd–0.083%), eugenol, coumarin (nd–0.004%)
Cinnamomum cassia	cinnamaldehyde (0.005–9.383%), cinnamic acid (0.001–0.191%), cinnamyl alcohol (0.001–0.177%) cinnamamyl acetate, cinnamon alcohol, eugenol, coumarin 0.001–1.218% (up to 5%)

Source: [81].

A number of compounds contained in cinnamon such as aldehydes, alcohols, acids, esters, terpenes and others cause that it is used as a flavoring agent in seasonings, sauces, pastries and sweets, drinks, meat dishes, cereals, chewing gums and fruit preserves. The cinnamaldehyde in cinnamon is responsible for its sweet taste, and the addition of sugar in food products further enhances this effect. Cinnamon oil obtained from bark is also a flavoring agent in perfumes and toilet waters, lotions, shampoos, soaps and other cosmetics [82].

In the food trade, cinnamon occurs in the form of bark fragments of various sizes, but more often in the form of ground powder. Ground cinnamon is sometimes falsified and instead of *C. verum* contains *C. cassia* or powder of other cinnamon trees, and on the labels of food products containing cinnamon there is no information about the origin of the raw material [83]. For this purpose, very specialized and advanced analytical methods are used to detect adulteration, which allow the identification of the raw material [51,84–87].

The identification of geographical origin is also important due to the presence of coumarin in cinnamon. Coumarin compounds with a benzopyrone structure are found in plants of the families *Leguminosae* (Legumes), *Rutaceae* (Ruthaceae), *Umbellifereae* (Apiaceae) (Umbellate), *Compositae* (Complex), *Gramineae* (Grassy), forming connections of a glycosidic nature. However, in 1954, when the FDA report on the hepatotoxic effects of coumarin was published, the European Commission directive 88/388/EEC (Annex II) recommended that the intake of coumarin from natural sources, used as a food additive, should not exceed 2 mg/kg of food.

In 2005, Chemisches and Veterinäruntersuchungsamt in Münster, Germany, determined a coumarin content of 22 mg/kg in a sample of Christmas cinnamon cookies, which led to a discussion on increasing the supervision of food products containing coumarin, as well as on compliance with the limits by food producers [88]. As a result of these studies and discussions, the EFSA (European Food Safety Authority) has set the Acceptable Daily Intake (ADI for animals at 0.1 mg/kg body weight based on the maximum non-harmful dose (NOAEL). In 2008, the EFSA established the same TDI for humans [89].

C. verum is mainly used for medicinal purposes, due to the fact that only this species of cinnamon contains small amounts of coumarin. Cinnamon for medicinal purposes has been used since ancient times [90] and the medicinal properties of cinnamon have been confirmed in numerous studies (Figure 4 and Table 4).

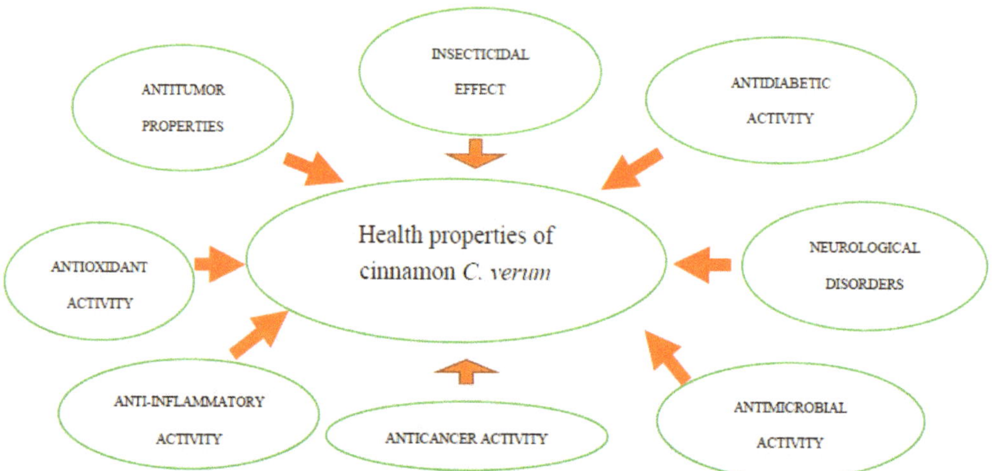

Figure 4. Health properties of cinnamon *C. verum*. *Source: own study.*

Table 4. Health properties of cinnamon based on other authors' studies.

Health Properties	Source
Analgetic activity	Pandey & Chandra 2015 [91]
Anticancer activity	Dutta & Chakraborty 2018; Thompson et al., 2019 [92,93]
Antidiabetes activity	Anderson et al., 2015; Zare et al., 2019; Shinjyo et al., 2020 [94–96]
Antiinflammatory activity	Han & Parker 2017; Schink et al., 2018; Shishehbor et al., 2018 [97–99]
Antimicrobial activity	Abd El-Hack et al., 2020; Arancibia et al., 2014; Atki et al., 2019 [100–102]
Antiobesity activity	Mnafgui et al., 2015 [103]
Antioxidant activity	Gulcin et al., 2019; Weerasekera et al., 2021 [76,104]
Cardiovascular protection	Jain et al., 2017; Mnafgui et al., 2015; Mousavi et al., 2020; Tarkhan et al., 2019 [103,105–107]
Insecticidal effect	Khan et al., 2020; Attia et al., 2020 [108,109]
Neurological disorders	Kang et al., 2016; Khasnavis & Pahan 2014 [110,111]

Source: own study.

3. Materials and Method

Samples (25) of black pepper (*Piper nigrum* L.) were obtained from the polish market. The test samples were representative of the entire Polish market. They came from a network of super and hyper stores present throughout the country, i.e., Auchan, Carrefour, Biedronka Jeronimo Martens, Kaufland, Lidl, Netto, Dino. These included both branded products and own brands of retail chains. All samples were encoded and analyzed. The piperine, essential oil content and antioxidant activity against DPPH radical and total polyphenols were investigated.

Material for the study consisted of 12 samples of ground cinnamon (*Cinnamomum* spp.) purchased in Poland. The coumarin content and antioxidant activity against DPPH radical and total polyphenols were investigated.

3.1. Used Chemicals

2,2-diphenyl-1-picrylhydrazyl (DPPH), Folin–Ciocalteu reagent, gallic acid, standard coumarin (purity $\geq$ 99%) were purchased from Sigma Aldrich GmbH (Steinheim, Germany), HPLC grade methanol and ethanol were obtained from POCH S.A. (Gliwice, Poland).

Five packages of each spice sample were taken from the market according to PN-ISO 948 [112], then milled in case of black pepper and cinnamon according to PN-ISO 2825 [113], and then 1% aqueous extracts were prepared.

3.2. Total Phenolic Content

The total phenolic content (TP) in water crude extracts was determined by the Folin–Ciocialteau method with modifications [114] 2.5 mL of 0.2 N FC reagent was added to tested solutions and mixed. After 5 min, 2 mL 75 g/L Na_2CO_3 solution was added. After 120 min incubation, the absorbance relative to that of a prepared blank was read at 760 nm using a spectrophotometer (Unicam UV2, Varian). The TP content is expressed in mg of gallic acid equivalents (mg GAE/g of product).

3.3. DPPH Free Radical Scavenging

Antioxidant activity was determined using the DPPH reagent and showed as DPPH radical scavenging percent. Free radical scavenging effect was determined using the free radical DPPH (2,2-diphenyl-1-picrylhydrazyl) reagent. 1 mL of the extract was added to 2 mL DPPH. The samples were gently mixed and left to stand in the darkness for 60 min. Absorbance was read at 517 nm using spectrophotometer. A control sample was prepared by mixing DPPH with distilled water. The ability of extracts to scavenge DPPH free radicals was calculated according to the following equation:

$$\text{Radical scavenging activity [\%] } AA[\%] = \frac{\text{Abs}_{contr} - \text{Abs}_{sample}}{\text{Abs}_{contr}} \times 100$$

The values are presented as the means of triplicate analyses.

3.4. Piperine Assay

Crushed black pepper was extracted with ethanol (97% or absolute), and the obtained extracts were determined spectrophotometrically at a wavelength $\lambda = 342$ nm using UV-VIS spectrophotometer Unicam. The determination was made in three repetitions [115].

3.5. Essential Oil Content

The determination of essential oil content in crushed black pepper was made by steam distillation in a Dering apparatus [116]. The essential oil content was expressed as mL/100 g d.m. Determinations were performed in triplicate.

3.6. Coumarin Content

The separation of coumarin was obtained in ground cinnamon by the HPLC method using RP-Nova Pack C18 (240 × 4.6 mm, 5 μm) column. A mobile phase was composed of acetonitrile water with a gradient elution at a flow rate of 1 mL/min. The identification of coumarin in the cinnamon samples was conducted based on their retention time compared with the retention time of the standard. The HPLC method was validated in term of linearity, limit of detection, limit of quantification, precision and accuracy.

3.7. Statistical ANALYSIS

Results were presented as the mean and standard deviation. The experimental designs and calculations were conducted using the Software Package Statistica 10.0 (StatSoft Inc., Tulsa, OK, USA).

The experiments were evaluated using analysis of variance (ANOVA) to find the impact of the type and quantity of spices on the evaluated parameters. Statistical hypotheses were verified at a significance level of *p* values < 0.05.

4. Results and Discussion

The analyses of selected bioactive compounds content and antioxidant activity were carried out in samples of black pepper and cinnamon available on the Polish retail market. These studies will allow to assess the amounts of bioactive substances which Polish consumers will find in these spices. The obtained results were compared with the other authors' studies.

4.1. Analysis of Black Pepper (Piper nigrum L.)

Black pepper samples (25) were encoded, the total content of polyphenols, antioxidant activity, the content of essential oil and piperine were determined. The analyses were performed in three repetitions and the results are presented in Table 5.

Table 5. Antioxidant activity (TP, AA), piperine and essential oil content in black pepper.

	n	TP M ± SD [mg GAE/g]	AA$_{DPPH}$ M [%]	Piperine M ± SD [%]	Essential Oil M ± SD [mL/100 g d.m.]
1	3	13.02 ± 1.56	84.42 ± 2.32	6.49 ± 0.07	1.09 ± 0.09
2	3	32.13 ± 0.24	84.51 ± 3.92	5.26 ± 0.03	1.29 ± 0.01
3	3	11.52 ± 0.30	83.93 ± 1.29	7.11 ± 0.05	1.77 ± 0.08
4	3	13.18 ± 0.35	85.42 ± 2.34	9.23 ± 0.05	2.05 ± 0.17
5	3	11.72 ± 0.85	84.86 ± 3.79	6.19 ± 0.05	1.12 ± 0.01
6	3	11.90 ± 0.29	82.67 ± 3.99	5.77 ± 0.04	1.48 ± 0.08
7	3	10.67 ± 1.30	85.32 ± 4.71	6.40 ± 0.05	1.31 ± 0.08
8	3	14.41 ± 0.73	81.49 ± 2.22	5.91 ± 0.00	2.08 ± 0.08
9	3	11.37 ± 0.18	82.01 ± 6.48	6.17 ± 0.04	1.43 ± 0.00
10	3	11.80 ± 1.13	80.64 ± 3.21	6.61 ± 0.02	1.03 ± 0.16
11	3	12.29 ± 0.78	84.27 ± 4.43	7.19 ± 0.05	2.18 ± 0.08
12	3	11.18 ± 0.06	80.85 ± 3.84	5.77 ± 0.04	1.75 ± 0.08
13	3	10.88 ± 0.37	85.13 ± 2.93	7.19 ± 0.05	1.55 ± 0.14
14	3	13.31 ± 0.12	81.73 ± 5.11	5.90 ± 0.06	1.32 ± 0.17
15	3	12.38 ± 0.54	78.07 ± 3.27	6.44 ± 0.03	1.86 ± 0.14
16	3	12.85 ± 1.02	84.45 ± 4.12	3.92 ± 0.35	1.61 ± 0.16
17	3	11.59 ± 0.27	71.29 ± 3.84	4.87 ± 0.02	1.48 ± 0.08
18	3	12.03 ± 0.32	81.83 ± 4.01	5.55 ± 0.02	1.42 ± 0.07
19	3	11.65 ± 0.70	84.60 ± 5.82	3.98 ± 0.06	2.17 ± 0.14
20	3	14.02 ± 1.08	85.34 ± 4.17	5.78 ± 0.64	1.47 ± 0.08
21	3	12.85 ± 1.02	84.43 ± 2.22	4.61 ± 0.12	1.76 ± 0.08
22	3	11.24 ± 0.53	84.59 ± 2.81	4.95 ± 0.17	1.28 ± 0.14
23	3	9.75 ± 0.96	73.88 ± 3.23	4.87 ± 0.02	0.89 ± 0.08
24	3	14.53 ± 0.61	81.48 ± 1.11	5.58 ± 0.01	1.48 ± 0.08
25	3	11.28 ± 0.74	63.93 ± 0.86	4.51 ± 0.22	2.19 ± 0.15

Source: own study.

4.1.1. Antioxidant Properties of Black Pepper (Piper nigrum L.)

The research showed different antioxidant activity of the tested black pepper samples available on the Polish market. The content of total polyphenols ranged from 9.75 to 32.13 mg GAE/g. The results of the statistical analysis confirmed that the pepper origin (brand) significantly influences the total content of polyphenols (KW-H(24) = 114.126, (p = 0.000)). The antiradical activity of the tested samples was at a high level, ranging from 63.93% to 85.42%. Statistical analysis showed that the brand of pepper influences the ability to scavenge DPPH free radicals (KW-H(24) = 114.63, p = 0.001).

The obtained results of our own research showed that the content of essential oils in pepper significantly influences the antioxidant activity measured with the use of DPPH radicals. A statistically significant, positive weak correlation between these parameters was found $r = 0.251$ ($p = 0.045$). It means that as the content of essential oil increases, the ability of the spice to scavenge DPPH radicals also increases. It should also be noted that pepper oil has a moderate ability to scavenge free radicals compared to other oils [117]. Gülçin points to the strong antioxidant properties of water and ethanolic black pepper extracts. The total content of polyphenols was determined at the level of 54.3 mg GAE in aqueous extracts [118]. The total content of polyphenols determined by Andradea and Ferreira, depending on the modification in the extraction, ranged from 14 to 27 mg GAE/g [119]. Ahmad determined the total content of polyphenols in the methanol extracts of black pepper purchased in Delhi, at the level of 172.8 mg GAE/100 g [120]. Nagy et al., determined the content of polyphenols in black pepper at the level of 338 ± 1.41 mg GAE/100 g, and the scavenging capacity of DPPH free radicals in methanolic solutions of pepper at the level of 13.28% [121].

The ability to scavenge free radicals of aqueous pepper extracts in the studies by Nahak and Sahu was for cubeb pepper (*Piper cubeba*) from 35.38 to 45.84% and for black pepper from 28.15 to 39.92%. In ethanol solutions it increased to 77.61% for cubeb pepper and 74.61% for black pepper [122]. The DPPH scavenging capacity of 35.20% was determined for an aqueous solution of black pepper by Gupta [123].

The polyphenol content of various pepper extracts was determined by Sruthi and Zachariah. Pepper extracts in chloroform and methanol showed the highest content of polyphenols compared to those in which n-hexane and water were used. In the case of black pepper, the total content of polyphenols in the aqueous solution was at the level of 3.84 µg GAE/g, and in the case of *P. longum* pepper, 2.16 µg GAE/g [124].

The high total polyphenol content in black pepper grown in Korea was determined by Lee et al., at the level of 1421.95 ± 22.35 mg GAE/100 g [125]. In another study, m, determined the total content of polyphenols in ethanol extracts of whole Korean peppercorns and after debarking, and it was 1046 ± 22 and 797 ± 28 mg GAE/100 g, respectively [26]. One of the lowest total polyphenols contents in black pepper from Bhubaneswar, India was shown by Mallick et al., and it was at the level of 11.9 ± 0.1 mg GAE/100 g [126].

In this context, the results obtained by Trifan et al., in different extracts are very interesting. The scavenging capacity of DPPH extracts with hexane, dichloromethane, 50% aqueous methanol and methanol was 18.77 ± 0.24, 19.56 ± 0.59, 45.41 ± 0.03, and 32.41 ± 0.07 mg Trolox equivalents (TE)/g, respectively. The presented research results indicate the importance of the analytical technique used in the extraction of the research material, and specifically in the type of polar and non-polar solvent used for the extraction of the spice [127].

In their research, Suchaj et al., showed a statistically significant effect of black pepper irradiation on the increase in antioxidant activity determined with the use of DPPH radicals after 2 months of storage. After 4 months of storage, these changes ranged from 4 to 9% [128].

4.1.2. Piperine Content in Black Pepper (*Piper nigrum* L.)

The content of piperine in the studied samples of pepper varies. Three samples of pepper did not meet the requirements of the PN-A-86965:1997, the piperine content in them being less than the required 4%, but they met the requirements of Codex Alimentarius. A piperine content of 3.5% in black pepper is mandatory according to ESA 2015 and Codex Alimentarius FAO/WHO [129,130]. The content of piperine of four manufacturers was at the limit of the requirements of the standard and ranged from 3.92 to 3.98%. The remaining samples met the requirements of the standard, with the highest average piperine content equal 6.61%. The results of the statistical analysis confirmed that the brand of pepper affected the content of piperine. The calculated statistic was KW-H(24) = 620.83431 ($p = 0.000$).

The content of piperine at a similar level (2–7.4%) was determined by Ravidran [131]. Hamrapurkar et al., determined the piperine content of black pepper by the HPLC method. The piperine content was 8.13% in *P. nigrum* and 4.32% in *P. longum*, and these levels were slightly higher when extracted with CO_2 gas (supercritical fluid) and methanol as a cosolvent and amounted to 8.76% and 4.96%, respectively [132]. A similar method was used by Rajopadhye et al., to determine the piperine content of various types of pepper purchased in Indian supermarkets. They obtained mixed results: 4.52% for *P. nigrum*, 3.71% for *P. longum*, and 1.19% for *P. cubeba* [133].

Zachariah et al., determined the piperine content at a much lower level from 2.8 to 3.8%. In a later study, the same authors determined piperine in black pepper from different locations in India: the lowest contents were found in pepper from Thevam (1.6%) and Neelamundi (2.0%), and the highest in samples from Perumkodi (9.5%) and Kuthiravally (8.7%) [134]. In addition to the type, the place of cultivation also affects the content of bioactive compounds in pepper. Sruthi et al., in the genus Panniyur-1, depending on the place of cultivation, determined the piperine content at the level of 2.13 to 4.49% [135].

The geographical origin also affects the varying levels of piperine content in black pepper. The high content of piperine in pepper from Malaysia (Malacca) was determined by HPLC by Rezvanian et al., and amounted to 5.85% [136]. Piperine in Indian and Malaysian pepper as determined by Jansz et al., was at the level of 2–7%, while in pepper from Sri Lanka above 7% [137]. Shango et al., found that the piperine content in black pepper depends on climatic factors such as air temperature, humidity, water availability and the amount of precipitation. In addition, they showed that the piperine content was influenced by the height at which the crop is located. Black pepper grown in Indonesia at an altitude of 650, 450 and 190 m above sea level contained 4.52%, 4.47% and 3.38% piperine, respectively [138]. The lowest piperine content in black pepper so far was determined by Ajaml [139]. The used reversed-phase liquid chromatography (RP-HPLC) method for the determination of piperine content in samples of pepper grown in various districts in Kerala, India, and it was in the range of 1.53 ± 0.002 to 1.78 ± 0.002 % *w/w*.

In turn, Lee et al. [125] determined piperine in samples of Korean pepper at the level of 2352.19 ± 68.88 mg/100 g. A similar content of piperine was determined by Shrestha et al. [140] in black pepper samples taken from various areas of Kathmandu in Nepal, and it ranged from 2.33% to 3.34% with an average of 2.75% and a standard deviation of 0.31%. In another study by Lee et al. [26] in Korean pepper samples, they determined piperine in the whole black pepper grain and in the grain with the peel stripped off, at 3728 ± 0.180 and $4035 \pm 0.108\%$, respectively. It turned out that when the peel was removed, the piperine content was 8.2% higher, and the removal of the outer skin led to a softer taste and greater bioactivity of the black pepper.

4.1.3. Essential Oil Content in Black Pepper (*Piper nigrum* L.)

While the content of piperine affects the sensory properties of pepper and determines its sharpness, the essential oil has a decisive influence on the aromatic properties. However, it should be remembered that both the content of piperine and essential oil are important factors influencing the health properties of black pepper, which is widely discussed in the theoretical part of this study. Moreover, the essential oil not only influences the aromatizing properties but also inhibits the growth of many microorganisms, including pathogenic microorganisms. The research of Nikoloć et al., showed that *P. niger* essential oil has a preservative effect on food [141].

The essential oil content of the 25 tested samples of black peppercorns was diversified. On average, it ranged from 0.89 to 2.19 mL/100 g of dry product. According to the requirements of the PN-A-86965 standard [117], the oil content should not be lower than 1.5 mL/100 g for black peppercorns and 1 mL/100 g for ground black pepper. The average content of essential oil in 50% of the tested samples did not meet the requirements of the above-mentioned standards, and was below 1.5 mL/100 g. According to Codex

Alimentarius, the content of essential oil in black pepper should not be less than 2% of the dry weight [129,142].

The Kruskal–Wallis test showed the existence of differences in the essential oil content between the tested samples (KW-H(25) = 62.547, p = 0.001).

Published works on the content of essential oil in black pepper are rare and concern various aspects related to it. The content of essential oil, apart from factors related to the product, depends on the extraction conditions. The amount of essential oil increases as the extraction temperature increases. The efficiency of the process increases for black peppercorns from 0.4% after 40 min at 100 °C to 2.6% at a temperature of 250 °C [143]. These amounts are similar to those presented by Pino et al., who determined them at the level of 1 to 3% [144]. Rezvanian found the content of essential oil in peppercorns from Malaysia also at a relatively low level, ranging from 0.76% for Jahor pepper to 1.06% for Malacca pepper [138]. A more varied amount of essential oil is given by Zachariach et al. It ranges from 0.6% for the genus Perunkodi to 6.0% for the genus Subhakara [145].

Kurian et al., observed the variability in the essential oil content in the range from 2.7 to 5.1%, they also believe that the classical hydrodistillation method is the best method of obtaining volatile oils compared to other techniques [146]. On the other hand, other researchers publish results showing significantly lower levels of essential oil, also obtained by hydrodistillation. Rmili et al., confirmed that the hydrodistillation method is one of the best methods for obtaining essential oil from black pepper, however, the amount of essential oil obtained by them from black pepper remained at a low level of 1.24% [147]. A slightly higher content of essential oil in the range of 1.60–2.80% was obtained by hydrodistillation from black pepper by Hussain et al. [148].

The results of other authors' studies confirm that the location of crops also significantly affects the content of essential oil in the same type of pepper. Sruthi determined the content of an oil of the Panniur-1 genus depending on the cultivation location, the determined amounts of the oil varied, and their content ranged from 1.6 to 3.2% [135]. Additionally, the studies carried by Chen et al., showed that the content of essential oil in pepper is influenced by the place of cultivation. They tested 25 samples of black and white pepper from different growing regions. The Investigated black pepper from the Yunnan Province in China contained 4.12%, and white pepper contained 3.01% of essential oil. On the other hand, the content of essential oil obtained from pepper originating in Indonesia and Vietnam was at the level of approximately 2% [25].

4.2. Analysis of Cinnamon (Cinnamonum sp.)

Table 6 presents the results of the determination of the content of bioactive substances and antioxidant properties in cinnamon samples purchased on the Polish retail market. They are referred to the other authors studies.

Table 6. Antioxidant activity (TP, AA) and coumarine content in cinnamon *Cinamomum* sp.

	n	TP M ± SD [mgGAE/g]	AA$_{DPPH}$ M [%]	Coumarine M ± SD [mg/kg]
1	3	94.71 ± 3.34	90.48 ± 4.13	1027.67 ± 50.36
2	3	92.32 ± 2.01	69.39 ± 2.88	3120.67 ± 81.25
3	3	85.77 ± 1.29	88.285 ± 1.49	2107.33 ± 103.24
4	3	74.58 ± 5.71	87.57 ± 2.11	3157.67 ± 59.77
5	3	73.38 ± 3.82	55.52 ± 7.56	2240.67 ± 57.01
6	3	72.33 ± 5.53	85.50 ± 2.65	3383.67 ± 57.07
7	3	71.22 ± 1.16	82.53 ± 1.99	2368.33 ± 37.58

Table 6. *Cont.*

	n	TP **M ± SD** [mgGAE/g]	AA$_{DPPH}$ **M** [%]	Coumarine **M ± SD** [mg/kg]
8	3	77.575 ± 1.59	72.03 ± 3.47	3111.12 ± 34.28
9	3	52.345 ± 0.96	68.19 ± 3.47	3564.67 ± 42.12
10	3	69.16 ± 1.24	76.12 ± 2.81	3066.33 ± 60.12
11	3	97.17 ± 2.18	91.87 ± 2.93	2725.32 ± 41.29
12	3	94.64 ± 2.36	72.64 ± 3.02	4012.00 ± 79.57
13	3	64.38 ± 1.67	90.48 ± 3.33	3198.71 ± 81.82
14	3	88.27 ± 1.11	79.94 ± 4.61	3798.00 ± 90.04
15	3	79.25 ± 2.13	91.11 ± 2,63	3812.24 ± 61.29
16	3	69.91 ± 1.18	87.51 ± 1.85	2603.77 ± 83.64

Source: own study

4.2.1. Antioxidant Properties of Cinnamon (*Cinnamonum* sp.)

Among the analyzed samples, the highest total content of phenolic compounds was found in the sample 11 at 97.17 mg GAE/g, and the lowest in the sample 9 at 52.345 mg GAE/g (Table 6). The highest capacity of scavenging DPPH radicals was observed in the samples of producer 11 (91.87% on average), and the lowest in sample number 5, for which it amounted to 55.52%. The performed statistical analysis confirmed that the origin (brand) of cinnamon significantly influences the total phenolic content (K-W, H(15) = 59.025, p = 0.001), and the ability to scavenge free radicals (K-W, H(15) = 69.582, p = 0.000).

The total content of polyphenols in cinnamon bark determined by Abraham was 289.0 ± 2.2 mg GAE/g of plant [149].

The highest content of polyphenols was found in cassia bark ethanol extracts (9.534 g GAE/100 g d.m.), in leaves (8.854 g GAE/100 g d.m.) and the lowest in buds (6.313 g GAE/100 g d.m.). Extracts obtained from extraction with a CO_2 supercritical fluid with metanol were characterized by a lower content of total polyphenols [150]. In addition, they confirmed the high DPPH radical scavenging capacity (over 80%) of *Cinnamomum cassia* ethanolic extracts. They obtained the highest values for cinnamon leaves.

Mathew and Abraham found a statistically significant decrease in the concentration of the DPPH radical along with an increase in the concentration of cinnamon bark extract from about 60% for a 3.125 µg/mL solution to about 5% for a 50 µg/mL solution [151]. Dragland found very high concentrations of antioxidants (>75 mmol/100 g) in the *Cinnamomi cortex* [152]. Additionally, other studies confirm the ability of plants from the *Cinnamomum* family to scavenge free radicals [153–155].

Prasad found a difference in the DPPH free radical scavenging ability between different grades of cinnamon. They decrease as follows: *C. zeylanica* > *C. cassia* > BHT > *C. pasiflorum* > *C. burmannii* > *C. tamala* [156]. The aqueous and alcoholic extracts of cinnamon (1:1) showed a significant ability to inhibit lipid oxidation in the in vitro lipid oxidation test [157].

Cinnamon has a higher oxidation inhibitory capacity than BHA, BHT and propyl gallate tested in the lipid peroxidation test [158]. Prakash et al. [159] found a high free radical scavenging capacity in solutions from cinnamon bark of *C. zeylanicum*, similar to the characteristics obtained by ascorbic acid.

Lin et al., assessed the antioxidant activity (using the DPPH radical) of 42 types of essential oils, including cinnamon oil. They showed the highest DPPH radical scavenging activity among the studied essential oils at the level of 91.4 ± 0.002% [160]. In another test, this spice was recognized as the best natural antioxidant, stronger than its synthetic counterparts (BHA, BHT). This is important to extend the shelf life of foodstuffs as oxidation is one of the most common chemical reactions responsible for food spoilage. It is likely that the high content of flavonoids is an essential basis for such a strong antioxidant activity [161].

Interesting, especially related to its use in cooking, is the fact that cinnamon's antioxidant properties increase with the extraction temperature. Shobana and Akhilender Naidu subjected that the extract in a temperature of 100 °C for 30 min, not only did not lose its antioxidant properties, but rather showed a significant increase in them. "Cold" cinnamon extract had an antioxidant activity equal to 21%, while when cooked it increased by as much as 35% [157].

Trifan et al., showed that the antioxidant properties of cinnamon depend on the extraction method. They extracted the cinnamon bark with hexane, dichloromethane, 50% aqueous methanol and methanol. The result was a total polyphenol content of 18.46 ± 0.27, 14.00 ± 0.15, 92.90 ± 0.46, and 63.68 ± 1.48 mg GAE/g, respectively. In the same study, the antioxidant activity (using the DPPH radical) was as follows: 6.96 ± 0.41, 9.39 ± 0.57, 473.74 ± 1.45, 178.42 ± 0.81%. These results clearly indicate a high relationship between antioxidant activity and the extraction technique, i.e., the polarity of the solvent, which influences the extraction of cinnamon essential oil [127].

4.2.2. Coumarin Content in Cinnamon (*Cinnamonum* spp.)

In 16 samples of cinnamon purchased on the Polish market, the varied content of coumarin was determined and ranged from 1027.67 ± 50.36 to 4012.00 ± 79.57 mg/kg. The performed statistical analysis allowed us to establish the existence of a relationship between the origin (brand) of cinnamon and the content of coumarin (K-W, $H(15) = 35.325$, $p = 0.0003$). The conducted research allows us to assume due to the high content of coumarin, but also a strong aroma and slightly sweet taste that we are dealing with cassia.

These results are also confirmed by other authors [162–164]. Authors examining the content of coumarin in cinnamon found that it is directly related to the species of cinnamon and the degree of processing (it differs for the bark of cinnamon and ground cinnamon). Woehrlin et al., report the coumarin content in cinnamon bark at the level of 1740 to 7670 mg/kg [165]. Its content in the bark found in Chinese studies was even over 12,000 mg/kg [166]. In ground cinnamon, Blahova and Svobodova [164] gave it at the level from 2650 to 7017 mg/kg in samples from the Czech market. A high level of coumarin was also determined by Ho et al.—29,400 mg/kg DS [167]. On the other hand, Lungarini found very low contents (<100 mg/kg) in ground *C. verum*, and up to 3094 mg/kg in cassia [168].

The higher content of coumarin in cassia than in Ceylon cinnamon is also confirmed by studies by other authors. Sproll et al. [163] found in the tested samples from the German market the absence of coumarin in Ceylon cinnamon, while in cassia, it was determined at the level of 2880–4820 mg/kg. In the samples of cinnamon of undefined origin, these amounts reached even 8790 mg/kg, and 85% of the 20 samples contained coumarin, so it can be presumed that 15% were samples of Ceylon cinnamon.

Among the food products containing cinnamon, an important source of coumarin were cinnamon cookies with up to 88 mg of coumarin/kg, and bread and breakfast cereals (up to 32 mg of coumarin/kg). This results in an amount exceeding the TDI by a child consuming just three to four cinnamon cookies weighing 5 g, and for an adult it is about ten pieces. Coumarin was present in small amounts in dairy products (up to 2 mg/k) and alcohols (up to 8 mg/kg) [163].

Research conducted in India has shown that cinnamon sourced directly from plantations has a low to moderate coumarin content, ranging from 12.3 to 143 mg/kg. On the other hand, a high content of coumarin was found in samples of ground cinnamon purchased in retail stores and it ranged from 819 to 3462 mg/kg. Only one market sample of cinnamon purchased from an authentic spice vendor showed low coumarin content (19.6 mg/kg). The authors of the study suggest that market cinnamon is very often adulterated with cheaper substitutes such as *C. cassia* and *C. burmanii*, because market cinnamon containing the least coumarin was three times more expensive than other market products [169].

In spice mixtures containing cinnamon, the coumarin content reaches 4308 mg/kg [170], and in plants from the *Lamiaceace* family it ranges from 14,300 to 276,900 mg/kg DS, *Lavandula*-lavender and *Salvia*-sage show its highest content [171].

Authors including Sproll [163] and Abraham [149] consider that regulations on the content of coumarin in food should be established, which was reflected in the Regulation of the European Parliament and of the Council (EC) No. 1334/2008 of 16 December 2008 [172].

Ground cinnamon and spice blends purchased in retail stores have led food regulatory authorities in many countries to increase the frequency of inspections [173,174]. The Norwegian Scientific Committee for Food Safety (Vitenskapskomiteen for mattrygghet, VKM), at the request of the Norwegian Food Safety Authority (Mattilsynet), conducted a risk assessment of coumarin consumption in the Norwegian population. As a result of these studies, it was found that a small or occasional exceeding of the TDI is not considered to increase the risk of adverse health effects. The consumption of coumarin may in some cases exceed the TDI seven to twenty times. Liver toxicity may occur soon after the initiation of coumarin exposure. Such large daily exceedances of TDI, even within a limited period of 1–2 weeks, raise concerns about adverse health effects [175].

5. Conclusions

The principal role of spices is to raise the quality of food to a higher level. Hence, spices are used primarily to provide consumers with a sensory experience and pleasure from consuming food [176]. Black pepper (*Piper nigrum* L.) and cinnamon (*Cinnamomum* spp.) are frequently consumed spices in Poland. The research discussed in this paper shows that these spices, apart from influencing the sensory value of food, may have a positive effect on the human body. The research carried out in this study shows that the quality of black pepper available to consumers in retail trade was at a good level, as measured by the results of the content of piperine and pepper oil. The content of piperine in the black pepper samples was in the range of 3.92 ± 0.35–$9.23 \pm 0.05\%$. The tested black pepper samples contained 0.89 ± 0.08–2.19 ± 0.15 mL/100 g d.m. essential oil. The coumarin content in the cinnamon samples remained in the range of 1027.67 ± 50.36–4012.00 ± 79.57 mg/kg. Taking into account the coumarin content, we suppose that the majority of cinnamon available in the Polish retail trade is *Cinnamomum cassia* (L.) J. Presl.

In the wider context of the research carried out, the question is what are the benefits of consuming spices for consumers? These benefits appear to depend primarily on the amount of spices consumed and their quality, measured by the content of the important bioactive substances. Therefore, the inspection services face a major task of ensuring that the quality of spices available on the market is as high as possible, because only then will these spices will have a high sensory value and will have an impact on human health. Consequently, there is still a need for more research into how spices affect the specific organs of the human body [177,178].

Author Contributions: Conceptualization, writing—original draft preparation, writing—review and editing, visualization, J.N.-G. and M.Ś.; methodology, investigation, J.N.-G. All authors have read and agreed to the published version of the manuscript.

Funding: This research received no external funding.

Data Availability Statement: The data are available from the corresponding author on request.

Conflicts of Interest: The authors declare no conflict of interest.

References

1. Embuscado, M.E. Spices and herbs: Natural sources of antioxidants—A mini review. *J. Funct. Foods* **2015**, *18*, 811–819. [CrossRef]
2. Yashin, A.; Yashin, Y.; Xia, X.; Nemzer, B. Antioxidant Activity of Spices and Their Impact on Human Health: A Review. *Antioxidants* **2017**, *6*, 70. [CrossRef]
3. Abdel-Aziz, S.M.; Abd El-Hack, M.E.; Alagawany, M.; Abdel-Moneim, A.-M.E.; Mohammed, N.G.; Khafaga, A.F.; Bin-Jumah Aeron, A.; Kahil, T.A. Health Benefits and Possible Risks off Herbal Medicine. In *Microbes in Food and Health*; Garg, N., Abdel-Aziz, S.M., Aeron, A., Eds.; Springer International Publishing: Cham, Switzerland, 2016.

4. Vázquez-Fresno, R.; Rosana, A.R.R.; Sajed, T.; Onookome-Okome, T.; Wishart, N.A.; Wishart, D.S. Herbs and Spices—Biomarkers of Intake Based on Human Intervention Studies—A Systematic Review. *Genes Nutr.* **2019**, *14*, 18. [CrossRef] [PubMed]

5. Directive 2004/24/EC of the European Parliament and of the Council of 31 March 2004. "Herbal Substances Are: All, Mainly Whole, Divided or Cut Plants, Parts of Plants, Algae, Fungi, Lichen in Un-Processed Form, Usually Dried, Sometimes Fresh". Available online: https://eur-lex.europa.eu/legal-content/EN/ALL/?uri=CELEX:32004L0024 (accessed on 4 May 2022).

6. Śmiechowska, M.; Newerli-Guz, J.; Skotnicka, M. Spices and seasoning mixes in European Union-innovations and ensuring safety. *Foods* **2021**, *10*, 2289. [CrossRef]

7. Chironi, S.; Bacarella, S.; Altamore, L.; Columba, P.; Ingrassia, M. Consumption of spices and ethnic contamination in the daily diet of Italians -consumers' preferences and modification of eating habits. *J. Ethn. Foods* **2021**, *8*, 6. [CrossRef]

8. Bortnowska, G.; Kałużna-Zajączkowska, J. Preferencje wyboru przypraw sypkich do potraw przez osoby pracujące zawodowo z uwzględnieniem innowacyjnych zmian w ich produkcji. *Rocz. Państw. Zakł. Hig.* **2011**, *62*, 445–452. [PubMed]

9. Krishnamoorthy, B.; Parthasarathy, V.A. Improvement of Black Pepper. *CAB Rev.* **2009**, *4*, 85. Available online: https://cabidigitallibrary.org/doi/10.1079/PAVSNNR20105003 (accessed on 4 May 2022). [CrossRef]

10. Korikanthimath, V.S.; Dhas, P.H.A. Processing and quality of black pepper—A review. *J. Spices Aromat. Crops* **2003**, *12*, 1–13.

11. Ganapathi, T.; Patil, S.V.; Rajakumar, G.R. Processing of black pepper (*Piper nigrum* L.) through solarisation. *Int. J. Agric. Sci.* **2018**, *14*, 211–214. [CrossRef]

12. Pickova, D.; Ostry, V.; Malir, J.; Toman, J.; Malir, F. A Review on Mycotoxins and Microfungi in Spices in the Light of the Last Five Years. *Toxins* **2020**, *12*, 789. [CrossRef]

13. Jalili, M.; Jinap, S.; Noranizam, M.A. Aflatoxins and ochratoxins a reduction in black and white pepper by gamma radiation. *Radiat. Phys. Chem.* **2012**, *81*, 1786–1788. [CrossRef]

14. Ee, K.P.; Shang, C.Y. Novel Farming Innovation for High Production of Black Pepper (*Piper nigrum* L.) Planting Materials. *J. Agric. Sci. Technol. B* **2017**, *7*, 301–308.

15. Sulok, K.M.T.; Ahmed, O.H.; Khew, C.Y.; Zehnder, J.A.M. Introducing Natural Farming in Black Pepper (*Piper nigrum* L.) Cultivation. *Int. J. Agron.* **2018**, *2018*, 9312537. [CrossRef]

16. Nguyen, S.D.; Trinh, T.H.T.; Tra, T.D.; Nguyen, T.V.; Chuyen, H.V.; Ngo, V.A.; Nguyen, A.D. Combined Application of Rhizosphere Bacteria with Endophytic Bacteria Suppresses Root Diseases and Increases Productivity of Black Pepper (*Piper nigrum* L.). *Agriculture* **2021**, *11*, 15. [CrossRef]

17. Karmawati, E.; Ardana, I.K.; Siswanto; Soetopo, D. Factors effecting pepper production and quality in several production center. *IOP Conf. Ser. Earth Environ. Sci.* **2020**, *418*, 012051. [CrossRef]

18. Hu, L.; Yin, C.; Ma, S.; Liu, Z. Assessing the authenticity of black pepper using diffuse reflectance midinfrared Fourier transform spectroscopy coupled with chemometrics. *Comput. Electron. Agric.* **2018**, *154*, 491–500. [CrossRef]

19. Rivera-Pérez, A.; Romero-González, R.; Frenich, A.G. Application of an innovative metabolomics approach to discriminate geographical origin and processing of black pepper by untargeted UHPLC-Q-Orbitrap-HRMS analysis and mid-level data Fusion. *Food Res. Int.* **2021**, *150*, 110722. [CrossRef]

20. Cantarelli, M.Á.; Moldes, C.A.; Marchevsky, E.J.; Azcarate, S.M.; Camiña, J.M. Low-cost analytic method for the identification of Cinnamon adulteration. *Microchem. J.* **2020**, *159*, 1055132020. [CrossRef]

21. Modupalli, N.; Naik, M.; Sunil, C.K.; Natarajan, V. Emerging non-destructive methods for quality and safety monitoring of spices. *Trends Food Sci. Technol.* **2021**, *108*, 133–147. [CrossRef]

22. Milenković, A.N.; Stanojević, L.P. Black pepper—Chemical composition and biological activities. *Adv. Technol.* **2021**, *10*, 40–50. [CrossRef]

23. Hammouti, B.; Dahmani, M.; Yahyi, A.; Ettouhami, A.; Messali, M.; Asehraou, A.; Bouyanzer, A.; Warad, I.; Touzani, R. Black Pepper, the "King of Spices": Chemical composition to applications. *Arab. J. Chem. Environ. Res.* **2019**, *6*, 12–56.

24. Bermawie, N.; Wahyuni, S.; Heryanto, R.; Darwati, I. Morphological characteristics, yield and quality of black pepper Ciinten variety in three agro ecological conditions. *IOP Conf. Ser. Earth Environ. Sci.* **2019**, *292*, 012065. [CrossRef]

25. Chen, S.-A.; Yang, K.; Xiang, J.-Y.; Kwaku, O.R.; Han, J.-X.; Zhu, X.-A.; Huang, Y.-T.; Liu, L.-J.; Shen, S.-B.; Li, H.-Z.; et al. Comparison of Chemical Compositions of the Pepper EOs from Different Cultivars and Their AChE Inhibitory Activity. *Nat. Prod. Commun.* **2020**, *15*, 1934578X20971469. [CrossRef]

26. Lee, J.-G.; Kim, D.-W.; Shin, Y.; Kim, Y.-J. Comparative study of the bioactive compounds, Lee flavours and minerals present in black pepper before and after removing the outer skin. *LW-Food Sci. Technol.* **2020**, *125*, 109356. [CrossRef]

27. Aswar, U.; Shintre, S.; Chepurwar, S.; Aswar, M. Antiallergic effect of piperine on ovalbumin-induced allergic rhinitis in mice. *Pharm. Biol.* **2015**, *53*, 1358–1366. [CrossRef]

28. Kim, S.-H.; Lee, Y.-C. Piperine inhibits eosinophil infiltration and airway hyperresponsiveness by suppressing T cell activity and Th2 cytokine production in the ovalbumin-induced asthma model. *J. Physiol. Pharmacol.* **2009**, *61*, 353–359. [CrossRef]

29. Bukhari, I.A.; Alhumayyd, M.S.; Mahesar, A.L.; Gilani, A.H. The analgesic and anticonvulsant effects of piperine in mice. *J. Physiol. Pharmacol.* **2013**, *64*, 789–794.

30. Jeena, K.; Liju, V.B.; Umadevi, N.P.; Kuttan, R. Antioxidant, anti-inflammatory and antinociceptive properties of black pepper essential oil (*Piper nigrum* L.). *J. Essent. Oil Bear. Plants* **2014**, *17*, 1–12. [CrossRef]

31. Tasleem, F.; Ashar, I.; Ali, S.N.; Perveen, S.; Mahmood, Z.A. Analgesic and anti-inflammatory activities of *Piper nigrum* L. *Asian Pac. J. Trop. Med.* **2014**, *7* (Suppl. S1), 461–468. [CrossRef]

32. Bang, J.S.; Oh, D.H.; Choi, H.M.; Sur, B.-J.; Lim, S.-J.; Kim, J.Y.; Yang, H.-I.; Yoo, M.C.; Hahm, D.-H.; Kim, K.S. Anti-inflammatory and antiarthritic effects of piperine in human interleukin 1β-stimulated fibroblast-like synoviocytes and in rat arthritis models. *Arthritis Res. Ther.* **2009**, *11*, R49. [CrossRef]

33. Umar, S.; Sarwar, A.H.M.G.; Umar, K.; Ahmad, N.; Sajad, M.M.; Ahmad, S.; Katiyar, C.K.; Khan, H.A. Piperine ameliorates oxidative stress, inflammation and histological out-come in collagen induced arthritis. *Cell. Immunol.* **2013**, *284*, 51–59. [CrossRef]

34. Banerjee, S.; Katiyar, P.; Kumar, V.; Saini, S.S.; Varshney, R.; Krishnan, V.; Sircar, D.; Roy, P. Black pepper and piperine induce anticancer effects on leukemia cell line. *Toxicol. Res.* **2021**, *10*, 232–248. [CrossRef]

35. Greenshields, A.L.; Doucette, C.D.; Sutton, K.M.; Madera, L.; Annan, H.; Yaffe, P.B.; Knickle, A.F.; Dong, Z.; Hoskin, D.W. Piperine inhibits the growth and motility of triple-negative breast cancer cells. *Cancer Lett.* **2015**, *357*, 129–140. [CrossRef] [PubMed]

36. Morsy, F.N.S.; Abd El-Salam, E.A. Antimicrobial and Antiproliferative Activities of Black Pepper (*Piper nigrum* L.) Essential Oil and Oleoresin. *J. Essent. Oil Bear. Plants* **2017**, *20*, 779–790. [CrossRef]

37. Prashant, A.; Rangaswamy, C.; Yadav, A.K.; Reddy, V.; Sowmya, M.N.; Madhunapantula, S. In vitro anticancer activity of ethanolic extracts of *Piper nigrum* against colorectal carcinoma cell lines. *Int. J. Appl. Basic Med. Res.* **2017**, *7*, 67–72. [CrossRef] [PubMed]

38. Atal, S.; Atal, S.; Vyas, S.; Phadnis, P. Bio-enhancing Effect of Piperine with Metformin on Lowering Blood Glucose Level in Alloxan Induced Diabetic Mice. *Pharmacogn. Res.* **2016**, *8*, 56–60. [CrossRef]

39. Kharbanda, C.; Alam, M.S.; Hamid, H.; Javed, K.; Bano, S.; Ali, Y.; Dhulap, A.; Alam, P.; Pasha, M.A.Q. Novel piperine derivatives with antidiabetic effect as PPAR-γ agonists. *Chem. Biol. Drug Des.* **2016**, *88*, 354–362. [CrossRef]

40. Oboh, G.; Ademosun, A.O.; Odubanjo, O.V.; Akinbola, I.A. Antioxidative Properties and Inhibition of Key Enzymes Relevant to Type-2 Diabetes and Hypertension by Essential Oils from Black Pepper. *Adv. Pharmacol. Sci.* **2013**, *2013*, 926047. [CrossRef] [PubMed]

41. Sarfraz, M.; Khaliq, T.; Hafizur, R.M.; Raza, S.A.; Ullah, H. Effect of black pepper, turmeric and ajwa date on the endocrine pancreas of the experimentally induced diabetes in wister albino rats: A histological and immunohistochemical study. *Endocr. Metab. Sci.* **2021**, *4*, 100098. [CrossRef]

42. Emon, N.U.; Alam, S.; Rudra, S.; Riya, S.R.; Paul, A.; Hossen, S.M.M.; Kulsum, U.; Ganguly, A. Antide-pressant, anxiolytic, antipyretic, and thrombolytic profiling of methanol extract of the aerial part of *Piper nigrum*: In vivo, in vitro, and in silico approaches. *Food Sci. Nutr.* **2020**, *9*, 833–846. [CrossRef]

43. Ghosh, S.; Kumar, A.; Sachan, N.; Chandra, P. Anxiolytic and antidepressant-like effects of essential oil from the fruits of *Piper nigrum* Linn. (Black pepper) in mice: Involvement of serotonergic but not GABAergic transmission system. *Heliyon* **2021**, *7*, e06884. [CrossRef]

44. Hritcu, L.; Noumedem, J.A.; Cioanca, O.; Hancianu, M.; Postu, P.; Mihasan, M. Anxiolytic and antidepressant profile of the methanolic extract of *Piper nigrum* fruits in beta-amyloid (1–42) rat model of Alzheimer's disease. *Behav. Brain Funct.* **2015**, *11*, 13. [CrossRef]

45. Mao, Q.-Q.; Huang, Z.; Zhong, X.-M.; Xian, Y.-F.; Ip, S.-P. Piperine reverses the effects of corti-costerone on behawior and hippocampal BDNF expression in mice. *Neurochem. Int.* **2014**, *74*, 36–41. [CrossRef] [PubMed]

46. Hlavačková, L.; Janegová, A.; Uličná, O.; Janega, P.; Černá, A.; Babál, P. Spice up the hypertension diet—curcumin and piperine prevent remodeling of aorta in experimental L-NAME induced hypertension. *Nutr. Metab.* **2011**, *8*, 72. [CrossRef] [PubMed]

47. Lee, K.P.; Lee, K.; Park, H.-W.; Kim, H.; Hong, H. Piperine inhibits platelet-derived growth factor-BB-induced proliferation and migration in vascular smooth muscle cells. *J. Med. Food* **2015**, *18*, 208–215. [CrossRef]

48. Taqvi, S.I.H.; Shah, A.J.; Gilani, A.H. Blood pressure lowering and vasomodulator effects of piperine. *J. Cardiovasc. Pharmacol.* **2008**, *52*, 452–458. [CrossRef]

49. Yu, S.; Liu, X.; Yu, D.C.E.; Yang, J. Piperine protects LPS-induced mastitis by inhibiting inflammatory response. *Int. Immunophar-macol.* **2020**, *87*, 106804. [CrossRef] [PubMed]

50. Bawazeer, S.; Khan, I.; Rauf, A.; Aljohani, A.S.M.; Alhumaydhi, F.A.; Khalil, A.A.; Qureshi, M.N.; Ahmad, L.; Khan, S.A. Black pepper (*Piper nigrum*) fruit-based gold nanoparticles (BP-AuNPs): Synthesis, characterization, biological activities, and catalytic applications—A green approach. *Green Process. Synth.* **2022**, *11*, 11–28. [CrossRef]

51. Chen, P.; Sun, J.; Ford, P. Differentiation of the Four Major Species of Cinnamons (*C. burmannii*, *C. verum*, *C. cassia*, and *C. loureiroi*) Using a Flow Injection Mass Spectrometric (FIMS) Fingerprinting Method. *J. Agric. Food Chem.* **2014**, *62*, 2516–2521. [CrossRef] [PubMed]

52. Daigham, G.E.; Mahfouz, A.Y. Assessment of Antibacterial, Antifungal, Antioxidant and Antiviral Activities of Black pepper Aqueous Seed Extract. *Al-Azhar Univ. J. Res. Stud.* **2020**, *2*, 1–12.

53. Hien, L.T.M.; Dao, D.T.A. Antibacterial Activity of Black Pepper Essential Oil Nanoemulsion Formulated by Emulsion Phase Inversion Method. *Curr. Res. Nutr. Food Sci.* **2022**, *10*, 311–320. [CrossRef]

54. Martinelli, L.; Rosa, J.M.; Fereira, C.S.B.; Nascimento, G.M.L.; Freitas, M.S.; Pizato, L.C.; Santos, W.O.; Pires, R.F.; Okura, M.H.; Malpass, G.R.P.; et al. Antimicrobial activity and chemical constituents of essential oils and oleoresins extracted from eight pepper species. *Ciênc. Rural* **2017**, *47*, e20160899. [CrossRef]

55. Zhang, J.; Ye, K.-P.; Zhang, X.; Pan, D.-D.; Sun, Y.-Y.; Cao, Y.-X. Antibacterial Activity and Mechanism of Action of Black Pepper Essential Oil on Meat-Borne *Escherichia coli*. *Front. Microbiol.* **2017**, *7*, 2094. [CrossRef]

56. Zou, L.; Hu, Y.-Y.; Chen, W.-X. Antibacterial mechanism and activities of black pepper chloroform extract. *J. Food Sci. Technol.* **2015**, *52*, 8196–8203. [CrossRef] [PubMed]

57. Chopathompikunrt, P.; Wattanathorn, J.; Muchimapura, S. Piperine, the main alkaloid of Thai black pepper, protects against neurodegeneration and cognitive impairment in animal model of cognitive deficit like condition of Alzheimer's disease. *Food Chem. Toxicol.* **2010**, *48*, 798–802. [CrossRef] [PubMed]

58. Elnaggar, Y.S.R.; Etman, S.M.; Abdelmonsif, D.A.; Abdallah, O.Y. Novel piperine-loaded Tween-integrated monoolein cubosomes as brain-targeted oral nanomedicine in Alzheimer's disease: Pharmaceutical, biological, and toxicological studies. *Int. J. Nanomed.* **2015**, *10*, 5459–5473. [CrossRef] [PubMed]

59. Etman, S.M.; Elnaggar, Y.S.R.; Abdelmonsif, D.A.; Abdallah, O.Y. Oral brain-targeted microemulsion for enhanced piperine delivery in Alzheimer's disease therapy: In vitro appraisal, in vivo activity, and nanotoxicity. *AAPS PharmSciTech* **2018**, *19*, 3698–3711. [CrossRef] [PubMed]

60. Du, Y.; Chen, Y.; Fu, X.; Gu, J.; Sun, Y.; Zhang, Z.; Xu, J.; Qin, L. Effects of piperine on lipid metabolism in high-fat diet induced obese mice. *J. Funct. Foods* **2020**, *71*, 104011. [CrossRef]

61. Lailiyah, N.N.; Ibrahim, M.D.; Fitriani, C.A.; Hermanto, F.E. Anti-Obesity Properties of Black Pepper (*Piper nigrum* L.): Completing Puzzle using Computational Analysis. *J. Smart Bioprospect. Technol.* **2021**, *2*, 101–106. [CrossRef]

62. Meriga, B.; Parim, B.; Chunduri, V.R.; Naik, R.R.; Nemani, H.; Suresh, P.; Ganapathy, S.; Wu, S.B. Antiobesity potential of Piperonal: Promising modulation of body composition, lipid profiles and obesogenic marker expression in HFD-induced obese rats. *Nutr. Metab.* **2017**, *14*, 72. [CrossRef] [PubMed]

63. Shah, S.S.; Shah, G.B.; Singh, S.D.; Gohil, P.V.; Chauhan, K.; Shah, K.A.; Chorawala, M. Effect of piperine in the regulation of obesity-induced dyslipidemia in high-fat diet rats. *Indian J. Pharmacol.* **2011**, *43*, 296–299. [CrossRef]

64. Srinivasan, K. Black Pepper and its Pungent Principle-Piperine: A Review of Diverse Physiological Effects. *Crit. Rev. Food Sci. Nutr.* **2007**, *47*, 735–748. [CrossRef] [PubMed]

65. Vijayakumar, R.S.; Nalini, N. Efficacy of piperine, an alkaloidal constituent from *Piper nigrum* on erythrocyte antioxidant status in high fat diet and antithyroid drug induced hyperlipidemic rats. *Cell Biochem. Funct.* **2006**, *24*, 491–498. [CrossRef]

66. Dutta, M.; Ghosh, A.K.; Mishra, P.; Jain, G.; Rangari, V.; Chattopadhyay, A.; Das, T.; Bhowmick, D.; Bandyopadhyay, D. Protective effects of piperine against copper ascorbate induced toxic injury to goat cardiac mitochondria in vitro. *Food Funct.* **2014**, *5*, 2252. [CrossRef] [PubMed]

67. Wang, D.; Zhang, L.; Huang, J.; Himabindu, K.; Tewari, D.; Horbańczuk, J.O.; Xu, S.; Chen, Z.; Atanasov, A.G. Cardiovascular protective effect of black pepper (*Piper nigrum* L.) and its major bioactive constituent piperine. *Trends Food Sci. Technol.* **2021**, *117*, 34–45. [CrossRef]

68. Mehmood, M.H.; Gilani, A.H. Pharmacological Basis for the Medicinal Use of Balck Pepper and Piperine in Gastrointestinal Disorders. *J. Med. Food* **2010**, *13*, 1086–1096. [CrossRef]

69. Shamkuwar, P.B.; Shahi, S.R.; Jadhav, S.T. Evaluation of antidiarrhoeal effect of Black pepper (*Piper nigrum* L.). *Asian J. Plant Sci. Res.* **2012**, *2*, 48–53.

70. Shamkuwar, P.B. Mechanisms of Antidiarrhoeal Effect of *Piper nigrum*. *Int. J. PharmTech Res.* **2013**, *5*, 1138–1141.

71. Bae, G.-S.; Kim, M.-S.; Jeong, J.; Lee, H.-Y.; Park, K.-C.; Koo, B.C.; Kim, B.-J.; Kim, T.-H.; Lee, S.H.; Hwang, S.-Y.; et al. Piperine ameliorates the severity of cerulein-induced acute pancrea-titis by inhibiting the activation of mitogen activated protein kinases. *Biochem. Biophys. Res. Commun.* **2011**, *410*, 382–388. [CrossRef]

72. Christina, A.J.M.; Saraswathy, G.R.; Heison, R.S.J.; Kothai, R.; Chidambaranatha, N.; Nalini, G.; Therasal, R.L. Inhibition of CCl4-induced liver fibrosis by *Piper longum*. *Phytomedicine* **2006**, *13*, 196–198. [CrossRef] [PubMed]

73. Gurumurthy, P.; Vijayalatha, S.; Sumathy, A.; Asokan, M.; Naseema, M. Hepatoprotective effect of aqueous extract of *Piper longum* and piperine when administered with anti-tubercular drugs. *Bioscan* **2012**, *7*, 661–663.

74. Matsuda, H.; Ninomiya, K.; Morikawa, T.; Yasuda, D.; Yamaguchi, I. Protective effects of amide constit-uents from the fruit of Piper chaba on D-galactosamine/TNF-alpha-induced cell death in mouse hepatocytes. *Bioorg. Med. Chem. Lett.* **2008**, *18*, 2038–2042. [CrossRef] [PubMed]

75. Nirwane, A.M.; Bapat, A.R. Effect of methanolic extract of *Piper nigrum* fruits in Ethanol-CCl4 induced hepatotoxicity in Wistar rats. *Pharm. Lett.* **2012**, *4*, 795–802.

76. Weerasekera, A.C.; Samarasinghe, K.; Sameera de Zoysa, H.K.; Bamunuarachchige, T.C.; Waisundara, V.Y. Cinnamomum zeylanicum: Morphology. In *Antioxidant Properties and Bioactive Compounds: Antioxidants*; Intech Open: London, UK, 2021.

77. Al-Dhubiab, B.E. Pharmaceutical applications and phytochemical profile of Cinnamomum burmannii. *Pharmacogn. Rev.* **2012**, *6*, 125–131. [CrossRef] [PubMed]

78. Tiwari, S.; Talreja, S. Importance of Cinnamomum Tamala in the Treatment of Various Diseases. *Pharmacogn. J.* **2020**, *12*, 1792–1796. [CrossRef]

79. Kumar, S.; Kumari, R.; Mishra, S. Pharmacological properties and their medicinal uses of Cinnamomum: A review. *J. Pharm. Pharmacol.* **2019**, *71*, 1735–1761. [CrossRef]

80. Anonymous. Information Cassia and Cinnamon. Available online: https://citeseerx.ist.psu.edu/viewdoc/similar?doi=10.1.1.409.6671&type=ab (accessed on 4 May 2022).

81. Suriyagoda, L.; Mohotti, A.J.; Vidanarachchi, J.K.; Kodithuwakku, S.P.; Chathurika, M.; Bandaranayake, P.C.G.; Hetherington, A.M.; Beneragama, C.K. Ceylon cinnamon: Much more than just a spice. *Plants People Planet* **2021**, *3*, 319–336. [CrossRef]

82. Gunia-Krzyżak, A.; Słoczyńska, K.; Popiół, J.; Marona, H.; Pękala, E. Cinnamic acid derivatives in cosmetics: Current use and future prospects. *Int. J. Cosmet. Sci.* **2018**, *40*, 356–366. [CrossRef]

83. De Silva, D.A.M.; Jeewanthi, R.K.C.; Rajapaksha, R.H.N.; Weddagala, W.M.T.B.; Hirotsu, N.; Shimizu, B.-I.; Munasinghe, M.A.J.P. Clean vs. dirty labels: Transparency and authenticity of the labels of Ceylon cinnamon. *PLoS ONE* **2021**, *16*, e0260474. [CrossRef] [PubMed]

84. Avula, B.; Smillie, T.J.; Wang, J.-H.; Zweigenbaum, J.; Khan, I.A. Authentication of true cinnamon (*Cinnamon verum*) utilising direct analysis in real time (DART)-QToF-MS. *Food Addit. Contam. Part A* **2015**, *32*, 1–8. [CrossRef]

85. Rana, P.; Liaw, S.-Y.; Lee, M.-S.; Sheu, S.-C. Discrimination of Four Cinnamomum Species with Physico-Functional Properties and Chemometric Techniques: Application of PCA and MDA Models. *Foods* **2021**, *10*, 2871. [CrossRef] [PubMed]

86. Zhou, W.; Liang, Z.; Li, P.; Zhao, Z.; Chen, J. Tissue-specific chemical profiling and quantitative analysis of bioactive components of *Cinnamomum cassia* by combining laser-microdissection with UPLC-Q/TOF–MS. *Chem. Cent. J.* **2018**, *12*, 71. [CrossRef] [PubMed]

87. Lixourgioti, P.; Goggin, K.A.; Zhao, X.; Murphy, D.J.; Van Ruth, S.; Koidis, A. Authentication of cinnamon spice samples using FT-IR spectroscopy and chemometric classification. *LWT-Food Sci. Technol.* **2022**, *154*, 112760. [CrossRef]

88. Lacy, A.; O'Kennedy, R. Studies on Coumarins and Coumarin-Related Compounds to Determine their Therapeutic Role in the Treatment of Cancer. *Curr. Pharm. Des.* **2004**, *10*, 3797–3811. [CrossRef] [PubMed]

89. European Food Safety Authority (EFSA). Scientific Opinion of the Panel on Food Additives, Flavourings, Processing Aids and Materials in Contact with Food on a Request from the European Commission on Coumarinin Flavourings and Other Food Ingredients with Flavouring Properties. *EFSA J.* **2008**, *793*, 1–15.

90. Ribeiro-Santos, R.; Andrade, M.; Madella, D.; Martinazzo, A.P.; Moura, L.A.G.; Melo, N.R.; Sanches-Silva, A. Revisiting an ancient spice with medicinal purposes: Cinnamon. *Trends Food Sci. Technol.* **2017**, *62*, 154–169. [CrossRef]

91. Pandey, M.; Chandra, D.R. Evaluation of Ethanol and Aqueous extracts of *Cinnamomum verum* Leaf Galls for Potential Antioxidant and Analgesic activity. *Indian J. Pharm. Sci.* **2015**, *77*, 243–247.

92. Dutta, A.; Chakraborty, A. Cinnamon in Anticancer Armamentarium: A Molecular Approach. *J. Toxicol.* **2018**, *2018*, 8978731. [CrossRef]

93. Thompson, M.; Schmelz, E.M.; Bickford, L. Anti-Cancer Properties of Cinnamon Oil and its Active Component, TransCinnamaldehyde. *J. Nutr. Food Sci.* **2019**, *9*, 1000750.

94. Anderson, R.A.; Zhan, Z.; Luo, R.; Guo, X.; Guo, Q.; Zhou, J.; Kong, J.; Davis, P.A.; Stoecker, B.J. Cinnamon extract lowers glucose, insulin and cholesterol in people with elevated serum glucose. *J. Tradit. Complement. Med.* **2015**, *6*, 332–336. [CrossRef]

95. Zare, R.; Nadjarzadech, A.; Zarshenas, M.M.; Shams, M.; Heydari, M. Efficacy of cinnamon in patients with type II diabetes mellitus: A randomized controlled clinical trial. *Clin. Nutr.* **2019**, *38*, 549–556. [CrossRef]

96. Shinjyo, N.; Waddell, G.; Green, J. A tale of two cinnamons: A comparative review of the clinical evidence of *Cinnamomum verum* and *C. cassia* as diabetes interventions. *J. Herb. Med.* **2020**, *21*, 00342. [CrossRef]

97. Han, X.; Parker, T.L. Antiinflammatory Activity of Cinnamon (*Cinnamomum zeylanicum*) Bark Essential Oil in a Human Skin Disease Model. *Phytother. Res.* **2017**, *31*, 1034–1038. [CrossRef]

98. Schink, A.; Naumoska, K.; Kitanovski, Z.; Kampf, C.J.; Nowoisky, J.F.; Thines, F.; Pöschl, U.; Schuppan, D.; Lucas, K. Anti-inflammatory effects of cinnamon extract and identification of active compounds influencing the TLR2 and TLR4 signaling pathways. *Food Funct.* **2018**, *9*, 5950–5964. [CrossRef] [PubMed]

99. Shishehbor, F.; Safar, M.R.; Rajaei, E.; Haghighizadeh, M.H. Cinnamon Consumption Improves Clinical Symptoms and Inflammatory Markers in Women with Rheumatoid Arthritis. *J. Am. Coll. Nutr.* **2018**, *37*, 685–690. [CrossRef] [PubMed]

100. Abd El-Hack, M.E.; Alagawany, M.; Abdel-Moneim, A.-M.E.; Mohammed, N.G.; Khafaga, A.F.; Bin-Jumah, M.; Othman, S.I.; Allam, A.A.; Elnesr, S.S. Cinnamon (*Cinnamomum zeylanicum*) Oil as a Potential Alternative to Antibiotics in Poultry. *Antibiotics* **2020**, *9*, 210. [CrossRef]

101. Arancibia, M.; Giménez, B.; López-Caballero, M.E.; Gómez-Guillén, M.C.; Montero, P. Release of Cinnamon Essential Oil from Polysaccharide Bilayer Films and Its Use for Microbial Growth Inhibition in Chilled Shrimps. *LWT-Food Sci. Technol.* **2014**, *59*, 989–995. [CrossRef]

102. Atki, Y.E.; Aouam, I.; Kamari, F.E.; Taroq, A.; Nayme, K.; Timinouni, M.; Lyoussi, B.; Abdellaoui, A. Antibacterial activity of cinnamon essential oils and their synergistic potential with antibiotics. *J. Adv. Pharm. Technol. Res.* **2019**, *10*, 63–67. [CrossRef]

103. Mnafgui, K.; Derbali, A.; Sayadi, S.; Gharsallah, N.; Alfeki, A.; Allouche, N. Anti-obesity and cardioprotective effects of cinnamic acid in high fat diet- induced obese rats. *J. Food Sci. Technol.* **2015**, *52*, 4369–4377. [CrossRef]

104. Gulcin, I.; Kaya, R.; Goren, A.C.; Akincioglu, H.; Topal, M.; Bingol, Z.; Çakmak, K.C.; Sarikaya, S.B.O.; Durmaz, L.; Alwasel, S. Anticholinergic, antidiabetic and antioxidant activities of cinnamon (*cinnamomum verum*) bark extracts: Polyphenol contents analysis by LC-MS/MS. *Int. J. Food Prop.* **2019**, *22*, 1511–1526. [CrossRef]

105. Jain, S.G.; Puri, S.; Misra, A.; Gulati, S.; Mani, K. Effect of oral cinnamon intervention on metabolic profile and body composition of Asian Indians with metabolic syndrome: A randomized double -blind control trial. *Lipids Health Dis.* **2017**, *16*, 113. [CrossRef] [PubMed]

106. Mousavi, M.S.; Karimi, E.; Hajishafiee, M.; Milajerdi, A.; Amini, M.R.; Esmaillzadeh, A. Anti-hypertensive effects of cinnamon supplementation in adults: A systematic review and dose-response Meta-analysis of randomized controlled trials. *Crit. Rev. Food Sci. Nutr.* **2020**, *60*, 3144–3154. [CrossRef]

107. Tarkhan, M.M.; Balamsh, K.S.; El-Bassossy, H.M. Cinnamaldehyde protects from methylglyoxal-induced vascular damage: Effect on nitric oxide and advanced glycation end products. *J. Food Biochem.* **2019**, *43*, e12907. [CrossRef]

108. Khan, A.U.; Khan, A.U.; Khanal, S.; Gyawali, S. Insect pests and diseases of cinnamon (*Cinnamomum verum* Presi.) and their management in agroforestry system: A review. *Acta Entomol. Zool.* **2020**, *1*, 51–59. [CrossRef]

109. Attia, R.G.; Rizk, S.A.; Hussein, M.A.; Fatah, H.M.A.; Khalil, M.M.A.; Ma'moun, S.A.M. Effect of cinnamon oil encapsulated with silica nanoparticles on some biological and biochemical aspects of the rice moth, *Corcyra cephalonica* (Staint.) (Lepidoptera: Pyralidae). *Ann. Agric. Sci.* **2020**, *65*, 1–5. [CrossRef]

110. Kang, Y.J.; Seo, D.G.; Park, S.Y. Phenylpropanoids from cinnamon bark reduced beta-amyloid production by the inhibition of beta-secretase in Chinese hamster ovarian cells stably expressing amyloid precursor protein. *Nutr. Res.* **2016**, *36*, 1277–1284. [CrossRef] [PubMed]

111. Khasnavis, S.; Pahan, K. Cinnamon treatment upregulates neuroprotective proteins Parkin and DJ-1 and protects dopaminergic neurons in a mouse model of Parkinson's disease. *J. Neuroimmune Pharmacol.* **2014**, *9*, 569–581. [CrossRef]

112. *PN-ISO 948*; Spices Sampling. Polish Committee for Standardization: Warsaw, Poland, 1980.

113. *PN-ISO 2825*; Herbs and Spices, Preparation of the Ground Sample for Analysis. Polish Committee for Standardization: Warsaw, Poland, 1981.

114. Amin, I.; Norazaidah, Y.; Hainida, E.K.I. Antioxidant activity and phenolic content of raw and blanched Amaranthus species. *Food Chem.* **2006**, *94*, 47–52. [CrossRef]

115. *PN-A-86965:1997*; Herbal Spices—Black Pepper. Polish Committee for Standardization: Warsaw, Poland, 1997.

116. *PN-EN ISO 6571:2009/A1*; Spices and Herbs. Determination of Essential Oil Content (Hydrodistillation Method). Polish Committee for Standardization: Warsaw, Poland, 2009.

117. Misharina, T.A.; Terenina, M.B.; Krikunova, N.I. Antioxidant properties of essential oils. *Appl. Biochem. Microbiol.* **2009**, *45*, 642–647. [CrossRef]

118. Gülçin, I. The antioxidant and radical scavenging activities of black pepper (*Piper nigrum*) seeds. *Int. J. Food Sci. Nutr.* **2005**, *56*, 491–499. [CrossRef]

119. Andradea, K.S.; Ferreira, K.S. Antioxidant activity of black pepper (*Piper nigrum* L.) oil obtained by supercritical CO_2. In Proceedings of the III Iberoamerican Conference on Supercritical Fluids, Cartagena de Indias, Colombia, 1–5 April 2013.

120. Ahmad, A.; Husain, A.; Mujeeb, M.; Khan, S.A.; Alhadrami, H.A.A.; Bhandari, A. Quantification of total phenol, flavonoid content and pharmacognostical evaluation including HPTLC fingerprinting for the standardization of *Piper nigrum* Linn fruits. *Asian Pac. J. Trop. Biomed.* **2015**, *5*, 101–107. [CrossRef]

121. Nagy, M.; Socaci, S.A.; Tofana, M.; Pop, C.; Muresan, C.; Pop, A.V.; Salanta, L.; Rotar, A.M. Determination of total phenolics, antioxidant capacity and antimicrobial activity of selected aromatic spices. *Bull. UASVM Food Sci. Technol.* **2015**, *72*, 82–85. [CrossRef]

122. Nahak, G.; Sahu, R.K. Phytochemical evaluation and antioxidant activity of *Piper cubeba* and *Piper nigrum*. *J. Appl. Pharm. Sci.* **2011**, *1*, 153–157.

123. Gupta, R.K.; Chawla, P.; Tripathi, M.; Shukla, A.K.; Pandey, A. Synergistic antioxidant activity of tea with ginger, black pepper and tulusi. *Int. J. Pharm. Pharm. Sci.* **2014**, *6*, 477–479.

124. Sruthi, D.; Zachariah, T.J. In vitro antioxidant activity and cytotoxicity of sequential extracts from selected black pepper (*Piper nigrum* L.) varieties and Piper species. *Int. Food Res. J.* **2017**, *24*, 75–85.

125. Lee, J.-G.; Chae, Y.; Shin, Y.; Kim, Y.-J. Chemical composition and antioxidant capacity of black pepper pericarp. *Appl. Biol. Chem.* **2020**, *63*, 35. [CrossRef]

126. Mallick, M.; Bose, A.; Mukhi, S. Comparative Evaluation of the Antioxidant Activity of Some commonly used Spices. *Int. J. PharmTech Res.* **2016**, *9*, 1–8.

127. Trifan, A.; Zengin, G.; Brebu, M.; Skalicka-Woźniak, K.; Luca, S.V. Phytochemical Characteriza-tion and Evaluation of the Antioxidant and Anti-Enzymatic Activity of Five Common Spices: Focus on Their Essential Oils and Spent Material Extractives. *Plants* **2021**, *210*, 2692. [CrossRef]

128. Suhaj, M.; Rácová, J.; Polovka, M.; Brezová, V. Effect of c-irradiation on antioxidant activity of black pepper (*Piper nigrum* L.). *Food Chem.* **2006**, *97*, 696–704. [CrossRef]

129. Codex Alimentarius Standard for Black, White and Green Peppers. CXS 326-2017. Available online: https://www.fao.org/fao-who-codexalimentarius/sh-proxy/en/?lnk=1&url=https%253A%252F%252Fworkspace.fao.org%252Fsites%252Fcodex%252FStandards%252FCXS%2B326-2017%252FCXS_326e.pdf (accessed on 4 May 2022).

130. European Spice Association (ESA). *Quality Minima Document*; European Spice Association (ESA): Bonn, Germany, 2015.

131. Ravindran, P.N. Black Pepper, *Piper nigrum*, Medicinal and Aromatic Plants—Industrial Profiles. *Phytochemistry* **2001**, *58*, 827–829.

132. Hamrapurkar, P.D.; Jadhav, K.; Zine, S. Quantitative Estimation of Piperine in *Piper nigrum* and *Piper longum* Using High Performance Thin Layer Chromatography. *J. Appl. Pharm. Sci.* **2011**, *1*, 117–120.

133. Rajopadhye, A.A.; Upadhye, A.S.; Mujumdar, A.M. HPTLC Method for Analysis of Piperine in Fruits of Piper Species. *J. Planar Chromatogr.* **2011**, *24*, 57–59. [CrossRef]

134. Zachariah, T.J.; Mathew, P.A.; Gobinath, P. Chemical quality of berries from black pepper varieties grafted on *Piper colubrinum*. *J. Med. Aromat. Plant Sci.* **2005**, *27*, 39–42.

135. Sruthi, D.; Zachariah, T.J.; Leela, N.K.; Jayarajan, K. Correlation between chemical profiles of black pepper (*Piper nigrum* L.) var. Panniyur-1 collected from different locations. *J. Med. Plants Res.* **2013**, *7*, 2349–2357.

136. Rezvanian, M.; Ooi, Z.T.; Jamal, J.A.; Husain, K.; Jalil, J.; Yaacob, Z.; Ghani, A.A. Pharmacognostical and Chromatographic Analysis of Malaysian *Piper nigrum* Linn. Fruits. *Indian J. Pharm. Sci.* **2016**, *78*, 334–343.

137. Jansz, E.R.; Pathirana, I.C.; Packiyasothy, E.V. Determination of piperine in pepper (*Piper nigrum* L.). *J. Natl. Sci. Counc. Sri Lanka* **1983**, *11*, 129–138. [CrossRef]

138. Shango, A.J.; Mkojera, B.T.; Majubwa, R.O.; Mamiro, D.P.; Maerere, A.P. Pre- and postharvest factors affecting quality and safety of Pepper (*Piper nigrum* L.). *CABI Rev.* **2021**, *16*, 031. [CrossRef]

139. Ajmal, H. Isolation, identification and quantitative analysis of piperine from *Piper nigrum* Linn. of various regions of kerala by rp-hplc method. *World J. Pharm. Pharm. Sci.* **2018**, *7*, 1023–1049.

140. Shrestha, S.; Chaudhary, N.; Sah, R.; Malakar, N. Analysis of Piperine in Black Pepper by High Performance Liquid Chromatography. *J. Nepal Chem. Soc.* **2020**, *41*, 80–86. [CrossRef]

141. Nikolić, M.; Stojković, D.; Glamočlija, J.; Ćirić, A.; Marković, T.; Smiljković, M.; Soković, M. Could essential oils of green and black pepper be used as food preservatives? *J. Food Sci. Technol.* **2015**, *52*, 6565–6573. [CrossRef]

142. Codex Alimentarius. *CODEX STAN 192-1995—Codex General Standard for Food Additives. Adopted in 1995*; Revision 1997, 1999, 2001, 2003, 2004, 2005, 2006, 2007, 2008, 2009, 2010, 2011, 2012, 2013; FAO: Rome, Italy; WHO: Geneva, Switzerland, 1995.

143. Rouatbi, M.; Duquenoy, A.; Giampaoli, P. Extraction of the essential oil of thyme and black pepper by superheated steam. *J. Food Eng.* **2007**, *78*, 708–714. [CrossRef]

144. Pino, J.; Rodriguez-Feo, G.; Borges, P.; Rosado, A. Chemical and sensory properties of black pepper oil. *Nahrung* **1990**, *34*, 555–560. [CrossRef]

145. Zachariah, T.J.; Safeer, A.L.; Jayarajan, K.; Leela, N.K. Correlation of metabolites in the leaf and berries of selected black pepper varieties. *Sci. Hortic.* **2010**, *123*, 418–422. [CrossRef]

146. Kurian, P.S.; Backiyarani, S.; Josephrajkumar, A.; Murugan, M. Varietal evaluation of black pepper (*Piper nigrum* L.) for yield, quality and anthracnose disease resistance in Idukki District, Kerala. *J. Spices Aromat. Crops* **2002**, *11*, 22–24.

147. Rmili, R.; Ramdani, M.; Ghazi, Z.; Saidi, N.; El Mahi, B. Composition comparison of essential oils extracted by hydrodistillation and microwave-assisted hydrodistillation from *Piper nigrum* L. *J. Mater. Environ. Sci.* **2014**, *5*, 1560–1567.

148. Hussain, M.D.S.; Hegde, L.; Goudar, S.A.; Hegde, N.K.; Shantappa, T.; Gurumurthy, S.B.; Manju, M.J.; Shivakumar, K.M. Evaluation of local black pepper (*Piper nigrum* L.) genotypes for yield and quality under arecanut based cropping system. *Int. J. Pure Appl. Biosci.* **2017**, *5*, 1396–1400. [CrossRef]

149. Abraham, K. Cinnamon and coumarin—Clarification from the scientific and administrative angle. *Dtsch. Lebensm. Rundsch.* **2007**, *103*, 480–487.

150. Yang, C.C.H.; Li, R.X.; Chuang, L.Y. Antioxidant activity of various parts of *Cinnamomum cassia* extracted with different extraction methods. *Molecules* **2012**, *17*, 7294–7304. [CrossRef] [PubMed]

151. Mathew, S.; Abraham, T. Studies on the antioxidant activities of cinnamon (*Cinnamomum verum*) bark extracts, through various in vitro models. *Food Chem.* **2006**, *94*, 520–528. [CrossRef]

152. Dragland, S.; Senoo, H.; Wake, K.; Holte, K.; Blomhoff, R. Several culinary and medicinal herbs are important sources of dietary antioxidants. *J. Nutr.* **2003**, *133*, 1286–1290. [CrossRef]

153. Foti, M.; Piattelli, M.; Baratta, M.T.; Ruberto, G. Flavonoids, coumarins, and cinnamic acids as antioxidants in a micellar system structure-activity relationship. *J. Agric. Food Chem.* **1996**, *44*, 497–501. [CrossRef]

154. Lin, C.C.; Wu, S.J.; Chang, C.H.; Ng, L.K. Antioxidant activity of *Cinnamomum cassia*. *Phytother. Res.* **2003**, *17*, 726–730. [CrossRef] [PubMed]

155. Lee, H.S.; Kim, B.S.; Kim, M.K. Suppression effect of Cinnamomum barks-derived component on nitric oxide synthase. *J. Agric. Food Chem.* **2002**, *50*, 7700–7703. [CrossRef] [PubMed]

156. Prasad, K.N.; Yang, B.; Dong, X.; Jiang, G.; Zhang, H.; Xie, H.; Jiang, Y. Flavonoid contents and antioxidant activities from Cinnamomum species. *Innov. Food Sci. Emerg. Technol.* **2009**, *10*, 627–632. [CrossRef]

157. Shobana, S.; Naidu, K.A. Antioxidant activity of selected Indian spices. *Prostaglandins Leukot. Essent. Fat. Acids* **2000**, *62*, 107–110. [CrossRef]

158. Murcia, M.A.; Egea, I.; Romojaro, F.; Parras, P.; Jimenez, A.M.; Martinez-Tom, M. Antioxidant evaluation in dessert spices compared with common food additives. Influence of irradiation procedure. *J. Agric. Food Chem.* **2004**, *52*, 1872–1881. [CrossRef] [PubMed]

159. Prakash, D.; Suri, S.; Upadhyay, G.; Singh, B.N. Total phenol, antioxidant and free radical scavenging activities of some medicinal plants. *Int. J. Food Sci. Nutr.* **2007**, *58*, 18–28. [CrossRef] [PubMed]

160. Lin, C.C.; Yu, C.; Wu, S.J.; Yih, K. DPPH free-radical scavenging activity, total phenolic contents and chemical composition analysis of forty-two kinds of essential oils. *J. Food Drug Anal.* **2009**, *17*, 386–395. [CrossRef]

161. Lee, R.; Balic, M.J. Sweet wood—Cinnamon and its importance as a spice and medicine. *Ethnomedicine* **2005**, *1*, 61–64. [CrossRef]

162. Jayatilaka, A.; Poole, S.K.; Poole, C.F.; Chichila, T.M.P. Simultaneous micro steam distillation/solvent extraction for the isolation of semivolatile flavor compounds from cinnamon and their separation by series coupled-column gas chromatography. *Anal. Chim. Acta* **1995**, *302*, 147–162. [CrossRef]

163. Sproll, C.; Ruge, W.; Andlauer, C.; Godelmann, R.; Lachenmeier, D.W. HPLC analysis and safety assessment of coumarin in foods. *Food Chem.* **2008**, *109*, 462–469. [CrossRef] [PubMed]

164. Blahova, J.; Svobodova, Z. Assessment of coumarin levels in ground cinnamon available in the Czech retail market. *Sci. World J.* **2012**, *2012*, 263851. [CrossRef] [PubMed]
165. Woehrlin, F.; Fry, H.; Abraham, K.; Preiss-Weigert, A. Quantificion of Flavoring Constituents in Cinnamon High Variation of Coumarin in Cassia Bark from the German Retail Market and in Authentic Samples from Indonesia. *J. Agric. Food Chem.* **2010**, *58*, 10568–10575. [CrossRef] [PubMed]
166. He, Z.D.; Qiao, C.F.; Han, Q.B.; Cheng, C.L.; Xu, H.X.; Jiang, R.W.; But, P.P.H.; Shaw, P.C. Authentication and quantitative analysis on the chemical profile of cassia bark (cortex. *cinnamomum*) by high pressure liquid chromatography. *J. Agric. Food Chem.* **2005**, *53*, 2424–2428. [CrossRef] [PubMed]
167. Ho, S.-C.; Hang, K.-S.; Hang, P.-W. Inhibition of neuroinflammation by cinnamon and its main componenets. *Food Chem.* **2013**, *138*, 2275–2282. [CrossRef] [PubMed]
168. Lungarini, S.; Aureli, F.; Coni, E. Coumarin and cinnamaldehyde in cinnamon marketed in Italy: A natural chemical hazard? *Food Addit. Contam. Part A Chem. Anal. Control. Expo. Risk Assess.* **2008**, *25*, 1297–1305. [CrossRef] [PubMed]
169. Ananthakrishnan, R.; Chandra, P.; Kumar, B.; Rameshkumar, K.B. Quantification of coumarin and related phenolics in cinnamon samples from south India using UHPLC-ESI-QqQLIT-MS/MS method. *Int. J. Food Prop.* **2018**, *21*, 50–57. [CrossRef]
170. Raters, M.; Matissek, R. Analysis of coumarin in various foods using liquid chromatography with tandem mass spectrometric detection. *Eur. Food Res. Technol. J.* **2008**, *227*, 637–642. [CrossRef]
171. Lee, C.J.; Chen, L.G.; Chang, T.L.; Ke, W.M.; Lo, Y.F.; Wang, C.C. The correlation between skin-care effects and phytochemical contents in Lamiaceae plants. *Food Chem.* **2011**, *124*, 833–841. [CrossRef]
172. Regulation of the European Parliament and of the Council (EC) No. 1334/2008 of 16 December 2008 on Flavourings and Certain Food Ingredients with Flavouring Properties for Use in and on Foods and Amending Council Regulation (EEC) no 1601/91, Regulations (EC) no 2232/96 and (EC) no 110/2008 and Directive 2000/13/EC. *Off. J. Eur. Communities* **2008**, *12*, 34–50.
173. French Agency for Food, Environmental and Occupational Health & Safety. *Opinion of the French Agency for Food, Environmental and Occupational Health & Safety—Assessment of the Risk of Hepatotoxicity Associated with the Coumarin Content of Certain Plants That Can Be Consumed in Food Supplements or in Other Foodstuffs*; ANSES Opinion Request No. 2018-SA-0180; ANSES: Maisons-Alfort, France, 2021.
174. Balin, N.Z.; Sørensen, A.T. Coumarin content in cinnamon containing food products on the Danish market. *Food Control* **2014**, *38*, 198–203. [CrossRef]
175. Steffensen, I.L.; Alexander, J.; Binderup, M.-L.; Helkåas, K.; Dahl Granum, B.; Hetland, R.B.; Husøy, T.; Paulsen, J.E.; Sanner, T.; Thrane, V. Risk Assessment of Coumarin Intake in the Norwegian Population. *Eur. J. Nutr. Food Saf.* **2019**, *11*, 72–75. [CrossRef]
176. Schifferstein, H.N.J.; Kudrowitz, B.M.; Breuer, C. Food Perception and Aesthetics—Linking Sensory Science to Culinary Practice. *J. Culin. Sci. Technol.* **2020**, *20*, 293–335. [CrossRef]
177. Embuscado, M.E. Bioactives from culinary spices and herbs: A review. *J. Food Bioact.* **2019**, *6*, 68–99. [CrossRef]
178. Almatroodi, S.A.; Alashi, M.A.; Almatroodi, A.; Anwar, S.; Verma, A.K.; Dev, K.; Rahmani, A.H. Cinnamon and its active compounds: A potential candidate in disease and tumour management through modulating various genes activity. *Gene Rep.* **2020**, *21*, 100966. [CrossRef]

 *foods*

Article

Pyrrolizidine Alkaloids in the Food Chain: Is Horizontal Transfer of Natural Products of Relevance?

Mohammad Said Chmit [1], Gerd Horn [2], Arne Dübecke [3] and Till Beuerle [1,*]

[1] Institute of Pharmaceutical Biology, Technical University of Braunschweig, Mendelssohnstr. 1, 38106 Braunschweig, Germany; s.chmit@tu-bs.de

[2] Exsemine GmbH, Am Wehr 4, 06198 Salzatal, Germany; g.horn@exsemine.de

[3] Quality Services International GmbH, Flughafendamm 9a, 28199 Bremen, Germany; arne.duebecke@tentamus.com

* Correspondence: t.beuerle@tu-bs.de; Tel.: +49-(0)531-391-5690

Abstract: Recent studies have raised the question whether there is a potential threat by a horizontal transfer of toxic plant constituents such as pyrrolizidine alkaloids (PAs) between donor-PA-plants and acceptor non-PA-plants. This topic raised concerns about food and feed safety in the recent years. The purpose of the study described here was to investigate and evaluate horizontal transfer of PAs between donor and acceptor-plants by conducting a series of field trials using the PA-plant *Lappula squarrosa* as model and realistic agricultural conditions. Additionally, the effect of PA-plant residues recycling in the form of composts or press-cakes were investigated. The PA-transfer and the PA-content of soil, plants, and plant waste products was determined in form of a single sum parameter method using high-performance liquid chromatography mass spectroscopy (HPLC-ESI-MS/MS). PA-transfer from PA-donor to acceptor-plants was frequently observed at low rates during the vegetative growing phase especially in cases of close spatial proximity. However, at the time of harvest no PAs were detected in the relevant field products (grains). For all investigated agricultural scenarios, horizontal transfer of PAs is of no concern with regard to food or feed safety.

Keywords: biodegradation; pyrrolizidine alkaloid (PA); contamination; food chain; compost; plant waste; *Lappula squarrosa*; *Senecio jacobaea*; HPLC-ESI-MS/MS; horizontal alkaloid transfer

Citation: Chmit, M.S.; Horn, G.; Dübecke, A.; Beuerle, T. Pyrrolizidine Alkaloids in the Food Chain: Is Horizontal Transfer of Natural Products of Relevance?. *Foods* **2021**, *10*, 1827. https://doi.org/10.3390/foods10081827

Academic Editors: Andreas Eisenreich and Bernd Schaefer

Received: 23 July 2021
Accepted: 5 August 2021
Published: 7 August 2021

Publisher's Note: MDPI stays neutral with regard to jurisdictional claims in published maps and institutional affiliations.

1. Introduction

To date, many cases of human and livestock poisonings are recorded in literature as a result of food, phytomedicines, or feed contamination with secondary metabolites of toxic plants such as pyrrolizidine alkaloids (PAs), cardiac glycosides or other toxic compounds [1,2]. In the case of PAs, these toxins are introduced mainly by accidental co-harvesting of poisonous PA plants together with the crop of interest [3]. Pyrrolizidine alkaloids (PAs) are a class of phytotoxins occurring in an estimated 3% of the flowering plants worldwide [4–6]. Presently, more than 660 individual PAs and PA-*N*-oxides (PANOs) have been structurally characterized [4], and three plant families (Asteraceae, Boraginaceae, and Fabaceae) are by far the most important sources of these toxins [4,6]. In the case of the Asteraceae, PAs occur mainly in the tribes Senecioneae and Eupatorieae. Within the Fabaceae, the genus Crotalaria is well known for PAs, while PAs are common in many genera of the Boraginaceae family [4]. Most derived PAs from plants can be assigned to one of four main structural classes and in particular, 1,2-unsaturated ester-PAs (see Figure 1), have been associated with hepatotoxicity and carcinogenicity [7,8].

The increasing awareness of PA-toxicity and the elucidation of the corresponding mode of action of PA-toxicity over the past few years, has triggered studies on various routes on how these toxic plant metabolites may enter and contaminate food and feed chains, although (with only a few exceptions) PA-plants are not used for crop production [9–11].

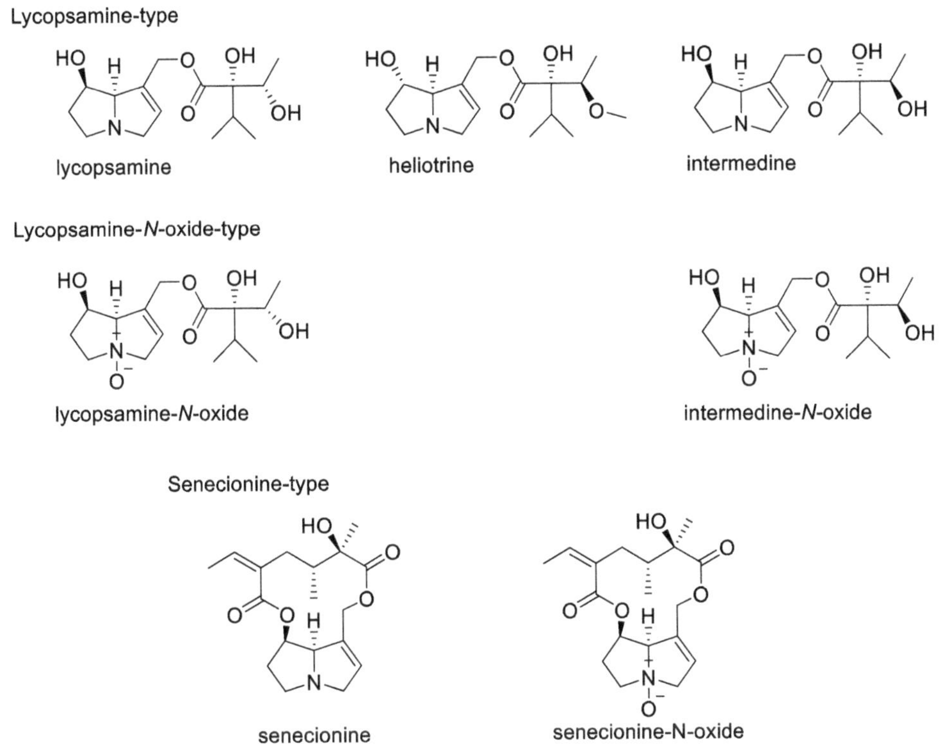

Figure 1. Exemplary chemical structures and features of some pyrrolizidine alkaloids (PAs).

So far, accidental or unwanted co-harvesting has been identified as the main source of PA-contamination potentially harmful to humans or livestock as a result of mixing leaves of plants that contain PAs with leaves of medicinal herbs or teas [12] or of mixing the seeds of PA-plants with wheat and barley crops [13]. Commercial fraud and mixing medicinal herbs with other plants to meet price, volume, or other market demands, might be another source for this contamination [2]. In addition, flowers, with their contents of nectar and pollen, are by now a well-known cause of food contamination with PAs [14]. Many studies have been conducted on beehive products (honey, royal jelly, pollen food supplements) and have demonstrated that these products could naturally be contaminated by PAs [14]. When tracing the sources of contamination, it was found that nectar of PA-producing plants was the main source of contamination of these beehive products which could additionally increase by the PA-load of pollen from those flowers [14,15]. Furthermore, animal products can be also contaminated with PAs [12]. As stated, eggs, milk, cheese, and at low levels, meat, were shown to contain PAs if the feed of animals contained PA-plants in the first place [10].

Recently, a new path via horizontal PA-transfer of PA-plants or decaying PA-plant material to neighboring non-PA-plants was proposed and discussed as an additional source of PA-contamination of non-PA-crops [16]. Similar observations are known for xenobiotica such as insecticides, polycyclic aromatic hydrocarbons (PAH), pentafluorophenol or pharmaceutical products/metabolites, which are transferred directly from contaminated soil to plants via the roots [17–19]. However, such transfers are less known and so far, less studied for natural compounds and plant-to-plant transfer. Since chronic intake of low amounts of PAs may be potentially hazardous to humans and livestock [20], the contribution of each possible route of PA-entry (e.g., honey, tea, spices, etc.) should be carefully investigated to evaluate the importance of its contribution to the overall risk of PAs in the food chain. So far, horizontal PA-transfer was demonstrated under laboratory or laboratory-like conditions only, demonstrating the possibility of a transfer of PAs from mulched PA-plants to non-PA-plants as well as, e.g., from *Senecio jacobaea* to various herbs such as mentha or chamomilla [21], or between living plants in close spatial proximity, e.g., from *S. jacobaea* to various herbs like parsley or melissa [22]. In addition, a recent published study reported the transfer from strongly PA-plant infested fallow land (*Chromolaena odorata*) after slash-and-burn and slash-and-mulch practice to the subsequently cultivated maize plants [23]. These studies have in common that the transfer is usually observed from a PA-weed to crops under a more or less extreme PA-plant excess (potted plant experiments or PA-plant invasions). These conditions are usually not found in Western agricultural economies; hence these findings hardly mirror the potential impact to evaluate a realistic risk of horizontal PA-transfer in agricultural practice. So far, the ratio of non-PA-plant to neighboring PA-plants or decaying PA-plant material was not reflecting a realistic agricultural practice, where PA-plants/PA-biomass (here: weeds) should be by far outnumbered by the cultivated non-PA-crop plants. In addition, climate, soil, and seasons should have a major impact on the transfer rate (e.g., in the majority of cases the crop is harvested before PA-plants start to accumulate or decay). Furthermore, while in previous model experiments higher PA-levels were detected in roots and leaves/stems, in many realistic scenarios, seeds would be the final crop/product, but so far only very little data is available here [23]. In order to gain better insight into how these above-mentioned second-degree contamination routes impact the PA-contamination of non-PA-crops, we decided to conduct field experiments for several seasons to investigate the scenario of horizontal PA-transfer under realistic, actual agricultural and climate conditions.

2. Materials and Methods

2.1. Field Experiments

Field trials were carried out in accordance with the general guidelines of field trials methodology [24] in the time period between 2017 to 2020, in the surroundings of Halle/Saale located in the rain shadow of the Harz Mountains range ("Mitteldeutsches Trockengebiet", Central German Dry Area). This consists of a dry but fertile landscape in the center of Germany close to Halle/Saale (Saxony-Anhalt, Germany). The altitude there is approximately 100 m above sea level. The mean for the long-term annual precipitation in this area is below 500 L/m^2, and the mean annual temperature is 9.0 °C [25].

However, comparing the vegetation periods between 2018 and 2020, 2018 was considerably drier (<400 L/m^2), whereas other locations showed even lower measurements (<300 L/m^2). All three periods were significantly warmer following the general trend of the climate changes that took place there (annual mean values of up to >11 °C during the trial periods).

All soils of the experiments had a medium to high credit rating (diluvial and loess sites, soil value indices, "Ackerzahl" [26]) of 54–98 and had a medium to elevated level of macro and micronutrient supply. All field experiments and the control plots were managed ecologically during these experiments and were in line with good agricultural practice methodology, while weed control was carried out exclusively by mechanical means. For the cultivation of *L. squarrosa* only the variety "Laira" was used. This is a biennial genotype, which is cultivated during autumn sowing of the previous year. The crops grown after *L. squarrosa* cultivations were chosen based on their importance for a rational crop rotation scheme for this region and for organic farming, constituting the idea that different possible plant families should be represented in these experiments. Regionally common varieties were used for these subsequent cultivations. Several scenarios of field experiments were conducted and the details are described below. In addition, Figure 2 in Results and Discussion illustrates and summarizes the chronology and rotation of field and crops.

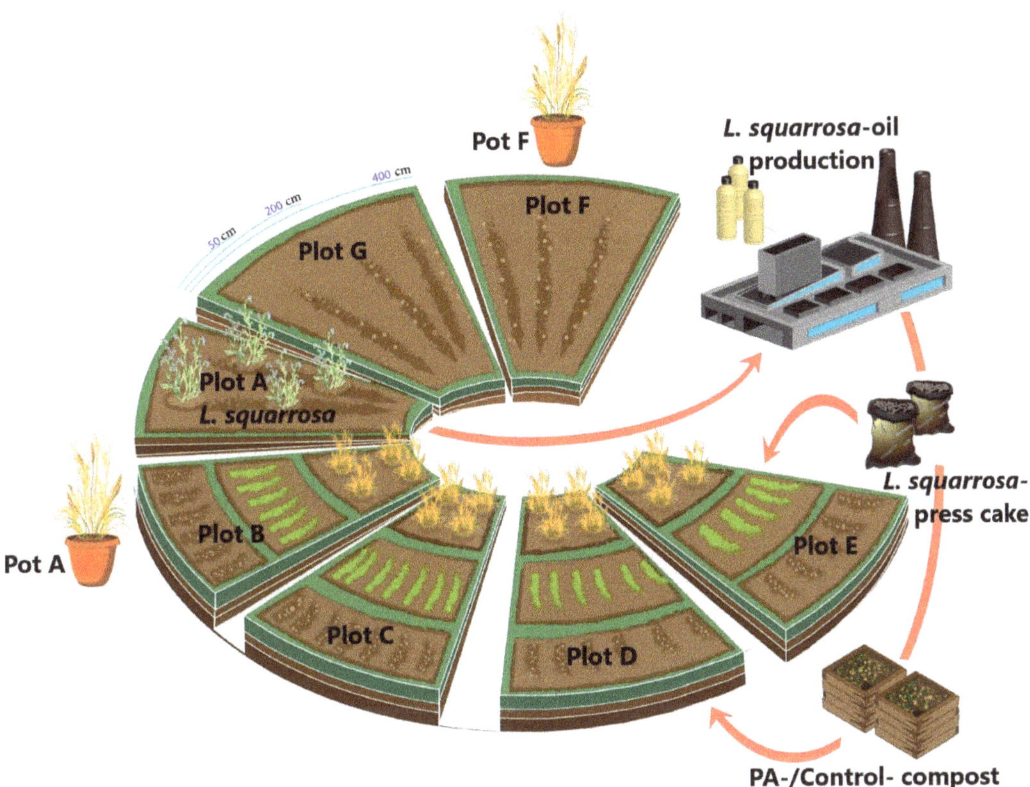

Figure 2. Experimental design to investigate possible horizontal PA-transfer in the context of PA-plant (*L. squarrosa*) cultivation.

2.2. Soil Improvement Measures

2.2.1. Composts

Composts were prepared in 2018 and 2019 (July through September) to yield four different model composts which were used to test the impact of PA-plant compost on PA-transfer to subsequent crops: (a) control-compost contained only compost stock; (b) Senecio-compost contained compost stock of which 32% (w/w) of fresh cut *S. jacobaea* material were included; (c) Lappula-compost contained compost stock of which 14% (w/w) *L. squarrosa* press-cake powder was included; (d) Senecio/Lappula-compost contained both 15% (w/w) fresh cut *S. jacobaea* and 9% (w/w) *L. squarrosa* press-cake powder in the composite.

Details on the compost making, PA-degradation and final PA-levels of the composts are given in Chmit et al. [27]. These four composts were analyzed for value-determining soil nutrients [27].

2.2.2. *L. squarrosa* Press-Cake

Dried *L. squarrosa* press-cake (residues of the pressed *L. squarrosa* seeds for oil production; harvest 2017) was used in homogenous milled form (3 mm) provided by Exsemine GmbH (Salzatal, Germany). Samples were taken to analyze the PA-content of the press-cake before it was used to set up the composts (2018 and 2019) and before the material was added directly to the soil for field plot preparation in 2019 and 2020.

2.3. PA-Transfer via PA-Plant Compost/Press-Cake Using Experimental Field Plots

Compost/press-cake experiments were performed in 2018/2019 in Etzdorf (Saxony-Anhalt, Germany) on a single field plot with a high-quality a high-quality loess-loam No.1 soil ("chernozem", soil value index 98). The area was divided in individual 1.7 m × 1.7 m plots with a 2 m distance to each other in all directions. Six different cultivation experiments were investigated: control-compost, Senecio-compost, Lappula-compost, Senecio/Lappula-compost, direct *L. squarrosa* press-cake, and mineral fertilizer as a control plot. Two crops were used as potential acceptor-plants: winter wheat (*Triticum aestivum*), variety "Wiwa" and summer barley (*Hordeum vulgare*), variety "Avalon". Four replicates of each treatment were conducted, resulting in 48 individual plots that were managed according to good agricultural practice (Figure 4). The corresponding amounts of compost/press-cakes were added to each plot to meet the nutrient requirements of the crop to be grown, based on total nitrogen content (*T. aestivum* 100 kg/ha and *H. vulgare* 50 kg/ha). A table showing the nutritional content and the added amount per plot is given in the Supplementary Materials file (Table S1).

The corresponding amounts of 2018 composts/press-cake quantities were added and worked in, on 18 October 2018 for both crops. This was done manually to avoid cross contamination between different plots. Winter wheat was sown on 19 October 2018 and spring barley in 4 March 2019, using a row distance of 25 cm. The mineral fertilization of the control plot was carried out for winter wheat on 26 April 2019, and for spring barley 23 May 2019 (at the two-node stage).

A second, very similar experiment was conducted in 2019/2020, in the area of Zappendorf (Saxony-Anhalt, Germany) near a high-quality, loess-loam No. 2 site (soil value index 85). The analog experiment had the same variants. However, due to technical/agricultural reasons, instead of 4 replicate plots for each treatment larger individual plots were used (3 m × 5 m) to allow similar numbers and distant sampling of soil and plants for analysis. This resulted in 12 individually treated plots.

2.4. One- and Two-Year Follow-Up Studies on PA-Transfer to Acceptor-Plants on Fields Previously Used for L. squarrosa Cultivation

These experiments were carried out on 50–400 m² plots in Zappendorf (Black loess No. 2/clay soil) after harvesting *L. squarrosa* crops.

The experiments were performed growing winter wheat and spring barley as subsequent crops in the first year after cultivation. For the second year, the plots were swapped, and spring barley followed winter wheat and vice versa. Controls using the same crop varieties were grown on a nearby comparable location, with the difference being that no PA-plants (e.g., *L. squarrosa*) were grown on these plots previously. The plots were sampled three times per season: (a) soil right before the start of the cultivation, (b) soil, roots and plantlets at the two-node stage and (c) soil, roots, above ground plant parts and grain/seeds/fruits of the cultivated crop. The subsequent crops were selected according to their importance for local agriculture. A total of three seasons were covered.

In addition, a more diverse range of follow-up crops of common regional varieties from other plant families were also included in these experiments, however, not under the strict procedure of three independent replicates. These other crops included *Brassica napus*, *Pisum sativum* and *Coriandrum sativum*. These experiments were conducted nearby on similar soil and climate conditions (diluvial site, soil value index 54, Halle/Saale, Martin Luther University, Halle-Wittenberg, Germany).

2.5. Investigations of Distance-Related Effects on PA-Transfer

To investigate the effects of distance on PA-transfer to non-PA-plants, field strips right next to the *L. squarrosa* cultivation were prepared. The location in 2019 was Kühnfeld (Martin Luther University, Halle-Wittenberg, Germany, soil value index 54), and in 2020 it was Zappendorf (soil value index 85, Saxony-Anhalt, Germany). These strips were planted with *Lolium multiflorum*. Plant and soil samples were taken at a distance of 50 cm, 200 cm and 400 cm away from the field of *L. squarrosa* cutivation. In addition, to monitor and minimize the PA-carry-over via soil, four Kick-Brauckmann vessels with a diameter of 22 cm and a height of 22 cm (0.038 m^2 soil) were buried up to the rim in these strips. Two vessels each were placed at a 50 cm and two vessels at 200 cm distance to the *L. squarrosa* cultivation. All four vessels per season also contained *L. multiflorum* as plants. To illustrate this experiment, we added a picture to the Supplementary Materials (Figure S1).

2.6. Controls and Pot Experiments

All experiments described above were accompanied through seasons with control experiments using plots that had no history of PA-plant cultivation (in particular, *L. squarrosa*). On these control patches *Brassica napus*, *Pisum sativum*, *Coriandrum sativum*, *T. aestivum*, *H. vulgare* and *L. multiflorum* were cultivated.

In addition to these field-controls, a one-season model experiment was conducted, using 20 Kick-Brauckmann pots (including replicates) to grow *T. aestivum* and *H. vulgare* separately, under controlled conditions. At the start of the experiment, the pots were individually filled using two kinds of soil: (a) commercial potting soil, and (b) top soil from a harvested *L. squarrosa* field. Shoot samples of both grain crops from both treatments were collected and analyzed for PAs at the two-node stage.

2.7. Sample Collection

In all experiments, samples of soil at the different growing stages (sowing stage, vegetation stage, two-node stage, and harvesting stage) were taken in the soil layer of the main root horizon (0–30 cm depth) using the boring stick method (8–14 impacts per variant depending on the crop acreage). All soil samples were frozen in plastic zip-lock bags and stored at −20 °C, then lyophilized and stored under dry and dark conditions until analysis. On the other hand, whole plants (including roots) at the vegetation stage (two-node stage) and crops at the stage of harvest were manually sampled using a spade (8–10 stakes) to collect about 20–50 plants per variant, depending on the species. In addition, individual plants of weeds growing in the plots were sampled as well. The plant material of the vegetation stage (two-node stage) and weeds were cleaned from adherent soil and separated into roots and shoot material. Harvest-ready plants were separated into roots, straw, and crop product (seeds, fruits); all different plant parts were collected in individual paper bags, air dried and stored at room temperature until analysis. Right before analysis all individual samples of one plot/field were well mixed resulting in an individual cross-section sample per data point, hence each value represents a biological average.

2.8. Chemicals and Reagents

All chemicals used were purchased from Roth (Karlsruhe, Germany) and Sigma-Aldrich (Seelze, Germany) and of HPLC grade purity or redistilled before use. Lithium aluminum hydride solution (1 M) in THF and pyridine both in AcroSeal quality were acquired from ACROS Organics (Fair Lawn, New Jersey, USA). Strong cation exchange solid phase extraction cartridges (SCX-SPE) were obtained from Phenomenex (Aschaffenburg, Germany). Isotopically labeled internal standard (IST) 7-*O*-9-*O*-dibutyroyl-[9,9−^{2}H$_2$]-retronecine, was synthesized in our laboratory [28].

2.9. Chemical Analysis of PAs

2.9.1. Sample Preparation and HPLC-ESI-MS/MS Quantification of the Total PA-Content

All analyzed samples were in homogenous form, lyophilized and homogenized in 50 mL conical centrifuge tubes using a mixer mill (MM 400; Retsch, Haan, Germany) for 2 min at 30 Hz. Three-hundred mg from each sample was soaked in a 15 mL polypropylene, conical centrifuge tube using 11 mL H$_2$SO$_4$ 0.05 M. Then, 40 µL of internal standard solution (2 µg/mL in methanol) was added to each tube and the tubes were thoroughly mixed and extracted overnight using a continuous tube rotator (Multi Bio RS-24, Biosan, Riga, Latvia) and the following settings: orbital: 21/01, reciprocal: 15/01, vibrio: 5/1, duration:

14 h. Sample preparation to analyze the total PA-content of the individual samples was carried out according to the methods of Cramer et al. [28] and Letsyo et al. [29]. During the course of sample preparation all 1,2-unsaturated retronecine- and heliotridine-type ester PAs/PANOs were converted into the corresponding core structures, i.e., retronecine and/or heliotridine. After derivatization, these analytes were analyzed by HPLC-ESI-MS/MS, generating a single signal which could be quantified by the use of the added isotopically labeled IST, allowing the calculation of the total content of all 1,2-unsaturated retronecine- and heliotridine-type ester PAs including all metabolites thereof, which still bear both necessary features of PA-toxicity (1,2-unsaturation and ester-PAs). Using an internal deuterated standard allowed the direct quantification across all matrices (matrix effects were covered and the IST was detected in all analyzed samples). A limit of detection (LoD) of 1 µg PA/kg and a limit of quantification (LoQ) of 5 µg PA/kg was obtained across all different matrices. A detailed description of this method and the sum parameter approach as well as the underlying calculations was added to the Supplementary Material or can be found at Cramer et al. [28]. Individual samples (positive and negative analytical results) were re-analyzed randomly, to confirm the integrity of the method. Data analysis and integration was achieved with Analyst 1.6.2 Software (Applied Biosystems MDS Sciex, Darmstadt, Germany). All analytical values are presented based on dry weight (d.w.).

2.9.2. Profiling of Individual PA-Patterns by HPLC-ESI-MS/MS

Pyrrolizidine alkaloid patterns of soil and plant materials were determined by an HPLC-ESI-MS/MS method, using an individual PA-detection approach according to the published BfR-method [30]. Extraction and profiling of the PAs in these products were conducted as a service by a commercial laboratory (QSI, Bremen, Germany; Chemisch-physikalische Analyze (#45183)) comprising 31 individual PAs and PANOs. The detailed results of this analysis can be found in the Supplementary Materials Table S2.

3. Results and Discussion

As outlined in the introduction, the transfer of toxic PAs from decaying PA-plant material as well as the transfer from nearby PA-plants to acceptor non-PA-plants was demonstrated in laboratory model experiments and its possible impact on food and feed safety was discussed [22,31]. In order to be able to derive realistic conclusions for a possible risk of horizontal PA-transfer from PA-plants to non-PA-plants, which might end up in food and feed, it was necessary to choose an appropriate experimental design. This means, in particular, a realistic ratio of PA-plants/PA-plant material (PA-donor) to non-PA-crop (PA-acceptor). Under given circumstances (Western agricultural economy) there should only be a very low number of PA-weeds compared to non-PA-crops. We know from several studies that these few so-called "accessory herbs" can cause PA-contamination via co-harvesting of the non-PA-crop [32,33], thus, it can most likely be ruled out that horizontal PA-transfer causes an additional risk in such a scenario. However, there is one realistic scenario which might cause problems in terms of horizontal PA-transfer that should be considered and investigated in more detail. A few PA-plant species are grown for commercial reasons. Hence, under this circumstance, there are exclusively PA-plants growing on an agricultural scale, generating potential PA-pressure directly on: (a) neighboring plants/cultivations, or indirectly via; (b) PA residues in the soil, which may impact subsequent cultivations on these plots. In addition, these PA-plant cultivations generate biomass waste that might return to fields as a form of organic fertilizer (composts, harvest residues or biogas residues) as it is promoted for circular bioeconomies. In all these scenarios there would be a potential risk that higher PA-loads might be accidently transferred to non-PA-crops, generating a significant accumulation of PAs in these acceptor-food-crops.

Currently, there are only a few PA-plants used commercially, mainly from the Boraginaceae plant family for the production of high-quality seed oils, such as *Borago officinalis* [34], *Buglossoides arvensis* [35], and *Echium plantagineum* [36]. In our experiments, we used *Lappula squarrosa*, another PA-producing Boraginaceae species currently being investigated for its potential as an alternative source for seed oils rich in stearidonic acid [37]. In addition to these cultivated PA-plants or production residues thereof, some organic fertilizers like composts might be produced containing, e.g., press-cake from Boraginaceae seed oil production or cuttings from other PA-plants for bio-recycling. In particular, it has recently been observed that some species like *S. jacobaea* (syn.: *Jacobaea vulgaris*), *Senecio aquaticus* [38,39], or *Senecio inaequidens* (a neophyte to European areas) [40], may mass-infest pasture areas or public land (parks, greenery, nature preserves, and ancillary areas of transport routes/roadsides) [41]. Those cuttings and green wastes naturally contain a high load of toxic PAs and if used as organic fertilizers in non-PA-crops, these PAs could potentially be passed on via horizontal PA-transfer. Instead of unrealistic laboratory experiments we designed a series of field experiments to investigate the possibility and the potential risks of horizontal PA-transfer. The experiments were conducted over a period of three seasons in order to trace as precisely and comprehensively as possible the impact of "worst case" scenarios by cultivating PA-plants and recycling their biomass, e.g., as compost. The experimental design is summarized in Figure 2.

Each year, the donor-PA-plant *L. squarrosa* was cultivated on different field plots (Figure 2, Plot A) and soil and plant samples (accessory herbs) were analyzed for PA-content. Plots B and C, respectively (Figure 2), represent two fields, where *L. squarrosa* grew before. While on plot B L. squarrosa grew there the season before, the second, plot C, was free of the donor-PA-plant for one season, resulting in a two-year follow-up study on plots growing *L. squarrosa*. Here, soil and plant samples (accessory herbs) were analyzed for PA-content three times and two times during each season, respectively. Plots D and E had no history of PA-plant cultivation before and were sub-divided into small plots where different crops were grown using additional organic fertilization (Plot D, PA-plant composts; Plot E, *L. squarrosa* press-cake from seed oil production residues). Plots F and G had no history of *L. squarrosa* cultivation and were used as controls. While Plot F was a pure control plot, Plot G was a strip next to the *L. squarrosa* cultivation (starting at a distance of 50 cm) to monitor possible distance effects on neighboring non-PA-plants. All experiments were repeated using new plots where necessary (e.g., new plots for controls, composts and so on) and rotating the *L. squarrosa* follow-up crops, which means the scheme illustrated in Figure 2 was performed twice, resulting in a two-year follow-up period for plots where *L. squarrosa* was grown in the past. Furthermore, in the period from March to August 2019, pot experiments were added as additional controls using *L. squarrosa* soil from the pre-season and commercial potting soil as substrates, and *T. aestivum* and *H. vulgare* as acceptor-plants (Figure 2; Pot A and Pot F)

3.1. PA-Content of Soil

The soil of all plots was sampled and was analyzed for PA-content before the cultivation started, then during the growth phase (two-node stage) as well as at the time of harvest. Since the applied analytical method was a sum parameter method, it comprised all toxicological relevant 1,2-unsaturated heliotrine- and retronecine-type PAs or even metabolized forms, which might occur due to biodegradation by soil microorganisms. Surprisingly, only some samples showed elevated PA-concentrations. Hence, the PA-positive soil results will be listed and discussed in the corresponding sections of the individual experiments below.

3.2. PA-Content of Non-PA-Plants Growing on L. squarrosa Plots (Plot A)

These experiments roughly reflect the experiments conducted so far to demonstrate the phenomena of horizontal PA-transfer [21,31]. In this case, a large number of PA-plants (*L. squarrosa*) were growing next to non-PA-plants (here common endemic weeds). For this scenario, elevated levels of PAs were observed generally in roots and in the above ground plant parts of those accessory herbs. Table 1 shows the level of PAs transferred to individual non-PA-plants via horizontal transfer.

Table 1. PA-content caused by horizontal PA-transfer to non-PA-accessory plants growing in *L. squarrosa* cultivations (Plot A, Figure 1).

Accessory Plant Species	Total Sum of PAs [µg PA/kg d.w.]	
	Root	Shoot
Chenopodium album	49.8	16.6
Convolvulus arvensis	512.9	4253.9
Echinochloa crus-galli 1	118.9	746.6
Echinochloa crus-galli 2	213.1	2944.2
Echinochloa crus-galli 3	359.2	988.5
Echinochloa crus-galli 4	214.4	710.1
Echinochloa crus-galli 5	1180.1	2184.8
Equisetum arvense	285.3	3691.9
Lactuca serriola	21.9	778.9
Lamium purpureum	1048.7	2975.4
Portulaca oleracea	trace	198.4
Atriplex patula	37.6	2386.6
Stellaria media	29.4	126.6
Urtica dioica	209.6	807.6
Veronica spp.	76.2	1850.1
Viola arvensis	2917.6	7254.7
L. squarrosa [1]	1,251,880.0	4,053,520.0

[1] PA-donor-plant.

The PA-levels in these plants ranged from traces (*P. oleracea*) up to 2917.6 µg PA/kg in the roots (average: 455 µg PA/kg) and 16.6–7254.7 µg PA/kg in the shoots (average: 1995 µg PA/kg). Besides the exemption of *C. album*, the PA-levels in the shoots always exceeded the levels in the roots, here by a factor of 15 on average (shoot/root factor: ranging from 0.3 (*C. album*) to 64 (*A. patula*)). The highest levels of PAs were recorded for the shoots and in the roots of *V. arvensis* (Table 1). This corresponds well to previous findings by Nowak et al. [21] and Letsyo et al. [23], where there was also the trend of higher PA-levels in the transpiration-active organs of the acceptor-plants, like shoots and leaves. In addition, the PA-levels observed in the acceptor-plants correspond to the results reported from the laboratory-like experiments, either by the transfer of mulching/decaying PA-plants ([21]: 50–500 µg PA/kg) or by co-cultivation of acceptor-plants with *S. jacobaea* ([31]: 100–1500 µg PA/kg). However, as interesting as these observations of horizontal natural-product transfers are, in our opinion it is absolutely essential to put these numbers into perspective and in direct comparison to the corresponding donor-plants to receive a realistic impression on how relevant this transfer is in terms of possible PA-contamination. Hence, for comparison reasons, the PA-levels of *L. squarrosa* were determined as well and are given in Table 1 (last row). As a result, the PA-levels of the donor-plant are two to three times the magnitude of the levels in the acceptor-plants. Our field experiments reflected the worst-case scenario possible for a realistic agricultural practice. Accessory plant/weeds (here: non-PA-plants) growing as close as naturally possible were by far outnumbered by PA-Plants. In this setting, a transfer rate of 0–0.23% and 0–0.18% compared to the donor-PA-plant could be observed. Hence, in all other possible agricultural scenarios these ratios are exactly the other way around (a non-PA-plant is the cultivated crop and by far outnumbers some individual PA-containing weeds). As a main result of this study, a low number of PA-weeds, together with the low observed transfer rate, will not lead to detectable contaminations of final crop as a whole via horizontal PA-transfer. Instead, co-harvesting of PA-weeds together with crops can cause significant PA-levels in food and feed [13,42,43] and should be considered the major route for the possible PA-contamination.

During our standard sum parameter quantitative analytical approach the structural information of the PAs is lost. Therefore, we analyzed the individual PAs of some selected samples to obtain the PA-profile using a second analytical method, established for the simultaneous determination of 17 PAs and 14 PANOs [30]. The typical PA-pattern observed for *L. squarrosa* in this study was; lycopsamine (39.7%), lycopsamine-*N*-Ox (26.1%), intermedine-*N*-Ox (20.7%) and intermedine-*N*-Ox (13.6%) (Figure 1), all members of the 1,2-unsaturated lycopsamine-type family PAs, typical for Boraginaceae plant spp. The results of this individual PA-analysis confirmed that *C. arvensis*, *E. crus-galli*, *A. patula*, and *V. arvensis* had similar PA-patterns as observed in *L. squarrosa*, where lycopsamine ranged between 48.3–67.6%, intermedine 23.6–32.4%, lycopsamine-*N*-Ox 0–21.3%, and intermedine-*N*-Ox 0–6.9%, all *L. squarrosa* typical open-chain retronecine-monoester-type of PAs. No other PAs/PANOs e.g., from the Senecionine-type PAs (closed-ring, diester-type) were detected (Figure 1). These findings confirmed that *L. squarrosa* was always the source of the PAs found in the accessory herbs.

3.3. Distance Effect of PA-Transfer (Plot G)

The effect of distance on the transfer of PAs was also investigated in this study (Table 2). These plots (Plot G, Figure 2) were located directly next to *L. squarrosa* cultivations (Plot A, Figure 2). The soil PA-content before the start of the experiment was below the limit of detection. As a model, *L. multiflorum*, an additional grass variety (compared to Plots B and C), was chosen for these experiments. The PA-levels in relation to the distance from the *L. squarrosa* cultivation were measured at the full-blooming stage of the nearby *L. squarrosa* cultivation (Figure S2). Two setups were investigated and sampled: (a) regular growing grass at a certain distance; and (b) grass samples growing in an extra pot, which was buried in the soil before the start to see how a barrier in the soil (closed system in terms of soil-transfers) would affect the PA-transfer.

Table 2. Effect of distance on the PA-content of *L. multiflorum* (Plot G, Figure 1) growing next to a plot of *L. squarrosa* (Plot A, Figure 1).

Distance between *L. multiflorum* and *L. squarrosa*	Total Sum of PAs [μg PA/kg d.w.]	
	Root	**Shoot**
50 cm without vessel	19.2	335.1
50 cm with vessel	11.9	38.1
200 cm without vessel	Trace	21.4
200 cm with vessel	Trace	8.6
400 cm without vessel	Trace	8.3

The results showed significantly reduced PA-transitions compared to the data obtained for the accessory herbs growing in Plot A (Figure 1). While accessory herbs in Plot A showed average values of 455 and 1995 μg PA/kg on average for roots and shoots, respectively, a distance of 50 cm to *L. squarrosa* lowered the PA-levels significantly to 19 and 335 μg PA/kg for roots and shoots, respectively. In addition, greater distances resulted in steadily decreasing PA-levels.

In all cases analyzed, the PA-profiles detected in *L. multiflorum* were matching the PA-profile of *L. squarrosa* (see Section 3.2.), with a domination of lycopsamine (66.03−77.78%) and intermidine (20.15−27.91%), which points to *L. squarrosa* as the sole PA-source.

Surprisingly, the potted samples also still showed some low levels of PAs, mainly in the shoots. It is expected that the horizontal PA-transfer takes place via the roots; however, there seems to be the possibility that a small fraction of this transfer might occur differently. At this point in time, we assume that maybe air-borne particles (pollen or dust of the nearby *L. squarrosa* plants) or rainwater flowing on the soil surface and over the rims of the buried pots and carrying a PA-load from the neighboring *L. squarrosa* cultivation [44,45] might cause these low PA-transfers.

As a result, for agricultural practice of growing PA-plants, a distance of four meters should be an adequate isolation distance to reduce PA-contamination of neighboring cultivations and reduce PA-transfers to a minimum.

3.4. Crops on Fields Used for L. squarrosa Cultivation Before

In another series of experiments, we wanted to monitor the possible transfer of PAs on fields which were previously used for *L. squarrosa* cultivation, to crops that grew on these soils in the following seasons. Two different follow-up crops (two types of cereals grains) usually recommended as follow-up crops for *L. squarrosa* were used to cover for a broader picture.

3.4.1. One-Year Follow-Up Studies on PA-Transfer to Acceptor-Plants on Fields Previously Used for L. *squarrosa* Cultivation (Plot B)

Several growth stages were observed. At the beginning, only soil at the stage of sowing was sampled. Later on, soil and plants at the two-node stage and just before harvest, including the crop fruits (here: cereal grains), were sampled and analyzed for total PA-content. As shown in Table 3, soil of such plots might contain low levels of PAs due to the preceding *L. squarrosa* cultivation. This PA-load can also be transferred to the next generation of crops on these plots, as it can be seen by the PA-values of the two-node stage (roots and shoots, 8.2 and 37.1 µg PA/kg, respectively), higher values were again and consistently observed in above ground parts of the plants (Table 3).

Table 3. PA-content of follow-up crops (*T. aestivum* and *H. vulgare*) on plots used for the cultivation of *L. squarrosa* the year before.

Sample		Total Sum of PAs [µg PA/kg d.w.]			
		T. aestivum		*H. vulgare*	
		2019	2020	2019	2020
sowing stage	soil	20.8	17.4	<LoD [1]	<LoD
two-node stage	soil	<LoD	<LoD	<LoD	<LoD
	roots	8.1	10	7.4	7.3
	shoots	60.8	14.4	34.4	38.7
time of harvest	soil	<LoD	<LoD	<LoD	<LoD
	roots	<LoD	<LoD	<LoD	<LoD
	straw	<LoD	<LoD	<LoD	<LoD
	caryopses	<LoD	<LoD	<LoD	<LoD

[1] Limit of detection.

This result for the vegetative part of the season is in accordance with Letsyo et al. [23] for *Zea mays*; there it was reported that PAs passed through the roots and accumulated at low levels in the leaves (16.3–21.1 µg/kg). Moreover, Hama et al. [44] demonstrated that during winter, soil contains lower amounts of PAs due to low temperature and the leaching of PAs into deeper soil layers out of the reach of the roots. However, at the point of harvest there were no longer PAs detected in the crops (*T. aestivum* and *H. vulgare*), suggesting that during the ripening of the grain and the die-off of the plant, the low PA-amounts of the growing phase in the green parts either vanish or get diluted below the detection limit by the increase of the above ground biomass. No PAs could be detected in the final harvest products, suggesting that fruits/grains are not a PA-sink in non-PA-plants and most importantly, no PAs would be transferred into the food chain under such circumstances.

Besides these elaborate multi-year experiments with grains (including numerous replicates of plots) individual cultivations of *C. sativum*, *P. sativum*, and *B. napus* were planted in the season right after *L. squarrosa* cultivation and monitored for PA-content. In these experiments, the soil samples at the sowing stage showed higher levels of PAs, 111–714.4 µg PA/kg (Table 4). However, while in some cases PA-soil levels were maintained at low levels throughout the season/experiment, this PA-contamination was not transferred to the later stages of the above ground plant parts nor to the harvested crops.

Table 4. PA-content of various follow-up crops on plots used for the cultivation of *L. squarrosa* the year before.

		Total Sum of PAs [µg PA/kg d.w.]		
		C. sativum	*P. sativum*	*B. napus*
sowing stage	soil	714.4	111.0	332.7
two-node stage	soil	27.4	11.3	<LoD [1]
	roots	<LoD	12.0	<LoD
	shoots	<LoD	<LoD	<LoD
time of harvest	soil	13.4	8.3	8.7
	roots	<LoD	8.3	<LoD
	straw	<LoD	<LoD	<LoD
	caryopses	<LoD	<LoD	<LoD

[1] Limit of detection.

The additional analyzed PA-profile for the positive samples confirmed *L. squarrosa* as the origin of the PA-transfer. The detailed analytical results are summarized in the Supplementary Materials (Table S2).

3.4.2. Second-Year Follow-Up Studies on PA-Transfer to Acceptor-Plants on Fields Previously Used for *L. squarrosa* Cultivation (Plot C)

The former *L. squarrosa* plots described above were monitored for an additional season (Plot C; Figure 2), switching the follow-up crops from *T. aestivum* to *H. vulgare* and vice versa. Except for some soil and a single root sample all of these second successor crop samples were PA-negative (Table 5).

Table 5. PA-content of follow-up crops (*T. aestivum* and *H. vulgare*) on plots used for the cultivation of *L. squarrosa* two years before.

Sample		Total Sum of PAs [µg PA/kg d.w.]			
		T. aestivum		*H. vulgare*	
		2019	2020	2019	2020
sowing stage	soil	<LoD [1]	15.4	<LoD	<LoD
two-node stage	soil	<LoD	<LoD	<LoD	<LoD
	root	<LoD	<LoD	8.1	32.6
	shoot	61.7	<LoD	<LoD	<LoD
time of harvest	soil	<LoD	15.7	<LoD	25.9
	roots	<LoD	<LoD	<LoD	44.8
	straw	<LoD	<LoD	<LoD	<LoD
	caryopses	<LoD	<LoD	<LoD	<LoD

[1] Limit of detection.

However, at the final stage of crop cultivation in this second season, a re-growth of new *L. squarrosa* plants appeared strictly between the rows of the sown crops (Figure 3).

These plants originated from twice hibernated *L. squarrosa* seeds, which were still present in the soil from the *L. squarrosa* cultivation two seasons ago and which were still able to germinate. Since this somewhat resembled the situation as observed in the co-cultivation of *L. squarrosa* and accessory herbs in Plot A (Figure 2), we addressed this phenomenon in more detail. To study a possible effect of this phenomenon, three adjacent plants of *T. aestivum* and of *H. vulgare*, located right next to germinating/developing *L. squarrosa* patches, were collected and analyzed.

Figure 3. *L. squarrosa* plantlets (red box) between rows of the follow-up crop *H. vulgare*, originating from hibernated seeds of *L. squarrosa* growing on this plot two years ago (Plot C).

As shown in Table 6, roots and straw of most of these crop plants next to the *L. squarrosa* patches were PA-positive, ranging between 9.8–397.7 µg PA/kg in the roots and 13.2–108 µg PA/kg in the straw. However, there was no PA-transfer to the caryopsis of these plants. As it could be confirmed in this and other studies, the simultaneous growth of PA-plants next to non-PA-plants results in the transfer of small amounts of secondary plant metabolites. Interestingly, in this case, germinating/developing PA-plants next to "adult" acceptor-plants generally resulted in higher PA-levels in the roots instead of the shoot parts, while in the experiments before, in adult PA-donor-plants in combination with young accessory plants, this was exactly the other way around.

Table 6. PA-content caused by horizontal PA-transfer to non-PA-accessory plants originating from re-germination of hibernated *L. squarrosa* seeds maintained in the soil after two years.

| | Total Sum of PAs [µg PA/kg d.w.] | | | | | |
| | *T. aestivum* | | | *H. vulgare* | | |
	Plant 1	Plant 2	Plant 3	Plant 1	Plant 2	Plant 3
root	70.6	<LoD [1]	20.9	<LoD	397.7	9.8
straw	66.1	<LoD	108	13.2	41.8	<LoD
caryopses	<LoD	<LoD	<LoD	<LoD	<LoD	<LoD

[1] Limit of detection.

The re-appearance/outgrowth of *L. squarrosa* seeds in these experiments can be attributed to the special experimental design in combination with the ecological agriculture practice applied. This led to extended row spacing allowing sunlight to reach these interspaces and promoting the development of the *L. squarrosa* plantlets. However, this does not seem a problem in general, since, due to economic reasons, the cultivation of *L. squarrosa* is intended for conventional cultivation and is very sensitive to the conventional herbicide groups and is easily outcompeted under conventional cultivation conditions with higher crop populations and denser row spacing. Hence, the possible germinating and possible carry-over of PAs seems only relevant for ecological agriculture practice and might need to be addressed there more specifically.

The analytical determined PA-profiling confirmed that *L. squarrosa* was the source of the PA-transfer.

3.5. PA-Plant Composts/L. squarrosa Press-Cake Experiments (Plots D and E)

In today's promoted circular bioeconomies, there should be no bio-waste accumulated; instead, it should be re-utilized, and harvest residues should stay in or return to the field. This is particularly tricky if those materials contain potentially toxic compounds, in our case, the PAs. Recently, we demonstrated that a plant derived PA-load added to the composting fermentation process is dramatically reduced by more than 99.9%, while a 91 to 99% reduction was observed in bio-gas fermentation [27]. However, despite the tremendous PA-reduction through such processing, there were still some residual PA-levels in those materials ranging from 0 to 26 µg PA/kg [27].

To understand the full picture of the impact of the cultivation of PA-producing Boraginaceae species, we conducted field experiments on the effects of returned *L. squarrosa* harvest materials to the field on subsequently planted crops. In particular, we investigated different methods for improving soil quality (mainly nitrogen content) by using PA-plant composts, including *L. squarrosa* press-cake compost (residue from seed-oil production) but also used the press-cake directly (Figure 4).

Figure 4. Picture showing the PA-plant composts/*L. squarrosa* press-cake experiments. *T. aestivum* (dark green) and *H. vulgare* (light green) plots are marked (red box). Each plot represents a different experimental variant or a replicate thereof.

In summary, all soil samples collected at sowing stage of *T. aestivum* and *H. vulgare* in 2018 and 2019 were below the limit of detection. In addition, the PA-content in soil, root, and shoot samples of the two-node vegetative stage of *T. aestivum* and of *H. vulgare* as well as all the samples at the stage of harvest, including the grains, also tested negative for PAs.

This result clearly showed that returning harvest wastes containing toxic PAs to the soil after composting or using *L. squarrosa* press-cake directly, do not lead to any risk of PA-transfers to the follow-up cultivated crops; moreover, plants, soil and farming economies benefit from this measure of returning harvest residues back to the fields. In our opinion, the so far published studies on horizontal transfer of natural products [21,31,46] do not consider common cultivation and farming conditions, instead describe a phenomenon at artificial/worst-case conditions using mulched poisonous plants (*S. jacobaea*, *C. odorata* and *C. autumnale*). It seems extremely unlikely that those conditions could be achieved under standard agricultural practice. However, it seems appropriate to reduce possible PA-loads through fermentation processes (composting, biogas fermentation) and incorporate treated plant material into the soil before the start of the next cultivation period. As demonstrated, this practice allows the safe handling and recycling of *L. squarrosa* harvest residues without any impact on follow-up cultivations.

3.6. Controls and Additional Pot Experiments (Plot F and Pot A and C)

As expected, in all control samples (soil, root, above ground parts and fruits) had no detectable PA-levels.

The control experiment using either fresh commercial potting soil or surface soil of a field which was used for *L. squarrosa* cultivation before, has confirmed the previous results. The analytical result showed that only those crops which grew (two-node stage) in PA-plant soil were contaminated with PAs in the shoots at levels ranging between 8.7–20.7 µg PA/kg, while the crops growing in potting soil were devoid of PAs. At the stage of harvest, on the other hand, significant levels of PAs could also be detected in the straw, but the caryopsis was PA-free. Table 7 summarizes the results obtained for PA-acceptor- plants *T. aestivum* and *H. vulgare* of pots containing PA-plant soil.

Table 7. PA-content of *T. aestivum* and *H. vulgare* growing in pots with surface soil from *L. squarrosa* fields of the previous year.

		Total Sum of PAs [µg PA/kg d.w.]	
		T. aestivum	*H. vulgare*
sowing stage	soil	<LoD [1]	<LoD
two-node stage	soil	<LoD	<LoD
	roots	<LoD	<LoD
	shoots	20.7	8.7
time of harvest	soil	79.1	179.5
	roots	71.5	84.2
	straw	132.4	2019.2
	caryopses	<LoD	<LoD

[1] Limit of detection.

Again, the reason for the transfer of PAs was the germinating of *L. squarrosa* plantlets in pots containing topsoil of a *L. squarrosa* field, where the seeds had fallen on the ground during harvest and germinated in the next season after hibernation. However, under this regime of forced closeness of the pot and the high abundance of *L. squarrosa* plantlets, a significant amount of PAs was found in the straw at the time of harvest (up to 2019.2 µg PA/kg). Hence, as a general preventive measure in the cultivation of PA-plants for seed oil production, the germination of remaining PA-plant seeds in the next season should be monitored and controlled, and if necessary reduced or prevented (e.g., use of herbicides).

4. Conclusions

In summary, the conducted experiments help to better understand or re-evaluate two different aspects. Firstly, under certain conditions, we could reproduce the so-called "horizontal transfer of natural products", however all our experiments did not show any marked difference to the well-known fact of the uptake of xenobiotics (organic pollutants/compounds) by plants [47]. The processes of uptake and distribution within plants are well known and mainly depend on physio/chemical characteristics of the compounds in the near environment (water, soil, air) of the acceptor-plant [47]. Whether this compound is of natural origin or synthetic seems of no relevance. Hence, the uptake is a completely neutral process and as we can extract from our data, the low transfer rates of 0.1% compared to concentrations in the "donor-plant" do not have any impact for the "acceptor", since it results in low overall concentrations which do not benefit the acceptor (no toxic or deterrent effects can be expected which would increase the fitness). In addition, there is also no great loss for the donor, which still largely maintains its metabolite concentration (not losing fitness/protection or wasting too much energy for the production of these compounds). Therefore, there is no real "transfer", since transfer somehow implies an intentional handover of a compound or a trait, at least from one partner. Under these circumstances of neutral and/or non-intentional ("it just happens"), using an already existing terminology like "uptake" seems to be appropriate.

Secondly, we can address the recently discussed possible negative effects of the uptake of natural products (plant toxins) and its possible impact of food and feed safety. To answer these questions, we tried to use field-trial setup spanning several seasons instead of laboratory-style experiments, using the worst-case scenario—cultivation of a PA-plant. Since the uptake only occurred at a low rate, we can define simple measures to eliminate the potential risk of PAs entering the food or feed chain. Boraginaceae seed oil plants can be safely produced by keeping distance to neighboring cultivations (e.g., using existing farm roads, ditches or the recently promoted flowering strips for biodiversity). Harvest residues can be efficiently re-used, however interposed fermenting methods (composts, biogas) are recommended, and the germinating of hibernated PA-plant seeds should be monitored and if necessary, contained appropriately. To meet or maintain quality in this specific area of PAs in food, feed or phytopharmaceuticals, the main issues for the future will still be the control and prevention of co-harvesting and processing of PA-plants together with the crop or plant of interest [3].

Supplementary Materials: The following are available online at https://www.mdpi.com/article/10.3390/foods10081827/s1. Figure S1. Setting up of compost-bins; A: Control-Compost, B: *S. jacobaea*-Compost, C: *S. jacobaea* and *L. squarrosa*-Compost, D: *L. squarrosa*-Compost. 1-compost starter, 2-*S. jacobaea* material, 3-bio starter, 4-compost stock, 5-*L. squarrosa* press cake. Table S1 Calculated amounts of compost/press-cakes for each plot to meet nutritional requirements. Table S2 The detailed analysis results of extraction and profiling of the PAs of plant samples as Chemisch-physikalische Analyse (#45183) comprising 31 individual PAs and PANOs by QSI (Bremen, Germany).Figure S2. Picture to illustrate investigations of distance-related effects on PA-transfer. Green: *L. squarrosa* field; Red: Excavated Kick-Brauckmann vessel (50 cm) next to its hole; Black: field strip of *L. multiflorum*; Blue: Still buried Kick-Brauckmann vessel (200 cm) including *L. multiflorum*.

Author Contributions: Conceptualization, T.B., G.H. and M.S.C.; methodology, T.B. and G.H.; formal analysis, M.S.C.; resources, T.B., A.D. and G.H.; data curation, M.S.C., A.D. and T.B.; writing—original draft preparation, M.S.C. and T.B.; writing—review and editing, M.S.C., T.B., G.H. and A.D.; visualization, M.S.C. and T.B.; supervision, T.B. and G.H.; project administration, T.B. and G.H. All authors have read and agreed to the published version of the manuscript.

Funding: This research received no external funding.

Acknowledgments: M.S.C. thanks the Niedersächsisches Ministerium für Wissenschaft und Kultur for the scholarship within the framework of Wissenschaft.Niedersachsen.Weltoffen. We gratefully acknowledge the support by the Open Access Publication Funds of Technische Universität Braunschweig.

Conflicts of Interest: The authors M.S.C., T.B. and A.D., declare no conflict of interest. The author G.H. breeds and conducts agricultural research on *L. squarrosa*. G.H. has contributed his expertise in this area to design and conduct the field trials, but had no role in the collection, analyses, or interpretation of data or in the decision to publish the results.

References

1. Malysheva, S.V.; Mulder, P.P.J.; Masquelier, J. Development and Validation of a UHPLC-ESI-MS/MS Method for Quantification of Oleandrin and Other Cardiac Glycosides and Evaluation of Their Levels in Herbs and Spices from the Belgian Market. *Toxins* **2020**, *12*, 243. [CrossRef] [PubMed]
2. Izcara, S.; Casado, N.; Morante-Zarcero, S.; Sierra, I. A Miniaturized QuEChERS Method Combined with Ultrahigh Liquid Chromatography Coupled to Tandem Mass Spectrometry for the Analysis of Pyrrolizidine Alkaloids in Oregano Samples. *Foods* **2020**, *9*, 1319. [CrossRef] [PubMed]
3. Steinhoff, B. Pyrrolizidine Alkaloid Contamination in Medicinal Plants: Regulatory Requirements and Their Impact on Production and Quality Control of Herbal Medicinal Products. *Planta Med.* **2021**, *10*, 34041723.
4. Hartmann, T.; Witte, L. Chemistry, biology and chemoecology of the pyrrolizidine alkaloids. In *Alkaloids: Chemical and Biological Perspectives*; Pelletier, S.W., Ed.; Pergamon Press: Oxford, UK, 1995; pp. 155–233.
5. WHO. Pyrrolizidine alkaloids. Environmental Health Criteria Geneva: World Health Organ. 1988. Available online: http://www.inchem.org/documents/ehc/ehc/ehc080.htm (accessed on 21 July 2021).
6. Schrenk, D.; Gao, L.; Lin, G.; Mahony, C.; Mulder, P.P.J.; Peijnenburg, A.; Pfuhler, S.; Rietjens, I.M.C.M.; Rutz, L.; Steinhoff, B.; et al. Pyrrolizidine alkaloids in food and phytomedicine: Occurrence, exposure, toxicity, mechanisms, and risk assessment—A review. *Food Chem. Toxicol.* **2020**, *136*, 111107. [CrossRef]
7. Xu, J.; Wang, W.; Yang, X.; Xiong, A.; Yang, L.; Wang, Z. Pyrrolizidine alkaloids: An update on their metabolism and hepatotoxicity mechanism. *Liver Res.* **2019**, *3*, 176–184. [CrossRef]
8. Fu, P.; Xia, Q.; Lin, G.; Chou, M. Pyrrolizidine alkaloids—Genotoxicity, metabolism enzymes, metabolic activation, and mechanisms. *Drug Metab. Rev.* **2004**, *36*, 1–55. [CrossRef]
9. Moreira, R.; Pereira, D.M.; Valentão, P.; Andrade, P.B. Pyrrolizidine Alkaloids: Chemistry, Pharmacology, Toxicology and Food Safety. *Int. J. Mol. Sci.* **2018**, *19*, 1668. [CrossRef]
10. (EFSA) European Food Safety Authority. Scientific opinion on pyrrolizidine alkaloids in food and feed. EFSA panel on contaminants in the food chain (CONTAM). Scientific opinion on pyrrolizidine alkaloids in food and feed. *EFSA J.* **2011**, *9*, 2406. [CrossRef]
11. (EFSA) European Food Safety Authority. Risks for Human Health Related to the Presence of Pyrrolizidine Alkaloids in Honey, Tea, Herbal Infusions and Food Supplements. EFSA Panel on Contaminants in the Food Chain (CONTAM) 2017. Available online: https://efsa.onlinelibrary.wiley.com/doi/epdf/10.2903/j.efsa.2017.4908 (accessed on 21 July 2021).
12. Mulder, P.P.J.; López Sánchez, P.L.; These, A.; Preiss-Weigert, A.; Castellari, M. Occurrence of Pyrrolizidine Alkaloids in food. *EFSA Support. Publ.* **2015**, *12*, 116. [CrossRef]
13. Kakar, F.; Akbarian, Z.; Leslie, T.; Mustafa, M.; Watson, J.; Egmond, H.; Omar, M.; Mofleh, J. An Outbreak of Hepatic Veno-Occlusive Disease in Western Afghanistan Associated with Exposure to Wheat Flour Contaminated with Pyrrolizidine Alkaloids. *J. Toxicol.* **2010**, *10*, 313280. [CrossRef]
14. Kempf, M.; Heil, S.; Haßlauer, I.; Schmidt, L.; von der Ohe, K.; Theuring, C.; Reinhard, A.; Schreier, P.; Beuerle, T. Pyrrolizidine alkaloids in pollen and pollen products. *Mol. Nutr. Food Res.* **2010**, *54*, 292–300. [CrossRef]
15. Matteo, A.; Lucchetti, M.A.; Glauser, G.; Kilchenmann, V.; Dübecke, A.; Beckh, G.; Praz, C.; Kast, C. Pyrrolizidine Alkaloids from *Echium vulgare* in Honey Originate Primarily from Floral Nectar. *J. Agric. Food Chem.* **2020**, *64*, 5267–5273.
16. Selmar, D.; Radwan, A.; Nowak, M. Horizontal Natural Product Transfer: A so far Unconsidered Source of Contamination of Plant-Derived Commodities. *Environ. Anal. Toxicol.* **2015**, *5*, 1–3.
17. Pullagurala, V.L.R.; Rawat, S.; Adisa, I.O.; Hernandez-Viezcas, J.A.; Peralta-Videa, J.R.; Gardea-Torresdey, J.L. Plant uptake and translocation of contaminants of emerging concern in soil. *Sci. Total Environ.* **2018**, *636*, 1585–1596. [CrossRef] [PubMed]
18. Eggen, T.; Heimstad, E.S.; Stuanes, A.O.; Norli, H.R. Uptake and translocation of organophosphates and other emerging contaminants in food and forage crops. *Environ. Sci. Pollut. Res.* **2013**, *20*, 4520–4531. [CrossRef] [PubMed]
19. Fismes, J.; Perrin-Ganier, C.; Empereur-Bissonnet, P.; Morel, J.L. Soil-to-Root Transfer and Translocation of Polycyclic Aromatic Hydrocarbons by Vegetables Grown on Industrial Contaminated Soils. *J. Environ. Qual.* **2002**, *31*, 1649–1656. [CrossRef]
20. Edgar, J.A.; Molyneux, R.J.; Colegate, S.M. Linking Dietary Exposure to 1,2-Dehydropyrrolizidine Alkaloids with Cancers and Chemotherapy-Induced Pulmonary and Hepatic Veno-Occlusive Diseases. *J. Agric. Food Chem.* **2020**, *68*, 5995–5997. [CrossRef]
21. Nowak, M.; Wittke, C.; Lederer, I.; Klier, B.; Kleinwächter, M.; Selmar, D. Interspecific transfer of pyrrolizidine alkaloids: A so far unconsidered source of contaminations of phytopharmaceuticals and plant derived commodities. *Food Chem.* **2016**, *213*, 163–168. [CrossRef] [PubMed]
22. Yahyazadeh, M.; Nowak, M.; Kima, H.; Selmar, D. Horizontal natural product transfer: A potential source of alkaloidal contaminants in phytopharmaceuticals. *Phytomedicine* **2017**, *34*, 21–25. [CrossRef]
23. Letsyo, E.; Adams, Z.; Dzikunoo, J.; Asante-Donyinah, D. Uptake and accumulation of pyrrolizidine alkaloids in the tissues of maize (*Zea mays* L.) plants from the soil of a 4-year-old *Chromolaena odorata* dominated fallow farmland. *Chemosphere* **2020**, *17*, 128669. [CrossRef]
24. Wagner, F.; Prediger, G.; Tiggemann, B.; Schmidt, I. *Der Feldversuch-Durchführung und Technik*; Selbstverlag Fritz Wagner: Bad Hersfeld, Germany, 2007.
25. Schliephake, W.; Garz, J.; Schmidt, L. *Exkursionsführer zu den Dauerversuchen auf dem Julius-Kühn-Versuchsfeld Halle*; Institut für Bodenkunde und Pflanzenernährung der Martin-Luther-Universität: Halle (Saale), Germany, 1999.

26. Wikipedia. Available online: https://de.wikipedia.org/wiki/Ackerzahl (accessed on 21 July 2021).

27. Chmit, M.S.; Müller, J.; Wiedow, D.; Horn, G.; Beuerle, T. Biodegradation and utilization of crop residues contaminated with poisonous pyrrolizidine alkaloids. *J. Environ. Manag.* **2021**, *290*, 112629. [CrossRef] [PubMed]

28. Cramer, L.; Schiebel, H.; Ernst, L.; Beuerle, T. Pyrrolizidine alkaloids in the food chain: Development, validation, and application of a new HPLC-ESI-MS/MS sum parameter method. *J. Agric. Food Chem.* **2013**, *61*, 11382–11391. [CrossRef]

29. Letsyo, E.; Jerz, G.; Winterhalter, P.; Lindigkeit, R.; Beuerle, T. Incidence of pyrrolizidine alkaloids in herbal medicines from German retail markets: Risk assessments and implications to consumers. *Phytother. Res.* **2017**, *31*, 1903–1909. [CrossRef]

30. BfR. Bestimmung von PA in Pflanzenmaterial Mittels SPE-LC-MS/MS. BfR-PA-Tee-2.0. 2014. Available online: https://www.bfr. bund.de/cm/343/bestimmung-von-pyrrolizidinalkaloiden.pdf (accessed on 21 July 2021).

31. Selmar, D.; Wittke, C.; Beck-von Wolffersdorff, I.; Klier, B.; Lewerenz, L.; Kleinwächter, M.; Nowak, M. Transfer of pyrrolizidine alkaloids between living plants: A disregarded source of contaminations. *Environ. Pollut.* **2019**, *248*, 456–461. [CrossRef]

32. Wiesner, J. Regulatory Perspectives of Pyrrolizidine Alkaloid Contamination in Herbal Medicinal Products. *Planta Med.* **2021**, in press. [CrossRef]

33. Van Wyk, B.; Stander, M.; Long, H. Senecio angustifolius as the major source of pyrrolizidine alkaloid contamination of rooibos tea (*Aspalathus linearis*). *S. Afr. J. Bot.* **2017**, *110*, 124–131. [CrossRef]

34. Navarro-Herrera, D.; Aranaz, P.; Eder-Azanza, L.; Zabala, M.; Romo-Hualde, A.; Hurtado, C.; Calavia, D.; Lopez-Yoldi, M.; Martinez, J.A.; Gonzalez-Navarro, C.J.; et al. *Borago officinalis* seed oil (BSO), a natural source of omega-6 fatty acids, attenuates fat accumulation by activating peroxisomal beta-oxidation both in *C. elegans* and in diet-induced obese rats. *Food Funct.* **2018**. [CrossRef] [PubMed]

35. Lefort, N.; LeBlanc, R.; Giroux, M.A.; Surette, M.E. Consumption of *Buglossoides arvensis* seed oil is safe and increases tissue long-chain n-3 fatty acid content more than flax seed oil-results of a phase I randomised clinical trial. *J. Nutr. Sci.* **2016**, *5*, e2. [CrossRef]

36. Kitessa, S.M.; Nichols, P.D.; Abeywardena, M. Purple Viper's Bugloss (*Echium plantagineum*) Seed Oil in Human Health. In *Nuts and Seeds in Health and Disease Prevention*, 1st ed.; Preedy, V.R., Watson, R.R., Patel, V.B., Eds.; Academic Press: Cambridge, MA, USA, 2011; Chapter 112; pp. 951–958.

37. Cramer, L.; Fleck, G.; Horn, G.; Beuerle, T. Process Development of *Lappula squarrosa* Oil Refinement: Monitoring of Pyrrolizidine Alkaloids in Boraginaceae Seed Oils. *J. Am. Oil Chem. Soc.* **2014**, *91*, 721–731. [CrossRef]

38. Chizzola, R.; Bassler-Binder, G.; Karrer, G.; Kriechbaum, M. Pyrrolizidine alkaloid production of *Jacobaea aquatica* and contamination of forage in meadows of Northern Austria. *Grass Forage Sci.* **2019**, *74*, 19–28. [CrossRef]

39. Zacharias, P. Offene Verwaltungsdaten zur Analyse des Befallspotenzials von Grünlandbeständen mit Schadpflanzen am Beispiel von Kreuzkräutern. *GIS-Z Geoinformatik* **2018**, *31*, 22–31.

40. Eller, A.; Chizzola, R. Seasonal variability in pyrrolizidine alkaloids in *Senecio inaequidens* from the val venosta (northern Italy). *Plant Biosyst.* **2016**, *150*, 1306–1312. [CrossRef]

41. Lattrell, B. PA-haltige Pflanzen auf nicht landwirtschaftlich genutzten Flächen: BFR-Forum Verbraucherschutz "Pyrrolizidinalkaloide -Herausforderungen an Landwirtschaft und Verbraucherschutz". Bundesinstitut für Risikobewertung 2015. Available online: https://www.bfr.bund.de/cm/343/pa-haltige-pflanzen-auf-nicht-landwirtschaftlich-genutzten-flaechen.pdf (accessed on 21 July 2021).

42. Azadbakht, M.; Talavaki, M. Qualitative and Quantitative Determination of Pyrrolizidine Alkaloids of Wheat and Flour Contaminated with Senecio in Mazandaran Province Farms. *Iran. J. Pharm. Sci.* **2003**, *2*, 179–183.

43. Kaltner, F.; Kukula, V.; Gottschalk, C. Screening of food supplements for toxic pyrrolizidine alkaloids. *J. Consum. Prot. Food Saf.* **2020**, *15*, 237–243. [CrossRef]

44. Hama, J.; Strobel, B. Occurrence of pyrrolizidine alkaloids in ragwort plants, soils and surface waters at the field scale in grassland. *Sci. Total Environ.* **2021**, *755*, 142822. [CrossRef]

45. Kisielius, V.; Hama, J.R.; Skrbic, N.; Hansen, H.C.B.; Strobel, B.W.; Rasmussen, L.H. The invasive butterbur contaminates stream and seepage water in groundwater wells with toxic pyrrolizidine alkaloids. *Sci. Rep.* **2020**, *10*, 19784. [CrossRef]

46. Grant, K.; Jilg, A.; Messner, J.; Elsäßer, M. Herbstzeitlose (*Colchicum autumnale*) in Mulchschicht führt zu leichter Kontamination mit Colchizin eines Pflanzenbestandes im Folgeaufwuchs. Deutsche Arbeitsbesprechung über Fragen der Unkrautbiologie und-bekämpfung, 3.-März in Braunschweig. *Open J. Syst.* **2020**, *464*, 85–89.

47. Verkleij, J.S.C.; Golan-Goldhirsh, A.; Antosiewisz, D.M.; Schwitzguébel, J.P.; Schröder, P. Dualities in plant tolerance to pollutants and their uptake and translocation to the upper plant parts. *Environ. Exp. Bot.* **2009**, *67*, 10–22. [CrossRef]

Article

The Food Contaminants Pyrrolizidine Alkaloids Disturb Bile Acid Homeostasis Structure-Dependently in the Human Hepatoma Cell Line HepaRG

Josephin Glück [1], Marcus Henricsson [2], Albert Braeuning [1] and Stefanie Hessel-Pras [1,*]

[1] Department of Food Safety, German Federal Institute for Risk Assessment, Max-Dohrn-Straße 8-10, 10589 Berlin, Germany; josephin.glueck@bfr.bund.de (J.G.); albert.braeuning@bfr.bund.de (A.B.)
[2] Wallenberg Laboratory, Department of Molecular and Clinical Medicine, Institute of Medicine, University of Gothenburg, 413 45 Gothenburg, Sweden; marcus.henricsson@wlab.gu.se
* Correspondence: stefanie.hessel-pras@bfr.bund.de; Tel.: +49-30-18412-25203

Abstract: Pyrrolizidine alkaloids (PAs) are a group of secondary plant metabolites being contained in various plant species. The consumption of contaminated food can lead to acute intoxications in humans and exert severe hepatotoxicity. The development of jaundice and elevated bile acid concentrations in blood have been reported in acute human PA intoxication, indicating a connection between PA exposure and the induction of cholestasis. Additionally, it is considered that differences in toxicity of individual PAs is based on their individual chemical structures. Therefore, we aimed to elucidate the structure-dependent disturbance of bile acid homeostasis by PAs in the human hepatoma cell line HepaRG. A set of 14 different PAs, including representatives of all major structural characteristics, namely, the four different necine bases retronecine, heliotridine, otonecine and platynecine and different grades of esterification, was analyzed in regard to the expression of genes involved in bile acid synthesis, metabolism and transport. Additionally, intra- and extracellular bile acid levels were analyzed after PA treatment. In summary, our data show significant structure-dependent effects of PAs on bile acid homeostasis. Especially PAs of diester type caused the strongest dysregulation of expression of genes associated with cholestasis and led to a strong decrease of intra- and extracellular bile acid concentrations.

Keywords: pyrrolizidine alkaloids; structure dependency; hepatotoxicity; cholestasis; bile acids

Citation: Glück, J.; Henricsson, M.; Braeuning, A.; Hessel-Pras, S. The Food Contaminants Pyrrolizidine Alkaloids Disturb Bile Acid Homeostasis Structure-Dependently in the Human Hepatoma Cell Line HepaRG. *Foods* **2021**, *10*, 1114. https://doi.org/10.3390/foods10051114

Academic Editor: Dirk W. Lachenmeier

Received: 22 April 2021
Accepted: 12 May 2021
Published: 18 May 2021

Publisher's Note: MDPI stays neutral with regard to jurisdictional claims in published maps and institutional affiliations.

1. Introduction

Secondary plant compounds have increasingly come into the focus of risk assessment as food contaminants in recent years. One group of these contaminants are the 1,2-unsaturated pyrrolizidine alkaloids (PAs). Humans are exposed to PAs mainly via the consumption of contaminated food. The Federal Institute for Risk Assessment identified in 2013 tea, herbal teas and honey as the main sources for human PA uptake in Western countries [1]. In addition, a consumption of contaminated salad mixes, herbs, flour or cereals can also lead to the uptake of substantial amounts of PA [2–5]. However, there are also some plants used as food that produce PA themselves such as borage. Chen et al. [6] showed that lycopsamine *N*-oxide, lycopsamine and acetyllycopsamine were the main PAs in the sample they studied. Dietary supplements based on plants containing PAs may also contribute to increased exposure to PAs. However, the total intake of PAs through their consumption cannot be estimated yet.

Although the levels of PA in most foodstuffs have been significantly reduced in recent years, exposure via highly contaminated dried herbs or herbal mixtures with PA levels of up to 3000 μg/kg seem to be possible [7].

PAs most commonly enter foods through mechanical harvesting processes, but contamination of honey by pollen from PA-containing plants or the use of PA-containing plants as herbal medicine is also a possible source of exposure [8–10].

Since PAs are primarily a random contamination of otherwise PA-free foods, it is not possible to predict which PAs are present in which foods. On the one hand, there are considerable differences in PA content between plant species, and on the other hand, factors such as soil conditions, climate and geographical origin can lead to considerable variations in the composition and quantity of PA within a plant species. Large intra-plant differences between different parts are also possible [11,12]. Therefore, exposure to PAs can only be estimated based on previous studies, but cannot be calculated accurately. Some studies have investigated the PA content in different foods and food supplements, and all the 1,2-unsaturated PAs used in this study were found in foodstuffs. For example, lasiocarpine and senecionine, two of the most toxic PAs examined in our study, were detected in various (herbal) teas and food supplements. Echimidine was also frequently detected [5,13].

After uptake, PAs can cause severe damage to humans and livestock after consumption of contaminated food or feed. Depending on the exposure, acute and chronic liver damage can result, such as liver hardening, ascites and the hepatic sinusoidal obstruction syndrome (HSOS), as well as liver cirrhosis, fibrosis, or liver cancer [14,15]. Due to their widespread distribution in more than 6000 plant species, especially in the *Asteraceae*, *Boraginaceae* and *Fabaceae* families, around the world, PA contamination is not locally limited [3,8,9,16,17].

Currently, more than 660 different PAs and their corresponding N-oxides are known. About half of them are considered to exhibit genotoxic and hepatotoxic properties [18]. Chemically, PAs consist of a necine base, which can be esterified with organic acids at the OH-groups at the C 7 and C 9 positions of the double-ring system. Based on their chemical structure, PAs can be classified into different groups. PAs can be divided according to their corresponding necine base (1-hydroxymethylpyrrolizidine) into platynecine-, heliotridine-, retronecine- and otonecine-type PAs. The different grades of esterification allow a further subdivision into free bases (no esterification), monoesters and open-chained or cyclic diesters (Figure 1) [19].

The PA parent substance and its N-oxide are not very reactive as such, and therefore do not directly induce toxic effects. Due to metabolic activation in the liver, reactive metabolites such as dehydropyrrolizidine alkaloids (DHP) are formed, resulting in DNA and protein adducts [20–22]. For the formation of these metabolites, a double bond at the C 1/C 2 position is necessary. Therefore, the 1,2-saturated platynecine-type PAs are considered to be non-toxic [23,24].

The molecular mode of action of PAs and their metabolites in the liver is not fully understood yet. In previous studies, effects on various intracellular pathways, such as the induction of apoptosis, DNA damage response, and prostanoid synthesis were investigated [25–29]. In a whole-genome microarray analysis by Luckert et al. [30] in primary human hepatocytes, evidence for PA-induced disturbance of bile acid homeostasis was found. In association with PA-induced HSOS, jaundice is often diagnosed. An accumulation of bilirubin is also an indication that the normal pathway of bilirubin degradation and bile flow may be impaired [15,31].

The bile, produced in the liver, is essential for efficient absorption of fats and lipophilic substances in the intestine, as well as for the excretion of metabolites and endogenous substances from the liver. It consists of bile acids, phospholipids, cholesterol, proteins, bilirubin, electrolytes and water. The first and rate-limiting step in the de novo synthesis of bile acids is the 7α-hydroxylation of cholesterol, catalyzed by CYP7A1. The primary bile acids cholic acid (CA) and chenodeoxycholic acid (CDCA) are conjugated very rapidly with the amino acids glycine and taurine after their synthesis. Secondary bile acids are formed by conversion by microorganisms in the intestine and enter the liver after absorption from the intestine [32–36].

A disturbance of bile acid homeostasis can have serious consequences for the liver. Reduced bile flow can cause accumulation of potentially cytotoxic bile acids in the liver, leading to serious cell damage [37–39].

Some recently published studies by Waizenegger et al. [40] and Hessel-Pras et al. [41] described the disturbance of bile homeostasis by selected PAs. Due to the small number of PAs investigated in these studies, predictions on the structure–activity relationship are not applicable. Therefore, in the present study, a set of 22 structurally different PAs was systematically investigated in relation to selected endpoints associated with the disturbance of bile acid homeostasis. The endpoints in the focus of this study include the induction of cytotoxicity, changes in the expression of cholestasis-associated genes and the influence on the levels of intra- and extracellular bile acids in the metabolically competent human hepatoma cell line HepaRG.

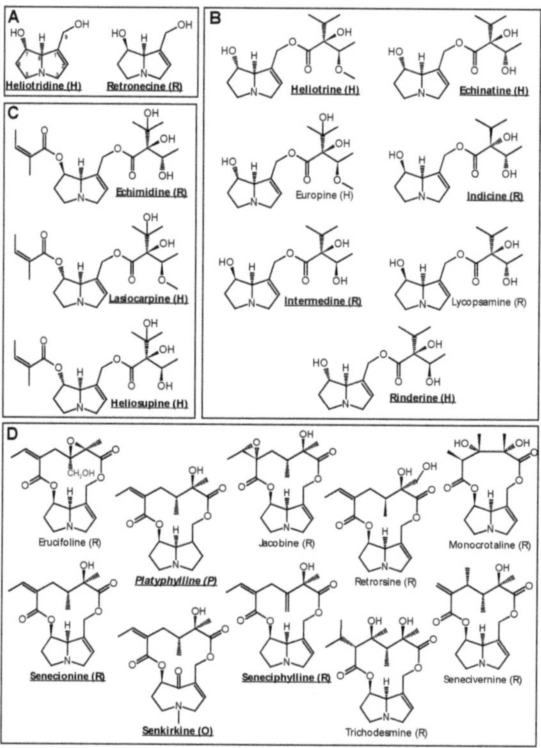

Figure 1. Overview of different PAs and their structural characteristics sorted by their grade of esterification: (**A**)—free bases; (**B**)—monoesters; (**C**)—open-chained diesters; (**D**)—cyclic diesters. The respective type of necine base is indicated in brackets: H—heliotrine (7S); O—otonecine (7R); P—platynecine (7R); R—retronecine (7R). The only 1,2-saturated PA, platyphylline, is indicated in italics. For cytotoxicity studies, all listed esters and the free bases heliotridine and retronecine were analyzed. The bold and underlined PAs represent the reduced test set for all subsequent experiments.

2. Materials and Methods

2.1. Chemicals

The PAs used in this study, except platyphylline, were purchased from Phytoplan Diehm & Neuberger GmbH (Heidelberg, Germany) with a purity of at least 95%. Platyphylline was obtained from BOC Sciences (New York, NY, USA). All PAs were dissolved in 50% acetonitrile (ACN, Sigma-Aldrich, Taufkirchen, Germany)/50% H_2O. The 5 mM stock solutions were stored at −20 °C. All other chemicals used in this study were obtained from Merck (Darmstadt, Germany) or Sigma-Aldrich (Taufkirchen, Germany) in the highest available purity.

2.2. HepaRG Cell Culture

The human hepatoma cell line HepaRG was purchased from Biopredic International (Saint-Gregoire, France). The cells were cultivated for two weeks in proliferation medium composed of William's Medium E with stable glutamine (PAN Biotech, Aidenbach, Germany), 10% fetal bovine serum (FBS Good forte, PAN Biotech, Aidenbach, Germany), 100 U/mL penicillin and 100 µg/mL streptomycin (Capricorn Scientific, Ebsdorfergrund, Germany), 5 µg/mL human insulin (PAN Biotech, Aidenbach, Germany) and 50 µM hydrocortisone hemisuccinate (Sigma-Aldrich, Taufkirchen, Germany). After the proliferation phase, differentiation was initiated by adding 1.7% dimethyl sulfoxide (DMSO, Merck, Darmstadt, Germany). The differentiation of HepaRG cells was completed after two weeks of cultivation in differentiation medium consisting of proliferation medium with 1.7% DMSO. All experiments with HepaRG were conducted with differentiated cells seeded at passages 16 to 20.

2.3. Cell Viability Assay

For cytotoxicity testing, 9000 HepaRG cells per well were seeded in the inner 60 wells of a 96-well plate. After proliferation and differentiation, FBS was set to 2% for 48 h before incubation. DMSO concentration of 1.7% was kept to reach the maximum expression of various CYP enzymes and transporters to ensure a strong metabolic activation of PAs [42]. The cells were treated with PAs in different concentrations as indicated in the figures. After 24 h of incubation, 10 µL of undiluted MTT reagent (3-(4,5-dimethylthiazol-2-yl)-2,5-diphenyltetrazolium bromide, Sigma-Aldrich, Taufkirchen, Germany) were added per well. After an incubation time of 30 min at 37 °C, the supernatant was removed and the formazan crystals were dissolved in 130 µL isopropanol with 0.7% sodium dodecyl sulfate (SDS, Merck, Darmstadt, Germany) for approximately 30 min under shaking and light protection. Absorption was measured at 570 nm, with 630 nm as reference wavelength [43]. Cell viability was calculated by subtracting the background and normalizing the treatment to solvent control. Solvent control was set to 100%. Four independent experiments with three replicates each were performed.

2.4. Isolation of Total RNA and Quantitative Real-Time Polymerase Chain Reaction (qPCR)

The effect of PAs on the expression of cholestasis-associated genes was analyzed using real-time qPCR. HepaRG cells were seeded at a density of 0.2×10^6 cells per well in 6-well plates. After proliferation and differentiation, DMSO and FBS were set to 0.5% and 2% for 48 h before incubation to ensure inducibility of gene expression. The cells were treated with PAs in different concentrations as indicated in the figures. After 24 h of incubation, the cells were washed two times with ice-cold phosphate-buffered saline (PBS). The total RNA was extracted using the RNeasy Mini Kit (Qiagen, Hilden, Germany) following the manufacturer's protocol, including the on-column DNase digestion step. Concentration and purity of the RNA was measured at 260 and 280 nm by a TecanM200Pro using a NanoQuant plate.

cDNA was synthesized by transcribing 1 µg of RNA, using the High Capacity cDNA Reverse Transcriptase Kit according the manufacturer's instructions (Applied Biosystems, Foster City, CA, USA). qPCR was conducted with Maxima SYBR Green/ROX qPCR Master Mix (Thermo Fisher Scientific, Waltham, MA, USA), 300 nM primers and 1 µL cDNA per sample in a total volume of 10 µL. The amplification protocol comprised the following steps:

- initial denaturation (15 min at 95 °C)
- 40 cycles of denaturation (30 s at 95 °C) and annealing/elongation (1 min at 60 °C)
- final elongation (10 min at 60 °C)
- dissociation curve

All qPCRs were performed on a 7900HT Fast Real-Time PCR System (Applied Biosystems, Foster City, CA, USA) using the 384-well format. Primer sequences used for the amplification are summarized in Table S2 in the supplementary material. The primers were obtained from Eurofins (Hamburg, Germany). The results were evaluated according to the

$2^{-\Delta\Delta Ct}$ method [44], normalized to the housekeeping gene GUSB (β-glucuronidase) and referred to the solvent control. Three replicates per sample were measured.

2.5. Bile Acid Quantification

The content of different bile acids was detected via UPLC-MS/MS in the cell culture supernatant and cell lysates. Thus, HepaRG cells were seeded at a density of 0.2×10^6 cells per well in 6-well plates. After the normal proliferation and differentiation period of four weeks, the FBS level in the medium was set to 2% for 48 h. DMSO concentration of 1.7% was kept to reach the maximum expression of various CYP enzymes and transporters to ensure a strong metabolic activation of PAs [42]. Before the incubation with PAs, the cells were washed once with PBS (room temperature) to remove the traces of FBS from cell culture medium. The incubation with PAs was conducted with FBS-free differentiation medium to reduce the background signal from bile acids contained in FBS, as described in the study by Sharanek et al. [39]. The volume of incubation medium (differentiation medium containing the respective amount of PAs) was reduced by 50% to 1 mL per well, to increase the concentration of excreted bile acids in the supernatant. Cells were incubated in triplicates to pool the medium and cells of three wells during harvesting, in order to facilitate analytical determination of bile acids present only at low levels. After incubation for 48 h, the medium was collected and stored at −80 °C. Cells were washed twice with PBS, trypsinized for 15 min at 37 °C and collected (Trypsin-EDTA, Capricorn Scientific, Ebsdorfergrund, Germany). After a second washing step with 1 mL PBS and centrifugation at $250 \times g$ for 5 min, the supernatant was discarded and the cell pellets were stored at −80 °C.

Bile acid analysis was conducted using UPLC coupled to tandem mass spectrometry as described previously [45]. Briefly, bile acids in cell lysates and medium were extracted by adding methanol containing deuterated internal standards. After vortexing and centrifugation, the methanol was evaporated under a stream of nitrogen, the samples were reconstituted in methanol:water [1:1 v/v] and injected onto the UPLC system (Infinity1290, Agilent Technologies, Palo Alto, CA, USA). Separation was made on a Kinetex C18 column (Phenomenex, Torrance, CA, USA) using H_2O and ACN as mobile phases. Detection was made in negative mode using a QTRAP 5500 mass spectrometer (Sciex, Concord, Canada). Three independent experiments were performed.

2.6. Statistical Analysis

For statistical analysis, SigmaPlot 14.0 software (Systat Software, Erkrath, Germany) was used. Statistically significant differences in a concentration series were calculated using a one-way ANOVA. Following the differences of the treated samples versus the respective solvent control were tested using Dunnett's post hoc analysis. Differences that were statistically significant are indicated by * $p < 0.05$, ** $p < 0.01$, *** $p < 0.001$.

3. Results

3.1. PA-Induced Cytotoxic Effects in HepaRG Cells

To elucidate the structure-dependent effects of PAs on HepaRG cells and to establish a suitable concentration range for further investigations, cell viability studies were performed. Metabolically competent HepaRG cells were treated with the 22 PAs for 24 h at six different concentrations ranging from 0.1 to 250 μM. The upper concentration limit resulted from the low PA solubility and the resulting high concentration of solvent in the incubation medium (max. 2.5% ACN). The high concentration of 1.7% DMSO in the medium was chosen to obtain the highest possible activity of xenobiotic-metabolizing CYP enzymes for effective bioactivation of PAs [42].

The results of the MTT assay are summarized in the heat map in Figure 2. For better overview, the PAs were sorted in descending order according to cell viability detected at the highest concentration. Clear differences in cell viability became obvious at a concentration of 250 μM, varying between 100 and 41%. Upon closer examination, it can be seen that among the less cytotoxic PAs (cell viability > 80% at 250 µM), the free bases heliotridine and retronecine, as well as monoesters, such as lycopsamine, indicine, intermdine, rinderine, echinatine and europine, were mainly represented. Moderate (cell viability between 80 and 60%) to strong toxicity (cell viability < 60%) was more likely to be induced by open-chained and cyclic diesters. Platyphylline, which is actually considered to be non-toxic, nevertheless led to a statistically significant reduced cell viability of 88% and 81% at the highest concentrations of 100 µM and 250 µM, respectively. Corresponding cytotoxicity data for HepaRG cells with a reduced DMSO content of only 0.5% have been published previously [25]. The levels of the effects here turned out to be somewhat smaller, but the order of PAs by strength of induced cytotoxicity is comparable.

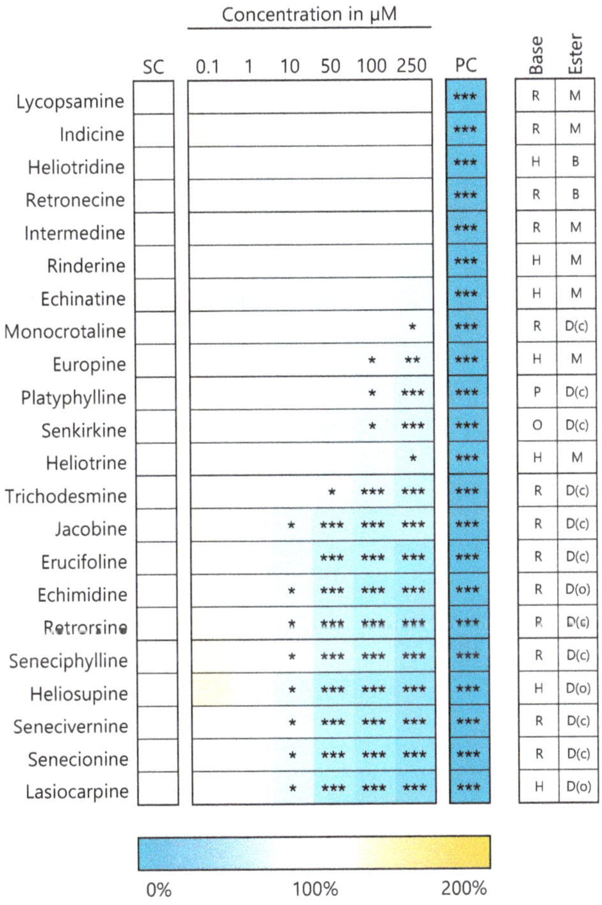

	SC	0.1	1	10	50	100	250	PC	Base	Ester
				Concentration in µM						
Lycopsamine								***	R	M
Indicine								***	R	M
Heliotridine								***	H	B
Retronecine								***	R	B
Intermedine								***	R	M
Rinderine								***	H	M
Echinatine								***	H	M
Monocrotaline							*	***	R	D(c)
Europine						*	**	***	H	M
Platyphylline						*	***	***	P	D(c)
Senkirkine						*	***	***	O	D(c)
Heliotrine							*	***	H	M
Trichodesmine					*	***	***	***	R	D(c)
Jacobine				*	***	***	***	***	R	D(c)
Erucifoline					***	***	***	***	R	D(c)
Echimidine				*	***	***	***	***	R	D(o)
Retrorsine				*	***	***	***	***	R	D(c)
Seneciphylline				*	***	***	***	***	R	D(c)
Heliosupine				*	***	***	***	***	H	D(o)
Senecivernine				*	***	***	***	***	R	D(c)
Senecionine				*	***	***	***	***	R	D(c)
Lasiocarpine				*	***	***	***	***	H	D(o)

0% — 100% — 200%

Figure 2. Decreasing viability of HepaRG cells 24 h after PA exposure. Cell viability was measured by the MTT assay. The mean values out of three biological replicates with three technical replicates each were normalized to the solvent control (SC, 2.5% ACN, 1.7% DMSO). Triton X-100 (0.05%) was used as positive control (PC). The heat map shows the cell viabilities in percent of the solvent control. Blue color indicates a decrease in cell viability and yellow indicates an increase. Statistics: * $p < 0.05$, ** $p < 0.01$, *** $p < 0.001$ (one-way ANOVA followed by Dunnett's post hoc test versus the solvent control). Mean values, standard deviations and p-values can be found in the supplemental material. Abbreviations for structural characteristics of the PAs: retronecine (R), heliotrine (H), otonecine (O) or platynecine (P) type; free base (B), monoester (M), open-chained diester (D(o)) or cyclic diester (D(c)).

3.2. PAs Affect Structure-Dependently Expression of Genes Involved in Bile Acid Homeostasis

Effects of PAs on the regulation of cholestasis-associated gene expression were elucidated with a test set of 14 selected PAs in which representatives of all structural groups were present. qPCR was used to investigate the expression of 45 genes associated with cholesterol and bile acid metabolism. HepaRG cells were treated with PAs at concentrations of 5, 21, and 35 μM for 24 h. These concentrations were chosen to induce no or only weak cytotoxic effects. For the analysis of gene expression, the DMSO concentration was lowered from 1.7% to 0.5% to ensure inducibility of gene expression.

In Figure 3 the induction of gene expression in percent of the solvent control is represented as the heat map. Further details such as means and standard deviations are summarized in the supplemental material (Table S3). For better comparability of the results for the different endpoints tested, the PAs were always sorted according to their cytotoxicity-inducing potential, as shown in Figure 2. The analyzed genes were classified related to the function of their corresponding proteins into the groups transport proteins, xenobiotic-metabolizing enzymes, transcription factors, and enzymes of cholesterol metabolism.

In the heat map, it can be clearly seen that the expression of all investigated genes was downregulated or unchanged without an exception. A significant upregulation of expression was not detected for any of the examined genes. In addition, it is noticeable that the effects, particularly in the case of the xenobiotic-metabolizing enzymes, were sometimes extremely strong with a reduction of the expression down to around 0.02% (*CYP7A1*, senecionine, 35 μM). Summarizing the gene expression data, it can additionally be concluded that PAs showing the strongest effects on cell viability also have more pronounced effects on the regulation of gene expression.

3.2.1. Transport Proteins

Within the group of transporters, the bile salt export pump (BSEP, ABCB11), the Na+/taurochlorate cotransporting polypeptide (NTCP, SLC10A1) and the bile acyl-CoA synthetase (SLC27A5) interact very specifically with bile acids and are thus directly involved in bile acid and cholesterol homeostasis. For these transporters, a very significant concentration-dependent downregulation of gene expression was observed after treatment with the three most cytotoxic PAs (lasiocarpine, senecionine, and heliosupine). Even after exposure of HepaRG cells to echimidine and seneciphylline, a weaker but still significant downregulation was evident at 21 and 35 μM of the PAs. Less or non-cytotoxic PAs exhibited no or only weak effects on the expression of genes encoding the above-mentioned transporters. A similar pattern was observed for the expression of the less bile-acid-specific transport proteins ABCB4, ABCC3, ABCC6, ABCG5, SLC22A7, SLC22A9, SLC51A, SLCO1B1 and SLCO2B1. For the genes encoding the transport proteins ABCB1, ABCC2, SLC51B, SLCO1B3, no substantial changes in gene expression were detected.

3.2.2. Enzymes of Cholesterol and Bile Acid Metabolism

While the expression of the genes for the enzymes of HMG-CoA synthesis (HMGCS1/2) was significantly downregulated, no change was detected in the gene expression of the rate-limiting HMGCR. The expression of INSIG1, responsible for the regulation of HMGCR expression via negative feedback was slightly downregulated (maximum reduction to 16% for 21 μM senecionine). BAAT, an enzyme for the amidation of bile acids with taurine and glycine, showed a significant reduced gene expression down to 5.5% of the solvent control. The decrease of the gene expression levels was stronger, the more cytotoxic and higher concentrated the respective PA was. The transcripts of CYP enzymes directly involved in the catabolism of cholesterol or the de novo synthesis of bile acids (CYP7A1, CYP8B1) were strongly downregulated in a structure-dependent manner, while the expression of *CYP27A1* and *CYP39A1*, relevant for the synthesis of secondary bile acids, were not affected at all.

3.2.3. Xenobiotic-Metabolizing Enzymes

The expression levels of all investigated genes of xenobiotic-metabolizing enzymes were affected in a structure- and concentration-dependent manner. For *CYP1A1* and *POR*, however, the decrease of mRNA content was rather weak. For these genes we could still detect around 20–30% of the transcript level of the solvent control after the exposure with the strongest regulating PAs. The transcription of the genes encoding the enzymes CYP1A2, CYP2B6, CYP2E1, CYP3A4, SULT2A1 and UGT2B4 was strongly downregulated.

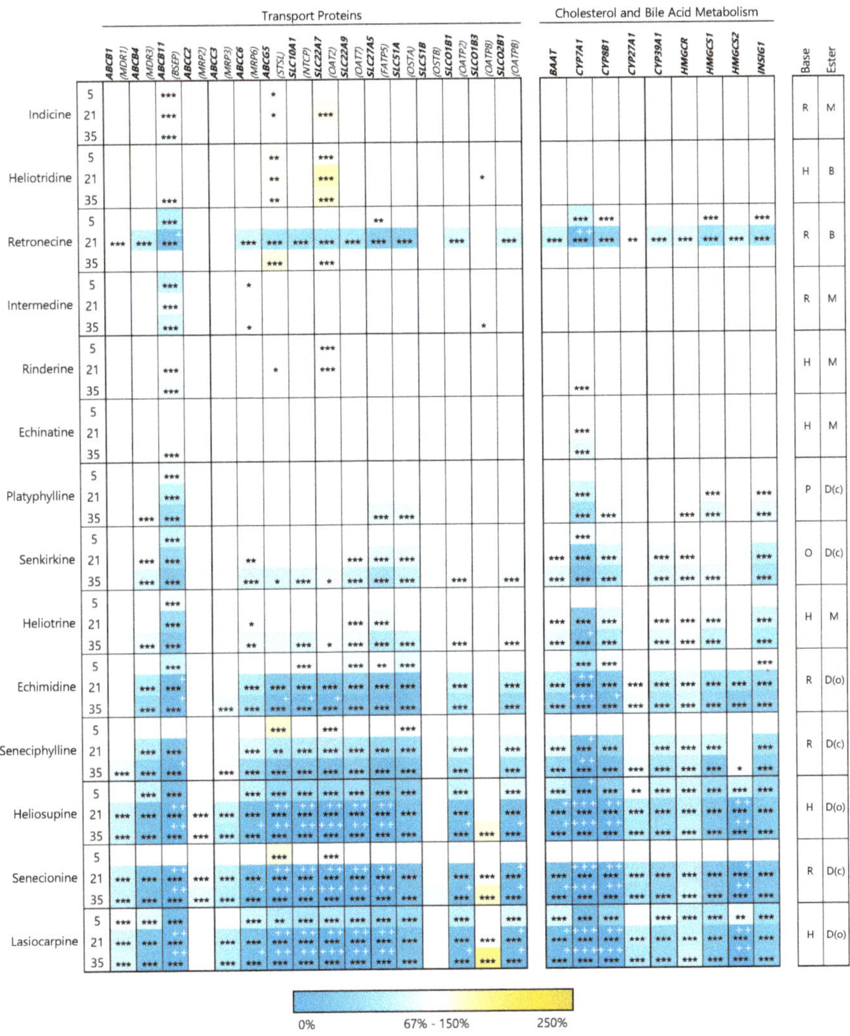

Figure 3. *Cont.*

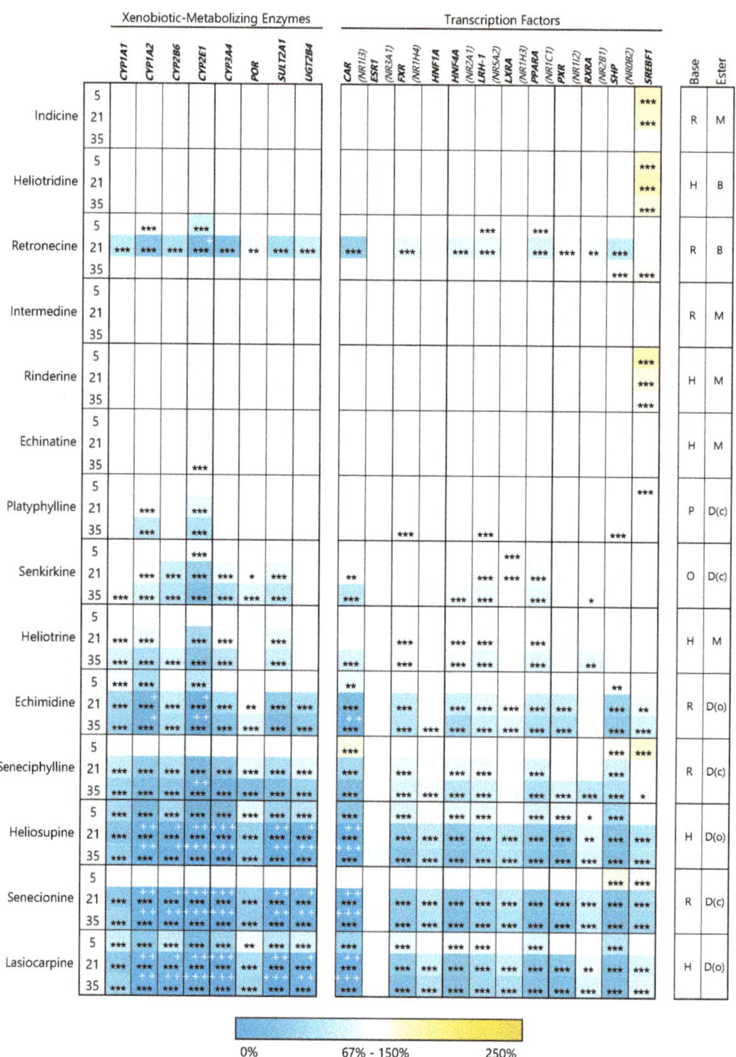

Figure 3. Changes in expression of cholestasis-associated genes after PA treatment of HepaRG cells for 24 h. Differentiated HepaRG were treated with PAs in concentrations of 5, 21 and 35 µM. The Ct-values were evaluated according to the $2^{-\Delta\Delta Ct}$ method by normalizing Ct-values of the respective gene to the housekeeping gene β-glucuronidase (GUSB) and by referring to solvent-treated cells (0.35% ACN and 0.5% DMSO). The cutoff for gene expression regulation was set from 67% to 150% of solvent control. Changes in gene expression within this range were considered not to be biologically relevant. The cells of the heat map show the changes in gene expression of the target genes in percent of the solvent control as means of three replicates. Blue color indicates a downregulation and yellow color an upregulation of gene expression. Expression levels below 10% of solvent control are additionally highlighted by white + (+ expression level below 10%; ++ expression level below 5%; +++ expression level below 1%). Statistics: * $p < 0.05$, ** $p < 0.01$, *** $p < 0.001$ (one-way ANOVA followed by Dunnett's post hoc analysis versus the respective solvent control). Mean values, standard deviations and *p*-values are summarized in the supplemental material (Table S3), as well as the sequences of used primers and the full-length names and synonyms of the detected genes (Table S2). Structural characteristics: retronecine (R), heliotrine (H), otonecine (O) or platynecine (P) type; free base (B), monoester (M), open-chained diester (D(o)) or cyclic diester (D(c)).

3.2.4. Transcription Factors

In general, gene expression of transcription factors was less affected by PA treatment in HepaRG cells. For transcription factors typically expressed in the liver (*FXR, HNF1A, HNF4A, LRH-1, LXR, PPARA*, and *PXR*), a downregulation of expression at the mRNA level down to 14% compared to untreated cells could be detected. No changes were detected for the nuclear receptors *ESR1* and *RXRA*. The gene expression of the constitutive androstane receptor (CAR) was strongly affected by PAs in a structure-dependent manner: a reduction of the CAR transcript down to 0.6% was detected in the treated HepaRG cells for the most cytotoxic PAs senecionine, heliosupine and lasiocarpine.

3.3. PAs Disturb Bile Acid Homeostasis Intra- and Extracellular

The amounts of the primary bile acids cholic acid (CA) and chenodeoxycholic acid (CDCA), as well as the respective glycine- or taurine-conjugated bile acids (GCA, TCA, GCDCA, TCDCA), were measured in the cell lysates and medium supernatants of PA-treated HepaRG cells by UPLC/MS after 48 h of incubation with PAs under serum-free conditions. Due to their formation by gut bacteria, secondary bile acids were not considered in our model.

The detected intra- and extracellular amounts of bile acids are summarized in Figure 4. PAs were sorted according to their potential to induce cytotoxicity in HepaRG cells (cp. Figure 2). The amounts of bile acids in the medium of the treated cells were normalized to the values obtained with medium of the respective solvent control (untreated cells, set to 100%). The amount of bile acids in the cell lysates was first normalized to the total protein content of the cells and then to the corresponding solvent control.

The pattern of the effects on bile acid homeostasis is obviously similar to the effects on gene regulation and cytotoxicity. The higher the cytotoxic potential of the PA, the stronger the effects on bile acid balance. Furthermore, the amount of bile acids in the medium and in the cells decreased significantly with increasing concentration of the respective PA. The effect was much more pronounced in the cell lysates than in the medium. Besides the reduced amounts of most detected bile acids, the conjugated bile acid TCDCA represents an exception due to a slightly increased level in the medium of PA-treated cells. Additionally, a slight but significant increase of intracellular CDCA could be detected after treatment with the non-cytotoxic PAs indicine and heliotridine.

Cyclosporine A, a drug known to induce cholestasis, was used as a positive control [46]. Exposure of HepaRG cells to 20 µM cyclosporine A for 48 h showed similar but much weaker effects than treatment with toxic PAs.

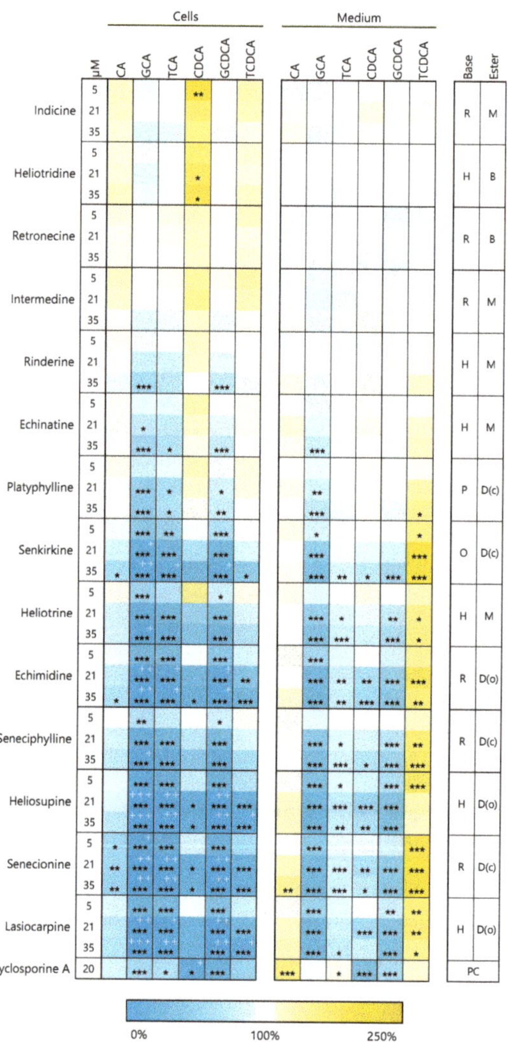

Figure 4. Changes in the intra- and extracellular bile acid concentration after PA treatment for 48 h. Differentiated HepaRG cells were treated with PAs in concentrations of 5, 21 and 35 µM under serum-free conditions. The cells of the heat map are colored according to the relative change of the respective bile acid as mean of three replicates compared to untreated cells (solvent control, 0.35% ACN and 1.7% DMSO). An increase is indicated by yellow color and blue filling indicates a decrease compared to the solvent control (100%). Cyclosporine A (20 µM) was used as positive control (PC) for the induction of cholestasis [46]. Bile acid levels below 10% of solvent control are additionally highlighted by white + (+ below 10%; ++ below 5%; +++ below 1%). Statistics: * $p < 0.05$, ** $p < 0.01$, *** $p < 0.001$ (one-way ANOVA followed by Dunnett's post hoc analysis versus the respective solvent control). Mean values, standard deviations and *p*-values are summarized in the supplemental material (Table S4). Structural characteristics: retronecine (R), heliotrine (H), otonecine (O) or platynecine (P) type; free base (B), monoester (M), open-chained diester (D(o)) or cyclic diester (D(c)). Abbreviations for measured bile acids: cholic acid (CA), glycocholic acid (GCA), taurocholic acid (TCA), chenodeoxycholic acid (CDCA), glycochenodeoxycholic acid (GCDCA), taurochenodeoxycholic acid (TCDCA).

4. Discussion

Due to their widespread distribution and their strong hepatotoxic properties, 1,2-unsaturated PAs are among the most important naturally occurring toxins. In current risk assessment, the general assumption that all PAs have equipotent hepatotoxic properties regardless of their structural characteristics has been followed [13]. However, many recent studies suggest structure-dependent effects of PAs on various endpoints like the induction of cytotoxicity and apoptosis, the occurrence of DNA double-strand breaks, and the formation of micronuclei and DHP-DNA adducts [26,47–52]. For all studies mentioned, a highly comparable order of PAs, sorted by their respective effect level, could be observed. The group of PAs showing the strongest effects, if used in the respective test set, always included the representatives of the open-chained diesters of the heliotridine type (lasiocarpine and heliosupine), and the cyclic diesters of the retronecine type (senecionine and seneciphylline). Echimidine (open-chained diester, retronecine type) also always showed a clear effect on the endpoint studied, despite some variations. This ranking has been observed independently of the test system or endpoint investigated and is comparable to the observations in this study. As discussed in detail in Glück et al. [25], it is becoming increasingly apparent that the representatives of the open-chained and cyclic diester groups of the heliotrine and retronecine type show the strongest toxic effects, whereas free bases and monoesters show no or only weak effects. This pattern of effect levels was also observed in the present study dealing with the disturbance of bile acid homeostasis.

A balanced bile acid level is important for normal liver function. Changes in the concentration of cytotoxic bile acids in the liver can lead to severe damage to hepatocytes. Jaundice and HSOS are frequently reported symptoms after PA intoxication, and indicate a connection between PA exposure and disturbance in bile flow [31]. Currently, there are very few studies dealing with an association between PA exposure and bile acid or cholesterol imbalance. Yan and Huxtable identified the first evidence for such a relation in the 1990s by studying changes in bile flow and bile composition in rat liver after PA administration [53,54]. They showed that detoxification of PAs occurs via glutathione conjugation. The resulting hydrophilic conjugates are then efficiently secreted via the bile. Taurine from liver cells and bile also appears to reduce PA toxicity [55]. Furthermore, Yan and Huxtable [56] found evidence that different PAs can induce different effects or effect levels, as retrorsine and senecionine stimulated bile flow and monocrotaline and trichodesmine did not.

Xiong et al. [57,58] investigated the effect of the PA senecionine as a single substance, or of extracts from PA-containing plants, on the expression of genes associated with the synthesis and transport of bile acids in two in vivo studies. In addition, the concentration of various bile acids was measured in the serum of rats after PA ingestion. In both studies, an increase in serum concentration was detected for all bile acids analyzed. In addition to the abovementioned studies, Hessel-Pras et al. [41] showed an induction of liver necrosis, inflammation and a disturbance of bile acid homeostasis in mouse liver after the exposure to senecionine, resulting in increased bile acid concentrations in serum.

Effects on bile acid homeostasis have also been described in several in vitro studies. Comparable to the information provided by the abovementioned in vivo studies, Luckert et al. [30] found indirect evidence for effects of PAs on the metabolism and transport of bile acids in primary human hepatocytes. After treatment with four different PAs, downregulation of the transcript level was observed for the liver transport proteins ABCB11 (BSEP), ABCC2, ABCC3, ABCC6, SLC10A1 (NTCP), SLC22A7, SLC22A9, SLCO1B1, SLCO1B3, SLCO2B1. Waizenegger et al. [40] examined the regulation of an extensive set of genes associated with bile acid homeostasis, with an experimental setup comparable to the design of this study. After treatment of HepaRG with four different PAs, the qPCR analyses and bile acid content measurements showed results very similar to the present study, including a strong reduction of gene expression of several enzymes involved in bile acid uptake, synthesis, metabolism, and excretion. Enzymes involved in de novo bile acid synthesis (CYP7A1, CYP8B1, BAAT) show decreased transcript levels in the in vitro experiment of this study. In rat liver, however, this change is only partially visible and rather weak. After exposure to senecionine, the expression of *Cyp7a1* and *Baat* was slightly decreased, whereas after ingestion of *Senecio vulgaris* extract, the transcript levels of *Cyp8b1* and *Baat* were reduced [57,58].

According to formerly published results from Waizenegger et al. [40], the analysis of the intra- and extracellular bile acid levels in the present study shows a strong reduction of their concentrations. The effects of the cholestasis-inducing drug cyclosporine A [46] were very similar to the PA-induced variations in bile acid balance, indicating a possible cholestasis-inducing potential of PAs in vivo. Overall, the changes of bile acid amounts were intracellularly stronger than extracellularly. This may be due to the fact that the cells are able to secrete the toxic bile acids very efficiently. The concentration changes for the primary bile acids CA and CDCA were much weaker than for their corresponding conjugates. However, since the unconjugated bile acids are rapidly converted into the glycine and taurine conjugates, their intracellular levels and effects on overall bile acid concentrations are very small compared to the conjugated bile acids.

Comparing gene expression data from the two in vivo studies in rat liver and the in vitro study in human HepaRG, many similarities can be found [40,57,58]. The genes of the nuclear receptors FXR and SHP are less expressed in both, in vitro in the human cell system and in vivo in the rat liver, after exposure to senecionine and the extract of *Senecio vulgaris*. The gene expression of the transporters SLC10A1 (NTCP) and SLCO1B1 is also significantly downregulated in all three studies. These transporters are responsible for the uptake of bile acids into hepatocytes from the blood [59–62]. Reduced expression of these transporters could result in lower bile acid concentrations in hepatocytes and thus counteract cholestasis. Differences are apparent in the regulation of the transporters ABCB11 (BSEP), SLC22A7, and ABCC3. While *Abcb11* (*Bsep*) and *Slc22a7* showed hardly any altered expression at the mRNA level in rat liver, their orthologs were significantly downregulated in human HepaRG cells after PA exposure. ABCB11 (BSEP) transports bile acids from the hepatocytes into the bile, whereas SLC22A7 imports organic anions (potentially bile acids) from the basolateral (blood) side into the hepatocytes [63]. *ABCC3*, on the other hand, showed only minimal downregulation in cell culture experiments, whereas gene expression was greatly increased in the rat. ABCC3 is responsible for exporting bile acids to the basolateral side of hepatocytes. This pathway is considered as an alternative route for the secretion of bile acids from cholestatic hepatocytes [64]. Thus, *Abcc3* upregulation in vivo fits very well with the increased serum concentration of bile acids that was also measured. However, this discrepancy of enhanced bile acid amounts in vivo to low levels observed in our in vitro model can probably be associated with the simplified 2D culture of HepaRG cells. Despite differentiation of HepaRG into hepatocyte-like and biliary epithelium-like cells, the culture model does not fully reflect the compartmentalization and polarization of the liver that may strongly affect the regulation of various processes like the efflux of bile acids. Nevertheless, the results of the cell culture experiment clearly show a PA-mediated impairment of bile acid homeostasis in human hepatocyte-like HepaRG cells.

5. Conclusions

In the present study, we have shown that PAs have a significant structure-dependent effect on bile acid homeostasis. We were able to demonstrate that especially PAs of the diester type caused strongest dysregulation of expression of several genes responsible for bile acid synthesis, uptake and secretion in HepaRG cells. Furthermore, the amounts of intra- as well as extracellular bile acids were strongly affected. The dramatic decreases of intra- and extracellular bile acid amounts were also predominantly detected for diester-type PAs showing a clear impairment of the bile acid balance, which may contribute to cholestatic liver disease in vivo. Therefore, our in vitro results support in vivo observations [36,53,54] that PAs could stimulate the formation of cholestasis. Our data show very impressively the structure dependence of this effect, although this correlation must be verified and further investigated in vivo.

Nevertheless, for a reliable risk assessment of PAs, some knowledge gaps need to be filled. Especially with regard to the classification of PAs according to their structural characteristics and the resulting differences in their toxicity, further refining studies are necessary. Important aspects for future studies should be the analysis of toxicokinetics and the associated metabolic reactions of toxification and detoxification. Further in vivo and in vitro studies are therefore essential for a more precise and reliable assessment of the risk to human health from PA contamination in food.

Supplementary Materials: The following are available online at https://www.mdpi.com/article/10.3390/foods10051114/s1, Table S1: Cytotoxicity MTT, Table S2: PCR Primer Sequences, Table S3: Gene expression PCR, Table S4: Bile acid quant. LC-MS.

Author Contributions: Conceptualization, J.G., A.B. and S.H.-P.; methodology, M.H.; investigation, J.G. and M.H.; writing—original draft preparation, J.G.; writing—review and editing, M.H., A.B. and S.H.-P.; visualization, J.G.; supervision, A.B. and S.H.-P.; project administration, S.H.-P.; funding acquisition, S.H.-P. All authors have read and agreed to the published version of the manuscript.

Funding: This research was funded by the German Research Foundation (Grant number LA1177/12-1) and by the German Federal Institute for Risk Assessment (grant numbers 1322-591 and 1322-624).

Institutional Review Board Statement: Not applicable.

Informed Consent Statement: Not applicable.

Data Availability Statement: The data presented in this study are available in the Supplementary Materials.

Conflicts of Interest: The authors declare no conflict of interest.

Abbreviations

ACN	Acetonitrile
BA	Bile acids
CA	Cholic acid
CDCA	Chenodeoxycholic acid
CYP	Cytochrome P450
DHP	Dehydropyrrolizidine alkaloid
DMSO	Dimethyl sulfoxide
FBS	Fetal bovine serum
GCA	Glycocholic acid
GCDCAHSOS	Glycochenodeoxycholic acidHepatic sinusoidal obstruction syndrome
MTT	3-(4,5-Dimethylthiazol-2-yl)-2,5-diphenyltetrazoliumbromid
PA(s)	Pyrrolizidine alkaloid(s)
PBS	Phosphate-buffered saline
qPCR	quantitative Polymerase chain reaction
SDS	Sodium dodecyl sulfate
TCA	Taurocholic acid
TCDCA	Taurochenodeoxycholic acid

References

1. BfR. Pyrrolizidine alkaloids in herbal teas and teas. *BfR Opin.* **2013**, *2013*, 1–31.
2. Edgar, J.A.; Roeder, E.; Molyneux, R.J. Honey from plants containing pyrrolizidine alkaloids: A potential threat to health. *J. Agric. Food Chem.* **2002**, *50*, 2719–2730. [CrossRef] [PubMed]
3. Kakar, F.; Akbarian, Z.; Leslie, T.; Mustafa, M.L.; Watson, J.; van Egmond, H.P.; Omar, M.F.; Mofleh, J. An outbreak of hepatic veno-occlusive disease in Western afghanistan associated with exposure to wheat flour contaminated with pyrrolizidine alkaloids. *J. Toxicol.* **2010**, *2010*, 313280. [CrossRef] [PubMed]

4. Bodi, D.; Ronczka, S.; Gottschalk, C.; Behr, N.; Skibba, A.; Wagner, M.; Lahrssen-Wiederholt, M.; Preiss-Weigert, A.; These, A. Determination of pyrrolizidine alkaloids in tea, herbal drugs and honey. *Food Addit. Contam. Part A* **2014**, *31*, 1886–1895. [CrossRef]

5. Mulder, P.P.J.; Lopez, P.; Castellari, M.; Bodi, D.; Ronczka, S.; Preiss-Weigert, A.; These, A. Occurrence of pyrrolizidine alkaloids in animal- and plant-derived food: Results of a survey across Europe. *Food Addit. Contam. Part A* **2018**, *35*, 118–133. [CrossRef]

6. Chen, L.; Mulder, P.P.J.; Peijnenburg, A.; Rietjens, I. Risk assessment of intake of pyrrolizidine alkaloids from herbal teas and medicines following realistic exposure scenarios. *Food Chem. Toxicol.* **2019**, *130*, 142–153. [CrossRef] [PubMed]

7. BfR. Updated risk assessment on levels of 1,2-unsaturated pyrrolizidine alkaloids (PAs) in foods. *BfR Opin.* **2020**, *2020*, 1–62. [CrossRef]

8. Roeder, E. Medicinal plants in Europe containing pyrrolizidine alkaloids. *Pharmazie* **1995**, *50*, 83–98. [PubMed]

9. Roeder, E. Medicinal plants in China containing pyrrolizidine alkaloids. *Pharmazie* **2000**, *55*, 711–726. [PubMed]

10. Steenkamp, V.; Stewart, M.J.; van der Merwe, S.; Zuckerman, M.; Crowther, N.J. The effect of Senecio latifolius a plant used as a South African traditional medicine, on a human hepatoma cell line. *J. Ethnopharmacol.* **2001**, *78*, 51–58. [CrossRef]

11. Allgaier, C.; Franz, S. Risk assessment on the use of herbal medicinal products containing pyrrolizidine alkaloids. *Regul. Toxicol. Pharmacol.* **2015**, *73*, 494–500. [CrossRef] [PubMed]

12. Kirk, H.; Vrieling, K.; Van Der Meijden, E.; Klinkhamer, P.G.L. Species by Environment Interactions Affect Pyrrolizidine Alkaloid Expression in Senecio jacobaea, Senecio aquaticus, and Their Hybrids. *J. Chem. Ecol.* **2010**, *36*, 378–387. [CrossRef] [PubMed]

13. EFSA. CONTAM Panel. Risks for human health related to the presence of pyrrolizidine alkaloids in honey, tea, herbal infusions and food supplements. *EFSA J.* **2017**, *15*, 4908.

14. Wiedenfeld, H.; Edgar, J. Toxicity of pyrrolizidine alkaloids to humans and ruminants. *Phytochem. Rev.* **2011**, *10*, 137–151. [CrossRef]

15. Chen, Z.; Huo, J.R. Hepatic veno-occlusive disease associated with toxicity of pyrrolizidine alkaloids in herbal preparations. *Neth. J. Med.* **2010**, *68*, 252–260.

16. Reimann, A.; Nurhayati, N.; Backenkohler, A.; Ober, D. Repeated evolution of the pyrrolizidine alkaloid-mediated defense system in separate angiosperm lineages. *Plant Cell* **2004**, *16*, 2772–2784. [CrossRef] [PubMed]

17. Chauvin, P.; Dillon, J.-C.; Moren, A.; Talbak, S.; Barakaev, S. Heliotrope poisoning in Tadjikistan. *Lancet* **1993**, *341*, 1663. [CrossRef]

18. Stegelmeier, B.L.; Edgar, J.A.; Colegate, S.M.; Gardner, D.R.; Schoch, T.K.; Coulombe, R.A.; Molyneux, R.J. Pyrrolizidine alkaloid plants, metabolism and toxicity. *J. Nat. Toxins* **1999**, *8*, 95–116.

19. Wiedenfeld, H.; Röder, E.; Bourauel, T.; Edgar, J.A.; Alkaloids, P. *Pyrrolizidine Alkaloids-Structure and Toxicity*; Bonn University Press: Bonn, Germany, 2008.

20. Fu, P.P.; Xia, Q.; Lin, G.; Chou, M.W. Pyrrolizidine alkaloids—Genotoxicity, metabolism enzymes, metabolic activation, and mechanisms. *Drug Metab. Rev.* **2004**, *36*, 1–55. [CrossRef]

21. Chen, T.; Mei, N.; Fu, P.P. Genotoxicity of pyrrolizidine alkaloids. *J. Appl. Toxicol.* **2010**, *30*, 183–196. [CrossRef]

22. Ruan, J.; Yang, M.; Fu, P.; Ye, Y.; Lin, G. Metabolic activation of pyrrolizidine alkaloids: Insights into the structural and enzymatic basis. *Chem. Res. Toxicol.* **2014**, *27*, 1030–1039. [CrossRef] [PubMed]

23. Ruan, J.; Liao, C.; Ye, Y.; Lin, G. Lack of metabolic activation and predominant formation of an excreted metabolite of nontoxic platynecine-type pyrrolizidine alkaloids. *Chem. Res. Toxicol.* **2014**, *27*, 7–16. [CrossRef] [PubMed]

24. Mattocks, A.R.; White, I.N. Pyrrolic metabolites from non-toxic pyrrolizidine alkaloids. *Nat. New Biol.* **1971**, *231*, 114–115. [CrossRef]

25. Gluck, J.; Waizenegger, J.; Braeuning, A.; Hessel-Pras, S. Pyrrolizidine Alkaloids Induce Cell Death in Human HepaRG Cells in a Structure-Dependent Manner. *Int. J. Mol. Sci.* **2020**, *22*, 202. [CrossRef] [PubMed]

26. Waizenegger, J.; Braeuning, A.; Templin, M.; Lampen, A.; Hessel-Pras, S. Structure-dependent induction of apoptosis by hepatotoxic pyrrolizidine alkaloids in the human hepatoma cell line HepaRG: Single versus repeated exposure. *Food Chem. Toxicol.* **2018**, *114*, 215–226. [CrossRef]

27. Ebmeyer, J.; Rasinger, J.D.; Hengstler, J.G.; Schaudien, D.; Creutzenberg, O.; Lampen, A.; Braeuning, A.; Hessel-Pras, S. Hepatotoxic pyrrolizidine alkaloids induce DNA damage response in rat liver in a 28-day feeding study. *Arch. Toxicol.* **2020**, *94*, 1739–1751. [CrossRef]

28. Ebmeyer, J.; Franz, L.; Lim, R.; Niemann, B.; Glatt, H.; Braeuning, A.; Lampen, A.; Hessel-Pras, S. Sensitization of Human Liver Cells Toward Fas-Mediated Apoptosis by the Metabolically Activated Pyrrolizidine Alkaloid Lasiocarpine. *Mol. Nutr. Food Res.* **2019**, *63*, e1801206. [CrossRef] [PubMed]

29. Ebmeyer, J.; Behrend, J.; Lorenz, M.; Gunther, G.; Reif, R.; Hengstler, J.G.; Braeuning, A.; Lampen, A.; Hessel-Pras, S. Pyrrolizidine alkaloid-induced alterations of prostanoid synthesis in human endothelial cells. *Chem. Biol. Interact.* **2019**, *298*, 104–111. [CrossRef]

30. Luckert, C.; Hessel, S.; Lenze, D.; Lampen, A. Disturbance of gene expression in primary human hepatocytes by hepatotoxic pyrrolizidine alkaloids: A whole genome transcriptome analysis. *Toxicol. In Vitro* **2015**, *29*, 1669–1682. [CrossRef]

31. Yang, X.Q.; Ye, J.; Li, X.; Li, Q.; Song, Y.H. Pyrrolizidine alkaloids-induced hepatic sinusoidal obstruction syndrome: Pathogenesis, clinical manifestations, diagnosis, treatment, and outcomes. *World J. Gastroenterol.* **2019**, *25*, 3753–3763. [CrossRef]

32. Marksteiner, J.; Blasko, I.; Kemmler, G.; Koal, T.; Humpel, C. Bile acid quantification of 20 plasma metabolites identifies lithocholic acid as a putative biomarker in Alzheimer's disease. *Metabolomics* **2018**, *14*, 1–10. [CrossRef]

33. Dawson, P.A. Chapter 12—Bile Acid Metabolism. In *Biochemistry of Lipids, Lipoproteins and Membranes*, 6th ed.; Ridgway, N.D., McLeod, R.S., Eds.; Elsevier: Boston, MA, USA, 2016; pp. 359–389. [CrossRef]
34. Dawson, P.A. Chapter 53—Bile Formation and the Enterohepatic Circulation. In *Physiology of the Gastrointestinal Tract*, 5th ed.; Johnson, L.R., Ghishan, F.K., Kaunitz, J.D., Merchant, J.L., Said, H.M., Wood, J.D., Eds.; Academic Press: Boston, MA, USA, 2012; pp. 1461–1484. [CrossRef]
35. García-Cañaveras, J.C.; Donato, M.T.; Castell, J.V.; Lahoz, A. Targeted profiling of circulating and hepatic bile acids in human, mouse, and rat using a UPLC-MRM-MS-validated method. *J. Lipid Res.* **2012**, *53*, 2231–2241. [CrossRef]
36. Hofmann, A.F.; Hagey, L.R. Bile Acids: Chemistry, Pathochemistry, Biology, Pathobiology, and Therapeutics. *Cell. Mol. Life Sci.* **2008**, *65*, 2461–2483. [CrossRef]
37. Trauner, M.; Meier, P.J.; Boyer, J.L. Molecular Pathogenesis of Cholestasis. *N. Engl. J. Med.* **1998**, *339*, 1217–1227. [CrossRef]
38. Anthérieu, S.; Azzi, P.B.-E.; Dumont, J.; Abdel-Razzak, Z.; Guguen-Guillouzo, C.; Fromenty, B.; Robin, M.-A.; Guillouzo, A. Oxidative stress plays a major role in chlorpromazine-induced cholestasis in human HepaRG cells. *Hepatology* **2013**, *57*, 1518–1529. [CrossRef]
39. Sharanek, A.; Burban, A.; Humbert, L.; Bachour-El Azzi, P.; Felix-Gomes, N.; Rainteau, D.; Guillouzo, A. Cellular Accumulation and Toxic Effects of Bile Acids in Cyclosporine A-Treated HepaRG Hepatocytes. *Toxicol. Sci.* **2015**, *147*, 573–587. [CrossRef] [PubMed]
40. Waizenegger, J.; Gluck, J.; Henricsson, M.; Luckert, C.; Braeuning, A.; Hessel-Pras, S. Pyrrolizidine Alkaloids Disturb Bile Acid Homeostasis in the Human Hepatoma Cell Line HepaRG. *Foods* **2021**, *10*, 161. [CrossRef]
41. Hessel-Pras, S.; Braeuning, A.; Guenther, G.; Adawy, A.; Enge, A.M.; Ebmeyer, J.; Henderson, C.J.; Hengstler, J.G.; Lampen, A.; Reif, R. The pyrrolizidine alkaloid senecionine induces CYP-dependent destruction of sinusoidal endothelial cells and cholestasis in mice. *Arch. Toxicol.* **2020**, *94*, 219–229. [CrossRef] [PubMed]
42. Kanebratt, K.P.; Andersson, T.B. Evaluation of HepaRG Cells as an in Vitro Model for Human Drug Metabolism Studies. *Drug Metab. Dispos.* **2008**, *36*, 1444–1452. [CrossRef] [PubMed]
43. Mosmann, T. Rapid colorimetric assay for cellular growth and survival: Application to proliferation and cytotoxicity assays. *J. Immunol. Methods* **1983**, *65*, 55–63. [CrossRef]
44. Livak, K.J.; Schmittgen, T.D. Analysis of Relative Gene Expression Data Using Real-Time Quantitative PCR and the $2^{-\Delta\Delta CT}$ Method. *Methods* **2001**, *25*, 402–408. [CrossRef]
45. Tremaroli, V.; Karlsson, F.; Werling, M.; Ståhlman, M.; Kovatcheva-Datchary, P.; Olbers, T.; Fändriks, L.; le Roux, C.W.; Nielsen, J.; Bäckhed, F. Roux-en-Y Gastric Bypass and Vertical Banded Gastroplasty Induce Long-Term Changes on the Human Gut Microbiome Contributing to Fat Mass Regulation. *Cell. Metab.* **2015**, *22*, 228–238. [CrossRef] [PubMed]
46. Sharanek, A.; Azzi, P.B.; Al-Attrache, H.; Savary, C.C.; Humbert, L.; Rainteau, D.; Guguen-Guillouzo, C.; Guillouzo, A. Different dose-dependent mechanisms are involved in early cyclosporine a-induced cholestatic effects in hepaRG cells. *Toxicol. Sci.* **2014**, *141*, 244–253. [CrossRef]
47. Louisse, J.; Rijkers, D.; Stoopen, G.; Holleboom, W.J.; Delagrange, M.; Molthof, E.; Mulder, P.P.J.; Hoogenboom, R.; Audebert, M.; Peijnenburg, A. Determination of genotoxic potencies of pyrrolizidine alkaloids in HepaRG cells using the γH2AX assay. *Food Chem. Toxicol.* **2019**, *131*, 110532. [CrossRef] [PubMed]
48. Merz, K.H.; Schrenk, D. Interim relative potency factors for the toxicological risk assessment of pyrrolizidine alkaloids in food and herbal medicines. *Toxicol. Lett.* **2016**, *263*, 44–57. [CrossRef] [PubMed]
49. Allemang, A.; Mahony, C.; Lester, C.; Pfuhler, S. Relative potency of fifteen pyrrolizidine alkaloids to induce DNA damage as measured by micronucleus induction in HepaRG human liver cells. *Food Chem. Toxicol.* **2018**, *121*, 72–81. [CrossRef] [PubMed]
50. Gao, L.; Rutz, L.; Schrenk, D. Structure-dependent hepato-cytotoxic potencies of selected pyrrolizidine alkaloids in primary rat hepatocyte culture. *Food Chem. Toxicol.* **2020**, *135*, 110923. [CrossRef]
51. Rutz, L.; Gao, L.; Kupper, J.H.; Schrenk, D. Structure-dependent genotoxic potencies of selected pyrrolizidine alkaloids in metabolically competent HepG2 cells. *Arch. Toxicol.* **2020**. [CrossRef]
52. Lester, C.; Troutman, J.; Obringer, C.; Wehmeyer, K.; Stoffolano, P.; Karb, M.; Xu, Y.; Roe, A.; Carr, G.; Blackburn, K.; et al. Intrinsic relative potency of a series of pyrrolizidine alkaloids characterized by rate and extent of metabolism. *Food Chem. Toxicol.* **2019**, *131*, 110523. [CrossRef]
53. Yan, C.C.; Huxtable, R.J. Effect of the pyrrolizidine alkaloid, monocrotaline, on bile composition of the isolated, perfused rat liver. *Life Sci.* **1995**, *57*, 617–626. [CrossRef]
54. Yan, C.C.; Huxtable, R.J. The effect of the hepatotoxic pyrrolizidine alkaloid, retrorsine, on bile composition in the rat in vivo. *Proc. West Pharmacol. Soc.* **1996**, *39*, 19–22. [PubMed]
55. Yan, C.C.; Huxtable, R.J. Effects of taurine and guanidinoethane sulfonate on toxicity of the pyrrolizidine alkaloid monocrotaline. *Biochem. Pharmacol.* **1996**, *51*, 321–329. [CrossRef]
56. Yan, C.C.; Cooper, R.A.; Huxtable, R.J. The comparative metabolism of the four pyrrolizidine alkaloids, seneciphylline, retrorsine, monocrotaline, and trichodesmine in the isolated, perfused rat liver. *Toxicol. Appl. Pharmacol.* **1995**, *133*, 277–284. [CrossRef]
57. Xiong, A.; Fang, L.; Yang, X.; Yang, F.; Qi, M.; Kang, H.; Yang, L.; Tsim, K.W.; Wang, Z. An application of target profiling analyses in the hepatotoxicity assessment of herbal medicines: Comparative characteristic fingerprint and bile acid profiling of Senecio vulgaris L. and Senecio scandens Buch.-Ham. *Anal. Bioanal. Chem.* **2014**, *406*, 7715–7727. [CrossRef] [PubMed]

58. Xiong, A.; Yang, F.; Fang, L.; Yang, L.; He, Y.; Wan, Y.J.; Xu, Y.; Qi, M.; Wang, X.; Yu, K.; et al. Metabolomic and genomic evidence for compromised bile acid homeostasis by senecionine, a hepatotoxic pyrrolizidine alkaloid. *Chem. Res. Toxicol.* **2014**, *27*, 775–786. [CrossRef] [PubMed]

59. Hagenbuch, B.; Meier, P.J. Molecular cloning, chromosomal localization, and functional characterization of a human liver Na+/bile acid cotransporter. *J. Clin. Invest.* **1994**, *93*, 1326–1331. [CrossRef]

60. Hagenbuch, B.; Stieger, B.; Foguet, M.; Lübbert, H.; Meier, P.J. Functional expression cloning and characterization of the hepatocyte Na+/bile acid cotransport system. *Proc. Natl. Acad. Sci. USA* **1991**, *88*, 10629–10633. [CrossRef]

61. Hsiang, B.; Zhu, Y.; Wang, Z.; Wu, Y.; Sasseville, V.; Yang, W.P.; Kirchgessner, T.G. A novel human hepatic organic anion transporting polypeptide (OATP2). Identification of a liver-specific human organic anion transporting polypeptide and identification of rat and human hydroxymethylglutaryl-CoA reductase inhibitor transporters. *J. Biol. Chem.* **1999**, *274*, 37161–37168. [CrossRef]

62. Meng, L.J.; Wang, P.; Wolkoff, A.W.; Kim, R.B.; Tirona, R.G.; Hofmann, A.F.; Pang, K.S. Transport of the sulfated, amidated bile acid, sulfolithocholyltaurine, into rat hepatocytes is mediated by Oatp1 and Oatp2. *Hepatology* **2002**, *35*, 1031–1040. [CrossRef]

63. Hayashi, H.; Takada, T.; Suzuki, H.; Onuki, R.; Hofmann, A.F.; Sugiyama, Y. Transport by vesicles of glycine- and taurine-conjugated bile salts and taurolithocholate 3-sulfate: A comparison of human BSEP with rat Bsep. *Biochim. Biophys. Acta* **2005**, *1738*, 54–62. [CrossRef]

64. Hirohashi, T.; Suzuki, H.; Takikawa, H.; Sugiyama, Y. ATP-dependent transport of bile salts by rat multidrug resistance-associated protein 3 (Mrp3). *J. Biol. Chem.* **2000**, *275*, 2905–2910. [CrossRef] [PubMed]

 *foods*

Article

Pyrrolizidine Alkaloids Disturb Bile Acid Homeostasis in the Human Hepatoma Cell Line HepaRG

Julia Waizenegger [1,2], Josephin Glück [1], Marcus Henricsson [3], Claudia Luckert [1], Albert Braeuning [1] and Stefanie Hessel-Pras [1,*]

[1] Department of Food Safety, German Federal Institute for Risk Assessment, Max-Dohrn-Straße 8-10, 10589 Berlin, Germany; julia.waizenegger@web.de (J.W.); josephin.glueck@bfr.bund.de (J.G.); claudia.luckert@bfr.bund.de (C.L.); albert.braeuning@bfr.bund.de (A.B.)

[2] German Nutrition Society, Godesberger Allee 18, 53175 Bonn, Germany

[3] Wallenberg Laboratory and Sahlgrenska Center for Cardiovascular and Metabolic Research, Institute of Medicine, University of Gothenburg, 413 45 Gothenburg, Sweden; marcus.henricsson@wlab.gu.se

* Correspondence: stefanie.hessel-pras@bfr.bund.de; Tel.: +49-30-18412-25203

Abstract: 1,2-unsaturated pyrrolizidine alkaloids (PAs) belong to a group of secondary plant metabolites. Exposure to PA-contaminated feed and food may cause severe hepatotoxicity. a pathway possibly involved in PA toxicity is the disturbance of bile acid homeostasis. Therefore, in this study, the influence of four structurally different PAs on bile acid homeostasis was investigated after single (24 h) and repeated (14 days) exposure using the human hepatoma cell line HepaRG. PAs induce a downregulation of gene expression of various hepatobiliary transporters, enzymes involved in bile acid synthesis, and conjugation, as well as several transcription regulators in HepaRG cells. This repression may lead to a progressive impairment of bile acid homeostasis, having the potential to accumulate toxic bile acids. However, a significant intracellular and extracellular decrease in bile acids was determined, pointing to an overall inhibition of bile acid synthesis and transport. In summary, our data clearly show that PAs structure-dependently impair bile acid homeostasis and secretion by inhibiting the expression of relevant genes involved in bile acid homeostasis. Furthermore, important biliary efflux mechanisms seem to be disturbed due to PA exposure. These mole-cular mechanisms may play an important role in the development of severe liver damage in PA-intoxicated humans.

Keywords: pyrrolizidine alkaloids; hepatotoxicity; HepaRG; cholestasis; bile acid

Citation: Waizenegger, J.; Glück, J.; Henricsson, M.; Luckert, C.; Braeuning, A.; Hessel-Pras, S. Pyrrolizidine Alkaloids Disturb Bile Acid Homeostasis in the Human Hepatoma Cell Line HepaRG. *Foods* **2021**, *10*, 161. https://doi.org/10.3390/foods10010161

Received: 7 December 2020
Accepted: 11 January 2021
Published: 14 January 2021

Publisher's Note: MDPI stays neutral with regard to jurisdictional claims in published maps and institutional affiliations.

1. Introduction

Pyrrolizidine alkaloids comprise a large group of secondary plant compounds occurring ubiquitously in the plant kingdom. They are constitutively produced in about 3% of the world's flowering plants as a protective mechanism against herbivores [1,2]. To date, over 660 structurally different pyrrolizidine alkaloids and their *N*-oxide derivates have been identified, and about half of them are considered to be toxic [3,4]. The PA-associated harmful effects on livestock and humans [5–8] are associated with a double bound in C1,2-position in the necine base. In the following, the abbreviation PA is always used for 1,2-unsaturated PAs. Intoxications in humans by PA-contaminated cereals, herbal teas, and herbal medicines were reported in the United States, India, Tajikistan, Afghanistan, and South Africa [9–14]. PAs can cause severe liver damage characterized by hemorrhagic liver necrosis, ascites, cirrhosis, and the development of the characteristic hepatic sinusoidal obstruction syndrome (HSOS) [6,14,15].

The molecular mechanisms of PA hepatotoxicity are not yet fully elucidated. a well-known molecular mechanism is the formation of adducts with DNA and proteins by reactive PA metabolites, which are formed by enzymatic conversion in the liver [7]. Further mechanisms at the molecular level, such as interactions of PAs with specific signaling and metabolic pathways, are not completely understood. Recently, a genome-wide expression study in primary human hepatocytes provided evidence for PA-mediated induction of apoptosis and impairment of bile acid homeostasis [16], suggesting a partial involvement of these mechanisms in PA-induced hepatotoxicity. Additionally, increased bile acid levels in blood, as well as disturbance of bile acid secretion, were shown in mice following an acute toxic dose of senecionine [17].

The bile produced in the liver is essential for the absorption and digestion of lipids from the intestinal lumen, as well as for the elimination of xenobiotics, endogenous compounds, and metabolic products (e.g., cholesterol, bilirubin, and hormones) [18,19]. It consists predominantly of water (82%), followed by bile acids, phospholipids, cholesterol, proteins, bilirubin, and electrolytes. Primary bile acids are synthesized from cholesterol in the hepatocytes by various enzymes (including cytochrome P450 monooxygenase 7A1 (CYP7A1), CYP8B1, and CYP27A1) [20,21]. Under physiological conditions, the primary bile acids are conjugated with the amino acids glycine or taurine, followed by secretion by hepatocytes into the bile canaliculi, along with the remaining bile constituents, and storage and concentration in the gallbladder. After release into the duodenum and subsequent absorption and digestion of the food constituents, about 95% of the bile acids are reabsorbed into the liver via blood circulation, thus entering enterohepatic circulation [20,22]. Therefore, hepatic transport proteins are essential for the formation of bile and maintaining bile flow. Sinusoidal (basolateral) transporters are responsible for the uptake of endogenous and exogenous substances into the hepatocytes, and canalicular (apical) transporters mediate the biliary secretion into the bile canaliculi [23,24]. If the secretion of bile acids and consequently the bile flow are impaired, potentially toxic bile components, such as bile acids, bilirubin, and cholesterol, can accumulate, and thus may lead to the damage of hepatic cells [25,26]. In 2013, Vinken et al. proposed an adverse outcome pathway (AOP) for drug-mediated cholestasis [26]. The proposed AOP connects as the primary molecular initiating event the inhibition of the canalicular transporter ATP-binding cassette subfamily B member 11 (ABCB11) with various key events (e.g., bile acid accumulation, regulation of nuclear receptors, induction of inflammation, and oxidative stress) and intermediate steps (e.g., induction of apoptosis/necrosis and gene expression changes) to the adverse outcome, cholestasis.

Based on the proposed AOP for cholestatic liver disease [26] and data suggesting that an impairment of bile acid homeostasis may contribute to the PA-induced hepatotoxicity [16,17], the present study aims to systematically investigate the effect of PA treatment on bile acid homeostasis in the human hepatoma cell line HepaRG. In this regard, possible structure-dependent effects were examined using the four structurally different PAs echimidine, heliotrine, senecionine, and senkirkine (for chemical structures see Figure 1). In addition, two different exposure scenarios were comparatively investigated, namely a single treatment for 24 h and a repeated treatment for 14 days. The human hepatoma cell line HepaRG was used due to its hepatocyte-like morphology, high metabolic activity, and its suitability to investigate bile acid homeostasis [21,27]. Furthermore, HepaRG cells can be cultivated under long-term conditions, enabling a continuous treatment with test substances [28,29].

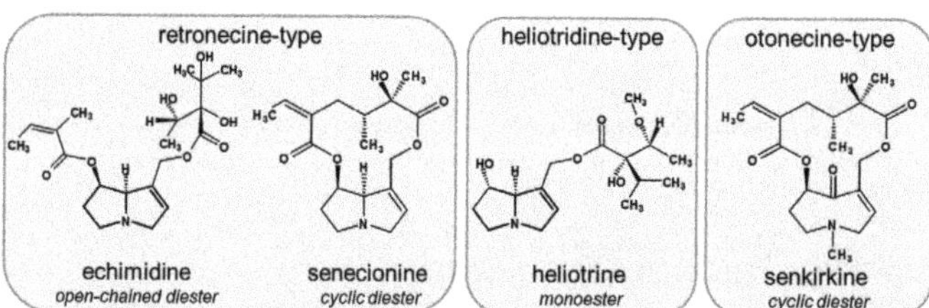

Figure 1. Chemical structure of PAs used in this study. Selected PAs represent the main occurring basic structures of necine base types and structures of esters: heliotrine is a monoester of the heliotridine-type, and senecionine and senkirkine are both cyclic diesters representing retronecine- and otonecine-type PAs, respectively. The open-chained diester echimidine belongs to the retronecine-type group.

2. Materials and Methods

2.1. Chemicals

Heliotrine was purchased from Latoxan SAS (Valence, France). Echimidine, senecionine, and senkirkine were purchased from PhytoLab GmbH & Co. KG (Vestenbergsgreuth, Germany). All other chemicals were purchased from Sigma-Aldrich (Taufkirchen, Germany) in the highest purity available.

2.2. Plasmids

The plasmid pGL4.14-CYP7A1-Prom was constructed as already described [30]. Briefly, the CYP7A1 promoter region was cloned into the vector pGL4.14 (Promega, Madison, WI, USA) upstream of the firefly luciferase reporter gene by means of sequence and ligation-independent cloning [31,32]. The CYP7A1 promoter (-2014 to -1 bp from translation start site) was amplified from human genomic DNA using the primers 5′-CGG TAC CTG AGC TCG CTA GCC AGG AAA GAA CTG CAC CCA TAA T-3′ and 5′-CAG ATC TTG ATA TCC TCG AGT TTG CAA ATC TAG GCC AAA ATC T-3′ and subsequently inserted between the NheI/XhoI site of pGL4.14. The construction of the Renilla luciferase expression plasmid pcDNA3-Rluc was described elsewhere [33].

2.3. Cell Culture

HepaRG cells were obtained from Biopredic International (Saint-Gregoire, France) and cultured in William's E medium supplemented with 10% (*v/v*) fetal bovine serum (FBS) (both from Pan-Biotech GmbH, Aidenbach, Germany), 100 U/mL of penicillin, 100 µg/mL of streptomycin (Capricorn Scientific GmbH, Ebsdorfergrund, Germany), 5 µg/mL of insulin (Pan-Biotech GmbH, Aidenbach, Germany), and 5×10^{-5} M of hydrocortisone hemisuccinate (Sigma-Aldrich, Taufkirchen, Germany) at 37 °C in a humidified atmosphere containing 5% CO_2. For all experiments, HepaRG cells were used in passages between 17 and 20. After seeding, HepaRG cells were cultivated in the medium for two weeks. To initiate differentiation of the cells, the HepaRG cells were then cultivated with a medium containing 1% dimethyl sulfoxide (DMSO) for two days, followed by a medium with 1.7% DMSO (Merck KGaA, Darmstadt, Germany) for a further 12 days. For investigating PA-induced hepatotoxic effects in HepaRG cells, two different incubation scenarios were performed: a single incubation for 24 h and a repeated incubation for 14 days comprising a total of seven incubations every two days. The exact cultivation and incubation schemes have recently been published [29].

The human hepatocarcinoma cell line HepG2 was obtained from the European Collection of Cell Cultures (ECACC, Porton Down, UK). The cells were grown in high-glucose Dulbecco's modified Eagle's medium (DMEM, PAN-Biotech GmbH, Aidenbach, Germany) supplemented with 10% (*v/v*) FBS (Capricorn Scientific GmbH, Ebsdorfergrund, Germany), 100 U/mL of penicillin, and 100 µg/mL of streptomycin (both from Capricorn Scientific GmbH, Ebsdorfergrund, Germany) at 37 °C in a humidified atmosphere containing 5% CO_2. At a confluence of about 80–90%, the cells were passaged and plated at a density of 3 to 5×10^4 cells/cm^2. Cells at passages up to 12 were used for all experiments.

2.4. Preparation of RNA and Quantitative Real-Time PCR Analysis (qRT-PCR)

For gene expression analysis, HepaRG cells were seeded in six-well plates at a density of 2×10^5 cells/well. After differentiation, cells were incubated with 5, 35, or 70 µM of PA or the solvent (1.7% DMSO, 0.7% acetonitril (ACN)) for either 24 h or 14 days. To investigate the effects of PAs on CYP7A1 gene expression in HepG2 cells, 7.5×10^5 cells/well were seeded in six-well plates and incubated on the next day with the three abovementioned concentrations of PA for 24 h. Following incubation, HepaRG cells were harvested directly into an RLT buffer from the RNeasy Kit (Qiagen, Hilden, Germany) containing 1% β-mercaptoethanol, while HepG2 cells were harvested into phosphate-buffered saline (PBS) and centrifuged for 5 min at 300× *g*. After centrifugation, the supernatant was discarded, and the HepG2 cell pellet was dissolved in an RLT buffer containing 1% β-mercaptoethanol. The RNA of HepG2 and HepaRG cells was isolated using the RNeasy Mini Kit (Qiagen, Hilden, Germany). a total of 1 μg of RNA was reverse transcribed into single-stranded cDNA using the High Capacity cDNA Reverse Transcription Kit (Applied Biosystems, Foster City, CA, USA) according to the manufacturer's protocol. Quantitative real-time PCR was performed on a 7900HT Fast Real-Time PCR System (Applied Biosystems, Darmstadt, Germany) using Maxima SYBR Green/ROX qPCR Master Mix (2×) (Thermo Fisher Scientific, Waltham, MA, USA). Thermal cycling conditions have been described elsewhere [16]. Relative quantification of gene expression was calculated according to the $2^{-\Delta\Delta Ct}$ method [34] by normalizing the Ct values of the PA-treated samples to that of the reference gene ACTB (encoding β-actin) and the solvent-treated samples. an upregulation of gene expression was represented by relative expression values > 1, while a downregulation was reflected by values in the range of $0 < x < 1$. For allowing a dimension-matched expression for the up- and downregulation, the reciprocals of the values for downregulation were calculated and are therefore shown as values < -1.

2.5. Transcriptional Activation of CYP7A1 by Dual Luciferase Reporter Gene Assay

For the CYP7A1 reporter gene assay, HepG2 cells were seeded in 96-well plates at a density of 1.8×10^4 cells/well. After 18–24 h, the cells were transiently transfected using the TransIT-LT1 transfection reagent (Mirus Bio, Madison, WI, USA) with 80 ng of the plasmid pGL4.14-CYP7A1-Prom and 1 ng of the Renilla luciferase expression plasmid (pcDNA3-Rluc) used as internal control for normalization. Four to 6 h after transfection, the cells were treated with four different concentrations of the PAs echimidine, heliotrine, senecionine, and senkirkine (5, 35, 70, or 250 μM) or the solvent (2.5% ACN) for 24 h. The known CYP7A1 promotor activity inhibitor phorbol 12-myristate 13-acetate (PMA, 5 μM) was used as a positive control [35]. After 24 h, the cells were lysed by adding 50 μL of lysis buffer to each well (100 mM of potassium phosphate with 0.2% (*v/v*) Triton X-100, pH = 7) and incubated for 15 min on a plate shaker. After centrifugation, luciferase activity was analyzed as previously described [36]. Firefly luciferase values were normalized to Renilla luciferase values and expressed as fold-induction referred to solvent control.

2.6. Staining of Bile Canaliculi to Assay Canalicular Efflux

The fluorescent dye CDFDA was used to investigate canalicular efflux in HepaRG cells. The membrane-permeable CDFDA is metabolized by intracellular esterases to CDF, a substrate of ABCC2, resulting in fluorescence labeling of bile canaliculi. Therefore, 5.5×10^4 HepaRG cells/well were seeded in 24-well plates. After differentiation, the cells were treated with a PA (5 or 35 μM), the solvent (1.7% DMSO, 0.35% ACN), or the positive control (10 mM of acetaminophen (APAP)) for 24 h or 14 days. Cells were washed three times with Hank's balanced salt solution buffer (HBSS; 5.4 mM of KCl, 0.44 mM of KH_2PO_4, 0.5 mM of $MgCl_2 \times 6\ H_2O$, 0.41 mM of $MgSO_4 \times 7\ H_2O$, 137 mM of NaCl, 4.2 mM of $NaHCO_3$, 0.34 mM of Na_2HPO_4, 25 mM of D-glucose, and 10 mM of 4-(2-hydroxyethyl)-1-piperazineethanesulfonic acid (HEPES, pH = 7.4)) prior to incubation with 5 μM of CDFDA for 30 min at 37 °C. Supernatants were discarded and cells were washed again three times with an HBSS buffer. The distribution of CDF was analyzed using the fluorescence microscope Axio Observer.D1 (objective EC Plan-Neofluar 5x/0.16 Ph 1) at $\lambda_{ex/em} = 470/525$ nm.

2.7. Staining of the Tight Junction Protein Zonula Occludens-1 (ZO-1) Combined with Nuclei Staining

HepaRG cells were seeded in 24-well plates on gelatin-coated cover slips (Ø 13 mm) at a density of 5.5×10^4 cells/well to determine effects of PAs on tight junction proteins. Subsequent to the treatment with either 5 or 35 μM of PA (echimidine, heliotrine, senecionine, and senkirkine), the solvent (1.7% DMSO, 0.35% ACN), or the positive control APAP (10 mM) for 24 h, cells were washed twice with PBS and fixed by treatment with ice-cold methanol for 20 min. After two washing steps with PBS, nuclei were stained by using SYTOX™ Orange Nucleic Acid Stain (1 μM) (Thermo Fisher Scientific, Braunschweig,

Germany), and cells were washed again twice with PBS. Subsequently, cells were blocked with 10% bovine serum albumin (BSA) in PBS for 1 h at room temperature. The cells were washed twice with PBS before incubation with the primary antibody ZO-1 (1:400 in blocking solution) (Cell Signaling Technology Europe, B.V., Leiden, Netherlands) for 4 h at room temperature. Following primary antibody incubation, the cells were washed twice in PBS and afterwards incubated with the secondary antibody AlexaFlour488-conjugated Goat Anti-Rabbit IgG Alexa Fluor 488 secondary antibody (1:400 in blocking solution; Thermo Fisher Scientific, Waltham, MA, USA) for 1 h in the dark at room temperature. Coverslips were washed twice in PBS and once in H_2O and then mounted onto slides using Vecta Shield HardSet Mounting Medium (Vector Laboratories, Burlingame, CA, USA). Imaging of SYTOX Orange ($\lambda_{ex/em}$ = 547/570) and ZO-1/Alexa Fluor 488 ($\lambda_{ex/em}$ = 488/519) was performed using the confocal laser scanning microscope Leica TCS SP5 (objective HCX PL APO 63x/1.40 OIL PH3 CS).

2.8. Analysis of Bile Acid Content

The amount of different bile acids was detected via ultra-performance liquid chromatography-tandem mass spectrometry (UPLC-MS/MS) in cell culture supernatants and cell lysates. Therefore, HepaRG cells were seeded at a density of 0.2×10^5 cells/well in six-well plates. After the proliferation and differentiation period of four weeks, the FBS level in the medium was reduced to 2% for 48 h. The cells were then washed once with pre-warmed PBS and incubated with either 5, 21, or 35 µM of PA, the solvent (1.7% DMSO, 0.35% ACN), or the positive control (20 µM of cyclosporine A) in an FBS-free medium to avoid interactions with bovine bile acids occurring in FBS [21]. The amount of cell culture medium was reduced to 1 mL per well to yield a higher concentration of secreted bile acids. Supernatants were collected after incubation for 48 h. Cells were washed twice with PBS, trypsinized for 15 min at 37 °C, and collected by adding 1 mL of PBS. After centrifugation at $250 \times g$ for 5 min, the supernatant was discarded. To increase the amount of bile acids, the cells and medium of three wells were pooled.

Bile acids were quantified using UPLC-MS/MS as described previously [37]. Briefly, bile acids in cell lysates and cell culture supernatants were extracted by adding methanol containing deuterated internal standards. After intensive mixing and centrifugation, methanol was evaporated under a stream of nitrogen. The samples were resuspended in a 1:1 (*v/v*) methanol:water mixture and injected onto the UPLC system (Infinity1290, Agilent Technologies, Santa Clara, CA, USA). Separation was performed on a Kinetex C18 column (Phenomenex, Torrance, CA, USA) using water and acetonitrile as mobile phases. Detection was operated in negative mode using a QTRAP 5500 mass spectrometer (Sciex, Concord, ON, Canada).

2.9. Statistical Analysis

Statistical analysis was performed with SigmaPlot 13.0 software. Statistical differences between the mean values of the treatments and the solvent control were determined by one-way ANOVA, followed by Dunnett's post hoc test. Statistically significant differences were assumed at $p < 0.05$.

3. Results

3.1. PA-Dependent Alterations of Gene Expression of Transporters, Enzymes, and Transcription Regulators Involved in Bile Acid Homeostasis

Considering the microarray data of Luckert et al. (2015) [16] and the proposed mechanisms for the development of cholestasis according to the AOP for cholestatic liver diseases [26], 32 target genes were selected to examine possible effects on their expression after 24 h and 14 days of treatment with the four structurally different PAs echimidine, heliotrine, senecionine, and senkirkine via qRT-PCR. These 32 target genes comprise 14 hepatobiliary transporters, eight enzymes, and ten transcriptional regulators involved in bile acid homeostasis. Data of cell viability were recently summarized in Waizenegger et al.

(2018) [29]. Based on these results, the following three concentrations were chosen: 5 μM (non-cytotoxic), 35 μM (non- to slightly cytotoxic), and 70 μM (cytotoxic). The qRT-PCR results showed that only the higher concentrations (35 and 70 μM) affected gene expression, whereas no significant changes occurred after treatment with the lowest concentration (5 μM). Furthermore, the regulatory effects after 24 h of PA treatment seemed to be higher than after 14 days. Additionally, the strongest regulatory effects were found after treatment with the retronecine-type PA echimidine and senecionine, while the weakest were detected for the heliotridine-type PA heliotrine.

The gene expression data showed a significant downregulation of five ABC transporters (*ABCB4, ABCB11, ABCC2, ABCC3, ABCC6*) and six solute carrier (SLC) transporters (*SLC10A1, SLC22A7, SLC22A9, SLC51A, SLCO1B1,* and *SLCO2B1*). After 24 h of PA treatment (Figure 2A), the most pronounced effects were found for the three SLC transporters *SLC22A7, SLC22A9,* and *SLC51A,* followed by the two SLC transporters *SLC10A1* and *SLCO2B1.* a complete list of the gene expression values is provided in Table S1 in the Supplemental Materials. a lower but significantly reduced gene expression was observed for the three ABC transporters *ABCB4, ABCB11,* and *ABCC6.* The weakest significant regulatory effects were detected for the two ABC transporters *ABCC2* and *ABCC3.* For the two SLC transporters *SLC51B* and *SLCO1B3,* no significant changes in gene expression were observed, except for the treatment with 35 and 70 μM of senecionine, as well as 70 μM of heliotrine. In comparison, after 14 days of treatment, the strongest decrease in gene expression was also observed for the SLC transporters *SLC22A7* and *SLC51A,* while the regulatory effect on *SLC22A9,* as well as *SLCO1B1* and *SLCO2B1,* was significantly lower compared to the 24 h treatment (Figure 2A). In contrast, *ABCB11* was more downregulated after 14 days of PA treatment than after 24 h. The weakest significant downregulation was detected for the ABC transporters *ABCB4, ABCC2,* and *ABCC6* and the SLC transporter *SLC10A1,* whereas no significant effects were found for the three transporters *ABCC3, SLC51B,* and *SLCO1B3* after continuous PA treatment. Finally, for the transporter *ABCB1,* a significant downregulation of gene expression was only observed after 14 days of treatment with the retronecine-type PAs echimidine and senecionine.

Concerning the effects of PAs on several enzymes involved in bile acid homeostasis, gene expression data revealed the strongest downregulation for the three CYP monooxygenases *CYP3A4, CYP7A1,* and *CYP8B1* after 24 h of PA treatment, with the most prominent repression for *CYP7A1,* the rate-limiting enzyme in bile acid formation (see Figure 2B). Furthermore, a significant downregulation was found for the phase II enzymes *SULT2A1* (encoding sulfotransferase 2A1) and *UGT2B4* (encoding UDP-glucuronosyltransferase 2B4), as well as for *BAAT* (encoding bile acid-CoA:amino acid *N*-acyltransferase). All three enzymes are required for bile acid conjugation. Compared to the aforementioned enzymes, gene expression of the two CYP monooxygenases *CYP27A1* and *CYP39A1,* also involved in bile acid synthesis, was only slightly decreased after PA treatment. In line with the 24 h treatment, the 14-day treatment also led to the strongest decrease in gene expression of *CYP7A1,* followed by *CYP3A4* and *CYP8B1.* However, the downregulation of *BAAT, SULT2A1,* and *UGT2B4* was less pronounced after 14 days. In contrast to the 24 h PA treatment, no effect was found on the expression of *CYP27A1* and *CYP39A1* after 14 days.

Additionally, for both PA treatment schemes, gene expression data showed a downregulation of all 10 investigated transcription regulators (see Figure 2C). After 24 h, the strongest downregulation was observed for *NR1I3* (encoding the constitutive androstane receptor), followed by *NR1I2* (encoding the pregnane X receptor) and further on *HNF4A* (encoding the hepatocyte nuclear factor 4 alpha), *PPARA* (encoding the peroxisome proliferator activated receptor alpha), and *NR0B2* (encoding small heterodimer partner). The weakest downregulation was found for the five remaining transcription regulators *ESR1* (encoding the estrogen receptor alpha), *NR1H4* (encoding the farnesoid X receptor), *HNF1A* (encoding the hepatocyte nuclear factor 1 alpha), *INSIG2* (encoding the insulin induced gene 2), and *SREBF1* (encoding the sterol regulatory element-binding transcription factor 1). After 14 days of PA treatment, the strongest downregulation was also detected for

NR1I3, followed by *NR0B2*, *NR1I2*, and *SREBF1*. In line with the 24 h treatment, *NR1H4*, *HNF4A*, and *INSIG2* were only slightly downregulated after 14 days of PA treatment, whereas no regulation of gene expression was observed for *ESR1*, *HNF1A*, and *PPARA*.

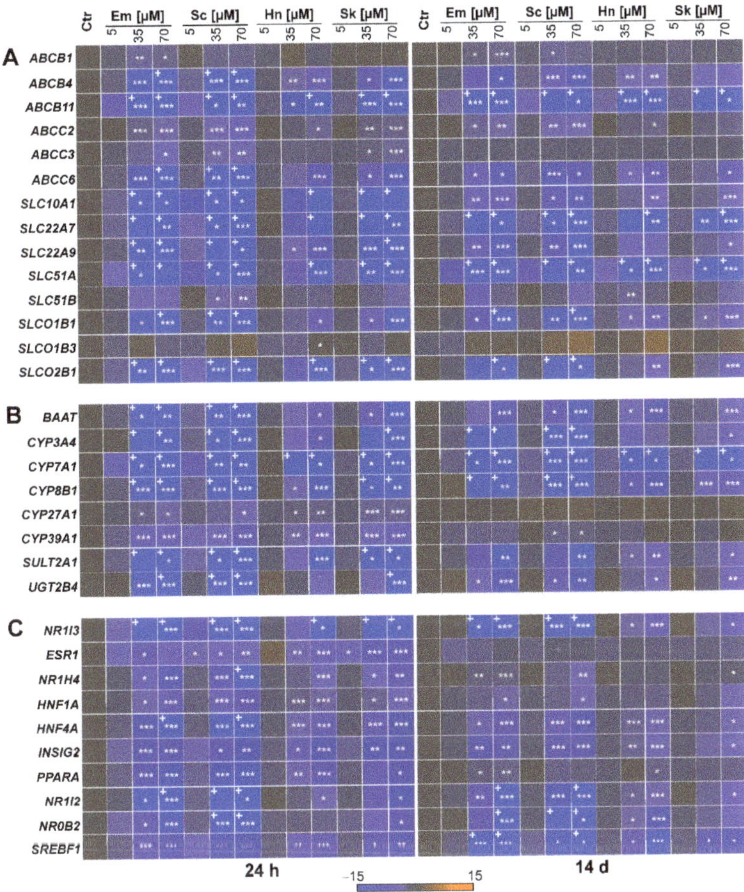

Figure 2. PA-dependent alterations of bile acid homeostasis-associated gene expressions ((**A**): transporters, (**B**): enzymes, (**C**): nuclear receptors) in HepaRG cells. HepaRG cells were seeded and cultivated as described in the material and methods section. Subsequently, the cells were incubated for 24 h or 14 days with 5, 35, or 70 µM of PA or the solvent (Ctr; 1.7% DMSO and 0.7% ACN). The total RNA was isolated and transcribed into cDNA. Expression analysis was performed by qRT-PCR. Gene expression of the target gene was normalized to *ACTB* and referred to solvent control to obtain relative expression ($2^{-\Delta\Delta Ct}$ method). The heat map shows the relative expression values for up- and downregulation as positive and negative fold changes (means of three independent experiments with three replicates each). For better comparability, only the fold change range between -15 and 15 is shown. Values exceeding this range are marked with +. Statistical differences were evaluated using one-way ANOVA followed by Dunnett's test: * $p < 0.05$, ** $p < 0.005$, *** $p < 0.001$. Em, echimidine; Sc, senecionine; Hn, heliotrine; Sk, senkirkine.

3.2. PA-Dependent Inhibition of CYP7A1 Promoter Activity and Gene Expression

Since PA treatment in HepaRG cells led to a strong decrease in *CYP7A1* gene expression (see Section 3.1), possible inhibitory effects of PAs on the transcriptional activity of *CYP7A1* were investigated using a reporter gene assay. Therefore, HepG2 cells were transfected with the reporter gene plasmid pGL4.14-CYP7A1-Prom and the control plasmid pcDNA3-RLuc.

Following 24 h of PA treatment (5, 35, 70, or 250 µM), the firefly and Renilla luciferase activities were determined. In HepG2 cells, even the highest PA concentration did not induce cytotoxicity. This might be due to the low expression level of phase I xenobiotic-metabolizing enzymes. As shown in Figure 3, no concentration-dependent inhibition of *CYP7A1* promoter activity was observed for the four PAs. Although there was a significant decrease in firefly luciferase activity after exposure to the PAs echimidine and heliotrine, this inhibition was relatively constant across all four investigated concentrations (~1.5- and 1.9-fold for 5 µM and 250 µM of echimidine, respectively, and ~1.6- and 1.8-fold for 5 µM and 250 µM of heliotrine, respectively).

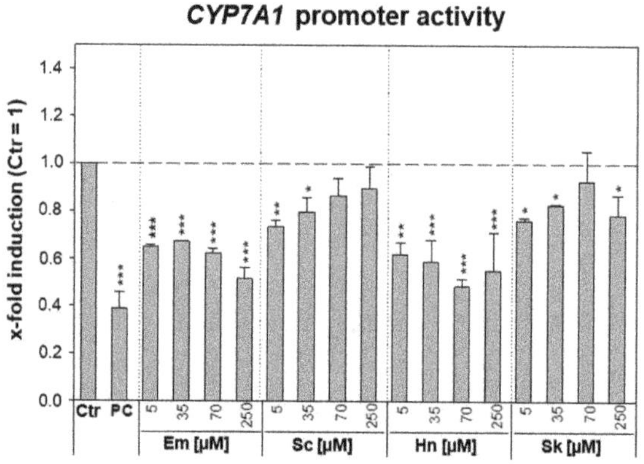

Figure 3. Interaction of PAs with *CYP7A1* promoter activity in HepG2 cells. HepG2 cells were cultivated and seeded as described in the materials and methods section. Cells were transfected with the reporter gene plasmid pGL4.14-CYP7A1-Prom (80 ng) and the control plasmid pcDNA3-Rluc (1 ng) for 6 h and subsequently treated with 5, 35, 70, or 250 µM of PA, solvent (Ctr; 2.5% ACN), or positive control (PC, 5 µM of PMA). After an incubation period of 24 h, the cells were lysed, and the firefly and Renilla luciferase activity was detected. The activity of the firefly luciferase was normalized to the activity of the Renilla luciferase and referred to solvent control (= 1) to obtain the x-fold induction. Shown are means ± standard deviations of three independent experiments with three replicates each. Statistical differences were evaluated using one-way ANOVA followed by Dunnett's test: * $p < 0.05$, ** $p < 0.005$, *** $p < 0.001$. Em, echimidine; Sc, senecionine; Hn, heliotrine; Sk, senkirkine.

The treatment with the known inhibitor of *CYP7A1* promoter activity PMA resulted in a significant inhibition of transcriptional activity by about 2.5-fold. Furthermore, this substance showed a concentration-dependent repression. To investigate whether the effect on *CYP7A1* repression by PAs is dependent on the metabolism of the PA, the influence of the four PAs on *CYP7A1* gene expression in HepG2 cells was investigated. Treatment for 24 h with the retronecine-type PAs echimidine and senecionine did not lead to a significant change in *CYP7A1* gene expression in HepG2 cells, whereas treatment with 70 µM of the PAs heliotrine and senkirkine resulted in a weak but significant downregulation of 1.9- or 1.5-fold, respectively (data depicted in Supplemental Materials, Figure S1).

3.3. Effect of PAs on ABCC2-Driven Canalicular Efflux

In order to examine whether PA treatment for 24 h or 14 days affects bile flow, HepaRG cells were incubated with CDFDA and analyzed by fluorescence microscopy. Since CDFDA is intracellularly converted by esterases to the fluorescent ABCC2 substrate CDF, ABCC2-mediated export of CDF into the bile canaliculi leads to their fluorescence labeling [38]. As shown in Figure 4 and Figure S2 in the Supplemental Materials, after treatment with the solvent control and the low concentration (5 µM) of the retronecine-type PAs echimidine and senecionine, CDF fluorescence was visible as clear, distinct green dots revealing an intact ABCC2 transport system. In contrast, treatment with the higher concentration of echimidine and senecionine (35 μM) for 24 h and 14 days led to a substantial effect on the transport and distribution of CDF (Figure 4B). The staining of bile canaliculi appeared rather diffuse and indistinct, with more extension across the area, indicating an accumulation of the dye in the bile canaliculi. After treatment with the heliotridine-type PA heliotrine and the otonecine-type PA senkirkine for 24 h, no effects on CDF transport were observed (Figure S3 in the Supplemental Materials), and the staining of the bile canaliculi was comparable to the solvent control. However, after 14 days of treatment, heliotrine (35 μM) showed a slight impairment of CDF transport, whereas senkirkine had no effect after prolonged exposure (Figure S4 in the Supplemental Materials). In summary, in particular, the retronecine-type PA seem to affect the transport of CDF and thus possibly the bile acid flow.

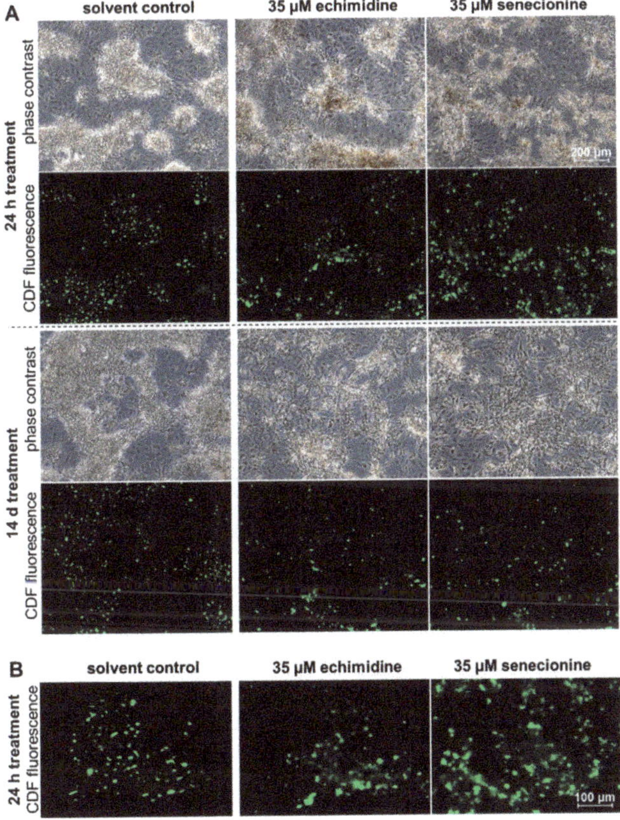

Figure 4. PA-dependent disturbance of ABCC2 driven efflux in HepaRG cells. (**A**) Differentiated HepaRG cells were incubated for 24 h or 14 days with 35 μM of echimidine, senecionine, or the solvent (1.7% DMSO and 0.35% ACN). To localize the bile canaliculi, the cells were incubated with 5 μM of 5(6)-carboxy-2′,7′-dichloro-fluorescein diacetate (CDFDA) for 30 min at 37 °C, and then analyzed on the fluorescence microscope Axio Observer.D1 (objective EC Plan-Neofluar 5x/0.16 Ph 1) under transmitted light and after excitation with 470 nm at 525 nm. The membrane-bound, non-fluorescent CDFDA is intracellularly converted by esterases into the green fluorescent ABCC2 substrate 5(6)-carboxy-2′,7′-dichlorofluorescein (CDF). By ABCC2-mediated transport, CDF enters the bile ducts. Representative sections are shown. The indicated scale applies to all images. (**B**) Exemplarily enlarged images of CDF fluorescence after treatment of HepaRG cells for 24 h with the solvent, 35 μM of echimidine, or 35 μM of senecionine. The indicated scale applies to all images of this enlargement.

3.4. Influence of PAs on the Tight Junction Protein ZO-1

According to the proposed AOP by Vinken et al. (2013), the impairment of tight junctions is one of the secondary molecular events that may contribute to the development of cholestatic liver disease [26]. The localization of the tight junction protein ZO-1 was investigated in more detail using immunofluorescence microscopy to check a possible influence of PA treatment on cell–cell contacts. As depicted in Figure 5, only the high concentration of the retronecine-type PA senecionine (35 µM) showed slight changes in immunofluorescence of the ZO-1 protein after 24 h compared to the solvent control, highlighted by discontinuities in the grid structure and a decrease in the intensity of the green immunofluorescence similar to the positive control (10 mM of APAP). No influence on the tight junction protein ZO-1 was observed for the treatments with lower concentration (5 µM) or for the treatment with 35 µM of the other three PAs echimidine, heliotrine, and senkirkine.

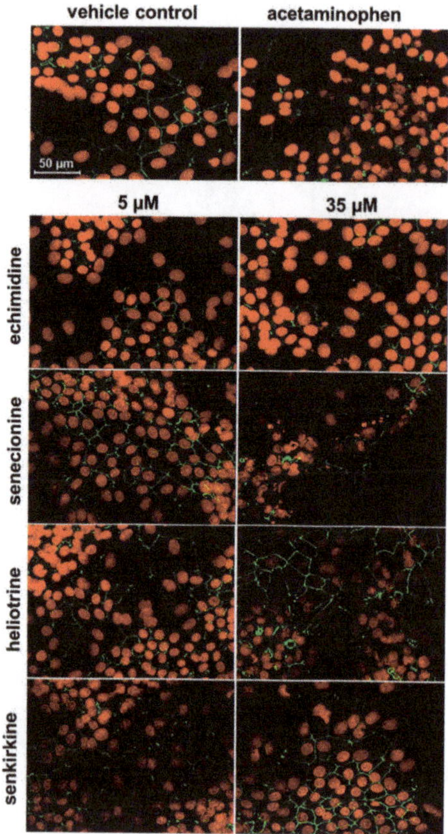

Figure 5. Effects of PA treatment on the tight junction protein ZO-1 in HepaRG cells. HepaRG cells were cultivated and seeded as described in the materials and methods section. After 14 days of further cultivation, the cells were treated with 5 or 35 µM of PA, solvent (1.7% DMSO and 0.35% ACN), or the positive control (10 mM of acetaminophen) for 24 h. Subsequently, the F-actin cytoskeleton was stained with ActinGreen 488 (green) and ZO-1 with an anti-ZO-1 antibody (1:500 in blocking solution) for 4 h. Afterwards, cells were treated with the secondary antibody Alexa Fluor 488 (anti-Rabbit IgG, 1:400 in blocking solution) for 1 h. Staining was determined using confocal fluorescence microscopy (Leica TCS SP5 with a 63× objective). Representative sections are shown. Indicated scale applies to all images.

3.5. Effects of PAs on Bile Acid Content

Amounts of primary and conjugated bile acids were measured in HepaRG cells and supernatants after treatment for 48 h with 5, 21, or 35 µM of PA. The primary bile acids synthesized in the liver are cholic and chenodeoxycholic acid, which are further conjugated with glycine or taurine. Within the UPLC/MS method used, the following bile acids were measured: primary bile acids, such as cholic acid and chenodeoxycholic acid, as well as conjugated bile acids, such as taurohyocholic acid, taurocholic acid, taurochenodeoxycholic acid, taurodeoxycholic acid, glycohyocholic acid, glycocholic acid, glycochenodeoxycholic acid, and glycodeoxycholic acid.

For a better comparison of the intracellular and extracellular effects, total bile acid amounts were referred to the solvent control (0.35% ACN), which was set as 100%. The sum of the determined bile acids in each treatment group is depicted in Figure 6. an overall concentration-dependent decrease of bile acids was observed intra- and extracellularly after PA incubation. However, the strongest decrease occurred for the retronecine-type PAs senecionine and echimidine. The treatment of HepaRG cells with the known cholestasis inducer cyclosporine a also resulted in a decrease to 50% and 20% of the extra- and intracellular bile acid content, respectively.

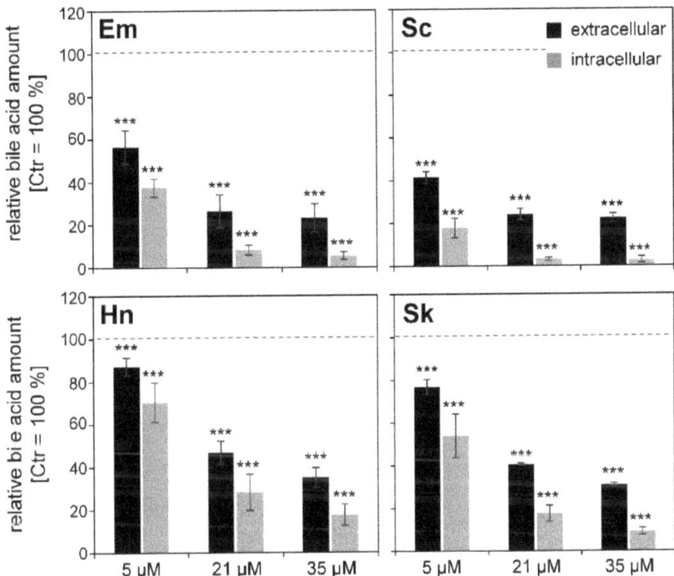

Figure 6. Sum of intra- and extracellular bile acid content after PA treatment. Differentiated HepaRG cells were treated under serum-free conditions for 48 h with PAs, as indicated in the figure. Bile acids were quantified as described in the material and methods section. The sum of bile acids was calculated and normalized to solvent control (1.7% DMSO, 0.35% ACN) set as 100%. As the control, cells were treated with 20 µM of known cholestasis inducer cyclosporine A, resulting in a decrease of bile acid amounts to 50% ± 15 in the medium and 20% ± 12 in the cells. Statistical differences were evaluated using one-way ANOVA followed by Dunnett's test: *** $p < 0.001$.

4. Discussion

A balanced bile acid homeostasis is a basic requirement for healthy normal liver function, including enzymes involved in the production and detoxification of bile acids, as well as sinusoidal and canalicular transporters mediating the bile acid flow. Recent research results point to an involvement of impaired bile acid homeostasis in PA-induced hepatotoxicity [16,17]. Therefore, the present study initially investigated the influence of PA treatment on the gene expression of 14 hepatobiliary transporters by qRT-PCR. This revealed a significant downregulation of gene expression of five ABC transporters and six sinusoidal SLC transporters. Additionally, the expression of several enzymes involved in bile acid homeostasis, synthesis, and detoxification was analyzed, also showing a significant PA-dependent downregulation. Furthermore, the effects of PA treatment on the gene expression of various transcriptional regulators were determined. a clear repression of the transcription regulators was shown, especially for *NR1I3* and *NR1I2*, followed by *HNF4A* and *NR0B2*. In general, these effects were more pronounced after 24 h of PA treatment than after 14 days of PA treatment. In regard to the chemical structure of the PA, the strongest effects were observed for the retronecine-type PAs echimidine and senecionine, whereas heliotridine-type PA heliotrine exerted the weakest effects. Similar results regarding the influence on gene expression of hepatobiliary transporters and also of enzymes involved in bile acid metabolism could be observed by Luckert et al. (2015) in primary human hepatocytes [16]. After treatment with 100 µM of the PAs echimidine, heliotrine, senecionine, and senkirkine, a downregulation of the gene expression of *SLC10A1*, *SLC22A7*, *SLCO1B1*, and *SLCO2B1* was observed. In accordance with the results presented here, Luckert et al. (2015) also observed a PA-related downregulation of the canalicular transporters *ABCC2* and *ABCB11* in primary human hepatocytes in their microarray analysis. The authors also observed a PA-induced downregulation for the enzymes *BAAT*, *CYP7A1*, *CYP8B1*, *CYP27A1*, and *SULT2A4* in primary human hepatocytes. Furthermore, the bioinformatic analysis of the microarray data performed by Luckert et al. (2015) predicted an inhibition of the transcriptional regulators NR1I3 and HNF4A, whereas for NR0B2, an activation was predicted. In agreement with the results of the present study, Xiong et al. [39] observed a reduction in the gene expression of *Slc10a1* and *Slco1b2* in rat livers after treatment with senecionine (35 mg per kg body weight, 36 h). In contrast, no effect on the expression of *ABCB11* was observed in rat livers. However, after treatment with *Senecio vulgaris*, a PA-producing plant, Xiong et al. (2014) also observed a reduced gene expression of *Abcb11* and the SLC transporters *Slc10a1*, *Slco1a2*, and *Slco1b2* in rat livers [40]. a reduced expression of *Cyp7a1* and *Baat* was detected in rat livers after treatment with senecionine, whereas the expression of *Cyp8b1*, in contrast to the results presented here, was not affected [39]. After treatment with *Senecio vulgaris*, reduced gene expression of *Baat*, but not of *Cyp7a1* and *Cyp8b1*, was observed in rat livers [40]. Consistent with the results of the present study, the two in vivo studies mentioned above also described a reduced gene expression of *NR0B2* in rat livers after treatment with senecionine or *Senecio vulgaris* [39,40].

The reduced expression of the transporters *SLC10A1*, *SLC22A7*, *SLC22A9*, *SLC51A*, *SLCO1B1*, and *SLCO2B1*, determined in the present study, may possibly lead to a reduced uptake of bile acids, since these transporters are located in the basolateral membrane of hepatocytes and are responsible for the uptake of bile acids and other organic anions from sinusoidal blood. In contrast, the transporters ABCC2 and ABCB11 are located on the canalicular side, and are responsible for the secretion of bile acids into the bile ducts. Therefore, a reduced gene expression of these transporters may result in a reduced secretion of bile acids. In this context, in the present study, an impairment of hepatobiliary transport was detected in vitro via fluorescence microcopy. Similar effects were also observed in vivo in mice [17]. Furthermore, inhibition of ABCB11, in particular by accumulation of bile acids in the hepatocytes, has already been described for the development of cholestasis [25,41,42]. Regarding the PA-induced downregulation of various transcription factors, PAs may also affect bile acid, cholesterol homeostasis, and associated signaling pathways via interaction with transcriptional regulators. Furthermore, the observed PA-related downregulation of the enzymes *CYP27A1*, *CYP39A1*, *CYP7A1*, and *CYP8B1* may inhibit bile acid synthesis, which could be due to a possible feedback regulation induced by accumulating toxic bile acids. In addition, the decreased gene expression of *CYP3A4*, *BAAT*, *SULT2A1*, and *UGT2B4* indicates a potential decrease in bile acid metabolism and conjugation, which in turn would lead to a reduced detoxification of bile acids.

With regard to the downregulation of *CYP7A1*, the rate-limiting enzyme of bile acid synthesis, the present study used a reporter gene assay to investigate the inhibition of transcriptional activity of the *CYP7A1* promoter in the hepatocyte cell line HepG2. The observed inhibition of the *CYP7A1* promoter activity after PA treatment was not concentration-dependent. Additionally, the effect of PA treatment on the gene expression of *CYP7A1* in HepG2 cells was investigated. Whereas in the metabolically highly active HepaRG cells PA treatment caused a significantly reduced *CYP7A1* expression, in HepG2 cells, only a slight effect on the gene expression of *CYP7A1* was observed after 24 h of PA exposure. Since HepG2 cells generally exert low metabolic activity, the results suggest that the PA metabolites may be responsible for the reduced *CYP7A1* gene expression shown in HepaRG cells.

Impairment of tight junctions, changes in the cytoskeleton, oxidative stress, and induction of apoptosis/necrosis are also key events associated with the development of cholestasis [20,24–26]. The induction of apoptotic processes by PA treatment, both intrinsic and sensitized to extrinsic apoptosis, has been shown in various studies [29,43]. Damage to tight junctions can lead to an increase in paracellular permeability, reflux of bile constituents into the sinusoidal space of Disse and plasma, and a reduction of the osmotic gradient, which is normally the driving force for bile secretion [24,44,45]. With regard to a possible influence on the tight junction protein ZO-1, protein location was investigated by immunofluorescence staining. ZO-1 showed discontinuities in the staining only after 24 h of treatment with senecionine (35 µM), indicating a potential damage of the tight junctions. For the other three PAs echimidine, heliotrine, and senkirkine, no influence on ZO-1 could be detected, suggesting that this mechanism might only be marginally involved in a potential PA-associated cholestasis. To the best of our knowledge, no further studies investigating a PA-related effect on hepatocellular tight junction proteins are available.

A key event in the development of cholestasis is the accumulation of bile acids, leading to the induction of apoptosis/necrosis, oxidative stress, and a regulation of nuclear receptors and gene expression. Various in vivo studies have shown elevated bile acid levels in rat serum after administration of both senecionine and *Senecio vulgaris* [17,39,40]. The authors concluded that PA treatment resulted in an excess accumulation of bile acids in hepatocytes, which led to an adaptive response to avoid bile acid overload in hepatocytes. This adaptive response includes suppression of bile acid de novo synthesis (reduced gene expression of e.g., *Cyp7a1*), reduced sinusoidal reabsorption (reduced expression of e.g., *Slc10a1*), and reduction of bile acid accumulation by activation of alternative basolateral export pumps (Abcc3 and Abcc4). In the present in vitro study, a potential suppression of bile acid synthesis and sinusoidal uptake by regulation of corresponding enzymes and transporters (e.g., *SLC10A1* and *CYP7A1*) was shown in HepaRG cells at the mRNA level. However, no increased mRNA expression of alternative export pumps was detected, but, on the contrary, a downregulation of *ABCC3* was detected. In accordance with the studies of Xiong et al. (2014) [39,40], further studies showed an increase in bile acids in the serum and an increase in liver toxic enzyme markers and bilirubin in horses, as well as in calves, after ingestion of Senecio species or *Cynoglossum officinale* [46–49]. All of these data suggest that PA treatment leads to increased serum bile acid levels. However, in the present in vitro study, reduced bile acid levels were measured intra- and extracellularly in PA-treated cells compared to solvent-treated cells. The strength of the reduction is associated with the previously observed cytotoxicity [29] and the described potency of the tested PA [29,50]. a possible explanation for the observed reduced bile acid levels may be that PAs intracellularly disturb bile acid metabolism in HepaRG cells so dramatically that significantly less bile acids are formed and thus secreted.

In summary, we observed a strong PA-mediated impairment of bile acid homeostasis in human HepaRG cells, comprising a downregulation of numerous genes involved in bile acid uptake, synthesis, detoxification, and secretion. Furthermore, bile acid secretion seems to be disturbed, and thus might contribute to the development of cholestatic liver disease. Generally, the most pronounced effects were observed for senecionine and echimidine, representing PAs of the retronecine type. Using the HepaRG cell model and the molecular initiating and key events addressed in the AOP for cholestasis, we could confirm the evidence from in vivo studies for the development of cholestasis in our in vitro cell model.

Supplementary Materials: The following are available online at https://www.mdpi.com/2304-815 8/10/1/161/s1, Figure S1: Influence of PA on CYP7A1 gene expression in HepG2 cells, Figure S2: PA-dependent disturbance of ABCC-2 driven efflux in HepaRG cells (5 µM echimidine or senecionine), Figure S3: PA-dependent disturbance of ABCC-2 driven efflux in HepaRG cells (5 µM heliotrine or senkirkine), Figure S4: PA-dependent disturbance of ABCC-2 driven efflux in HepaRG cells (35 µM heliotrine or senkirkine), Table S1: Alterations of gene expressions of hepatobiliary transport proteins after 24 h PA treatment in HepaRG cells, Table S2: Alterations of gene expressions of hepatobiliary transport proteins after 14 days PA treatment in HepaRG cells, Table S3: Alterations of gene expressions of enzymes involved in bile acid/cholesterol metabolism after 24 h PA treatment in HepaRG cells, Table S4: Alterations of gene expressions of enzymes involved in bile acid/cholesterol metabolism after 14 days PA treatment in HepaRG cells, Table S5: Alterations of gene expressions of transcription regulators involved in bile acid/cholesterol metabolism after 24 h PA treatment in HepaRG cells, Table S6: Alterations of gene expressions of transcription regulators involved in bile acid/cholesterol metabolism after 14 days PA treatment in HepaRG cells.

Author Contributions: Conceptualization, J.W., A.B., and S.H.-P.; methodology and investigation, J.W., J.G., M.H., and C.L.; software, J.W. and S.H.-P.; writing—original draft preparation, J.W. and S.H.-P.; writing—review and editing, M.H., C.L., and A.B.; visualization, J.W. and S.H.-P.; supervision, A.B. and S.H.-P.; project administration, S.H.-P.; funding acquisition, S.H.-P. All authors have read and agreed to the published version of the manuscript.

Funding: This research was funded by the German Research Foundation (Grant number LA1177/12-1) and by the German Federal Institute for Risk Assessment (grant numbers 1322-591 and 1322-624).

Data Availability Statement: The data presented in this study are available in the Supplementary Materials.

Conflicts of Interest: The authors declare no conflict of interest.

References

1. Hartmann, T.; Witte, L. Chapter Four—Chemistry, Biology and Chemoecology of the Pyrrolizidine Alkaloids. In *Alkaloids: Chemical and Biological Perspectives*; Pelletier, S.W., Ed.; Elsevier: Amsterdam, The Netherlands, 1995; pp. 155–233.
2. Smith, L.W.; Culvenor, C.C.J. Plant Sources of Hepatotoxic Pyrrolizidine Alkaloids. *J. Nat. Prod.* **1981**, *44*, 129–152. [CrossRef]
3. Roeder, E. Medicinal plants in China containing pyrrolizidine alkaloids. *Die Pharm.* **2000**, *55*, 711–726.
4. Stegelmeier, B.L.; Edgar, J.A.; Colegate, S.M.; Gardner, D.R.; Schoch, T.K.; Coulombe, R.A.; Molyneux, R.J. Pyrrolizidine alkaloid plants, metabolism and toxicity. *J. Nat. Toxins* **1999**, *8*, 95–116. [PubMed]
5. Cheeke, P.R. Toxicity and Metabolism of Pyrrolizidine Alkaloids. *J. Anim. Sci.* **1988**, *66*, 2343–2350. [CrossRef] [PubMed]
6. Chen, Z.; Huo, J.R. Hepatic veno-occlusive disease associated with toxicity of pyrrolizidine alkaloids in herbal prepara-tions. *Neth. J. Med.* **2010**, *68*, 252–260. [PubMed]
7. Fu, P.P.; Xia, Q.; Lin, G.; Chou, M.W. Pyrrolizidine Alkaloids—Genotoxicity, Metabolism Enzymes, Metabolic Activation, and Mechanisms. *Drug Metab. Rev.* **2004**, *36*, 1–55. [CrossRef] [PubMed]
8. Huxtable, R.J. Herbal Teas and Toxins: Novel Aspects of Pyrrolizidine Poisoning in the United States. *Perspect. Biol. Med.* **1980**, *24*, 1–14. [CrossRef]
9. Chauvin, P.; Dillon, J.C.; Moren, A. an outbreak of Heliotrope food poisoning, Tadjikistan, November 1992–March 1993. *Sante* **1994**, *4*, 263–268.
10. Kakar, F.; Akbarian, Z.; Leslie, T.; Mustafa, M.L.; Watson, J.; Van Egmond, H.P.; Omar, M.F.; Mofleh, J. an Outbreak of Hepatic Veno-Occlusive Disease in Western Afghanistan Associated with Exposure to Wheat Flour Contaminated with Pyrrolizidine Alkaloids. *J. Toxicol.* **2010**, *2010*, 1–7. [CrossRef]
11. Krishnamachari, K.A.; Bhat, R.V.; Krishnamurthi, D.; Krishnaswamy, K.; Nagarajan, V. Aetiopathogenesis of endemic ascites in Surguja district of Madhya Pradesh. *Indian J. Med Res.* **1977**, *65*, 672–678.
12. Steenkamp, V.; Stewart, M.J.; Van der Merwe, S.; Zuckerman, M.; Crowther, N.J. The effect of Senecio latifolius a plant used as a South African traditional medicine, on a human hepatoma cell line. *J. Ethnopharmacol.* **2001**, *78*, 51–58. [CrossRef]
13. Tandon, H.D.; Tandon, B.N.; Mattocks, A.R. an epidemic of veno-occlusive disease of the liver in Afghanistan. Pathologic features. *Am. J. Gastroenterol.* **1978**, *70*, 607–613. [PubMed]
14. Wiedenfeld, H.; Edgar, J. Toxicity of pyrrolizidine alkaloids to humans and ruminants. *Phytochem. Rev.* **2011**, *10*, 137–151. [CrossRef]

15. EFSA. Scientific Opinion on Pyrrolizidine alkaloids in food and feed—EFSA Panel on Contaminants in the Food Chain (CONTAM). *EFSA J.* **2011**, *9*, 2406–2540. [CrossRef]

16. Luckert, C.; Hessel-Pras, S.; Lenze, D.; Lampen, A. Disturbance of gene expression in primary human hepatocytes by hepatotoxic pyrrolizidine alkaloids: a whole genome transcriptome analysis. *Toxicol. In Vitro* **2015**, *29*, 1669–1682. [CrossRef] [PubMed]

17. Hessel-Pras, S.; Braeuning, A.; Guenther, G.; Adawy, A.; Enge, A.-M.; Ebmeyer, J.; Henderson, C.J.; Hengstler, J.G.; Lampen, A.; Reif, R. The pyrrolizidine alkaloid senecionine induces CYP-dependent destruction of sinusoidal endothelial cells and cholestasis in mice. *Arch. Toxicol.* **2019**, *94*, 219–229. [CrossRef] [PubMed]

18. Arrese, M.; Ananthananarayanan, M.; Suchy, F.J. Hepatobiliary Transport: Molecular Mechanisms of Development and Cholestasis. *Pediatr. Res.* **1998**, *44*, 141–147. [CrossRef]

19. Trauner, M.; Boyer, J.L. Bile Salt Transporters: Molecular Characterization, Function, and Regulation. *Physiol. Rev.* **2003**, *83*, 633–671. [CrossRef]

20. Cañaveras, J.C.G.; Donato, M.T.; Castell, J.V.; Lahoz, A. Targeted profiling of circulating and hepatic bile acids in human, mouse, and rat using a UPLC-MRM-MS-validated method. *J. Lipid Res.* **2012**, *53*, 2231–2241. [CrossRef]

21. Sharanek, A.; Burban, A.; Humbert, L.; Azzi, P.B.-E.; Felix-Gomes, N.; Rainteau, M.; Guillouzo, A. Cellular Accumulation and Toxic Effects of Bile Acids in Cyclosporine A-Treated HepaRG Hepatocytes. *Toxicol. Sci.* **2015**, *147*, 573–587. [CrossRef]

22. Hofmann, A.F.; Hagey, L.R. Bile acids: Chemistry, pathochemistry, biology, pathobiology, and therapeutics. *Cell. Mol. Life Sci.* **2008**, *65*, 2461–2483. [CrossRef]

23. Kullak-Ublick, G.A.; Beuers, U.; Paumgartner, G. Hepatobiliary transport. *J. Hepatol.* **2000**, *32*, 3–18. [CrossRef]

24. Trauner, M.; Meier, P.J.; Boyer, J.L. Molecular Pathogenesis of Cholestasis. *N. Engl. J. Med.* **1998**, *339*, 1217–1227. [CrossRef] [PubMed]

25. Antherieu, S.; Azzi, P.B.-E.; Dumont, J.; Fromenty, B.; Robin, M.-A.; Guillouzo, A.; Abdel-Razzak, Z.; Guguen-Guillouzo, C. Oxidative stress plays a major role in chlorpromazine-induced cholestasis in human HepaRG cells. *Hepatology* **2013**, *57*, 1518–1529. [CrossRef] [PubMed]

26. Vinken, M.; Landesmann, B.; Goumenou, M.; Vinken, S.; Shah, I.; Jaeschke, H.; Willett, C.; Whelan, M.; Rogiers, V. Development of an Adverse Outcome Pathway From Drug-Mediated Bile Salt Export Pump Inhibition to Cholestatic Liver Injury. *Toxicol. Sci.* **2013**, *136*, 97–106. [CrossRef] [PubMed]

27. Woolbright, B.L.; McGill, M.R.; Yan, H.; Jaeschke, H. Bile Acid-Induced Toxicity in HepaRG Cells Recapitulates the Response in Primary Human Hepatocytes. *Basic Clin. Pharmacol. Toxicol.* **2016**, *118*, 160–167. [CrossRef] [PubMed]

28. Josse, R.; Aninat, C.; Glaise, D.; Dumont, J.; Fessard, V.; Morel, F.; Poul, J.-M.; Guguen-Guillouzo, C.; Guillouzo, A. Long-Term Functional Stability of Human HepaRG Hepatocytes and Use for Chronic Toxicity and Genotoxicity Studies. *Drug Metab. Dispos.* **2008**, *36*, 1111–1118. [CrossRef] [PubMed]

29. Waizenegger, J.; Braeuning, A.; Templin, M.; Lampen, A.; Hessel-Pras, S. Structure-dependent induction of apoptosis by hepatotoxic pyrrolizidine alkaloids in the human hepatoma cell line HepaRG: Single versus repeated exposure. *Food Chem. Toxicol.* **2018**, *114*, 215–226. [CrossRef]

30. Behr, A.-C.; Kwiatkowski, A.; Ståhlman, M.; Schmidt, F.F.; Luckert, C.; Braeuning, A.; Buhrke, T. Impairment of bile acid metabolism by perfluorooctanoic acid (PFOA) and perfluorooctanesulfonic acid (PFOS) in human HepaRG hepatoma cells. *Arch. Toxicol.* **2020**, *94*, 1673–1686. [CrossRef]

31. Jeong, J.-Y.; Yim, H.-S.; Ryu, J.-Y.; Lee, H.S.; Lee, J.-H.; Seen, D.-S.; Kang, S.G. One-Step Sequence- and Ligation-Independent Cloning as a Rapid and Versatile Cloning Method for Functional Genomics Studies. *Appl. Environ. Microbiol.* **2012**, *78*, 5440–5443. [CrossRef]

32. Li, M.Z.; Elledge, S.J. Harnessing homologous recombination in vitro to generate recombinant DNA via SLIC. *Nat. Methods* **2007**, *4*, 251–256. [CrossRef] [PubMed]

33. Luckert, C.; Ehlers, A.; Buhrke, T.; Seidel, A.; Lampen, A.; Hessel, S. Polycyclic aromatic hydrocarbons stimulate human CYP3A4 promoter activity via PXR. *Toxicol. Lett.* **2013**, *222*, 180–188. [CrossRef]

34. Livak, K.J.; Schmittgen, T.D. Analysis of relative gene expression data using real-time quantitative PCR and the 2− $\Delta\Delta$CT method. *Methods* **2001**, *25*, 402–408. [CrossRef] [PubMed]

35. Crestani, M.; Stroup, D.; Chiang, J.Y. Hormonal regulation of the cholesterol 7 alpha-hydroxylase gene (CYP7). *J. Lipid Res.* **1995**, *36*, 2419–2432. [CrossRef]

36. Hampf, M.; Gossen, M. a protocol for combined Photinus and Renilla luciferase quantification compatible with protein assays. *Anal. Biochem.* **2006**, *356*, 94–99. [CrossRef]

37. Tremaroli, V.; Karlsson, F.; Werling, M.; Ståhlman, M.; Kovatcheva-Datchary, P.; Olbers, T.; Fändriks, L.; Le Roux, C.W.; Nielsen, J.; Bäckhed, F. Roux-en-Y Gastric Bypass and Vertical Banded Gastroplasty Induce Long-Term Changes on the Human Gut Microbiome Contributing to Fat Mass Regulation. *Cell Metab.* **2015**, *22*, 228–238. [CrossRef]

38. Zamek-Gliszczynski, M.J.; Xiong, H.; Patel, N.J.; Turncliff, R.Z.; Pollack, G.M.; Brouwer, K.L.; Leslie, E.M.; Bowers, R.J.; Deeley, R.G.; Cole, S.P.C. Pharmacokinetics of 5 (and 6)-Carboxy-2′,7′-Dichlorofluorescein and Its Diacetate Promoiety in the Liver. *J. Pharmacol. Exp. Ther.* **2003**, *304*, 801–809. [CrossRef]

39. Xiong, A.; Yang, F.; Fang, L.; Yang, L.; He, Y.; Wan, Y.Y.J.; Xu, Y.; Qi, M.; Wang, X.; Yu, K.; et al. Metabolomic and genomic evidence for compromised bile acid homeostasis by senecionine, a hepatotoxic pyrrolizidine alkaloid. *Chem. Res. Toxicol.* **2014**, *27*, 775–786. [CrossRef]

40. Xiong, A.; Fang, L.; Yang, X.; Yang, F.; Qi, M.; Kang, H.; Yang, L.; Tsim, K.W.K.; Wang, Z. an application of target profiling analyses in the hepatotoxicity assessment of herbal medicines: Comparative characteristic fingerprint and bile acid profiling of *Senecio vulgaris* L. and *Senecio scandens* Buch.-Ham. *Anal. Bioanal. Chem.* **2014**, *406*, 7715–7727. [CrossRef]

41. Pauli-Magnus, C.; Meier, P.J. Hepatobiliary transporters and drug-induced cholestasis. *Hepatology* **2006**, *44*, 778–787. [CrossRef]

42. Paumgartner, G. Medical treatment of cholestatic liver diseases: From pathobiology to pharmacological targets. *World J. Gastroenterol.* **2006**, *12*, 4445–4451. [CrossRef] [PubMed]

43. Ebmeyer, J.; Franz, L.; Lim, R.; Niemann, B.; Glatt, H.; Braeuning, A.; Lampen, A.; Hessel-Pras, S. Sensitization of Human Liver Cells Toward Fas-Mediated Apoptosis by the Metabolically Activated Pyrrolizidine Alkaloid Lasiocarpine. *Mol. Nutr. Food Res.* **2019**, *63*, e1801206. [CrossRef] [PubMed]

44. Anderson, J.M.; Glade, J.L.; Stevenson, B.R.; Boyer, J.L.; Mooseker, M.S. Hepatic immunohistochemical localization of the tight junction protein ZO-1 in rat models of cholestasis. *Am. J. Pathol.* **1989**, *134*, 1055–1062. [PubMed]

45. Fallon, M.B.; Mennone, A.; Anderson, J.M. Altered expression and localization of the tight junction protein ZO-1 after common bile duct ligation. *Am. J. Physiol. Physiol.* **1993**, *264 Pt 1*, C1439–C1447. [CrossRef]

46. Baker, D.C.; Pfister, J.A.; Molyneux, R.J.; Kechele, P. Cynoglossum officinale toxicity in calves. *J. Comp. Pathol.* **1991**, *104*, 403–410. [CrossRef]

47. Craig, A.; Pearson, E.; Meyer, C.; Schmitz, J. Clinicopathologic studies of tansy ragwort toxicosis in ponies: Sequential serum and histopathological changes. *J. Equine Veter. Sci.* **1991**, *11*, 261–271. [CrossRef]

48. Mendel, V.E.; Witt, M.R.; Gitchell, B.S.; Gribble, D.N.; Rogers, Q.R.; Segall, H.J.; Knight, H.D. Pyrrolizidine alkaloid-induced liver disease in horses: an early diagnosis. *Am. J. Veter. Res.* **1988**, *49*, 572–578.

49. Stegelmeier, B.L.; Gardner, D.R.; James, L.F.; Molyneux, R.J. Pyrrole detection and the pathologic progression of Cynoglossum officinale (houndstongue) poisoning in horses. *J. Vet. Diagn. Investig.* **1996**, *8*, 81–90. [CrossRef]

50. Merz, K.-H.; Schrenk, D. Interim relative potency factors for the toxicological risk assessment of pyrrolizidine alkaloids in food and herbal medicines. *Toxicol. Lett.* **2016**, *263*, 44–57. [CrossRef]

Article

Identification of microRNAs Implicated in Modulating Senecionine-Induced Liver Toxicity in HepaRG Cells

Anne-Margarethe Enge, Heike Sprenger , Albert Braeuning and Stefanie Hessel-Pras *

Department of Food Safety, German Federal Institute for Risk Assessment, Max-Dohrn-Str. 8-10,
10589 Berlin, Germany; anne-margarethe.enge@bfr.bund.de (A.-M.E.); heike.sprenger@bfr.bund.de (H.S.);
albert.braeuning@bfr.bund.de (A.B.)
* Correspondence: stefanie.hessel-pras@bfr.bund.de; Tel.: +49-30-18412-25203

Abstract: 1,2-unsaturated Pyrrolizidine Alkaloids (PAs) are secondary plant metabolites that occur as food contaminants. Upon consumption, they can cause severe liver damage. PAs have been shown to induce apoptosis, to have cytotoxic and genotoxic effects, and to impair bile acid homeostasis in the human hepatoma cell line HepaRG. The major mode of action of PAs is DNA- and protein-adduct formation. Beyond that, nuclear receptor activation has only been observed for one receptor and two PAs, yielding the possibility that other cellular mediators are involved in PA-mediated toxicity. Here, the mode of action of Senecionine (Sc), a prominent and ubiquitous representative of hepatotoxic PAs, was investigated by analyzing 7 hepatic microRNAs (miRNAs) in HepaRG cells. Ultimately, 11 target genes that were predicted with Ingenuity Pathway Analysis software (IPA) were found to be significantly downregulated, while their assigned miRNAs showed significant upregulation of gene expression. According to IPA, these targets are positively correlated with apoptosis and cellular death and are involved in diseases such as hepatocellular carcinoma. Subsequent antagomiR-inhibition analysis revealed a significant correlation between PA-induced *miRNA-4434* induction and P21-Activated Kinase-1 (*PAK1*) downregulation. *PAK1* downregulation is usually associated with cell cycle arrest, suggesting a new function of Sc-mediated toxicity in human liver cells.

Keywords: cell cycle; IPA; microRNA; *PAK1*; pyrrolizidine alkaloids; senecionine

Citation: Enge, A.-M.; Sprenger, H.; Braeuning, A.; Hessel-Pras, S. Identification of microRNAs Implicated in Modulating Senecionine-Induced Liver Toxicity in HepaRG Cells. *Foods* **2022**, *11*, 532. https://doi.org/10.3390/foods11040532

Academic Editor: Junsoo Lee

Received: 18 January 2022
Accepted: 8 February 2022
Published: 12 February 2022

Publisher's Note: MDPI stays neutral with regard to jurisdictional claims in published maps and institutional affiliations.

1. Introduction

Pyrrolizidine alkaloids are the most common natural toxins with over 660 different chemical structures. They are present in more than 6000 plant species worldwide and can occur as contaminants in food and feed, posing a possible health risk for humans and livestock [1]. For example, they can be found in products such as tea, honey, culinary herbs, dietary supplements or herbal preparations such as traditional medicine [2–5]. It must be stressed that the exposure to pyrrolizidine alkaloids can occur via a wide variety of foodstuff, which increases the likelihood of ingestion. Because 1,2-unsaturated Pyrrolizidine Alkaloids (PAs) have genotoxic properties and are considered to be carcinogenic, no safe threshold value such as the TDI (Tolerable Daily Intake) can be derived. Therefore, it is recommended to apply the ALARA principle (As Low As Reasonably Achievable) to prevent unnecessary exposure [6].

PAs are known to require metabolic activation in the liver, usually provided by members of the cytochrome P450 superfamily [7]. The hepatotoxic metabolites may cause acute toxic effects such as Hepatic Sinusoidal Obstruction Syndrome (HSOS), resulting in ascites, hepatomegaly, and ultimately death. Chronic intoxication may lead to megalocytosis, liver cirrhosis and cancer [7–16]. To a lesser extent, PAs are also pneumotoxic, most likely due to systemic distribution of reactive metabolites to the lung [17].

Pyrrolizidine alkaloids consist of one so-called necine base, which is esterified with one or two aliphatic mono- or dicarboxylic acids (necic acids). Accordingly, they are classified into different structure types, depending on the structure of their necine base (retronecine-, heliotridine-, otonecine-, and platynecine-type) and further into monoesters, open-chained diesters and cyclic diesters [18].

Until today, many in vitro studies have been conducted showing that PAs cause apoptosis [19,20], cytotoxicity [21], and genotoxicity [22] in a structure-dependent manner. Moreover, they were shown to disturb bile acid homeostasis in the human hepatoma cell line HepaRG [23,24]. Upon PA treatment, it is consistently observed that the expression pattern of genes relevant for apoptosis, bile acid homeostasis, and metabolism is highly deregulated in liver cells in vivo and in vitro [21,23,25]. Nonetheless, some ambiguities remain regarding the underlying molecular mechanisms of PA-mediated hepatotoxicity. The major molecular mode of action is the binding of PA metabolites to DNA and proteins to form adducts via alkylation [7]. Nevertheless, interactions between PAs and specific regulatory elements may pose additional molecular mechanisms leading to deregulation of cellular signaling and disturbance of metabolic pathways. Therefore, PA-mediated nuclear receptor activation was investigated, mainly because nuclear receptors are typical targets in xenobiotic-induced toxicity. However, no PA-induced activation could be observed, with the exception of Pregnane X Receptor (PXR), which was exclusively activated by the PAs echimidine and Lasiocarpine (Lc), resulting in the activation of its target CYtochrome P450 monooxygenase (CYP) 3A4 [26]. Therefore, it appears likely that cellular signaling pathways and mediators other than nuclear receptor interference are involved in mediating the effects of PAs. miRNAs are small, non-coding, 21- to 25-nucleotide-long RNA molecules that regulate a wide variety of physiological processes such as cell growth, development, apoptosis, differentiation, and carcinogenesis at the post-transcriptional level [27,28]. In animals, miRNAs bind their target mRNAs in the 3'-UnTranslated Region (UTR). Depending on the complementarity of base pairing between the miRNA and the mRNA target, the binding of the miRNA either leads to mRNA cleavage or translational repression. It is postulated that translational repression is the predominant mechanism by which metazoan miRNAs negatively regulate their targets [29]. Many miRNAs are crucial regulators of bile acid homeostasis, lipid and glucose metabolism, inflammation, apoptosis, and proliferation [30–32], for example, and are deregulated in many liver diseases [32,33]. Notably, CYP7A1, the rate-limiting enzyme in bile acid synthesis, is regulated by two miRNAs, with one of them being the most abundant miRNA in the liver [34]. Interestingly, a correlation between chronic exposure to the PA riddelliine and an altered miRNA expression pattern in the liver has been observed in a 12-week feeding study with rats [35]. Moreover, an integrative analysis studying the miRNA-mRNA interaction after acute incubation with the PA Monocrotaline (Mc) in high doses found the phagosome signaling pathway to be a relevant molecular mechanism of PA-induced HSOS in mice [36]. Some miRNAs are solely expressed in a tissue-specific manner; others circulate in body fluids such as blood or urine and have been proposed as non-invasive biomarkers for disease prediction and progression. For example, circulating miRNAs in blood samples from patients suffering from PA-induced HSOS were positively correlated with the severity of liver injury and progression of HSOS [37]. Additionally, patients suffering from intrahepatic cholestasis showed elevated miRNA levels in blood and liver tissue [33]. Therefore, miRNA signatures are currently discussed as early-stage biomarkers for HSOS and other liver-specific injuries [28,38].

The present study aimed to examine the role of hepatic miRNAs in the regulation of PA-induced hepatotoxicity in HepaRG cells. For this purpose, relevant hepatic miRNAs were discovered that are sensible to PA treatment. With quantitative analysis and Ingenuity Pathway Analysis (IPA), an evidence-based tool for target and pathway prediction, a selection of target genes was further subjected to analysis to identify the potential engaged molecular functions and biological pathways involved in PA-mediated toxicity in human liver cells.

2. Materials and Methods

2.1. Chemicals

The PAs Lc and senecionine (Sc) (>95% purity) were obtained from PHYTOPLAN Diehm & und Neuberger GmbH (Heidelberg, Germany). In order to get 5 mM stock solutions, the PAs were dissolved in 50% acetonitrile (ACN) and 50% water (*v/v*). All other chemicals were purchased from Merck KGaA (Darmstadt, Germany) and Thermo Fisher Scientific (Waltham, MA, USA).

2.2. Cell Culture

The human hepatoma cell line HepaRG was purchased from Biopredic International (Saint-Grégoire, France). The cells were cultivated at 37 °C in a humidified atmosphere of 5% CO_2 [39]. After seeding, cells were cultivated in William's Medium E with stable glutamine (PAN-Biotech, Aidenbach, Germany) for proliferation. The medium was supplemented with 10% Fetal Bovine Serum (FBS; PAN-Biotech, Aidenbach, Germany), 50 µM hydrocortisone hemisuccinate (Merck KGaA, Darmstadt, Germany), 100 U/mL penicillin and 100 µg/mL streptomycin (Capricorn Scientific, Ebsdorfergrund, Germany) and 5 µg/mL human insulin (PAN-Biotech, Aidenbach, Germany). After two weeks of proliferation, 1.7% of DiMethyl SulfOxide (DMSO; Merck KGaA, Darmstadt, Germany) was added to initiate differentiation and HepaRG cells were cultivated for another two weeks to fully differentiate before being used for experiments.

2.3. Liver-Specific miRNA Array

For liver-specific miRNA expression screening, HepaRG cells were seeded at a density of 0.2×10^6 cells per well in 6-well plates and cultivated as described in Section 2.2. After four weeks of cultivation and 48 h prior to incubation, fully differentiated cells were adapted to treatment medium containing 1.7% DMSO and a reduced FBS concentration of 2%. The cells were incubated with 10, 35 and 250 µM of the most potent Lc for 8 h or with 2.5, 10 and 35 µM Lc for 24 h, respectively. The miRNA was isolated using the miRNeasy Mini Kit (Qiagen, Hilden, Germany) according to the manufacturer's protocol, including a purification step of the miRNA-enriched fraction using the RNeasy MinElute Cleanup Kit (Qiagen, Hilden, Germany). Initially, cells were lysed with 700 µL QIAzol lysis reagent. As a miRNA isolation efficiency control, a spike-in control consisting of three synthetic templates in different concentrations (Qiagen, Hilden, Germany) was added to the lysis buffer. The subsequent procedure was as follows: incubation for 5 min at Room Temperature (RT), addition of 140 µL chloroform per sample and thoroughly mixing, incubation for 3 min at RT, centrifugation for 15 min at 12,000× *g* at 4 °C, transfer of the upper aqueous phase intro a new tube, addition of one volume of 70% ethanol (usually 350 µL) and thorough mixing, transfer of the sample into an RNeasy Mini spin column, and centrifugation for 15 s at 8000× *g* at RT. The RNeasy Mini spin columns containing the mRNA fractions were discarded. The flow-through containing the smaller miRNA fraction was purified as follows: addition of 450 µL 100% ethanol, thorough mixing, transfer of the sample into an RNeasy MinElute spin column and centrifugation for 15 s at 8000× *g* at RT, washing of column with 700 µL RWT buffer and 500 µL RPE buffer, respectively, for 15 s at 8000× *g* at RT, final washing with 500 µL 80% ethanol with subsequent centrifugation for 2 min at 8000× *g* at RT and subsequent drying by centrifugation for 5 min at 8000× *g* at RT. Afterwards, the miRNA fractions were eluted in 14 µL water and quantified at 260 and 280 nm on a TecanM200Pro spectrometer (Tecan Group Lt., Männedorf, Switzerland). Next, 10 ng miRNA (naturally occurring as non-polyadenylated) were polyadenylated by a poly(A) polymerase and reverse transcribed into cDNA using the miRCURY LNA RT Kit (Qiagen, Hilden, Germany) according to the manufacturer's instructions. Additionally, as a reverse transcription efficiency control, a spike-in control consisting of a synthetic RNA target (Qiagen, Hilden, Germany) was added to the reaction buffer. The reverse transcription step lasted for 60 min at 42 °C, followed by an inactivation step for 5 min at 95 °C. Quantitative real-time Polymerase Chain

Reaction (qPCR) on pre-designed PCR panels with Locked Nucleic Acids (LNA)-enhanced oligonucleotides (miRCURY LNA miRNA custom PCR array of 84 liver-specific miRNAs (see Supplementary Materials Table S1 for a list of all miRNAs investigated)) was performed from 62.5 pg cDNA on an ABI 7900HT Fast Real-Time PCR system (Applied Biosystems, Foster City, CA, USA) in 384-well format using miRCURY LNA SYBR Green Master Mix and ROX reference dye (20×) (Qiagen, Hilden, Germany). The thermal profile was as follows: initiation (2 min, 95 °C), denaturation (10 s, 95 °C, 40 cycles), annealing/elongation (1 min, 56 °C, 40 cycles) and dissociation curve analysis. Results were analyzed using the QIAGEN GeneGlobe miRCURY LNA miRNA PCR Data analysis software. That is, C_t values were evaluated in consideration of an interplate calibrator, RNA isolation efficiency spike-in controls and reverse transcription efficiency spike-in control, and subsequently normalized to the expression of a set of housekeeping genes such as the miRNAs *miR-103a-3p, -16-5p, -191-5p, -423-5p* and *-23a-3p* and a small nucleolar RNA (*snoRD38D*) and referred to the values of the solvent control (2.5% ACN for 8 h and 0.35% ACN for 24 h), which is in accordance with the $2^{-\Delta\Delta Ct}$ method [40].

2.4. RNA Isolation and qPCR Analysis

To investigate the deregulation of gene expression of the selected liver-specific miR-NAs, HepaRG cells were seeded, cultivated and adapted to treatment medium as described in Section 2.3. The cells were incubated with 35 μM Sc for 2, 4, 6, 8 and 24 h, respectively. The mRNA and miRNA were isolated using the miRNeasy Mini Kit (Qiagen, Hilden, Germany) according to the manufacturer's protocol, including a purification step of miRNA-enriched fractions using the RNeasy MinElute Cleanup Kit (Qiagen, Hilden, Germany). Cell lysis and total RNA isolation occurred as mentioned in Section 2.3. with the exception that no spike-in control was added. Additionally, the RNeasy Mini spin columns containing the mRNA fractions were not discarded, but stored at 4 °C until further isolation. The flow-through containing the smaller miRNA fractions was purified and miRNA eluted as mentioned in Section 2.3. Afterwards, mRNA isolation occurred as follows: washing of RNeasy Mini columns once with 700 μL RWT buffer and twice with 500 μL RPE buffer with centrifugation for 15 s at 8000× *g* at RT between each step, respectively, and drying by centrifugation for 1 min at full speed at RT. The samples were eluted in 30 μL water. Both the miRNA enriched samples and the total RNA samples were quantified at 260 and 280 nm on a TecanM200Pro spectrometer (Tecan Group Lt., Männedorf, Switzerland).

Five hundred ng miRNA were polyadenylated and reverse transcribed into cDNA using the miScript II RT Kit (Qiagen, Hilden, Germany) according to the manufacturer's protocol. The cDNA was synthesized by using a poly(T) primer with a 5′ universal tag, enabling the alignment with a universal primer during qPCR amplification. qPCR was performed from 227.3 pg cDNA on an ABI 7900HT Fast Real-Time PCR system in 384-well format using miScript SYBR Green PCR Kit (Qiagen, Hilden, Germany) with 10x universal primer and 10x miRNA-specific miScript Primer Assay (Qiagen, Hilden, Germany). The thermal profile was as follows: initial activation (15 min, 95 °C), 3-step cycling (40 cycles in total) with denaturation (15 s, 94 °C), annealing (30 s, 55 °C), elongation (34 s, 70 °C), and dissociation curve analysis. C_t values were evaluated according to the $2^{-\Delta\Delta Ct}$ method [40] by normalizing the respective C_t-values to the expression level of miR-103a and referring to the solvent control (0.35% ACN).

One microgram of total RNA was reverse transcribed into cDNA using the High-Capacity cDNA Reverse Transcriptase Kit (Applied Biosystems, Foster City, CA, USA) according to the manufacturer's protocol. Using 20 ng cDNA, qPCR was conducted on an ABI 7900HT Fast Real-Time PCR system in 384-well format with Maxima SYBR Green/ROX qPCR Mastermix (Thermo Fisher Scientific, Waltham, MA, USA) and 300 nM of each primer (synthesized at Eurofins Genomics, Ebersberg, Germany, see Table 1). The thermal profile was as follows: initial denaturation (15 min, 95 °C), 3-step cycling (40 cycles in total) with denaturation (30 s, 95 °C), annealing and elongation (1 min, 60 °C), final elongation

(10 min, 60 °C), and dissociation curve analysis. C_t values were evaluated according to the $2^{-\Delta\Delta Ct}$ method and normalized to the housekeeping gene *GUSB* (β-glucuronidase).

Table 1. Sequences of primers used for qPCR analysis of target genes.

Gene	Ensembl ID	Forward Primer (5′–3′)	Reverse Primer (5′–3′)
ACVR2A	ENSG00000121989	TGTTTTGGGCACAGGTTATTGT	CAAGGTGGGGTTTTATGGGGA
AOX1	ENSG00000138356	TCCCTGCCATCTGTGACATG	CCGACTCTCCCAGACCCTTA
ATP6V1H	ENSG00000047249	CCAAAACTGTGGCCATGCTA	TGCTGGTGAGAGGCTTCAGA
ATP7A	ENSG00000165240	CAAAACCAGGCGTCTCAACC	TCGTCCACTGCTGTTTTCGG
BOK	ENSG00000176720	CACACACAGCCTTCCCTTGA	GCCTGTATCTCCTGAGTGCC
CASP2	ENSG00000106144	GAATTCCACCGGTGCAAGGA	TGGATGATGGGGAGGTGACA
FBXW7	ENSG00000109670	TATACTCCCTGCCCTTCCCC	CCAACATCCTGCACCACTGA
FOXO1	ENSG00000150907	CAGGGGTGGCCATGTAAGTC	GGAACAAGAACGTGGAATCTGC
GAS2	ENSG00000148935	GTGCCGAGATTTAGGGGTGG	GCAATCCGGCCAAGCTCTAG
GUSB	ENSG00000169919	CACCAGGGACCATCCAATAC	ATGTAGGTGGTGGGTGTCGT
IFIH1	ENSG00000115267	CAAAAGAAGTGTGCCGACTATCA	TGCACCATCATTGTTCCCCA
LEPROT	ENSG00000213625	ATTGAGTTCCCAGGCCAAGC	AGACCCAAAACTCAGGCAGG
PAK1	ENSG00000149269	GCCTGACATGATACCCTGCC	AGCAAACATCCCCAACACCC
PATZ1	ENSG00000100105	AGTCTGGTCAGGGAAGTAGGG	ACACAATGTCCCTACCTGCC
PRKAR1A	ENSG00000108946	ACTTTGCTGGAGTGGTGGTG	TGGTCTTGAACTCCTGGGCT
RACGAP1	ENSG00000161800	CTGTCCCCTTCCCTGCATTC	TGCACAACAATGGAGGGGAT
REL	ENSG00000162924	CTTCATGCCCCTTCCCAGTC	ACGTTGACAACCCAGCTGTT
SFXN1	ENSG00000164466	AACAGGGTCATGCTTGGATCA	GGCTTCTTGAGGTCTCTGGC
SLC25A29	ENSG00000197119	GCCCACACTGTAGAGTCACG	AGAAAGGGGCTGGAGTGTCT
SLC4A4	ENSG00000080493	CAAACATTGCAACTCAGGGCT	TTCACATTGTAGGACTGGGACA
STX17	ENSG00000136874	TGTTAGCAAGGGTCAGCACG	TCAAGCACCACTCAGCAATGT
ST6GAL1	ENSG00000073849	AGCCCACTTTCCCTCTCCAT	TGCCTCTCTCACTGAACCGT

2.5. Ingenuity Pathway Analysis

The target genes of miRNAs with significantly altered expression values in HepaRG cells after treatment with 35 µM Sc for 24 h were predicted with Ingenuity Pathway Analysis software (IPA, version 70750971, Qiagen Bioinformatics, Redwood City, CA, USA). IPA uses experimentally validated targeting interactions from TarBase [41], miRecords [42] and Ingenuity expert findings, and predicted miRNA–mRNA interactions from TargetScan [43]. The predicted target genes were compared to the gene expression dataset from a whole genome microarray conducted with primary human hepatocytes incubated with 100 µM Sc for 24 h [44]. The matched targets were prioritized according to their miRNA and mRNA relationship. This included opposing expression pairing and "highly predicted" and/or "experimentally observed" confidence of miRNA-target gene correlation. Subsequently, IPA core analysis and expression analysis with the matched target genes were performed to predict affected diseases and functions and to determine possible signaling pathways involved in PA toxicity.

2.6. AntagomiR Experiments

In order to identify the targets of the miRNA-4434 (also known as human miRNA-4516) and to assess if this specific miRNA expression is sensitive to PA exposure in HepaRG cells, so-called antagomiR experiments were conducted. For this purpose, HepaRG cells were seeded, cultivated and adapted to treatment medium as described in Section 2.3. The cells were incubated with 35 µM Sc and transfected with Lipofectamine RNAiMAX (Thermo Fisher Scientific Waltham, MA, USA) reagent with either miRCURY LNA miRNA Power Inhibitor (antagomiR)-4434 (5′-3′ sequence: TTCTACTTTACTTCTCCT, matches miRNA-4434 in its seed region GGAGAAG), miScript Inhibitor Negative Control (5′-3′ sequence: TAACACGTCTATACGCCCA; both from Qiagen, Hilden, Germany) as an antagomiR-Negative Control (antagomiR-NC) or water as a negative treatment control. RNAiMAX and antagomiR-4434, antagomiR-NC or water were prepared separately in Opti-MEM reduced serum medium before being mixed 1:1 and incubated for 5 min at RT. Afterwards, the antagomiR-lipid complexes were added to the cells, resulting in a final concentration of 100 pmol (50 nM) or 150 pmol (75 nM) antagomiR-4434 or anagomiR-NC per 6-well, respectively. According to the manufacturer's protocol, antisense effects are usually assessed 24–72 h after transfection. Therefore, in this case, total RNA was isolated 48 h after transfection and isolation was conducted with the RNeasy Mini Kit (Qiagen, Hilden, Germany) according to the manufacturer's instructions, as described elsewhere [45]. Quantity and purity of isolated RNA were measured at 260 and 280 nm on a TecanM200Pro spectrometer. Then, 1 µg total RNA was reverse transcribed into cDNA and qPCR was performed from 20 ng cDNA as described in Section 2.4.

2.7. Cell Viability Assay

In order to exclude antagomiR-mediated cytotoxicity on PA-incubated HepaRG cells, cells were seeded at a density of 9000 cells per well in the inner 60 wells of a 96-well plate and cultivated as described in Section 2.2. After differentiation, cells were incubated with 35 µM Sc and transfected with either antagomiR-4434 or antagomiR-NC at concentrations of 25, 50, 75 and 100 nM, respectively, as described in Section 2.5. Transfection reagent with water was used as a negative control and 0.01% Triton-X-100 served as positive control for cytotoxicity. The Neutral Red Uptake (NRU) assay was applied for assessing the cell viability. Briefly, 120 µL 3-amino-7-dimethylamino-2-methylphenazine hydrochloride (neutral red) reagent dissolved in cell culture medium (40 µg/mL) was added per well and incubated for 2 h at 37 °C. Afterwards, the supernatant was discarded, cells were washed twice with Phosphate Buffered Saline (PBS) and dye crystals dissolved in 150 µL desorption solution (1% acetic acid in 50% ethanol). After shaking for 10 min (under protection from light), fluorescence was measured at λ_{ex} = 530 nm and λ_{em} = 645 nm on a TecanM200Pro spectrometer.

2.8. Statistical Analysis

Statistical analysis was performed using SigmaPlot 14.0 software (Systat Software, Erkrath, Germany). All qPCR data were subjected to logarithmic tranformation and statistically significant differences were calculated by Student's *t*-test in the case of single comparison analyses. If applicable, *p*-values were subjected to False Discovery Rate (FDR) correction [46]. Results were considered as significant at $p < 0.05$ and are indicated in the graphs by * $p < 0.05$, ** $p < 0.01$, *** $p < 0.001$.

3. Results

3.1. Sc Induces miRNA Expression in HepaRG Cells

First, the miRNA expression profile in HepaRG cells was analyzed with a customized liver-specific miCURY LNA Array covering 84 miRNAs. Compared to the control group, the number of deregulated miRNAs above 1 $\log_2$ Fold Change (FC) or below 1 $\log_2$ FC tended to be highest at 250 µM after 8 h and at 35 µM after 24 h, respectively (see Supplementary Figure S1). Among the top deregulated miRNAs, a selection of five downregulated and four upregulated miRNAs were chosen to be investigated within senecionine (Sc)-treated HepaRG cells. Sc was used for qPCR in order to have greater certainty for general PA effects and to exclude very specific effects of one single PA. To this end, HepaRG cells were treated with 35 µM of Sc and miRNA expression was analyzed after five different time points. Seven out of nine miRNAs proved to be differentially expressed after Sc treatment. The results are summarized in a heat map in Figure 1. Interestingly, all but one miRNA showed an increased expression compared to the solvent control in a time-dependent manner and for three miRNAs, the upregulation was highest after 8 h. *miRNA-122-3p* (alias *miRNA-122**) was the only one which showed significant downregulation after 24 h.

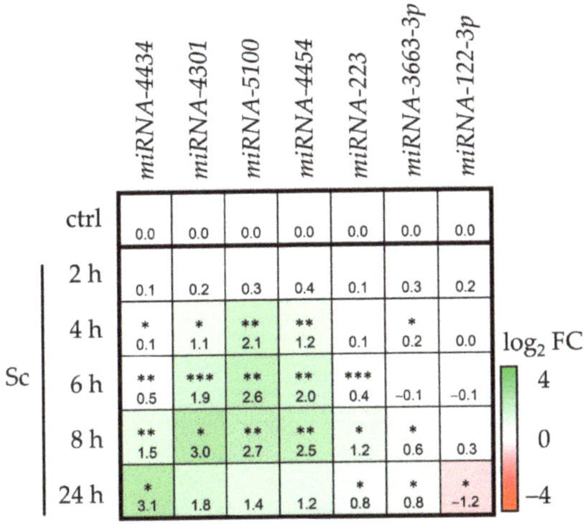

		miRNA-4434	miRNA-4301	miRNA-5100	miRNA-4454	miRNA-223	miRNA-3663-3p	miRNA-122-3p
	ctrl	0.0	0.0	0.0	0.0	0.0	0.0	0.0
Sc	2 h	0.1	0.2	0.3	0.4	0.1	0.3	0.2
	4 h	* 0.1	* 1.1	** 2.1	** 1.2	0.1	* 0.2	0.0
	6 h	** 0.5	*** 1.9	** 2.6	** 2.0	*** 0.4	−0.1	−0.1
	8 h	** 1.5	* 3.0	** 2.7	** 2.5	* 1.2	* 0.6	0.3
	24 h	* 3.1	1.8	1.4	1.2	* 0.8	* 0.8	* −1.2

$\log_2$ FC: 4 / 0 / −4

Figure 1. miRNA expression in differentiated HepaRG cells after exposure to 35 µM Sc at 5 different time-points. The results were evaluated using the $2^{-\Delta\Delta Ct}$ method [40]. Results are shown as $\log_2$ fold changes ($\log_2$ FC) and as mean of three independent biological replicates. Gene expression values were referred to the solvent control (ctrl) of the respective time point but, for clarity, only the ctrl values of the 2 h time point are depicted here. Up- or downregulation of gene expression is indicated in green or red, respectively. The higher the values, the stronger the coloration. Mean values and standard deviations (SD) are summarized in the Supplementary Materials Table S2. Statistical analysis was performed using Student's *t*-test as this is a case of single comparison analysis. That is, each miRNA expression value of one time point was compared to the ctrl of this respective time point. Subsequently, *p*-values were subjected to FDR correction. Significant differences are depicted as follows: * $p < 0.05$, ** $p < 0.01$, *** $p < 0.001$.

3.2. Prediction of Biological Consequences of PA-Mediated miRNA Level Alterations

Subsequently, the target genes of seven miRNAs that were identified to have significantly altered expression $\log_2$ FC values in HepaRG cells (see Figure 1) were predicted with IPA software (see Figure 2 for workflow).

The predicted target genes (773 in total) were compared to the gene expression dataset from a whole genome microarray conducted with primary human hepatocytes incubated with 100 µM Sc for 24 h [44]. Matched and prioritized targets (143 target genes) were submitted to a so-called IPA expression analysis to predict affected diseases and functions. In Figure 3 the top "diseases and bio functions" of the deregulated target genes of seven selected miRNAs predicted in Sc-treated primary human hepatocytes are shown. "Cancer" and "organismal injury and abnormalities" are the two main diseases and disorders that are regulated by the highest number of predicted targets. Among molecular and cellular functions, "cell death and survival" has the highest number of target genes with the most pronounced statistical significance. "Liver hyperproliferation" is the top hepatotoxicity function with 57 target genes involved, followed by "hepatocellular carcinoma" with 16 target genes.

Based on the expression analysis, 21 target genes involved in cellular growth and development, apoptosis and inflammation were chosen to be further investigated by qPCR in HepaRG cells incubated with 35 µM Sc for five different time points. The results are depicted in Figure 4 as a heat map. A total of 11 out of 21 genes showed an opposing expression pairing (indicated by superscript number 1); that is, significant downregulation, while their attributed miRNAs were upregulated. For the miRNA-4434 in particular, interaction with 5 out of 11 targets was predicted. These five targets comprised Growth Arrest-Specific protein 2 (*GAS2*), P21-Activated Kinase-1 (*PAK1*), LEPtin Receptor Overlapping Transcript (*LEPROT*), POZ/BTB and AT hook containing Zinc finger 1 (*PATZ1*) and

ST6 beta-GALactoside alpha-2,6-sialyltransferase 1 (*ST6GAL1*). Therefore, this miRNA was chosen for further experiments.

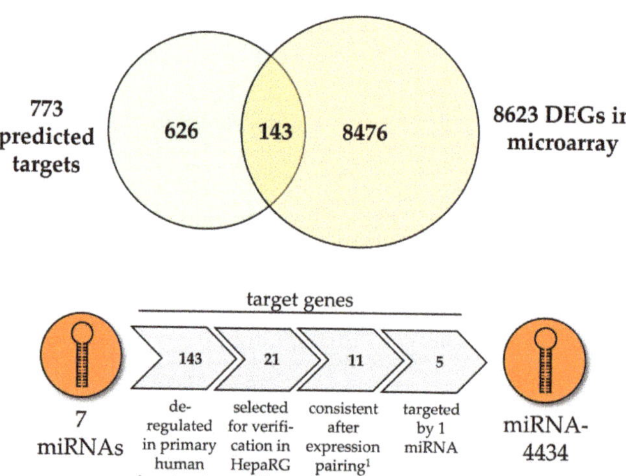

Figure 2. Workflow of miRNA target gene prediction with IPA. The target genes of 7 differentially and significantly expressed miRNAs in HepaRG cells were predicted with IPA (773 predicted targets) and compared to the dataset of a whole genome microarray conducted with primary human hepatocytes [44]. Among the 8623 Differentially Expressed Genes (DEGs) of the primary human hepatocytes after Sc-treatment (100 µM, 24 h), 143 targets were found to overlap with the predicted miRNA target genes. Out of these, 21 targets that simultaneously showed both a significant deregulation in the microarray as well as a high (predicted) confidence of miRNA-target interaction, were selected to be further investigated in HepaRG cells exposed to 35 µM of Sc for 5 different time points. A total of 11 out of 21 target genes showed an opposing expression pairing to their annotated miRNAs (indicated by superscript number 1), as verified by qPCR in HepaRG cells. IPA analysis revealed that 5 out of 11 targets were regulated by one miRNA (miRNA-4434).

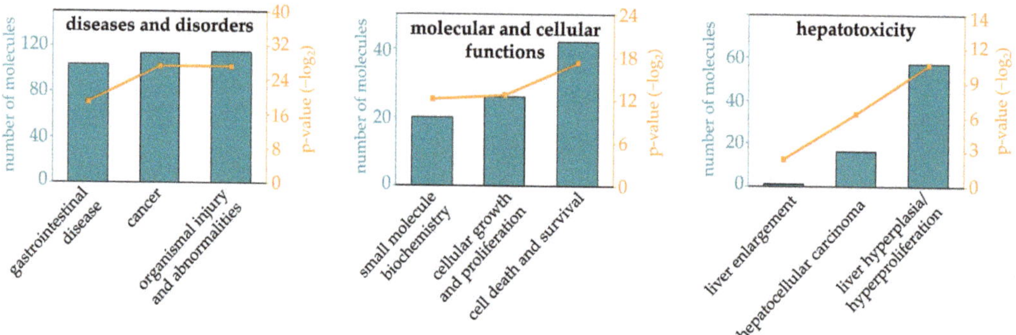

Figure 3. Top 3 pathways in 3 "diseases and bio functions" of deregulated target genes in primary human hepatocytes listed in IPA after expression analysis with 7 miRNAs. IPA target gene prediction of 7 significantly deregulated miRNAs in HepaRG cells treated with 35 µM of Sc for 24 h was compared to the gene expression data from Sc-treated primary human hepatocytes (100 µM, 24 h) [44]. The numbers of deregulated target genes (number of molecules) predicted to be involved in diseases and bio functions are depicted as bar charts and their lower range of the *p*-value is shown in −log₂ FC (orange graph).

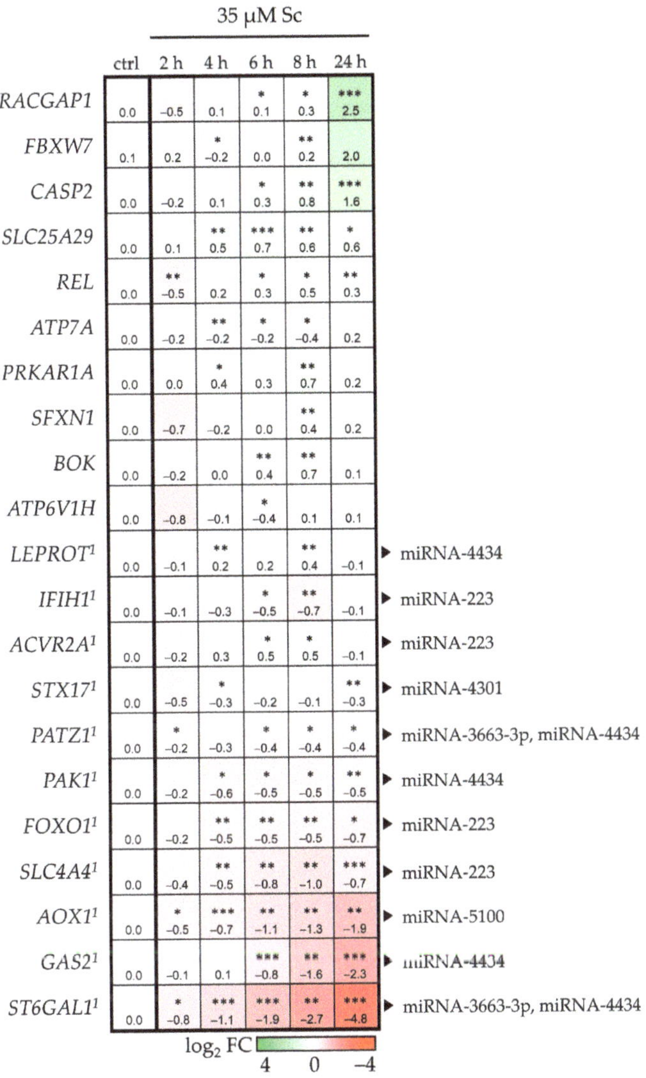

Figure 4. Target gene expression in differentiated HepaRG cells after exposure to 35 µM Sc at 5 different time points. The results were evaluated using the $2^{-\Delta\Delta Ct}$ method [40]. Results are shown as $\log_2$ FC and as mean of three biological replicates. Gene expression values were referred to the solvent control (ctrl) of the respective time point but, for clarity, only the ctrl values of the 2 h time point are depicted here. Up- or downregulation of gene expression is indicated in green or red, respectively. The higher the values, the stronger the coloration. Mean values and standard deviations are summarized in the Supplementary Materials Table S3. Superscript number 1 indicates the target genes that show an opposite regulation to their annotated miRNAs and the black arrows indicate the predicted regulatory miRNAs, respectively. Statistical analysis was performed using Student's *t*-test as this is a case of single-comparison analysis. That is, each target gene expression value of one time point was compared to the solvent control of this respective time point. Subsequently, *p*-values were subjected to FDR correction. Significant differences are depicted as follows: * $p < 0.05$, ** $p < 0.01$, *** $p < 0.001$.

3.3. MiR-4434 Negatively Regulates PAK1 Gene Expression in HepaRG Cells

In order to investigate whether it is indeed the PA-mediated induction of miR-4434 levels that leads to the downregulation of the pre-selected target genes, differentiated HepaRG cells were treated with 35 μM of the PA Sc and, subsequently, transiently transfected either with the synthetic miRNA antagonist antagomiR-4434 or with an antagomiR-NC (both at 50 and 75 nM concentrations) or with transfection reagent and water only, as a negative treatment control. The miRNA inhibitor does not degrade its target, but forms a stable complex with it, resulting in an inhibition of miRNA-4434 function. The antagomiR-NC is non-homologous to any mammalian gene and was used to examine if the results of the antagomiR-4434-mediated inhibition were specific. That is, results achieved with the antagomiR-NC should be similar to results from negative treatment control (transfection reagent and water only). It was ensured that the combined incubation with antagomiR and Sc was non-cytotoxic (see Supplementary Materials Figure S2). The gene expression of all 21 target genes was assessed after 48 h and the values of the Sc-treated cells were always referred to their respective solvent control (antagomiR-4434, antagomiR-NC, or negative treatment control) (see Supplementary Materials Figure S3). When comparing antagomiR-4434-transfected cells with the negative control (antagomiR-NC) within the Sc-treatment group, the antisense effect of the antagomiR and its subsequent effect on miR-4434 target gene expression becomes evident. Here, the inhibition of miR-4434 led to significantly attenuated downregulation of *PAK1* gene expression, with increased significance at the higher antagomiR concentration (Figure 5).

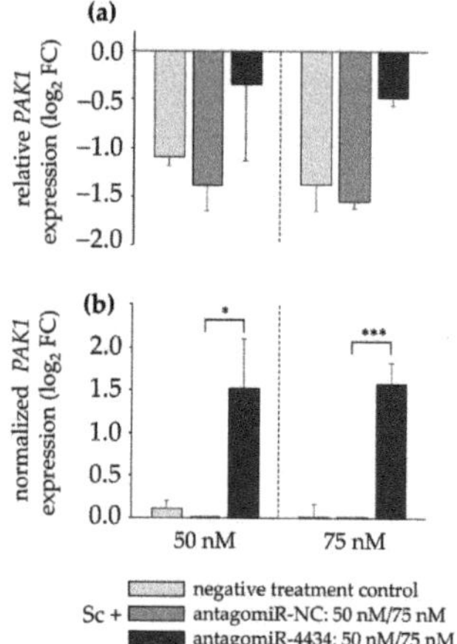

Figure 5. *PAK1* gene expression in differentiated HepaRG cells after exposure to 35 μM Sc and antagomiR-mediated inhibition of miR-4434 after 48 h. The results were evaluated according to the $2^{-\Delta\Delta Ct}$ method [40]. (a) After housekeeper normalization, the values of the Sc-treated cells were referred to the respective solvent control (0.35% ACN). Noteworthy, the different treatments (antagomiR-4434, antagomiR-NC and negative treatment control) were referred to their respective solvent controls. AntagomiR/antagomiR-NC treatment was applied in the two concentrations 50 and 75 nM. Results are shown as log₂ FC and as mean of three replicates. Mean values and standard deviations are summarized in the Supplementary Materials Table S4. (b) For normalized *PAK1* expression analysis within the Sc-treated cells, antagomiR-4434 treatment was referred to antagomiR-NC treatment. Again, antagomiR/antagomiR-NC treatment was applied in the two concentrations 50 and 75 nM. Additionally, the negative treatment control was referred to antagomiR-NC to ensure specific antagomiR-mediated inhibition. Statistical analysis was performed using Student's *t*-test as this is a case of single-comparison analysis. That is, the value of *PAK1* gene expression with antagomiR-4434 treatment (50 and 75 nM, respectively) was compared to the value of *PAK1* gene expression with antagomiR-NC treatment (50 and 75 nM, respectively). Significant differences are depicted as follows: * $p < 0.05$, *** $p < 0.001$.

4. Discussion

This study aimed to investigate the PA-induced effects on miRNA expression. Furthermore, miRNA-mediated effects on potential target genes were considered as an additional, however indirect, molecular regulator in PA-mediated hepatotoxicity. In the last years, miRNAs have emerged as very important players in the regulation of a variety of biological processes in many cell types, including those in the liver [32]. miRNA function ensures fine tuning of target gene expression, and subsequently protein abundance and protein distribution under constantly changing cellular conditions [47]. For example, in the liver intracellular miRNA levels regulate lipid and glucose metabolism [48], inflammation [49–51], apoptosis and necrosis [52–54], cell cycle and proliferation [55], as well as epithelial–mesenchymal transition during liver regeneration [56] and liver fibrosis [57,58]. Alterations in physiological miRNA levels correlate with various liver diseases such as viral hepatitis, alcoholic and nonalcoholic steatohepatitis, drug-induced liver injury, and autoimmune liver disease [32]. In metazoan cells, it is generally assumed that translational repression, presumably occurring during translation initiation, is the predominant mechanism by which miRNAs negatively regulate their target genes [29]. Therefore, it could be concluded that protein analysis is the investigative method of choice. However, transcript degradation of the target mRNA is an unfailing secondary effect triggered by the initial event of translational repression [59–61]. Therefore, qPCR analysis of target mRNA expression is a reliable and widely used method to assess the effects of miRNA-mediated gene silencing and was also applied here.

To elucidate the regulatory effects of miRNAs on their target genes is rather challenging because of its sheer complexity. For example, one miRNA can regulate many targets, and one target can be regulated by many different miRNAs in turn. Furthermore, miRNAs can directly bind their target gene, or they can indirectly regulate a target gene via binding to regulatory molecules like transcription factors, which have their own mode of action [62]. Thus, miRNAs can actually decrease, increase or not change the expression levels of their target genes [63]. This study aimed to achieve the first insights into the mechanistic of miRNA expression in HepaRG cells to understand its implication in PA-mediated hepatotoxicity. Therefore, for simplicity we only considered target genes that showed reverse correlation in gene expression in relation to their assigned miRNAs. Thus, target genes potentially showing a positive correlation or no correlation at all could have been overlooked. In general, miRNA upregulation upon PA-treatment has been described before. For example, *miRNA-34a*, which is considered to be a biomarker for exposure to genotoxic compounds, showed significant upregulation after chronic PA ingestion in a feeding study with rats [35]. Furthermore, blood samples from patients with HSOS showed elevated levels of miRNA-148a-3p, miRNA-362-5p, and miRNA-194-5p, which could be correlated to the severity of the PA-induced liver injury [37].

Upon Sc-incubation, seven liver-specific miRNAs could be verified to be significantly deregulated in HepaRG cells. All but one showed increased gene expression in a time-dependent manner. For *miRNA-4301, miRNA-5100* and *miRNA-4454*, upregulation was highest after 8 h, showing the early responsiveness of miRNA levels. miRNA-4301 and miRNA-5100 have been observed to regulate proliferation and apoptosis in lung and breast cancer cells [64–67], but according to our assessment there are no reports on their implication in hepatic diseases. For *miRNA-4454*, it was reported that upregulation positively enhances hepatic carcinoma progression [68], miR-223 is a common regulator in various liver diseases [69], and *miRNA-3663-3p* was shown to be downregulated in hepatocellular carcinoma cells, thus positively regulating cell proliferation of cancer cells [70]. The involvement of these three miRNAs in liver diseases might pose an interface between their PA-induced upregulation and possible carcinogenic properties that have been described for PAs [71–73]. *miRNA-4434* had the highest upregulation of all miRNAs investigated. This miRNA plays a role in different tumors, either as promoter or inhibitor of proliferation [74–76]. However, reports on its relevance in liver function and disease are rare. One study observed miRNA-4434 to be inhibited by a long non-coding RNA (lncRNA) called Long Stress Induced Non-Coding Transcripts 5 (LSINCT5) in hepato-cellular carcinoma progression, potentially resulting in inhibited miRNA-4434 induced apoptosis [77]. *miRNA-122-3p* (alias miRNA-122*, derived from the antisense strand of the precursor (pre)-miRNA; [78]) was the only miRNA that showed a significant downregulation only after 24 h. Noteworthy, *miRNA-122-3p* downregulation has been observed during early and advanced liver fibrosis [79]. As fibrosis is a disease which is also observed after chronic PA-intoxication, this could point towards a connection between downregulated *miRNA-122-3p* levels and PA-mediated toxicity resulting in fibrosis. The most abundant miRNA in the liver, *miRNA-122-5p* (alias miRNA-122a, derived from the sense strand of the pre-miRNA), did not show a statistically relevant deregulation in gene expression after Sc-treatment at any time-point and was therefore excluded from further analysis.

The comparison analysis between IPA-predicted mRNA targets and differentially expressed genes obtained from a whole genome microarray [44] revealed a target match of 143 genes that are involved in many diseases and disorders such as cancer in general and hepatocellular carcinoma in particular, with cellular growth and proliferation as underlying molecular and cellular functions. Of course, this analysis is a rather generalized evaluation. Therefore, out of these targets, a set of 21 genes regulating processes in cellular growth, development, apoptosis and inflammation was chosen to be verified in Sc-treated HepaRG cells and subjected to further IPA analysis. For cAMP-dependent PRotein Kinase type I-Alpha Regulatory subunit (*PRKAR1A*), a miRNA-dependent regulation was also observed in Mc-treated mice for type II-alpha subunit (*Prkar2a*). Moreover, in the same study, target gene V-type proton ATPase subunit e 2 (*Atp6v0e2*) also showed a significant upregulation in a miRNA-dependent way [36]. Here, another subunit type (*ATP6V1H*) was selected. Both *PRKAR1A* and *ATP6V1H* showed upregulation of gene expression, as was observed in the study mentioned above. A small subgroup of 11 genes showed reverse correlation in gene expression in comparison to their assigned miRNAs. These five miRNAs out of the initial set of seven were miRNA-223, miRNA-3663-3p, miRNA-4301, miRNA-4434, and miRNA-5100. According to IPA analysis, the 11 dysregulated genes generally promote apoptosis and necrosis and decrease cellular survival. Interestingly, miRNA-4434 was predicted to be an upstream regulator of 5 out of the 11 genes, which was the highest number of annotated targets for one miRNA. Surprisingly, it was not a well-established hepatic miRNA such as miRNA-223 or miRNA-122-5p, but the rather less-known miRNA-4434. The five target genes are involved in processes such as cell cycle progression (*GAS2* and *PAK1*), growth hormone signaling (*LEPROT*), apoptosis (*PATZ1*) and immunity (*ST6GAL1*), suggesting a correlation between PA-induced upregulation of miRNAs and subsequent downregulation of these five target genes, presumably resulting in disturbed cellular function. Ultimately, antagomiR-mediated inhibition of miRNA-4434 resulted in significantly altered gene expression pattern of the target gene *PAK1*, strongly indicating a biological connection. The *PAK1* gene encodes one family member of the serine/threonine-specific intracellular protein kinases that are involved in a number of cellular functions including cell cycle regulation, apoptosis, and cytoskeletal motility, usually through substrate phosphorylation [80]. PAK1 is a regulator in key signaling pathways which are relevant for cell cycle progression and proliferation. On the one hand, *PAK1* is overexpressed in many cancers and positively correlates with promotion of cell survival, invasion and metastasis, and drug resistance [81]. In the context of tumor therapy, for example, PAK1 triggers DNA repair caused by genotoxic therapeutic agents [82]. On the other hand, *PAK1* hinders cell cycle progression upon inhibition [83]. Additionally, miRNA interaction with *PAK1* expression has been described previously: in hepatocellular carcinoma development, miRNA-485-5p was observed to suppress *PAK1* levels, and lncRNA-mediated binding of miRNA-485-5p resulted in the upregulation of *PAK1* during hepatocellular carcinoma progression [84]. Here, a PA-induced upregulation of miRNA-4434 can be assumed to negatively regulate *PAK1* expression, presumably resulting in cell-cycle arrest. Notably, an implication of PAs in cell cycle regulation, which is closely linked to DNA damage, has been observed before [85]. This effect was observed to be higher for the more toxic PAs in comparison to the less toxic PAs. In conclusion, we suggest that the identification of the regulatory mechanism of miRNA-4434-initiated and *PAK1*-induced dysregulation of cell cycle signaling may help to understand the molecular mode of action of some hepatotoxic and carcinogenic effects of Sc.

Supplementary Materials: The following supporting information can be downloaded at: https://www.mdpi.com/article/10.3390/foods11040532/s1: Table S1: miRNAs investigated with miR-CURY LNA miRNA custom PCR array, including 84 liver-specific miRNAs, 7 housekeeping genes, 1 interplate calibrator, 3 RNA isolation efficiency spike-in controls and 1 reverse transcription efficiency spike-in control; Figure S1: miRNA expression profile in differentiated HepaRG cells after exposure to Lc and determined with miRCURY LNA miRNA Array; Table S2: miRNA expression in differentiated HepaRG cells after exposure to 35 μM Sc at 5 different time-points; Table S3: Target gene expression in differentiated HepaRG cells after exposure to 35 μM of Sc for 5 time points; Figure S2: Cytotoxicity of differentiated HepaRG cells after antagomiR/antagomiR-NC- and Sc-incubation; Figure S3: Target gene expression in differentiated HepaRG cells after exposure to 35 μM Sc and antagomiR-mediated inhibition of miR-4434 after 48 h; Table S4: Target gene expression in differentiated HepaRG cells after exposure to 35 μM Sc and antagomiR-mediated inhibition of miR-4434 after 48 h.

Author Contributions: A.-M.E. conceptualization, investigation, visualization, writing. H.S. conceptualization, evaluation, writing—review and editing. A.B. conceptualization, supervision, writing—review and editing. S.H.-P. conceptualization, supervision, funding acquisition, project administration, writing—review and editing. All authors have read and agreed to the published version of the manuscript.

Funding: This work was supported by the GERMAN FEDERAL INSTITUTE FOR RISK ASSESSMENT (grant number 1322-624).

Institutional Review Board Statement: Not applicable.

Informed Consent Statement: Not applicable.

Data Availability Statement: Data is contained within the article (or Supplementary Material).

Acknowledgments: We thank Claudia Luckert for graciously providing additional documents and Beatrice Rosskopp for her excellent technical work.

Conflicts of Interest: The authors declare no conflict of interest.

Abbreviations

ALARA	As Low As Reasonably Achievable
ATP6V1H/0E2	V-type proton ATPase subunit H/E 2
DEG	Differentially Expressed Gene
DMSO	DiMethyl SulfOxide
FBS	Fetal Bovine Serum
FC	Fold Change
FDR	False Discovery Rate
GAS2	Growth Arrest-Specific protein 2
GUSB	β-glucuronidase
HSOS	Hepatic Sinusoidal Obstruction Syndrome
IPA	Ingenuity Pathway Analysis
Lc	Lasiocarpine
lncRNA	long non-coding RNA
LSINCT5	Long Stress-Induced Non-Coding Transcripts 5
LEPROT	LEPtin Receptor Overlapping Transcript
Mc	Monocrotaline
miRNA	microRNA
NRU	Neutral Red Uptake
PA	1,2-unsaturated PyrrolizidineAalkaloid
PAK1	P21-Activated Kinase-1

PATZ1	POZ/BTB and AT hook containing Zinc finger 1
PRKAR1A/2A	cAMP-dependent PRotein Kinase type I (II)-Alpha Regulatory subunit
PXR	Pregnane X Receptor
qPCR	quantitative real-time Polymerase Chain Reaction
RT	Room Temperature
Sc	Senecionine
ST6GAL1	ST6 beta-GALactoside alpha-2,6-sialyltransferase 1
TDI	Tolerable Daily Intake
UTR	3′-UnTranslated Region

References

1. Mattocks, A. *Chemistry and Toxicology of Pyrrolizidine Alkaloids*; Academic Press: London, UK, 1986.
2. Mulder, P.P.J.; López, P.; Castellari, M.; Bodi, D.; Ronczka, S.; Preiss-Weigert, A.; These, A. Occurrence of pyrrolizidine alkaloids in animal- and plant-derived food: Results of a survey across Europe. *Food Addit. Contam. Part A Chem. Anal. Control Expo. Risk Assess.* **2018**, *35*, 118–133. [CrossRef] [PubMed]
3. Bodi, D.; Ronczka, S.; Gottschalk, C.; Behr, N.; Skibba, A.; Wagner, M.; Lahrssen-Wiederholt, M.; Preiss-Weigert, A.; These, A. Determination of pyrrolizidine alkaloids in tea, herbal drugs and honey. *Food Addit. Contam. Part A Chem. Anal. Control Expo. Risk Assess.* **2014**, *31*, 1886–1895. [CrossRef] [PubMed]
4. Edgar, J.A.; Lin, H.J.; Kumana, C.R.; Ng, M.M. Pyrrolizidine alkaloid composition of three Chinese medicinal herbs, *Eupatorium cannabinum*, *E. japonicum* and *Crotalaria assamica*. *Am. J. Chin. Med.* **1992**, *20*, 281–288. [CrossRef] [PubMed]
5. Kaltner, F.; Rychlik, M.; Gareis, M.; Gottschalk, C. Occurrence and Risk Assessment of Pyrrolizidine Alkaloids in Spices and Culinary Herbs from Various Geographical Origins. *Toxins* **2020**, *12*, 155. [CrossRef]
6. BfR. *Pyrrolizidine Alkaloids in Herbal Teas and Teas*; BfR Opinion No. 018/2013; German Federal Institute for Risk Assessment: Berlin, Germany, 2013; pp. 1–31.
7. Ruan, J.; Yang, M.; Fu, P.; Ye, Y.; Lin, G. Metabolic activation of pyrrolizidine alkaloids: Insights into the structural and enzymatic basis. *Chem. Res. Toxicol.* **2014**, *27*, 1030–1039. [CrossRef]
8. Bach, N.; Thung, S.N.; Schaffner, F. Comfrey herb tea-induced hepatic veno-occlusive disease. *Am. J. Med.* **1989**, *87*, 97–99. [CrossRef]
9. Ridker, P.M.; McDermott, W.V. Comfrey herb tea and hepatic veno-occlusive disease. *Lancet* **1989**, *1*, 657–658. [CrossRef]
10. Ridker, P.M.; Ohkuma, S.; McDermott, W.V.; Trey, C.; Huxtable, R.J. Hepatic venocclusive disease associated with the consumption of pyrrolizidine-containing dietary supplements. *Gastroenterology* **1985**, *88*, 1050–1054. [CrossRef]
11. Kumana, C.R.; Ng, M.; Lin, H.J.; Ko, W.; Wu, P.C.; Todd, D. Herbal tea induced hepatic veno-occlusive disease: Quantification of toxic alkaloid exposure in adults. *Gut* **1985**, *26*, 101–104. [CrossRef]
12. Roitman, J.N. Comfrey and liver damage. *Lancet* **1981**, *1*, 944. [CrossRef]
13. Edgar, J.A.; Molyneux, R.J.; Colegate, S.M. Pyrrolizidine Alkaloids: Potential Role in the Etiology of Cancers, Pulmonary Hypertension, Congenital Anomalies, and Liver Disease. *Chem. Res. Toxicol.* **2015**, *28*, 4–20. [CrossRef] [PubMed]
14. Edgar, J.A.; Roeder, E.; Molyneux, R.J. Honey from plants containing pyrrolizidine alkaloids: A potential threat to health. *J. Agric. Food Chem.* **2002**, *50*, 2719–2730. [CrossRef]
15. Svoboda, D.; Reddy, J.; Bunyaratvej, S. Hepatic megalocytosis in chronic lasiocarpine poisoning. Some functional studies. *Am. J. Pathol* **1971**, *65*, 399–409.
16. Jago, M.V. The development of the hepatic megalocytosis of chronic pyrrolizidine alkaloid poisoning. *Am. J. Pathol.* **1969**, *56*, 405–421.
17. Molteni, A.; Ward, W.F.; Ts'Ao, C.-H.; Port, C.D.; Solliday, N.H. Monocrotaline-Induced Pulmonary Endothelial Dysfunction in Rats. *Proc. Soc. Exp. Biol. Med.* **1984**, *176*, 88–94. [CrossRef]
18. Stegelmeier, B.L.; Edgar, J.A.; Colegate, S.M.; Gardner, D.R.; Schoch, T.K.; Coulombe, R.A.; Molyneux, R.J. Pyrrolizidine alkaloid plants, metabolism and toxicity. *J. Nat. Toxins* **1999**, *8*, 95–116. [PubMed]
19. Waizenegger, J.; Braeuning, A.; Templin, M.; Lampen, A.; Hessel-Pras, S. Structure-dependent induction of apoptosis by hepatotoxic pyrrolizidine alkaloids in the human hepatoma cell line HepaRG: Single versus repeated exposure. *Food Chem. Toxicol.* **2018**, *114*, 215–226. [CrossRef] [PubMed]
20. Ebmeyer, J.; Braeuning, A.; Glatt, H.; These, A.; Hessel-Pras, S.; Lampen, A. Human CYP3A4-mediated toxification of the pyrrolizidine alkaloid lasiocarpine. *Food Chem. Toxicol.* **2019**, *130*, 79–88. [CrossRef]
21. Glück, J.; Waizenegger, J.; Braeuning, A.; Hessel-Pras, S. Pyrrolizidine Alkaloids Induce Cell Death in Human HepaRG Cells in a Structure-Dependent Manner. *Int. J. Mol. Sci.* **2021**, *22*, 202. [CrossRef] [PubMed]
22. Allemang, A.; Mahony, C.; Lester, C.; Pfuhler, S. Relative potency of fifteen pyrrolizidine alkaloids to induce DNA damage as measured by micronucleus induction in HepaRG human liver cells. *Food Chem. Toxicol.* **2018**, *121*, 72–81. [CrossRef]
23. Glück, J.; Henricsson, M.; Braeuning, A.; Hessel-Pras, S. The Food Contaminants Pyrrolizidine Alkaloids Disturb Bile Acid Homeostasis Structure-Dependently in the Human Hepatoma Cell Line HepaRG. *Foods* **2021**, *10*, 1114. [CrossRef] [PubMed]

24. Waizenegger, J.; Glück, J.; Henricsson, M.; Luckert, C.; Braeuning, A.; Hessel-Pras, S. Pyrrolizidine Alkaloids Disturb Bile Acid Homeostasis in the Human Hepatoma Cell Line HepaRG. *Foods* **2021**, *10*, 161. [CrossRef] [PubMed]

25. Hessel-Pras, S.; Braeuning, A.; Guenther, G.; Adawy, A.; Enge, A.-M.; Ebmeyer, J.; Henderson, C.J.; Hengstler, J.G.; Lampen, A.; Reif, R. The pyrrolizidine alkaloid senecionine induces CYP-dependent destruction of sinusoidal endothelial cells and cholestasis in mice. *Arch. Toxicol.* **2020**, *94*, 219–229. [CrossRef]

26. Luckert, C.; Braeuning, A.; Lampen, A.; Hessel-Pras, S. PXR: Structure-specific activation by hepatotoxic pyrrolizidine alkaloids. *Chem. Biol. Interact.* **2018**, *288*, 38–48. [CrossRef] [PubMed]

27. He, L.; Hannon, G.J. MicroRNAs: Small RNAs with a big role in gene regulation. *Nat. Rev. Genet.* **2004**, *5*, 522–531. [CrossRef] [PubMed]

28. Takeuchi, M.; Oda, S.; Tsuneyama, K.; Yokoi, T. Comprehensive analysis of serum microRNAs in hepatic sinusoidal obstruction syndrome (SOS) in rats: Implication as early phase biomarkers for SOS. *Arch. Toxicol.* **2018**, *92*, 2947–2962. [CrossRef]

29. Bartel, D.P. MicroRNAs: Genomics, biogenesis, mechanism, and function. *Cell* **2004**, *116*, 281–297. [CrossRef]

30. Munoz-Garrido, P.; García-Fernández de Barrena, M.; Hijona, E.; Carracedo, M.; Marín, J.J.G.; Bujanda, L.; Banales, J.M. MicroRNAs in biliary diseases. *World J. Gastroenterol.* **2012**, *18*, 6189–6196. [CrossRef]

31. Garzon, R.; Marcucci, G. Potential of microRNAs for cancer diagnostics, prognostication and therapy. *Curr. Opin. Oncol.* **2012**, *24*, 655–659. [CrossRef] [PubMed]

32. Szabo, G.; Bala, S. MicroRNAs in liver disease. *Nat. Rev. Gastroenterol. Hepatol.* **2013**, *10*, 542–552. [CrossRef]

33. Marin, J.J.; Bujanda, L.; Banales, J.M. MicroRNAs and cholestatic liver diseases. *Curr. Opin. Gastroenterol.* **2014**, *30*, 303–309. [CrossRef]

34. Song, K.H.; Li, T.; Owsley, E.; Chiang, J.Y. A putative role of micro RNA in regulation of cholesterol 7alpha-hydroxylase expression in human hepatocytes. *J. Lipid Res.* **2010**, *51*, 2223–2233. [CrossRef] [PubMed]

35. Chen, T.; Li, Z.; Yan, J.; Yang, X.; Salminen, W. MicroRNA expression profiles distinguish the carcinogenic effects of riddelliine in rat liver. *Mutagenesis* **2012**, *27*, 59–66. [CrossRef] [PubMed]

36. Huang, Z.; Chen, M.; Zhang, J.; Sheng, Y.; Ji, L. Integrative analysis of hepatic microRNA and mRNA to identify potential biological pathways associated with monocrotaline-induced liver injury in mice. *Toxicol. Appl. Pharmacol.* **2017**, *333*, 35–42. [CrossRef]

37. Wang, X.; Zhang, W.; Yang, Y.; Chen, Y.; Zhuge, Y.; Xiong, A.; Yang, L.; Wang, Z. Blood microRNA Signatures Serve as Potential Diagnostic Biomarkers for Hepatic Sinusoidal Obstruction Syndrome Caused by *Gynura japonica* Containing Pyrrolizidine Alkaloids. *Front. Pharmacol.* **2021**, *12*, 627126. [CrossRef]

38. Oda, S.; Takeuchi, M.; Akai, S.; Shirai, Y.; Tsuneyama, K.; Yokoi, T. miRNA in Rat Liver Sinusoidal Endothelial Cells and Hepatocytes and Application to Circulating Biomarkers that Discern Pathogenesis of Liver Injuries. *Am. J. Pathol.* **2018**, *188*, 916–928. [CrossRef] [PubMed]

39. Gripon, P.; Rumin, S.; Urban, S.; Le Seyec, J.; Glaise, D.; Cannie, I.; Guyomard, C.; Lucas, J.; Trepo, C.; Guguen-Guillouzo, C. Infection of a human hepatoma cell line by hepatitis B virus. *Proc. Natl. Acad. Sci. USA* **2002**, *99*, 15655–15660. [CrossRef]

40. Livak, K.J.; Schmittgen, T.D. Analysis of relative gene expression data using real-time quantitative PCR and the $2^{-\Delta\Delta CT}$ Method. *Methods* **2001**, *25*, 402–408. [CrossRef]

41. Karagkouni, D.; Paraskevopoulou, M.D.; Chatzopoulos, S.; Vlachos, I.S.; Tastsoglou, S.; Kanellos, I.; Papadimitriou, D.; Kavakiotis, I.; Maniou, S.; Skoufos, G.; et al. DIANA-TarBase v8: A decade-long collection of experimentally supported miRNA–gene interactions. *Nucleic Acids Res.* **2018**, *46*, D239–D245. [CrossRef]

42. Xiao, F.; Zuo, Z.; Cai, G.; Kang, S.; Gao, X.; Li, T. miRecords: An integrated resource for microRNA–target interactions. *Nucleic Acids Res.* **2009**, *37*, D105–D110. [CrossRef]

43. Agarwal, V.; Bell, G.W.; Nam, J.-W.; Bartel, D.P. Predicting effective microRNA target sites in mammalian mRNAs. *eLife* **2015**, *4*, e05005. [CrossRef] [PubMed]

44. Luckert, C.; Hessel, S.; Lenze, D.; Lampen, A. Disturbance of gene expression in primary human hepatocytes by hepatotoxic pyrrolizidine alkaloids: A whole genome transcriptome analysis. *Toxicol. Vitr.* **2015**, *29*, 1669–1682. [CrossRef]

45. Enge, A.M.; Kaltner, F.; Gottschalk, C.; Kin, A.; Kirstgen, M.; Geyer, J.; These, A.; Hammer, H.; Pötz, O.; Braeuning, A.; et al. Organic Cation Transporter I and Na⁺/taurocholate Co-Transporting Polypeptide are Involved in Retrorsine- and Senecionine-Induced Hepatotoxicity in HepaRG cells. *Mol. Nutr. Food Res.* **2021**, e2100800. [CrossRef] [PubMed]

46. Benjamini, Y.; Hochberg, Y. Controlling the False Discovery Rate: A Practical and Powerful Approach to Multiple Testing. *J. R. Stat. Soc. Ser. B* **1995**, *57*, 289–300. [CrossRef]

47. Bartel, D.P. MicroRNAs: Target recognition and regulatory functions. *Cell* **2009**, *136*, 215–233. [CrossRef] [PubMed]

48. Rottiers, V.; Näär, A.M. MicroRNAs in metabolism and metabolic disorders. *Nat. Rev. Mol. Cell Biol.* **2012**, *13*, 239–250. [CrossRef]

49. O'Neill, L.A.; Sheedy, F.J.; McCoy, C.E. MicroRNAs: The fine-tuners of Toll-like receptor signalling. *Nat. Rev. Immunol.* **2011**, *11*, 163–175. [CrossRef]

50. Bala, S.; Szabo, G. MicroRNA Signature in Alcoholic Liver Disease. *Int. J. Hepatol.* **2012**, *2012*, 498232. [CrossRef]

51. Thai, T.H.; Calado, D.P.; Casola, S.; Ansel, K.M.; Xiao, C.; Xue, Y.; Murphy, A.; Frendewey, D.; Valenzuela, D.; Kutok, J.L.; et al. Regulation of the germinal center response by microRNA-155. *Science* **2007**, *316*, 604–608. [CrossRef]

52. An, F.; Gong, B.; Wang, H.; Yu, D.; Zhao, G.; Lin, L.; Tang, W.; Yu, H.; Bao, S.; Xie, Q. miR-15b and miR-16 regulate TNF mediated hepatocyte apoptosis via BCL2 in acute liver failure. *Apoptosis* **2012**, *17*, 702–716. [CrossRef]

53. Yu, D.S.; An, F.M.; Gong, B.D.; Xiang, X.G.; Lin, L.Y.; Wang, H.; Xie, Q. The regulatory role of microRNA-1187 in TNF-α-mediated hepatocyte apoptosis in acute liver failure. *Int. J. Mol. Med.* **2012**, *29*, 663–668. [CrossRef] [PubMed]

54. Sharma, A.D.; Narain, N.; Händel, E.M.; Iken, M.; Singhal, N.; Cathomen, T.; Manns, M.P.; Schöler, H.R.; Ott, M.; Cantz, T. MicroRNA-221 regulates FAS-induced fulminant liver failure. *Hepatology* **2011**, *53*, 1651–1661. [CrossRef] [PubMed]

55. Yuan, Q.; Loya, K.; Rani, B.; Möbus, S.; Balakrishnan, A.; Lamle, J.; Cathomen, T.; Vogel, A.; Manns, M.P.; Ott, M.; et al. MicroRNA-221 overexpression accelerates hepatocyte proliferation during liver regeneration. *Hepatology* **2013**, *57*, 299–310. [CrossRef] [PubMed]

56. Kalluri, R.; Weinberg, R.A. The basics of epithelial-mesenchymal transition. *J. Clin. Investig.* **2009**, *119*, 1420–1428. [CrossRef]

57. Noetel, A.; Kwiecinski, M.; Elfimova, N.; Huang, J.; Odenthal, M. microRNA are Central Players in Anti- and Profibrotic Gene Regulation during Liver Fibrosis. *Front. Physiol.* **2012**, *3*, 49. [CrossRef]

58. He, Y.; Huang, C.; Zhang, S.P.; Sun, X.; Long, X.R.; Li, J. The potential of microRNAs in liver fibrosis. *Cell Signal* **2012**, *24*, 2268–2272. [CrossRef]

59. Hu, W.; Coller, J. What comes first: Translational repression or mRNA degradation? The deepening mystery of microRNA function. *Cell Res.* **2012**, *22*, 1322–1324. [CrossRef]

60. Djuranovic, S.; Nahvi, A.; Green, R. miRNA-mediated gene silencing by translational repression followed by mRNA deadenylation and decay. *Science* **2012**, *336*, 237–240. [CrossRef]

61. Bazzini, A.A.; Lee, M.T.; Giraldez, A.J. Ribosome profiling shows that miR-430 reduces translation before causing mRNA decay in zebrafish. *Science* **2012**, *336*, 233–237. [CrossRef]

62. Krek, A.; Grün, D.; Poy, M.N.; Wolf, R.; Rosenberg, L.; Epstein, E.J.; MacMenamin, P.; da Piedade, I.; Gunsalus, K.C.; Stoffel, M.; et al. Combinatorial microRNA target predictions. *Nat. Genet.* **2005**, *37*, 495–500. [CrossRef]

63. Valinezhad Orang, A.; Safaralizadeh, R.; Kazemzadeh-Bavili, M. Mechanisms of miRNA-Mediated Gene Regulation from Common Downregulation to mRNA-Specific Upregulation. *Int. J. Genom.* **2014**, *2014*, 970607. [CrossRef]

64. Gholipour, N.; Ohradanova-Repic, A.; Ahangari, G. A novel report of MiR-4301 induces cell apoptosis by negatively regulating DRD2 expression in human breast cancer cells. *J. Cell Biochem.* **2018**, *119*, 6408–6417. [CrossRef] [PubMed]

65. Avval, A.J.; Majd, A.; Gholipour, N.; Noghabi, K.A.; Ohradanova-Repic, A.; Ahangari, G. An Inventive Report of Inducing Apoptosis in Non-Small Cell Lung Cancer (NSCLC) Cell Lines by Transfection of MiR-4301. *Anticancer Agents Med. Chem.* **2019**, *19*, 1609–1617. [CrossRef] [PubMed]

66. Zhang, H.M.; Li, H.; Wang, G.X.; Wang, J.; Xiang, Y.; Huang, Y.; Shen, C.; Dai, Z.T.; Li, J.P.; Zhang, T.C.; et al. MKL1/miR-5100/CAAP1 loop regulates autophagy and apoptosis in gastric cancer cells. *Neoplasia* **2020**, *22*, 220–230. [CrossRef]

67. Yang, L.; Lin, Z.; Wang, Y.; Gao, S.; Li, Q.; Li, C.; Xu, W.; Chen, J.; Liu, T.; Song, Z.; et al. MiR-5100 increases the cisplatin resistance of the lung cancer stem cells by inhibiting the Rab6. *Mol. Carcinog.* **2018**, *57*, 419–428. [CrossRef]

68. Lin, H.; Zhang, R.; Wu, W.; Lei, L. miR-4454 Promotes Hepatic Carcinoma Progression by Targeting Vps4A and Rab27A. *Oxid. Med. Cell Longev.* **2021**, *2021*, 9230435. [CrossRef]

69. Ye, D.; Zhang, T.; Lou, G.; Liu, Y. Role of miR-223 in the pathophysiology of liver diseases. *Exp. Mol. Med.* **2018**, *50*, 1–12. [CrossRef]

70. Tian, J.; Li, J.; Bie, B.; Sun, J.; Mu, Y.; Shi, M.; Zhang, S.; Kong, G.; Li, Z.; Guo, Y. MiR-3663-3p participates in the anti-hepatocellular carcinoma proliferation activity of baicalein by targeting SH3GL1 and negatively regulating EGFR/ERK/NF-κB signaling. *Toxicol. Appl. Pharmacol.* **2021**, *420*, 115522. [CrossRef] [PubMed]

71. National Toxicology Program. Bioassay of lasiocarpine for possible carcinogenicity. *Natl. Cancer Inst. Carcinog. Tech. Rep. Ser.* **1978**, *39*, 1–66.

72. National Toxicology Program. Toxicology and carcinogenesis studies of riddelliine (CAS No. 23246-96-0) in F344/N rats and B6C3F1 mice (gavage studies). *Natl. Toxicol. Program Tech. Rep. Ser.* **2003**, *508*, 1–280.

73. Chen, T.; Mei, N.; Fu, P.P. Genotoxicity of pyrrolizidine alkaloids. *J. Appl. Toxicol.* **2010**, *30*, 183–196. [CrossRef]

74. Jin, X.H.; Lu, S.; Wang, A.F. Expression and clinical significance of miR-4516 and miR-21-5p in serum of patients with colorectal cancer. *BMC Cancer* **2020**, *20*, 241. [CrossRef] [PubMed]

75. Cui, T.; Bell, E.H.; McElroy, J.; Becker, A.P.; Gulati, P.M.; Geurts, M.; Mladkova, N.; Gray, A.; Liu, K.; Yang, L.; et al. miR-4516 predicts poor prognosis and functions as a novel oncogene via targeting PTPN14 in human glioblastoma. *Oncogene* **2019**, *38*, 2923–2936. [CrossRef] [PubMed]

76. Chen, S.; Xu, M.; Zhao, J.; Shen, J.; Li, J.; Liu, Y.; Cao, G.; Ma, J.; He, W.; Chen, X.; et al. MicroRNA-4516 suppresses pancreatic cancer development via negatively regulating orthodenticle homeobox 1. *Int. J. Biol. Sci.* **2020**, *16*, 2159–2169. [CrossRef]

77. Li, O.; Li, Z.; Tang, Q.; Li, Y.; Yuan, S.; Shen, Y.; Zhang, Z.; Li, N.; Chu, K.; Lei, G. Long Stress Induced Non-Coding Transcripts 5 (LSINCT5) Promotes Hepatocellular Carcinoma Progression Through Interaction with High-Mobility Group AT-hook 2 and MiR-4516. *Med. Sci. Monit.* **2018**, *24*, 8510–8523. [CrossRef]

78. Ambros, V.; Bartel, B.; Bartel, D.P.; Burge, C.B.; Carrington, J.C.; Chen, X.; Dreyfuss, G.; Eddy, S.R.; Griffiths-Jones, S.; Marshall, M.; et al. A uniform system for microRNA annotation. *RNA* **2003**, *9*, 277–279. [CrossRef]

79. Van Keuren-Jensen, K.R.; Malenica, I.; Courtright, A.L.; Ghaffari, L.T.; Starr, A.P.; Metpally, R.P.; Beecroft, T.A.; Carlson, E.W.; Kiefer, J.A.; Pockros, P.J.; et al. microRNA changes in liver tissue associated with fibrosis progression in patients with hepatitis C. *Liver Int.* **2016**, *36*, 334–343. [CrossRef]

80. Rane, C.K.; Minden, A. P21 activated kinase signaling in cancer. *Semin. Cancer Biol.* **2019**, *54*, 40–49. [CrossRef]

81. Radu, M.; Semenova, G.; Kosoff, R.; Chernoff, J. PAK signalling during the development and progression of cancer. *Nat. Rev. Cancer* **2014**, *14*, 13–25. [CrossRef]
82. Advani, S.J.; Camargo, M.F.; Seguin, L.; Mielgo, A.; Anand, S.; Hicks, A.M.; Aguilera, J.; Franovic, A.; Weis, S.M.; Cheresh, D.A. Kinase-independent role for CRAF-driving tumour radioresistance via CHK2. *Nat. Commun.* **2015**, *6*, 8154. [CrossRef]
83. Arias-Romero, L.E.; Chernoff, J. A tale of two Paks. *Biol. Cell* **2008**, *100*, 97–108. [CrossRef]
84. Tu, J.; Zhao, Z.; Xu, M.; Chen, M.; Weng, Q.; Ji, J. LINC00460 promotes hepatocellular carcinoma development through sponging miR-485-5p to up-regulate PAK1. *Biomed. Pharmacother.* **2019**, *118*, 109213. [CrossRef]
85. Abdelfatah, S.; Naß, J.; Knorz, C.; Klauck, S.M.; Küpper, J.H.; Efferth, T. Pyrrolizidine alkaloids cause cell cycle and DNA damage repair defects as analyzed by transcriptomics in cytochrome P450 3A4-overexpressing HepG2 clone 9 cells. *Cell Biol. Toxicol.* **2021**, 1–21. [CrossRef]

Review

Pathways Affected by Falcarinol-Type Polyacetylenes and Implications for Their Anti-Inflammatory Function and Potential in Cancer Chemoprevention

Ruyuf Alfurayhi [1,2,*], Lei Huang [3] and Kirsten Brandt [1]

[1] Human Nutrition & Exercise Research Centre, Population Health Sciences Institute, Faculty of Medical Sciences, Newcastle University, Newcastle upon Tyne NE1 7RU, UK

[2] Department of Food Science and Human Nutrition, College of Agriculture and Veterinary Medicine, Qassim University, Qassim, Buraydah 52571, Saudi Arabia

[3] Immunity and Inflammation Research Theme, Translational and Clinical Research Institute, Faculty of Medical Sciences, Newcastle University, Newcastle upon Tyne NE1 7RU, UK

* Correspondence: r.s.a.alfurayhi2@newcastle.ac.uk; Tel.: +966-564944111

Abstract: Polyacetylene phytochemicals are emerging as potentially responsible for the chemoprotective effects of consuming apiaceous vegetables. There is some evidence suggesting that polyacetylenes (PAs) impact carcinogenesis by influencing a wide variety of signalling pathways, which are important in regulating inflammation, apoptosis, cell cycle regulation, etc. Studies have shown a correlation between human dietary intake of PA-rich vegetables with a reduced risk of inflammation and cancer. PA supplementation can influence cell growth, gene expression and immunological responses, and has been shown to reduce the tumour number in rat and mouse models. Cancer chemoprevention by dietary PAs involves several mechanisms, including effects on inflammatory cytokines, the NF-κB pathway, antioxidant response elements, unfolded protein response (UPR) pathway, growth factor signalling, cell cycle progression and apoptosis. This review summarises the published research on falcarinol-type PA compounds and their mechanisms of action regarding cancer chemoprevention and also identifies some gaps in our current understanding of the health benefits of these PAs.

Keywords: polyacetylenes; phytochemicals; anti-inflammatory; anticancer

Citation: Alfurayhi, R.; Huang, L.; Brandt, K. Pathways Affected by Falcarinol-Type Polyacetylenes and Implications for Their Anti-Inflammatory Function and Potential in Cancer Chemoprevention. *Foods* **2023**, *12*, 1192. https://doi.org/10.3390/foods12061192

Academic Editor: Carla Gentile

Received: 25 January 2023
Revised: 5 March 2023
Accepted: 8 March 2023
Published: 11 March 2023

1. Introduction

Studies have indicated the beneficial impact of eating vegetables and fruits on human health in preventing chronic diseases including cancer, which is one of the major causes of death around the world [1]. Polyacetylenes (PAs) are a class of chemicals defined by the presence of two or more carbon–carbon triple bonds in the carbonic skeleton [2]. Falcarinol-type PAs are biologically active compounds that are widely found in plants in the Apiaceae family, such as carrots, celery and parsley, and the Araliaceae family, such as ginseng. Carrot is the main dietary source of polyacetylenic oxylipins, including falcarinol (FaOH), falcarindiol (FaDOH) and falcarindiol 3-acetate (FaDOH3Ac) (Figure 1), with FaOH serving as the intermediate metabolite of PA, from which the other forms are generated [3–5]. Carrots have been studied for their nutritional value, in addition to their disease-curative effects, for almost 90 years [6]. Carrot is a rich source of the vitamin A precursor β-carotene and also provides some potentially beneficial dietary fibre. Carrot also contains other potentially bioactive phytochemicals including carotenoids, phenolics, PAs, isocoumarins, terpenes and sesquiterpenes, many of which have been extensively investigated for potential therapeutic properties against a wide range of diseases including cancer, cardiovascular disease, diabetes, anaemia, colitis, ocular diseases and obesity [7]. Ginseng is also rich in PAs; in addition to FaOH (also called panaxynol), they include panaxydiol and panaxydol (Figure 1) [8], which have similar properties to FaOH [9–11]. Despite the extensive research on the analytical and biochemical identification and characterization of plant PAs, as well as the large number of papers on their putative biological functions, little is known about the structures and functions of the enzymes involved in PA biosynthesis [12]. Furthermore, the molecular genetic principles underlying PA production in various plant tissues are poorly understood, and little is known about the genetics and inheritance of specific PA patterns and concentrations in (crop) plants [3].

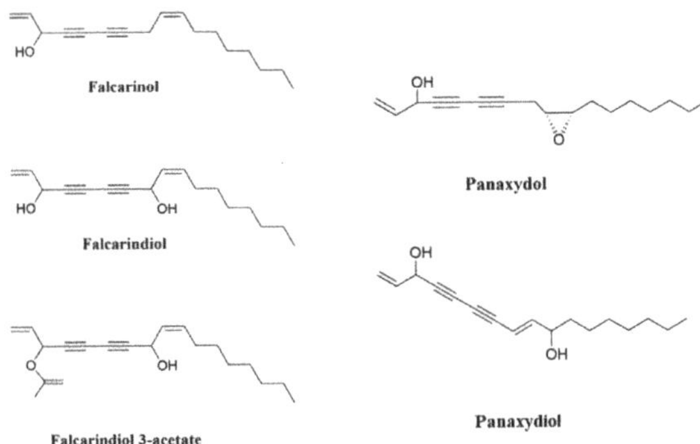

Figure 1. Chemical structures of FaOH, (3R)-falcarinol (also known as panaxynol); FaDOH, (3R,8S)-falcarindiol; FaDOH3Ac, (3R,8S)-falcarindiol-3-acetate; panaxydol; and panaxydiol.

β-carotene was initially believed to be protective against multiple chronic diseases, particularly cardiovascular disease and cancer, due to observations of associations of reduced risk of these diseases with a high dietary intake (with carrots as the primary dietary source of β-carotene) [13–16]. However, meta-analyses of subsequent intervention studies ruled out a role of β-carotene in suppressing non-communicable diseases affecting lifespan [17]. for example, dietary supplementation with purified β-carotene showed a dose-dependent increased risk of lung cancer for intakes higher than what can be obtained from food [18]. Thus, it was suggested that other bioactive substances such as PAs (FaOH and FaDOH) could be responsible for the health benefits of carrot [19]. Phytochemicals have been used as the main sources of the primary structures for conventional drugs used for curing cancer. Natural products historically have been essential in the development of new treatments for cancer and infectious diseases [20–23]. This review focuses on the possible mechanisms of action of PAs in inflammation and cancer, with emphasis placed on the various pathways involved including growth factor signalling, inflammatory processes, oxidative stress, cell cycle progression and apoptosis. We will also discuss the potential for PAs to modulate one or more of these pathways to contribute toward the treatment or prevention of inflammation and cancers.

2. Polyacetylenes and Inflammation
2.1. Chronic Inflammation Disease and Cancer

Chronic inflammation is recognized as a leading promoting factor of diseases including carcinogenesis [24], which continues to be the leading cause of mortality and disability around the world [25–28]. Tumour-promoting inflammation is recognised as an enabling hallmark of cancer [29]. Cancer and inflammation are linked by intrinsic and extrinsic pathways. Intrinsically, oncogenes regulate the inflammatory microenvironment, whereas extrinsically, the inflammatory microenvironment promotes the growth and spread of cancer [30]. Various cell types involved in chronic inflammation can be found in tumours, both in the surrounding stroma and within the tumour itself. Neoplasms, including some of epithelial origin, contain a significant inflammatory cell component [31]. Multiple studies on human clinical samples have revealed that inflammation influences epithelial cell turnover [32,33]. Significantly, human susceptibility to breast, liver, large bowel, bladder, prostate, gastric mucosa, ovary and skin carcinoma is increased when proliferation occurs in the context of chronic inflammation [32–37].

Chronic inflammation is linked to approximately 25% of all human cancers and increases cancer risk [38] by stimulating angiogenesis and cell proliferation, inducing gene mutations and/or inhibiting apoptosis [38]. Chronic inflammation can develop from acute inflammation if the irritant persists, although in most cases the response is chronic from the start. Chronic inflammation is characterized by the infiltration of injured tissue by mononuclear cells such as macrophages, lymphocytes and plasma cells, as well as tissue destruction and attempts at repair [31]. *Helicobacter pylori* infections in gastric cancer, human papillomavirus infections in cervical cancer, hepatitis B or C infections in hepatocellular carcinoma and inflammatory bowel disease in colorectal cancer (CRC) are common causes of chronic inflammation associated with cancer development [39,40]. Inflammation also causes epigenetic changes that are linked to cancer development.

Natural PAs from diverse food and medicinal plants and their derivatives exert multiple bioactivities, including anti-inflammatory properties [41]. PAs can impact inflammation through known and unknown pathways. Evidence supports that PA compounds improve human health by stimulating anti-cancer and anti-inflammatory mechanisms [3]. These PAs contain triple bonds that functionality convert them into highly alkylating compounds that are reactive to proteins and other biomolecules. This unique molecular structure might be the key to understanding the beneficial effects of PAs such as their anti-inflammatory and cytoprotective function [41]. Recent research has suggested that the anti-cancer role of certain foods might be attributed to their anti-inflammatory function. Root vegetables, and particularly carrots, are promising natural sources in this respect thanks to their rich content of PAs [3,41–43]. The anti-inflammatory properties of purple carrots have been suggested to be due to the high levels of anthocyanin pigments [44]; however, another study showed that PAs, not anthocyanins, are responsible for the anti-inflammatory bioactivity of purple carrots [45]. In vitro and in vivo studies have demonstrated that the health-benefitting effects of carrots and other root vegetables might be attributed to PAs, such as FaOH and FaDOH [46]. Other dietary compounds, including several different phytochemicals, have been examined in the context of cancer chemoprevention; however, until now the measured effects [47–50] have been quite small and inconsistent compared with those found for PAs.

2.2. Inhibition of Nuclear Factor Kappa B (NF-κB) Pathways

NF-κB is a transcription factor that regulates the expression of many genes involved in the regulation of inflammation and autoimmune diseases [51,52]. Moreover, NF-κB plays a significant role in inflammation-induced cancers, as NF-κB is one of the major inflammatory pathways that are triggered by, for example, infections causing chronic inflammation [39,40,53]. Cellular immunity, inflammation and stress are all regulated by NF-κB signalling, as are cell differentiation, proliferation and apoptosis (Figure 2) [54,55]. Both solid and hematologic malignancies frequently modify the NF-κB pathway in ways that promote tumour cell proliferation and survival [56–58].

NF-κB, a key factor in the inflammatory process, provides a mechanistic link between inflammation and cancer, and the components of this pathway are targets for chemoprevention, particularly in CRC [59]. There are two major signalling pathways for NF-κB activation, namely the canonical and the non-canonical NF-κB signalling pathways. The canonical pathway activates NF-κB1 p50, RELA and c-REL, which are also called canonical NF-κB family members. The non-canonical NF-κB pathway, on the other hand, selectively activates p100-sequestered NF-κB members, mostly NF-κB2 p52 and RELB, which are also called non-canonical NF-κB family members [60]. LPS and proinflammatory cytokines, among other pathogenic substances, activate NF-κB through degrading inhibitors of κB (IκBs) [61] to release the common subunit P65 (RELA). In order to trigger the transcription of these genes, activated NF-κB travels into the nucleus and attaches to its associated DNA motifs. When activated, the NF-κB p65 subunit binds to the promoter regions of genes involved in inflammation, leading to the production of *IL-6*, *IL-1β* and *TNF-α* [62].

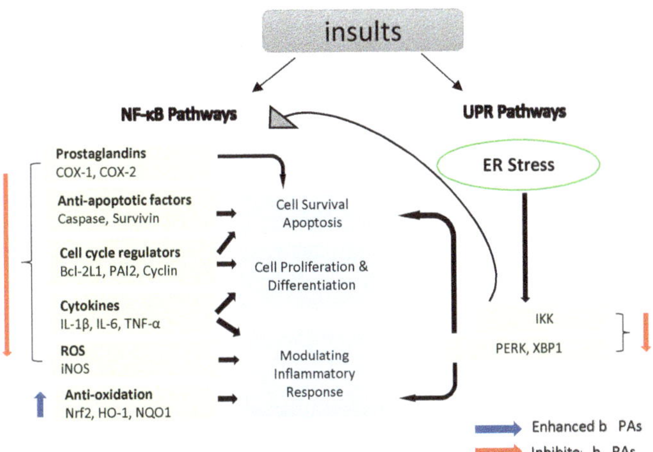

Figure 2. NF-κB target genes implicated in the onset and progression of inflammation. NF-κB is a transcription factor that is inducible. After activation, it can regulate inflammation by activating the transcription of several genes. NF-κB regulates cell proliferation, apoptosis and differentiation in addition to promoting the production of pro-inflammatory cytokines and chemokines. Endoplasmic reticulum (ER) stress results in an inflammatory unfolded protein response (UPR). Stress on the ER induces apoptosis by activating inflammation. This can be accomplished by stimulating IKK complex (element of the NF-κB) or (X-box binding protein 1) XBP1 and (protein kinase R-like ER kinase) PERK through a mediator. These trigger the release of pro-inflammatory molecules, hence accelerating cell death.

Carrot PAs, particularly FaOH and FaDOH, were studied for their anti-inflammatory properties [3,63], in part by inhibiting the transcription factor NF-κB [64]; however, their exact mechanism of action is still unknown. Mice fed a diet containing FaOH were less likely to develop severe inflammation after being exposed to LPS [5]. FaOH from *Saposhnikovia divaricata (S. divaricata)* significantly reduced the levels of LPS-induced *TNF-α* and *IL-6* in cultured BV-2 microglia cells and murine serum [61]. FaOH and FaDOH purified from carrots were demonstrated for their preventative effects on colorectal precancerous lesions in azoxymethane (AOM)-induced rats. Biopsies of neoplastic tissue were analysed for gene expression, and the results showed that FaOH and FaDOH inhibited NF-κB and the downstream inflammatory markers *TNF-α, IL-6 and COX-2* [46]. FaOH from Radix Saposhnikoviae (dried roots of *S. divaricata*, Apiaceae) inhibited LPS-induced NF-κB p65 activation and IκB-α phosphorylation in BV-2 microglia cells [61]. Treatment using FaOH from the roots of *Heracleum moellendorffii (H. moellendorffii)* inhibited LPS-induced NF-κB signalling activation by inhibiting IκB-α degradation and nuclear accumulation of p65 [65] in RAW264.7 cells. In addition, FaDOH reduced the level of LPS/IFNγ-induced NF-κB, IKK-α and IKK-β activation in rat primary astrocytes [64].

Prostaglandin (PG) synthesis is a hallmark of inflammation. Two enzymes, cyclooxygenase (COX) 1 and 2, catalyse the first step of PG synthesis, but *COX-2* is the major one that responds to inflammatory signals to produce PG at inflammatory sites [66]. However, *COX-2* can be suppressed by inhibiting the NF-κB translocation pathway (Figure 2) [67]. *COX-2* expressions in healthy tissues are low, but they can quickly increase in response to growth factors, cytokines and signals promoting tumour invasion, metastasis, aberrant proliferation and angiogenesis [68]. Malignancies, including colorectal [69], bladder [70], breast [71], lung [72], pancreatic [73], prostate [73] and head and neck cancer [74], tend to be associated with elevated levels of *COX-2*. Mechanistically, *COX-2* promotes carcinogenesis through the creation of prostaglandins (PGs), which suppress apoptosis and stimulate the development of blood vessels in tumour tissue, which helps in sustaining tumour cell viability and growth [39,75], suggesting that anti-inflammatory drugs targeting *COX-2* might be beneficial in the treatment of different types of cancer.

PAs modulate inflammation via suppressing *COX-2* expression, which depends on NF-κB activation by inflammation [76]. FaOH inhibited LPS-induced *COX-2* expression in RAW264.7 cells, thus blocking PGE2 overproduction [65]. FaOH isolated from American ginseng (*P. quinquefolius*) effectively reduced the severity of colitis in mice treated by dextran sulphate sodium (DSS) induced for a week before FaOH treatment. FaOH reduced the number of CD11b+ macrophages in the lamina propria and the inflammation hallmark protein *COX-2*. These data suggest that macrophages expressing *COX-2* might be an essential factor for colitis development. Interestingly, FaOH treatment prior to DSS did not prevent colitis or reduce colitis severity in mice [8]. Quiescent macrophages in the lamina propria in a healthy mouse might offer protection against colitis induction, as depleting macrophages prior to induction of colitis may exacerbate DSS-induced colitis [77]. However, when colitis develops, there is an increase in the number of activated macrophages that secrete pro-inflammatory cytokines to boost the inflammatory response, thus exacerbating colitis. At this stage, an overactive macrophage response to enteric microbiota greatly contributes to the pathogenesis of colitis [78]. Treatment with FaOH to target macrophages was shown to be highly effective in suppressing colitis at this stage, highlighting the utility of FaOH in the treatment of a hyper-inflammatory disease (Figure 3) [79]. In an azoxymethane (AOM)-induced rat colorectal cancer model, FaOH and FaDOH downregulated *COX-2* in precancerous lesions of CRC [46] and also reduced the number of malignant tumour foci.

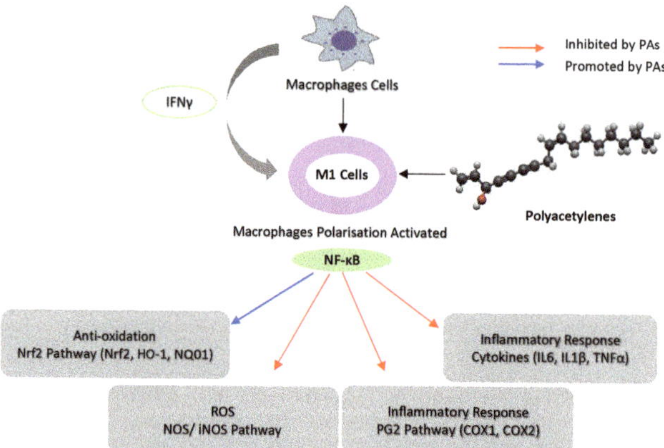

Figure 3. Schematic representation of the possible mechanism of immunoregulation activity of poly-acetylenes (PAs) in macrophages. (Interferon-γ) IFN-γ activates macrophage cells to M1, and PAs downregulate NF-κB activities in M1 macrophages by inhibiting *iNOS*, *COX-1* and *COX-2*. PAs suppress the inflammatory response by inhibiting cytokines (*IL-16*, *IL-1β* and *TNF-α*) and upregulating Nrf2 pathway (*HO-1* and *NQO1*) in macrophages. PAs structure modified from Mplanine (2022) https://www.wikiwand.com/en/Falcarinol#Media/File:Falcarinol_3D_BS.png (accessed on 1 January 2023).

2.3. Oxidative Stress

2.3.1. Inhibition of Nitric Oxide Synthase (NOS) and Pro-Inflammatory Cytokine Pathways

Nitric oxide (NO) is essential in a number of physiological functions, such as host defence, where it prevents the spread of disease-causing microbes within cells by stifling their reproduction [80]. The upregulation of NO expression in response to cytokines or pathogen-derived chemicals is a crucial part of the host's defence against different types of intracellular pathogens. Different cell types produce the enzyme NOS, which catalyses the synthesis of NO, at high levels in a number of different tumour types [81]. Inflammation induces a specific form of NOS, i.e., the inducible isoform of nitric oxide synthase (iNOS), via activating *iNOS* gene transcription (Figure 2) [82]. iNOS is involved in complex immunomodulatory and antitumor mechanisms, which have a role in eliminating bacteria, viruses and parasites [83].

A considerable number of studies have been published on the role of PAs in *iNOS* expression in inflammation. Studies have demonstrated that FaOH extracted from *P. quinquefolius* inhibited *iNOS* expression in ANA-1 mΦ macrophage cells that were polarized to M1 [8] and LPS-induced iNOS expression in macrophages [84,85], leading to colitis suppression [8]. Moreover, FaDOH was tested on rat primary astrocytes for its impact on LPS/IFN-γ-induced *iNOS* expression. FaDOH blocked 80% of LPS/IFN-γ-induced *iNOS* by reducing *iNOS* protein and mRNA in a dose-dependent manner. FaDOH was shown to suppress *iNOS* expression, and it inhibited *IKK, JAK, NF-κB* and *Stat1* (Figures 2 and 3) [64].

Another study showed a dose-dependent reduction in nitric oxide production in macrophage cells, where treatment with an extract of purple carrots containing PAs significantly reduced iNOS activity and *iNOS* expression in macrophage cells [45]. PAs reduced nitric oxide production in macrophage cells without cytotoxicity [45]. In vivo, purple carrots also inhibited inflammation in colitic mice and reduced colonic mRNA expression of *iNOS* [44]. FaOH from *H. moellendorffii* roots inhibits the LPS-induced overexpression of *iNOS* in RAW264.7 cells [65]. FaOH and other PAs from *P. quinquefolius* such as panaxydiol have a suppressive effect on *iNOS* expression in macrophages treated with LPS [85].

2.3.2. Reactive Oxygen Species (ROS) Pathways

Oxidative stress is described as an imbalance between the generation of free radicals and reactive metabolites, also known as oxidants or reactive oxygen species (ROS), and their removal by protective mechanisms, also known as antioxidants. Electron transfer is involved in oxidative and antioxidative processes, which influence the redox state of cells and the organism. The altered redox state stimulates or inhibits the activities of various signal proteins, which have an effect on cell fate [86,87]. PAs promote apoptosis preferentially in cancer cells mediated ROS stress. A study has analysed ROS production in MCF-7 cells after treating with panaxydol. The increase in the ROS levels started at 10–20 min after the panaxydol treatment. The role of NADPH oxidase was investigated in order to determine the source of ROS after panaxydol treatment. The creation of reactive oxygen species (ROS) by NADPH oxidase appeared to take precedence, while ROS production in the mitochondria was secondary but also necessary, suggesting that NADPH oxidase generates ROS in the presence of panaxydol. Panaxydol was tested on different cell lines to investigate whether the induction of apoptosis occurred preferentially in cancer cells. In this study, panaxydol induced apoptosis only in cancer cells [88].

FaOH and FaDOH from carrot were tested for their effects on the oxidative stress responses of primary myotube cultures. The effects of 100 μM of H_2O_2 on the intracellular formation of ROS, the transcription of the antioxidative enzyme, cytosolic glutathione peroxidase (cGPx), and the heat shock proteins (HSP) HSP70 and heme oxygenase 1 (*HO-1*) were studied after 24 h treatment with FaOH and FaDOH at a wide range of concentrations. At intermediate concentrations, under which only moderate cytotoxicity was shown, intracellular ROS formation was slightly enhanced by PAs. In addition, PAs increased the transcription of cGPx and decreased the transcription of HSP70 and *HO-1*. The enhanced cGPx transcription may have decreased the need for the protective properties of HSPs as an adaptive response to the elevated ROS production. However, ROS production was significantly reduced with higher doses of PAs (causing substantial cytoxicity), and the transcription of HSP70 and *HO-1* decreased to a lesser extent, while the induction of cGPx was marginally reduced, showing a necessity for the protective and repairing functions of HSPs [89].

Transcription factor *Nrf2* (also known as nuclear factor erythroid 2-like 2) regulates the expression of various antioxidant, anti-inflammatory and cytoprotective factors, including heme oxygenase-1 (*Hmox1, HO-1*) and NADPH:quinone oxidoreductase-1 (*NQO1*). S-alkylation of the protein Keap1, which normally inhibits *Nrf2*, is induced by FaDOH extracted from *Notopterygium incisum* (*N. incisum*), as reported in [90]. Moreover, nuclear accumulation of *Nrf2* and expression of *HO-1* were both enhanced in LPS-stimulated RAW264.7 cells by FaOH from *H. moellendorffii* roots [65]. FaOH also inhibited the inflammatory factor level and reduced nitric oxide production in BV-2 microglia. FaOH also reduced the levels of LPS-induced oxidative stress in BV-2 microglia [61]. In addition, FaOH inhibited inflammation in macrophages by activating *Nrf2* [85]. HO-1 is linked to redox-regulated gene expression. Chemical and physical stimuli that produce ROS either directly or indirectly cause the expression of *HO-1* to respond [91]. A one-week study looked at the effects of FaOH (5 mg/kg twice per day in CB57BL/6 mice) pre-treatment on acute intestinal and systemic inflammation. FaOH effectively increased *HO-1* mRNA and protein levels in both the mouse liver and intestine and reduced the levels of the plasma chemokine eotaxin and the myeloid inflammatory cell growth factor GM-CSF, both of which are involved in the recruitment and maintenance of first-responder immune cells [92].

3. Unfolded Protein Response (UPR) Pathways

The endoplasmic reticulum (ER) stress response, also known as unfolded protein response (UPR), is a cellular process that is activated by a number of conditions that disrupt protein folding in the ER. The UPR is an evolutionarily conserved adaptive mechanism in eukaryotic cells that aims to clear unfolded proteins and restore ER protein homeostasis. When ER stress is irreversible, cellular functions deteriorate, often leading to cell death (Figure 2) [93]. There is mounting evidence that ER stress plays a significant role in the development and progression of varied diseases, including cancer and inflammation [93,94]. FaDOH-induced cell death is mediated via ER stress induction and the activation of the UPR.

Reducing the extent of ER stress by overexpressing the ER chaperone protein glucose-regulated protein 78 (GRP78) or by knocking down components of the UPR pathway decreased FaDOH-induced apoptosis. In contrast, raising the level of ER stress by inhibiting GRP78 enhanced the apoptosis triggered by FaDOH extracted from *Oplopanax horridus* (*O. horridus*) [95]. In addition, ER stress mediated panaxydol-induced apoptosis in MCF-7 cells [96]. Another study investigated the effect of a sub-toxic dose of 5 µM of FaDOH in a series of experiments and found that it increased the lipid content and number of lipid droplets (LDs) in human mesenchymal stem cells (hMSCs) and enhanced *PPARγ2* expression in human colon adenocarcinoma cells. The activation of *PPARγ* can enhance *ABCA1* expression [97]. FaDOH treatment showed an upregulation of *ABCA1* in colon neoplastic rat tissue, suggesting a function for this transporter in the redistribution of lipids and the enhanced creation of LDs in cancer cells, which may result in endoplasmic reticulum (ER) stress and cancer cell death [97].

4. Cancer

4.1. In Vitro

4.1.1. Anti-Proliferative Activity

PAs derived from different plants exhibit potent cytotoxicity against a variety of cancer cells. These biologically active molecules engage directly or indirectly in biological processes, including cellular cycle arrest, HIF-1 (hypoxia-inducible factor-1 alpha) activation and signal transducer and transcriptional factor 3 (*STAT3*) suppression.

The anti-proliferative effects of FaOH isolated from carrots was initially shown in 2003. In addition, FaOH-type PAs show toxicity against human pancreatic carcinoma cells, but not against normal pancreatic cells, in vitro by modulating the expression of the genes involved in apoptosis, cell cycle, stress response and death receptors [42,98].

Treatment of leukaemia cell lines with carrot extract or isolated FaOH or FaDOH inhibited cell cycle progression, suggesting that carrots cause cell cycle arrest (G0/G1) in leukaemia cell lines [99]. Moreover, the cytotoxicity of FaOH, FaDOH and panaxydiol isolated from the dichloromethane extract of root celery was tested for its potential impact in a number of human cancer and leukaemia cell lines. All PAs examined exhibited moderate cytotoxicity against leukaemia, lymphoma and myeloma cell lines, although FaOH had significantly more activity against the lymphocytic leukaemia cells than FaDOH and panaxydiol [100]. In other studies, FaDOH also had less cytotoxic activity than FaOH and FaDOH3Ac [99,101].

4.1.2. Pro-Apoptosis Activity

Cancer prevention and treatment depend on the use of a variety of bioactive compounds that inhibit the early stages of cellular transformation required for the development of the neoplastic phenotype, such as initiating autophagy, apoptosis or other forms of cell death such as oncosis or necrosis [102]. Apoptosis dysfunction is a major contributor to cancer development and progression. Tumour cells' ability to avoid apoptosis can play a significant role in their resistance to traditional therapies.

One study investigated the effects of FaOH on human pancreatic ductal adenocarcinoma cell lines compared with normal pancreatic cells. FaOH regulated the genes related to pro-apoptosis, anti-apoptosis, apoptosis, cell cycle, stress and death receptors in adenocarcinoma cells more preferentially than in normal pancreatic cells [98]. FaOH suppressed pro-apoptosis genes (*BAD* and *HTRA-2*), anti-apoptosis genes (*Livin* and *XIAP*), a cell cycle controller (Phospho-p53 at amino acids serine 15, 46 and 392 (S15, S46 and S392), stress-related genes (*Clusterin* and *Hsp60*) and death receptor genes (*TNFR1* and *TNFSF1A*). In addition, FaOH increased cell cycle checkpoint phosphorylation (Phospho-Rad17 (S635)) and induced stress-related genes (*HO-1, HMXO1, HP32 and Hsp27*). Furthermore, FaOH-type and other PAs are potent inhibitors of pancreatic cancer cell proliferation [98].

Tumour recurrence and drug resistance are both facilitated by cancer stem-like cells (CSCs) [103]. *Hsp90* is known to enhance cancer cell survival and their ability to acquire anti-cancer drug resistance; its overexpression has been linked to a poor prognosis in human malignancies [104,105]. An in vivo study showed that orally administered FaOH significantly suppressed the proliferation of lung cancer in a mouse model without overt symptoms of toxicity at concentrations of 50 mg/kg body weight [103], which would correspond to a human dose of 4 mg/kg [106]. FaOH selectively inhibited carcinogenesis cells but not normal cells both in vitro and in vivo by inhibiting the function and viability of cancer stem-like cells of non-small-cell lung cancer by triggering apoptosis without enhancing *Hsp70* expression. Moreover, FaOH, at a low dose of 1 μM, induced apoptosis in cancer stem-like cells [103]. The pro-apoptotic function of panaxydol from *P. ginseng* was also tested on different cell lines to check whether the induction of apoptosis occurred preferentially in cancer cells. Indeed, panaxydol selectively induced apoptosis in malignant cancer cells [85].

4.1.3. Gut Microbiota Composition

A study aimed to investigate whether the antibacterial effects of FaOH and FaDOH may be a mechanism of action in the antineoplastic properties of FaOH and FaDOH. They tested the effect of FaOH and FaDOH on gut microbiota composition in an AOM-induced rat colorectal cancer model. Rats treated with AOM were fed either a normal rat diet or a diet enriched with FaOH and FaDOH. Analysis of cecum faecal samples revealed a significant change in the gut microbiota among the groups. FaOH and FaDOH, which suppressed the growth of neoplastic tumours in the colon in a rat colon cancer model, modified the composition of low-abundance gut microbiota GM members, such as *Lactobacillus reuteri*, and high-prevalence *Turicibacter*, which was also correlated with a reduction in the number of macroscopic sites of neoplasms. Thus, this study demonstrated that modifications in the gut microbiota may play a significant role in the colon-protective action of FaOH and FaDOH against neoplastic transformation [107].

4.1.4. Other Effects

FaOH stimulated the differentiation of primary mammalian cells at concentrations as low as 0.004 to 0.4 µM, whereas cytotoxic effects were observed at concentrations of >4 µM [108]. Moreover, one study evaluated PAs (FaOH and FaDOH) isolated from carrots in non-cancerous human intestinal epithelial cells (FHs 74 Int. cells) and intestinal cancer cells (CaCo-2). The growth–inhibition response was seen in concentrations above 1 µg/mL (~4 µM), with maximum inhibition at 20 µg/mL (~80 µM). The FaOH showed a higher inhibitory potency compared with FaDOH. In addition, cancer cells treated with combinations of FaOH and FaDOH showed a synergistic response for the inhibition of cell growth [109]. FaOH purified from carrots inhibited caspase-3 expression to prevent cell death and reduced basal DNA strand breakage in CaCo-2 cells. Thus, FaOH is either pro-survival or pro-death in a concentration-dependent manner in CaCo-2 cells. The effects of FaOH on CaCo-2 cells appear to be biphasic, with low and high concentrations of falcarinol inducing pro-proliferative and apoptotic characteristics, respectively [110].

PAs have other effects relevant for cancer. PAs can be used to heal or relieve symptoms by interacting with other foods or drugs. Cisplatin, which has nephrotoxicity as a side effect, is a therapeutic drug for various solid tumours. FaDOH attenuates cisplatin-induced injury and down-regulates mRNA levels of *TNF-α, IL-1β* and the protein expression of p-NF-κB p65 in mice [111].

Another study demonstrated the effects of FaOH, FaDOH, FaDOH3Ac and falcarindiol 3,8-diacetate on breast cancer multidrug resistance protein (BCRP/ABCG2), a xenobiotic efflux transporter that causes chemotherapy resistance in cancer. PAs inhibited mitox-antrone efflux (an ABCG2 substrate) in HEK293 cells overexpressing ABCG2. In a vesicular transport assay, a concentration-dependent inhibition of methotrexate (another ABCG2 substrate) uptake into ABCG2-overexpressing Sf9 insect cell membrane vesicles was observed. PAs also inhibited baseline and sulfasalazine-stimulated vanadate-sensitive ATPase activities in ABCG2-overexpressing Sf9 insect cell membrane vesicles. This study suggested that PAs might mitigate multidrug resistance in chemotherapy. As ABCG2 may play a role in the absorption and disposition of PAs, there may be food–drug interactions between PA-rich foods and ABCG2 substrate drugs [112].

4.2. In Vivo

There is not yet any direct in vivo evidence supporting an anti-cancer role of PA in humans. Observational human studies have reported that carrot consumption was associated with a reduced risk of several cancer types. For example, in a prospective cohort study including 57,053 Danes, an intake of 2–4 or more raw carrots each week (>32 g/day) was associated with a 17% reduction in the risk of colorectal cancer [16], pancreatic cancer and leukaemia [113] compared to individuals with no intake of raw carrots. One experimental study with carrot juice (500 mL) containing approximately 18 mg FaOH reduced the level of inflammatory cytokines *IL-1* and *IL-16* significantly in LPS-stimulated human blood an hour after intake compared with before the intake of carrot juice [114]. More detailed studies were performed using animal models. The consumption of carrot powder reduced the growth of intestinal tumours in an Apc$^{Min/+}$ mouse colon cancer model [115,116]. A study examined colon preneoplastic lesions in AOM-treated rats that were fed carrots (10% freeze-dried carrot with a natural concentration of FaOH at 35 µg/g), FaOH (purified FaOH mixed at 3.5 µg/g in food) or a control for 18 weeks. The number and size of lesions decreased significantly in the rats that received either one of the two experimental treatments compared to the control group, indicating that carrots and FaOH slowed the growth of aberrant crypt foci (ACF) and tumours [117]. In a similar study, again using AOM-treated rats as a colon cancer model, feed containing a mixture of FaOH and FaDOH at concentrations four times higher than the previous study significantly reduced the number of tumours >1 mm, from 21 in controls to 12 in PA-treated rats [118]. An inverse correlation was found between a higher intake of a combination of FaOH and FaDOH with the multiplicity of colorectal neoplastic lesions [46]. These studies support the hypothesis that diets rich in FaOH and FaDOH can be a preventive treatment of colorectal cancer. A human dose of PAs corresponding to a 2017 rat experiment would be 24 mg per day for a 70 kg person, which could be provided by consuming 260 g per day of the cultivated carrot cultivar 'Nantes Empire' [118].

PAs from ginseng have shown selective tumour reduction activities similar to chemotherapeutic agents. Panaxydol isolated from *Panax ginseng* (*P. ginseng*) inhibited tumour growth in mouse tumour models, including PC3 human prostate cancer xenograft and mouse renal carcinoma (Renca) cells. BALB/c nude mice bearing PC3 or Renca cell tumours were injected with panaxydol every two days for a course of three weeks. Panaxydol inhibited the growth of the PC3 xenograft dose-dependently, with complete suppression at 100 mg/kg. Panaxydol also reduced Renca tumour size in dose-dependent manner, demonstrating an in vivo anticancer effect in this model [96].

5. Polyacetylene Toxicology and Pharmacokinetics

5.1. Toxicology of PAs

PAs in high concentrations have toxic effects that depend on cell sensitivity. In in vitro studies, FaOH has shown cytotoxic activity against intestinal cell lines at concentrations of 4 μM [108] and 10 μM [109]. In addition, FaDOH has shown a toxic activity in human colon adenocarcinoma (HT-29) cells in concentrations >50 μM, while it exhibited a toxic effect on human mesenchymal stem (hMSC) cells in concentrations > 20 μM [97]. In another study, FaDOH and panaxydol showed toxicity at concentrations of 40 μM [109]. In vivo, FaOH showed neurotoxic effects at a high concentration (LD50 = 100 mg/kg) when injected into mice [119], whereas FaDOH had no neurotoxic effects in rats when injected with similar concentrations (LD50 > 200 mg/kg) [120]. However, by inhibiting the Foxo–Notch axis, FaDOH can disrupt the maintenance of normal neural stem cells and modify the balance between the self-renewal and differentiation of neural stem cells with negative consequences [121]. There was also a report on the modulation of $GABA_A$ receptors by FaOH, which may underlie a sedative effect [122]. Mammals have not been observed to be poisoned after the voluntary consumption of FaOH-type PAs in natural sources; this is probably related with the bitter taste of PAs, in particular FaDOH, which causes a bitter or burning sensation when occurring in concentrations > 40 μM [123], thus preventing the eating of unsafe amounts of vegetables with too high levels of these PAs. This contrasts with other types of polyacetylenes such as oenanthotoxin, which is found in the neurotoxic plant hemlock water dropwort [124]. However, while FaOH-type PAs can also cause neurotoxic symptoms, this requires much higher concentrations than those that occur in edible plants, and therefore their presence in food plants is deemed harmless [125]. FaOH from ivy (*Hedera helix* L.) has a moderate allergic potential on human skin. Repeated direct contact with ivy or other plants containing FaOH can cause sensitization in susceptible individuals and subsequently lead to allergic contact dermatitis after long-term frequent exposure to skin [126]. Most patients with this rare condition become sensitised in occupational settings, e.g., plant nursery workers handling ornamental plants [127]; only very few cases of PA-related contact dermatitis from Apiaceous vegetables have been reported [128]. This contact dermatitis may be related to falcarinol selectively alkylating the anandamide binding site in the CB1 receptor [129].

As far as we are aware, no research has been published on the safety and side effects of falcarinol-type PAs in other contexts than food safety or contact dermatitis. Therefore, it is necessary to carry out research on the toxicological evaluation and potential toxicity mechanisms of PAs and to establish scientifically justified safe doses and applications as a prerequisite for their use in therapeutic and clinical applications.

5.2. Pharmacokinetics of PAs

The study of pharmacokinetics determines the fate of drugs supplied externally to a living body [42]. Pharmacokinetic research, such as pharmacodynamic and toxicological research, has become an essential part of drug preclinical and clinical research. It is critical in the development of new drugs, the improvement of dosage forms and the study of dosage form mechanisms [130].

Two studies have reported on the bioavailability of serum FaOH concentrations in humans after the consumption of carrot juice. One study used three doses of fresh carrot juice providing 19, 33 or 49 µmol FaOH and demonstrated the dose-dependent bioavailability of this compound, reaching 0.010 µM (4.0 ng/mL) after the highest dose [19]. A recent study prepared juice from 30 g of freeze-dried carrot powder containing approximately 18 mg FaOH, corresponding to 300 g raw carrot (two–three carrots). The powder was mixed with water to a total of 500 mL and given to participants. Serum FaOH concentrations reached their peak at 1 h after consumption, and the peak concentration was 0.9–4.0 ng/mL. FaOH had a half-life of 1.5 h in human serum [131]. Another study reported measuring the pharmacokinetics of FaOH in mice. FaOH was intravenously (IV) administered to mice at 5 mg/kg and orally administrated at 20 mg/kg to determine the pharmacokinetic parameters in the plasma and tissue using LC-MS/MS. FaOH reached a peak plasma concentration of 8.24 µg/mL after IV administration and then declined in a multiphasic manner. A plasma analysis of FaOH after oral administration showed that the compound's concentration quickly reached a maximum of 1.72 µg/mL in 1 h. The plasma concentration then decreased in a multiphasic manner, reaching a final measurable concentration of 32.2 ng/mL after 24 h. FaOH had a half-life of 1.5 h when IV injected and 5.9 h when administered orally, with a bioavailability of 50.4%. The mice did not show any toxicity up to 300 mg/kg orally [132]. When mice were orally given 20 mg/kg of FaOH, the FaOH concentrations in the colon tissue were the highest at 2 h after treatment at 121 ng/mL [132]. The very high murine values contrast remarkably with the similarity of the maximal plasma concentration ranges reported in the above two human studies. However, the lack of detail in the description of the LC-MS methodology in the mouse paper opens the possibility that this may reflect issues with the methodology rather than substantial differences in bioavailability between the species; additional bioavailability studies in mice would be useful to resolve this important question.

6. Conclusions

Food plants of the Apiaceae and Araliaceae families rich in PAs have important potential regarding cancer prevention. The findings reviewed here (Table 1) consistently support that PAs are anti-neoplastic natural phytochemicals with the potential for advancement into multiple applications in cancer prevention and treatment and as leading compounds in the discovery of new anticancer drugs. The mechanisms of the action attributed to PAs are similar to those of many other anticancer drugs, which include triggering cell cycle arrest, apoptosis, UPR and reducing inflammation but potentially with lower toxic side effects. The PA concentrations in widely consumed vegetables such as carrots are sufficiently high to potentially provide substantial chemo-preventive effects within the recommended vegetable and fruit intake of 400 g per day while at the same time being sufficiently low to exclude concerns about toxicity from these dietary sources. PAs have significant inhibitory effects on multiple cancer cell pathways, indicating anti-proliferative and anti-tumorigenic properties.

Nevertheless, much work still needs to be carried out to assess and possibly develop the medical development of PAs. Despite being widely distributed in plants, PAs are instable and present in relatively small amounts, which makes it challenging to isolate abundant polyacetylenes from natural sources for large-scale experiments. Improved isolation methods or the development of improved stereospecific chemical synthesis procedures will provide better feasibility for further study. Future studies are needed to determine the safe doses of PAs in humans, which is a prerequisite for any non-food application, whether as food supplements or drugs. In this regard, additional studies are needed on animals and humans for a better determination of the toxicological effects of PAs. Pharmacokinetic studies on PAs are limited, and the pharmacological actions of many PAs are unknown. Advancing this research will provide a scientific foundation for assessments of their potential for clinical applications and new drug development. NF-κB pathways are influenced by PAs, indicating their significance in terms of not only cancer prevention and treatment but also various other biological processes. However, future studies should focus on investigating the exact mechanism of action of PAs, particularly on NF-κB pathways, to distinguish the influence of PAs on gene transcription, translation or post-translation functions, such as enzymatic activity. Animal experiments, clinical studies and human intervention studies must be conducted to investigate and compare health benefits of PAs in whole foods or isolated PA metabolites using biomarkers indicating inflammation and other cancer-related processes to guide the optimisation of the implementation of affordable food-based cancer prevention programmes.

Table 1. Studies of polyacetylene compounds extracted from or contained in natural food and herbs and their applications, doses, time and effect or pathway assessed.

Compound/Source	Dose/Route	Time	Model	Effect/Pathway Investigated	Ref.
FaOH/carrot	Orally 10 mg/kg	1 weeks	Mice	Activate Nrf2 pathway	[5,92]
FaOH/*P. quinquefolius*	0.5–100 μM/mL	12 h	Primary Macrophages	Induce DNA damage/apoptosis	[8]
FaOH/*P. quinquefolius*	Orally 0.01–1 mg/kg	2 weeks	Mice	Reduce inflammation/Induce apoptosis	[8]
PAs/purple carrot	6.6 or 13.3 μg/mL	16 h	RAW 264.7 cells	Reduce secretion of the proinflammatory cytokines (IL-6, IL-1β, TNF-α)	[45]
FaDOH/*P. quinquefolius*	50 μM/mL	30–60 min	Astrocytes cells	Modulate NF-κB Pathway	[64]
FaOH/*S. divaricata*	Orally 0.1–1 mg/kg	1 weeks	Mice	Modulate NF-κB pathway/proinflammatory	[61]
FaOH/*S. divaricata*	1–10 μM/mL	3.5–24 h	Microglia cells	Reduce nitric oxide secretion/NF-κB pathway	[61]
FaOH, FaDOH/carrot	Orally 0.16–35 μg/g feed	20 weeks	Rats	Reduce tumour growth in colon/NF-κB pathway	[46]
FaOH/*H. moellendorffii*	6.25–50 μg/mL	20 h	RAW 264.7 cells	Activate ROS/Nrf2/HO-1 signaling pathway/Modulate NF-κB pathway	[65]
FaOH/*P. quinquefolius*	0.5 μM/mL	6 h	Macrophages cell lines	Activate Nrf2 pathway/NF-κB pathway	[85]
Panaxydol/*P. ginseng*	50 μg/mL	6 h	MCF-7 cells	Induce ROS generation, induce apoptosis/activate Nrf2 pathway	[88]
FaOH, FaDOH/carrot	6.25–50 μM/mL	24 h	Myotube cells	Modulate ROS pathway	[89]
FaDOH/*N. incisum*	2.5 μM/mL	24 h	HEK293 cells	Modulate Nrf2/ARE pathway	[90]
FaDOH/*O. horridus*	10 μM/mL	8 h	Colon cancer cells	Modulate ER/UPR pathway	[95]
Panaxydol/*P. ginseng*	20 μg/mL	2–4 h	MCF-7 cells	Modulate ER/EGFR pathway/induce apoptosis	[96]
FaDOH/carrot	5 μg/mL	24 h	HT-29/hMSCs cells	Modulate PPARγ pathway	[97]
FaOH/*O. horridus*	0.3 μg/mL	48 h	PANC-1 cells	Inhibit human pancreatic cancer cells/Induce apoptosis	[98]
FaOH, FaDOH/carrot	25–100 μM/mL	24 h	Leukaemia cells	Induced apoptosis in leukaemia cells/arrest of cell cycle	[99]
FaOH/*P. ginseng*	Orally 50–100 mg/kg mouse	8 weeks	Mice	Reduced lung tumorigenesis	[103]
FaOH/*P. ginseng*	1 μM/mL	-	NSCLC cells	Induce apoptosis	[103]
FaOH, FaDOH/carrot	Orally 7 μg/g feed	20 weeks	Rats	Alter gut microbiota, *Lactobacillus reuteri, Turicibacter*	[107]
FaOH/carrot	0.5–100 μM/mL	72 h	CaCo-2 Cell	Induce proliferation/apoptosis	[110]
FaDOH	IP injection/50–100 mg/kg	4 days	Mice	Activate Nrf2 pathway/NF-κB pathway	[111]
FaOH, FaDOH, FaDOH3Ac/carrot	20 μM/mL	105 min	HEK293 cells	Modulate ABCG2 pathway	[112]
FaOH in purple carrot juice	Orally approx. 18 mg in 500 mL	1 h	Human	Reduce secretion of the proinflammatory cytokines IL-1α and IL-16	[114]
PAs in carrot powder	Orally approx. 20 μg/g feed	12 weeks	Mice	Reduced tumours in intestine	[116]
PAs in carrot powder	Orally approx. 20 μg/g feed	10 weeks	Mice	Reduced tumours in intestine	[115]
Panaxydol/*P. ginseng*	IP injection/50–100 mg/kg	30 days	Mice	Reduced syngeneic and xenogeneic tumours	[96]
FaOH/carrot	Orally 3.5 μg/g feed	18 weeks	Rats	Reduced tumours in colon	[117]
FaOH+FaDOH/carrot	Orally 7 + 7 μg/g feed	18 weeks	Rats	Reduced tumours in colon	[118]

Author Contributions: Investigation, figure design, writing—original draft preparation, writing—review and editing, R.A.; supervision, writing—review and editing, L.H.; supervision, writing—review and editing, K.B. All authors have read and agreed to the published version of the manuscript.

Funding: This work was funded by Qassim University, Qassim, KSA. The funder had no role in data collection, data interpretation, or writing of the review.

Data Availability Statement: Data are contained within the article.

Acknowledgments: We thank M. Kobæk-Larsen for dose data supplementary to [114].

Conflicts of Interest: R.A. and K.B. declare no conflict of interest. L.H. is a consultant and scientific advisor for Bioperfectus Technologies, and a director of Bioperfectus U.K. Limited.

References

1. Wang, X.; Ouyang, Y.; Liu, J.; Zhu, M.; Zhao, G.; Bao, W.; Hu, F.B. Fruit and vegetable consumption and mortality from all causes, cardiovascular disease, and cancer: Systematic review and dose-response meta-analysis of prospective cohort studies. *Postgrad. Med. J.* **2014**, *349*, g4490. [CrossRef]
2. Negri, R. Polyacetylenes from terrestrial plants and fungi: Recent phytochemical and biological advances. *Fitoterapia* **2015**, *106*, 92–109. [CrossRef]
3. Dawid, C.; Dunemann, F.; Schwab, W.; Nothnagel, T.; Hofmann, T. Bioactive C17-polyacetylenes in carrots (*Daucus carota* L.): Current knowledge and future perspectives. *J. Agric. Food Chem.* **2015**, *63*, 9211–9222. [CrossRef]
4. Hansen, L.; Boll, P.M. Polyacetylenes in Araliaceae: Their chemistry, biosynthesis and biological significance. *Phytochemistry* **1986**, *25*, 285–293. [CrossRef]
5. Stefanson, A.; Bakovic, M. Dietary polyacetylene falcarinol upregulated intestinal heme oxygenase-1 and modified plasma cytokine profile in late phase lipopolysaccharide-induced acute inflammation in CB57BL/6 mice. *Nutr. Res.* **2020**, *80*, 89–105. [CrossRef] [PubMed]
6. Moore, T. Vitamin A and carotene: The association of vitamin A activity with carotene in the carrot root. *Biochem. J.* **1929**, *23*, 803. [CrossRef] [PubMed]
7. Butnariu, M. Therapeutic Properties of Vegetable. *J. Bioequivalence Bioavailab.* **2014**, *6*, e55. [CrossRef]
8. Chaparala, A.; Poudyal, D.; Tashkandi, H.; Witalison, E.E.; Chumanevich, A.A.; Hofseth, J.L.; Nguyen, I.; Hardy, O.; Pittman, D.L.; Wyatt, M.D. Panaxynol, a bioactive component of American ginseng, targets macrophages and suppresses colitis in mice. *Oncotarget* **2020**, *11*, 2026. [CrossRef]
9. Cambria, C.; Sabir, S.; Shorter, I.C. *Ginseng*; StatPearls Publishing: Treasure Island, FL, USA, 2019.
10. Hong, H.; Baatar, D.; Hwang, S.G. Anticancer activities of ginsenosides, the main active components of ginseng. *Evid. -Based Complement. Altern. Med.* **2021**, *2021*, 8858006. [CrossRef] [PubMed]
11. Yang, S.; Li, F.; Lu, S.; Ren, L.; Bian, S.; Liu, M.; Zhao, D.; Wang, S.; Wang, J. Ginseng root extract attenuates inflammation by inhibiting the MAPK/NF-κB signaling pathway and activating autophagy and p62-Nrf2-Keap1 signaling in vitro and in vivo. *J. Ethnopharmacol.* **2022**, *283*, 114739. [CrossRef]
12. Jeon, J.E.; Kim, J.-G.; Fischer, C.R.; Mehta, N.; Dufour-Schroif, C.; Wemmer, K.; Mudgett, M.B.; Sattely, E. A pathogen-responsive gene cluster for highly modified fatty acids in tomato. *Cell* **2020**, *180*, 176–187.e119. [CrossRef]
13. Mares-Perlman, J.A.; Millen, A.E.; Ficek, T.L.; Hankinson, S.E. The body of evidence to support a protective role for lutein and zeaxanthin in delaying chronic disease. Overview. *J. Nutr.* **2002**, *132*, 518S–524S. [CrossRef] [PubMed]
14. Sommer, A.; Vyas, K.S. A global clinical view on vitamin A and carotenoids. *Am. J. Clin. Nutr.* **2012**, *96*, 1204S–1206S. [CrossRef]
15. Tapiero, H.; Townsend, D.M.; Tew, K.D. The role of carotenoids in the prevention of human pathologies. *Biomed. Pharmacother.* **2004**, *58*, 100–110. [CrossRef]
16. Deding, U.; Baatrup, G.; Christensen, L.P.; Kobaek-Larsen, M. Carrot intake and risk of colorectal cancer: A prospective cohort study of 57,053 Danes. *Nutrients* **2020**, *12*, 332. [CrossRef] [PubMed]
17. Bjelakovic, G.; Nikolova, D.; Gluud, C. Antioxidant Supplements to Prevent Mortality. *JAMA-J. Am. Med. Assoc.* **2013**, *310*, 1178–1179. [CrossRef] [PubMed]
18. Kordiak, J.; Bielec, F.; Jabłoński, S.; Pastuszak-Lewandoska, D. Role of Beta-Carotene in Lung Cancer Primary Chemoprevention: A Systematic Review with Meta-Analysis and Meta-Regression. *Nutrients* **2022**, *14*, 1361. [CrossRef]
19. Brandt, K.; Christensen, L.P.; Hansen-Møller, J.; Hansen, S.; Haraldsdottir, J.; Jespersen, L.; Purup, S.; Kharazmi, A.; Barkholt, V.; Frøkiær, H. Health promoting compounds in vegetables and fruits: A systematic approach for identifying plant components with impact on human health. *Trends Food Sci. Technol.* **2004**, *15*, 384–393. [CrossRef]
20. Atanasov, A.G.; Waltenberger, B.; Pferschy-Wenzig, E.-M.; Linder, T.; Wawrosch, C.; Uhrin, P.; Temml, V.; Wang, L.; Schwaiger, S.; Heiss, E.H. Discovery and resupply of pharmacologically active plant-derived natural products: A review. *Biotechnol. Adv.* **2015**, *33*, 1582–1614. [CrossRef]
21. Harvey, A.L.; Edrada-Ebel, R.; Quinn, R.J. The re-emergence of natural products for drug discovery in the genomics era. *Nat. Rev. Drug Discov.* **2015**, *14*, 111–129. [CrossRef]
22. Atanasov, A.G.; Zotchev, S.B.; Dirsch, V.M.; Supuran, C.T. Natural products in drug discovery: Advances and opportunities. *Nat. Rev. Drug Discov.* **2021**, *20*, 200–216. [CrossRef]
23. Newman, D.J.; Cragg, G.M. Natural products as sources of new drugs over the nearly four decades from 01/1981 to 09/2019. *J. Nat. Prod.* **2020**, *83*, 770–803. [CrossRef] [PubMed]
24. Singh, N.; Baby, D.; Rajguru, J.P.; Patil, P.B.; Thakkannavar, S.S.; Pujari, V.B. Inflammation and cancer. *Ann. Afr. Med.* **2019**, *18*, 121. [CrossRef] [PubMed]
25. Furman, D.; Campisi, J.; Verdin, E.; Carrera-Bastos, P.; Targ, S.; Franceschi, C.; Ferrucci, L.; Gilroy, D.W.; Fasano, A.; Miller, G.W. Chronic inflammation in the etiology of disease across the life span. *Nat. Med.* **2019**, *25*, 1822–1832. [CrossRef] [PubMed]
26. Ju, T.; Hernandez, L.; Mohsin, N.; Labib, A.; Frech, F.; Nouri, K. Evaluation of risk in chronic cutaneous inflammatory conditions for malignant transformation. *J. Eur. Acad. Dermatol. Venereol.* **2023**, *37*, 231–242. [CrossRef]
27. Park, J.M.; Kim, J.; Lee, Y.J.; Bae, S.U.; Lee, H.W. Inflammatory bowel disease–associated intestinal fibrosis. *J. Pathol. Transl. Med.* **2023**, *57*, 60–66. [CrossRef]

28. Lirhus, S.S. Inflammatory Bowel Disease and Health Registry Data: Estimates of Incidence, Prevalence and Regional Treatment Variation. Ph.D. Thesis, University of Oslo, Oslo, Norway, January 2022.
29. Hanahan, D. Hallmarks of Cancer: New Dimensions. *Cancer Discov* **2022**, *12*, 31–46. [CrossRef] [PubMed]
30. Raposo, T.; Beirão, B.; Pang, L.; Queiroga, F.; Argyle, D. Inflammation and cancer: Till death tears them apart. *Vet. J.* **2015**, *205*, 161–174. [CrossRef]
31. Macarthur, M.; Hold, G.L.; El-Omar, E.M. Inflammation and Cancer II. Role of chronic inflammation and cytokine gene polymorphisms in the pathogenesis of gastrointestinal malignancy. *Am. J. Physiol. Gastrointest. Liver Physiol.* **2004**, *286*, G515–G520. [CrossRef]
32. De Marzo, A.M.; Marchi, V.L.; Epstein, J.I.; Nelson, W.G. Proliferative inflammatory atrophy of the prostate: Implications for prostatic carcinogenesis. *Am. J. Pathol.* **1999**, *155*, 1985–1992. [CrossRef] [PubMed]
33. Kuper, H.; Adami, H.O.; Trichopoulos, D. Infections as a major preventable cause of human cancer. *J. Intern. Med.* **2001**, *249*, 61–74. [CrossRef]
34. Scholl, S.; Pallud, C.; Beuvon, F.; Hacene, K.; Stanley, E.; Rohrschneider, L.; Tang, R.; Pouillart, P.; Lidereau, R. Anti-colony-stimulating factor-1 antibody staining in primary breast adenocarcinomas correlates with marked inflammatory cell infiltrates and prognosis. *J. Natl. Cancer Inst.* **1994**, *86*, 120–126. [CrossRef]
35. Ernst, P.B.; Gold, B.D. The disease spectrum of Helicobacter pylori: The immunopathogenesis of gastroduodenal ulcer and gastric cancer. *Annu. Rev. Microbiol.* **2000**, *54*, 615. [CrossRef] [PubMed]
36. Ness, R.B.; Cottreau, C. Possible role of ovarian epithelial inflammation in ovarian cancer. *J. Natl. Cancer Inst.* **1999**, *91*, 1459–1467. [CrossRef] [PubMed]
37. Bröcker, E.; Zwadlo, G.; Holzmann, B.; Macher, E.; Sorg, C. Inflammatory cell infiltrates in human melanoma at different stages of tumor progression. *Int. J. Cancer* **1988**, *41*, 562–567. [CrossRef] [PubMed]
38. Kundu, J.K.; Surh, Y.-J. Inflammation: Gearing the journey to cancer. *Mutat. Res./Rev. Mutat. Res.* **2008**, *659*, 15–30. [CrossRef]
39. Coussens, L.M.; Werb, Z. Inflammation and cancer. *Nature* **2002**, *420*, 860–867. [CrossRef] [PubMed]
40. Multhoff, G.; Molls, M.; Radons, J. Chronic inflammation in cancer development. *Front. Immunol.* **2012**, *2*, 98. [CrossRef]
41. Christensen, L.P. Bioactive C17 and C18 acetylenic oxylipins from terrestrial plants as potential lead compounds for anticancer drug development. *Molecules* **2020**, *25*, 2568. [CrossRef]
42. Xie, Q.; Wang, C. Polyacetylenes in herbal medicine: A comprehensive review of its occurrence, pharmacology, toxicology, and pharmacokinetics (2014–2021). *Phytochemistry* **2022**, 113288. [CrossRef]
43. Ahmad, T.; Cawood, M.; Iqbal, Q.; Ariño, A.; Batool, A.; Tariq, R.M.S.; Azam, M.; Akhtar, S. Phytochemicals in Daucus carota and their health benefits. *Foods* **2019**, *8*, 424. [CrossRef] [PubMed]
44. Kim, Y.-J.; Ju, J.; Song, J.-L.; Yang, S.-g.; Park, K.-Y. Anti-colitic effect of purple carrot on dextran sulfate sodium (DSS)-induced colitis in C57BL/6J Mice. *Prev. Nutr. Food Sci.* **2018**, *23*, 77. [CrossRef]
45. Metzger, B.T.; Barnes, D.M.; Reed, J.D. Purple carrot (Daucus carota L.) polyacetylenes decrease lipopolysaccharide-induced expression of inflammatory proteins in macrophage and endothelial cells. *J. Agric. Food Chem.* **2008**, *56*, 3554–3560. [CrossRef] [PubMed]
46. Kobaek-Larsen, M.; Baatrup, G.; Notabi, M.K.; El-Houri, R.B.; Pipó-Ollé, E.; Christensen Arnspang, E.; Christensen, L.P. Dietary polyacetylenic oxylipins falcarinol and falcarindiol prevent inflammation and colorectal neoplastic transformation: A mechanistic and dose-response study in a rat model. *Nutrients* **2019**, *11*, 2223. [CrossRef]
47. George, B.P.; Chandran, R.; Abrahamse, H. Role of phytochemicals in cancer chemoprevention: Insights. *Antioxidants* **2021**, *10*, 1455. [CrossRef] [PubMed]
48. Islam, S.U.; Ahmed, M.B.; Ahsan, H.; Islam, M.; Shehzad, A.; Sonn, J.K.; Lee, Y.S. An update on the role of dietary phytochemicals in human skin cancer: New insights into molecular mechanisms. *Antioxidants* **2020**, *9*, 916. [CrossRef] [PubMed]
49. Rahman, M.A.; Rahman, M.H.; Hossain, M.S.; Biswas, P.; Islam, R.; Uddin, M.J.; Rahman, M.H.; Rhim, H. Molecular insights into the multifunctional role of natural compounds: Autophagy modulation and cancer prevention. *Biomedicines* **2020**, *8*, 517. [CrossRef]
50. Kumar, S.; Gupta, S. Dietary phytochemicals and their role in cancer chemoprevention. *J. Cancer Metastasis Treat.* **2021**, *7*, 51. [CrossRef]
51. Spooner, R.; Yilmaz, Ö. The role of reactive-oxygen-species in microbial persistence and inflammation. *Int. J. Mol. Sci.* **2011**, *12*, 334–352. [CrossRef]
52. Kauppinen, A.; Suuronen, T.; Ojala, J.; Kaarniranta, K.; Salminen, A. Antagonistic crosstalk between NF-κB and SIRT1 in the regulation of inflammation and metabolic disorders. *Cell. Signal.* **2013**, *25*, 1939–1948. [CrossRef]
53. Crusz, S.M.; Balkwill, F.R. Inflammation and cancer: Advances and new agents. *Nat. Rev. Clin. Ocol.* **2015**, *12*, 584–596. [CrossRef]
54. Song, W.; Mazzieri, R.; Yang, T.; Gobe, G.C. Translational significance for tumor metastasis of tumor-associated macrophages and epithelial–mesenchymal transition. *Front. Immunol.* **2017**, *8*, 1106. [CrossRef] [PubMed]
55. Kwon, H.-J.; Choi, G.-E.; Ryu, S.; Kwon, S.J.; Kim, S.C.; Booth, C.; Nichols, K.E.; Kim, H.S. Stepwise phosphorylation of p65 promotes NF-κB activation and NK cell responses during target cell recognition. *Nat. Commun.* **2016**, *7*, 11686. [CrossRef] [PubMed]
56. Hayden, M.S.; Ghosh, S. Shared principles in NF-κB signaling. *Cell* **2008**, *132*, 344–362. [CrossRef]

57. Sau, A.; Lau, R.; Cabrita, M.A.; Nolan, E.; Crooks, P.A.; Visvader, J.E.; Pratt, M.C. Persistent activation of NF-κB in BRCA1-deficient mammary progenitors drives aberrant proliferation and accumulation of DNA damage. *Cell Stem Cell* **2016**, *19*, 52–65. [CrossRef]
58. Salazar, L.; Kashiwada, T.; Krejci, P.; Meyer, A.N.; Casale, M.; Hallowell, M.; Wilcox, W.R.; Donoghue, D.J.; Thompson, L.M. Fibroblast growth factor receptor 3 interacts with and activates TGFβ-activated kinase 1 tyrosine phosphorylation and NFκB signaling in multiple myeloma and bladder cancer. *PLoS ONE* **2014**, *9*, e86470. [CrossRef]
59. Burstein, E.; Fearon, E.R. Colitis and cancer: A tale of inflammatory cells and their cytokines. *J. Clin. Investig.* **2008**, *118*, 464–467. [CrossRef]
60. Sun, S.-C. The non-canonical NF-κB pathway in immunity and inflammation. *Nat. Rev. Immunol.* **2017**, *17*, 545–558. [CrossRef] [PubMed]
61. Zhao, Y.; Sun, X.; Zhang, T.; Liu, S.; Cai, E.; Zhu, H. Study on the antidepressant effect of panaxynol through the IκB-α/NF-κB signaling pathway to inhibit the excessive activation of BV-2 microglia. *Biomed. Pharmacother.* **2021**, *138*, 111387. [CrossRef]
62. Chen, T.; Mou, Y.; Tan, J.; Wei, L.; Qiao, Y.; Wei, T.; Xiang, P.; Peng, S.; Zhang, Y.; Huang, Z. The protective effect of CDDO-Me on lipopolysaccharide-induced acute lung injury in mice. *Int. Immunopharmacol.* **2015**, *25*, 55–64. [CrossRef]
63. Kang, H.; Bang, T.-S.; Lee, J.-W.; Lew, J.-H.; Eom, S.H.; Lee, K.; Choi, H.-Y. Protective effect of the methanol extract from Cryptotaenia japonica Hassk. against lipopolysaccharide-induced inflammation in vitro and in vivo. *BMC Complement. Altern. Med.* **2012**, *12*, 199. [CrossRef]
64. Shiao, Y.-J.; Lin, Y.-L.; Sun, Y.-H.; Chi, C.-W.; Chen, C.-F.; Wang, C.-N. Falcarindiol impairs the expression of inducible nitric oxide synthase by abrogating the activation of IKK and JAK in rat primary astrocytes. *Br. J. Pharmacol.* **2005**, *144*, 42. [CrossRef] [PubMed]
65. Kim, H.N.; Kim, J.D.; Yeo, J.H.; Son, H.-J.; Park, S.B.; Park, G.H.; Eo, H.J.; Jeong, J.B. Heracleum moellendorffii roots inhibit the production of pro-inflammatory mediators through the inhibition of NF-κB and MAPK signaling, and activation of ROS/Nrf2/HO-1 signaling in LPS-stimulated RAW264.7 cells. *BMC Complement. Altern. Med.* **2019**, *19*, 310. [CrossRef]
66. Yao, C.; Narumiya, S. Prostaglandin-cytokine crosstalk in chronic inflammation. *Br. J. Pharmacol.* **2019**, *176*, 337–354. [CrossRef]
67. Lee, S.; Shin, S.; Kim, H.; Han, S.; Kim, K.; Kwon, J.; Kwak, J.-H.; Lee, C.-K.; Ha, N.-J.; Yim, D. Anti-inflammatory function of arctiin by inhibiting COX-2 expression via NF-κB pathways. *J. Inflamm.* **2011**, *8*, 16. [CrossRef] [PubMed]
68. Nagaraju, G.P.; El-Rayes, B.F. Cyclooxygenase-2 in gastrointestinal malignancies. *Cancer* **2019**, *125*, 1221–1227. [CrossRef]
69. Ghosh, N.; Chaki, R.; Mandal, V.; Mandal, S.C. COX-2 as a target for cancer chemotherapy. *Pharmacol. Rep.* **2010**, *62*, 233–244. [CrossRef] [PubMed]
70. Agrawal, U.; Kumari, N.; Vasudeva, P.; Mohanty, N.K.; Saxena, S. Overexpression of COX2 indicates poor survival in urothelial bladder cancer. *Ann. Diagn. Pathol.* **2018**, *34*, 50–55. [CrossRef]
71. Harris, R.E.; Casto, B.C.; Harris, Z.M. Cyclooxygenase-2 and the inflammogenesis of breast cancer. *World J. Clin. Oncol.* **2014**, *5*, 677. [CrossRef] [PubMed]
72. Petkova, D.; Clelland, C.; Ronan, J.; Pang, L.; Coulson, J.; Lewis, S.; Knox, A. Overexpression of cyclooxygenase-2 in non-small cell lung cancer. *Respir. Med.* **2004**, *98*, 164–172. [CrossRef]
73. Yip-Schneider, M.T.; Barnard, D.S.; Billings, S.D.; Cheng, L.; Heilman, D.K.; Lin, A.; Marshall, S.J.; Crowell, P.L.; Marshall, M.S.; Sweeney, C.J. Cyclooxygenase-2 expression in human pancreatic adenocarcinomas. *Carcinogenesis* **2000**, *21*, 139–146. [CrossRef] [PubMed]
74. Saba, N.F.; Choi, M.; Muller, S.; Shin, H.J.C.; Tighiouart, M.; Papadimitrakopoulou, V.A.; El-Naggar, A.K.; Khuri, F.R.; Chen, Z.G.; Shin, D.M. Role of Cyclooxygenase-2 in Tumor Progression and Survival of Head and Neck Squamous Cell CarcinomaRole of COX-2 in Head and Neck Cancer. *Cancer Prev. Res.* **2009**, *2*, 823–829. [CrossRef] [PubMed]
75. Masferrer, J.L.; Leahy, K.M.; Koki, A.T.; Zweifel, B.S.; Settle, S.L.; Woerner, B.M.; Edwards, D.A.; Flickinger, A.G.; Moore, R.J.; Seibert, K. Antiangiogenic and Antitumor Activities of cyclooxygenase-2 inhibitors. *Cancer Res.* **2000**, *60*, 1306–1311. [PubMed]
76. Shi, G.; Li, D.; Fu, J.; Sun, Y.; Li, Y.; Qu, R.; Jin, X.; Li, D. Upregulation of cyclooxygenase-2 is associated with activation of the alternative nuclear factor kappa B signaling pathway in colonic adenocarcinoma. *Am J Transl Res* **2015**, *7*, 1612–1620. [PubMed]
77. Qualls, J.E.; Kaplan, A.M.; Van Rooijen, N.; Cohen, D.A. Suppression of experimental colitis by intestinal mononuclear phagocytes. *J. Leukoc. Biol.* **2006**, *80*, 802–815. [CrossRef]
78. Xavier, R.J.; Podolsky, D.K. Unravelling the pathogenesis of inflammatory bowel disease. *Nature* **2007**, *448*, 427–434. [CrossRef]
79. Chen, S. Natural products triggering biological targets-a review of the anti-inflammatory phytochemicals targeting the arachidonic acid pathway in allergy asthma and rheumatoid arthritis. *Curr. Drug Targets* **2011**, *12*, 288–301. [CrossRef] [PubMed]
80. Bogdan, C.; Röllinghoff, M.; Diefenbach, A. The role of nitric oxide in innate immunity. *Immunol. Rev.* **2000**, *173*, 17–26. [CrossRef] [PubMed]
81. Ekmekcioglu, S.; Grimm, E.A.; Roszik, J. Targeting iNOS to increase efficacy of immunotherapies. *Hum. Vaccines Immunother.* **2017**, *13*, 1105–1108. [CrossRef] [PubMed]
82. Hausel, P.; Latado, H.; Courjault-Gautier, F.; Felley-Bosco, E. Src-mediated phosphorylation regulates subcellular distribution and activity of human inducible nitric oxide synthase. *Oncogene* **2006**, *25*, 198–206. [CrossRef]
83. Pautz, A.; Art, J.; Hahn, S.; Nowag, S.; Voss, C.; Kleinert, H. Regulation of the expression of inducible nitric oxide synthase. *Nitric Oxide* **2010**, *23*, 75–93. [CrossRef] [PubMed]
84. Ichikawa, T.; Li, J.; Nagarkatti, P.; Nagarkatti, M.; Hofseth, L.J.; Windust, A.; Cui, T. American ginseng preferentially suppresses STAT/iNOS signaling in activated macrophages. *J. Ethnopharmacol.* **2009**, *125*, 145–150. [CrossRef]

85. Qu, C.; Li, B.; Lai, Y.; Li, H.; Windust, A.; Hofseth, L.J.; Nagarkatti, M.; Nagarkatti, P.; Wang, X.L.; Tang, D. Identifying panaxynol, a natural activator of nuclear factor erythroid-2 related factor 2 (Nrf2) from American ginseng as a suppressor of inflamed macrophage-induced cardiomyocyte hypertrophy. *J. Ethnopharmacol.* **2015**, *168*, 326–336. [CrossRef]

86. Ďuračková, Z. Some current insights into oxidative stress. *Physiol. Res.* **2010**, *59*, 459–469. [CrossRef] [PubMed]

87. Scialo, F.; Sanz, A. Coenzyme Q redox signalling and longevity. *Free Radic. Biol. Med.* **2021**, *164*, 187–205. [CrossRef] [PubMed]

88. Kim, J.Y.; Yu, S.-J.; Oh, H.J.; Lee, J.Y.; Kim, Y.; Sohn, J. Panaxydol induces apoptosis through an increased intracellular calcium level, activation of JNK and p38 MAPK and NADPH oxidase-dependent generation of reactive oxygen species. *Apoptosis* **2011**, *16*, 347–358. [CrossRef] [PubMed]

89. Young, J.F.; Christensen, L.P.; Theil, P.K.; Oksbjerg, N. The polyacetylenes falcarinol and falcarindiol affect stress responses in myotube cultures in a biphasic manner. *Dose-Response* **2008**, *6*, 239–251. [CrossRef]

90. Ohnuma, T.; Nakayama, S.; Anan, E.; Nishiyama, T.; Ogura, K.; Hiratsuka, A. Activation of the Nrf2/ARE pathway via S-alkylation of cysteine 151 in the chemopreventive agent-sensor Keap1 protein by falcarindiol, a conjugated diacetylene compound. *Toxicol. Appl. Pharmacol.* **2010**, *244*, 27–36. [CrossRef] [PubMed]

91. Chiang, S.K.; Chen, S.E.; Chang, L.C. A Dual Role of Heme Oxygenase-1 in Cancer Cells. *Int J Mol Sci* **2018**, *20*, 39. [CrossRef]

92. Stefanson, A.L.; Bakovic, M. Falcarinol is a potent inducer of heme oxygenase-1 and was more effective than sulforaphane in attenuating intestinal inflammation at diet-achievable doses. *Oxidative Med. Cell. Longev.* **2018**, *2018*, 3153527. [CrossRef] [PubMed]

93. Sano, R.; Reed, J.C. ER stress-induced cell death mechanisms. *Biochim. Et Biophys. Acta (BBA)-Mol. Cell Res.* **2013**, *1833*, 3460–3470. [CrossRef] [PubMed]

94. Li, Y.; Lu, L.; Zhang, G.; Ji, G.; Xu, H. The role and therapeutic implication of endoplasmic reticulum stress in inflammatory cancer transformation. *Am. J. Cancer Res.* **2022**, *12*, 2277–2292. [PubMed]

95. Jin, H.; Zhao, J.; Zhang, Z.; Liao, Y.; Wang, C.; Huang, W.; Li, S.; He, T.; Yuan, C.; Du, W. The antitumor natural compound falcarindiol promotes cancer cell death by inducing endoplasmic reticulum stress. *Cell Death Dis.* **2012**, *3*, e376. [CrossRef]

96. Kim, H.S.; Lim, J.M.; Kim, J.Y.; Kim, Y.; Park, S.; Sohn, J. Panaxydol, a component of P anax ginseng, induces apoptosis in cancer cells through EGFR activation and ER stress and inhibits tumor growth in mouse models. *Int. J. Cancer* **2016**, *138*, 1432–1441. [CrossRef]

97. Andersen, C.B.; Runge Walther, A.; Pipó-Ollé, E.; Notabi, M.K.; Juul, S.; Eriksen, M.H.; Lovatt, A.L.; Cowie, R.; Linnet, J.; Kobaek-Larsen, M. Falcarindiol Purified From Carrots Leads to Elevated Levels of Lipid Droplets and Upregulation of Peroxisome Proliferator-Activated Receptor-γ Gene Expression in Cellular Models. *Front. Pharmacol.* **2020**, *11*, 565524. [CrossRef]

98. Cheung, S.S.; Hasman, D.; Khelifi, D.; Tai, J.; Smith, R.W.; Warnock, G.L. Devil's Club falcarinol-type polyacetylenes inhibit pancreatic cancer cell proliferation. *Nutr. Cancer* **2019**, *71*, 301–311. [CrossRef] [PubMed]

99. Zaini, R.; Brandt, K.; Clench, M.; Maitre, C. Effects of bioactive compounds from carrots (*Daucus carota* L.), polyacetylenes, beta-carotene and lutein on human lymphoid leukaemia cells. *Anti-Cancer Agents Med. Chem. (Former. Curr. Med. Chem. -Anti-Cancer Agents)* **2012**, *12*, 640–652.

100. Zidorn, C.; Jöhrer, K.; Ganzera, M.; Schubert, B.; Sigmund, E.M.; Mader, J.; Greil, R.; Ellmerer, E.P.; Stuppner, H. Polyacetylenes from the Apiaceae vegetables carrot, celery, fennel, parsley, and parsnip and their cytotoxic activities. *J. Agric. Food Chem.* **2005**, *53*, 2518–2523. [CrossRef]

101. Bernart, M.W.; Cardellina, J.H.; Balaschak, M.S.; Alexander, M.R.; Shoemaker, R.H.; Boyd, M.R. Cytotoxic falcarinol oxylipins from Dendropanax arboreus. *J. Nat. Prod.* **1996**, *59*, 748–753. [CrossRef]

102. Sapienza, C.; Issa, J.-P. Diet, nutrition, and cancer epigenetics. *Annu. Rev. Nutr.* **2016**, *36*, 665–681. [CrossRef]

103. Le, H.T.; Nguyen, H.T.; Min, H.-Y.; Hyun, S.Y.; Kwon, S.; Lee, Y.; Van Le, T.H.; Lee, J.; Park, J.H.; Lee, H.-Y. Panaxynol, a natural Hsp90 inhibitor, effectively targets both lung cancer stem and non-stem cells. *Cancer Lett.* **2018**, *412*, 297–307. [CrossRef]

104. Chatterjee, S.; Burns, T.F. Targeting heat shock proteins in cancer: A promising therapeutic approach. *Int. J. Mol. Sci.* **2017**, *18*, 1978. [CrossRef]

105. Neckers, L.; Workman, P. Hsp90 molecular chaperone inhibitors: Are we there yet? *Clin. Cancer Res.* **2012**, *18*, 64–76. [CrossRef]

106. Nair, A.; Morsy, M.A.; Jacob, S. Dose translation between laboratory animals and human in preclinical and clinical phases of drug development. *Drug Dev. Res.* **2018**, *79*, 373–382. [CrossRef]

107. Kobaek-Larsen, M.; Nielsen, D.S.; Kot, W.; Krych, Ł.; Christensen, L.P.; Baatrup, G. Effect of the dietary polyacetylenes falcarinol and falcarindiol on the gut microbiota composition in a rat model of colorectal cancer. *BMC Res. Notes* **2018**, *11*, 411. [CrossRef]

108. Hansen, S.L.; Purup, S.; Christensen, L.P. Bioactivity of falcarinol and the influenceof processing and storage on its content in carrots (*Daucus carota* L). *J. Sci. Food Agric.* **2003**, *83*, 1010–1017. [CrossRef]

109. Purup, S.; Larsen, E.; Christensen, L.P. Differential effects of falcarinol and related aliphatic C17-polyacetylenes on intestinal cell proliferation. *J. Agric. Food Chem.* **2009**, *57*, 8290–8296. [CrossRef] [PubMed]

110. Young, J.F.; Duthie, S.J.; Milne, L.; Christensen, L.P.; Duthie, G.G.; Bestwick, C.S. Biphasic effect of falcarinol on CaCo-2 cell proliferation, DNA damage, and apoptosis. *J. Agric. Food Chem.* **2007**, *55*, 618–623. [CrossRef] [PubMed]

111. Dolatpanah, M.; Rashtchizadeh, N.; Abbasi, M.M.; Nazari, S.; Mohammadian, J.; Roshangar, L.; Argani, H.; Ghorbanihaghjo, A. Falcarindiol attenuates cisplatin-induced nephrotoxicity through the modulation of NF-kB and Nrf2 signaling pathways in mice. *Res. Sq.* **2022**. [CrossRef]

112. Tan, K.W.; Killeen, D.P.; Li, Y.; Paxton, J.W.; Birch, N.P.; Scheepens, A. Dietary polyacetylenes of the falcarinol type are inhibitors of breast cancer resistance protein (BCRP/ABCG2). *Eur. J. Pharmacol.* **2014**, *723*, 346–352. [CrossRef]

113. Deding, U.; Baatrup, G.; Kaalby, L.; Kobaek-Larsen, M. Carrot Intake and Risk of Developing Cancer: A Prospective Cohort Study. *Nutrients* **2023**, *15*, 678. [CrossRef] [PubMed]

114. Deding, U.; Clausen, B.H.; Al-Najami, I.; Baatrup, G.; Jensen, B.L.; Kobaek-Larsen, M. Effect of Oral Intake of Carrot Juice on Cyclooxygenases and Cytokines in Healthy Human Blood Stimulated by Lipopolysaccharide. *Nutrients* **2023**, *15*, 632. [CrossRef] [PubMed]

115. Almqbel, M.; Seal, C.; Brandt, K. Effects of carrot powder intake after weaning on tumours in APCMin mice. *Proc. Nutr. Soc.* **2017**, *76*. [CrossRef]

116. Saleh, H.; Garti, H.; Carroll, M.; Brandt, K. Effect of carrot feeding to APC(Min) mouse on intestinal tumours. *Proc. Nutr. Soc.* **2013**, *72*, E183. [CrossRef]

117. Kobæk-Larsen, M.; Christensen, L.P.; Vach, W.; Ritskes-Hoitinga, J.; Brandt, K. Inhibitory effects of feeding with carrots or (−)-falcarinol on development of azoxymethane-induced preneoplastic lesions in the rat colon. *J. Agric. Food Chem.* **2005**, *53*, 1823–1827. [CrossRef]

118. Kobaek-Larsen, M.; El-Houri, R.B.; Christensen, L.P.; Al-Najami, I.; Fretté, X.; Baatrup, G. Dietary polyacetylenes, falcarinol and falcarindiol, isolated from carrots prevents the formation of neoplastic lesions in the colon of azoxymethane-induced rats. *Food Funct.* **2017**, *8*, 964–974. [CrossRef]

119. Crosby, D.; Aharonson, N. The structure of carotatoxin, a natural toxicant from carrot. *Tetrahedron* **1967**, *23*, 465–472. [CrossRef] [PubMed]

120. Uwai, K.; Ohashi, K.; Takaya, Y.; Ohta, T.; Tadano, T.; Kisara, K.; Shibusawa, K.; Sakakibara, R.; Oshima, Y. Exploring the structural basis of neurotoxicity in C17-polyacetylenes isolated from water hemlock. *J. Med. Chem.* **2000**, *43*, 4508–4515. [CrossRef] [PubMed]

121. Kim, T.-J.; Kwon, H.-S.; Kang, M.; Leem, H.H.; Lee, K.-H.; Kim, D.-Y. The antitumor natural compound falcarindiol disrupts neural stem cell homeostasis by suppressing Notch pathway. *Int. J. Mol. Sci.* **2018**, *19*, 3432. [CrossRef] [PubMed]

122. Czyzewska, M.M.; Chrobok, L.; Kania, A.; Jatczak, M.; Pollastro, F.; Appendino, G.; Mozrzymas, J.W. Dietary acetylenic oxylipin falcarinol differentially modulates GABAA receptors. *J Nat Prod* **2014**, *77*, 2671–2677. [CrossRef]

123. Czepa, A.; Hofmann, T. Structural and sensory characterization of compounds contributing to the bitter off-taste of carrots (Daucus carota L.) and carrot puree. *J. Agric. Food Chem.* **2003**, *51*, 3865–3873. [CrossRef] [PubMed]

124. Downs, C.; Phillips, J.; Ranger, A.; Farrell, L. A hemlock water dropwort curry: A case of multiple poisoning. *Emerg. Med. J.* **2002**, *19*, 472–473. [CrossRef] [PubMed]

125. Christensen, L. *Bioactivity of Polyacetylenes in Food Plants: Bioactive Foods in Promoting Health (Chapter 20)*; Academic Press: Cambridge, MA, USA, 2010.

126. Ozdemir, C.; Schneider, L.; Hinrichs, R.; Staib, G.; Weber, L.; Weiss, J.; Scharffetter-Kochanek, K. Allergic contact dermatitis to common ivy (Hedera helix L.). *Der Hautarzt; Z. Fur Dermatol. Venerol. Und Verwandte Geb.* **2003**, *54*, 966–969.

127. Paulsen, E.; Christensen, L.; Andersen, K. Dermatitis from common ivy (Hedera helix L. subsp helix) in Europe: Past, present, and future. *Contact Dermat.* **2010**, *62*, 201–209. [CrossRef] [PubMed]

128. Machado, S.; Silva, E.; Massa, A. Occupational allergic contact dermatitis from falcarinol. *Contact Dermat.* **2002**, *47*, 113–114. [CrossRef]

129. Leonti, M.; Casu, L.; Raduner, S.; Cottiglia, F.; Floris, C.; Altmann, K.; Gertsch, J. Falcarinol is a covalent cannabinoid CB1 receptor antagonist and induces pro-allergic effects in skin. *Biochem. Pharmacol.* **2010**, *79*, 1815–1826. [CrossRef] [PubMed]

130. Li, Y.; Wang, Y.; Tai, W.; Yang, L.; Chen, Y.; Chen, C.; Liu, C. Challenges and solutions of pharmacokinetics for efficacy and safety of traditional Chinese medicine. *Curr. Drug Metab.* **2015**, *16*, 756–776. [CrossRef]

131. Jakobsen, U.; Kobæk-Larsen, M.; Kjøller, K.D.; Antonsen, S.; Baatrup, G.; Trelle, M.B. Quantification of the anti-neoplastic polyacetylene falcarinol from carrots in human serum by LC-MS/MS. *J. Chromatogr. B* **2022**, *1210*, 123440. [CrossRef]

132. Tashkandi, H.; Chaparala, A.; Peng, S.; Nagarkatti, M.; Nagarkatti, P.; Chumanevich, A.A.; Hofseth, L.J. Pharmacokinetics of Panaxynol in Mice. *J. Cancer Sci. Clin. Ther.* **2020**, *4*, 133. [CrossRef]

 *foods*

MDPI

Review

Biologically Active Compounds in Mustard Seeds: A Toxicological Perspective

Julika Lietzow

Department of Food Safety, German Federal Institute for Risk Assessment, Max-Dohrn-Straße 8-10, 10589 Berlin, Germany; julika.lietzow@bfr.bund.de

Abstract: Mustard plants have been widely cultivated and used as spice, medicine and as source of edible oils. Currently, the use of the seeds of the mustard species *Sinapis alba* (white mustard or yellow mustard), *Brassica juncea* (brown mustard) and *Brassica nigra* (black mustard) in the food and beverage industry is immensely growing due to their nutritional and functional properties. The seeds serve as a source for a wide range of biologically active components including isothiocyanates that are responsible for the specific flavor of mustard, and tend to reveal conflicting results regarding possible health effects. Other potentially undesirable or toxic compounds, such as bisphenol F, erucic acid or allergens, may also occur in the seeds and in mustard products intended for human consumption. The aim of this article is to provide comprehensive information about potentially harmful compounds in mustard seeds and to evaluate potential health risks as an increasing use of mustard seeds is expected in the upcoming years.

Keywords: mustard; *Sinapis alba*; *Brassica nigra*; *Brassica juncea*; glucosinolates; isothiocyanates; bisphenol F; erucic acid; allergy; toxicology

Citation: Lietzow, J. Biologically Active Compounds in Mustard Seeds: A Toxicological Perspective. *Foods* **2021**, *10*, 2089. https://doi.org/10.3390/foods10092089

Academic Editor: Verica Dragović-Uzelac

Received: 9 August 2021
Accepted: 31 August 2021
Published: 3 September 2021

Publisher's Note: MDPI stays neutral with regard to jurisdictional claims in published maps and institutional affiliations.

1. Introduction

Mustard has been known as one of the oldest condiments ever and has been considered as one of the most widely grown and multifunctional plant in the world for thousands of years. The first known cultivars and use of mustard plants dating back to 3000 B.C. [1]. The following section gives a short overview about mustard plants, their taxonomic classification and the economic relevance of mustard seeds.

Mustard plants belong to the commonly known mustard family Brassicaceae of the order Brassicales (previously denoted as Capparales) including over 330 genera and over 3700 species that are distributed worldwide [2]. Notable characteristics of this family are the four sepals in median position of the flowers followed by four alternating petals that are arranged in a crossform referring to the old family name Cruciferae. The presence of organosulphur compounds is also a unique characteristic of this plant family.

The most cultivated varieties within this family almost all belong to the six species *Brassica rapa*, *Brassica juncea*, *Brassica nigra*, *Brassica carinata*, *Brassica oleracea* and *Brassica napus* in which also two of three of the most important mustard species are included (*B. juncea* and *B. nigra*). Based on cytogenetic and hybridization experiments of the Korean botanist U Nagaharu (1935), a well-known model named "U's triangle" have demonstrated the genetic relationship among these different species. The triangle is based on the theory that the genomes of three ancestral diploid species of Brassica combined to give rise to three common tetraploid vegetables and oilseed crop species. Figure 1 illustrates the triangle and shows the main forms *B. rapa*, *B. nigra* and *B. oleracea* with their own elementary genome A, B and C. The three amphidiploid species *B. juncea* (AB), *B. napus* (AC) and *B. carinata* (BC) are natural hybrids of the diploids. A genetic relationship exists also between *B. nigra* and the genera *Sinapis* L. in which the species *Sinapis alba*, belonging to one of the important mustard species, is included [1,3–5].

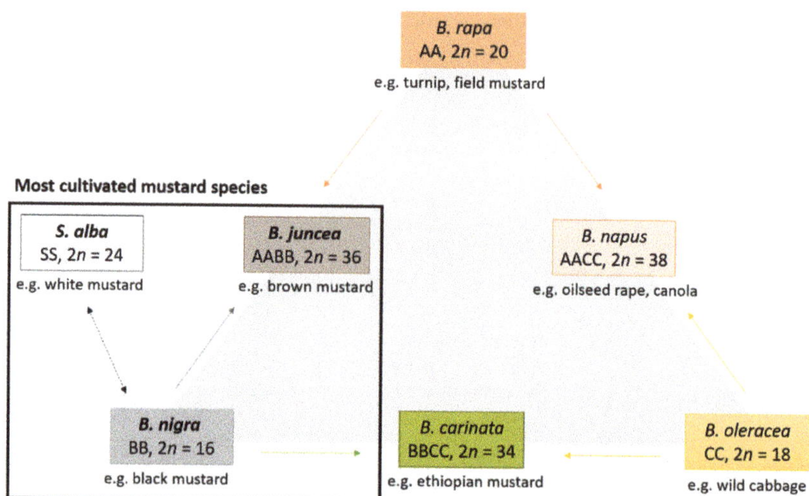

Figure 1. Taxa, genomes, chromosome number (n) and genetic relationship among Brassica species represented by U's triangle. The most cultivated mustard species that are subject of this article are framed in black.

Amongst the approximately 40 species of mustard plants in the Brassicaceae family, three types, derived from the genera *Sinapis* and *Brassica*, are mainly commercially cultivated throughout the world and used to make the mustard condiment.

The genus *Sinapis* L. contains several different types of mustard species, including *Sinapis alba* (L.) (white mustard, syn. *B. hirta* Moench or *B. alba* Linn) or *Sinapis arvensis* (charlock mustard). As mentioned above the cultivated form *S. alba* L. is phylogenetically close to Brassica species and occasionally referred to as *Brassica alba* or *B. hirta*. The seeds of *S. alba* are light straw yellow-colored and are slightly larger than the seeds of *B. nigra* and *juncea*. White mustard), believed to be native to the Mediterranean region, is the most used mustard species in Europe [1,6]. In warm regions, the mustard plant is primarily cultivated for its seeds, as a spice and for its high oil content, whilst in cold regions it is grown as catch crop, green fodder crop or as green manure. Varieties of mustard blends made from white mustard are available on the market. For example the American ballpark-style mustard is made from the white seeds, blended with vinegar and spices and usually coloured with turmeric [7]. Characteristic of the white mustard seeds is the glucosinolate sinalbin, which is enzymatically hydrolysed to 4-hydroxybenzyl isothiocyanate contributing significantly to the irritating odour and the sharp flavour of the mustard. However, white mustard is milder in taste than other mustard species. The seeds are often used for pickling gherkins or mixed pickles or making sausages [7]. Several varieties of yellow mustard exist with different qualities such as earlier maturing, nematode-resistance or altered content of glucosinolates or erucic acid. According to the catalogue of varieties of agricultural plant species more than 100 varieties of *S. alba* L. are listed [8].

Brassica nigra (L.) W.D.J. Koch (black mustard, syn.: *S. nigra, B. sinapioides*) is widely cultivated for its blackish brown-red seeds which are slightly bitter and more pungent than the seeds of the white (*S. alba*) or brown (*B. juncea*) mustard. Sinigrin is the major glucosinolate in the seeds of black mustard and can be hydrolysed to allyl-isothiocyanate (AITC) giving the characteristic of a pungent irritating odour [9]. Black mustard is commonly used as spice, potherb and as source of oil, however in the Unites States (U.S.) and Europe difficulties in harvesting reduce its popularity [7]. Likewise, the plant plays an important role in traditional medicine since eternal times, either used internally or externally. Black mustard oil is also utilized for production of soap and for medicinal remedies [10]. When black mustard seeds are grounded to meal and mixed with vinegar, a pungent condiment occurs. Among the different mustard varieties, the traditional Dijon mustard originated in the city of Dijon in France, is very popular and made from black or brown mustard seeds blended with wine (vinegar) and/or verjuice and seasonings. If vinegar is added to a ground mixture of seeds of black and white mustard a milder blend is produced (e.g., German mustard) [7].

Plants of *Brassica juncea* (L.) Czern. and Coss. (brown mustard, syn. *B. integrifolia*) are grown worldwide, in North America and Europe mainly for the seeds to be used as condiment, in the Indian subcontinent for oil production, and in Asiatic countries like China and Japan the mustard plant is used as a root and leaf vegetable [1,6]. *B. juncea* plants are shorter and have larger seeds than its relative *B. nigra*. *B. juncea* is more suitable for its agricultural cultivation and mechanical harvesting than *B. nigra* due to its higher growth and its pods, which are less prone to bursting. Therefore, brown mustard enables less labour-intensive and cost-effective production and increasingly replaces cultivation of black mustard [9]. *B. juncea* is a hybrid form derived from interspecific crosses between *B. nigra* and *B. rapa* giving it the characteristics of rapid growth from *B. rapa* and the mustard oil of *B. nigra*. Several subspecies of *B. juncea* exist with different morphologies and characteristics. Besides the oilseed types, vegetable and root type forms are cultivated for its edible leaves, inflorescences, stems and roots [1,11]. Due to its close relative to *B. napus* (oilseed rape, canola) a canola variety known as *Brassica juncea var. juncea canola* was developed from *B. juncea* resulting in low erucic acid and glucosinolate content and comparable oil and meal quality as canola (rapeseed) species [12,13]. The most important constituent in brown mustard is the natural glucosinolate sinigrin which on hydrolysis yields up to 1.4% allyl-isothiocyanate, known as volatile oil responsible for the pungent taste of brown mustard. *B. juncea* is also used as fodder and in traditional medicine, e.g., as diuretic or stimulant [10].

The world's largest producer of mustard seeds in the last years was Nepal, accounting for more than 32% of the global production in 2019, followed by Russia with 25% and Canada with 21%. In the recent past, the largest importers of mustard seeds across the globe were the U.S. and Germany, followed by France, however the import values strongly vary depending on the different mustard species intended to be market. In this case, North America is the major market for yellow mustard used for the condiment industry, whereas Europe is important for brown mustard for use as a condiment, often in the form of specialty mustards. The Asia–Pacific region is the major market for brown mustard used as spicy cooking oil and condiments [14]. According to data provided by FAOSTAT (Food and Agriculture Organization of the United Nations), the countries UK (221 K tonnes), Germany (202 K tonnes) and Italy (190 K tonnes) had the highest volumes of prepared mustard consumption in 2019, together comprising 44% of total consumption [15]. In 2019, the average consumption of mustard per capita in Germany was around 900 g [16]. So far, the average daily consumption of mustard might be in the range from 1 to 2 g and it is very common to consume around 10–20 g mustard (as condiment) during a meal.

Mustard plants, especially the seeds, are used as or in food in many different forms and due to different functional properties. Processed foods are often mixed with natural mustard seeds, such as in pickled gherkins or small white onions. Mustard is also used as an ingredient in many ready-cooked dishes like crackers, appetizers, various flours and dehydrated products for soups. Moreover, various spicy sauces, vinaigrettes and mayonnaises often contain mustard condiment. Mustard may even be present in baby food. The risk of contamination is high, e.g., in fast food restaurants or snack stands where several products with mustard are handled [9,10].

To put it briefly, the seeds of mustard plants are characterized by various nutritional and functional properties and the use in the food and beverage industry is constantly growing. As mustard plants contain several bioactive compounds, the purpose of this review is to present a comprehensive overview about potentially harmful substances in the seeds from a toxicological perspective and to evaluate possible risks to human health.

2. Biologically Active Compounds in Mustard Seeds

Long ago, mustard was considered as medicinal plant rather than as culinary one pointing out the existence of biologically active compounds in the plants and their seeds. In general, these so-called "bioactive compounds" are typically present in various natural-based sources characterized by specific functionalities. In recent years, the research interest in these phytochemicals increased exponentially in various commercial industries. Moreover, scientific and (bio)technologic developments have led to immense nutritional discoveries and product developments resulting in increasing numbers of food products with potential medical and health benefits. Due to its broad use as nutraceutical or functional constituents for several health purposes, biologically active compounds have entered the market in many different forms (e.g., as food supplements), partially without reliable scientific basis with respect to efficacy or safety [17,18].

The growing consumer interest towards different taste preferences and healthy eating habits increases the use of mustard seeds and products made thereof. Due to its different functional and technological properties, mustard is gradually utilized in a wide-range by the food and beverage industry as well as in the cosmetic and pharmaceutical industry Moreover, the use of mustard plants has gained increasing interest for several non-food uses. For instance, seed meal of yellow mustard (*S. alba*) was shown to be efficient in controlling weeds, and oriental mustard (*B. juncea*) seed meal has been used as a broad-spectrum pesticide to control nematodes, insects, and fungi [19,20]. Current and future studies will undoubtedly focus on the health benefits of mustard plants and seeds whereas information on undesirable and antinutritional compounds in these products intended for human consumption will tend to fade into the background. Therefore, this article underscores and evaluates those compounds in mustard seeds potentially exerting undesired effects in humans.

2.1. Glucosinolates

The unique properties of glucosinolates and their breakdown products isothiocyanates, so called mustard oils, providing the typical flavour and (bitter) taste of mustard plants. Especially their seeds have been known for hundreds of years [21]. Since then, up to 200 different glucosinolates have been reported, predominantly in the plant families of the order Brassicales, but also in plants of the genus Drypetes [22,23].

Glucosinolates may be defined as amino acid-derived sulphur-containing glycosides sharing a generic chemical structure consisting of thiohydroximates carrying an S-linked b-glucopyranosyl moiety and an O-linked sulfate residue, and with a variable side chain (R). Until now, more than 200 side-groups (alkyl, alkenyl, aryl, indolyl) have been identified. Occasionally, further substituents are attached to O, S or N atoms of the side chain or glucosyl moiety [22–24].

Glucosinolates can be divided into three chemical groups including aliphatic, indole and aromatic glucosinolates depending on the amino acid precursor. Aliphatic glucosinolates are derived from alanine, leucine, isoleucine, valine and methionine. Indolic group are derived from tryptophan and aromatic glucosinolates are derived from phenylalanine and tyrosine, respectively. The majority of those are the aliphatic glucosinolates derived from chain-elongated methionine [25–27].

Most of the hydrophilic glucosinolates are chemically and thermically stable. However, they can be converted into diverse breakdown products by specific ß-thioglucosidases, referred to as myrosinases, which are present in glucosinolate-producing plants but also in fungi and in bacteria commonly associated with gut microflora [28,29] (Figure 2).

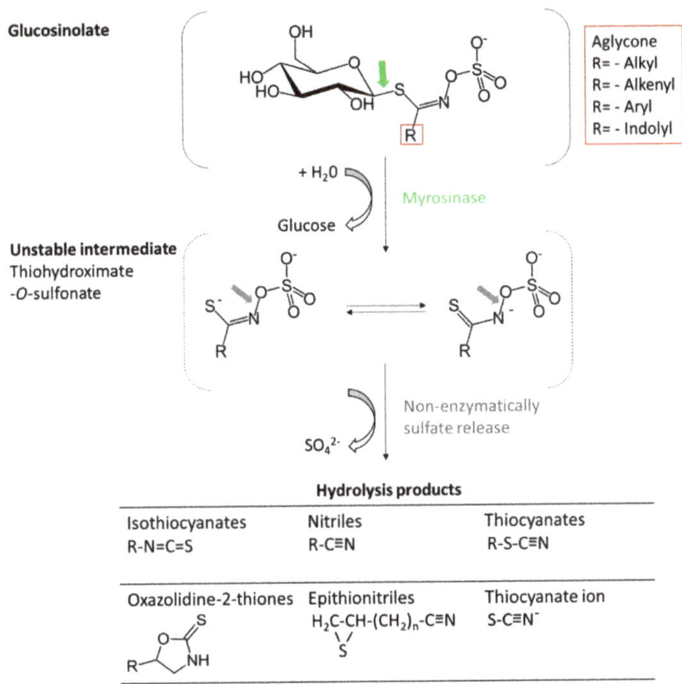

Figure 2. Enzymatic processing of glucosinolates by myrosinase into hydrolysis products.

In that regard it is of interest, that significant amounts of metabolites of glucosinolates (largely dithiocarbamates) were excreted via the urine of healthy human volunteers after eating brassica vegetables, even when myrosinase has been completely heat-inactivated. The excretion falls to negligible levels when the microbiota was disturbed, e.g., due to antibiotic treatment [30]. Therefore, bacterial myrosinase activity can cause the breakdown of glucosinolates in the distal gut, leading to release of several glucosinolate metabolites, e.g., isothiocyanates into the faecal stream and/or the urine.

In intact plants, myrosinases are separated from its substrates, e.g., in different cells or in different intracellular compartments. Damage of the plant tissue in case of preparing or chewing results in enzyme release. In the presence of water myrosinase initiates upon catalyzation by ascorbic acid hydrolytic cleavage of the β-glucosyl moity forming glucose, hydrogen ion and an unstable aglycone (thiohydroxamate-O-sulfonate). It is of note, that myrosinases are activated to various degrees by ascorbic acid, and in some instances the enzyme is almost inactive in its absence [22]. After the non-enzymatically release of the sulfate residue various volatile and non-volatile compounds derive, which exert different biological activities such as plant defence against insects or phytopathogens, or as attractants, and also give the specific flavour and pungency of mustard [31,32].

The hydrolysis products include isothiocyanates, previously known as "thioglucosides" or "mustard oils", thiocyanates, nitriles, epithionitriles, oxazolidine-2-thiones (5-vinyl-2-oxazolidinethione and 5-vinyl-1,3 oxazolidine-2-thione), and indol derivatives depending on the aglycone structure and reaction conditions (e.g., pH value, ferrous ion concentration, specifier proteins) (Figure 2). Considering plant biology, these alternative breakdown products provide an additional level of plant defence [33,34].

Under physiological conditions, isothiocyanates contribute to 60–90% of the total glucosinolate breakdown products. After release of glucose, the thiohydroximate-O-sulfonate is formed, which can be spontaneously degraded by a Lossen-like rearrangement to the relatively stable isothiocyanates.

Isothiocyanates are very reactive and toxic for microbes, nematodes, fungi and insects, however many species of the Brassicaceae are able to promote alternative activation pathways by the action of specifier proteins. In the presence of these proteins, formation of isothiocyanates is reduced in favour of alternative breakdown products [35].

For instance, the presence of nitrile specifier proteins (NSPs) leads to an increased formation of nitriles [36,37], whereas, the occurrence and activity of the epithiospecifier protein (ESP) leads to the generation of epithionitriles from alkenyl-glucosinolate aglycons as well as nitriles from non-alkenyl-glucosinolate-aglycons. Moreover, it was shown that at physiological pH, isothiocyanates are the major products, whereas nitriles are formed at more acid pH [38].

Each type of Brassica vegetables including mustard plants show a characteristic glucosinolate composition and most species contain a limited number of glucosinolates [22]. The amount and composition vary among the different plant organs such as roots, leaves, stems and seeds, mainly with the highest concentrations often found in the reproductive tissues (florets, flowers and seeds). Therefore, for quantification of glucosinolates and their corresponding breakdown products, the seeds seem to be the best bulk source. The glucosinolate profile also depends on various environmental and ecophysiological factors such as plant nutrition, water availability, plant age and plant cycle [39].

Tables 1 and 2 give an overview of glucosinolates and their corresponding desulfated breakdown products identified in the mainly cultivated mustard species *S. alba, B. nigra, B. juncea* [22,24,29,40]. The mustard species *S. alba, B. nigra* and *B. juncea* show large ranges between the highest and lowest levels of total glucosinolate content as well as different content of predominant glucosinolates, rise up to 200 µmol/g seed.

Table 1. Distribution of glucosinolates among mustard plant species.

Mustard Species	Common Names	Glucosinolates Identified	Systematic Name
		Trivial Name	**(a-Glycone = R Side Chain)**
S. alba	white, yellow mustard	Gluconapin [a]	3-Butenyl
		Progoitrin [a]	2R-2-Hydroxy-3-butenyl
		Glucobrassicanapin [a]	Pent-4-enyl
		# (Gluco-)Sinalbin [b]	4-Hydroxybenzyl
		Glucotropaeolin [b]	Benzyl
		Gluconasturtiin [b]	2-Phenylethyl
		Glucoerucin [c]	4-Methylthiobutyl
		Glucoibe(rve)rin [c]	3-Methylthiopropyl
		Glucoiberin [c]	3-Methylsulphinylpropyl
		2-Methylpropyl Isobutyl [d]	
		Glucobrassicin [e]	3-Indolylmethyl
		Neoglucobrassicin [e]	N-Methoxy-3-indolylmethyl
B. nigra	wlack, shortpod mustard, moutarde noire	# (Gluco-)Sinigrin [a]	2-Propenyl
B. juncea	brown, indian, asiatic, chinese, sarepta mustard	# (Gluco-)Sinigrin [a]	2-Propenyl
		# Gluconapin [a]	3-Butenyl
		Progoitrin [a]	2R-2-Hydroxy-3-butenyl
		Epiprogoitrin [a]	2S-2-Hydroxy-3-butenyl
		Glucosinalbin [b]	4-Hydroxybenzyl

predominant glucosinolates in mustard seeds; [a] aliphatic olefin, [b] aromatic aryl, [c] S-containing, [d] aliphatic branched chain, [e] aromatic indol.

Table 2. Glucosinolate content in mustard species.

Mustard Species	Total Content of Glucosinolates	Content of Predominant Glucosinolates	Reference
S. alba		Sinalbin (seed): 250 μmol/g sinalbin	[23]
		Sinalbin (seed): 165 ± 3 μmol/g Sinigrin (seed): 155 ± 4 μmol/g	[41]
	Seed: 45.4–61.9 g/kg FW	Sinalbin (seed): 16.6–46.2 g/kg FW Progoitrin (seed): 0.8–1.3 g/kg FW Glucoibervirin/Glucobrassicin/4-hydroxglucobrassicin (seed): <0.2 g/kg FW	[42]
B. nigra	Seed: 207.2–229.0 μmol/g FW	Sinigrin (seed): 202–226.7 μmol/g FW	[43]
	Seed: 26–62.4 g/kg FW	Sinigrin (seed): 24.5–61.2 g/kg FW Gluconapin/Progoitrin/Glucoibervirin (seed): <0.3 g/kg FW Sinalbin/4-hydroxglucobrassicin (seed): <0.9 g/kg FW	[42]
B. juncea	Intact seed: 15.7–127.6 μmol/g DW Leaf: 4.3–129.9 μmol/g DW	Gluconapin: (leaf): 0.6–70.1 μmol/g DW (seed): 2.3–90.8 μmol/g Sinalbin (leaf): 0.1–64.9 μmol/g DW (seed): 0.2–35.2 μmol/g Epiprogoitrin (leaf): 0.7–16.4 μmol/g DW	[44]
	Seed: 94.4–169.2 μmol/g FW	Sinigrin (seed): 43.79–145.5 μmol/g FW Gluconapin (seed): 0–110.58 μmol/g FW	[43]
	Seed: 68–153 μmol/g FW	Sinigrin (seed): 0–134.2 μmol/g FW Gluconapin: 68–153 μmol/g FW	[45]
	Defatted seed meal: 58.83–132.26 μmol/g		[46]
	Oil free seed meal: 123.8–152.1 μmol/g		[47]
		Sinigrin (seed): 155 ± 4 μmol/g	[41]
	Seed: 18.1–61.8 g/kg FW	Sinigrin (seed): 16.6–46.2 g/kg FW Gluconapin/Progoitrin/Glucoibervirin/Glucoerucin/ Sinalbin/Gluconasturiin (seed): <1 g/kg FW 4-hydroxyglucobrassicin (seed): 0.6–1.6 g/kg FW	[42]

DW dry weight, FW fresh weight.

Sinalbin and sinigrin whose chemical structure was elucidated by Ettlinger and Lundeen [48,49] belong to the main glucosinolates in mustard seeds. In brown mustard (*B. juncea*) and black mustard (*B. nigra*) seeds sinigrin is the predominant compound degrading to allyl-isothiocyanate (AITC) after hydrolysis, whereas, the main glucosinolate of white mustard (*S. alba*) is sinalbin yielding 4-hydroxybenzyl isothiocyanate (29) (Figure 3).

The concentration of glucosinolates and their breakdown products in mustard plants, especially in the seeds, is largely dependent on the activity of myrosinase or related bacterial enzymes or further chemical transformation prior to consumption [26].

In case of mustard as food one has to distinguish between the intake of whole not-processed seeds as seasoning in several dishes, and processed seeds, for example as mustard flour made from ground, peeled and non-degreased seeds or as oil made by pressing the seeds. Prepared mustard is mostly composed of a mixture of ground mustard seed and/or mustard flour and/or mustard cake [9,10].

In general, glucosinolates are relatively stable in the seeds until cell damage results in chemical degradation involving myrosinase-catalysed hydrolysis. Thermal treatment leads to inactivation of myrosinase via denaturation [50] and preserve most of the glucosinolates. Van Eylen et al. showed that the optimum temperature of myrosinase activity of mustard seeds from *S. alba* was at 60 °C, however the activity decreased when the temperature of the experimental system was higher and was influenced by several other factors such as pH or presence of ascorbic acid [51].

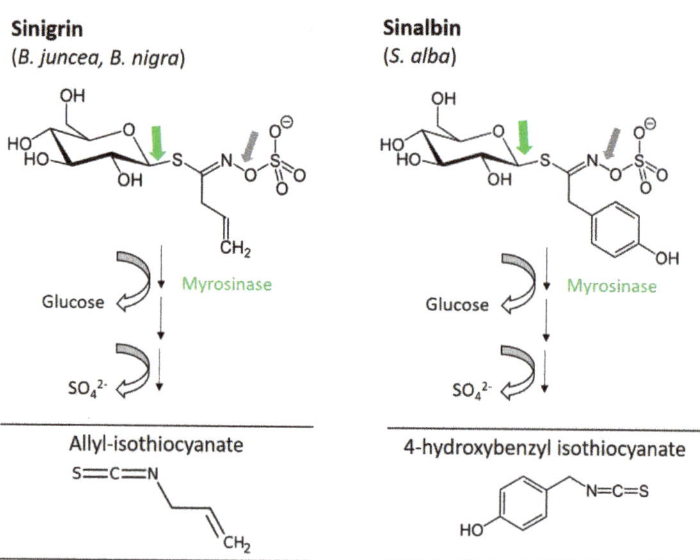

Figure 3. Main glucosinolates and corresponding hydrolysis products in brown (*B. juncea*), black (*B. nigra*) and white (*S. alba*) mustard seeds.

This occurs for example when whole mustard seeds are added as spice to dishes which are cooked, like soups or stews. In this case, intact glucosinolates might reach the colon, however a certain amount of the glucosinolates seem to be hydrolysed in the stomach or can be absorbed via passive transport or via diffusion in the small intestine. In the colon, several bacteria strains are able to hydrolyse glucosinolates to form isothiocyanates, amines or nitriles depending on the type of bacterial myrosinase-like activity. Usually, cooking reduces the concentration of glucosinolates, partly through thermal breakdown and partly to leaching of the intact glucosinolates and their derivatives into the cooking liquid. Besides, heat treatment inhibits the activity of myrosinase through denaturation of the enzyme [52,53]. Therefore, the processing method can make a very large difference, both to the intake of glucosinolates, and to the bioavailability of their breakdown products.

The pattern of intact glucosinolates and breakdown products upon heat treatment is influenced by the thermal stability of the corresponding molecules. For instance, indol glucosinolates—one of the best-known representatives is progoitrin—seem to be less stable than aliphatic ones such as sinigrin, which is the predominant glucosinolate in black and brown mustard seeds. However, sinigrin totally disappears upon boiling, while 5-vinyloxazolidine-2-thione (derived from progoitrin) and 3-methyl-sulphinylpropylisothiocyanate (derived from glucoiberin) may partly escape decomposition [54].

Comparison between different mustard seed preparations showed that the seed alone had higher sinigrin and lower content of the breakdown product isothiocyanate compared to mustard preparations, whereas the wholegrain-style mustard contained lower sinigrin and higher isothiocyanate levels. This suggests that during mustard preparation, the enzyme myrosinase can break down sinigrin in the presence of high salinity and acidity from added ingredients, such as salt or vinegar [55]. Isothiocyanates are volatile and water-soluble. It can be expected that they largely disappear after processing steps using high temperatures and soaking in water, but those processes are commonly not included during mustard preparation.

Interestingly, in a thesis, it was shown that beside allyl-isothiocyanate derived from sinigrin, sulforaphane derived from glucoraphanin also corresponds to the most abundant breakdown product detected in different samples of table mustard. In this case, concentration of allyl-isothiocyanates ranged from 4 to above 200 mg/100 g fresh weight whereas sulforaphane ranges from 24 to above 188 mg/kg. These unexpected results of sulforaphane quantification may be explained by the fact that table mustard is a mixture and probably also contain glucoraphanin-containing ingredients [56].

Consumption of mustard seeds, either whole or after preparation, may lead to the intake of intact glucosinolates as well as their different breakdown products. One has to note, that the pungent taste of different prepared mustard samples strongly depends on levels of allyl-isothiocyanates [57]. The dietary intake of glucosinolates in general have been reported ranging from 2 to 29 mg/day depending on population groups and countries [58]. However, dietary intake of glucosinolates through the consumption of mustard as well as exposure data on different types of glucosinolates, especially of sinigrin and sinalbin, and their corresponding breakdown products such as allyl-isothiocyanates, are not available.

2.1.1. Toxicological and Antinutritional Effects

Many plants of the Brassicaceae family containing glucosinolates are used for human and animal nutrition. Glucosinolates themselves are biologically inactive, however various breakdown products exert a variety of toxicological and antinutritional effects in animals whereas the information on adverse effects in humans is limited.

Numerous studies confirm the negative effects that occur in animals after feeding of glucosinolate-containing rapeseed cake (de-oiled seeds). After oil extraction from oilseed crops, such as rapeseed (*B. napus)* or white mustard (*S. alba)*, the hydrophilic glucosinolates remain in the seed meal fraction (by-product) used in animal diets [25]. The presence of the glucosinolates sinigrin and progoitrin in the diet is strongly associated with bitter taste and is responsible for reduced palatability corresponding to a lower intake of glucosinolate containing animal diets [26,59].

Although progoitrin is a non-bitter glucosinolate, it produces more profound bitter taste compared to sinigrin due its degradation to the extremely bitter compound goitrin by myrosinase or by heat treatment [60]. In Annex I of Directive 2002/32/EC, vinyl thiooxazolidone (5-vinyloxazolidine-2-thione (5-VOT), known as goitrin) and volatile mustard oil expressed as allyl-isothiocyanate are listed as undesirable substances with maximal amounts for animal feed. Humans are also sensitive to the strong flavours of glucosinolate breakdown products, and these compounds are therefore important determinants of intentionally or unintentionally flavour [61].

The main toxic effects in animals can be described as decreased feed intake, growth depression, enlargement of liver and kidneys, structural and functional alteration of the thyroid gland, and reprotoxic effects such as embryo mortality in mammals and reduced egg production in birds, amongst which the adverse effects on thyroid metabolism are the most thoroughly studied. Even though the extent of the detrimental effects both on productivity and on health vary within the animal species and the amount of glucosinolates, it is suggested that the formation of isothiocyanates, thiocyanates, oxazolidinethiones and nitriles from glucosinolates strongly contribute to the observed toxicological effects [62,63].

2.1.2. Goitrogenic Effects

The reduction of the livestock feeding quality of seed meal following oil extraction is largely due to the presence of thiocyanate ion (SCN-), isothiocyanates and oxazolidine-2-thiones, which all have been shown to be goitrogenic [64].

At neutral pH, myrosinase-initiated hydrolysis of glucosinolates mainly yield isothiocyanates which are relatively stable in aqueous solutions. In comparison and especially at higher pH, isothiocyanates from glucobrassicin and neoglucobrassicin, which belong to indolylglucosinolates, as well as sinalbin may degrade to free thiocyanate ion and further metabolites [65]. McGregor et al. reported in oil-free seed meal from four cultivars of *B. juncea* thiocyanate ion concentration of 5.4 to 6.9 µmol/g whereas the amount of volatile isothiocyanates correspond to 116.9 to 145.9 µmol/g [47]. Mustard seeds of *S. alba* contain the glucosinolate sinalbin, which under physiological condition (pH 5–7) is decomposed mainly to 4-hydroxybenzylalcohol and thiocyanate ion. Under acidic conditions other degradation compounds were identified, such as 4-hydroxybenzyl cyanide or 4-hydroxybenzylnitrile (2-(4-hydroxyphenyl)acetonitrile) [66]. Paunovic et al. investigated sinalbin degradation products in ground yellow mustard seeds and paste. They identified 2-(4-hydroxyphenyl)acetonitrile as the major degradation product in ground seeds, whereas the most abundant sinalbin degradation product in mustard paste was 4-(hydroxymethyl)phenol. Other degradation compounds identified were 4-methyl phenol, 4-ethyl phenol, 4-(2-hydroxyethyl)phenol and 2-(4-hydroxyphenyl) ethanoic acid [67].

The thiocyanate ion formed by degradation of glucosinolates is known to be a competitive inhibitor of the sodium/iodide (NIS) symporter located on the basolateral membrane of the thyroid follicular cell. Consequently, the iodide uptake by the thyroid gland is impaired which can lead to reduced synthesis of thyroid hormones [68]. The effect is only evident in iodine-deficient diets and increased iodine intake can prevent the adverse effects.

This is not the case for goitrin. Isothiocyanates that are derived from progoitrin (2-hydroxy-3-butenyl glucosinolate) cyclize to produce 5-vinyloxazolidine-2-thione, known as goitrin. Goitrin is a potent inhibitor of the thyroid peroxidase, which plays a central role in the thyroid hormone biosynthesis by regulating organification, iodination and coupling reactions to form thyroid hormones (T3 and T4) in the thyroid gland. This leads to suppressed thyroxine secretion and reduced serum tetraiodothyronine (T4) concentration and can result in a compensatory increase in thyroid glandular mass (known as goiter-hyperplasia and hypertrophy). This has been shown in several animal species. Adverse effects in animals were observed with different concentrations of progoitrin/goitrin in the diet, ranging from 71 to 755 μmol/100 g depending on species investigated. In general, ruminants seem to be less sensitive to dietary glucosinolates, unlike pigs, which are severely affected by dietary glucosinolates compared to rabbit, poultry, and fish [63,69].

Besides isothiocyanates, nitriles have been reported to exert goitrogenic effects [70] however only sparse descriptions exist about the mode of action of nitriles on the thyroid gland. More reliable data report adverse effects on liver and kidney functions of this substance class [62]. The formation of nitriles from glucosinolates is favoured under acid conditions or in the presence of Fe(II) ions, whereas epithiospecific protein (ESP) promotes the formation of epithionitriles.

It was shown that nitriles were the dominant glucosinolate products from hydrolysis of aliphatic glucosinolates in cabbage presumably due to the effect of endogenous Fe(II) ions [71]. However, limited data exist regarding the amount or formation of nitriles consuming mustard seeds. Cole et al. showed negligible concentrations of nitriles in seeds of *B. juncea* or *B. nigra* [72] and Choi et al. detected only very small amounts of organic nitriles in serum of rats treated with sinigrin, the predominant glucosinolate in mustard seeds [73].

In principle, exposure of isothiocyanates and goitrin derived from glucosinolates, predominantly sinigrin and gluconapin occurring in mustard seeds, might also be capable of inducing goitrogenic effects in humans. However, no reliable studies exist regarding exposure of goitrogens due to the consumption of mustard and mustard products and higher risk of goiter or related pathologies.

Studies in human volunteers indicate that exposure of goitrogens (e.g., progoitrin) through the intake of realistic amounts of Brassica vegetables does not alter thyroid function. However, if consumed in considerable quantities it may contribute to development of goiter [74,75].

The absence of epidemiological evidence regarding goitrogenic effects after consumption of Brassica vegetables containing high amount of these toxic compounds, such as goitrin from Brussels sprouts, is partly due to the fact that cooking inactivates myrosinase responsible for activation of progoitrin to goitrin, thus reduces the biological availability of goitrogenic breakdown products to sub-clinical levels.

It needs to be emphasized, that mustard seeds or mustard condiment are generally not heat-treated and damage of the seeds might form high amounts of goitrogens whose exposition should still be considered and monitored, especially in iodine-deficient population groups or in subjects where thyroid hormone production is already impaired. Additionally, even if myrosinase activity is diminished, the ability of intestinal microorganisms to readily convert inactive precursors, such as progoitrin, into goitrogenic compounds, should also be considered.

2.1.3. Genotoxic and Carcinogenic Effects

Certain glucosinolate breakdown products have been under observation as possible human carcinogens. Several of these compounds are electrophilically reactive, for example, due to the highly electrophilic central carbon in the isothiocyanate group (-N=C=S), and possibly undergo addition as well as substitution reactions with nucleophiles.

In the past, several isothiocyanates from Brassica vegetables including broccoli tested as freshly prepared juice were investigated using in vitro models and partially demonstrated dose-dependent genotoxic effects in bacterial and mammalian cells [76–80].

Musk et al. observed significant clastogenic activity (chromosome aberration and sister chromatid exchanges) in in vitro cell lines for phenyl isothiocyanate (PITC), phenethyl isothiocyanate (PEITC) and benzyl isothiocyanate (BITC), which is one of the major breakdown products of the glucosinolate sinalbin contained in white (*S. alba*) mustard seeds. No mutagenic effects were observed for allyl-isothiocyanate (AITC), known as the predominant hydrolysis product of sinigrin and the major flavour constituent of brown (*B. juncea*) and black mustard (*B. nigra*) seeds [81,82].

With regard to AITC, the EFSA Panel on Food Additives and Nutrient Sources added to Food (ANS) summarized in its scientific opinion on the safety of AITC used as a food preservative genotoxicity tests in in vitro and in vivo studies (summary of the studies in [83]). Positive as well as negative results have been observed in genotoxicity tests with AITC in bacterial and mammalian cells in vitro when no metabolic activation occurred. Mutagenic effects mostly occurred when cytotoxic concentrations of AITC were used. In the presence of metabolic activation, the results were generally negative, even in modified Ames tests with extended preincubation time suggesting that cytochrome P450 isoenzymes seem to be able to reduce markedly the occurrence of mutagenic derivatives. According to the EFSA Panel, reactive oxygen species might be involved in the genotoxic effects of AITC in vitro. On the other hand, no convincing evidence on genotoxicity was shown in in vivo studies performed on different genetic endpoints. Therefore, the Panel did not raise concern regarding genotoxic effects of AITC [83].

Moreover, AITC administered via gavage was not carcinogenic for B6C3F1 mice of either sex, however an increased incidence of transitional cell papillomas of the urinary bladder of F344/N male rats was reported [84]. Possible explanations for the carcinogenic effects of AITC at repeated high doses in male rats consider the accumulation of the corresponding mercapturic acid conjugate of AITC. The AITC conjugate *N*-acetyl-S-(*N*-allylthiocarbamoyl)-L-cysteine is the major urinary metabolite in humans and rats whereby in rats the AITC clearance is slower compared to humans. High concentrations of the N-*acetyl*-cysteine conjugate in the bladder might therefore be a direct irritant on the bladder epithelium or dissociate into free AITC likewise act as irritating substance resulting in regenerative hyperplasia and subsequent benign papillomas. Considering that no genotoxicity in vivo was observed, a threshold mechanism (high-dose response) of AITC underlying these effects in the urinary bladder was assumed [83].

Several aliphatic, aromatic and indole glucosinolates were also tested by Baasanjav-Geber et al. for their mutagenic potential using Ames test and two different *S. typhimurium* strains. Mutagenicity was observed only when myrosinase was added to the test compounds demonstrating the predominant role of the glucosinolate breakdown products with respect to the mutagenic potency. The indole glucosinolate neoglucobrassicin, mainly occurring in broccoli and kale, had the highest increases in the number of revertants in both strains. Sinalbin, which is the main glucosinolate in white mustard (*S. alba*) seeds and hydrolysed to 4-benzyl isothiocyanate (BITC), likewise had a high number of revertants. However, the mutagenic potency of sinalbin was much lower and similar to other glucosinolates tested [85].

In summary, high concentrations of several glucosinolates, especially indolic glucosinolates and their degradation products, isothiocyanates have been shown to be mutagenic in vitro and the results might give some cause for concern. However, clear genotoxic or carcinogenic effects in humans have not been reported [86,87]. In this context, it has to be emphasized that under in vivo test conditions isothiocyanates may be efficiently detoxified [26,76,78].

2.1.4. Evaluation of Toxicological Effects

The characteristic hot and pungent flavour of mustard is based predominantly on isothiocyanates produced from the parent compounds sinigrin and sinalbin, though various types of other glucosinolate-derived products, such as goitrin, may also occur but in much smaller quantities. The described toxic and antinutritional effects of these biologically active breakdown products seem predominantly occur in animal species rather than in humans and with the administration of high doses.

In 2008, EFSA considered the Joint FAO/WHO Expert Committee on Food Additives (JECFA) evaluations of miscellaneous nitrogen-containing substances including allyl-isothiocyanate and benzyl-isothiocyanate. The panel agreed, based on a Maximised Survey-derived Daily Intake (MSDI) approach, with the JECFA conclusion that "No safety concern at estimated levels of intake as flavouring substances" is made, which also applied to allyl-isothiocyanate and benzyl-isothiocyanate, and that "the available data on genotoxicity and carcinogenicity do not preclude the evaluation of the flavouring substances through the Procedure" [88,89].

Except for AITC [83], no comprehensive toxicological evaluation and derivation of health-based guidance values of glucosinolates or their degradation products predominantly occurring in mustard seeds have been identified. In its opinion dealing with the safety of AITC for the proposed use as food additive, EFSA established an acceptable daily intake (ADI) of 20 µg/kg body weight. AITC concentrations in mustard seeds and products vary widely depending on mustard species and process conditions [10,57,83,90]. Assuming an average mustard consumption of 1 g/day containing 1000 mg AITC per kg mustard product, the intake of AITC would be 1 mg per day and would not exceed the ADI of 1.4 mg for an adult weighing 70 kg. However, if the body weight is lower or the AITC concentration is higher, up to 15,000 mg/kg mustard, as it is documented by EFSA [83], would result in exceedance of the ADI, e.g., for children or adolescents.

Numerous factors, such as the activity of myrosinase and specifier proteins as well as processing conditions (temperature, pH, acidic environment) may influence the conversion rate of glucosinolates to isothiocyanates and the formation of more or less, partially not yet known, toxic derivatives.

Early experimental data have indicated that isothiocyanates might be mutagenic and genotoxic in various assays, however evidence of carcinogenicity in experimental animals is inconclusive and human studies are limited. On the opposite, research has recently focused on the anti-carcinogenic properties of constituents of Brassica vegetables, especially of isothiocyanates. Anti-carcinogenic effects have been attributed to the inhibition of cytochrome P450 enzymes, and a concomitant induction of phase II detoxifying enzymes, thereby preventing the activation of pro-carcinogens and improving their conjugation and elimination [91–93].

Therefore, it is not surprising that glucosinolates and their hydrolysis products are already on the market as ingredients in dietary supplements and concentrated herbal preparations. However, most evidence concerning anti-carcinogenic effects and the mechanism of action has come from in vitro or animal studies.

Despite the huge number of epidemiological studies reporting on decreased cancer risk in humans along with a higher intake of Brassica vegetables, an increasing (therapeutic) use of high-dose, partially isolated glucosinolates (isothiocyanates) should be questioned when clear dose–effect relationships are not clinically proven and the tissue bioavailability is largely unknown. It is certainly possible that some glucosinolates have properties that are both carcinogenic and chemopreventive [94].

Furthermore, it has to be noticed that volatile mustard oils including AITC and goitrin are listed as undesirable substances in animal feed [61]. Pointing out the harmfulness of these compounds and the need of monitoring including exposure assessments when used by humans in concentrated forms is necessary because human studies focusing on adverse effects are still scarce. However, the intake of biologically active glucosinolate breakdown products from mustard seeds is comparatively low and could be considered to be tolerated when for example the ADI established for allyl isothiocyanates from whatever source is not exceeded.

2.2. Bisphenol F

Bisphenol F is an aromatic compound related to bisphenol A through its basic structure containing two phenol groups. The two aromatic rings of bisphenol F are linked through methylene. Generally, bisphenol F comprises several isomers such as 2,2′-, 2,4′- and 4,4′-dihydroxydiphenylmethane. Recently, bisphenol F was detected in several mustard products containing seeds of white mustard (*S. alba*) as well as in various plants of the orchid family (Orchidaceae) (*Coeloglossum viride* var. *bracteatum* (rhizome), *Galeola faberi* (rhizome), *Gastrodia elata* (rhizome), *Tropidia curculioides* (root)) and in the seeds of *Xanthium strumarium*, which are partly used as herbal remedies [95] (Table 3).

Table 3. Occurrence of bisphenol F in foods related to mustard.

Food Items	Bisphenol F Content (in mg/kg)	Reference
Condiments	2.4	[96]
Commercial table mustard (n = 61) thereof mild mustards (n = 19)	mean 1.84 (max 8.4) mean 3.2	[97,98]
Commercial table mustard Medium hot mustard (n = 39) Sweet mustard (n = 7) Hot mustard (n = 10) Quince mustard (n = 1) Self-made mustard products Mustard flour Table mustard from *B. nigra* Table mustard from *B. juncea* Table mustard from *S. alba*	<5 (n = 37); >9 (n = 2) <3 <3 <1 n.d. n.d. n.d. 0.18–0.94	[99]
Mustard dressing	1.13	[100]

n.d. not detected.

Bisphenol F is structurally similar to bisphenol A and belongs to one of the main substitutes replacing bisphenol A in the production of epoxy resins and polycarbonates used in the manufacture of adhesives, plastics, coatings and other applications [101]. Thus, the occurrence of bisphenol F is constantly increasing in products leading to consumer exposure from a broad number of sources.

In contrast to bisphenol A, bisphenol F in mustard and other plant species is not derived from synthetic materials, packaging or other sources of contamination, but is most likely formed as a breakdown product of glucosinolates [97].

Evaluation of Possible Health Risks

The occurrence of bisphenol F in mustard products is obviously due to the formation from a natural ingredient in the seeds during mustard production. Zoller et al. excludes the origin of bisphenol F from contamination through epoxy resin or other sources like food packaging. In support of this, the 4,4′-isomer was nearly exclusively detected in the samples. A contamination with bisphenol F would most likely lead to the occurrence of all isomers which are commercially used as a mixture of the three isomers, respectively [97].

It was observed that the highest levels of bisphenol F were detected in mild mustard samples compared to spicy ones. Additionally, bisphenol F was detected in mustard produced from seeds of *S. alba* but not from *B. juncea* or *B. nigra* or in mustard flour [97,99]. Therefore, based on differences in composition of the mustard species it is considered that bisphenol F is formed from the glucosinolate glucosinalbin, because it is predominantly found in the seeds of white mustard (*S. alba*) but not in *B. juncea* or *B. nigra*. This possibly can also explain the higher levels of bisphenol F in mild mustard types mainly contain *S. alba* which is milder in taste than other mustard species.

Recent findings from Zoller et al. verified the myrosinase-catalysed formation of 4-hydroxybenzyl isothiocyanate from 4-hydroxy benzyl glucosinolate (sinalbin). In the presence of water, the isothiocyanate is hydrolysed to the intermediate 4-hydroxybenzylalcohol. Subsequently, formation of a carbocation from the isothiocyanate by release of thiocyanate or from the 4-hydroxybenzyl alcohol by protonation and release of water is suggested. A further 4-hydroxybenzyl alcohol molecule could result in dimerization and formation of the 4,4′-isomer of bisphenol F [97]. Additionally, it was pointed out that 4,4′-bisphenol F could also be formed from 4-hydroxybenzylalcohol in the stomach due to its acidic conditions as it was also shown for other glucosinolate breakdown products [102].

After absorption bisphenol F is distributed throughout the body, also including the reproductive organs and the foetus by crossing the placental barrier. As glucuronide or sulfate conjugate, bisphenol F is rapidly excreted mainly via urine and to a lesser extent in faeces, however residues of bisphenol F were still detectable in tissues 96 h after a single dose was administered to rats [103].

So far, no comprehensive evaluation of possible health risks due to the exposure of bisphenol F in humans is available. Subchronic and chronic toxicity studies as well as studies on reproductive and developmental toxicity are still lacking. However, the structural similarity of bisphenol F to bisphenol A suggests comparable biological (adverse) effects and the potential to disrupt the human endocrine system [104–108].

In contrast to bisphenol A [104], no health-based guidance values for bisphenol F or legal regulations for foodstuff exist. However, some authorities evaluated the possible health risks to consumers from the natural occurrence of bisphenol F in mustard products. The Swiss authority FSVO (Federal Food Safety and Veterinary Office) carried out a risk assessment and evaluated the ratio between the lowest observed adverse effect level (LOAEL) and the dietary intake of bisphenol F from mustard products. In a very rudimentary approach, it was concluded that a daily intake of 0.67 mg of bisphenol F may be tolerable for a person of 60 kg body weight corresponding to a tolerable daily intake of 11 µg/kg body weight/day. The calculation was based on a LOAEL (Lowest Observed Adverse Effect Level) of 20 mg/kg body weight/day which was the lowest dose tested in a repeated 28-day toxicity study in rats leading to reduced body weight and several changes of blood parameters. Including several extrapolation factors (from LOAEL to NOAEL (No Observed Adverse Effect Level); from subacute to chronic toxicity) and an uncertainty factor of 100, it was assumed that consumption of 80 g mustard with the highest detected level of bisphenol F (8.4 mg/kg) would pass the margin of safety of 1800 which was postulated by the authority [98].

The German Federal Institute for Risk Assessment (BfR) used for its assessment the tolerable daily intake (TDI) of bisphenol A of 4 µg/kg body weight/day which was established by the European Food Safety Authority (EFSA) in 2015 [104] and is currently being re-evaluated [109]. The German authority concluded that even in the case of consuming high amount of mustard (4 g/day) with a maximum bisphenol F content of 6 mg/kg, the daily mean intake of bisphenol F would not exceed the assumed TDI of 4 µg/kg body weight/day. In the case of "normal consumers", the estimated intake quantity is 100 times lower [110]. Therefore, adverse health effects due to the intake of bisphenol F from mustard are not to be expected, although further toxicological studies are needed for a final evaluation. It should be underlined that bisphenol A substitutes appear to have similar metabolism, potencies and action and may pose similar potential health risks as bisphenol A. Thus, consumption of mustard may be an important source of bisphenol F and continues monitoring of this compound and its relevant sources, either natural or technical-induced, and thorough investigations on its health effects in humans remain important [100,111,112].

2.3. Erucic Acid

Mustard seeds are regarded as an important source of edible oil, which are used due to its nutty and pungent flavour and its high smoke point (250 °C), especially in Eastern and North-Western India [113]. Oil from Brassica plants differ from other vegetable oils mainly due to their significant proportion of long-chain monoenoic fatty acids, eicosenoic and erucic acids. Wendlinger et al. detected up to 19 fatty acids in mustard oils, of which oleic acid (18:1n-9), linoleic acid (18:2n-6), alpha-linolenic acid (18:3n-3), eicosanoic acid (20:1n-9) and erucic acid (22:1n-9) belong to the major ones [114]. Erucic acid is a typical example of a very-long chain mono-unsaturated fatty acid (VLCMFA) with 22 carbon atoms and one double bond between C13 and C14, also referred to as docosenoic acid (C22:1). Further, it belongs to the predominant fatty acid in seed oils from the three commercial cultivated mustard species *B. nigra*, *B. juncea* and *S. alba* containing levels over 30% erucic acid of total fatty acids (Table 4).

Table 4. Erucic acid concentration in mustard products.

Type		Origin	Amount of Erucic Acid (of Total FA)	Reference
Mustard seed	*B. juncea* mustard (54 varieties)	India	35.7–51.4%	[113]
	B. juncea varieties	Bangladesh	43.7–48.6%	[115]
	B. juncea (51 collections)	India	22.5–53.4%	[116]
	B. nigra (3 collections)		35.9–40.1%	
	S. alba (4 collections)		39.1–47.2%	
	B. nigra (n = 5)	Germany	20–40%	[117]
	B. juncea (n = 5)			
	S. alba (n = 5)			
	B. juncea (n = 12)	India	39.4–51.8%	[46]
	B. juncea (n = 3)		0.9–1.6%	
	B. nigra (n = 5)	Egypt	33.3–45/32–42.9 %[a]	[118]
	B. juncea (n = 5)		21.9–40/24.2–42.7 %[a]	
Mustard seed oil	Mustard oil (n = 3)	Australia	43.8–45.8%	[114]
	Mustard oil (n = 6)	Germany	0.3–50.8%	
	S. alba mustard oil	Egypt	37.90%	[90]
	B. juncea mustard oil		23.90%	
	Mustard oil	India		[119]
	Variety 1 (n = 3)		0.9–4.7%	
	Variety 2 (n = 2)		14.7–34.1%	
	Mustard oil	Bangladesh	41.80%	[120]
	Mustard oil (Erucic free)		20.10%	
	Mustard oil (n = 59)	India	48.5–54.02%	[121]
	Mustard oil	Turkey	11.40%	[122]
	Mustard oil	Malaysia	18.50%	[123]

Table 4. *Cont.*

Type		Origin	Amount of Erucic Acid (of Total FA)	Reference
	S. alba mustard oil			
	Cultivar 1	Poland	22.20%	[124]
	Cultivar 2 (Erucic free)		3.80%	
Other mustard products	Mustard samples (n = 15)	Germany	14–33%	[114]
	Mustard sauces (n = 5)		<5%	
	Mustard condiment	Germany	1.6 g/100 g condiment [b]	[125]
	Mustard seeds		11.5 g/100 g seeds [c]	
	Mustard sauces	Malaysia	33.10%	[123]
	Mustard sauces		32.20%	
	Mustard powder		25.50%	

[a] 2 crop years, [b] total fat content 4%, [c] total fat content 28.8%.

The fatty acid profile and erucic acid content of mustard seeds depend on several factors such as breeding techniques, total oil content, climatic conditions as well as morphological and physiological determinants [116]. Recently, breeding programs has successfully produced canola quality (low, zero erucic acid content) of *B. juncea* and low erucic acid content genotypes are cultivated in few countries [126]. Nevertheless, mustard seeds and mustard oils are one of the food products with the highest erucic acid content that can be found on the market. Based on the low consumption rate of mustard oil in the European Union, mustard seeds and mustard condiment or table mustard remain here the major sources regarding erucic acid intake [127].

2.3.1. Toxicological Effects

Like other long-chain fatty acids, erucic acid is transported to the tissues either bound to serum albumin or in esterified form incorporated in lipoproteins. The heart and skeletal muscles primarily utilize fatty acids as energy-providing substrates mainly through mitochondrial ß-oxidation. However, the capacity for mitochondrial ß-oxidation is reduced for long chain fatty acids (>18 carbon atoms). In this case, erucic acid is likewise sparsely oxidised by the mitochondrial β–oxidation system, probably due to the poor utilization of erucoyl-CoA as substrate by the mitochondrial acyl-CoA dehydrogenase. Furthermore, erucic acid also appears to inhibit the overall rate of fatty acid oxidation, by the mitochondria. It is worth noting, that both the capacity for ß-oxidation and inhibition of the tricarboxylic acid cycle differs between species. Rats compared to pigs seem to have a lower capacity for ß-oxidation and higher ability to inhibit oxidation of tricarboxylic acid-cycle intermediates [128,129].

Therefore, high exposure of erucic acid leads to accumulation of triacylglycerols in heart and other tissues. It has been shown that the lipidosis is reversible and transient during prolonged exposure, probably due to the increased peroxisomal chain shortening mediated by peroxisome proliferator-activated receptors. However, myocardial lipidosis is the most common and sensitive effect associated with short-term, and to a lesser extent, sub-chronic exposure of erucic acid in all animal species examined. With respect to liver tissue, the presence of erucic acid appears to induce the peroxisomal β–oxidation system, resulting in no accumulation of erucic acid and in reduced inhibition of mitochondrial ß-oxidation. Consequently, the liver is able to export erucic acid as very low-density lipoprotein (VLDL) [130–133].

Given what is known about the metabolism of erucic acid, it seems reasonable to expect that humans would also be susceptible to myocardial lipidosis following exposure to high levels of erucic acid. Clouet et al. observed a very low capacity in human heart mitochondria for the direct utilization of erucic acid as a substrate for energy requirements. Furthermore, activation of fatty acids due to the transfer of acyl groups from Coenzyme A (CoA) to carnitine, which belongs to the preliminary step of their beta-oxidation, was also reduced with high erucic acid exposure [134].

Imamura et al. showed that higher levels of docosenoic (22:1) and nervonic (24:1) acids in plasma phospholipids from diverse dietary sources, were associated with higher incidence of congestive heart failure in two independent cohorts, assuming possible cardiac toxicity of long-chain monounsaturated fatty acids [135]. In contrast, Matsumoto et al. described a lower incidence of coronary heart disease with increased levels of erucic acid in erythrocytes [136].

According to EFSA, treatment with Lorenzo's oil, which is a mixture of omega-9 fatty acids (oleic and erucic) to normalize the accumulation of very long chain fatty acids in adrenal and cerebral tissues in patients with adrenoleukodystrophy, led to undesirable effects on the hematopoietic system at doses of 100 mg/kg body weight per day [127].

So far, no reliable information is available regarding the development of myocardial lipidosis after high intake of erucic acid in humans. Epidemiological studies show no clear association between cardiac disease in humans and a diet high in erucic acid. One has to notice, that erucic acid metabolism is relatively complex in respect of differences in organ dependent metabolism of long chain fatty acids including lipid incorporation, chain shortening and elongation.

The EFSA Panel on Contaminants in the Food Chain (CONTAM) noted, that erucic acid-induced myocardial lipidosis observed in several animal species may also be relevant in humans, especially with the background that myocardial lipidosis is associated with cardiac insufficiency [127,137]. Therefore, oils, including high erucic acid content, are considered undesirable for human consumption. In 2016, EFSA conducted a comprehensive toxicological review and risk assessment of erucic acid in food and feed and established a tolerable daily intake (TDI) for erucic acid of 7 mg/kg body weight. The TDI is based on the observed development of myocardial lipidosis in juvenile rats which were treated for 7 days with 1 g erucic acid/kg body weight, and in neonatal piglets that received 1.1 g/kg erucic acid for 14 days [131,132]. Higher doses of erucic acid resulted in adverse effects on liver, kidney, skeletal muscles and caused changes in body and testis weight. Moreover, higher erucic acid intake was accompanied by mitochondrial damage and disorganisation of myofibrils as well as higher incidence of myocardial necrosis and fibrosis. No conclusion on genotoxicity and carcinogenicity could be made by EFSA due to limited data available. A single generation reproductive study was performed in rats and guinea pigs where doses of erucic acid up to 7500 mg/kg body weight/day were not associated with any adverse reproductive or developmental effects [127].

2.3.2. Evaluation of Toxicological Effects

Since it was reported that consumption of oils rich in erucic acid are accompanied with the onset of myocardial lipidosis and heart lesions in a number of species, erucic acid content in edible oils consumed by humans was restricted to certain levels by various regulatory agencies.

The Australia New Zealand Food Standards Code (FSANZ) has considered erucic acid as natural toxicant and set a maximum level of 20 g/kg (2%) in edible oils [138]. This also applies for the European Union. Commission Regulation (EU) 2019/1870 sets a maximum limit of 20 g/kg (2%) for erucic acid in vegetable oils and fats placed on the market for the final consumer or for use as an ingredient in food, whereas the maximum permitted level for mustard oil is 50 g/kg (5%).

In its scientific opinion, EFSA concludes that the 95th percentile of dietary exposure levels of erucic acid is especially high in infants and other children. For highly exposed children, this may pose an elevated health risk [127]. It has to be noted, that stricter levels apply to infant formula and follow-on formula (Commission Delegated Regulation (EU) 2019/828). The use of mustard (or the oil) as ingredient in baby food is not usual, however due to its numerous functional properties the utilization of small amounts cannot be excluded.

For mustard as condiment a maximum level of 35 g/kg (3.5%) was established, however "with acceptance from the competent authority, the maximum level does not apply to mustard oil locally produced and consumed". In the European Union level, no maximum permitted level for erucic acid is set for mustard seeds or powder, which may contribute to a high exposure of the undesirable substance. In contrast to Asiatic countries, use of mustard oil in Europe's cuisines is rare and mustard is mainly consumed in the form of prepared (table) mustard.

The U.S. Food and Drug Administration (FDA) banned mustard oil in pure form by publishing an Import Alert specifically stating that "Expressed mustard oil is not permitted for use as a vegetable oil. It may contain 20 to 40% erucic acid, which has been shown to cause nutritional deficiencies and cardiac lesions in test animals" [139].

According to the literature search on erucic acid content in mustard seeds and products derived from (Table 4), the findings indicate that both seeds and oils represent relevant sources for human intake. Wendlinger et al. showed in 2014 that most of the mustard oil samples analysed were well above the European Union limit of 5% erucic acid [114]. Recently, the same group from Germany addresses the intake of erucic acid from different sources including table mustard. According to the authors, teenagers and adults may exceed the tolerable daily intake of 7 mg/kg body weight/day by consuming one serving of mustard (10 g) depending on the contribution of erucic acid to the total fatty acids (>40%) of the seeds and the lipid content of the prepared product (>10%) [117].

It seems obvious that mustard consumption may contribute considerably to the intake of erucic acid. Especially in India and Pakistan, the use of mustard oil as cooking oil is widely distributed and still contain high amount of erucic acid. Several experts from India arguing that the health risks associated with mustard oil, also called the "Olive oil of India" were not observed in the Indian population, probably due to the high alpha-linolenic acid content which might compensate for the erucic acid [140]. Nevertheless, further development of mustard breeds low in erucic acid and a careful choice of the mustard cultivar used in food products will be in favour of consumer protection.

2.4. Allergens

Mustard seeds have a relatively high protein content, up to 36% depending on the mustard species. The mustard seed storage proteins 2S albumin, referred to as napin, are highly abundant in mustard seeds, and have been identified as major mustard allergens with significantly higher response in allergy tests compared to other storage proteins [141,142].

Allergenic 2S albumins have been characterized from white mustard (*S. alba*) [143,144] and brown mustard (*B. juncea*) [145]. According to the WHO/IUIS Allergen Nomenclature Subcommittee [146] the proteins are termed as Bra j 1 and Sin a 1 which show a close sequence homology [147] which increases the probability that people allergic to one mustard species are also sensitive for the other one [145]. For black mustard (*B. nigra*), one of the progenitor species of *B. juncea*, no information on allergens have been provided by the WHO/IUIS Allergen Nomenclature Subcommittee, however it can be assumed that seeds likewise contain allergenic 2 s albumins [148].

Beside the seed storage protein from the 2S albumin, three more allergens have been identified in *S. alba* including Sin a 2 which belongs to the seed storage 11S globulin and known as cruciferin; Sin a 3 corresponding to a non-specific lipid transfer protein, and Sin a 4, named as profilin (Table 5) [149].

Table 5. Mustard allergens.

Mustard Species	List of Allergens	Type	Source [a]
S. alba	Sin a 1	2S albumin	http://www.allergen.org
	Sin a 2	11S globulin (legumin-like) seed storage protein	http://www.allergen.org
	Sin a 3	Non-specific lipid transfer protein type 1 (ns-LTP)	http://www.allergen.org
	Sin a 4	Profilin	http://www.allergen.org
B. nigra	Bra j 1	2S albumin seed storage protein	http://www.uniprot.org *
B. juncea	Bra j 1	2S albumin seed storage protein	http://www.allergen.org

[a] accessed on 8 June 2021, * not listed according to the WHO/IUIS Allergen Nomenclature Subcommittee.

Recently, Hocine et al. assessed protein profiles and immunoglobulin E (IgE)-binding patterns of selected mustard varieti es (*S. alba* and *B. juncea*). In addition to proven allergens, the authors identified other new IgE-binding protein bands from *S. alba* and *B. juncea* varieties [150].

At present, no effective preventive treatments exist for mustard allergy. The allergenic proteins are highly resistant to heat treatment. Thermal denaturation for napin and cruciferin occurs upon 80 °C depending on the pH-value [151]. Moreover, mustard allergens are poorly digestible. Sin a 1 was shown to be resistant to digestion by trypsin and other proteolytic enzymes [141,152]. Therefore, the allergenic potential cannot be effectively reduced during mustard processing or preparation of table mustard.

Previous studies suggest that allyl-isothiocyanate (AITC), which derives from sinigrin mainly contained in *B. nigra* and *B. juncea* could have the potential to cause allergic contact dermatitis in humans [153,154]. EFSA evaluated the safety of AITC when used as food preservative and concluded that AITC may cause contact hypersensitivity which is an immunologically mediated adverse reaction but mechanistically different from food allergy, although it might be extremely unlikely that AITC acts as a direct food allergen [83]. In the scientific opinion on the evaluation of allergenic foods and food ingredients for labelling purposes, EFSA additionally pointed out that mustard may contain a number of further irritants triggering non-immune mediated reactions mimicking allergic reactions [155].

2.4.1. Clinical Indication

Food allergy can be IgE-mediated or non-IgE-mediated. IgE-mediated reactions have a rapid onset, affecting skin, respiratory and gastrointestinal tract, and in some cases can lead to systemic anaphylaxis, whereas non-IgE-mediated food allergy usually is delayed and affects mainly the skin and the gastrointestinal tract. The allergic symptoms after consuming mustard or mustard-containing products are comparable to other IgE-mediated food allergies. This includes allergic or atopic dermatitis resulting in skin reactions such as rashes or hives, urticaria, swelling of lips and tongue, facial flushing and oedema, chest tightness and respiratory problems, rhinitis, asthma, nausea and vomiting, dizziness and anaphylactic shock. Mustard allergy symptoms were also observed in children under the age of 3 years indicating a primary sensitization to mustard in at least some food allergies [148,156–158].

It is suggested that approximately 50% of patients allergic to mustard are also sensitized to mugwort pollen (*Artemisia vulgaris*) and several vegetable foods, mainly from the Rosaceae family, but also from tree nuts, peanuts and legumes [159]. Vereda et al. showed that 21 out of 34 subjects sensitized to mustard had cross-reactivity with fruits from the Rosaceae family (peach, apple, pear, apricot, plum, cherries and strawberries, excluding almond) and 20 suffered from allergy to one or more nuts. Moreover, allergy to mugwort was especially reported [160]. Sin a 1 was the most prevalent allergen and highly correlated with specific IgE levels. Therefore, Sin a 1 is considered the major allergen of white mustard (*S. alba*) and the most suitable marker for a precise diagnostic screening of mustard sensitization.

Figueroa et al. reported that all of the 38 patients with mustard IgE-mediated hypersensitivity showed associated sensitization to other members of Brassicaceae family, and cross-reactivity among them was confirmed. However only 40% of these had symptomatic reactions, mainly to cabbage, cauliflower and broccoli [161].

The reported cross-reactivities with pollens or with members of the Brassicaceae family may influence the positivity of specific IgE and skin-prick tests. This may result in overestimated prevalence of sensitization to mustard, and may possibly influence the occurrence of oral allergy syndrome-like symptoms elicited by mustard. Another factor contributing to a possible overestimation of mustard sensitization and allergy may be the presence of irritants in mustard preparations or mustard containing products. For instance, the neuropeptide-active agent capsaicin may affect the release of substance P causing a non-IgE mediated mast cell degranulation [156,162]. Therefore, irritants such as capsaicin or isothiocyanates may trigger non-immune reactions mimicking allergic reactions which may lead to false positive allergy-like reactions. For example, Morisset et al. have shown that only 23.3% of patients with positive skin prick tests are truly allergic to mustard based on positive oral food challenges [148].

2.4.2. Evaluation of the Allergic Potential

Mustard is one of the priority food allergens declared under several international food allergen labelling regulations, for instance in the European Union or Canada but not in the U.S. [163]. According to Regulation (EU) No 1169/2011 mustard belongs to one of the 14 major allergens that shall be indicated in the list of ingredients. The inclusion of mustard was based on the view that mustard allergy is a serious health problem in certain countries, although limited data are available according to the number of people affected. Mustard allergy is recognized as one of the most frequent spice allergies with a credible cause–effect relationship confirmed by single and double-blind placebo-controlled food challenges (DBPCFC). Several clinical studies and case reports documented severe systemic reactions, including anaphylactic reaction immediately after the ingestion of mustard, even after small quantities [164–166]. Only a small number of DBPCFC or oral food challenges (OFC) studies exist in the literature, which possibly can be justified by technical difficulties of masking the strong taste of mustard limiting the attempts to perform these studies. Previously, it was suggested that mustard allergy accounts for about 1% of food allergies in children and up to 7% of total food allergies based on estimated prevalence in France [152,155,167].

The amount of mustard required to elicit an allergic reaction may be very small. Estimations of the eliciting dose range from a mean cumulative reactive dose of mustard sauce of 891.4 ± 855.2 mg, equivalent to 124.8 ± 119.7 mg of mustard, published by Figueroa et al. to 40 mg of mustard seasoning (roughly equivalent to 0.8 mg of protein) reported by Morisset et al. [148,161]. In 36 children with a positive mustard skin-prick test, the cumulative reactive dose by open challenge or double-blind placebo controlled food challenge (SBPCFC) varied from 1 to 936 mg [156]. Most of the case reports indicate that allergic reactions come from mustard sauce or mustard hidden in other products, such as mayonnaise, dips, ketchup sauces, where mustard is present in small or trace amounts [164].

Using data from clinical studies and mathematical calculations the VITAL Scientific Expert Panel (VSEP) recently published an update on reference doses including mustard allergens. For mustard, 0.05 mg protein has been derived as a reference dose for 'ED01', or 0.4 mg protein for 'ED05' corresponding to a dose where 99% or 95% of people affected by a mustard allergy are protected from developing objectively measurable allergic reactions [168]. This is in line with EFSA's conclusion that 1 mg mustard protein may trigger allergic reactions in patients allergic to mustard [155].

Mustard is widely used in foods due to its sensory attributes, its high protein content and its numerous functional properties. Therefore, foods formulated with mustard are expected to rise in future and the risk of masked allergens in modern food products is increasing. In regions with high consumption, such as France, mustard might be among the most important food allergen for children. Hence, it deserves special attention, particularly as a potent "hidden allergen" provoking unexpected allergic reactions. The seriousness of the allergic reactions argues for informative labelling in countries where mustard is not considered as food allergen, such as the U.S.

3. Conclusions

Mustard seeds are widely used in foods due to its sensory attributes, nutritional values, and its numerous functional properties. The intake of mustard products and foods formulated with mustard is expected to rise in future and may contribute considerably to the exposure to several biologically active compounds that were objects of this work. Mustard products and their seeds are consumed in very small quantities primarily explainable by the characteristic hot and pungent flavour. Adverse effects of bioactive compounds such as erucic acid or glucosinolates breakdown products have been mainly observed in in vitro or in vivo animal studies rather than in humans and especially when the compounds were administered in concentrated or isolated form, and given in high doses. When consumed in average amounts typically found in a normal diet, it can be expected, that the intake of mustard seeds or products made thereof do not pose a direct health risk, except for consumers allergic to mustard proteins. However, it cannot be excluded that high intake levels of mustard seeds or products made thereof, such as mustard seed oil, cause health problems. Nowadays, research is basically focusing on the potential beneficial effects of the multifunctional mustard plant and its ingredients, however reliable data on potentially toxic compounds and possible health risks should likewise be considered.

Funding: This research received no external funding.

Institutional Review Board Statement: Not applicable.

Informed Consent Statement: Not applicable.

Data Availability Statement: Not applicable.

Acknowledgments: The author expresses profound thanks to Bernd Schäfer and Andreas Eisenreich (German Federal Institute for Risk Assessment, Department of Food Safety) for their constructive comments and support in the preparation of this manuscript.

Conflicts of Interest: The author declares no conflict of interest.

References

1. Dixon, G. Origins and diversity of Brassica and its relatives. In *Vegetable Brassicas and Related Crucifers*; CABI: Wallingford, UK, 2006.
2. Warwick, S.I.; Francis, A.; Al-Shehbaz, I.A. Brassicaceae: Species checklist and database on CD-Rom. *Plant Syst. Evol.* **2006**, *259*, 249–258. [CrossRef]
3. Cheng, F.; Wu, J.; Wang, X. Genome triplication drove the diversification of Brassica plants. *Hortic. Res.* **2014**, *1*, 14024. [CrossRef]
4. Warwick, S.I.; Black, L.D. Molecular systematics of Brassica and allied genera (Subtribe Brassicinae, Brassiceae)—Chloroplast genome and cytodeme congruence. *Theor. Appl. Genet.* **1991**, *82*, 81–92. [CrossRef]
5. Xue, J.-Y.; Wang, Y.; Chen, M.; Dong, S.; Shao, Z.-Q.; Liu, Y. Maternal Inheritance of U's Triangle and Evolutionary Process of Brassica Mitochondrial Genomes. *Front. Plant Sci.* **2020**, *11*, 805. [CrossRef]
6. Vaughan, J.G. A Multidisciplinary Study of the Taxonomy and Origin of Brassica Crops. *BioScience* **1977**, *27*, 35–40. [CrossRef]
7. Thomas, J.; Kuruvilla, K.M.; Hrideek, T.K. 21—Mustard. In *Handbook of Herbs and Spices*, 2nd ed.; Peter, K.V., Ed.; Woodhead Publishing: Cambridge, UK, 2012; pp. 388–398. [CrossRef]
8. EC (European Commission). *Common Catalogue of Varieties of Agricultural Plant Species—Version of 1 December 2020; European Commission*. 2020. Available online: https://ec.europa.eu/food/plants/plant-reproductive-material/plant-variety-catalogues-databases-information-systems.de/ (accessed on 26 July 2021).
9. Divakaran, M.; Babu, K.N. Mustard. In *Encyclopedia of Food and Health*; Caballero, B., Finglas, P.M., Toldrá, F., Eds.; Academic Press: Oxford, UK, 2016; pp. 9–19. [CrossRef]
10. Rahman, M.; Khatun, A.; Liu, L.; Barkla, B.J. Brassicaceae Mustards: Traditional and Agronomic Uses in Australia and New Zealand. *Molecules* **2018**, *23*, 231. [CrossRef]

11. Rakow, G. Species Origin and Economic Importance of Brassica. In *Brassica*; Pua, E.C., Douglas, C.J., Eds.; Springer: Berlin/Heidelberg, Germany, 2004; Volume 54.

12. McVetty, P.B.E.; Duncan, R.W. Canola, Rapeseed, and Mustard: For Biofuels and Bioproducts. In *Industrial Crops: Handbook of Plant Breeding 9*; Cruz, V.M.V., Dierig, D.A., Eds.; Springer: Berlin/Heidelberg, Germany, 2015; pp. 133–156. [CrossRef]

13. Government of Canada. Guidance Document Repository (GDR). The Biology of *Brassica juncea* (Canola/Mustard). Biology Document BIO2007-01: A Companion Document to the Directive 94-08 (Dir94-08), Assessment Criteria for Determining Environmental Safety of Plants with Novel Traits . 2007. Available online: https://inspection.canada.ca/plant-varieties/plants-with-novel-traits/applicants/directive-94-08/biology-documents/brassica-juncea/eng/1330727837568/1330727899677 (accessed on 26 July 2021).

14. Tridge (Global Sourcing Hub of Food & Agriculture). Available online: https://www.tridge.com/intelligences/mustard-seed/ (accessed on 26 July 2021).

15. FAOSTAT (Food and Agriculture Organization of the United Nations). Available online: http://www.fao.org/faostat/en/#search/mustard%20seed (accessed on 26 July 2021).

16. Statista (Statistisches Bundesamt). ID 620, Pro-Kopf-Verbrauch von Senf in Deutschland in den Jahren 1983 bis 2019 (in Gramm). Available online: https://de.statista.com/statistik/daten/studie/620/umfrage/pro-kopf-verzehr-von-senf-in-deutschland/ (accessed on 26 July 2021).

17. Santini, A.; Cicero, N. Development of Food Chemistry, Natural Products, and Nutrition Research: Targeting New Frontiers. *Foods* **2020**, *9*, 482. [CrossRef] [PubMed]

18. Durazzo, A.; D'Addezio, L.; Camilli, E.; Piccinelli, R.; Turrini, A.; Marletta, L.; Marconi, S.; Lucarini, M.; Lisciani, S.; Gabrielli, P.; et al. From Plant Compounds to Botanicals and Back: A Current Snapshot. *Molecules* **2018**, *23*, 1844. [CrossRef] [PubMed]

19. ReasearchAndMarkets.com. *Global Mustard Market—Forecasts from 2020 to 2025*; Knowledge Sourcing Intelligence LLP: Noida, India, 2020.

20. Kayaçetin, F. Botanical characteristics, potential uses, and cultivation possibilities of mustards in turkey: A review. *Turk. J. Bot.* **2020**, *44*, 101–127. [CrossRef]

21. Robiquet, P.J.; Bussy, A. Notice sur l'huile volatile de moutarde. *J. Pharm. Sci.* **1840**, *26*, 116–120.

22. Fahey, J.W.; Zalcmann, A.T.; Talalay, P. The chemical diversity and distribution of glucosinolates and isothiocyanates among plants. *Phytochemistry* **2001**, *56*, 5–51. [CrossRef]

23. Clarke, D.B. Glucosinolates, structures and analysis in food. *Anal. Methods* **2010**, *2*, 310–325. [CrossRef]

24. Agerbirk, N.; Olsen, C.E. Glucosinolate structures in evolution. *Phytochemistry* **2012**, *77*, 16–45. [CrossRef] [PubMed]

25. Griffiths, D.W.; Birch, A.N.E.; Hillman, J.R. Antinutritional compounds in the Brassicaceae: Analysis, biosynthesis, chemistry and dietary effects. *J. Hortic. Sci. Biotechnol.* **1998**, *73*, 1–18. [CrossRef]

26. Mithen, R.F. Glucosinolates and their degradation products. *Adv. Bot. Res.* **2001**, *35*, 213–232.

27. Sikorska-Zimny, K.; Beneduce, L. The glucosinolates and their bioactive derivatives in Brassica: A review on classification, biosynthesis and content in plant tissues, fate during and after processing, effect on the human organism and interaction with the gut microbiota. *Crit. Rev. Food Sci. Nutr.* **2021**, *61*, 2544–2571. [CrossRef]

28. Bhat, R.; Vyas, D. Myrosinase: Insights on structural, catalytic, regulatory, and environmental interactions. *Crit. Rev. Biotechnol.* **2019**, *39*, 508–523. [CrossRef]

29. Nguyen, V.P.T.; Stewart, J.; Lopez, M.; Ioannou, I.; Allais, F. Glucosinolates: Natural Occurrence, Biosynthesis, Accessibility, Isolation, Structures, and Biological Activities. *Molecules* **2020**, *25*, 4537. [CrossRef]

30. Shapiro, T.A.; Fahey, J.W.; Wade, K.L.; Stephenson, K.K.; Talalay, P. Human metabolism and excretion of cancer chemoprotective glucosinolates and isothiocyanates of cruciferous vegetables. *Cancer Epidemiol. Biomark. Prev.* **1998**, *7*, 1091–1100.

31. Bones, A.M.; Rossiter, J.T. The myrosinase-glucosinolate system, its organisation and biochemistry. *Physiol. Plant.* **1996**, *97*, 194–208. [CrossRef]

32. Halkier, B.A.; Gershenzon, J. Biology and Biochemistry of Glucosinolates. *Annu. Rev. Plant Biol.* **2006**, *57*, 303–333. [CrossRef]

33. Bones, A.M.; Rossiter, J.T. The enzymic and chemically induced decomposition of glucosinolates. *Phytochemistry* **2006**, *67*, 1053–1067. [CrossRef] [PubMed]

34. Grubb, C.D.; Abel, S. Glucosinolate metabolism and its control. *Trends Plant Sci.* **2006**, *11*, 89–100. [CrossRef]

35. Agrawal, S.; Yallatikar, T.; Gurjar, P. Brassica Nigra: Ethopharmacological Review of a Routinely Used Condiment. *Curr. Drug Discov. Technol.* **2019**, *16*, 40–47. [CrossRef]

36. Kuchernig, J.C.; Burow, M.; Wittstock, U. Evolution of specifier proteins in glucosinolate-containing plants. *BMC Evol. Biol.* **2012**, *12*, 127. [CrossRef]

37. Kissen, R.; Bones, A.M. Nitrile-specifier proteins involved in glucosinolate hydrolysis in Arabidopsis thaliana. *J. Biol. Chem.* **2009**, *284*, 12057–12070. [CrossRef]

38. Halkier, B.A.; Du, L. The biosynthesis of glucosinolates. *Trends Plant Sci.* **1997**, *2*, 425–431. [CrossRef]

39. Rangkadilok, N.; Nicolas, M.; Bennett, R.; Premier, R.; Eagling, D.; Taylor, P. Developmental changes of sinigrin and glucoraphanin in three Brassica species (Brassica nigra, Brassica juncea and Brassica oleracea var. italica). *Sci. Hortic.* **2002**, *96*, 11–26. [CrossRef]

40. Daxenbichler, M.E.; Spencer, G.F.; Carlson, D.G.; Rose, G.B.; Brinker, A.M.; Powell, R.G. Glucosinolate composition of seeds from 297 species of wild plants. *Phytochemistry* **1991**, *30*, 2623–2638. [CrossRef]

41. Popova, I.E.; Morra, M.J. Simultaneous Quantification of Sinigrin, Sinalbin, and Anionic Glucosinolate Hydrolysis Products in *Brassica juncea* and *Sinapis alba* Seed Extracts Using Ion Chromatography. *J. Agric. Food Chem.* **2014**, *62*, 10687–10693. [CrossRef]
42. Velíšek, J.; Mikulcová, R.; Míková, K.; Woldie, K.B.; Link, J.; Davídek, J. Chemometric investigation of mustard seed. *Lebensm. Wiss. Und-Technol. Food Sci. Technol.* **1995**, *28*, 620–624. [CrossRef]
43. Velasco, L.; Becker, H.C. Variability for seed glucosinolates in a germplasm collection of the genus Brassica. *Genet. Resour. Crop Evol.* **2000**, *47*, 231–238. [CrossRef]
44. Gupta, S.; Sangha, M.K.; Kaur, G.; Atwal, A.K.; Banga, S.; Banga, S.S. Variability for Leaf and Seed Glucosinolate Contents and Profiles in a Germplasm Collection of the Brassica juncea. *Biochem. Anal. Biochem.* **2012**, *1*. [CrossRef]
45. Merah, O. Genetic Variability in Glucosinolates in Seed of Brassica juncea: Interest in Mustard Condiment. *J. Chem.* **2015**, *2015*, 606142. [CrossRef]
46. Saikia, S.; Kumar, G.; Jammu, J.; Kashmir; Gyanendra, C.; Rai, K.; Rai, G.; Salgotra, R.; Rai, S.K.; Singh, M.; et al. Erucic acid and glucosinolate variability in *Brassica juncea* L. *Int. J. Chem. Stud.* **2018**, *6*, 1223–1226.
47. McGregor, D.L. Thiocyanate ion, a hydrolysis product of glucosinolates from rape and mustard seed. *Can. J. Plant Sci.* **1978**, *58*, 795–800. [CrossRef]
48. Ettlinger, M.G.; Lundeen, A.J. The structures of sinigrin and sinalbin; an enzymatic rearrangement. *J. Am. Chem. Soc.* **1956**, *78*, 4172–4173. [CrossRef]
49. Ettlinger, M.G.; Lundeen, A.J. First synthesis of a mustard oil glucoside, the enzymatic lossen rearrangement. *J. Am. Chem. Soc.* **1957**, *79*, 1764–1765. [CrossRef]
50. Nugrahedi, P.Y.; Verkerk, R.; Widianarko, B.; Dekker, M. A Mechanistic Perspective on Process-Induced Changes in Glucosinolate Content in Brassica Vegetables: A Review. *Crit. Rev. Food Sci. Nutr.* **2015**, *55*, 823–838. [CrossRef]
51. Van Eylen, D.; Oey, I.; Hendrickx, M.; Van Loey, A. Behavior of mustard seed (*Sinapis alba* L.) myrosinase during temperature/pressure treatments: A case study on enzyme activity and stability. *Eur. Food Res. Technol.* **2008**, *226*, 545–553. [CrossRef]
52. Srisangnam, C.; Salunkhe, K.; Dull, G. Quality of cabbage II. Physical, Chemical and biochemical modification in processing treatments to improve flavour of blanched cabbage (*Brassica oleracea* L.). *J. Food Qual.* **2007**, *3*, 233–250. [CrossRef]
53. Prieto, M.A.; López, C.J.; Simal-Gandara, J. Chapter Six—Glucosinolates: Molecular structure, breakdown, genetic, bioavailability, properties and healthy and adverse effects. In *Advances in Food and Nutrition Research*; Ferreira, I.C.F.R., Barros, L., Eds.; Academic Press: Cambridge, MA, USA, 2019; Volume 90, pp. 305–350.
54. de Vos, R.H.; Blijleven, W.G.H. The effect of processing conditions on glucosinolates in cruciferous vegetables. *Z. Lebensm. Unters. Forsch.* **1988**, *187*, 525–529. [CrossRef]
55. Cools, K.; Terry, L.A. The effect of processing on the glucosinolate profile in mustard seed. *Food Chem.* **2018**, *252*, 343–348. [CrossRef]
56. Kübler, K. *Dissertation: Analyse von Glucosinolaten und Isothiocyanaten mittels Flüssigkeitschromatographie- bzw. Gaschromatographie-Massenspektrometrie*; Justus-Liebig-University: Gießen, Germany, 2010. Available online: http://geb.uni-giessen.de/geb/volltexte/2010/7884/ (accessed on 9 June 2021).
57. Eib, S.; Schneider, D.; Hensel, O.; Seuss-Baum, I. Relationship between mustard pungency and allyl-isothiocyanate content: A comparison of sensory and chemical evaluations. *J. Food Sci.* **2020**, *85*, 2728–2736. [CrossRef]
58. EFSA (European Food Safety Authority). Safety of rapeseed powder from *Brassica rapa* L. and *Brassica napus* L. as a Novel food pursuant to Regulation (EU) 2015/2283. *EFSA J.* **2020**, *18*, e06197. [CrossRef]
59. Fenwick, G.R.; Heaney, R.K. Glucosinolates and their Breakdown Products in Cruciferous Crops, Foods and Feedingstuffs. *Food Chem.* **1983**, *11*, 249–271. [CrossRef]
60. van Doorn, H.E.; van der Kruk, G.C.; van Holst, G.-J.; Raaijmakers-Ruijs, N.C.M.E.; Postma, E.; Groeneweg, B.; Jongen, W.H.F. The glucosinolates sinigrin and progoitrin are important determinants for taste preference and bitterness of Brussels sprouts. *J. Sci. Food Agric.* **1998**, *78*, 30–38. [CrossRef]
61. EP (European Parlament and the Council). Directive 2002/32/EC of the European Parlament and of the Council of 7 May 2002 on Undesirable Substances in Animal Feed. 2002. Available online: http://data.europa.eu/eli/dir/2002/32/2019-11-28/ (accessed on 9 June 2021).
62. Mawson, R.; Heaney, R.K.; Zdunczyk, Z.; Kozlowska, H. Rapeseed meal-glucosinolates and their antinutritional effects Part 4. Goitrogenicity and internal organs abnormalities in animals. *Die Nahr.* **1994**, *38*, 178–191. [CrossRef]
63. Tripathi, M.K.; Mishra, A.S. Glucosinolates in animal nutrition: A review. *Anim. Feed Sci. Technol.* **2007**, *132*, 1–27. [CrossRef]
64. Langer, P.; Greer, M.A. *Antithyroid Substances and Naturally Occurring Goitrogens*; S. Karger AG: Basel, Switzerland, 1977.
65. McDanell, R.; McLean, A.E.M.; Hanley, A.B.; Heaney, R.K.; Fenwick, G.R. Chemical and biological properties of indole glucosinolates (glucobrassicins): A review. *Food Chem. Toxicol.* **1988**, *26*, 59–70. [CrossRef]
66. Kawakishi, S.; Namiki, M.; Watanabe, H.; Muramatsu, K. Studies on the Decomposition of Sinalbin. *Agric. Biol. Chem.* **1967**, *31*, 823–830. [CrossRef]
67. Paunovic, D.; Solevic Knudsen, T.; Krivokapic, M.; Zlatković, B.; Antić, M. Sinalbin degradation products in mild yellow mustard paste. *Hem. Ind.* **2012**, *66*, 29–32. [CrossRef]
68. Mitchell, M.L.; O'Rourke, M.E. Response of the thyroid gland to thiocyanate and thyrotropin. *J. Clin. Endocrinol. Metab.* **1960**, *20*, 47–56. [CrossRef]

69. Felker, P.; Bunch, R.; Leung, A.M. Concentrations of thiocyanate and goitrin in human plasma, their precursor concentrations in brassica vegetables, and associated potential risk for hypothyroidism. *Nutr. Rev.* **2016**, *74*, 248–258. [CrossRef]

70. Nishie, K.; Daxenbichler, M.E. Toxicology of glucosinolates, related compounds (nitriles, R-goitrin, isothiocyanates) and vitamin U found in cruciferae. *Food Cosmet. Toxicol.* **1980**, *18*, 159–172. [CrossRef]

71. Daxenbichler, M.E.; Van Etten, C.H.; Spencer, G.F. Glucosinolates and derived products in cruciferous vegetables. Identification of organic nitriles from cabbage. *J. Agric. Food Chem.* **1977**, *25*, 121–124. [CrossRef]

72. Cole, R.A. Isothiocyanates, nitriles and thiocyanates as products of autolysis of glucosinolates in Cruciferae. *Phytochemistry* **1976**, *15*, 759–762. [CrossRef]

73. Choi, E.-J.; Zhang, P.; Kwon, H. Determination of goitrogenic metabolites in the serum of male wistar rat fed structurally different glucosinolates. *Toxicol. Res.* **2014**, *30*, 109–116. [CrossRef] [PubMed]

74. Dahlberg, P.A.; Bergmark, A.; Björck, L.; Bruce, A.; Hambraeus, L.; Claesson, O. Intake of thiocyanate by way of milk and its possible effect on thyroid function. *Am. J. Clin. Nutr.* **1984**, *39*, 416–420. [CrossRef] [PubMed]

75. Langer, P.; Michajlovskij, N.; Sedlák, J.; Kueka, M. Studies on the antithyroid activity of naturally occurring L-5-vinyl-2-thiooxazolidone in man. *Endokrinologie* **1971**, *57*, 225–229. [PubMed]

76. Kassie, F.; Knasmüller, S. Genotoxic effects of allyl isothiocyanate (AITC) and phenethyl isothiocyanate (PEITC). *Chem. Biol. Interact.* **2000**, *127*, 163–180. [CrossRef]

77. Kassie, F.; Parzefall, W.; Musk, S.; Johnson, I.; Lamprecht, G.; Sontag, G.; Knasmüller, S. Genotoxic effects of crude juices from Brassica vegetables and juices and extracts from phytopharmaceutical preparations and spices of cruciferous plants origin in bacterial and mammalian cells. *Chem. Biol. Interact.* **1996**, *102*, 1–16. [CrossRef]

78. Kassie, F.; Pool-Zobel, B.; Parzefall, W.; Knasmüller, S. Genotoxic effects of benzyl isothiocyanate, a natural chemopreventive agent. *Mutagenesis* **1999**, *14*, 595–604. [CrossRef]

79. Baasanjav-Gerber, C.; Hollnagel, H.M.; Brauchmann, J.; Iori, R.; Glatt, H. Detection of genotoxicants in Brassicales using endogenous DNA as a surrogate target and adducts determined by 32P-postlabelling as an experimental end point. *Mutagenesis* **2010**, *26*, 407–413. [CrossRef]

80. Glatt, H.; Baasanjav-Gerber, C.; Schumacher, F.; Monien, B.H.; Schreiner, M.; Frank, H.; Seidel, A.; Engst, W. 1-Methoxy-3-indolylmethyl glucosinolate; a potent genotoxicant in bacterial and mammalian cells: Mechanisms of bioactivation. *Chem. Biol. Interact.* **2011**, *192*, 81–86. [CrossRef]

81. Musk, S.R.; Johnson, I.T. The clastogenic effects of isothiocyanates. *Mutat. Res.* **1993**, *300*, 111–117. [CrossRef]

82. Musk, S.R.; Smith, T.K.; Johnson, I.T. On the cytotoxicity and genotoxicity of allyl and phenethyl isothiocyanates and their parent glucosinolates sinigrin and gluconasturtiin. *Mutat. Res.* **1995**, *348*, 19–23. [CrossRef]

83. EFSA (European Food Safety Authority). Scientific Opinion on the safety of allyl isothiocyanate for the proposed uses as a food additive. *EFSA J.* **2010**, *8*, 1943. [CrossRef]

84. NTP (National Toxicology Program). Carcinogenesis Bioassay of Allyl Isothiocyanate (CAS No. 57-06-7) in F344/N Rats and B6C3F1 Mice (Gavage Study). *Natl. Toxicol. Prpgram Tech. Rep. Ser.* **1982**, *234*, 1–142.

85. Baasanjav-Gerber, C.; Monien, B.H.; Mewis, I.; Schreiner, M.; Barillari, J.; Iori, R.; Glatt, H. Identification of glucosinolate congeners able to form DNA adducts and to induce mutations upon activation by myrosinase. *Mol. Nutr. Food Res.* **2011**, *55*, 783–792. [CrossRef]

86. Kołodziejski, D.; Piekarska, A.; Hanschen, F.S.; Pilipczuk, T.; Tietz, F.; Kusznierewicz, B.; Bartoszek, A. Relationship between conversion rate of glucosinolates to isothiocyanates/indoles and genotoxicity of individual parts of Brassica vegetables. *Eur. Food Res. Technol.* **2019**, *245*, 383–400. [CrossRef]

87. Latté, K.P.; Appel, K.-E.; Lampen, A. Health benefits and possible risks of broccoli—An overview. *Food Chem. Toxicol.* **2011**, *49*, 3287–3309. [CrossRef]

88. EFSA (European Food Safety Authority). GE 85: Consideration of miscellaneous nitrogen-containing substances evaluated by JECFA (65th meeting). *EFSA J.* **2008**, *804*, 1–30.

89. JECFA (Joint FAO/WHO Expert Committee on Food Additives). Safety evaluation of certain food additives I prepared by the sixty-fifth meeting of the Joint FAO/WHO Expert Committee on Food Additives (JEFCA). *WHO Food Addit. Ser.* **2007**, *56*. Available online: https://apps.who.int/iris/handle/10665/43407 (accessed on 9 June 2021).

90. Aboulfadl, M.; El-Badry, N.; Ammar, M. Nutritional and Chemical Evaluation for Two Different Varieties of Mustard Seeds. *World Appl. Sci. J.* **2011**, *15*, 1225–1233.

91. Holst, B.; Williamson, G. A Critical Review of the Bioavailability of Glucosinolates and Related Compounds. *Nat. Prod. Rep.* **2004**, *21*, 425–447. [CrossRef]

92. Mandrich, L.; Caputo, E. Brassicaceae-Derived Anticancer Agents: Towards a Green Approach to Beat Cancer. *Nutrients* **2020**, *12*, 868. [CrossRef]

93. Verhoeven, D.T.; Goldbohm, R.A.; van Poppel, G.; Verhagen, H.; van den Brandt, P.A. Epidemiological studies on brassica vegetables and cancer risk. *Cancer Epidemiol. Biomark. Prev.* **1996**, *5*, 733–748.

94. Lamy, E.; Crössmann, C.; Saeed, A.; Schreiner, P.R.; Kotke, M.; Mersch-Sundermann, V. Three structurally homologous isothiocyanates exert "Janus" characteristics in human HepG2 cells. *Environ. Mol. Mutagenesis* **2009**, *50*, 164–170. [CrossRef] [PubMed]

95. Huang, T.; Danaher, L.A.; Brüschweiler, B.J.; Kass, G.E.N.; Merten, C. Naturally occurring bisphenol F in plants used in traditional medicine. *Arch. Toxicol.* **2019**, *93*, 1485–1490. [CrossRef]

96. Cao, X.-L.; Kosarac, I.; Popovic, S.; Zhou, S.; Smith, D.; Dabeka, R. LC-MS/MS analysis of bisphenol S and five other bisphenols in total diet food samples. *Food Addit. Contam. Part A* **2019**, *36*, 1740–1747. [CrossRef]

97. Zoller, O.; Brüschweiler, B.J.; Magnin, R.; Reinhard, H.; Rhyn, P.; Rupp, H.; Zeltner, S.; Felleisen, R. Natural occurrence of bisphenol F in mustard. *Food Addit. Contam. Part A* **2016**, *33*, 137–146. [CrossRef]

98. BLV (Bundesamt für Lebensmittelsicherheit und Veterinärwesen, Schweiz) Bisphenol F in Senf—Fakten und Risikobewertung des BLV. Bericht. 2015. Available online: https://www.blv.admin.ch/dam/blv/de/dokumente/lebensmittel-und-ernaehrung/lebensmittelsicherheit/stoffe-im-fokus/bisphenol-f-senf-fakten-risikobewertung-blv.pdf.download.pdf/Bisphenol_Risikobewertung_BLV_DE.pdf/ (accessed on 9 June 2021).

99. Reger, D.; Pavlovic, M.; Pietschmann-Keck, M.; Klinger, R. Bisphenol F in Senf: Aktueller Wissensstand und Nachweis mittels LC-MS/MS. *J. Consum. Prot. Food Saf.* **2017**, *12*, 131–137. [CrossRef]

100. Liao, C.; Kannan, K. Concentrations and profiles of bisphenol A and other bisphenol analogues in foodstuffs from the United States and their implications for human exposure. *J. Agric. Food Chem.* **2013**, *61*, 4655–4662. [CrossRef]

101. Karak, N. *Vegetable Oil-Based Polymers Properties, Processing and Applications*; Woodhead Publishing: Cambridge, UK, 2012. [CrossRef]

102. De Kruif, C.A.; Marsman, J.W.; Venekamp, J.C.; Falke, H.E.; Noordhoek, J.; Blaauboer, B.J.; Wortelboer, H.M. Structure elucidation of acid reaction products of indole-3-carbinol: Detection in vivo and enzyme induction in vitro. *Chem. Biol. Interact.* **1991**, *80*, 303–315. [CrossRef]

103. Cabaton, N.; Chagnon, M.C.; Lhuguenot, J.C.; Cravedi, J.P.; Zalko, D. Disposition and metabolic profiling of bisphenol F in pregnant and nonpregnant rats. *J. Agric. Food Chem.* **2006**, *54*, 10307–10314. [CrossRef] [PubMed]

104. EFSA (European Food Safety Authority). Scientific Opinion on the risks to public health related to the presence of bisphenol A (BPA) in foodstuffs. *EFSA J.* **2015**, *13*, 3978. [CrossRef]

105. EFSA (European Food Safety Authority). Scientific Opinion of the Panel on food contact materials, enzymes, flavourings and processing aids (CEF) on 24th list of substances for food contact materials. *EFSA J.* **2009**, *1157–1163*, 1–27.

106. Rochester, J.R.; Bolden, A.L. Bisphenol S and F: A Systematic Review and Comparison of the Hormonal Activity of Bisphenol A Substitutes. *Environ. Health Perspect.* **2015**, *123*, 643–650. [CrossRef]

107. Dietrich, D.R.; Hengstler, J.G. From bisphenol A to bisphenol F and a ban of mustard due to chronic low-dose exposures? *Arch. Toxicol.* **2016**, *90*, 489–491. [CrossRef] [PubMed]

108. Gramec Skledar, D.; Peterlin Mašič, L. Bisphenol A and its analogs: Do their metabolites have endocrine activity? *Environ. Toxicol. Pharmacol.* **2016**, *47*, 182–199. [CrossRef] [PubMed]

109. EFSA (European Food Safety Authority). In Proceedings of the Minutes of the 54th Meeting of the Working Group on BPA Evaluation, 12–13 July 2021. Available online: https://www.efsa.europa.eu/sites/default/files/wgs/food-ingredients-and-packaging/wg-BPA-re-evaluation-m.pdf (accessed on 26 July 2021).

110. BfR (Bundesinstitut für Risikobewertung). Bisphenol F in Mustard: Adverse Effects on Health Due to the Measured BPF Concentrations Are Unlikely. *Opinion No. 044/2015*. Available online: www.bfr.bund.de.BfR (accessed on 17 June 2021).

111. Liu, J.; Wattar, N.; Field, C.J.; Dinu, I.; Dewey, D.; Martin, J.W. Exposure and dietary sources of bisphenol A (BPA) and BPA-alternatives among mothers in the APrON cohort study. *Environ. Int.* **2018**, *119*, 319–326. [CrossRef]

112. Lehmler, H.-J.; Liu, B.; Gadogbe, M.; Bao, W. Exposure to Bisphenol A, Bisphenol F, and Bisphenol S in U.S. Adults and Children: The National Health and Nutrition Examination Survey 2013–2014. *Am. Chem. Soc. (AMS) Omega* **2018**, *3*, 6523–6532. [CrossRef] [PubMed]

113. Chauhan, J.; Bhadauria, V.P.S.; Singh, V.; Singh, M.; Singh, K.; Kumar, A. Quality characteristics and their interrelationships in Indian rapeseed-mustard (Brassica sp.) varieties. *Indian J. Agric. Sci.* **2007**, *77*, 616–620.

114. Wendlinger, C.; Hammann, S.; Vetter, W. Various concentrations of erucic acid in mustard oil and mustard. *Food Chem.* **2014**, *153*, 393–397. [CrossRef]

115. Mortuza, M.; Dutta, P. Erucic acid content in some rapeseed/mustard cultivars developed in Bangladesh. *J. Sci. Food Agric.* **2005**, *86*, 135–139. [CrossRef]

116. Mandal, S.; Yadav, S.; Singh, R.; Begum, G.; Suneja, P.; Singh, M. Correlation studies on oil content and fatty acid profile of some Cruciferous species. *Genet. Resour. Crop Evol.* **2002**, *49*, 551–556. [CrossRef]

117. Vetter, W.; Darwisch, V.; Lehnert, K. Erucic acid in Brassicaceae and salmon—An evaluation of the new proposed limits of erucic acid in food. *NFS J.* **2020**, *19*, 9–15. [CrossRef]

118. Mahmoud, K.; Tahoun, A.H.; Ghazy, I.; Amin, E.A. Oil and erucic acid. Contents of five Brassica species grown for four successive years in Egypt. In Proceedings of the 10th International Rapeseed Congress, Canberra, Australia, 26–29 September 1999.

119. Chowdhury, K.; Banu, L.A.; Khan, S.; Latif, A. Studies on the Fatty Acid Composition of Edible Oil. *Bangladesh J. Sci. Ind. Res.* **2007**, *42*, 311–316. [CrossRef]

120. Rahman, D.M.H.; Sarwar, M.; Rahman, M.; Raza, M.; Rouf, S.; Rahman, M. Determination of Erucic acid content in traditional and commercial mustard oils of Bangladesh by Gas- Liquid Chromatography. *Adv. Biochem.* **2014**, *2*, 9–13.

121. Dorni, C.; Sharma, P.; Saikia, G.; Longvah, T. Fatty acid profile of edible oils and fats consumed in India. *Food Chem.* **2018**, *238*, 9–15. [CrossRef]

122. Konuskan, D.B.; Arslan, M.; Oksuz, A. Physicochemical properties of cold pressed sunflower, peanut, rapeseed, mustard and olive oils grown in the Eastern Mediterranean region. *Saudi J. Biol. Sci.* **2019**, *26*, 340–344. [CrossRef] [PubMed]

123. Kok, W.-M.; Mainal, A.; Chuah, C.-H.; Cheng, S.-F. Content of Erucic Acid in Edible Oils and Mustard by Quantitative 13C NMR. *Eur. J. Lipid Sci. Technol.* **2018**, *120*, 1700230. [CrossRef]

124. Sawicka, B.; Kotiuk, E.; Kiełtyka-Dadasiewicz, A.; Krochmal-Marczak, B. Fatty Acids Composition of Mustard Oil from Two Cultivars and Physico-chemical Characteristics of the Seeds. *J. Oleo Sci.* **2020**, *69*, 207–217. [CrossRef] [PubMed]

125. BLS (Bundeslebensmittelschlüssel). Version 3.02. Max-Rubner Institut (MRI) 2005–2021, Standort Karlsruhe. Available online: https://blsdb.de/ (accessed on 17 June 2021).

126. Burton, W.; Salisbury, P.; Potts, D. The potential of canola quality Brassica juncea as an oilseed crop for Australia. In Proceedings of the 13th Australian Research Assembly on Brassicas, Tamworth, NSW, Australia, 8–12 September 2003.

127. EFSA (European Food Safety Authority). Erucic acid in feed and food. *EFSA J.* **2016**, *14*, e04593. [CrossRef]

128. Osmundsen, H.; Bremer, J. Comparative biochemistry of beta-oxidation. An investigation into the abilities of isolated heart mitochondria of various animal species to oxidize long-chain fatty acids, including the C22:1 monoenes. *Biochem. J.* **1978**, *174*, 379–386. [CrossRef]

129. Buddecke, E.; Filipović, I.; Wortberg, B.; Seher, A. Wirkungsmechanismus langkettiger Monoenfettsäuren im Energiestoffwechsel des Herzens. *Fette Seifen Anstrichm.* **1976**, *78*, 196–200. [CrossRef]

130. Chien, K.R.; Bellary, A.; Nicar, M.; Mukherjee, A.; Buja, L.M. Induction of a reversible cardiac lipidosis by a dietary long-chain fatty acid (erucic acid). Relationship to lipid accumulation in border zones of myocardial infarcts. *Am. J. Pathol.* **1983**, *112*, 68–77.

131. Kramer, J.K.; Farnworth, E.R.; Johnston, K.M.; Wolynetz, M.S.; Modler, H.W.; Sauer, F.D. Myocardial changes in newborn piglets fed sow milk or milk replacer diets containing different levels of erucic acid. *Lipids* **1990**, *25*, 729–737. [CrossRef]

132. Kramer, J.K.G.; Sauer, F.D.; Wolynetz, M.S.; Farnworth, E.R.; Johnston, K.M. Effects of dietary saturated fat on erucic acid induced myocardial lipidosis in rats. *Lipids* **1992**, *27*, 619–623. [CrossRef]

133. Beare-Rogers, J.L. Docosenoic acids in dietary fats. *Prog. Chem. Fats Other Lipids* **1977**, *15*, 29–56. [CrossRef]

134. Clouet, P.; Bézard, J. Comparative oxidation of erucic and oleic acids by mitochondria isolated from heart auricle of living man. *C. R. L'académie Sci.* **1979**, *288*, 1683–1686.

135. Imamura, F.; Lemaitre, R.N.; King, I.B.; Song, X.; Steffen, L.M.; Folsom, A.R.; Siscovick, D.S.; Mozaffarian, D. Long-chain monounsaturated Fatty acids and incidence of congestive heart failure in 2 prospective cohorts. *Circulation* **2013**, *127*, 1512–1521. [CrossRef] [PubMed]

136. Matsumoto, C.; Matthan, N.R.; Lichtenstein, A.H.; Gaziano, J.M.; Djoussé, L. Red blood cell MUFAs and risk of coronary artery disease in the Physicians' Health Study. *Am. J. Clin. Nutr.* **2013**, *98*, 749–754. [CrossRef]

137. Schulze, P.C. Myocardial lipid accumulation and lipotoxicity in heart failure. *J Lipid Res* **2009**, *50*, 2137–2138. [CrossRef]

138. FSANZ (Australia New Zealand Food Standards Code). Maximum Levels of Contaminants and Natural Toxicants. 2016. Available online: https://www.legislation.gov.au/Details/F2016C00167 (accessed on 17 June 2021).

139. FDA (U.S. Food and Drug Administration). Import Alert 26-04 "Detention without Physical Examination of Expressed Mustard Oil". 2016. Available online: https://www.accessdata.fda.gov/cms_ia/ind._26.html (accessed on 17 June 2021).

140. Rastogi, T.; Reddy, K.S.; Vaz, M.; Spiegelman, D.; Prabhakaran, D.; Willett, W.C.; Stampfer, M.J.; Ascherio, A. Diet and risk of ischemic heart disease in India. *Am. J. Clin. Nutr.* **2004**, *79*, 582–592. [CrossRef]

141. Wanasundara, J.P. Proteins of Brassicaceae oilseeds and their potential as a plant protein source. *Crit. Rev. Food Science Nutr.* **2011**, *51*, 635–677. [CrossRef]

142. Wanasundara, J.P.D.; Abeysekara, S.J.; McIntosh, T.C.; Falk, K.C. Solubility Differences of Major Storage Proteins of Brassicaceae Oilseeds. *J. Am. Oil Chem. Soc.* **2012**, *89*, 869–881. [CrossRef]

143. Menéndez-Arias, L.; Moneo, I.; Domínguez, J.; Rodríguez, R. Primary structure of the major allergen of yellow mustard (*Sinapis alba* L.) seed, Sin a I. *Eur. J. Biochem.* **1988**, *177*, 159–166. [CrossRef] [PubMed]

144. Menéndez-Arias, L.; Domínguez, J.; Moneo, I.; Rodríguez, R. Epitope mapping of the major allergen from yellow mustard seeds, Sin a I. *Mol. Immunol.* **1990**, *27*, 143–150. [CrossRef]

145. Gonzalez de la Peña, M.A.; Menéndez-Arias, L.; Monsalve, R.I.; Rodríguez, R. Isolation and Characterization of a Major Allergen from Oriental Mustard Seeds, Bra j I. *Int. Arch. Allergy Immunol.* **1991**, *96*, 263–270. [CrossRef]

146. WHO/IUIS Allergen Nomenclature Sub-Committee. Available online: http://www.allergen.org/ (accessed on 26 July 2021).

147. Moreno, F.J.; Clemente, A. 2S Albumin Storage Proteins: What Makes them Food Allergens? *Open Biochem. J.* **2008**, *2*, 16–28. [CrossRef]

148. Morisset, M.; Moneret-Vautrin, D.-A.; Maadi, F.; Frémont, S.; Guénard, L.; Croizier, A.; Kanny, G. Prospective study of mustard allergy: First study with double-blind placebo-controlled food challenge trials (24 cases). *Allergy* **2003**, *58*, 295–299. [CrossRef] [PubMed]

149. Sirvent, S.; Palomares, O.; Vereda, A.; Villalba, M.; Cuesta-Herranz, J.; Rodríguez, R. nsLTP and profilin are allergens in mustard seeds: Cloning, sequencing and recombinant production of Sin a 3 and Sin a 4. *Clin. Exp. Allergy* **2009**, *39*, 1929–1936. [CrossRef] [PubMed]

150. L'Hocine, L.; Pitre, M.; Achouri, A. Detection and identification of allergens from Canadian mustard varieties of *Sinapis alba* and *Brassica juncea*. *Biomolecules* **2019**, *9*, 489. [CrossRef] [PubMed]

151. Perera, S.P.; McIntosh, T.C.; Wanasundara, J.P.D. Structural Properties of Cruciferin and Napin of Brassica napus (Canola) Show Distinct Responses to Changes in pH and Temperature. *Plants* **2016**, *5*, 36. [CrossRef] [PubMed]
152. EFSA (European Food Safety Authority). Opinion of the Scientific Panel on Dietetic products, nutrition and allergies [NDA] on a request from the Commission relating to the evaluation of allergenic foods for labelling purposes. *EFSA J.* **2004**, *2*, 32. [CrossRef]
153. Maruyama, T.; Iizuka, H.; Tobisawa, Y.; Shiba, T.; Matsuda, T.; Kurohane, K.; Imai, Y. Influence of Local Treatments with Capsaicin or Allyl Isothiocyanate in the Sensitization Phase of a Fluorescein-Isothiocyanate-Induced Contact Sensitivity Model. *Int. Arch. Allergy Immunol.* **2007**, *143*, 144–154. [CrossRef] [PubMed]
154. Pasricha, J.S.; Gupta, R.; Gupta, S.K. Contact Hwersensitaity to Mustard Khal and Mustard Oil. *Indian J. Dermatol. Venereol. Leprol.* **1985**, *51*, 108–110.
155. EFSA (European Food Safety Authority). Scientific Opinion on the evaluation of allergenic foods and food ingredients for labelling purposes. *EFSA J.* **2014**, *12*, 3894.
156. Rancé, F.; Dutau, G.; Abbal, M. Mustard allergy in children. *Allergy* **2000**, *55*, 496–500. [CrossRef]
157. Rancé, F.; Abbal, M.; Dutau, G. Mustard allergy in children. *Pediatr. Pulmonol.* **2001**, *23*, 44–45. [CrossRef] [PubMed]
158. Rancé, F. Mustard allergy as a new food allergy. *Allergy* **2003**, *58*, 287–288. [CrossRef]
159. Sirvent, S.; Palomares, O.; Cuesta-Herranz, J.; Villalba, M.; Rodríguez, R. Analysis of the structural and immunological stability of 2S albumin, nonspecific lipid transfer protein, and profilin allergens from mustard seeds. *J. Agric. Food Chem.* **2012**, *60*, 6011–6018. [CrossRef]
160. Vereda, A.; Sirvent, S.; Villalba, M.; Rodríguez, R.; Cuesta-Herranz, J.; Palomares, O. Improvement of mustard (*Sinapis alba*) allergy diagnosis and management by linking clinical features and component-resolved approaches. *J. Allergy Clin. Immunol.* **2011**, *127*, 1304–1307. [CrossRef]
161. Figueroa, J.; Blanco, C.; Dumpiérrez, A.G.; Almeida, L.; Ortega, N.; Castillo, R.; Navarro, L.; Pérez, E.; Gallego, M.D.; Carrillo, T. Mustard allergy confirmed by double-blind placebo-controlled food challenges: Clinical features and cross-reactivity with mugwort pollen and plant-derived foods. *Allergy* **2005**, *60*, 48–55. [CrossRef] [PubMed]
162. Kulka, M.; Sheen, C.H.; Tancowny, B.P.; Grammer, L.C.; Schleimer, R.P. Neuropeptides activate human mast cell degranulation and chemokine production. *Immunology* **2008**, *123*, 398–410. [CrossRef]
163. FARRP (Food Allergy Research and Resource Program). Food Allergens—International Regulatory Chart. Available online: https://farrp.unl.edu/IRChart/ (accessed on 26 July 2021).
164. Monsalve, R.; Villalba, M.; Rodriguez, R. Allergy to mustard seeds: The importance of 2S albumins as food allergens. *Internet Symp. Food Allerg.* **2001**, *3*, 57–69.
165. Sharma, A.; Verma, A.K.; Gupta, R.K.; Neelabh; Dwivedi, P.D. A Comprehensive Review on Mustard-Induced Allergy and Implications for Human Health. *Clin. Rev. Allergy Immunol.* **2019**, *57*, 39–54. [CrossRef]
166. Pałgan, K.; Żbikowska-Gotz, M.; Bartuzi, Z. Dangerous anaphylactic reaction to mustard. *Arch. Med. Sci.* **2018**, *14*, 477–479. [CrossRef]
167. Moneret-Vautrin, A. Epidemiology of food allergies and relative prevalence of trophallergens. *Cah. Nutr. Diet.* **2001**, *36*, 247–252.
168. BfR (Bundesinstitut für Risikobewertung). 'VITAL 3.0': New and updated proposals for reference doses of food allergens. *BfR Opin.* **2020**. Available online: https://www.bfr.bund.de/cm/349/vital-30-new-and-updated-proposals-for-reference-doses-of-food-allergens.pdf/ (accessed on 26 July 2021). [CrossRef]

Article

Spices in the Apiaceae Family Represent the Healthiest Fatty Acid Profile: A Systematic Comparison of 34 Widely Used Spices and Herbs

Ramesh Kumar Saini [1], Awraris Derbie Assefa [2] and Young-Soo Keum [1,*]

[1] Department of Crop Science, Konkuk University, Seoul 05029, Korea; saini1997@konkuk.ac.kr
[2] National Agrobiodiversity Center, National Institute of Agricultural Sciences, Rural Development Administration, Jeonju 54874, Korea; awraris@korea.kr
* Correspondence: rational@konkuk.ac.kr

Abstract: Spices and herbs are well-known for being rich in healthy bioactive metabolites. In recent years, interest in the fatty acid composition of different foods has greatly increased. Thus, the present study was designed to characterize the fatty acid composition of 34 widely used spices and herbs. Utilizing gas chromatography (GC) flame ionization detection (FID) and GC mass spectrometry (MS), we identified and quantified 18 fatty acids. This showed a significant variation among the studied spices and herbs. In general, oleic and linoleic acid dominate in seed spices, whereas palmitic, stearic, oleic, linoleic, and α-linolenic acids are the major constituents of herbs. Among the studied spices and herbs, the ratio of $n-6/n-3$ polyunsaturated fatty acids (PUFAs) was recorded to be in the range of 0.36 (oregano) to 85.99 (cumin), whereas the ratio of PUFAs/saturated fatty acids (SFAs) ranged from 0.17 (nutmeg) to 4.90 (cumin). Cumin, coriander, fennel, and dill seeds represent the healthiest fatty acid profile, based upon fat quality indices such as the ratio of hypocholesterolemic/hypercholesterolemic (h/H) fatty acids, the atherogenic index (AI), and the thrombogenic index (TI). All these seed spices belong to the Apiaceae family of plants, which are an exceptionally rich source of monounsaturated fatty acids (MUFAs) in the form of petroselinic acid (C18:1n12), with a very small amount of SFAs.

Keywords: polyunsaturated fatty acids (PUFAs); erucic acid; petroselinic acid; fat quality indices; hypocholesterolemic fatty acids; atherogenic index (AI)

Citation: Saini, R.K.; Assefa, A.D.; Keum, Y.-S. Spices in the Apiaceae Family Represent the Healthiest Fatty Acid Profile: A Systematic Comparison of 34 Widely Used Spices and Herbs. *Foods* **2021**, *10*, 854. https://doi.org/10.3390/foods10040854

Academic Editors: Andreas Eisenreich and Bernd Schaefer

Received: 8 March 2021
Accepted: 12 April 2021
Published: 14 April 2021

Publisher's Note: MDPI stays neutral with regard to jurisdictional claims in published maps and institutional affiliations.

1. Introduction

Spices and herbs are a vital part of human nutrition around the world, especially in India, China, and southeastern Asian countries [1]. Spices and herbs are food adjuncts, traditionally used as flavoring, seasoning, coloring, and as a food preservative agent [1,2]. Moreover, spices and herbs are an exceptionally rich source of nutritionally important phenolic compounds [3]. These phenolic compounds are primarily responsible for the potent antioxidative, digestive stimulative, hypolipidemic, antibacterial, anti-inflammatory, antiviral, and anticancer properties of spices and herbs [4–6].

In general, the terms herbs and spices have more than one meaning. However, the most widely used are those that consider herbs to be derived from the green parts of a plant, such as a stem and leaves used in small amounts to impart flavor, whereas spices are obtained from seeds, buds, fruits, roots, or even the bark of the plants [2].

Fatty acids are the primary nutritional components found in edible seed oils [7]. Seed oils provide essential polyunsaturated fatty acids, linoleic acid ($\omega-6$ or $n-6$), and α-linolenic acid ($n-3$) to humans and other higher animals. In the human body, linoleic acid give rise to $n-6$ very long-chain (VLC)-PUFA arachidonic acid, and α-linolenic acid converts to $n-3$ VLC-PUFA eicosapentaenoic acid (EPA), and docosahexaenoic acid (DHA,

$n-3$). These $n-6$ and $n-3$ VLC-PUFAs plays key distinct roles in regulating body homeostasis. In general, $n-6$ VLC-PUFAs gives rise to proinflammatory mediators (eicosanoids) whereas $n-3$ VLC-PUFAs give rise to anti-inflammatory mediators. Thus, a higher amount of $n-3$ VLC-PUFAs in the body may protect from chronic diseases, including cancer, inflammatory, or cardiovascular diseases (CVD) [8]. Moreover, a diet with a high proportion of $n-6$ PUFAs (high ratio of $n-6/n-3$ PUFAs) cannot be considered beneficial to health, as $n-3$ PUFAs to $n-3$ VLC-PUFAs conversion occurs at a very low rate (e.g., 8% for EPA and less than 1% for DHA), and conversion is largely dependent upon the ratio of ingested $n-6$ (linoleic acid) and $n-3$ (α-linolenic) PUFAs [9]. In human hepatoma cells, this conversion is highest when these $n-6$ and $n-3$ acids are provided at a 1:1 ratio. Thus, the consumption of an appropriate amount of fats with a 1:1 $n-6/n-3$ PUFAs ratio, which was probably followed by our ancestors [10], may be considered beneficial.

Similar to the consumption of fats with a balanced ratio of $n-6/n-3$ PUFAs, growing evidence suggests that replacing saturated fatty acids (SFAs) with monounsaturated fatty acids (MUFAs) from plant sources may decrease the risk of CVD [11]. And with the health benefits associated with consumption of $n-3$ PUFAs and MUFAs, consumer interest is shifting towards foods with a low proportion of SFAs, a high proportion of MUFAs, and balanced $n-6/n-3$ PUFAs. Given this, it is necessary to characterize all the major and minor components of the diet to acquire a better estimate of the fatty acid composition of our food.

Spices and herbs are not a significant source of fatty acids, as they form a small part of the diet. However, a detailed and comparative study of the fatty acid composition of various spices and herbs may be useful to identify those with health-beneficial fatty acids. Considering these facts, this study aims to investigate the fatty acid composition of commercially available major spices and herbs utilizing gas chromatography-flame ionization detection and GC-mass spectrometry analysis. We used fatty acid composition data to study spices and herbs to determine their fat quality indices. We anticipate the results contained herein will contribute significantly to the identification of spices with a healthy fatty acid profile.

2. Materials and Methods

2.1. Plant Material, Reagents, and Standards

A total of 34 commercially packed spices and herbs (Table 1; 200–500 g each spice and herb from at least three different brands) were obtained from retail outlets in Seoul, Korea. The spice and herb samples of different brands were mixed in equal proportions (200–300 g total) to make a representative sample, ground into a fine powder using a 7010HG laboratory blender (Waring Commercial, Torrington, CT, USA), placed into an airtight container, and stored at room temperature. The fatty acid standard mix (37 Component FAME Mix, CRM47885) was obtained from Merck Ltd., Seoul, Korea. The organic solvents used for the extraction of lipids were of high-pressure liquid chromatography (HPLC) grade, obtained from Samchun Chemical Co., Ltd., Seoul, Korea.

2.2. Extraction of Crude Lipid Compounds

The crude lipids were extracted by using the previous method [12,13] with minor modification. Briefly, 0.6 g dehydrated and powdered spices and herb samples were precisely weighed and transferred to a 50 mL glass tube. In each tube, 150 mg sodium ascorbate and 22 mL (isopropyl alcohol/cyclohexane, 10:12, v/v) containing 0.075% butylated hydroxytoluene (BHT: w/v; antioxidant) were added, and the contents were subjected to bath sonication (JAC-2010; 300 w, 60 Hz, for 12 min) for efficient disintegration and complete extraction, followed by 15 h shaking (200 RPM at 22 °C) in a rotary shaker. Contents were centrifuged at $7000\times g$ (12 min at 4 °C). The supernatant was collected, and pellets were extracted again with 30 mL cyclohexane. Supernatants from both extractions were pooled (total volume of ~50 mL) and partitioned with an equal volume of 1 M of sodium chloride (NaCl). The upper cyclohexane phase containing crude lipids were collected, filtered over anhydrous sodium sulfate, transferred to a 250-mL round-bottom flask, and vacuum-dried in a rotary evaporator at 30 °C. The crude lipids were recovered into 3 mL methanol/dichloromethane (DCM) (1:3, *v/v*) containing 0.1% BHT, transferred to a 5 mL glass vial fitted with a Teflon-lined screw cap, and stored at −20 °C. One milliliter of sample was used to prepare fatty acid methyl esters (FAMEs).

Table 1. List of spices and herbs used in the present investigation (arranged according to botanical name).

Sample No.	Common Name	Botanical Name	Family	Part
1	Galangal root	*Alpinia galanga* (L.) Willd.	Zingiberaceae	Rhizomes
2	Dill	*Anethum graveolens* L.	Apiaceae	Seeds
3	Celery	*Apium graveolens* L.	Apiaceae	Seeds
4	Tarragon	*Artemisia dracunculus* L.	Asteraceae	Leaves
5	Cayenne pepper	*Capsicum annuum* L.	Solanaceae	Pods
6	Pepperoncini	*Capsicum annuum* L. *var. annuum*	Solanaceae	Pods
7	Hot chili pepper	*Capsicum frutescens* L.	Solanaceae	Pods
8	Caraway	*Carum carvi* L.	Apiaceae	Fruits
9	Cinnamon	*Cinnamomum verum* J.Presl	Lauraceae	Bark
10	Coriander seed	*Coriandrum sativum* L.	Apiaceae	Seeds
11	Cumin	*Cuminum cyminum* L.	Apiaceae	Seeds
12	Turmeric	*Curcuma longa* L.	Zingiberaceae	Rhizomes
13	Lemongrass	*Cymbopogon microstachys* (J. D. Hooker) Soenarko	Poaceae	Leaves
14	Cardamom	*Elettaria cardamomum* (L.) Maton	Zingiberaceae	Pods
15	Fennel	*Foeniculum vulgare* Mill.	Apiaceae	Seeds
16	Star anise	*Illicium verum* Hook.f.	Schisandraceae	Fruits
17	Allspice	*Pimenta dioica* (L.) Merr.	Myrtaceae	Fruits
18	Juniper berry	*Juniperus communis* L.	Cupressaceae	Fruits
19	Bay leaf	*Laurus nobilis* L.	Lauraceae	Leaves
20	Nutmeg	*Myristica fragrans* Houtt.	Myrtaceae	Seeds
21	Mace	*Myristica fragrans* Houtt.	Myrtaceae	Aril
22	Basil	*Ocimum basilicum* L.	Lamiaceae	Leaves
23	Marjoram	*Origanum majorana* L.	Lamiaceae	Leaves
24	Oregano	*Origanum vulgare* L.	Lamiaceae	Leaves
25	Parsley	*Petroselinum crispum* (Mill.) Fuss	Apiaceae	Leaves
26	Black pepper	*Piper nigrum* L.	Piperaceae	Fruits (unripe)
27	White pepper	*Piper nigrum* L.	Piperaceae	Seeds
28	Rosemary	*Rosmarinus officinalis* L.	Lamiaceae	Leaves
29	Sage	*Salvia officinalis* L.	Lamiaceae	Leaves
30	Pink peppercorn	*Schinus mole* L.	Anacardiaceae	Fruits
31	White mustard	*Sinapis alba* L.	Brassicaceae	Seeds
32	Clove	*Syzygium aromaticum* (L.) Merr. and L. M. Perry	Myrtaceae	Flower buds
33	Thyme	*Thymus vulgaris* L.	Lamiaceae	Leaves
34	Ginger	*Zingiber officinale* Roscoe	Zingiberaceae	Rhizomes

2.3. Preparation of Fatty Acid Methyl Esters (FAMEs)

The crude lipids extracted from the spices and herb samples were used to prepare the FAMEs, following the previously optimized method [14] with minor modification. Briefly, 1 mL of a crude lipids sample was transferred into a 5 mL glass vial fitted with a Teflon-lined screw cap. Contents were evaporated to dryness using a rotary evaporator at 30 °C. After evaporation, 3 mL of anhydrous methanolic-HCl (methanol/acetyl chloride, 95:5, *v/v*) was added and incubated for 2 h at 55 °C in a heat block. Samples were cooled in ice, and FAMEs were sequentially washed with 1M NaCl and 2% sodium bicarbonate (NaHCO$_3$) and recovered in 4 mL hexane. A pinch of anhydrous sodium sulfate (Na$_2$SO$_4$) was added to the recovered sample (hexane) to absorb the traces of water. One milliliter of sample was filtered through a 0.45 μm PTFE syringe filter and transferred to a 1.5 mL autosampler vial for GC-FID and GC-MS analysis.

2.4. GC-FID and GC-MS Analysis of FAMEs

FAMEs were quantitatively analyzed with GC (Agilent 7890B, Agilent Technologies Canada, Inc., Mississauga, ON, Canada) equipped with an autoinjector, an FID, and an SP-2560 capillary column (100 m, 0.20 μm film thickness, 0.25 mm ID; Merck KGaA, Darmstadt, Germany). The injector and the detectors were maintained at 250 °C and 260 °C, respectively. The inlet flow was 2 mL/min with a constant pressure of 54 psi. The FID parameters of hydrogen (H_2) fuel flow, airflow, and make flow (nitrogen, N_2) were set to 30, 400, and 25 mL/min, respectively. The column oven temperature was kept at 140 °C for 5 min, then progressively increased to 240 °C for 25 min (linear temperature program 4 °C/min and held at 240 °C for 15 min [15]. The FAMEs were precisely identified by comparing them with the retention time with authentic standards. For a more accurate qualitative analysis, the mass spectra were also recorded using a GC-MS system (QP2010 SE; Shimadzu, Kyoto, Japan), following the optimized GC-FID analysis thermal program. The identity of FAMEs was confirmed by comparing their fragmentation pattern with authentic standards, and also by using the National Institute of Standards and Technology (NIST; U.S. Department of Commerce, Gaithersburg, MD, USA) mass spectrum database (NIST08 and NIST08s).

2.5. Calculation of Fat Quality Indices

We used the spice and herbs fatty acid profile to determine several nutritional parameters of lipids, including the ratios of PUFAs/monounsaturated fatty acids (MUFAs), PUFAs/saturated fatty acids (SFAs), the ratio of hypocholesterolemic/hypercholesterolemic (h/H) fatty acids, atherogenic index (AI), and thrombogenic index (TI) [16]. The ratio of h/H fatty acids, AI, and TI was calculated with the following equations [16]:

$$h/H = \frac{cis\,C18:1 + \sum MUFAs + \sum PUFAs}{C12:0 + C14:0 + C16:0}$$

$$AI = \frac{C12:0 + (4 \times C14:0) + C16:0}{\sum MUFAs + \sum PUFAs}$$

$$TI = \frac{C14:0 + C16:0 + C18:0}{(0.5 \times \sum MUFAs) + (0.5 \times \sum n{-}6\,PUFAs) + (3 \times \sum n{-}3\,PUFAs) + \left(\frac{\sum n{-}3\,PUFAs}{\sum n{-}6\,PUFAs}\right)}$$

2.6. Statistical Analysis and Quality Control

We performed a total of six replicate extractions and analyses from each representative sample. The data were analyzed by one-way analysis of variance (ANOVA), and homogenous subsets (mean separation) were determined using Turkey HSD with a significance level of $p < 0.05$, utilizing the IBM statistical 25.0 software.

The method used for GC-FID quantification of FAMEs was validated recently [15].

3. Results and Discussion

3.1. Fatty Acids Composition

In the present study, 18 fatty acids were identified and quantified, utilizing GC-FID and GC-MS analyses (Table 2). The results, given in Table 2, show that oleic (C18:1n9) and linoleic acid (C18:2n6) are dominated in seed spices, and palmitic (C16:0), stearic, oleic, linoleic, and α-linolenic acid (C18:3n3) are the major constituents of herbs. An exception was myristic (C14:0) acid, which was 60.8% of total fatty acids in *Myristica fragrans* (nutmeg) seeds (Figure 1A,B). Surprisingly, myristic acid was just 1.59% of the total fatty acids in the *M. fragrans* (mace; Figure 1C) seed arils. The highest proportions of oleic acid (41.64–41.85%) were recorded in cardamon pods/capsules (Figure S1) and white pepper seeds (Table 2). The data of the fatty acid composition of cardamom pods and white pepper seeds are scarce. However, 40.6–49.2% of oleic acid has been reportedly extracted from cold-pressed cardamom seeds [17,18], which agrees with data obtained in the present study from whole cardamon pods.

Table 2. Fatty acid composition of spices and herbs.

Peak No	1	2	3	4	5	6	7	8	9	10	11	12	13	14	15	16	17	18
Component (FAME)	C12:0 (Lauric)	C14:0 (Myristic)	C15:0 (Pentade-canoic)	C16:0 (Palmitic)	C16:1 (Palmi-toleic)	C17:0 (Heptade-canoic)	C18:0 (Stearic)	C18:1n12c (Pet-roselinic)	C18:1n9c (Oleic)	C16:3n3 (Hexade-catrienoic)	C18:2n6c (Linoleic)	C20:0 (Arachidic)	20:1n9 (Eicosenoic)	C18:3n3 (α-Linolenic)	C22:0 (Be-henic)	C22:1n9 (Erucic)	C24:0 (Ligno-ceric)	C24:1n9 (Ner-vonic)
RT	13.44	16.78	18.58	20.42	21.74	22.20	23.96	25.03	25.10	25.20	26.81	27.29	28.36	28.60	30.38	31.48	33.46	34.57
S1	0.62	0.90	1.18 [a]	37.65	nd	nd	13.06	nd	28.21	nd	13.59	nd	nd	2.45	1.01	nd	1.33	nd
S2	1.20	0.90	nd	8.48	0.34	nd	1.54	50.37	15.26	nd	20.49	0.23	nd	0.85	0.23	nd	0.12	nd
S3	nd	0.44	nd	11.09	0.21	nd	1.91	49.42	7.48	nd	27.55	0.23	nd	1.32	0.21	nd	0.15	nd
S4	6.95	2.90	nd	24.86	nd	nd	3.03	nd	2.23	nd	24.10	2.21	nd	31.84	1.16	nd	0.72	nd
S5	0.27	0.95	nd	18.25	0.64	0.13	2.73	nd	12.31	nd	61.46	0.38	nd	2.53	0.22	nd	0.12	nd
S6	0.27	1.19	nd	18.71	0.86	nd	3.03	nd	12.68	nd	59.65	0.45	nd	2.75	0.27	nd	0.16	nd
S7	1.18	3.22	nd	21.98	0.85	nd	3.87	21.76	17.43	nd	27.99	0.48	nd	0.68	0.36	nd	0.20	nd
S8	nd	10.62	nd	5.55	0.13	nd	2.95	34.09	11.70	nd	33.70	0.50	nd	0.47	0.21	nd	0.07	nd
S9	Nd	1.77	nd	33.97	nd	nd	7.49	nd	23.25	nd	26.99	nd	nd	4.04	1.13	nd	1.36	nd
S10	0.10	0.45	0.14	6.91	0.28	nd	1.36	62.07	7.13	nd	20.85	nd	nd	0.35	0.23	nd	0.12	nd
S11	0.08	0.10	0.16	5.28	0.29	nd	1.08	49.89	9.21	nd	33.34	nd	nd	0.39	0.12	nd	0.07	nd
S12	2.81	0.40	nd	10.51	0.76	nd	2.67	nd	4.48	nd	72.86 [a]	nd	nd	4.50	0.32	nd	0.67	nd
S13	5.68	2.74	nd	47.82 [a]	nd	nd	9.46	nd	6.20	nd	12.29	3.89	nd	5.70	3.64 [a]	nd	2.57	nd
S14	nd	0.75	nd	32.84	1.58 [a]	nd	3.19	nd	41.81 [a]	nd	15.07	0.67	nd	3.44	0.26	nd	0.38	nd
S15	0.46	0.17	nd	7.25	0.21	nd	1.24	63.33 [a]	6.88	nd	19.60	0.16	nd	0.53	0.10	nd	0.07	nd
S16	1.47	0.26	nd	20.26	0.17	nd	3.74	nd	33.75	nd	39.27	nd	nd	0.71	0.29	nd	0.08	nd
S17	nd	0.29	nd	14.27	nd	0.21 [a]	27.16 [a]	nd	13.90	nd	39.58	1.39	nd	2.53	0.47	nd	0.20	nd
S18	0.61	0.92	nd	26.01	nd	nd	12.37	nd	22.41	nd	26.98	4.40 [a]	nd	3.56	1.50	nd	1.24	nd
S19	4.88	8.60	nd	37.43	nd	nd	5.43	nd	20.33	nd	9.55	1.49	nd	7.19	1.51	nd	3.58 [a]	nd
S20	2.19	60.81 [a]	nd	8.94	0.39	nd	1.26	nd	13.36	nd	11.94	0.08	nd	0.76	0.14	nd	0.13	nd
S21	0.08	1.59	nd	30.63	1.36	0.14	3.29	nd	28.00	nd	33.72	0.14	nd	0.81	0.13	nd	0.11	nd
S22	17.47 [a]	1.97	nd	29.85	nd	nd	7.81	nd	7.56	nd	10.18	4.18	nd	18.57	1.63	nd	0.79	nd
S23	1.20	1.33	nd	21.60	nd	nd	5.39	nd	33.65	nd	13.64	2.63	nd	18.23	1.33	nd	0.99	nd
S24	4.91	2.63	nd	26.85	nd	nd	7.89	nd	6.42	nd	12.49	1.83	nd	35.08 [a]	1.04	nd	0.87	nd
S25	1.15	0.72	nd	15.01	nd	nd	3.38	nd	2.14	17.74 [a]	24.49	0.24	nd	33.40	0.64	nd	1.09	nd
S26	4.71	1.93	nd	28.57	nd	nd	11.35	nd	14.95	nd	26.61	0.44	nd	9.32	0.81	nd	1.31	nd
S27	2.62	0.95	nd	22.55	nd	nd	11.25	nd	41.64 [a]	nd	17.59	nd	nd	1.49	0.81	nd	1.11	nd
S28	nd	2.93	nd	47.85 [a]	nd	nd	11.39	nd	15.62	nd	7.74	3.65	nd	5.31	3.05	nd	2.45	nd
S29	nd	1.59	nd	42.71	nd	nd	10.67	nd	14.51	nd	10.90	3.09	nd	12.62	2.76	nd	1.14	nd
S30	nd	0.35	nd	14.29	0.63	nd	4.59	nd	21.11	nd	56.56	0.43	nd	1.54	0.26	nd	0.24	nd
S31	0.12	0.14	nd	5.53	0.30	nd	1.77	nd	32.97	nd	15.78	0.43	7.96 [a]	16.59	0.23	17.28 [a]	0.09	0.82 [a]
S32	7.95	3.44	nd	27.93	nd	nd	12.60	nd	4.47	nd	28.69	2.29	nd	5.99	2.39	nd	4.26 [a]	nd
S33	0.91	1.83	nd	29.78	nd	nd	10.11	nd	11.03	nd	14.62	1.52	nd	28.15	1.18	nd	0.87	nd
S34	3.85	1.79	nd	31.47	nd	nd	10.01	nd	7.49	nd	32.47	nd	nd	11.10	0.75	nd	1.07	nd

Values (% of total fatty acids) are the mean of six determinations from each representative sample. RT: retention time. The letter "a" within a column represents the highest significant ($p < 0.05$) values. nd: not detected. Sample numbers (S1–S34) correspond to Table 1.

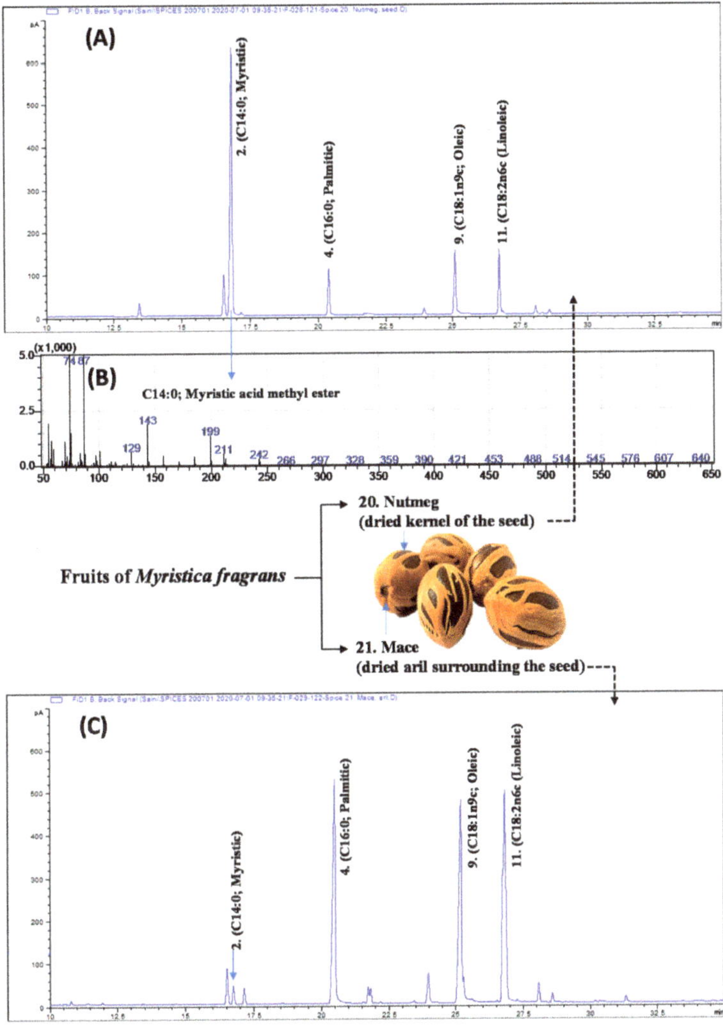

Figure 1. (**A**) The gas chromatography (GC)-flame ionization detection (FID) profiles of fatty acid methyl esters (FAMEs) of nutmeg. (**B**) The GC-mass spectrum of dominating fatty acid (myristic acid) from nutmeg. (**C**) The GC-FID profiles of FAMEs of mace. The numbers, 2, 4, 6, and 8 correspond to peak numbers illustrated in Table 1.

In the present study, a substantial amount of erucic (C22:1n9; 17.3%) and eicosenoic (20:1n9; gondoic acid; 8%) acids were exclusively recorded in white mustard (Sinapis alba; syn Brassica alba) seeds. Similarly, a significant amount of petroselinic acid (C18:1n12c; an isomer of oleic acid) was recorded only in Apiaceae family seeds.

Among the studied 34 spices and herbs, total fatty acids were recorded to be in the range of 2.3 (galangal root) to 130.32 mg/g (mace). The odd chain fatty acid, pentadecanoic (C15:0) acid, was recorded as being a minor constituent (1.18%) in the galangal root. Similarly, heptadecanoic (C17:0) was recorded at only 0.13–0.14% in cayenne pepper, allspice, and mace. In nutmeg (*Myristica fragrans*) seed hexane extract, Anaduaka et al. [19] reported a significant amount of (27%) heptadecanoic (C17:0; margaric) acid. However, in the present study, heptadecanoic acid is not detected in nutmeg seeds.

3.2. Black Pepper and White Pepper

Black pepper and white pepper are prepared from the fruits of *Piper nigrum* L., according to the harvesting time and inclusion of the outer skin. Black pepper is the dried immature but fully developed fruit, whereas white pepper consists of the mature fruit lacking the outer skin [20]. The fatty acid composition data of black and white pepper is scarce. In the present study, 28.57%, 14.95%, 26.61%, and 9.32% of palmitic, oleic, linoleic, and α-linolenic acid were recorded being in black pepper. In contrast, 22.55%, 41.64%, 17.19%, and 1.49% of palmitic, oleic, linoleic, and α-linolenic was reported as being in white pepper (Table 2). These observations show that oleic acid increases significantly, whereas the palmitic, linoleic, and α-linolenic acids decrease significantly during the maturation of pepper fruits.

3.3. Nutmeg and Mace

Nutmeg and mace spices are obtained from different parts of the same fruit of the nutmeg (*Myristica fragrans*; Myristicaceae) tree. Nutmeg is the dried kernel of the seed, whereas mace is the dried aril surrounding the seed [21]. Myristic acid's name is derived from *Myristica fragrans*, from which it was first isolated [22]. In the present study, myristic acid was 60.8% of total fatty acids in nutmeg, followed by oleic (C18:1n9c; 13.4%), linoleic (C18:2n6c; 11.9%), and palmitic (C16:0; 8.94%) (Figure 1A). Surprisingly, in mace, linoleic acid was 33.7% of total fatty acids, followed by palmitic (30.6%) and oleic (28.0%). Myristic acid was only 1.59% of the total fatty acids (Figure 1C, Table 2). In the investigations of Al-Khatib et al. [23], myristic acid was recorded as being 79.7% of the total fatty acids in nutmeg. Kozłowska et al. [24] analyzed the fatty acids composition of plant seeds, including anise, coriander, caraway, white mustard, and nutmeg. They reported dominance of oleic (56.5%), palmitic (18.29%), and linoleic (13.6%) acids in nutmeg. These contrasting observations probably arose as these authors reported only above C16 fatty acids. Myristic acid is widely used in the food industry as a flavor ingredient. It is approved as a pharmaceutical excipient by the Food and Drug Administration (FDA) and declared generally recognized as safe (GRAS) by various regulators [25].

3.4. Erucic Acid in White Mustard

Mustard (*Sinapis alba*; syn *Brassica alba*) seeds are well known for the occurrence of a substantial amount of erucic and eicosenoic acid [24]. In the present study, white mustard seeds were found containing 17.3% and 8.0% of erucic and eicosenoic acid, respectively (Figure 2A, Table 2). High intake of erucic acid is considered harmful for cardiac health [26]. The panel on contaminants in the food chain established a tolerable daily intake (TDI) of 7 mg/kg body weight (BW) for erucic acid based on a no-observed adverse effect level (NOAEL) for myocardial lipidosis in rats and pigs [26]. Considering the 43 mg of total fatty acids/g of white mustard seeds, consumption of 100 g of seeds may provide 7.31 mg of erucic acid. The intake of erucic acid from white mustard used as food condiments in daily food preparations is far below the TDI and is safe for consumption.

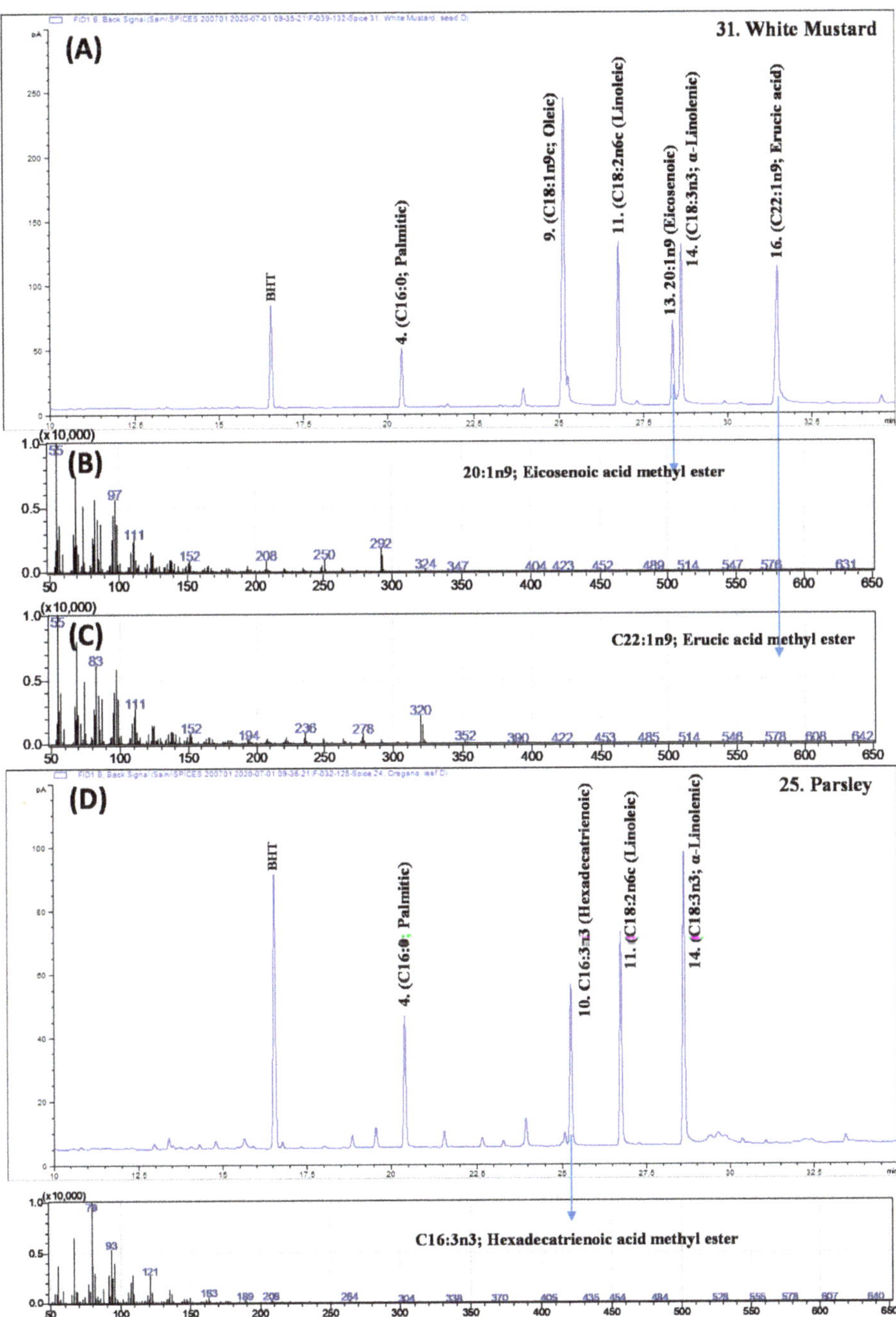

Figure 2. (**A**) The gas chromatography (GC)-flame ionization detection (FID) profiles of fatty acid methyl esters (FAMEs) of white mustard seeds. (**B,C**) The GC-mass spectrum of eicosenoic acid and erucic acid from white mustard seeds. (**D**) The GC-FID profiles of FAMEs of parsley leaves. The numbers, 4, 9, 10, 13, 14, and 16 correspond to peak numbers illustrated in Table 1. BHT: Butylated hydroxytoluene (A synthetic antioxidant used during lipid extraction).

Petroselinic acid (C18:1n12c; an isomer of oleic acid) is the major component of the lipid constituent of Apiaceae family seeds [27,28]. In a previous study [27] of dill (*Anethum graveolens*) seeds, 87.2% of total fatty acids were composed of petroselinic acid. Similarly, in celery (*Apium graveolens*), coriander seeds (*Coriandrum sativum*), and fennel seeds (*Foeniculum vulgare*), petroselinic acid was recorded as being 56.1%, 72.8%, and 31.32% of total fatty acids. In agreement with the present study, we have also recorded the 50.4%, 49.4%, 62.1%, and 63.3% of petroselinic acid in dill, coriander celery, and fennel seeds, respectively (Table 2). And a similar high amount of petroselinic acid was reported to be in the seeds of other Apiaceae family plants, such as caraway (*Carum carvi*, 34.1%) and cumin (*Cuminum cyminum*; 49.9%). In seeds of different varieties of caraway, Reiter et al. [28] recorded 33.5–42.5% of petroselinic acid, which is in agreement with the present study. Petroselinic acid possesses potent anti-inflammatory and antiaging properties by reducing the metabolites of arachidonic acid [29]. And owing to its anti-aging properties, petroselinic acid is widely used in cosmetics or dermatological compositions [29]. Surprisingly, petroselinic acid was not detected in herbs (leaves) of the Apiaceae family member parsley (*Petroselinum crispum*). In the parsley herb, hexadecatrienoic (C16:3n3) was reported to be 17.7% of the total fatty acids (Figure 2D), whereas no other spices were found to contain this fatty acid. Parsley has been previously classified as a "16:3" plant owing to the presence of a significant amount of hexadecatrienoic acid in photosynthetic tissues, which is part of primitive lipid metabolism [30].

3.5. Fat Quality Indices

The present study is based on the fatty acid composition of 34 spices and herbs. We evaluated them for fat quality indices, including the n–6/n–3 ratio, AI, TI, and h/H fatty acid ratios (Table 3). Among the studied spices and food condiments, the ratio of n–6/n–3 PUFAs was found to be in the range of 0.36 (oregano) to 85.99 (cumin). In view of health benefits associated with the consumption of n–6/n–3 PUFAs ratio of 0.5–2.0 (nearest to 1:1), lipids obtained from leaf spices, including tarragon (0.76), bay leaf (1.33), basil (0.55), marjoram (0.75), parsley (0.48), white mustard (0.95), sage (0.86), and thyme (0.52) can be considered to be beneficial. In general, the high occurrence of α-linolenic acids compared to linoleic acid is responsible for the low n–6/n–3 ratio in leaves (photosynthetic tissue).

In view of the high risk of CVD and other chronic diseases that are associated with the dietary intake of SFAs [11], fats with a PUFAs/SFAs ratio lower than 0.45 are not advised for diet [31]. In the present study, PUFAs/SFAs ratios ranged from 0.17 (nutmeg) to 4.90 (cumin). Low PUFAs/SFAs ratios of 0.17 in nutmeg lipids are the result of the dominance of myristic acid (an SFA; Figure 1A), whereas in the case of cumin, linoleic acid is dominant over SFAs. In addition to the nutmeg, low PUFAs/SFAs ratios (<0.44) were recorded from galangal root (0.29), lemongrass (0.24), rosemary (0.28), and sage (0.38) because of the occurrence of a substantial amount of palmitic acid (Figure S2).

Fats with lower AI and TI and higher ratios of h/H fatty acids are recommended for minimizing the risk of CVD [32]. In the present study, a significant difference was recorded for AI, TI values as well as h/H fatty acids among the studied spices and herbs. The lowest significant values of the AI (0.06) and the highest ratios of h/H fatty acids (17.0) were obtained from cumin seeds (Table 3, Figure 3), because of the presence of a low amount of atherogenic lauric, myristic, and palmitic fatty acids, and high amounts of hypocholesterolemic C18:1 MUFAs and PUFAs. Whereas the lowest significant values of TI (0.08) were recorded in white mustard, due to the low contents SFAs and high content of PUFAs.

Overall, based on a higher ratio of h/H fatty acids and their lower AI and TI values, cumin, coriander, fennel, and dill spices have the healthiest fatty acid profiles (Figure 3). These spices belong to the Apiaceae family. White mustard also represents a higher ratio of h/H fatty acids and lower values of AI and TI. However, it contains a substantial amount of erucic acid.

In Figure 3, cumin, coriander, fennel, and dill spices top the fat quality indices, the ratio of h/H fatty acids, AI, and TI. However, the occurrence of a very low proportion of α-linolenic acid (a $n-3$ PUFA; 0.35–0.85%) and a fairly good amount of linoleic acid (a $n-6$ PUFA; 19.60–33.34%) in these spices, give rise to the high ratio of $n-6/n-3$ PUFAs (24.02–85.99), which is substantially higher than the recommended ratio of 1:1. Considering this, the culinary use of these spices can be recommended with $n-3$ PUFA rich components to obtain the overall $n-6/n-3$ PUFAs ratio of 1:1.

Previously, we had analyzed the total phenolic contents (TPC) and antioxidant activities of 39 spices and herbs (including the 34 spices and herbs investigated in the present study) and found that cloves possess the highest antioxidant activities, followed by allspice, cinnamon, oregano, and marjoram [33]. The high antioxidant activities of these spices and herbs were probably the results of the richness of phenolic compounds, as the antioxidant activities showed a good correlation (0.835–0.966) with TPC. In contrast, in the present study, cumin, coriander, fennel, and dill spices showed the healthiest fatty acid profile among the 34 spices and herbs. These observations show that the selection of healthy spices and herbs may vary with nutrient requirements. Thus, in the present study, cumin, coriander, fennel, and dill spices are the recommendations based on the fatty acid profile. However, other spices and herbs might be richer in other health-beneficial dietary components.

Table 3. The fat quality indices of lipids of spices and herbs.

Sample No	Total FA (mg/g DW)	Total SFAs	Total MUFAs	Total PUFAs	PUFAs: SFAs	PUFAs: MUFAs	$n-3$ PUFA	$n-6$ PUFA	$n-6/n-3$	h/H	AI	TI
S1	2.30[b]	55.75	28.21	16.04	0.29	0.57	2.45	13.59	5.55	1.13	0.95	1.82
S2	23.30	12.70	65.97	21.34	1.68	0.32 [b]	0.85	20.49	24.02	8.22	0.15	0.24
S3	27.60	14.02	57.11	28.88	2.06	0.51	1.32	27.55	20.82	7.45	0.15	0.29
S4	8.30	41.83	2.23[b]	55.94	1.34	25.13	31.84	24.10	0.76 [b]	1.68	0.75	0.28
S5	120.83	23.06	12.96	63.99	2.78	4.94	2.53	61.46	24.30	3.92	0.29	0.49
S6	48.72	24.07	13.54	62.40	2.59	4.61	2.75	59.65	21.71	3.72	0.31	0.51
S7	42.00	31.29	40.04	28.67	0.92	0.72	0.68	27.99	41.12	2.57	0.52	0.81
S8	103.38	19.90	45.93	34.17	1.72	0.74	0.47 [b]	33.70	71.19	4.94	0.60	0.46
S9	2.78	45.73	23.25	31.03	0.68	1.33	4.04	26.99	6.68	1.52	0.76	1.16
S10	57.06	9.30	69.49 [a]	21.21	2.28	0.31 [b]	0.35 [b]	20.85	58.96	12.12	0.10	0.19
S11	36.37	6.88 [b]	59.39	33.73	4.90 [a]	0.57	0.39 [b]	33.34	85.99 [a]	17.01 [a]	0.06 [b]	0.14
S12	7.01	17.39	5.24	77.36 [a]	4.45	14.75	4.50	72.86 [a]	16.20	5.96	0.18	0.26
S13	2.75	75.80 [a]	6.20	17.99	0.24	2.90	5.70	12.29	2.16	0.43 [b]	2.66	2.24
S14	8.87	38.10	43.39	18.50	0.49	0.43 [b]	3.44	15.07	4.38	1.80	0.58	0.92
S15	53.18	9.45	70.42 [a]	20.13	2.13	0.29 [b]	0.53 [h]	19.60	37.30	11.47	0.09	0.19
S16	29.96	26.11	33.92	39.98	1.53	1.18	0.71	39.27	55.10	3.35	0.31	0.63
S17	22.49	43.98	13.90	42.11	0.96	3.03	2.53	39.58	15.62	3.85	0.28	1.21
S18	5.98	47.05	22.41	30.54	0.65	1.36	3.56	26.98	7.58	1.92	0.57	1.11
S19	5.39	62.92	20.33	16.74	0.27	0.82	7.19	9.55	1.33	0.73	2.07	1.38
S20	61.04	73.56	13.74	12.70 [b]	0.17 [b]	0.92	0.76	11.94	15.73	0.36 [b]	9.62 [a]	4.68 [a]
S21	130.32 [a]	36.12	29.36	34.53	0.96	1.18	0.81	33.72	41.69	1.94	0.58	1.04
S22	6.23	63.69	7.56	28.75	0.45	3.80	18.57	10.18	0.55 [b]	0.74	1.52	0.60
S23	8.09	34.48	33.65	31.87	0.92	0.95	18.23	13.64	0.75 [b]	2.72	0.43	0.36
S24	8.84	46.01	6.42	47.57	1.03	7.41	35.08	12.49	0.36 [b]	1.57	0.24	0.32
S25	14.05	22.23	2.14[b]	75.63	3.40	35.37 [a]	51.14 [a]	24.49	0.48 [b]	4.61	0.78	0.11
S26	5.38	49.12	14.95	35.93	0.73	2.40	9.32	26.61	2.86	1.45	0.81	0.85
S27	3.75	39.28	41.64	19.08	0.49	0.46	1.49	17.59	11.80	2.32	0.48	1.02
S28	4.38	71.32	15.62	13.06 [b]	0.18 [b]	0.84	5.31	7.74 [b]	1.46	0.56	2.08	2.20
S29	6.74	61.96	14.51	23.53	0.38	1.62	12.62	10.90	0.86 [b]	0.86	1.29	1.06
S30	32.31	20.16	21.74	58.10	2.88	2.67	1.54	56.56	36.78	5.41	0.20	0.44
S31	43.00	8.30	59.33	32.37	3.90	0.55	16.59	15.78	0.95 [b]	11.29	0.07 [b]	0.08 [b]
S32	7.13	60.86	4.47	34.68	0.57	7.76	5.99	28.69	4.79	1.00	1.27	1.27
S33	5.95	46.20	11.03	42.77	0.93	3.88	28.15	14.62	0.52 [b]	1.65	0.71	0.42
S34	5.26	48.94	7.49	43.57	0.89	5.81	11.10	32.47	2.93	1.38	0.83	0.81

Values are the mean of six determinations. PUFAs: total polyunsaturated fatty acids; MUFAs: total monounsaturated fatty acids; SFAs: total saturated fatty acids; AI: atherogenic index; and TI: thrombogenic index; h/H: ratios of hypocholesterolemic (h)/hypercholesterolemic (H) fatty acids. The letters a and b within a column represent the highest and lowest significant ($p < 0.05$) values, respectively. Sample numbers (S1–S34) correspond to the Table 1.

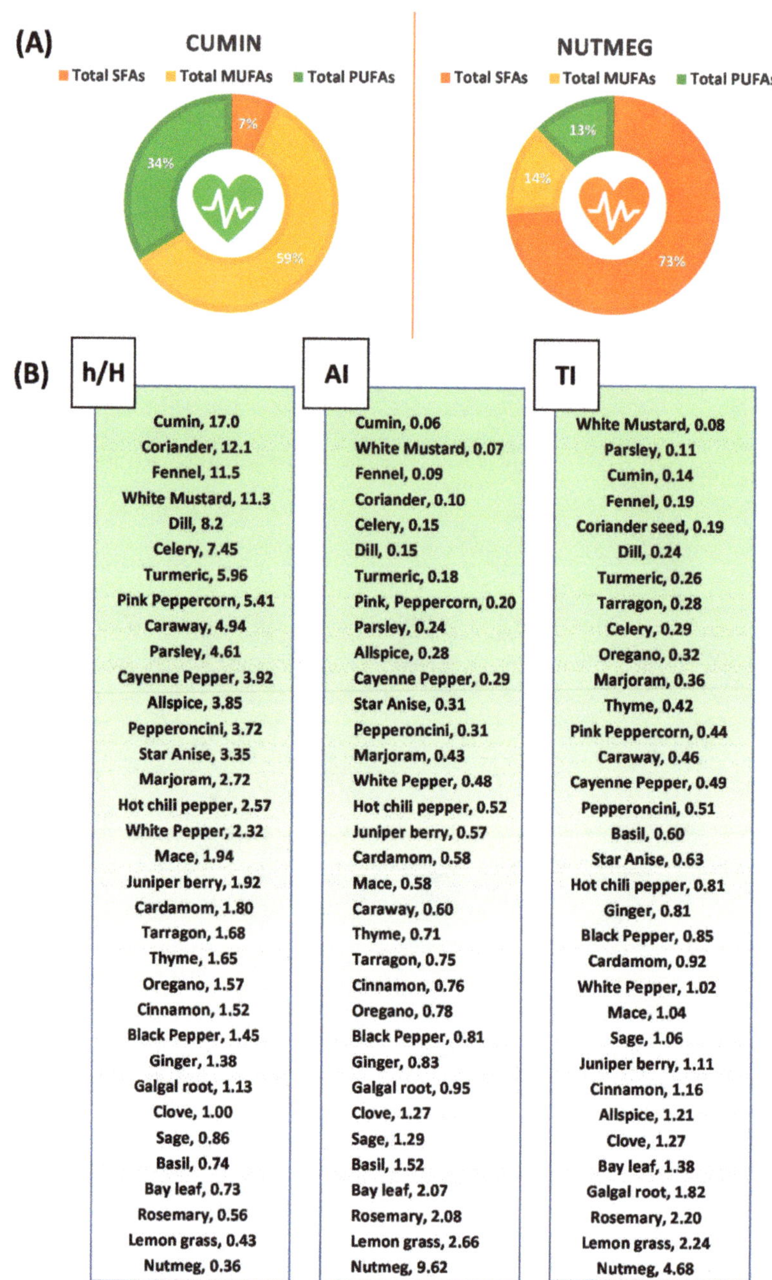

Figure 3. (**A**) Illustrations showing the high content of healthy monounsaturated (MUFAs) and polyunsaturated fatty acids (PUFAs) in cumin, compared to low contents of MUFAs and PUFAs, and high contents of saturated fatty acids (SFAs) in nutmeg. (**B**) Arrangements of studied spices and herbs in ascending/descending order according to the fat quality indices of the ratio of hypocholesterolemic (h)/hypercholesterolemic fatty acids, atherogenic index (AI), and thrombogenic index (TI).

4. Conclusions

Spices belonging to Apiaceae family plants (cumin, coriander, fennel, and dill) are an exceptionally rich source of monounsaturated fatty acids (MUFAs) in the form of petroselinic acid, a good amount of polyunsaturated fatty acids (PUFAs; linoleic acid), and a small amount of saturated fatty acids. And, with high proportions of MUFAs and PUFAs, the Apiaceae family spices top the fat quality indices, particularly in terms of a higher ratio of hypocholesterolemic/hypercholesterolemic fatty acids, and lower values of the atherogenic index and the thrombogenic index (Figure 3).

Supplementary Materials: The following are available online at https://www.mdpi.com/article/10.3390/foods10040854/s1, Figure S1: (A) The gas chromatography (GC)-flame ionization detection (FID) profiles of fatty acid methyl esters (FAMEs) of cardamom. (B) The GC-mass spectrum of dominating fatty acid (Palmitic acid); Figure S2. (A–C) The gas chromatography (GC)-flame ionization detection (FID) profiles of fatty acid methyl esters (FAMEs) of lemongrass, rosemary, and Sage. The GC-mass spectrum of dominating fatty acid (Palmitic acid). The numbers, 4, 7, 9, 11, and 14 correspond to peak numbers illustrated in Table 1. BHT: Butylated hydroxytoluene (A synthetic antioxidant used during lipid extraction).

Author Contributions: Conceptualization, R.K.S. and A.D.A.; methodology, R.K.S. and A.D.A.; software, R.K.S. and A.D.A.; validation, R.K.S. and A.D.A. and Y.-S.K.; formal analysis, R.K.S.; investigation, R.K.S.; resources, Y.-S.K.; data curation, R.K.S. and A.D.A.; writing—original draft preparation, R.K.S.; writing—review and editing, A.D.A. and Y.-S.K.; visualization, Y.-S.K.; supervision, Y.-S.K.; project administration, R.K.S.; funding acquisition, R.K.S. All authors have read and agreed to the published version of the manuscript.

Funding: This paper was supported by the KU Research Professor Program of Konkuk University, Seoul, Republic of Korea and "The APC was supported by Konkuk University research fund (2021A0190061)".

Acknowledgments: This paper was supported by the KU Research Professor Program of Konkuk University, Seoul, Korea.

Conflicts of Interest: The authors declare no conflict of interest.

References

1. Rubió, L.; Motilva, M.-J.; Romero, M.-P. Recent Advances in Biologically Active Compounds in Herbs and Spices: A Review of the Most Effective Antioxidant and Anti-Inflammatory Active Principles. *Crit. Rev. Food Sci. Nutr.* **2013**, *53*, 943–953. [CrossRef] [PubMed]
2. Martínez-Graciá, C.; Gonzalez-Bermúdez, C.A.; Cabellero-Valcárcel, A.M.; Santaella-Pascual, M.; Frontela Saseta, C. Use of herbs and spices for food preservation: Advantages and limitations. *Curr. Opin. Food Sci.* **2015**, *6*, 38–43. [CrossRef]
3. Shahidi, F.; Ambigaipalan, P. Phenolics and polyphenolics in foods, beverages and spices: Antioxidant activity and health effects—A review. *J. Funct. Foods* **2015**, *18*, 820–897. [CrossRef]
4. Viuda-Martos, M.; Ruiz-Navajas, Y.; Fernández-López, J.; Pérez-Álvarez, J. Spices as functional foods. *Crit. Rev. Food Sci. Nutr.* **2010**, *51*, 13–28. [CrossRef]
5. Jiang, T.A. Health Benefits of Culinary Herbs and Spices. *J. AOAC Int.* **2019**, *102*, 395–411. [CrossRef]
6. Alasvand, S.; Bridges, W.; Haley-Zitlin, V. Efficacy of Common Spices on Improving Serum Lipids in Individuals with Type 2 Diabetes: Systematic Review and Meta-Analysis of Clinical Trials. *Curr. Dev. Nutr.* **2020**, *4*, 1. [CrossRef]
7. Yu, L.; Choe, U.; Li, Y.; Zhang, Y. Oils from fruit, spice, and herb seeds. *Bailey Ind. Oil Fat Prod.* **2005**, 1–35. [CrossRef]
8. Saini, R.K.; Keum, Y.S. Omega-3 and omega-6 polyunsaturated fatty acids: Dietary sources, metabolism, and significance—A review. *Life Sci.* **2018**, *203*, 255–267. [CrossRef]
9. Harnack, K.; Andersen, G.; Somoza, V. Quantitation of alpha-linolenic acid elongation to eicosapentaenoic and docosahexaenoic acid as affected by the ratio of n6/n3 fatty acids. *Nutr. Metab.* **2009**, *6*, 8. [CrossRef]
10. Simopoulos, A.P. Evolutionary aspects of the dietary omega-6/omega-3 fatty acid Ratio: Medical implications. In *Evolutionary Thinking in Medicine*; Springer: Berlin/Heidelberg, Germany, 2016; pp. 119–134.
11. Briggs, M.; Petersen, K.; Kris-Etherton, P. Saturated Fatty Acids and Cardiovascular Disease: Replacements for Saturated Fat to Reduce Cardiovascular Risk. *Healthcare* **2017**, *5*, 29. [CrossRef]
12. Cruz, R.; Casal, S.; Mendes, E.; Costa, A.; Santos, C.; Morais, S. Validation of a Single-Extraction Procedure for Sequential Analysis of Vitamin E, Cholesterol, Fatty Acids, and Total Fat in Seafood. *Food Anal. Methods* **2013**, *6*, 1196–1204. [CrossRef]
13. Saini, R.K.; Mahomoodally, M.F.; Sadeer, N.B.; Keum, Y.S.; Rr Rengasamy, K. Characterization of nutritionally important lipophilic constituents from brown kelp Ecklonia radiata (C. Ag.) J. Agardh. *Food Chem.* **2021**, *340*. [CrossRef]

14. Saini, R.K.; Rengasamy, K.R.R.; Ko, E.Y.; Kim, J.T.; Keum, Y.S. Korean Maize Hybrids Present Significant Diversity in Fatty Acid Composition: An Investigation to Identify PUFA-Rich Hybrids for a Healthy Diet. *Front. Nutr.* **2020**, *7*. [CrossRef]

15. Saini, R.K.; Keum, Y.-S.; Rengasamy, K.R. Profiling of nutritionally important metabolites in green/red and green perilla (Perilla frutescens Britt.) cultivars: A comparative study. *Ind. Crop Prod.* **2020**, *151*, 112441. [CrossRef]

16. Chen, J.; Liu, H. Nutritional Indices for Assessing Fatty Acids: A Mini-Review. *Int. J. Mol. Sci.* **2020**, *21*, 5695. [CrossRef]

17. Parry, J.W.; Cheng, Z.; Moore, J.; Yu, L.L. Fatty Acid Composition, Antioxidant Properties, and Antiproliferative Capacity of Selected Cold-Pressed Seed Flours. *J. Am. Oil Chem. Soc.* **2008**, *85*, 457–464. [CrossRef]

18. Parry, J.; Hao, Z.; Luther, M.; Su, L.; Zhou, K.; Yu, L. Characterization of cold-pressed onion, parsley, cardamom, mullein, roasted pumpkin, and milk thistle seed oils. *J. Am. Oil Chem. Soc.* **2006**, *83*, 847–854. [CrossRef]

19. Anaduaka, E.G.; Uchendu, N.O.; Ezeanyika, L.U.S. Mineral, amino acid and fatty acid evaluations of Myristica fragrans seeds extracts. *Sci. Afr.* **2020**, *10*, e00567. [CrossRef]

20. Lee, J.-G.; Chae, Y.; Shin, Y.; Kim, Y.-J. Chemical composition and antioxidant capacity of black pepper pericarp. *Appl. Biol. Chem.* **2020**, *63*. [CrossRef]

21. Rema, J.; Krishnamoorthy, B. 22—Nutmeg and mace. In *Handbook of Herbs and Spices*, 2nd ed.; Peter, K.V., Ed.; Woodhead Publishing: Abington, UK, 2012; pp. 399–416. [CrossRef]

22. Playfair, L. XX. On a new fat acid in the butter of nutmegs. *Lond. Edinb. Dublin Philos. Mag. J. Sci.* **1841**, *18*, 102–113. [CrossRef]

23. Al-Khatib, I.M.H.; Hanifa Moursi, S.A.; Mehdi, A.W.R.; Al-Shabibi, M.M. Gas-liquid chromatographic determination of fatty acids and sterols of selected Iraqi foods. *J. Food Compos. Anal.* **1987**, *1*, 59–64. [CrossRef]

24. Kozłowska, M.; Gruczyńska, E.; Ścibisz, I.; Rudzińska, M. Fatty acids and sterols composition, and antioxidant activity of oils extracted from plant seeds. *Food Chem.* **2016**, *213*, 450–456. [CrossRef]

25. Burdock, G.A.; Carabin, I.G. Safety assessment of myristic acid as a food ingredient. *Food Chem. Toxicol.* **2007**, *45*, 517–529. [CrossRef]

26. Knutsen, H.K.; Alexander, J.; Barregård, L.; Bignami, M.; Brüschweiler, B.; Ceccatelli, S.; Dinovi, M.; Edler, L.; Grasl-Kraupp, B.; Hogstrand, C.; et al. Erucic acid in feed and food. *EFSA J.* **2016**, *14*. [CrossRef]

27. Ksouda, G.; Hajji, M.; Sellimi, S.; Merlier, F.; Falcimaigne-Cordin, A.; Nasri, M.; Thomasset, B. A systematic comparison of 25 Tunisian plant species based on oil and phenolic contents, fatty acid composition and antioxidant activity. *Ind. Crop Prod.* **2018**, *123*, 768–778. [CrossRef]

28. Reiter, B.; Lechner, M.; Lorbeer, E. The fatty acid profiles—Including petroselinic and cis-vaccenic acid-of different Umbelliferae seed oils. *Lipid/Fett* **1998**, *100*, 498–502. [CrossRef]

29. Alaluf, S.; Green, M.; Powell, J.; Rogers, J.; Watkinson, A.; Cain, F.; Hu, H.; Rawlings, A. Petroselinic Acid and Its Use in Food. U.S. Patent No. 6,365,175 B1, 2 April 2002.

30. Mongrand, S.; Bessoule, J.-J.; Cabantous, F.; Cassagne, C. The C16:3\C18:3 fatty acid balance in photosynthetic tissues from 468 plant species. *Phytochemistry* **1998**, *49*, 1049–1064. [CrossRef]

31. Wołoszyn, J.; Haraf, G.; Okruszek, A.; Wereńska, M.; Goluch, Z.; Teleszko, M. Fatty acid profiles and health lipid indices in the breast muscles of local Polish goose varieties. *Poult. Sci.* **2020**, *99*, 1216–1224. [CrossRef]

32. Ulbricht, T.L.V.; Southgate, D.A.T. Coronary heart disease: Seven dietary factors. *Lancet* **1991**, *338*, 985–992. [CrossRef]

33. Assefa, A.D.; Keum, Y.S.; Saini, R.K. A comprehensive study of polyphenols contents and antioxidant potential of 39 widely used spices and food condiments. *J. Food Meas. Charact.* **2018**, *12*, 1548–1555. [CrossRef]

 **foods**

MDPI

Article

Variation in the Chemical Composition of Five Varieties of *Curcuma longa* Rhizome Essential Oils Cultivated in North Alabama

William N. Setzer [1,2,*, Lam Duong [3], Ambika Poudel [2] and Srinivasa Rao Mentreddy [3,***

[1] Department of Chemistry, University of Alabama in Huntsville, Huntsville, AL 35899, USA
[2] Aromatic Plant Research Center, 230 N 1200 E, Suite 100, Lehi, UT 84043, USA; apoudel@aromaticplant.org
[3] Department of Biological and Environmental Sciences, Alabama A&M University, Normal, AL 35762, USA; lamduongvn@gmail.com
* Correspondence: wsetzer@chemistry.uah.edu (W.N.S.); srinivasa.mentreddy@aamu.edu (S.R.M.)

Abstract: Turmeric (*Curcuma longa* L.) is an important spice, particularly is Asian cuisine, and is also used in traditional herbal medicine. Curcuminoids are the main bioactive agents in turmeric, but turmeric essential oils also contain health benefits. Turmeric is a tropical crop and is cultivated in warm humid environments worldwide. The southeastern United States also possesses a warm humid climate with a growing demand for locally sourced herbs and spices. In this study, five different varieties of *C. longa* were cultivated in north Alabama, the rhizome essential oils obtained by hydrodistillation, and the essential oils were analyzed by gas chromatographic techniques. The major components in the essential oils were α-phellandrene (3.7–11.8%), 1,8-cineole (2.6–11.7%), α-zingiberene (0.8–12.5%), β-sesquiphellandrene (0.7–8.0%), *ar*-turmerone (6.8–32.5%), α-turmerone (13.6–31.5%), and β-turmerone (4.8–18.4%). The essential oil yields and chemical profiles of several of the varieties are comparable with those from tropical regions, suggesting that these should be considered for cultivation and commercialization in the southeastern United States.

Keywords: turmeric; essential oil composition; α-turmerone; β-turmerone; *ar*-turmerone

Citation: Setzer, W.N.; Duong, L.; Poudel, A.; Mentreddy, S.R. Variation in the Chemical Composition of Five Varieties of *Curcuma longa* Rhizome Essential Oils Cultivated in North Alabama. *Foods* **2021**, *10*, 212. https://doi.org/10.3390/foods 10020212

Academic Editor: Severino De Alencara
Received: 4 December 2020
Accepted: 18 January 2021
Published: 21 January 2021

Publisher's Note: MDPI stays neutral with regard to jurisdictional claims in published maps and institutional affiliations.

1. Introduction

Turmeric (*Curcuma longa* L.), belonging to Zingiberaceae, is a rhizomatous plant native to Southeast Asia, but is extensively cultivated worldwide, particularly in tropical countries (e.g., India, Pakistan, Bangladesh, China, Taiwan, Thailand, Sri Lanka, East Indies, Burma, Indonesia, northern Australia, Costa Rica, Haiti, Jamaica, Peru, and Brazil) [1–4]. Turmeric is well known for its use as a culinary ingredient and a traditional herbal medicine [5]. It is extensively used in Asian cuisine and is one of the key ingredients in curry powders [6]. Turmeric, either fresh or in dried form, has a long history of medicinal use, dating back 4000 years [7]. Due to its bright yellow/orange color, turmeric is often referred to as the "Indian saffron" or "golden spice". Curcuminoids are considered the main bioactive components of turmeric. Turmeric's medical properties are credited mainly to the curcuminoids, which are abundant in turmeric rhizome [8]. The total curcumin (sum of all curcuminoids) was proven to have significant health benefits along with the potential to prevent various diseases, including Alzheimer's, coronary heart disease, and cancer [9]. Due to a plethora of scientific articles on the health benefits of turmeric, the demand for turmeric is steadily increasing in the US and now represents an estimated US$36 million per annum. The US imports 90% of its requirements from various countries [10]. Due to recent inconsistencies associated with quality and production methods of raw materials imported from Asian countries, many US manufacturers of herbal products are seeking domestically-produced materials that meet their standards and requirements. Turmeric (*Curcuma longa*) is one such crop. The rhizome from which the curcumins are derived is tuberous, with a rough and segmented skin. The primary rhizome is called the "mother rhizome" or bulb, which is pear-shaped in the center (Figure 1). The branches of mother rhizomes are the secondary rhizomes, called lateral or "finger rhizomes" [11]. The mother rhizomes are more matured than finger rhizomes, therefore containing higher curcuminoid concentrations and perhaps higher essential contents than finger rhizomes. However, the curcumin yield from finger rhizomes is higher than that from mother rhizomes [12].

Figure 1. Underground parts of *Curcuma longa* showing the rhizomes and roots.

One of the important components of turmeric is its volatile oil. The role of turmeric oil in the treatment of a wide variety of diseases in animals and humans were reviewed in detail [4,7]. Thus, curcuma oil appears to be a promising agent for the treatment of simple dermatitis, cerebral stroke, and other disorders associated with oxidative stress [13]. The essential oils of turmeric are relatively complex, with hundreds of components. The major components, however, are α-turmerone (12.6–44.5%), β-turmerone (9.1–37.8%), *ar*-turmerone (12.2–36.6%), β-sesquiphellandrene (5.0–14.6%), α-zingiberene (5.0–12.8%), germacrone (10.3–11.1%), terpinolene (10.0–10.2%), *ar*-curcumene (5.5–9.8%), and α-phellandrene (5.0–6.7%) [14].

Turmeric is considered a viable cash crop with a ready market in Alabama and in the US. Similar to any essential oil crop, turmeric's essential oil also varies with variety, soil type, and environmental conditions [15,16]. Hence, evaluating different varieties for essential content and composition is an important consideration for determining a variety for cultivation in a particular location. Turmeric is a tropical crop and grows well in warm and humid environments with mean air temperatures between 20 and 30 °C. It can be planted in all soil types, but it does best in well-drained clay loam or sandy loam soils rich in humus or organic matter with a soil pH of about 5.5 to 6.5. It grows in a wide range of climatic conditions but requires about 100 to 200 cm of rainfall a year. Thus, Alabama's hot, humid, and rainy summer season is suited for turmeric production in the southeastern US. Furthermore, turmeric is potentially suited for catering to the herbal products industry, which prefers locally sourced materials. The purpose of the present study was to determine the variation in the essential oil chemistry of five *C. longa* genotypes that could potentially be cultivated for commercial purposes in north Alabama and to note any differences between the mother rhizome and the lateral rhizomes of each cultivar.

2. Materials and Methods

2.1. Curcuma longa Varieties

The five varieties used in this study were selected out of fourteen varieties according to three criteria: high yield but low curcuminoid content (varieties, CL5, CL3), low yield but high curcumin content (CL10), and high yield and high curcumin content, thus, high curcumin yield (CL9, CL11), based on studies at Auburn University and Alabama A&M University. Thus, CL3 and CL5 may be considered for the fresh rhizome market, CL 10 for the high curcumin dry rhizome herbal products market, and CL9 and CL11, which have attractive, orange-colored rhizomes, could serve both fresh and dry herbal produce markets. The two varieties CL3 and CL9 were consistent in their performance over three years in both south and north Alabama. A knowledge of their relative oil yield and composition could help in value-addition for either fresh rhizomes or dry herbal markets.

2.2. Cultivation of Curcuma longa

The rhizomes of five turmeric varieties belonging to *C. longa* (CL3, CL5, CL9, CL10, and CL11) obtained from various sources (Table 1) were planted in seed germination trays filled with a soilless potting mix (Pro-Mix) on 3 April 2019. After planting, the trays were placed in a glass greenhouse at Alabama A&M University, Normal, AL (natural daylight increasing from 11 h in mid-March to about 14.5 h in early June; mean air temperature maintained at 26 °C) for 70 days for sprouting and plant development. The 10-week-old plants were then transplanted onto raised beds (60 cm wide, 15 cm high, 25 m long, 2 m apart, covered with black plastic mulch with drip irrigation tubing underneath the plastic) on 25 June 2019 at the Alabama A&M Winfred Thomas Agricultural Research Station located in Hazelgreen, AL (Latitude 34°89′ N and longitude 86°56′ W). Soil at the experimental site was a Decatur silt loam (fine, kaolinitic, thermic Rhodic Paleudult). The plants were grown using organic production methods and irrigation was provided as and when needed by the drip method. Prior to making the raised beds, the soil was plowed with a rototiller, and a mixture of composted cow manure, poultry litter, and vermicompost was incorporated into the soil at a rate equivalent to 45.5 kg of N/ha. Once the crop was established, a fish emulsion-based organic soluble fertilizer, Neptune's Harvest™ (Seven Springs Organic Farming and Gardening Supplies, Check, VA, USA), was applied through the irrigation system at 3-week intervals. Three plants from the middle of each row were harvested in mid-February 2020 by digging the plants, separating the rhizomes from the shoot, and washing clean of soil and debris with forced water jets. The mother and lateral rhizomes (Figure 1) were separated and placed in mesh trays and dried of excess water using fans at room temperature. The rhizomes were then placed in a cooler box with ice and carried to the chemistry department at the University of Alabama in Huntsville for oil extraction and profiling.

2.3. Essential Oils

The fresh rhizomes, both mother and lateral rhizomes (Figure 1), were chopped and hydrodistilled for 4 h using a Likens–Nickerson apparatus with continuous extraction of the distillate with dichloromethane. Evaporation of the dichloromethane gave pale yellow to yellow rhizome essential oils (Table 1), which were stored at −20 °C until analysis.

Table 1. Hydrodistillation details for *Curcuma longa* rhizome essential oils cultivated in north Alabama.

C. longa Cultivar (Rhizome)	Source (Origin) of Rhizome	Mass of Rhizome (g)	Mass of Essential Oil (g)	% Yield	Color
CL3 (mother)	Horizon Herbs, 3350 Cedar Flat Road,	236.2	0.5935	0.251	pale yellow
CL3 (lateral)	Williams, Oregon (Hawaii)	268.7	0.6797	0.253	pale yellow
CL5 (mother)	James Simon, Rutgers University,	214.0	1.3120	0.613	yellow
CL5 (lateral)	New Jersey (India)	281.0	1.5540	0.553	pale yellow
CL9 (mother)	Lam T. Duong	207.1	1.3645	0.659	yellow
CL9 (lateral)	(Dak Lak Province, Vietnam)	188.3	0.8050	0.428	yellow
CL10 (mother)	International farmers Market, Chamblee,	232.0	0.6715	0.289	pale yellow
CL10 (lateral)	Georgia (unknown)	282.7	0.5763	0.204	pale yellow
CL11 (mother)	Dr. Anand Yadav, Fort Valley State	215.0	1.0760	0.500	pale yellow
CL11 (lateral)	University, Georgia (unknown)	264.6	1.4126	0.534	pale yellow

2.4. Gas Chromatographic–Mass Spectral (GC–MS) Analysis

Gas chromatography–mass spectrometry was carried out as previously described [14]: Shimadzu GCMS-QP2010 Ultra instrument, electron impact (EI) mode (electron energy = 70 eV), scan range = 40–400 atomic mass units, scan rate = 3.0 scans/s, and GC-MS solution software (Shimadzu Scientific Instruments, Columbia, MD, USA); ZB-5 fused silica capillary column, 30 m length, 0.25 mm internal diameter, 0.25 μm film thickness (Phenomenex, Torrance, CA, USA); He carrier gas, head pressure = 552 kPa, flow rate = 1.37 mL/min; injector temperature = 250 °C, ion source temperature = 200 °C, oven temperature program = 50 °C start, increased by 2 °C/min to 260 °C; 7% w/v solutions, 0.1 μL injections, split mode (30:1). Essential oil components were identified based on both their retention indices, which were determined by reference to a homologous *n*-alkane series, and their mass spectral fragmentation patterns available from the databases [17–20].

2.5. Hierarchical Cluster Analysis

Agglomerative hierarchical cluster (AHC) analysis was carried out using the chemical compositions of the *C. longa* rhizome essential oils. The compositions were treated as operational taxonomic units, with the percentages of the 16 most abundant components (α-turmerone, *ar*-turmerone, β-turmerone, α-phellandrene, 1,8-cineole, α-zingiberene, β-sesquiphellandrene, terpinolene, (6*S*,7*R*)-bisabolone, *p*-cymene, zingiberenol, *ar*-curcumene, β-bisabolene, 7-*epi-trans*-sesquisabinene hydrate, limonene, and *ar*-tumerol), using XLSTAT Premium, version 2018.1.1.62926. Euclidean distance was used to determine dissimilarity, and Ward's method was used to define the clusters.

3. Results and Discussion

The fresh mother and lateral rhizomes (Figure 1) were chopped and hydrodistilled to give pale yellow to yellow essential oils in yields ranging from 0.204% to 0.695% (Table 1). Varieties CL5, CL9, and CL11 gave better essential oil yields (0.443–0.659%) than CL3 or CL10 (<0.3%). The total oil content of CL5, CL9, and CL11 were similar to those reported for Indian chemotypes of *C. longa* [21,22]. In CL5, CL9, and CL10, the mother rhizomes had higher oil yields than the lateral rhizomes. A similar trend was reported for curcumin concentration of turmeric varieties grown in south–central AL [12].

The chemical compositions of the *C. longa* rhizome essential oils are compiled in Table 2. The major components in the essential oils were α-phellandrene (3.7–11.8%), 1,8-cineole (2.6–11.7%), α-zingiberene (0.8–12.5%), β-sesquiphellandrene (0.7–8.0%), *ar*-turmerone (6.8–32.5%), α-turmerone (13.6–31.5%), and β-turmerone (4.8–18.4%). A hierarchical cluster analysis of the *C. longa* rhizome essential oils in this study was carried out based on the concentrations of the 16 most abundant essential oil components (α-turmerone, *ar*-turmerone, β-turmerone, α-phellandrene, 1,8-cineole, α-zingiberene, β-sesquiphellandrene, terpinolene, (6*S*,7*R*)-bisabolone, *p*-cymene, zingiberenol, *ar*-curcumene, β-bisabolene, 7-*epi-trans*-sesquisabinene hydrate, limonene, and *ar*-tumerol) (Figure 2).

Table 2. Chemical compositions of "mother" and "lateral" rhizome essential oils of *Curcuma longa* varieties growing in north Alabama.

RI$_{calc}$	RI$_{db}$	Compound	CL3-M	CL3-L	CL5-M	CL5-L	CL9-M	CL9-L	CL10-M	CL10-L	CL11-M	CL11-L
924	925	α-Thujene	tr	tr	tr	tr	tr	tr	tr	tr	tr	tr
931	933	α-Pinene	0.1	0.2	0.2	0.3	0.1	0.1	0.3	0.3	0.1	0.1
971	972	Sabinene	0.1	0.1	0.1	tr	tr	tr	0.1	0.1	tr	tr
976	978	β-Pinene	tr	tr	tr	tr	tr	tr	0.1	0.1	tr	tr
988	991	Myrcene	0.4	0.6	0.3	0.4	0.2	0.3	0.6	0.7	0.2	0.2
999	1000	δ-2-Carene	tr	tr	tr	tr	tr	tr	tr	tr	tr	tr
1007	1007	α-Phellandrene	5.9	9.1	7.3	11.8	3.7	6.7	8.7	7.8	4.4	5.8
1009	1009	δ-3-Carene	0.1	0.2	0.1	0.2	0.1	0.1	0.2	0.2	0.1	0.1
1016	1017	α-Terpinene	0.2	0.3	0.1	0.2	0.1	0.1	0.3	0.2	0.2	0.2
1019	1022	*m*-Cymene	—	—	—	—	—	—	—	tr	—	—
1024	1025	*p*-Cymene	0.9	2.2	1.0	1.2	0.5	1.2	1.7	3.0	0.6	0.7
1029	1030	Limonene	0.5	0.7	0.5	0.6	0.3	0.4	0.9	0.8	0.4	0.4
1033	1032	1,8-Cineole	9.2	5.1	6.1	3.5	3.4	2.6	11.7	6.2	8.5	3.6
1034	1034	(*Z*)-β-Ocimene	—	tr	tr	tr	tr	tr	tr	tr	—	—
1045	1045	(*E*)-β-Ocimene	tr	tr	tr	tr	tr	tr	tr	tr	tr	tr
1057	1058	γ-Terpinene	0.3	0.4	0.3	0.5	0.2	0.3	0.4	0.4	0.3	0.3
1069	1069	*cis*-Sabinene hydrate	tr	tr	tr	tr	tr	tr	tr	tr	tr	tr
1085	1086	Terpinolene	4.0	4.4	0.7	0.7	0.5	0.7	3.6	3.0	2.0	1.8
1090	1093	2-Nonanone	0.1	—	—	—	—	—	0.1	—	—	—
1090	1093	*p*-Cymenene	—	tr	tr	tr	tr	tr	—	tr	tr	tr
1099	1099	Linalool	0.1	0.1	tr	tr	tr	tr	0.1	tr	tr	tr
1100	1101	*trans*-Sabinene hydrate	—	—	tr	tr	tr	tr	—	—	tr	tr
1101	1101	2-Nonanol	0.2	0.1	—	—	—	—	0.2	0.2	—	—
1124	1124	*cis-p*-Menth-2-en-1-ol	0.1	0.1	tr	0.1	tr	tr	0.1	0.1	tr	tr
1141	1146	Ipsdienol	—	—	—	—	—	—	—	—	tr	tr
1142	1142	*trans-p*-Menth-2-en-1-ol	0.1	0.1	0.1	0.1	tr	0.1	0.1	0.1	tr	tr
1145	1146	Myrcenone	tr	0.1	0.1	0.1	tr	0.1	—	—	0.1	0.1
1149	1146	*trans*-Limonene oxide	tr	0.3	tr	tr	tr	tr	0.2	0.3	tr	tr
1167	1169	*trans*-β-Terpineol	tr	tr	tr	tr	tr	tr	tr	tr	tr	tr
1170	1170	δ-Terpineol	0.1	0.1	0.1	tr	tr	tr	0.1	0.1	0.1	tr
1171	1165	*iso*-Borneol	tr	tr	—	—	—	—	tr	tr	—	—
1171	1171	*p*-Mentha-1,5-dien-8-ol	—	—	tr	tr	tr	tr	—	—	tr	tr
1173	1173	Borneol	tr	tr	tr	tr	tr	tr	tr	0.1	—	—
1180	1180	Terpinen-4-ol	0.4	0.1	0.2	0.1	0.1	0.1	0.4	0.2	0.3	0.1
1187	1188	*p*-Cymen-8-ol	0.1	0.2	tr	tr	tr	tr	0.1	0.2	0.1	tr

Table 2. *Cont.*

RI$_{calc}$	RI$_{db}$	Compound	CL3-M	CL3-L	CL5-M	CL5-L	CL9-M	CL9-L	CL10-M	CL10-L	CL11-M	CL11-L
1195	1195	α-Terpineol	0.7	0.3	0.3	0.2	0.2	0.2	0.7	0.4	0.6	0.2
1196	1196	*cis*-Piperitol	tr	tr	tr	tr	tr	tr	tr	tr	tr	tr
1203	1202	*cis*-Sabinol	0.1	0.2	0.1	0.1	tr	0.1	0.2	0.3	0.1	tr
1203	1203	*p*-Cumenol	0.1	0.1	—	—	—	—	—	—	tr	tr
1208	1209	*trans*-Piperitol	tr	tr	tr	tr	tr	tr	tr	tr	tr	tr
1289	1289	Thymol	0.1	0.2	0.2	0.2	0.1	0.2	—	—	0.1	0.1
1292	1293	2-Undecanone	tr	tr	—	—	—	—	tr	tr	—	—
1297	1300	Carvacrol	tr	tr	tr	tr	tr	tr	tr	tr	tr	tr
1309	1312	Livescone	—	—	—	—	—	—	—	—	tr	tr
1319	1318	3-Hydroxycineole	0.1	0.1	tr	tr	tr	tr	0.1	0.3	tr	tr
1400	1405	Sesquithujene	0.1	0.1	—	—	—	—	0.2	0.2	—	—
1417	1417	(*E*)-Caryophyllene	0.1	0.2	0.1	0.2	0.1	0.1	0.1	0.2	0.1	0.2
1431	1432	*trans*-α-Bergamotene	0.1	tr	—	—	—	—	0.1	0.1	—	—
1442	1439	(*Z*)-β-Farnesene	tr	tr	tr	tr	tr	0.1	—	—	0.1	0.1
1450	1452	(*E*)-β-Farnesene	0.3	0.2	tr	tr	tr	tr	0.3	0.3	tr	0.1
1453	1454	α-Humulene	tr	tr	tr	tr	tr	tr	—	—	tr	tr
1476	1482	γ-Curcumene	0.1	tr	tr	tr	tr	tr	0.1	0.1	tr	tr
1480	1482	*ar*-Curcumene	1.1	1.5	0.3	0.3	0.4	0.5	1.9	3.2	0.3	0.5
1482	1483	*trans*-β-Bergamotene	0.1	tr	tr	tr	tr	tr	0.1	0.1	tr	tr
1495	1494	α-Zingiberene	9.2	7.8	0.9	0.9	1.2	0.9	12.5	10.4	0.9	1.0
1507	1508	β-Bisabolene	1.3	1.2	0.2	0.2	0.2	0.2	2.0	2.2	0.2	0.2
1508	1511	β-Curcumene	tr	tr	tr	tr	tr	tr	—	—	tr	tr
1524	1523	β-Sesquiphellandrene	5.5	4.9	0.7	0.8	0.9	0.9	7.7	8.0	0.8	1.0
1526	1528	(*E*)-γ-Bisabolene	0.1	0.2	tr	0.1	tr	tr	0.1	0.1	tr	tr
1553	1555	7-*epi*-*cis*-Sesquisabinene hydrate	0.4	0.4	0.2	0.2	0.2	0.2	0.5	0.5	0.2	0.2
1559	1561	(*E*)-Nerolidol	0.1	0.1	0.1	0.1	0.1	0.1	0.1	0.1	0.1	0.1
1577	1578	*ar*-Tumerol	0.4	0.5	0.5	0.5	0.6	0.8	0.3	0.3	0.5	0.9
1589	1590	7-*epi*-*trans*-Sesquisabinene hydrate	0.9	0.8	0.5	0.4	0.5	0.4	1.1	1.2	0.4	0.4
1600	1594	*anti*-*anti*-*anti*-Helifolen-12-al B	0.2	0.2	0.3	0.4	0.4	0.4	0.2	0.2	0.2	0.4

Table 2. *Cont.*

RI$_{calc}$	RI$_{db}$	Compound	CL3-M	CL3-L	CL5-M	CL5-L	CL9-M	CL9-L	CL10-M	CL10-L	CL11-M	CL11-L
1615	1615	Zingiberenol	2.0	1.7	0.3	0.5	0.6	0.5	2.8	3.0	0.5	0.6
1620	1620	*anti-syn-syn*-Helifolen-12-al C	0.2	0.3	0.3	0.3	0.5	0.3	—	—	—	0.6
1623	1624	10-*epi*-γ-Eudesmol	0.2	0.2	—	0.2	—	0.4	0.2	0.4	—	0.1
1643	1647	Camphenone	—	—	—	—	—	—	—	0.3	—	—
1670	1668	*ar*-Turmerone	15.4	15.3	18.3	15.5	26.3	32.5	6.8	9.5	21.8	27.0
1675	1668	α-Turmerone	17.6	18.7	30.1	31.5	24.9	18.9	15.4	13.6	27.8	24.8
1687	1687	Himachal-4-en-1β-ol	0.7	0.7	—	—	—	—	0.9	1.0	—	—
1688	1687	α-Bisabolol	0.5	0.3	—	—	—	—	0.5	0.8	—	—
1695	1695	(2Z,6Z)-Farnesol	0.2	0.3	—	—	—	—	0.4	0.6	—	—
1702	1699	β-Turmerone (= Curlone)	8.9	8.8	17.1	16.8	18.4	17.0	5.0	4.8	17.9	17.5
1712	1712	Curcuphenol	0.2	0.2	0.2	0.2	0.2	0.2	0.2	0.2	0.1	0.2
1743	1742	(6S,7R)-Bisabolone	2.0	1.7	0.6	0.7	0.8	0.7	3.1	3.4	0.7	0.7
1773	1775	*trans*-α-Atlantone	0.3	0.2	0.2	0.4	0.4	0.4	0.2	0.2	0.3	0.4
1807	1807	Eudesm-11-en-4α,6α-diol	0.4	0.4	—	—	—	—	—	—	—	—
1985	1983	Methyl-β-(*E*)-ionyl tiglate	0.5	0.4	0.1	0.1	0.1	0.2	0.2	0.5	0.1	0.1
		Monoterpene hydrocarbons	12.6	18.3	10.6	15.9	5.6	9.9	17.0	16.6	8.3	9.5
		Oxygenated monoterpenoids	2.0	2.0	0.9	0.8	0.5	0.7	2.1	1.9	1.4	0.6
		Sesquiterpene hydrocarbons	17.9	16.1	2.2	2.5	2.8	2.7	25.0	24.8	2.3	2.9
		Oxygenated sesquiterpenoids	51.1	51.1	68.9	67.8	74.0	73.0	37.9	40.5	70.5	73.8
		Others	0.3	0.1	0.0	0.0	0.0	0.0	0.3	0.2	0.0	0.0
		Total Identified	83.8	87.6	82.7	86.9	82.9	86.2	82.2	84.0	82.4	86.8

RI$_{calc}$: Retention indices determined with respect to a homologous series of *n*-alkanes on a ZB-5 column. RI$_{db}$: Retention indices from databases [17–20]. tr: "trace" (<0.05%).

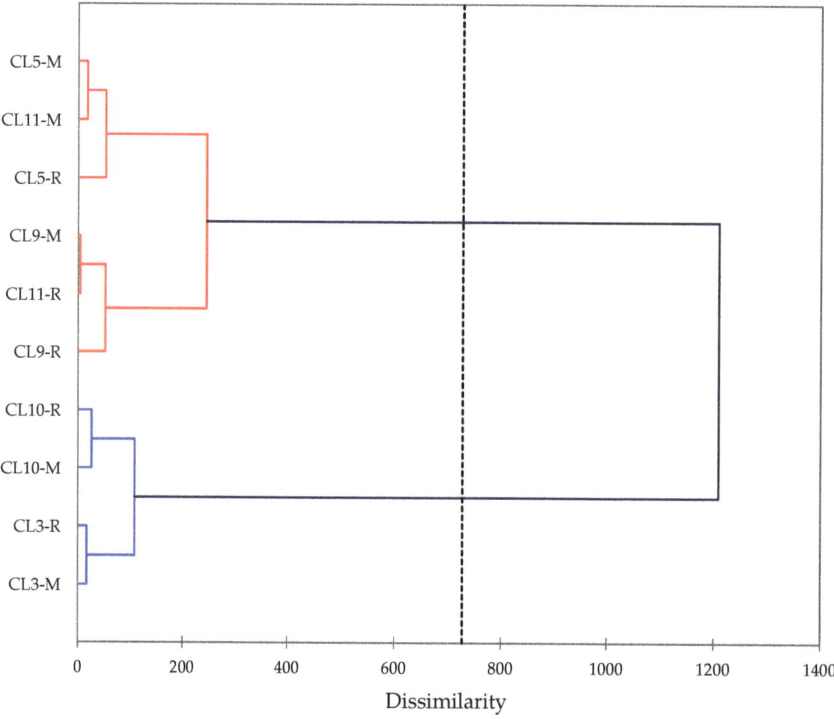

Figure 2. Dendrogram obtained from agglomerative hierarchical cluster analysis of the rhizome essential oils from varieties of *Curcuma longa* cultivated in north Alabama.

The cluster analysis of the *C. longa* varieties in this work revealed two well-defined groups. One group (varieties CL5, CL9, and CL11) was dominated by turmerones (α-turmerone, *ar*-turmerone, and β-turmerone). The second group demonstrated lower concentrations of turmerones, but higher concentrations of other components (e.g., α-zingiberene and β-phellandrene). Previous examination of *C. longa* rhizome essential oils from India as well as other geographical locations showed there to be four clusters based on the relative concentrations of the turmerones [14]: (1) dominated by turmerones, but with relatively large concentrations of other components; (2) dominated by turmerones, especially *ar*-turmerone; (3) dominated by turmerones, especially α-turmerone; and (4) very large concentrations of *ar*-turmerone. The chemical compositions of varieties CL5, CL9, and CL11 placed them into the cluster dominated by turmerones (i.e., cluster 2 of [14]), while varieties CL3 and CL10 had relatively lower concentrations of turmerones with relatively larger concentrations of components other than turmerones (i.e., cluster 1 of [14]). Thus, the essential oil compositions of these turmeric varieties adaptable to the Alabama summer growing season fit in well with essential oil compositions of turmeric varieties cultivated in Asia. The *ar*-turmerone and α-turmerone levels were similar to (CL3) or greater than (CL5, CL9, and CL11) those reported for turmeric grown in a tropical country, Brazil [23]. The *ar*-turmerone and α-turmerone were substantially greater than those reported for turmeric grown in India [24]. The β-turmerone levels were generally lower than those reported for turmeric grown in tropical countries [22–24], but similar to those reported for turmeric grown in Korea [25]. The cluster analysis also showed very little dissimilarity between the mother rhizome essential oil and the lateral rhizome essential oils for each of the varieties.

Xu and coworkers examined the extracts of 160 samples of *C. longa* from China [26]. Gas chromatographic analysis of the volatiles from the extracts revealed three volatile profile types, while high-performance liquid chromatographic (HPLC) analysis showed three types based on curcuminoid content. Unfortunately, Xu et al. identified only 10 volatile components, whereas 79 components were identified in our essential oil work. Furthermore, percent compositions were not reported and only "representative" chromatograms were presented. Nevertheless, although comparison is tenuous, based on the chromatograms, the volatile profile types identified seem to be analogous to essential oil types in our work. Importantly, volatile profile types "a" and "b" correspond to high-curcuminoid type "B" [26]. Therefore, we conclude that high turmerone concentrations are desirable qualities in turmeric essential oil.

The turmerones are responsible for the turmeric-like odor of *C. longa* [27]. In addition to their pungent scent, turmerones are important, biologically active constituents of *C. longa* rhizome essential oils [28], showing in vitro cytotoxic activities against several human tumor cell lines [29–32], anti-inflammatory activity through attenuated expression of proinflammatory cytokines [33,34], antibacterial activity against Gram-positive organisms [35], antifungal activity against phytopathogenic [36] and dermatopathogenic [37] fungi, mosquito larvicidal activity against *Anopheles gambiae* [38] and *Culex pipiens* [39], and insecticidal activity against the agricultural pests *Sitophilus zeamais* and *Spodoptera frugiperda* [40].

4. Conclusions

The chemical profiles of varieties CL5, CL9, and CL11 tested in this study in north Alabama were comparable to those growing in tropical regions of the world, suggesting that these varieties are suitable for commercialization in this region. However, CL3 and CL10 gave relatively poor essential oil yields and essential oils with lower concentrations of the turmerones. There is a growing market for *Curcuma longa* essential oils, with several varieties showing promise for development in the southeastern United States.

Author Contributions: Conceptualization, S.R.M.; methodology, S.R.M. and W.N.S.; formal analysis, W.N.S.; investigation, W.N.S., L.D. and A.P.; resources, S.R.M.; data curation, W.N.S.; writing—original draft preparation, W.N.S. and S.R.M.; writing—review and editing, all authors; supervision, S.R.M.; project administration, S.R.M.; funding acquisition, S.R.M. All authors have read and agreed to the published version of the manuscript.

Funding: This research and the APC were funded by USDA/National Institute of Food and Agriculture (NIFA)-Agriculture and Food Research Initiative (AFRI) Project, grant number 2016-68006-24785.

Institutional Review Board Statement: Not applicable.

Informed Consent Statement: Not applicable.

Data Availability Statement: Data are available from the corresponding authors.

Acknowledgments: W.N.S. and A.P. participated in this work as part of the activities of the Aromatic Plant Research Center (APRC, https://aromaticplant.org/). S.R.M. and L.D. thank Andrea Barr, Mounika Pudota, Jasmine Arnold, and Suresh Kumar for their assistance with field trials.

Conflicts of Interest: The authors declare no conflict of interest.

References

1. Ferreira, F.D.; Kemmelmeier, C.; Arrotéia, C.C.; Da Costa, C.L.; Mallmann, C.A.; Janeiro, V.; Ferreira, F.M.D.; Mossini, S.A.G.; Silva, E.L.; Machinski, M. Inhibitory effect of the essential oil of *Curcuma longa* L. and curcumin on aflatoxin production by *Aspergillus flavus* Link. *Food Chem.* **2013**, *136*, 789–793. [CrossRef] [PubMed]

2. Awasthi, P.K.; Dixit, S.C. Chemical composition of *Curcuma longa* leaves and rhizome oil from the plains of northern India. *J. Young Pharm.* **2009**, *1*, 312–316.

3. Dahal, K.R.; Idris, S. *Curcuma longa* L. In *Plant Resources of South-East Asia No 13: Spices*; de Guzman, C.C., Siemonsma, J.S., Eds.; Backhuys Publishers: Leiden, The Netherlands, 1999; pp. 111–116.

4. Dosoky, N.S.; Setzer, W.N. Chemical composition and biological activities of essential oils of *Curcuma* species. *Nutrients* **2018**, *10*, 1196. [CrossRef]

5. Helen, C.F.; Su, R.H.; Ghulam, J. Isolation, purification and characterization of insect repellents from *Curcuma longa* L. *J. Agric. Food Chem.* **1982**, *30*, 290–292.

6. Govindarajan, V.S. Turmeric- chemistry, technology and quality. *Crit. Rev. Food Sci. Nutr.* **1980**, *12*, 199–301. [CrossRef] [PubMed]

7. Prasad, S.; Aggarwal, B.B. Turmeric, the golden spice: From traditional medicine to modern medicine. In *Herbal Medicine: Biomolecular and Clinical Aspects*; Benzie, I.F.F., Wachtel-Galor, S., Eds.; CRC Press: Boca Raton, FL, USA, 2011; pp. 263–288, ISBN 978-1439807132.

8. Aggarwal, B.B.; Harikumar, K.B. Potential therapeutic effects of curcumin, the anti-inflammatory agent, against neurodegenerative, cardiovascular, pulmonary, metabolic, autoimmune and neoplastic diseases. *Int. J. Biochem. Cell Biol.* **2009**, *41*, 40–59. [CrossRef] [PubMed]

9. Hamaguchi, T.; Ono, K.; Yamada, M. Curcumin and Alzheimer's disease. *CNS Neurosci. Ther.* **2010**, *16*, 285–297. [CrossRef]

10. Shahbandeh, M. Trade Value of Turmeric Imported to the United States in 2019, by Country of Origin. Available online: https://www.statista.com/statistics/798301/us-turmeric-imports-by-country/ (accessed on 5 October 2020).

11. Hossain, M.A. Effects of harvest time on shoot biomass and yield of turmeric (*Curcuma longa* L.) in Okinawa, Japan. *Plant Prod. Sci.* **2015**, *13*, 97–103. [CrossRef]

12. Shannon, D.A.; van Santen, E.; Salmasi, S.Z.; Murray, T.J.; Duong, L.T.; Greenfield, J.T.; Gonzales, T.; Foshee, W. Shade, establishment method, and varietal effects on rhizome yield and curcumin content in turmeric in Alabama. *Crop Sci.* **2019**, *10*, 1–10. [CrossRef]

13. Rathore, P.; Dohare, P.; Varma, S.; Ray, A.; Sharma, U.; Jaganathanan, N.R.; Ray, M. *Curcuma* oil: Reduces early accumulation of oxidative product and is anti-apoptogenic in transient focal ischemia in rat brain. *Neurochem. Res.* **2008**, *33*, 1672–1682. [CrossRef]

14. Dosoky, N.S.; Satyal, P.; Setzer, W.N. Variations in the volatile compositions of *Curcuma* species. *Foods* **2019**, *8*, 53. [CrossRef] [PubMed]

15. Zachariah, T.J.; Sasikumar, B.; Nirmal Babu, K. Variation for quality components in ginger and turmeric and their interaction with environments. *Biotechnol. Plant. Crop.* **1999**, *3*, 230–251.

16. Sandeep, I.S.; Sanghamitra, N.; Sujata, M. Differential effect of soil and environment on metabolic expression of turmeric (*Curcuma longa* cv. Roma). *Indian J. Exp. Biol.* **2015**, *53*, 406–411. [PubMed]

17. Adams, R.P. *Identification of Essential Oil Components by Gas Chromatography/Mass Spectrometry*, 4th ed.; Allured Publishing: Carol Stream, IL, USA, 2007; ISBN 978-1-932633-21-4.

18. Mondello, L. *FFNSC 3*; Shimadzu Scientific Instruments: Columbia, MD, USA, 2016.

19. National Institute of Standards and Technology. *NIST17*; National Institute of Standards and Technology: Gaithersburg, MD, USA, 2017.

20. Satyal, P. Development of GC-MS Database of Essential Oil Components by the Analysis of Natural Essential Oils and Synthetic Compounds and Discovery of Biologically Active Novel Chemotypes in Essential Oils. Ph.D. Dissertation, University of Alabama in Huntsville, Huntsville, AL, USA, 2015.

21. Garg, S.N.; Bansal, R.P.; Gupta, M.M.; Kumar, S. Variation in the rhizome essential oil and curcumin contents and oil quality in the land races of turmeric *Curcuma longa* of North Indian plains. *Flavour Fragr. J.* **1999**, *14*, 315–318. [CrossRef]

22. Raina, V.K.; Srivastava, S.K.; Syamasundar, K.V. Rhizome and leaf oil composition of *Curcuma longa* from the lower Himalayan region of northern India. *J. Essent. Oil Res.* **2005**, *17*, 556–559. [CrossRef]

23. Avanço, G.B.; Ferreira, F.D.; Bomfim, N.S.; Peralta, R.M.; Brugnari, T.; Mallmann, C.A.; de Abreu Filho, B.A.; Mikcha, J.M.G.; Machinski, M., Jr. *Curcuma longa* L. essential oil composition, antioxidant effect, and effect on *Fusarium verticillioides* and fumonisin production. *Food Control* **2017**, *73*, 806–813. [CrossRef]

24. Leela, N.K.; Tava, A.; Shafi, P.M.; John, S.P.; Chempakam, B. Chemical composition of essential oils of turmeric (*Curcuma longa* L.). *Acta Pharm.* **2002**, *52*, 137–141.

25. Hwang, K.-W.; Son, D.; Jo, H.-W.; Kim, C.H.; Seong, K.C.; Moon, J.-K. Levels of curcuminoid and essential oil compositions in turmerics (*Curcuma longa* L.) grown in Korea. *Appl. Biol. Chem.* **2016**, *59*, 209–215. [CrossRef]

26. Xu, L.; Shang, Z.; Lu, Y.; Li, P.; Sun, L.; Guo, Q.; Bo, T.; Le, Z.; Bai, Z.; Zhang, X.; et al. Analysis of curcuminoids and volatile components in 160 batches of turmeric samples in China by high-performance liquid chromatography and gas chromatography mass spectrometry. *J. Pharm. Biomed. Anal.* **2020**, *188*, 113465. [CrossRef]

27. Hasegawa, T.; Nakatani, K.; Fujihara, T.; Yamada, H. Aroma of turmeric: Dependence on the combination of groups of several odor constituents. *Nat. Prod. Commun.* **2015**, *10*, 1047–1050. [CrossRef]

28. Aggarwal, B.B.; Yuan, W.; Li, S.; Gupta, S.C. Curcumin-free turmeric exhibits anti-inflammatory and anticancer activities: Identification of novel components of turmeric. *Mol. Nutr. Food Res.* **2013**, *57*, 1529–1542. [CrossRef] [PubMed]

29. Ji, M.; Choi, J.; Lee, J.; Lee, Y. Induction of apoptosis by *ar*-turmerone on various cell lines. *Int. J. Mol. Med.* **2004**, *14*, 253–256. [CrossRef] [PubMed]

30. Cheng, S.-B.; Wu, L.-C.; Hsieh, Y.-C.; Wu, C.-H.; Chan, Y.-J.; Chang, L.-H.; Chang, C.-M.J.; Hsu, S.-L.; Teng, C.-L.; Wu, C.-C. Supercritical carbon dioxide extraction of aromatic turmerone from *Curcuma longa* Linn. induces apoptosis through reactive oxygen species-triggered intrinsic and extrinsic pathways in human hepatocellular carcinoma HepG2 cells. *J. Agric. Food Chem.* **2012**, *60*, 9620–9630. [CrossRef] [PubMed]

31. Park, S.Y.; Kim, Y.H.; Kim, Y.; Lee, S.-J. Aromatic-turmerone attenuates invasion and expression of MMP-9 and COX-2 through inhibition of NF-κB in TPA-induced breast cancer cells. *J. Cell. Biochem.* **2012**, *113*, 3653–3662. [CrossRef]

32. Nair, A.; Amalraj, A.; Jacob, J.; Kunnumakkara, A.B.; Gopi, S. Non-curcuminoids from turmeric and their potential in cancer therapy and anticancer drug delivery formulations. *Biomolecules* **2019**, *9*, 13. [CrossRef] [PubMed]

33. Park, S.Y.; Jin, M.L.; Kim, Y.H.; Kim, Y.; Lee, S.J. Anti-inflammatory effects of aromatic-turmerone through blocking of NF-κB, JNK, and p38 MAPK signaling pathways in amyloid β-stimulated microglia. *Int. Immunopharmacol.* **2012**, *14*, 13–20. [CrossRef]

34. Li, Y.-L.; Du, Z.-Y.; Li, P.-H.; Yan, L.; Zhou, W.; Tang, Y.-D.; Liu, G.-R.; Fang, Y.-X.; Zhang, K.; Dong, C.-Z.; et al. Aromatic-turmerone ameliorates imiquimod-induced psoriasis-like inflammation of BALB/c mice. *Int. Immunopharmacol.* **2018**, *64*, 319–325. [CrossRef]

35. Negi, P.S.; Jayaprakasha, G.K.; Jagan Mohan Rao, L.; Sakariah, K.K. Antibacterial activity of turmeric oil: A byproduct from curcumin manufacture. *J. Agric. Food Chem.* **1999**, *47*, 4297–4300. [CrossRef]

36. Jayaprakasha, G.K.; Negi, P.S.; Anandharamakrishnan, C.; Sakariah, K.K. Chemical composition of turmeric oil—A byproduct from turmeric oleoresin industry and its inhibitory activity against different fungi. *Zeitschrift fur Naturforsch. Sect. C J. Biosci.* **2001**, *56*, 40–44. [CrossRef]

37. Jankasem, M.; Wuthi-udomlert, M.; Gritsanapan, W. Antidermatophytic properties of *ar*-turmerone, turmeric oil, and *Curcuma longa* preparations. *ISRN Dermatol.* **2013**, *2013*, 250597. [CrossRef]

38. Ajaiyeoba, E.O.; Sama, W.; Essien, E.E.; Olayemi, J.O.; Ekundayo, O.; Walker, T.M.; Setzer, W.N. Larvicidal activity of turmerone-rich essential oils of *Curcuma longa* leaf and rhizome from Nigeria on *Anopheles gambiae*. *Pharm. Biol.* **2008**, *46*, 279–282. [CrossRef]

39. Abdelgaleil, S.A.M.; Zoghroban, A.A.M.; Kassem, S.M.I. Insecticidal and antifungal activities of crude extracts and pure compounds from rhizomes of *Curcuma longa* L. (Zingiberaceae). *J. Agric. Sci. Technol.* **2019**, *21*, 1049–1061.

40. de Tavares, W.S.; de Freitas, S.S.; Grazziotti, G.H.; Parente, L.M.L.; Lião, L.M.; Zanuncio, J.C. *Ar*-turmerone from *Curcuma longa* (Zingiberaceae) rhizomes and effects on *Sitophilus zeamais* (Coleoptera: Curculionidae) and *Spodoptera frugiperda* (Lepidoptera: Noctuidae). *Ind. Crops Prod.* **2013**, *46*, 158–164. [CrossRef]

MDPI
St. Alban-Anlage 66
4052 Basel
Switzerland
www.mdpi.com

Foods Editorial Office
E-mail: foods@mdpi.com
www.mdpi.com/journal/foods